DYNAMICS IN ENGINEERING PRACTICE

TENTH EDITION

CRC Series in
COMPUTATIONAL MECHANICS and APPLIED ANALYSIS

Series Editor: J. N. Reddy
Texas A&M University

Published Titles

ADVANCED THERMODYNAMICS ENGINEERING
Kalyan Annamalai and Ishwar K. Puri

APPLIED FUNCTIONAL ANALYSIS
J. Tinsley Oden and Leszek F. Demkowicz

COMBUSTION SCIENCE AND ENGINEERING
Kalyan Annamalai and Ishwar K. Puri

CONTINUUM MECHANICS FOR ENGINEERS, Third Edition
Thomas Mase, Ronald E. Smelser, and George E. Mase

DYNAMICS IN ENGINEERING PRACTICE, Tenth Edition
Dara W. Childs

EXACT SOLUTIONS FOR BUCKLING OF STRUCTURAL MEMBERS
C. M. Wang, C. Y. Wang, and J. N. Reddy

THE FINITE ELEMENT METHOD IN HEAT TRANSFER AND FLUID DYNAMICS, Third Edition
J. N. Reddy and D. K. Gartling

MECHANICS OF LAMINATED COMPOSITE PLATES AND SHELLS: THEORY AND ANALYSIS, Second Edition
J. N. Reddy

NUMERICAL AND ANALYTICAL METHODS WITH MATLAB®
William Bober, Chi-Tay Tsai, and Oren Masory

PRACTICAL ANALYSIS OF COMPOSITE LAMINATES
J. N. Reddy and Antonio Miravete

SOLVING ORDINARY and PARTIAL BOUNDARY VALUE PROBLEMS in SCIENCE and ENGINEERING
Karel Rektorys

STRESSES IN BEAMS, PLATES, AND SHELLS, Third Edition
Ansel C. Ugural

DYNAMICS IN ENGINEERING PRACTICE

TENTH EDITION

DARA W. CHILDS

CRC Press
Taylor & Francis Group
Boca Raton London New York

CRC Press is an imprint of the
Taylor & Francis Group, an **informa** business

CRC Press
Taylor & Francis Group
6000 Broken Sound Parkway NW, Suite 300
Boca Raton, FL 33487-2742

Printed in the United States of America on acid-free paper
10 9 8 7 6 5 4 3 2 1

International Standard Book Number: 978-1-4398-3125-0 (Hardback)

Visit the Taylor & Francis Web site at
http://www.taylorandfrancis.com

and the CRC Press Web site at
http://www.crcpress.com

This book is dedicated to the students who have taken courses in dynamics from me since 1968,

particularly the hardworking students

at Texas A&M University in College Station, Texas

Contents

Preface

Dynamics concerns the motion of particles and rigid bodies and, as traditionally presented, has proven to be a "killer" course for both students who study engineering and professors who try to teach it. Students generally fail to grasp the pervasive nature of dynamics in engineering practice and view the subject matter as a collection of "tricks," rather than a unified body of knowledge based on an extremely limited number of basic principles. Professors grow tired of teaching courses that have a high level of failure and attrition, and professors in successor courses tend to complain that students "don't know how to derive equations of motions (EOMs) or do free-body diagrams."

This book's premise is that many of these difficulties arise because of the inappropriate nature of currently available dynamics texts. These books have a peculiar disconnect from either the requirements of subsequent course work or the practice of dynamics in an engineering career. The practice of engineering involves the development of (1) *general* kinematic equations, that is, equations that tell us the position, velocity, and acceleration of particles and/or rigid bodies without regard to the forces acting on the bodies; and (2) *general* EOM that tell us how the motion of particles and rigid bodies evolves as a function of time under the action of forces. Dynamics describes the *continuous* evolution of motion. It is not a series of "snapshots" at a few discrete times; yet many current textbooks treat it as such.

The natural language of dynamics is differential equations, and most of today's dynamics students have either studied differential equations or are concurrently enrolled. Nonetheless, many dynamics books carry on an extended conspiracy to avoid differential equations. Given that most current texts regularly use algebraic equations, for example, $f = ma$ and $T = I\alpha$, rather than the differential equation statements $\Sigma f = m\ddot{r}$ and $T = I\ddot{\theta}$, students regularly have trouble with follow-on courses that require the development and solution of governing differential EOM.

An additional shortcoming of most current texts is the treatment of computer skills. Most engineering freshmen have an introduction to at least one programming language and then frequently suffer at least 1 year of neglect before starting over with a computer-based numerical methods course. To be direct, modern dynamics cannot be fully appreciated without recourse to computer solutions. This book provides enough computer example problems and tools for students to grasp the subject.

In the past, many engineering curricula covered dynamics twice. The "survivors" of the algebra-based first course proceeded to a differential-equation-based second course. The *essential* modeling skills, involving the development of general kinematics equations and general EOM, were covered in the second course. The continuing contraction of engineering curricula is forcing an end to this practice. Many curricula developers are also struggling to find space for coverage of traditional vibration topics. Engineering educators should now be prepared to consider the *real* requirements of dynamics both in curricula and engineering practice and develop courses that meet these requirements. The contents of this book were developed to teach students how to model general dynamic systems. The writer's experience has been that most students have an easier time learning general skills, where effectiveness can be demonstrated for a wide range of problems, than a collection of problem-specific tasks.

Many current dynamics books have potentially superb problems; however, the wrong questions are being asked. In my experience, companies hiring engineers will never ask them to work the problems in these books, or even apply the tricks the students are taught to solve these problems. Engineers are employed to derive *general* kinematics equations and EOM. These general equations can be used to solve for maximum accelerations, loads, velocities, etc., and answer real engineering design problems. Evidence that our students understand this situation is provided by the massive resale of such books on the secondhand book market. Students regularly keep books on thermodynamics, fluid mechanics, strength of materials, controls, etc., but rush to dispose of their undergraduate dynamics texts.

Dynamics books that are written with a conscious avoidance of differential equations tend to either omit elements of mechanical vibrations or to provide a separate (late) chapter covering the topic. Given the differential-equation orientation of this book, elements of mechanical vibrations are integrated, from the outset, in the particle and planar-dynamics material. "Real-world" systems of particles and rigid bodies inevitably include springs and energy-dissipation mechanisms, including viscous, Coulomb, and aerodynamic damping. One- and two-degree-of-freedom problems are introduced in the application of particle and planar rigid-body kinetics. Damped and undamped natural frequencies and linear damping factors are introduced and defined for one-degree-of-freedom (1DOF) problems. Two-degree-of-freedom (2DOF) problems are used to define system natural frequencies (eigenvalues) and mode shapes (eigenvectors), and to demonstrate modal uncoupling of multi-degree-of-freedom (MDOF) vibration problems. The 2DOF material can be skipped without affecting the book's order.

The idea of an equilibrium position or positions defined by a differential EOM is introduced, leading to linearization of nonlinear EOM for small motion about an equilibrium position. Nonlinear terms are linearized by using Taylor-series expansions about an equilibrium position. Stable and unstable static equilibrium positions are demonstrated.

In considering his campaign, President George H.W. Bush referred to "The Vision Thing" (Wikipedia). In fact, vision is the missing element in current dynamics books and courses. The "vision" of dynamics presented in this book is *active* students, *actively* developing *general* kinematics equations and EOM, actively analyzing the systems to the extent possible, and then actively setting up and solving these equations via current computer techniques.

This book combines subjects that would normally be covered in both an introductory and an intermediate or advanced dynamics course. It is intended to be of value to students through their first course and subsequently in engineering practice, and was also written for working engineers who are trying to analyze real dynamic systems.

For MATLAB® and Simulink® product information, please contact

The MathWorks, Inc.
3 Apple Hill Drive
Natick, MA, 01760-2098 USA
Tel: 508-647-7000
Fax: 508-647-7001
E-mail: info@mathworks.com
Web: www.mathworks.com

Book Outline

The outline followed in achieving the book's vision is traditional. Chapter 1 covers some fundamental requirements of dynamics, including units, force, mass, etc., and provides a brief history of the development of dynamics. This material is intended to be a quick review of material covered earlier in physics.

Chapter 2 covers kinematics of a particle, including displacement, velocity, and acceleration in one and two dimensions. Cartesian, polar, and path coordinates are introduced, and coordinate transformations of the components of a vector in a plane are emphasized. Alternative kinematic statements of the same example using Cartesian, polar, and path coordinates are emphasized, using coordinate transformations to move between the coordinate representations.

Chapter 3 deals with the kinetics of a particle. It begins with Newton's laws for rectilinear motion. The rectilinear examples introduce physical modeling, covering forces due to gravity and spring forces, as well as linear, quadratic, and Coulomb damping. 1DOF vibration problems are covered, introducing the concepts of damped and undamped natural frequencies, damping factors, resonance, etc. Solutions are developed for free and forced motion and steady-state response due to harmonic excitation. The examples emphasize the development of the EOM from $\Sigma f = m\ddot{r}$ and the solution of the equation via either (1) direct time integration, or (2) the energy-integral substitution, $\ddot{x} = d(\dot{x}^2/2)/dx$. Next, the notion of degrees of freedom and kinematic equations of constraint are introduced, using masses connected by pulleys as examples. These examples include several 1DOF vibration problems.

Planar motion in two dimensions is introduced next, with applications involving Cartesian, polar, and path coordinates. The Cartesian examples include trajectory motion with and without quadratic drag and motion on a plane, including Coulomb friction forces. Polar coordinate applications include the simple pendulum, and this example is used as the initial demonstration of linearization of an EOM. EOM are developed for the pendulum, including viscous and quadratic damping. The path-coordinate applications tend to be traditional with beads sliding on wires or along a specified surface.

The fact that dynamics problems regularly have MDOF is obscured by many introductory dynamics texts that only present 1DOF examples. 2DOF problems are introduced in Section 3.5 with simple spring–mass systems. An approach is presented for developing EOM for masses connected by linear springs and dampers. Matrix statements are presented for linear 2DOF vibration problems with coordinate coupling provided by the stiffness and damping matrices. A double pendulum is introduced and analyzed as a second 2DOF example. Nonlinear EOM are developed for the two particles in the double pendulum from $\Sigma f = m\ddot{r}$. These equations are subsequently linearized to yield a matrix model with coordinate coupling in the inertia matrix entries.

Analysis procedures for 2DOF vibration problems are introduced by analyzing the free motion of a spring–mass system, leading to a quadratic characteristic equation to define the eigenvalues and natural frequencies. Solutions for the eigenvectors from calculated eigenvalues are presented next. Uncoupling of the coupled matrix vibration equations via the matrix of eigenvectors is also explained, leading to uncoupled modal differential equations. Solutions are presented for free and forced motion, proceeding from solutions of the modal differential equations. Forced motion resulting from harmonic excitation is also presented. 2DOF vibration problems are provided to show students that such problems exist, and methods to derive and analyze the EOM that model the systems are presented. This is done to help students understand how these systems behave, with multiple natural frequencies and resonance possibilities. The vibration coverage in Chapter 3 is a reasonable introduction to aspects of vibration but is not a substitute for a good engineering-vibrations text.

Work–energy applications for 1DOF examples are covered in Section 3.6. Kinetic and potential energies are introduced, and the general work–energy expression, $\text{Work}_{n.c} = \Delta(T + V)$, is developed. Work due to potential forces is not covered. Such forces are modeled by potential–energy functions from the outset. The energy-integral substitution, $d(\dot{x}^2/2)dx = \ddot{x}$, is employed to show the direct relation between the work–energy equation and the original $\Sigma f = m\ddot{r}$ differential EOM. A considerable amount of material is presented on the derivation of EOM for 1DOF examples starting from the work–energy equation.

Chapter 3 concludes with the coverage of linear momentum and moment-of-momentum topics. These are largely algebraic in nature and are customarily covered adequately by prior physics courses. The coefficient of restitution for particle collisions is covered in Section 3.7.2.

Chapter 4 covers planar kinematics of rigid bodies. Many engineering dynamics problems involve planar dynamics, that is, motion of rigid bodies in a plane (such as the plane of this page) versus three-dimensional vector kinematics that would be required to keep track of the position and orientation of an airplane or satellite. The kinematic relationships of Chapter 2 are adequate for the tasks presented in this chapter. The traditional approach for teaching the analysis of planar mechanisms involves vector velocity and acceleration equations for Cartesian motion that require a great deal of vector algebra with cross products. This traditional approach is presented in Chapter 4; however, an alternate approach using geometry is also presented for finding relationships between variables. (My students have found the geometric approach to be much easier, and are able to develop general kinematic equations for mechanisms quickly.) A considerable effort is expended at the start of this chapter to cover the rolling-without-slipping kinematic condition, given that students traditionally have trouble grasping this concept. Multiple representative examples including several commonly occurring mechanisms are analyzed using both traditional and geometric approaches.

Chapter 5 covers planar kinetics of rigid bodies, starting with inertia properties and including the mass moment of inertia, radius of gyration, and the parallel-axis formula. For planar motion of a rigid body, governing force and moment equations are developed in Section 5.3, and the kinetic energy definition is developed in Section 5.4. In Chapter 3, Newtonian and energy approaches are considered sequentially. In this chapter, students are assumed to have experience and confidence in developing the EOM for 1DOF examples using either Newtonian (free-body) approaches or the energy. Hence, for most 1DOF planar-kinetics examples, the EOM is derived using both approaches.

Fixed-axis-rotation examples provide the initial application of the governing equations. In Section 5.6, EOM are developed for various compound-pendulum geometries. EOM are developed, pivot reactions are defined in terms of the pendulum's rotation angle, and stability is examined for small motion about equilibrium positions. Linear damping is also considered. A bar supported by linear spring and damper supports is investigated, with considerations of motion about equilibrium. The base acceleration of a compound pendulum's pivot-point is also investigated.

Systems in general planar motion that require both force and moment equations are introduced in Section 5.7.1, including rolling-without-slipping examples. Various 1DOF examples are modeled in Section 5.7.2. In Section 5.7.3, EOM developments are carried out using Newtonian and energy approaches for one-body examples that include external forces. Generalized forces are introduced and defined for the energy approach. In Section 5.7.3, multi-body, 1DOF examples are introduced, demonstrating the advantages of the energy approach in deriving the EOM for this type of problem.

Several MDOF examples are introduced in Section 5.7.4, including a double compound pendulum. Torsional vibrations and systems for which beams are used as springs are also considered. Section 5.7.5 covers models for various planar mechanisms, including developing the EOM from Newtonian approaches as well as the kinematic-constraint equations that were developed in Chapter 4. Section 5.8 concludes the chapter with the development and application of moment-of-momentum equations for the planar motion of a rigid body.

1

Introduction and Fundamentals

1.1 Introduction

This book was written for students or working engineers who have completed prerequisites in physics, statics, possibly an introduction to dynamics, and (at a minimum) are concurrently enrolled in differential equations. You are assumed to have some basic familiarity with the fundamental concepts of dynamics, that is, forces, kinetic energy, potential energy, velocity, acceleration, etc. Accordingly, this is a short chapter. Following a brief history of dynamics, the SI and U.S. customary unit systems are discussed.

1.2 A Short History of Dynamics

Dynamics is generally (including in this book) broken into kinematics and kinetics. Kinematics is a geometrical study of motion without regard for the forces that either cause motion or result from motion. At the simplest level, kinematics tells us the position of particles (points in space) with respect to a specified coordinate system. Think about a simple street map or a GPS system that tells you how to move from your current location to a desired location. Kinematics also defines the orientation of rigid bodies with respect to a fixed reference. The leaning tower of Pisa is a notable landmark because of its inclination with respect to the vertical line. You could think about drawing an arc from a vertical line and specifying the angle of rotation of the tower's axis from that line. Kinematics of rigid bodies normally involves keeping track of positions and displacements; for example, the task of defining the position and orientation of a car traveling around a track could be accomplished by measuring and recording the coordinates of a point on the car with respect to an agreed-upon coordinate system, plus keeping track of the car's pointing angle with respect to a fixed axis. We have casually used the terms "particle" and "rigid body" that actually have specific definitions to be discussed below.

Kinetics is the study of motion of particles or bodies due to applied loads. Kinetics can also involve the definition of reaction loads due to motion. Since kinematics is based on geometry, its origins date back to prehistorical land measurements and surveying. Greek scholars began the formal study of geometry about 600 BC, culminating in Euclid's book, *The Elements*, in about 300 BC. Euclid founded the first school of mathematics in Alexandria, Egypt.

Most engineering analysis procedures have been developed to answer specific questions concerning the analysis (and frequently) failure of physical systems. Issues involving statics (equilibrium) and strength-of-material problems have been present since the beginning of civilization. Buildings that do not properly account for static forces regularly collapse; hence, the reaction forces that apply loads to elements must be calculated (statics), and an analysis is necessary to determine whether a structural element can withstand a given load (strength of materials). Our ancestors were able to build marvelous structures that are still present today without recourse to analysis; however, multiple undocumented (and nonsurviving) failures were normally the basis of these successes.

There was no comparable urgency in developing dynamics, and it started with astronomy, concerning the motion of heavenly bodies. Ancient mankind spent lots of time under the stars, and considerable effort was expended to understand the observed motion. Simply getting a workable calendar that fit the seasons and the solar year was a considerable effort. Modern astronomy started with the work of a Polish researcher, Nicolaus Copernicus (1473–1543), who proposed that the earth moved around the sun, versus the prior view that the earth was the center of the universe with a revolving sun. A German mathematician, Johannes Kepler (1571–1630), demonstrated that the path traced out by a planet's motion around the sun is elliptical, and the Italian astronomer, Galileo (1564–1642), made telescopic observations that supported the Copernican view of the solar system.

All of these results were based on careful observations by quite intelligent researchers and thinkers. However, their predictions of motion were entirely descriptive and could not be developed from any fundamental "laws of motion." Isaac Newton (1642–1727) radically changed these circumstances by discovering the universal law of gravitation. This law defines the mutual attraction force between two bodies and is stated as follows:

$$f_g = K_g \frac{m_1 m_2}{R^2}, \quad (1.1)$$

where

f_g is the attraction force
m_1 and m_2 are the masses of two bodies
$K_g = 6.673 \times 10^{-11}$ N m^2/kg^2 is the universal law of gravitation
R is the distance between the bodies' centers

Of considerably more interest in the study of engineering dynamics are the three laws of motion for a particle postulated by Newton:

Law 1. A particle remains at rest or continues to move in a straight line with constant velocity unless acted on by an external force.

Law 2. A particle's acceleration is proportional to the resultant force acting on the particle and is in the same direction as the resultant force.

Law 3. The forces of action and reaction between two contacting particles are equal and act in opposite directions.

Newton's laws apply to particles. You probably have some general sense of the definition of a particle. The formal dynamic sense follows: *Particle*—A particle refers to an idealized body having vanishing dimensions. Think of a sphere with a radius approaching zero, and you have a particle. In dynamics, circumstances dictate whether or not a body can be thought of as a particle. For example, most people would agree that a grain of sand being propelled by wind could normally be considered to be a particle versus a book, a car, or a house that have clear dimensions of length, width, depth, etc. However, the earth, which we obviously consider to be massive, can be considered to be a particle in considering its motion on a solar-system scale.

Together with kinematics, Newton's three laws of motion form the basis for almost all of engineering dynamics. They also pose a paradox that was only resolved in the twentieth century by "modern physics." Specifically, Newton's second law of motion is a vector differential equation that states

$$\Sigma f = m\ddot{r}, \quad (1.2)$$

where

Σf is the resultant force acting on the particle
m is the proportionality constant called mass
$\ddot{r}$ is the particle's acceleration with respect to *an inertial coordinate system*

However, the answer to the question, "What is an inertial coordinate system?" is, "An inertial coordinate system is a coordinate system for which $\Sigma f = m\ddot{r}$ is correct." Until Einstein (1879–1955), the argument was made that a reference, nonaccelerating, inertial coordinate system existed at some distant point in space. Einstein's theory of special and general relativity corrected Newton's laws of motion and are beyond the scope of this book. Newton's laws are entirely valid for almost all engineering applications, with the surface of the earth used as the inertial coordinate system (despite the earth's daily rotation about its own axis and annual rotation about the sun). Newton's laws break down at a particle physics level, and the earth's surface is an inadequate inertial system for the planet's motion around the sun. A nonrotating system centered within the sun is required for the study of planetary motion.

The brilliance and scope of Newton's developments are scarcely believable. In a comparatively short time he produced the universal law of gravitation, his three laws of motion, and invented differential and integral calculus. The only ready parallels for his productive genius would be Mozart and Bach in music and Einstein in physics. Genius is neither predictable nor explicable.

Newton's laws apply for particles not rigid bodies. The Swiss mathematician, Leonhard Euler (1707–1783), extended Newton's laws to the motion of rigid bodies. Again, you probably have a sense of a "rigid body." The formal definition is as follows: *Rigid body*—A rigid body is a collection of particles that are fixed in position relative to each other and has finite dimensions, that is, length, width, depth, etc. The *rigid* in rigid body is an idealization in that any body can be deflected or deformed due to loads. For example, the ball in a ball bearing has elastic deflections due to the static and dynamic loads that it experiences.

Most fields of engineering mechanics can be approached from either a Newtonian (force or moment equations leading directly to differential equations of motion) or a variational approach based on work and energy. For example, in statics, equilibrium can be defined from the requirement that the external forces and external moments be zero. Alternatively, equilibrium can be defined as the requirement that a system's potential energy be minimized. In dynamics, the mathematician, Joseph Lagrange (1736–1813), developed "Lagrange's equations of motion," from a variational viewpoint. Lagrange's equations provide an entirely separate approach for developing equations of motion for particles and rigid bodies.

The general application of dynamics awaited the Industrial Revolution with steam engines and then steam turbines at the beginning of the twentieth century. The design of high-speed machinery required dynamics and its offspring vibrations. Dimarogonas (1995), provides a recent and very thorough historical review of the development of dynamics and vibrations. Motion and speed from automobiles to spacecraft characterized the twentieth century and stimulated the development and application of dynamics.

The ability to derive governing differential equations of motion from either a Newtonian or variational viewpoint was a significant accomplishment; however, solving the equations was another matter. The ability to solve nonlinear differential equations or large systems of differential equations analytically (pencil and paper) was (and remains) exceptionally limited. A study of advanced dynamics texts over many years shows the same set of limited examples for which analytical solutions can be developed, reflecting, in many cases, exceptional inventiveness and physical insight.

Modern dynamics, involving the derivation, analysis, *and direct solution* of governing nonlinear differential equations of motion, began with the computer era. In the early 1960s, analog computers tended to be the computer of choice for solving nonlinear differential equations, supplanted rapidly by modern digital computers. Newton, Euler, and Lagrange could not have dreamed of the systems of nonlinear differential equations that are readily solved today. Classical solutions continue to be of value as the basis for understanding the expected nature of solutions and to provide checks for numerical solutions. Again, modern dynamics involves deriving *and solving* (generally numerically) equations of motion.

1.3 Units

With the notable exception of the United States, all engineers use the SI system of units involving the meter, newton, and kilogram, respectively, for length, force, and mass. The metric system, which preceded the SI system, was legalized for commerce by an act of the U.S. Congress in 1866. The act of 1866 reads in part, Mechtly (1969),

> It shall be lawful throughout the United States of America to employ the weights and measures of the metric system; and no contract or dealing or pleading in any court, shall be deemed invalid or liable to objection because the weights or measures referred to therein are weights or measures of the metric system.

Nonetheless, in the twenty-first century, U.S. engineers, manufacturers, and the general public continue to use the foot and pound as standard units for length and force. Both the SI and U.S. systems use the second as a unit of time. The U.S. customary system of units began in England and continues to be referred to in the United States as the "English System" of units. However, Great Britain adopted and has used the SI system for many years. A reasonable person would expect the United States to eventually and officially give up, adopt, and use the SI system.

Both the U.S. and SI systems use the same standard symbols for exponents of 10 base units. A partial list of these symbols is provided in Table 1.1. Only the "m = −3" (mm = millimeter = 10^{-3} m) and "k = +3" (km = kilometer = 10^3 m) exponent symbols are used to any great extent in this book.

Newton's second law of motion $\Sigma \boldsymbol{f} = m\ddot{\boldsymbol{r}}$ ties the units of time, length, mass, and force together. Using Equation 1.1 to define the force of gravity, and stating $\Sigma \boldsymbol{f} = m\ddot{\boldsymbol{r}}$ for a particle of mass m on the earth's surface gives

$$m\ddot{Y} = \sum f_Y = f_g = K_g \frac{m m_e}{R^2}, \tag{1.3}$$

where

m_e is the earth's mass
R is the earth's nominal (spherical) radius

Canceling m and solving for the resultant acceleration gives

$$\begin{aligned}\ddot{Y} = g(\text{ideal}) &= \frac{K_g m_e}{R^2} \\ &= \frac{6.673\times10^{-11}\,\text{N m}^2/\text{kg}^2\times \text{s}^5\,0.976\times10^{24}\,\text{kg}}{(6.37\times10^6\,\text{m})^2} \\ &= 9.824\,\frac{\text{N}}{\text{kg}} = 9.824\,\frac{\text{m}}{\text{s}^2}\end{aligned}$$

for the ideal "acceleration of gravity." The unit conversion N/kg = m/s^2 used in this equation follows from $\Sigma \boldsymbol{f} = m\ddot{\boldsymbol{r}}$ and will be discussed further below. Because

TABLE 1.1

Standard Symbols for Exponents of 10

Factor by Which Unit Is Multiplied	Prefix	Symbol
10^9	Giga	G
10^6	Mega	M
10^3	Kilo	k
10^2	Hecto	h
10	Deka	da
10^{-1}	Deci	d
10^{-2}	Centi	c
10^{-3}	Milli	m
10^{-6}	Micro	μ
10^{-9}	Nano	n

the earth rotates and is an oblate spheroid (not a sphere), measured values for g on the earth's surface differ from this result. The earth's rotation is accounted for via the international gravity formula definition

$$g = 9.784049(1 + 0.0052884\sin^2\gamma - 0.0000059\sin^2 2\gamma)\ \text{m/s}^2, \tag{1.4}$$

where γ is the latitude. In customary engineering usage, Equation 1.3 is stated

$$m\ddot{Y} = \sum f_Y = w = mg, \tag{1.5}$$

where

Y is pointed directly downward
w is the weight force due to gravity
g is the acceleration of gravity

In North America, at sea level, customary usage for the acceleration of gravity is

$$\begin{aligned} g &= 9.81\ \text{m/s}^2 = 9810\ \text{mm/s}^2 \\ &= 32.2\ \text{ft/s}^2 = 386\ \text{in/s}^2. \end{aligned} \tag{1.6}$$

In Australia, I am told that engineers use $g = 9.80\ \text{m/s}^2$.

We start our discussion of the connection between force and mass with the SI system, since it tends to be more rational (not based on the length of a man's foot or stride). The kilogram (mass) and meter (length) are fundamental units in the SI system, and the newton (force) is a derived unit. The formal definition of a newton is as follows: That force which gives to a mass of 1 kg an acceleration of 1 m/s/s. From Newton's second law as expressed in Equation 1.5, 9.81 N would be required to accelerate 1 kg at the constant acceleration rate of $g = 9.81\ \text{m/s}^2$, that is,

$$9.81\ \text{N} = 1\ \text{kg} \times 9.81\ \text{m/s}^2 = 9.81\ \text{kg m/s}^2.$$

Hence, the newton has derived dimensions of kg m/s^2.

From Equation 1.5, changing the length unit to the millimeter (mm) while retaining the kg as the (fundamental) mass unit gives $w = mg = m\ (\text{kg}) \times 9810\ \text{mm/s}^2$, which would imply a 1000-fold increase in the weight force; however, 1 N is still required to accelerate 1 kg at $g = 9.81\ \text{m/s}^2 = 9810\ \text{mm/s}^2$, and

$$m\ (\text{kg}) \times 9810\ \text{mm/s}^2 = w(10^{-3}\ \text{N}) = w\ (\text{mN}).$$

Hence, for a kg-mm-s system of units, the derived force unit is $10^{-3}\ \text{N} = 1\ \text{mN}$ (1 milli newton).

Another view of units and dimensions is provided by the natural-frequency (see Section 3.2.2) definition $\omega_n = \sqrt{K/M}$ of a mass M supported by a linear spring with spring coefficient K. Perturbing the mass from its equilibrium position causes harmonic motion at the frequency ω_n, and ω_n's dimension is rad/s, or s^{-1} (since the radian is dimensionless). Using the kg-m-s system for length, mass, and time, the dimensions for ω_n^2 follow from

$$\omega_n^2 = \frac{K\ (\text{N/m})}{M\ (\text{kg})} = K\left(\frac{\text{kg m}}{\text{s}^2} \times \frac{1}{\text{m}}\right) \times \frac{1}{M\ (\text{kg})} = \frac{K}{M}\ (\text{s}^{-2}) \tag{1.7}$$

confirming the expected dimensions. Note that we have used the derived units of kg m/s^2 for N in this development.

Shifting to mm for the length unit while continuing to use the newton as the force unit would change the dimensions of the stiffness coefficient K to N/mm and reduce K by a factor of 1000. Specifically, the force required to deflect the spring 1 mm should be smaller by a factor of 1000 than the force required to displace the same spring 1 m = 1000 mm. However, substituting K with dimensions of N/mm into Equation 1.5, while retaining M in kg would cause a decrease in the natural frequency by a factor of $\sqrt{1000}$. Obviously, changing the units should not change the natural frequency; hence, this proposed dimensional set is wrong. The correct answer follows by using mN as the derived unit for force. This choice gives the dimensions of mN/mm (*millinewton per millimeter*) for K, and leaves both K and ω_n unchanged. To confirm that K is unchanged (numerically) by this choice of units, suppose $K = 1000\ \text{N/m} = 1\ \text{N/mm} = 1000\ \text{mN/mm}$. The reaction force magnitude produced by the deflection $\delta = 1\ \text{mm} = 10^{-3}\ \text{m}$ is

$$\begin{aligned} f_s = K\delta &= 1000\ \frac{\text{N}}{\text{m}} \times 10^{-3}\ \text{m} = 1\ \text{N} \\ &= 1\ \frac{\text{N}}{\text{mm}} \times 1\ \text{mm} = 1\ \text{N} \\ &= 1000\ \frac{\text{mN}}{\text{mm}} \times 1\ \text{mm} = 10^3\ \text{mN} = 1\ \text{N} \end{aligned}$$

confirming that mN is the appropriate derived force unit for a kg-mm-s unit system. Table 1.2 summarizes the results of this discussion.

TABLE 1.2

SI Base and Derived Units

Base Mass Unit	Base Time Unit	Base Length Unit	Derived Force Unit	Derived Force Unit Dimensions
Kilogram (kg)	Second (s)	Meter (m)	Newton (N)	kg m/s^2
Kilogram (kg)	Second (s)	Millimeter (mm)	Millinewton (mN)	kg mm/s^2

Shifting to the U.S. customary unit system, the unit of pounds is the fundamental unit for force, and the mass dimension is to be derived. Applying Equation 1.5 to this situation gives

$$w = mg \Rightarrow m = \frac{w\ (\text{lb})}{32.2\ \text{ft/s}^2} = \frac{w}{32.2}\ \frac{\text{lb s}^2}{\text{ft}}.$$

The derived mass unit has dimensions of lb s^2/ft and is called a "slug." When acted on by a resultant force of 1 lb, a mass of one slug will accelerate at 32.2 ft/s^2. Alternatively, under standard conditions, a mass of one slug weighs 32.2 lbs. The U.S. customary unit system is also referred to as the foot-pound-second system.

If the inch-pound-second unit system is used for displacement, force, and time, respectively, Equation 1.5 gives

$$w = mg \Rightarrow m = \frac{w\ (\text{lb})}{386\ \text{in./s}^2} = \frac{w}{386}\ \frac{\text{lb s}^2}{\text{in.}},$$

and the mass has derived dimensions of lb s^2/in. Within the author's 1960s aerospace employer, a mass weighing one pound with the derived units of lb s^2/in. was called a "snail." To the author's knowledge, there is no commonly accepted name for this mass unit, so *snail* will be used in this discussion. When acted upon by a resultant 1 lb force, a mass of 1 snail will accelerate at 386 in./s^2, and under standard conditions, a snail weighs 386 lbs.

Returning to the natural-frequency discussion, from Equation 1.5 for a pound-ft-s system,

$$\omega_n^2 = \frac{K\ (\text{lb/ft})}{M\ (\text{lb s}^2/\text{ft})} = \frac{K}{M}\ (\text{s}^{-2}).$$

Switching to the inch-pound-second unit system gives

$$\omega_n^2 = \frac{K\ (\text{lb/in.})}{M\ (\text{lb s}^2/\text{in.})} = \frac{K}{M}\ (\text{s}^{-2}).$$

Table 1.3 summarizes the results of this discussion.

Many U.S. students and engineers use the pound mass (lbm) unit in thermodynamics, fluid mechanics, and heat transfer. A one pound mass weighs one pound under standard conditions. However, the lbm unit is not attractive in dynamics. If the pound mass is the fundamental *mass* unit, then a new force unit must developed. Inserting $m = 1$ lbm into $w = mg$ with $g = 32.2$ ft/s^2 will require a force unit equal to 32.2 lbs, called the *poundal*. This force unit is recognized, named, perfectly acceptable, and must be used as the derived force unit corresponding to the pound mass (as a fundamental mass unit). *Stated briefly, unless you are prepared to use the poundal force unit, the pound-mass unit should not be used in dynamics.*

TABLE 1.3

U.S. Customary Base and Derived Units

Base Force Unit	Base Time Unit	Base Length Unit	Derived Mass Unit	Derived Mass Unit Dimensions
Pound (lb)	Second (s)	Foot (ft)	Slug	lb s^2/ft
Pound (lb)	Second (s)	Inch (in.)	Snail	lb s^2/in.

From one viewpoint, applications of the U.S. system make more sense than the SI system in that a U.S. system scale states your weight (the force of gravity) in pounds, a force unit. A scale in SI units reports your weight in kilos (kilograms), the SI mass unit, rather than in newtons, the SI force unit. Another example of less than rational application of the SI system concerns pressure. The standard SI pressure unit is the Pascal (N/m^2); however, erroneous citations of pressures in units of kg/m^2 continue to appear.

Useful (and exact) conversion factors between the SI and U.S. systems are the following: 1 lbm = 0.45359237*kg, 1 in. = 0.0254*m, 1 ft = 0.3048*m, 1 lb = 4.4482216152605*N. The * in these definitions denotes internationally agreed-upon *exact* conversion factors. Inverting the lbm definition, this gives 1 kg = 2.2046226*lbm. Increasingly, references are being made to the "metric ton" as a measure of weight. A metric ton is actually 1000 kg or approximately 2200 lbm. It should be used as a mass unit instead of a weight unit, but its weight is fairly close to the U.S. customary force unit, 1 ton = 2000 lbf.

Conversions between SI and U.S. customary unit systems should be checked carefully. An article in the October 4, 1999, issue of *Aviation Week and Space Technology* states, "Engineers have discovered that use of English instead of metric units in a navigation software table contributed to, if not caused, the loss of Mars Climate Orbiter during orbit injection on Sept. 23." This press report covers a highly visible and public failure; however, less spectacular mistakes are regularly made in unit conversions. Actually, conversions need to be checked carefully in working in a single unit system. Recently, one of the author's research projects was delayed when the result of a calculation for the natural frequency of a test rotor supported by a hydrostatic bearing produced unreasonably low natural frequencies. The problem was traced to the mass output predictions from a separate computer model for the hydrostatic bearing. Working in the inch-pound-second system, the research assistant (reasonably) thought that the output units were in snails (lb s^2/in), while the actual output unit was lbm. Hence, the erroneous input for the bearing's added mass contributions was high by a factor of 386.

2

Planar Kinematics of Particles

2.1 Introduction

Dynamics generally involves motion of particles and rigid bodies and has been traditionally and usefully divided into "kinematics" and "kinetics." Kinematics is geometric in nature, considering motion without regard to the forces that either cause or result from motion. Kinetics *uses kinematics as a foundation* for studying the motion of particles or rigid bodies due to prescribed forces, or, conversely, examines reaction forces due to prescribed motion. By itself, kinematics is extremely useful in solving and analyzing engineering problems, and the mastery of kinematics is absolutely essential for the subsequent study of kinetics.

The notion of a particle was covered in Chapter 1 and basically implies an idealized situation where the physical dimensions of a body (length, height, width, etc.) can be neglected, with all of the body's mass concentrated at a single point. In conventional terms, one can think of a physically small body such as the head of a pin or a "beebee." However, the size of a body is, in most cases, a matter of perspective. For example, an aircraft at high altitude appears to be a dot in the sky. On a solar-system scale, the earth can be considered a particle. Kinematically, the nice thing about a particle is that we only need to keep track of its coordinates with respect to a reference system; for example, in a Cartesian-coordinate system, the coordinates (x, y, z) would tell us the position of a particle. Cartesian coordinates are named after the French philosopher, René Descartes (1596–1650), who invented analytical geometry. Descartes is known for the (philosophical) statement, "I think; therefore, I am."

An appreciation of the relative ease of dealing with particles, versus rigid bodies, can be had by thinking of the more complicated problem of defining the position *and orientation* of a rigid body. For example, an aircraft can be changing position with respect to ground while pitching, rolling, and yawing. Keeping track of the aircraft's position plus defining and keeping track of its orientation would obviously be a more complicated situation.

Thinking ahead, we will return to planar kinematics of a rigid body in Chapter 4. Three-dimensional kinematics of a rigid body is beyond the scope of this chapter, but is covered in books dealing with more advanced dynamics. A mastery of this chapter's contents is required throughout the book and will be used immediately in Chapter 3 on particle kinetics.

2.2 Motion in a Straight Line

Figure 2.1 illustrates a particle P moving along a straight line. The distance from O to the point P on the line varies with time and is denoted by $s(t)$. The positive direction for $s(t)$ is defined by the unit vector $\boldsymbol{e}$. The position *vector* from O to P is $\boldsymbol{r} = \boldsymbol{e}s$. A negative value for $s(t)$ indicates that $\boldsymbol{r}$ is pointed oppositely from $\boldsymbol{e}$. The velocity and acceleration of point P with respect to point O in the $\boldsymbol{e}$-direction are defined, respectively, by $\boldsymbol{v} = \boldsymbol{e}v$, and $\boldsymbol{a} = \boldsymbol{e}a$, where

$$v = \frac{ds}{dt} = \dot{s}, \quad a = \frac{d^2s}{dt^2} = \dot{v} = \ddot{s}. \tag{2.1}$$

Note that $s(t)$, $v(t)$, and $a(t)$ are *components*, respectively, of the *vectors*, $\boldsymbol{r} = \boldsymbol{e}s$, $\boldsymbol{v} = \boldsymbol{e}v$, and $\boldsymbol{a} = \boldsymbol{e}a$ and are scalar functions of time as with a temperature reading on a thermometer.

Equations 2.1 suggest that s is generally given as a function of time with v and a obtained by differentiation. However, it is also possible to start with $a(t)$ and integrate with given initial conditions; for example,

$$v(t) = v(t_0) + \int_{t_0}^{t} a(\tau)d\tau \tag{2.2a}$$

yields $v(t)$, and a second integration with respect to time gives

$$s(t) = s(t_0) + \int_{t_0}^{t} v(\tau)d\tau. \tag{2.2b}$$

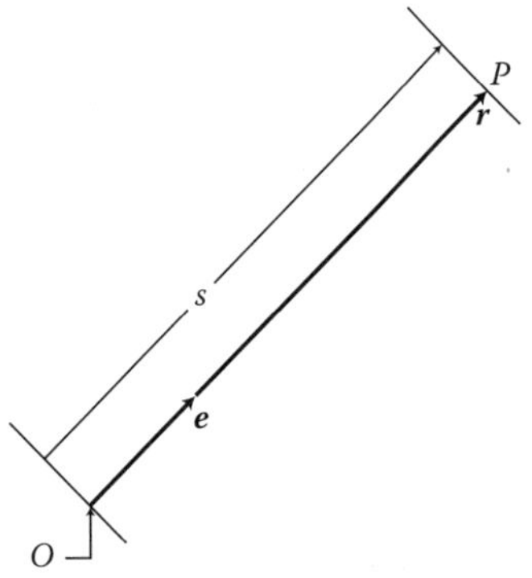

FIGURE 2.1
Position vector $\boldsymbol{r}$ along the unit vector $\boldsymbol{e}$.

Alternatively, one can start with a prescribed $v(t)$ and integrate to get

$$s(t) = s(t_0) + \int_{t_0}^{t} v(\tau)d\tau, \tag{2.3}$$

and then differentiate $v(t)$ to get

$$a(t) = \frac{dv}{dt} = \ddot{s}. \tag{2.4}$$

Figure 2.2 illustrates hypothetical motion of a particle moving along a straight line with s, v, and a given as functions of time. For any value of time t, v equals the slope of the $s(t)$ curve, ds/dt, while the slope of the $v(t)$ curve, dv/dt, equals $a(t)$, the acceleration. Conversely, s at t equals $s_0 = s(t_0)$ plus the area under the v–t curve from t_0 to t. Similarly, v at t is $v_0 = v(t_0)$ plus the area under the a–t curve from t_0 to t.

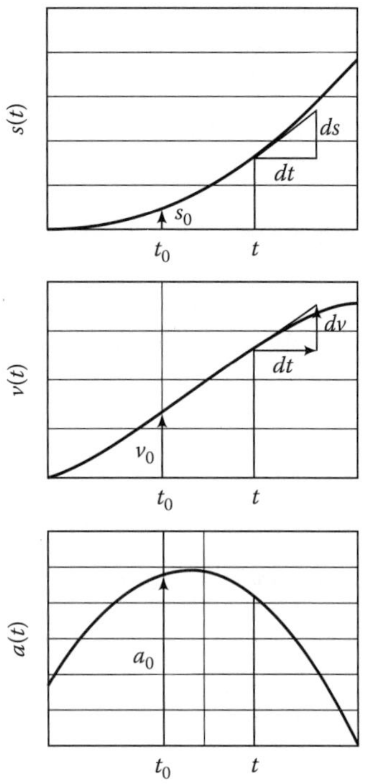

FIGURE 2.2
Position, velocity, and acceleration versus time.

A large class of *kinetics* problems arise from Newton's second law of motion for a particle moving along a straight line,

$$f = m\ddot{s} = ma, \tag{2.5}$$

where

m is the particle's mass
f is the force acting on the body in the $+s$ direction

If f is a function of time, the acceleration a is defined from

$$ma = ma(t) = f(t). \tag{2.6}$$

Integration with respect to time yields

$$mv(t) - mv(t_0) = \int_{t_0}^{t} f(\tau)d\tau \tag{2.7}$$

You may recall from physics that this equation is the linear, impulse–momentum integral of the equation of motion. On the left-hand side, we have the change in linear momentum, while on the right we have linear impulse. A second integration of Equation 2.7 yields $s(t)$. Most introductory kinematic problems, where the acceleration is given as a function of time, with integration required to find velocity and displacement, follow from Newton's second law of motion.

Example Problem 2.1

The acceleration of a particle is defined over a time span by

$$a = 10t - 3t^2 \text{ m/s}^2.$$

Task: Given that $v(0) = 5$ m/s, $s(0) = 10$ m, derive general expressions for $v(t)$ and $s(t)$. What is the maximum velocity?

Solution

The first part of this problem is solved by direct integration, following Equations 2.2. First,

$$v(t) = v(0) + \int_{0}^{t} (10\tau - 3\tau^2)d\tau \text{ m/s}$$

$$= 5 + \left. 5\tau^2 - \tau^3 \right|_{0}^{t} = 5 + 5t^2 - t^3 \text{ m/s}.$$

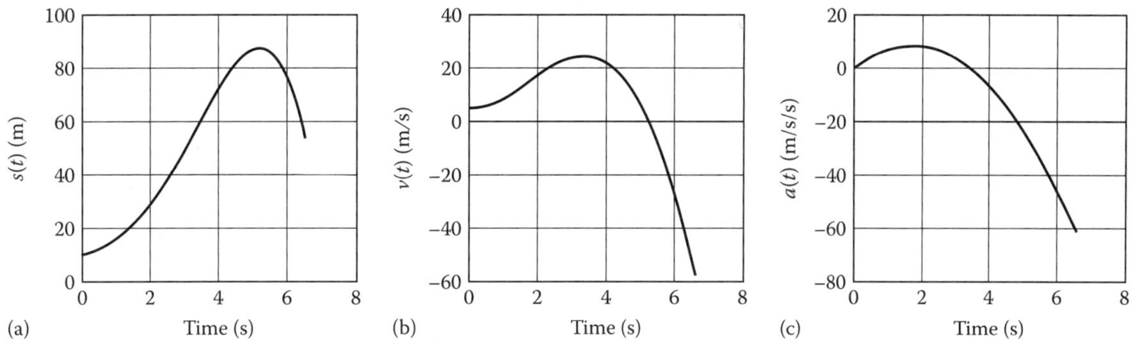

FIGURE XP2.1
(a) $s(t)$, (b) $v(t)$, (c) $a(t)$.

Next,

$$s(t) = s(0) + \int_0^t (5 + 5\tau^2 - \tau^3)d\tau \text{ m}$$

$$= 10 + \left| 5\tau + 5\frac{\tau^3}{3} - \frac{\tau^4}{4} \right|_0^t = 10 + 5t + 5\frac{t^3}{3} - \frac{t^4}{4} \text{ m.}$$

Plots of the solution are provided in Figure XP2.1. To find maximum values for $v(t)$ with respect to time, we set

$$\frac{dv}{dt} = a(t) = 10t - 3t^2 = 0,$$

and solve to obtain the following values for time: $t_1 = 0$, $t_2 = 10/3 = 3.33$ s.

The peak velocity is obtained for $t_2 = 3.33$ s;

$$v_{max} = 5 + 53.33^2 - 3.33^3 = 23.5 \text{ m/s.}$$

Example Problem 2.2

The motion of a particle is defined by

$$v(t) = 2\omega \cos \omega t \text{ m/s,}$$

where

$$\omega = 30 \text{ rad/s}, \quad s(0) = 5 \text{ m.}$$

Task: Find the acceleration and displacement functions.

Solution

The displacement function is found by integration; that is,

$$s(t) = s(0) + \int_0^t 2\omega \cos \omega\tau \, dt$$

$$= 5 + \left| 2 \sin \omega\tau \right|_0^t = 5 + 2 \sin \omega t$$

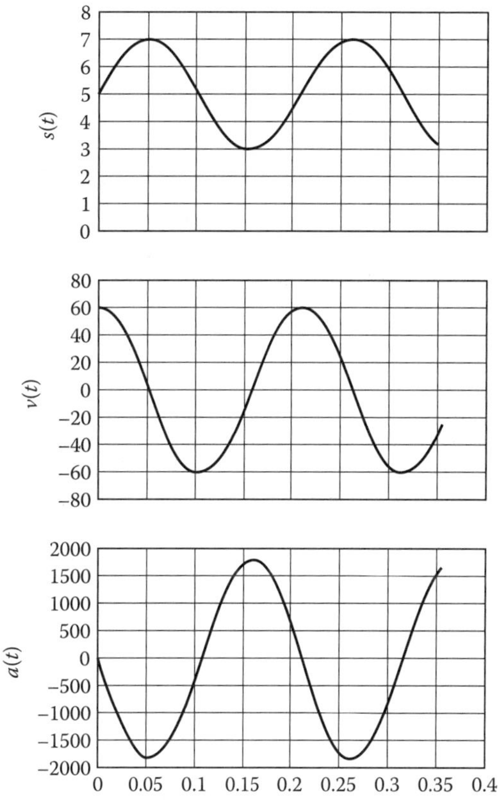

FIGURE XP2.2
Position, velocity, and acceleration versus time.

The acceleration function is found by differentiation, yielding

$$a(t) = \frac{dv}{dt} = -2\omega^2 \sin \omega t.$$

The functions $s(t)$, $v(t)$, and $a(t)$ are illustrated in Figure XP2.2.

2.2.1 Distance Traveled

For the two example problems provided above, the displacement function $s(t)$ answers the following question: Where is point P? Another question that could arise is as follows: How far has point P traveled to reach its present position? In Example Problem 2.1, suppose we were asked to answer the following specific question: How far has point P traveled from $t=0$ to $t=8$ s? Looking at the $s(t)$ function in Figure XP2.1a, we observe that a maximum is reached at about $t=5.19$ s, after which $s(t)$ becomes progressively smaller. The peak value for $s(t)$ coincides with the time t^* for which $v(t^*)=0$, changing sign from positive to negative (and defining the peak location for $s(t)$).

The distance traveled from time t_0 to t can formally be defined as

$$d = \int_{t_0}^{t} |v(\tau)| d\tau.$$

Incorporating the absolute velocity in the definition causes d to increase without regard to changes in the sign of v.

For Example Problem 2.1, the distance traveled from $t=0$–8 s is

$$\begin{aligned} d(0 \to 8) &= \int_0^{5.19} v(\tau)\, d\tau + \int_{5.19}^{8} -v(\tau)\, d\tau \text{ m} \\ &= \left| \left(5\tau + \frac{5\tau^3}{3} - \frac{\tau^4}{4}\right) \right|_0^{5.19} - \left| \left(5\tau + \frac{5\tau^3}{3} - \frac{\tau^4}{4}\right) \right|_{5.19}^{8} \text{ m} \\ &= 77.6 - (-130.7 - 77.6) = 285.9 \text{ m}. \end{aligned}$$

By contrast, the position at $t=8$ s is $s(8)=10-130.7=-120.7$ m.

2.3 Particle Motion in a Plane: Cartesian Coordinates

Figure 2.3 illustrates the position of a particle P in a Cartesian X, Y coordinate system.

The position vector locating P is

$$\boldsymbol{r} = \boldsymbol{I}r_X + \boldsymbol{J}r_Y, \tag{2.8}$$

with $\boldsymbol{I}$ and $\boldsymbol{J}$ being vectors of unit magnitudes pointed along the orthogonal X- and Y-axes. r_X and r_Y are *components* of the *vector* $\boldsymbol{r}$ in the X, Y coordinate system.

The velocity of point P with respect to the X, Y coordinate system is defined by

$$\boldsymbol{v} = \frac{d\boldsymbol{r}}{dt}\bigg|_{X,Y} = \dot{\boldsymbol{r}} = \boldsymbol{I}\dot{r}_X + \boldsymbol{J}\dot{r}_Y = \boldsymbol{I}v_X + \boldsymbol{J}v_Y. \tag{2.9}$$

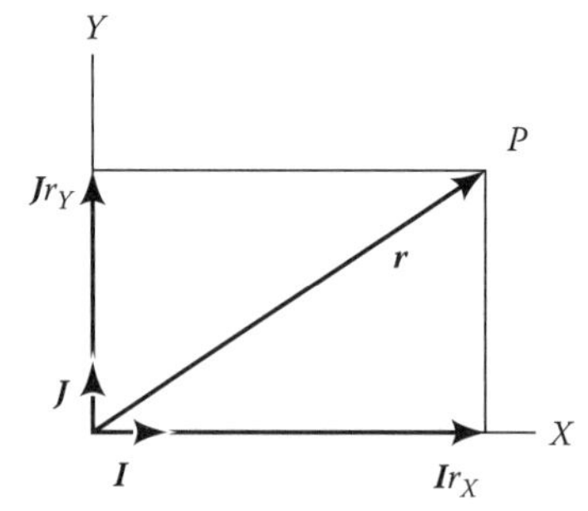

FIGURE 2.3
Particle P located in the X, Y system by the vector $\boldsymbol{r}$.

The magnitude of the velocity vector is

$$|v| = \left(v_X^2 + v_Y^2\right)^{1/2}. \tag{2.10}$$

Since there is only one coordinate system in Figure 2.3, the definition provided by Equation 2.9 would appear to be obvious; that is, the time derivative can only be performed with respect to the X, Y system. However, in kinematics one frequently has more than one coordinate system, and students regularly find themselves asking the following question: What do you mean, "Find a derivative of vector $\boldsymbol{B}$ with respect to a coordinate system?" A formal (and infallible) means to answer this question is in the following way: The time derivative of any vector $\boldsymbol{B}$ with respect to a specified coordinate system is found by writing $\boldsymbol{B}$ out in terms of its components in the specified coordinate system,

$$\boldsymbol{B} = \boldsymbol{I}B_X + \boldsymbol{J}B_Y,$$

and then differentiating with respect to time while holding the unit vectors constant.

$$\dot{\boldsymbol{B}} = \frac{d\boldsymbol{B}}{dt}\bigg|_{X,Y} = \boldsymbol{I}\dot{B}_X + \boldsymbol{J}\dot{B}_Y.$$

These are exactly the steps that are performed for $\boldsymbol{r}$ in Equations 2.8 and 2.9.

The acceleration of point P with respect to the X, Y system is defined by

$$\boldsymbol{a} = \frac{d\boldsymbol{v}}{dt}\bigg|_{X,Y} = \dot{\boldsymbol{v}} = \ddot{\boldsymbol{r}} = \boldsymbol{I}\dot{v}_X + \boldsymbol{J}\dot{v}_Y = \boldsymbol{I}a_X + \boldsymbol{J}a_Y. \tag{2.11}$$

The magnitude of the acceleration vector is

$$|a| = \left(a_X^2 + a_Y^2\right)^{1/2}.$$

Example Problem 2.3

Figure XP2.3 illustrates a particle that is moving in a circular path in the X, Y coordinate system. The center of the circle is located at coordinate $(a, 0)$, and the components of vector $\boldsymbol{r}$ are

$$r_X = a + b\cos\omega t \text{ m}, \quad r_Y = b\sin\omega t \text{ m}.$$

Tasks: Find the components of the velocity $\boldsymbol{v}$ and acceleration $\boldsymbol{a}$ of point P with respect to the X, Y system. Also, find their magnitudes

Solution

Following Equations 2.9 and 2.11, the components of $\boldsymbol{v}$ and $\boldsymbol{a}$ in the X, Y coordinate system are

$$v_X = \dot{r}_X = -b\omega \sin t\omega t \text{ m/s}, \quad v_Y = \dot{r}_Y = b\omega\cos\omega t \text{ m/s},$$

and

$$a_X = \dot{v}_X = \ddot{r}_X = -b\omega^2\cos\omega t \text{ m/s}^2$$
$$a_Y = \dot{v}_Y = \ddot{r}_Y = -b\omega^2\sin\omega t \text{ m/s}^2.$$

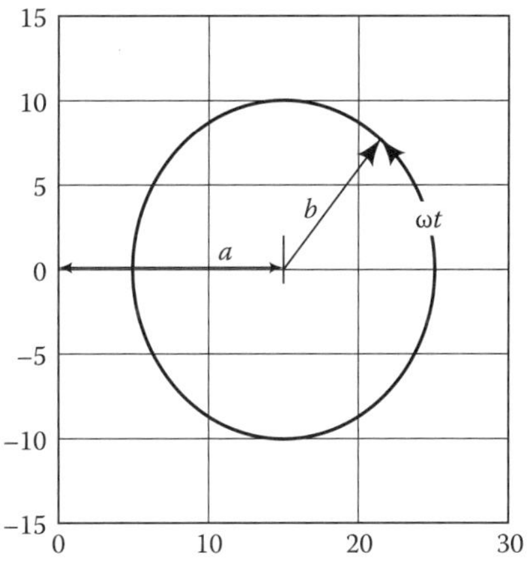

FIGURE XP2.3
Circular motion with $a = 15$ m, and $b = 10$ m.

Further, the velocity and acceleration magnitudes are

$$|v| = \left(v_X^2 + v_Y^2\right)^{1/2} = b\omega \text{ m/s}$$
$$|a| = \left(a_X^2 + a_Y^2\right)^{1/2} = b\omega^2 \text{ m/s}^2.$$

Example Problem 2.4

From Newton's second law of motion, the acceleration components of a particle that is falling under the influence of gravity (neglecting aerodynamic drag forces) are

$$a_X = \ddot{r}_X = 0, \quad a_Y = \ddot{r}_Y = -g. \qquad \text{(i)}$$

This definition of the acceleration components has the particle accelerating straight down in the $-Y$ direction.

Tasks: For the initial conditions illustrated in Figure XP2.4a, namely, $v_X(0) = v_0\sin\alpha$, $v_Y(0) = v_0\cos\alpha$ and $r_X(0) = r_Y(0) = 0$, find the components of the position and velocity vectors as a function of time.

Solution

Integrating Equations (i) once with respect to time yields the velocity components:

$$v_X = \dot{r}_X = v_X(0) = v_0\sin\alpha$$
$$v_Y = \dot{r}_Y = v_Y(0) - gt = v_0\cos\alpha - gt.$$

A second integration gives the displacement components:

$$r_X = r_Y(0) + tv_0\sin\alpha = tv_0\sin\alpha$$
$$r_Y = r_Y(0) + tv_0\cos\alpha - \frac{gt^2}{2} = tv_0\cos\alpha - \frac{gt^2}{2}.$$

For $v_0 = 100$ m/s, $\alpha = 30°$, and $g = 9.81$ m/s^2, Figure XP2.4b and c illustrate $r_X(t)$, $r_Y(t)$, and the trajectory r_Y versus r_X, respectively. The solution illustrated extends over a time span that results in negative r_Y values that might cause the particle to plow into a horizontal earth; however, the solution presented would be reasonable if the particle were launched over an inclined plane.

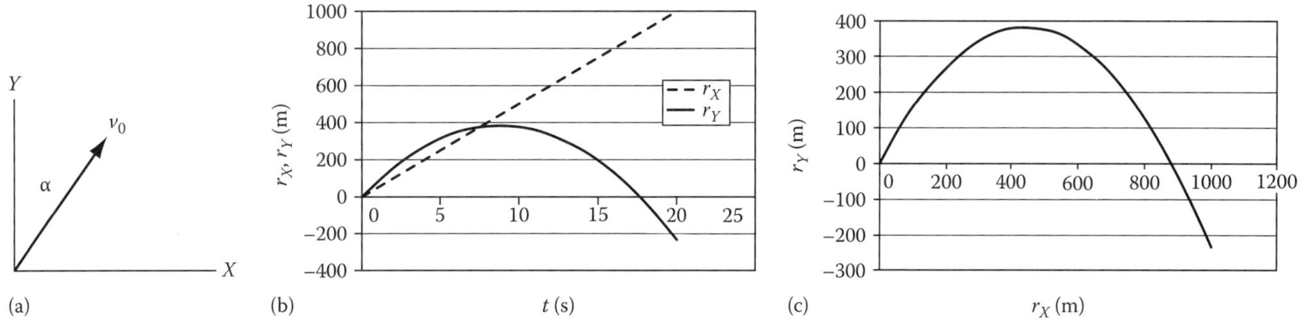

FIGURE XP2.4
(a) Particle with initial velocity v_0 launched at an angle α with respect to the vertical. (b) Displacement components $r_X(t)$, $r_Y(t)$. (c) Trajectory, r_Y versus r_X.

2.4 Coordinate Transformations: Relationships between Components of a Vector in Two Coordinate Systems

Figure 2.4a illustrates a vector $\boldsymbol{B}$ in the "original" X, Y coordinate system and a "new" X', Y' coordinate system. The orientation of the X', Y' system is defined with respect to the X, Y system by the *constant* angle α. Suppose that at a given time $\bar{t}$, the components of $\boldsymbol{B}$ are defined in the X, Y system by B_X, B_Y and "new" components are needed in the X', Y' system; that is, given B_X, B_Y and α, what are $B_{X'}$ and $B_{Y'}$?

Figure 2.4b illustrates an approach to the solution of this problem. Specifically, $\boldsymbol{I}B_X$ and $\boldsymbol{J}B_Y$ are projected (individually) into the $X' - Y'$ coordinate system, yielding

$$\begin{aligned} \boldsymbol{I}B_X &= \boldsymbol{I}'B_X\cos\alpha - \boldsymbol{J}'B_X\sin\alpha \\ \boldsymbol{J}B_Y &= \boldsymbol{I}'B_Y\sin\alpha + \boldsymbol{J}'B_Y\cos\alpha. \end{aligned} \tag{2.12}$$

Substituting these results into $\boldsymbol{B} = \boldsymbol{I}B_X + \boldsymbol{J}B_Y$ gives

$$\begin{aligned} \boldsymbol{B} &= \boldsymbol{I}'(B_X\cos\alpha + B_Y\sin\alpha) + \boldsymbol{J}'(-B_X\sin\alpha + B_Y\cos\alpha) \\ &= \boldsymbol{I}'B_{X'} + \boldsymbol{J}'B_{Y'}. \end{aligned} \tag{2.13}$$

Equating coefficients of $\boldsymbol{I}'$, and $\boldsymbol{J}'$ in Equations 2.13 yields

$$\begin{aligned} B_{X'} &= B_X\cos\alpha + B_Y\sin\alpha \\ B_{Y'} &= -B_X\sin\alpha + B_Y\cos\alpha. \end{aligned} \tag{2.14}$$

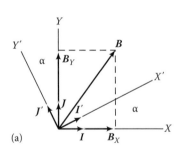

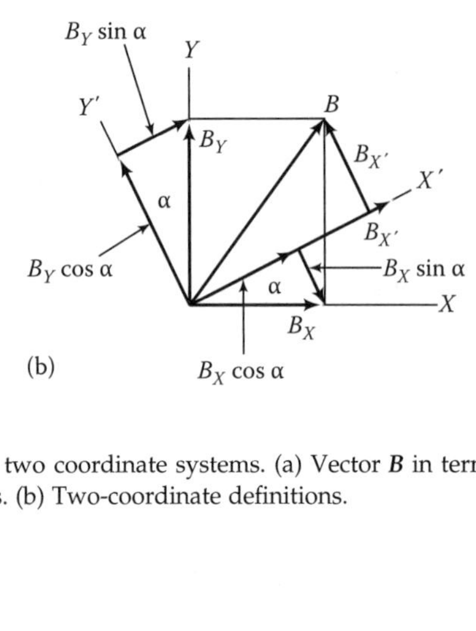

FIGURE 2.4
Vector $\boldsymbol{B}$ in two coordinate systems. (a) Vector $\boldsymbol{B}$ in terms of its X, Y components. (b) Two-coordinate definitions.

In matrix format, this equation becomes

$$\begin{Bmatrix} B_{X'} \\ B_{Y'} \end{Bmatrix} = \begin{bmatrix} \cos\alpha & \sin\alpha \\ -\sin\alpha & \cos\alpha \end{bmatrix} \begin{Bmatrix} B_X \\ B_Y \end{Bmatrix} \tag{2.15}$$

or, symbolically,

$$(B)_{I'} = [A](B)_I. \tag{2.16}$$

(Appendix A provides a refresher in matrix algebra.) $[A]$ is called the "direction-cosine matrix" and can be formally defined as

$$[A] = \begin{bmatrix} \cos(X', X)\cos(X', Y) \\ \cos(Y', X)\cos(Y', Y) \end{bmatrix}, \tag{2.17}$$

where
$\cos(X', X)$ is the cosine of the angle between X and X'
$\cos(X', Y)$ is the cosine of the angle between and X' and Y, etc.

Returning to Figure 2.4b,

$$\begin{aligned} &\cos(X', X) = \cos(Y', X) = \cos\alpha \\ &\cos(X', Y) = \cos\left(\frac{\pi}{2} - \alpha\right) = \cos\frac{\pi}{2}\cos\alpha + \sin\frac{\pi}{2}\sin\alpha = \sin\alpha \\ &\cos(Y', X) = \cos\left(\frac{\pi}{2} + \alpha\right) = \cos\frac{\pi}{2}\cos\alpha - \sin\frac{\pi}{2}\sin\alpha = -\sin\alpha. \end{aligned} \tag{2.18}$$

Substituting from Equations 2.18 into 2.17 gives

$$[A] = \begin{bmatrix} \cos\alpha & \sin\alpha \\ -\sin\alpha & \cos\alpha \end{bmatrix},$$

which is the same result provided earlier by Equation 2.15. With a little practice, the sketch done in Figure 2.4b can generally be used to develop the results of Equation 2.15 more rapidly than the formal definition of Equation 2.17.

Note that $[A]$ is orthogonal; that is, its inverse $[A]^{-1}$ is its transverse $[A]^T$. To confirm that $[A]^{-1} = [A]^T$,

$$\begin{aligned} [A]^T[A] &= \begin{bmatrix} \cos\alpha & -\sin\alpha \\ \sin\alpha & \cos\alpha \end{bmatrix} \begin{bmatrix} \cos\alpha & \sin\alpha \\ -\sin\alpha & \cos\alpha \end{bmatrix} \\ &= \begin{bmatrix} \cos^2\alpha + \sin^2\alpha & \cos\alpha\sin\alpha - \sin\alpha\cos\alpha \\ \cos\alpha\ \sin - \sin\alpha\cos\alpha & \sin^2\alpha + \cos^2\alpha \end{bmatrix} \\ &= \begin{bmatrix} 1 & 0 \\ 0 & 1 \end{bmatrix}. \end{aligned}$$

Hence, Equation 2.16 can be easily used to solve the inverse problem: Given $B_{X'}$ and $B_{Y'}$ what are B_X and B_Y? Premultiplying Equation 2.15 by $[A]^T$ gives

$$[A]^T(B)_{I'} = [A]^T[A](B)_I = (B)_I.$$

Hence,

$$(B)_I = [A]^T(B)_{I'},$$

or in expanded format

$$\begin{Bmatrix} B_X \\ B_Y \end{Bmatrix} = \begin{bmatrix} \cos\alpha & -\sin\alpha \\ \sin\alpha & \cos\alpha \end{bmatrix} \begin{Bmatrix} B_{X'} \\ B_{Y'} \end{Bmatrix}. \tag{2.19}$$

NOTE: We could have simply redrawn Figure 2.4, starting with the component definition $\boldsymbol{B}=\boldsymbol{I}'B_{X'}+\boldsymbol{J}'B_{Y'}$, found the components of $\boldsymbol{B}_{X'}$, $\boldsymbol{B}_{Y'}$ along the X- and Y-axes, and summed the result to find the X and Y coordinates of $\boldsymbol{B}$. In fact, later in this chapter, we will carry out exactly those "reverse" steps.

The importance of coordinate transformations in vector kinematics cannot be overstated. One regularly has results in terms of the components of a vector in one coordinate system and needs to express the answer in terms of another coordinate system. You have probably been told (correctly) that physical results are best stated in a "vector notation," since a vector equation is valid, *independent of coordinate systems*. However, most real problems must be worked in terms of components stated in specific coordinate systems, and Equations 2.15 and 2.19 allow you to shift the components of any vector back and forth between coordinate systems.

Example Problem 2.5

We return to the equations of motion for the free fall (no aerodynamic drag forces) of a particle. The acceleration components of a particle P in the X, Y coordinate system are given by

$$\ddot{r}_X = 0, \quad \ddot{r}_Y = -g,$$

where Y points vertically upward. As illustrated in Figure XP2.5a, the initial velocity vector is along the X'-axis that is oriented with respect to the horizontal X-axis by the angle α. Hence, the velocity initial conditions are

$$\dot{r}_X(0) = v_0\cos\alpha, \quad \dot{r}_Y(0) = v_0\sin\alpha$$
$$r_X(0) = r_Y(0) = 0.$$

Tasks: Find the velocity and position vectors of the particle P when the projectile crosses the X-axis, that is, when $r_Y=0$. Give your answer in terms of components in the X, Y and X', Y' coordinate systems.

Solution

Integrating the acceleration components to obtain the velocity-vector components is the first step in the solution, yielding

$$\begin{aligned} \dot{r}_X &= \dot{r}_X(0) = v_0\cos\alpha \\ \dot{r}_Y &= \dot{r}_Y(0) - gt = v_0\sin\alpha - gt. \end{aligned} \tag{i}$$

Integrating again to obtain the displacement-vector components gives

$$r_X = r_X(0) + tv_0\cos\alpha = tv_0\cos\alpha$$
$$r_Y = r_Y(0) + tv_0\sin\alpha - \frac{gt^2}{2} = tv_0\sin\alpha - \frac{gt^2}{2}.$$

To find the time at which the particle crosses the X-axis, we set

$$r_Y(\bar{t}) = 0 = \bar{t}\,v_0\sin\alpha - \frac{g\bar{t}^2}{2},$$

obtaining

$$\bar{t}_1 = 0, \quad \bar{t}_2 = \frac{2v_0\sin\alpha}{g}.$$

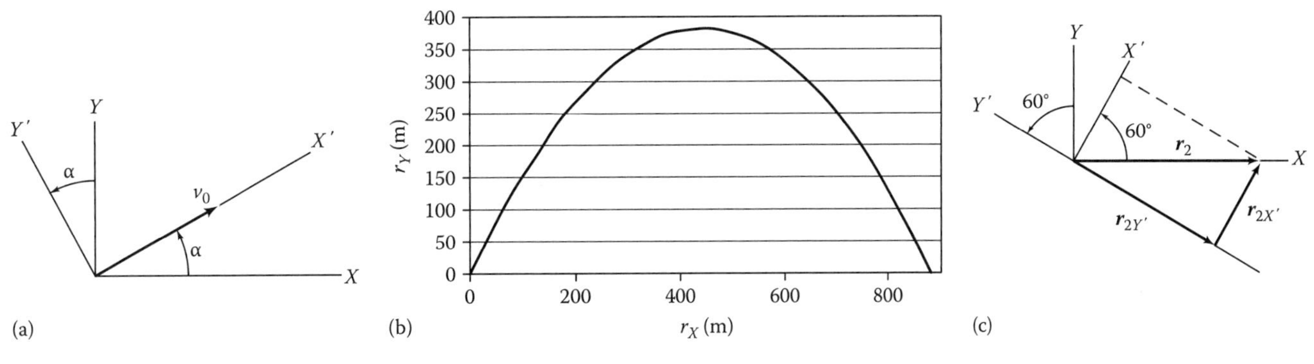

FIGURE XP2.5
(a) Initial velocity vector for a particle. (b) Trajectory solution, r_Y versus r_X. (c) Components of the final position vector of the particle $\boldsymbol{r}_2$ in the X, Y and X', Y' systems.

Substituting $\bar{t}_2$ into Equation (i), the velocity components in the X, Y system at t_2 are

$$\dot{r}_X(t_2) = v_0 \cos\alpha$$
$$\dot{r}_Y(t_2) = v_0 \sin\alpha - g \times \frac{2v_0 \sin\alpha}{g} = -v_0 \sin\alpha.$$

Figure XP2.5b illustrates the trajectory solution. An observer in the X, Y coordinate system would see the velocity $\boldsymbol{v}(\bar{t}_2)$ have the negative angle α when crossing the X-axis.

The velocity components in the $X'-Y'$ coordinate system at $\bar{t}_2$ can be obtained from Equation 2.15 as

$$\begin{Bmatrix} \dot{r}_{X'} & (\bar{t}_2) \\ \dot{r}_{Y'} & (\bar{t}_2) \end{Bmatrix} = \begin{bmatrix} \cos\alpha & \sin\alpha \\ -\sin\alpha & \cos\alpha \end{bmatrix} \begin{Bmatrix} \dot{r}_X & (\bar{t}_2) \\ \dot{r}_Y & (\bar{t}_2) \end{Bmatrix}$$
$$= \begin{bmatrix} \cos\alpha & \sin\alpha \\ -\sin\alpha & \cos\alpha \end{bmatrix} \begin{Bmatrix} v_0 & \cos\alpha \\ -v_0 & \sin\alpha \end{Bmatrix}$$
$$= v_0 \begin{Bmatrix} \cos^2\alpha & -\sin^2\alpha \\ -2\cos\alpha & \sin\alpha \end{Bmatrix} = v_0 \begin{Bmatrix} \cos 2\alpha \\ -\sin 2\alpha \end{Bmatrix}.$$

Hence, an observer in the X', Y' system would see the velocity $\boldsymbol{v}(\bar{t}_2)$ have an angle of -2α with respect to the X'-axis. In either case, the observed velocity magnitude is v_0. For $v_0 = 100$ m/s, and $\alpha = 60°$, the components are

$$\dot{r}_X(\bar{t}_2) = 50 \text{ m/s}, \quad \dot{r}_Y(\bar{t}_2) = -86.6 \text{ m/s}$$
$$\dot{r}_{X'}(\bar{t}_2) = -50 \text{ m/s}, \quad \dot{r}_{Y'}(\bar{t}_2) = -86.6 \text{ m/s}.$$

The components of $\boldsymbol{r}$ in the X, Y system at $\bar{t}_2$ are

$$r_X(\bar{t}_2) = v_0 \cos\alpha \cdot \frac{2v_0 \sin\alpha}{g} = \frac{v_0^2}{g} \sin 2\alpha$$
$$r_Y(\bar{t}_2) = 0.$$

Using the coordinate transformation, the components in the $X' - Y'$ system are

$$\begin{Bmatrix} r_{X'} & (\bar{t}_2) \\ r_{Y'} & (\bar{t}_2) \end{Bmatrix} = \begin{bmatrix} \cos\alpha & \sin\alpha \\ -\sin\alpha & \cos\alpha \end{bmatrix} \begin{Bmatrix} r_X & (\bar{t}_2) \\ r_Y & (\bar{t}_2) \end{Bmatrix}$$
$$= \begin{bmatrix} \cos\alpha & \sin\alpha \\ -\sin\alpha & \cos\alpha \end{bmatrix} \begin{Bmatrix} v_0^2/g & \sin 2\alpha \\ & 0 \end{Bmatrix}$$
$$= \frac{v_0^2}{g} \sin 2\alpha \begin{Bmatrix} \cos\alpha \\ -\sin\alpha \end{Bmatrix}.$$

For $v_0 = 100$ m/s, $\alpha = 60°$, and $g = 9.81$ m/s^2, $\boldsymbol{r}$'s components are

$$r_X(\bar{t}_2) = 883 \text{ m}, \quad r_Y(\bar{t}_2) = 0 \text{ m}$$
$$r_{X'}(\bar{t}_2) = 441 \text{ m}, \quad r_{Y'}(\bar{t}_2) = -764 \text{ m}.$$

Figure XP2.5c illustrates these two-component descriptions of the *same* vector $\boldsymbol{r}_2 = \boldsymbol{r}(\bar{t}_2)$ in the X, Y and $X' - Y'$ coordinate systems.

2.5 Particle Motion in a Plane: Polar Coordinates

Figure 2.5 illustrates a particle P located in the X, Y coordinate system by polar coordinates r, θ. You should be accustomed to locating a particle in a Cartesian-coordinate system using polar coordinates, versus the conventional coordinate pair X, Y. Many dynamics problems involve circular motion and are particularly well suited to polar coordinates. Developing useful expressions for velocity and acceleration using polar coordinates is our present task.

Returning to Figure 2.5a, note the two unit vectors: (1) $\boldsymbol{\varepsilon}_r$ shown parallel to r, and (2) $\boldsymbol{\varepsilon}_\theta$ drawn perpendicular to $\boldsymbol{\varepsilon}_r$. The unit vectors $\boldsymbol{\varepsilon}_r$, $\boldsymbol{\varepsilon}_\theta$ define a second coordinate system similar to the X'-, Y'-axes of the preceding section, but the situation is different in that θ, the angle between $\boldsymbol{I}$ and $\boldsymbol{\varepsilon}_r$ and between $\boldsymbol{J}$ and $\boldsymbol{\varepsilon}_\theta$, varies with time and is not constant.

In drawing Figure 2.5, a choice was made to have θ define counterclockwise rotation for $\boldsymbol{r}$. This is the conventional polar arrangement; however, a different convention defining positive clockwise rotations could be used. Note that the arc defining θ has a single arrow denoting a starting zero angle with the initial position for $\boldsymbol{\varepsilon}_r$ aligned with X. A positive value for θ indicates an $\boldsymbol{r}$ counterclockwise rotation; a negative value indicates a clockwise rotation direction. Figure 2.5b shows the $\boldsymbol{\varepsilon}_r$, $\boldsymbol{\varepsilon}_\theta$ unit vectors located at the end of the position vector $\boldsymbol{r}$ and emphasizes that the positive $\boldsymbol{\varepsilon}_\theta$-direction corresponds to the positive θ rotation direction.

Figure 2.5 introduces our first option in regard to differentiation of a vector with respect to two different coordinate systems; that is, we can differentiate $\boldsymbol{r}$ with respect to time in either the X, Y or $\boldsymbol{\varepsilon}_r$, $\boldsymbol{\varepsilon}_\theta$ systems. To illustrate this point, we write $\boldsymbol{r}$ in terms of the $\boldsymbol{\varepsilon}_r$, $\boldsymbol{\varepsilon}_\theta$ unit vectors as

$$\boldsymbol{r} = r\boldsymbol{\varepsilon}_r. \tag{2.20}$$

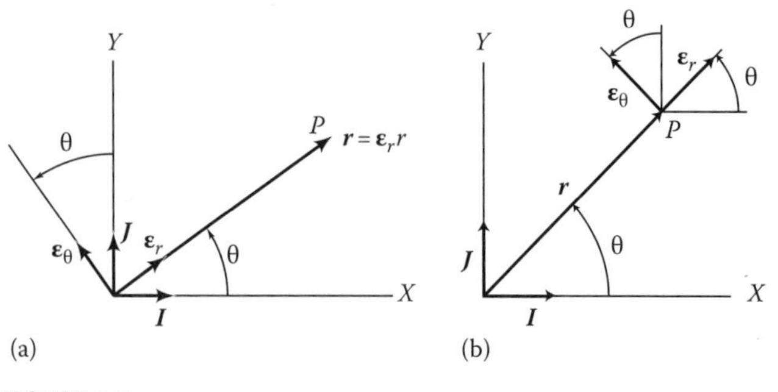

FIGURE 2.5
Polar-coordinate unit vectors.

The derivative of this vector with respect to time in the $\boldsymbol{\varepsilon}_r$, $\boldsymbol{\varepsilon}_\theta$ coordinate system is obtained by differentiating Equation 2.20 while holding $\boldsymbol{\varepsilon}_r$ constant, obtaining

$$\left.\frac{d\boldsymbol{r}}{dt}\right|\boldsymbol{\varepsilon}_r, \boldsymbol{\varepsilon}_\theta = \dot{r}\boldsymbol{\varepsilon}_r \tag{2.21}$$

Hence, an observer rotating with the $\boldsymbol{\varepsilon}_r$, $\boldsymbol{\varepsilon}_\theta$ coordinate system and looking at $\boldsymbol{r}$ would only see a change in $\boldsymbol{r}$'s magnitude, which is also its $\boldsymbol{\varepsilon}_r$ component.

Now consider the task of finding $\boldsymbol{r}$'s time derivative with respect to the X, Y system. Starting with Equation 2.20, differentiation with respect to time yields

$$\begin{aligned}\dot{\boldsymbol{r}} = \left.\frac{d\boldsymbol{r}}{dt}\right|X, Y &= \dot{r}\boldsymbol{\varepsilon}_r + r\left.\frac{d\boldsymbol{\varepsilon}_r}{dt}\right|X, Y \\ &= \dot{r}\boldsymbol{\varepsilon}_r + r\dot{\boldsymbol{\varepsilon}}_r.\end{aligned} \tag{2.22}$$

NOTE: This is exactly the same velocity definition provided by Equation 2.9 using Cartesian coordinates and the $\boldsymbol{I}$ and $\boldsymbol{J}$ unit vectors. Equation 2.22 is obtained by simply differentiating a product, but leaves open the following question: How can I obtain $\dot{\boldsymbol{\varepsilon}}_r$? Again, to obtain the time derivative of $\boldsymbol{\varepsilon}_r$ with respect to the X, Y system, we first state $\boldsymbol{\varepsilon}_r$ in terms of its components in the X, Y coordinate system:

$$\boldsymbol{\varepsilon}_r = \boldsymbol{I}\cos\theta + \boldsymbol{J}\sin\theta. \tag{2.23}$$

For future purposes we can also state $\boldsymbol{\varepsilon}_\theta$ in terms of its $\boldsymbol{I}$ and $\boldsymbol{J}$ components as

$$\boldsymbol{\varepsilon}_\theta = -\boldsymbol{I}\sin\theta + \boldsymbol{J}\cos\theta. \tag{2.24}$$

Differentiating $\boldsymbol{\varepsilon}_r$ in Equation 2.23 while holding $\boldsymbol{I}$ and $\boldsymbol{J}$ constant yields

$$\dot{\boldsymbol{\varepsilon}}_r = \left.\frac{d\boldsymbol{\varepsilon}_r}{dt}\right|X, Y = (-\boldsymbol{I}\sin\theta + \boldsymbol{J}\cos\theta)\dot{\theta}. \tag{2.25}$$

The correct units for $\dot{\theta}$ in Equation 2.25 are rad/s or s^{-1}. By comparison to Equation 2.24, this result can be written

$$\dot{\boldsymbol{\varepsilon}}_r = \dot{\theta}\boldsymbol{\varepsilon}_\theta. \tag{2.26}$$

Returning to Equation 2.22 then gives

$$\dot{\boldsymbol{r}} = \dot{r}\boldsymbol{\varepsilon}_r + r\dot{\theta}\boldsymbol{\varepsilon}_\theta = v_r\boldsymbol{\varepsilon}_r + v_\theta\boldsymbol{\varepsilon}_\theta. \tag{2.27}$$

This expression provides $\dot{\boldsymbol{r}}$, the time derivative of $\boldsymbol{r}$ with respect to the X, Y coordinate system, but the answer is given in terms of components in the $\boldsymbol{\varepsilon}_r$, $\boldsymbol{\varepsilon}_\theta$ coordinate system. Figure 2.6 illustrates the components of $\dot{\boldsymbol{r}}$ in the $\boldsymbol{\varepsilon}_r$- and $\boldsymbol{\varepsilon}_\theta$-directions. The "radial" component $\dot{\boldsymbol{r}}$ is aligned with $\boldsymbol{\varepsilon}_r$; the "circumferential" component $r\dot{\theta}$ is aligned with $\boldsymbol{\varepsilon}_\theta$.

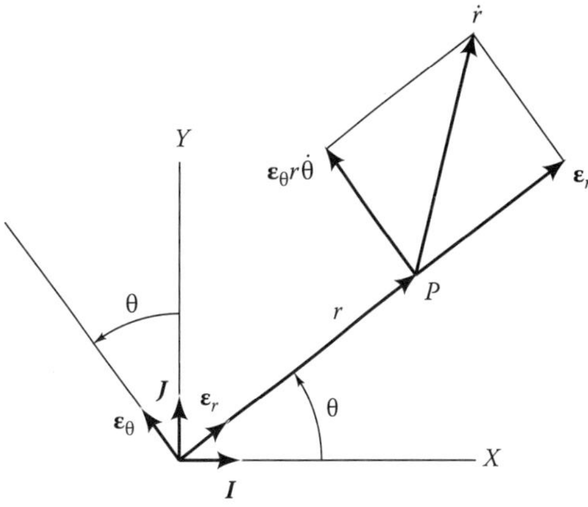

FIGURE 2.6
Components of $\dot{\boldsymbol{r}}$ in the $\boldsymbol{\varepsilon}_r$, $\boldsymbol{\varepsilon}_\theta$ system.

Having developed a polar component definition for the velocity of point P with respect to the X, Y coordinate system, a comparable definition for the acceleration $\ddot{\boldsymbol{r}}$ is needed. We begin by differentiating Equation 2.27 with respect to time in the X, Y coordinate system, while recalling that $\boldsymbol{\varepsilon}_\theta$ is also time-varying with respect to the X, Y coordinate system. The result is

$$\ddot{\boldsymbol{r}} = \left.\frac{d\dot{\boldsymbol{r}}}{dt}\right|X, Y = \ddot{r}\boldsymbol{\varepsilon}_r + \dot{r}\dot{\boldsymbol{\varepsilon}}_r + (\dot{r}\dot{\theta} + r\ddot{\theta})\boldsymbol{\varepsilon}_\theta + r\dot{\theta}\dot{\boldsymbol{\varepsilon}}_\theta. \tag{2.28}$$

We previously developed $\dot{\boldsymbol{\varepsilon}}_r$ as expressed in Equation 2.26; however, we now need a corresponding expression for $\dot{\boldsymbol{\varepsilon}}_\theta$, the time derivative of $\boldsymbol{\varepsilon}_\theta$ with respect to the X, Y system. Our end is achieved by differentiating Equation 2.24 with respect to time, holding $\boldsymbol{I}$ and $\boldsymbol{J}$ constant, to obtain

$$\dot{\boldsymbol{\varepsilon}}_\theta = \left.\frac{d\boldsymbol{\varepsilon}_\theta}{dt}\right|X, Y = -(\boldsymbol{I}\sin\theta + \boldsymbol{J}\cos\theta)\dot{\theta} = -\boldsymbol{\varepsilon}_r\dot{\theta}. \tag{2.29}$$

The substitution for $\boldsymbol{\varepsilon}_r$ in the last step of this equation follows from Equation 2.23. Substituting from Equations 2.26 and 2.29 into Equation 2.28 gives

$$\begin{aligned}\ddot{\boldsymbol{r}} &= \ddot{r}\boldsymbol{\varepsilon}_r + \dot{r}\dot{\theta}\boldsymbol{\varepsilon}_\theta + (\dot{r}\dot{\theta} + r\ddot{\theta})\boldsymbol{\varepsilon}_\theta - r\dot{\theta}^2\boldsymbol{\varepsilon}_r \\ &= (\ddot{r} - r\dot{\theta}^2)\boldsymbol{\varepsilon}_r + (r\ddot{\theta} + 2\dot{r}\dot{\theta})\boldsymbol{\varepsilon}_\theta \\ &= a_r\boldsymbol{\varepsilon}_r + a_\theta\boldsymbol{\varepsilon}_\theta.\end{aligned} \tag{2.30}$$

Equation 2.30 provides a definition for the acceleration of point P, with respect to the X, Y coordinate system, in

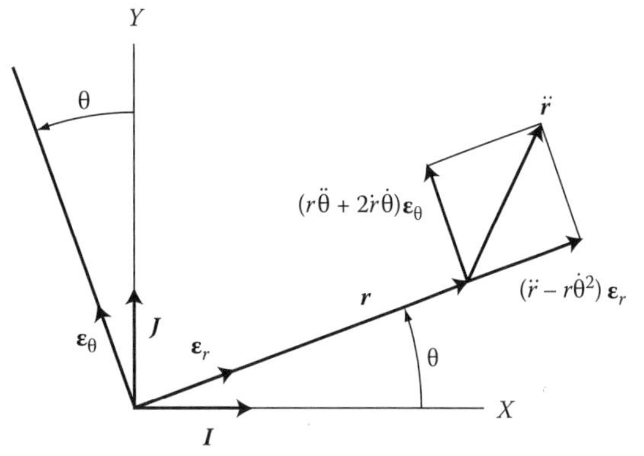

FIGURE 2.7
Components of $\ddot{\boldsymbol{r}}$ in the $\boldsymbol{\varepsilon}_r$, $\boldsymbol{\varepsilon}_\theta$ coordinate system.

terms of components in the $\boldsymbol{\varepsilon}_r$, $\boldsymbol{\varepsilon}_\theta$ coordinate system. The terms $-r\dot{\theta}^2$ and $r\ddot{\theta}$ are referred to, respectively, as centripetal and centrifugal acceleration components. The term $2\dot{r}\dot{\theta}$ is the Coriolis* acceleration component. Figure 2.7 illustrates the $\ddot{r}$ polar components.

Dynamics does not require much memorization; however, the component definition of velocity and acceleration in the $\boldsymbol{\varepsilon}_r$, $\boldsymbol{\varepsilon}_\theta$ coordinate systems must be committed to memory. In summary, they are

$$\begin{aligned} v_r &= \dot{r}, \quad v_\theta = r\dot{\theta} \\ a_r &= \ddot{r} - r\dot{\theta}^2, \quad a_\theta = r\ddot{\theta} + 2\dot{r}\dot{\theta}. \end{aligned} \tag{2.31}$$

After a little practice working kinematics problems, most of these terms will appear physically reasonable and consistent with your physical experiences. The Coriolis acceleration term tends to be an exception. The rotation of the $\boldsymbol{\varepsilon}_r$, $\boldsymbol{\varepsilon}_\theta$ unit vectors relative to the X, Y coordinate system causes this term, and it is not intuitively consistent with everyday experiences. You should satisfy yourself concerning the correctness of the derivation, and then use the definitions rigorously without "losing sleep" over a physical interpretation for the Coriolis term.

Example Problem 2.6

As illustrated in Figure XP2.6a, a mass is sliding freely along a bar that is rotating at a constant 50 cycles per minute (cpm). At $r = 0.38$ m and $\theta = 135°$, the additional conditions apply $\ddot{r} = 6.92\ \text{m/s}^2$, and $\dot{r} = 0.785$ m/s.

Tasks: Carry out the following steps:

a. Determine the velocity and acceleration components of the mass in the rotating $\boldsymbol{\varepsilon}_r$, $\boldsymbol{\varepsilon}_\theta$ coordinate system.
b. Determine the components of $\boldsymbol{v}$ and $\boldsymbol{a}$ in the stationary X, Y system.

Solution

In applying Equations 2.31 to find the polar components of velocity and acceleration, $\dot{\theta}$ is the angular velocity of the bar; however, we need to convert its given dimensions from cpm to radians per second via

$$\dot{\theta} = \frac{50\ \text{cycle}}{\text{min}} \times \frac{1\ \text{min}}{60\ \text{s}} \times \frac{2\pi\ \text{rad}}{1\ \text{cycle}} = 5.24\ \text{rad/s}$$

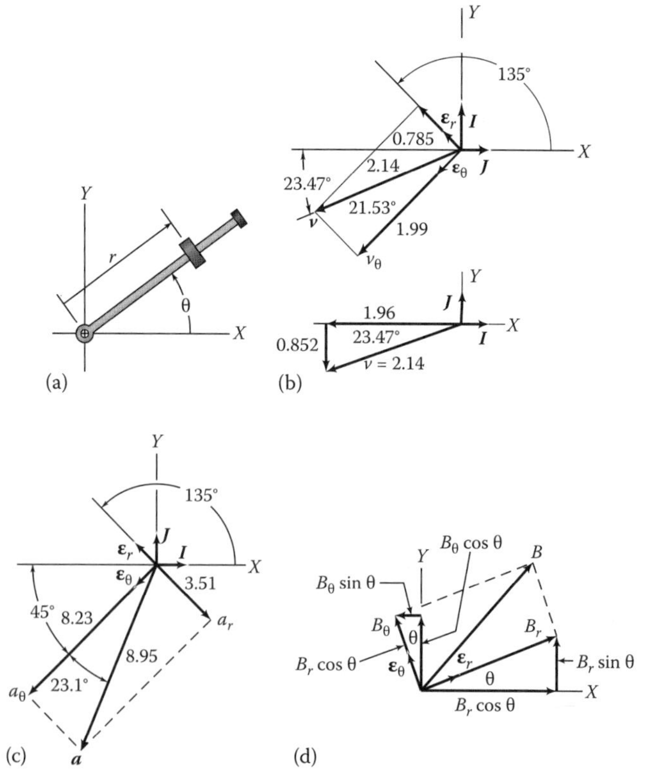

FIGURE XP2.6
(a) Mass sliding along a smooth rotating bar. (b) Velocity components in the $\boldsymbol{\varepsilon}_r$, $\boldsymbol{\varepsilon}_\theta$ and X, Y systems. (c) Acceleration components in the $\boldsymbol{\varepsilon}_r$, $\boldsymbol{\varepsilon}_\theta$ system. (d) Coordinate transformation development to move from the $\boldsymbol{\varepsilon}_r$, $\boldsymbol{\varepsilon}_\theta$ coordinate frame to the X, Y frame.

* Gustave Coriolis (1792–1843) was an assistant professor of mathematics at the Ecole Polytechnique, Paris from 1816 to 1838 and studied mechanics and engineering mathematics. He is best remembered for the Coriolis force which appears in the paper *Sur les équations du mouvement relatif des systèmes de corps* (1835). He showed that the laws of motion could be used in a rotating frame of reference if an extra "force term" incorporating the Coriolis acceleration is added to the equations of motion. Coriolis also introduced the terms "work" and "kinetic energy" with their present scientific meaning. (Web site, School of Mathematics and Statistics, Saint Andrews University, Scotland.)

Direct application of Equations 2.31 gives

$$v_r = \dot{r} = 0.785 \text{ m/s}$$
$$v_\theta = r\dot{\theta} = 0.38 \text{ m} \times 5.24 \text{ rad/s} = 1.99 \text{ m/s}$$
$$a_r = (\ddot{r} - r\dot{\theta}^2) = 6.92 \text{ m/s}^2 - 0.38 \text{ m} \times 5.24^2 \text{ rad}^2/\text{s}^2 = -3.51 \text{ m/s}^2$$
$$a_\theta = (r\ddot{\theta} + 2\dot{r}\dot{\theta}) = 0.38 \text{ m} \times 0 + 2 \times 0.785 \text{ m/s} \times 5.24 \text{ rad/s} = 8.23 \text{ m/s}^2, \quad \text{(i)}$$

and concludes Task a.

Figure XP2.6b illustrates the velocity components in the rotating $\boldsymbol{\varepsilon}_r$, $\boldsymbol{\varepsilon}_\theta$ coordinate system. There are several ways to develop the X and Y components of $\boldsymbol{v}$ and $\boldsymbol{a}$. We will first consider a development, based directly on geometry and then use coordinate transformations.

The velocity vector magnitude is $v = (0.785^2 + 1.99^2)^{1/2} = 2.1$ m/s, and $\boldsymbol{v}$ is oriented at the angle $21.53° = \tan^{-1}(0.785/1.99)$ with respect to $\boldsymbol{\varepsilon}_\theta$. Since $\boldsymbol{\varepsilon}_\theta$ is directed 45° below the $-X$-axis, $\boldsymbol{v}$ is at $45° - 21.53° = 23.47°$ below the $-X$-axis, and

$$v_X = -2.14 \cos 23.47° = -1.96 \text{ m/s}, \quad v_Y = -2.14 \sin 23.47° = -0.852 \text{ m/s}. \quad \text{(ii)}$$

Moving along to the acceleration components, Figure XP2.6c illustrates the a_r and a_θ components in the rotating $\boldsymbol{\varepsilon}_r$, $\boldsymbol{\varepsilon}_\theta$ coordinate system. Note that $a_r = -3.51$ m/s^2 is directed in the $-\boldsymbol{\varepsilon}_r$ direction. The acceleration magnitude is $a = (8.23^2 + 3.51^2)^{1/2} = 8.95$ m/s^2. The acceleration vector $\boldsymbol{a}$ is rotated $23.1° = \tan^{-1}(3.51/8.23)$ counterclockwise from $\boldsymbol{\varepsilon}_\theta$; hence, $\boldsymbol{a}$ is directed at $23.1° + 45° = 68.1°$ below the $-X$-axis. Accordingly,

$$a_X = -8.95 \cos 68.1° = -3.34 \text{ m/s}^2, \quad a_Y = -8.95 \sin 68.1° = -8.30 \text{ m/s}^2. \quad \text{(iii)}$$

These velocity and acceleration component results can also be obtained via a coordinate transformation. Figure XP2.6d illustrates a vector $\boldsymbol{B}$ defined in terms of its components B_r, B_θ in the $\boldsymbol{\varepsilon}_r$, $\boldsymbol{\varepsilon}_\theta$ coordinate system. Projecting these components into the X, Y system and adding the results along the X- and Y-axes gives

$$B_X = B_r \cos\theta - B_\theta \sin\theta, \quad B_Y = B_r \sin\theta + B_\theta \cos\theta.$$

In matrix notation, these results become

$$\begin{Bmatrix} B_X \\ B_Y \end{Bmatrix} = \begin{bmatrix} \cos\theta & -\sin\theta \\ \sin\theta & \cos\theta \end{bmatrix} \begin{Bmatrix} B_r \\ B_\theta \end{Bmatrix}. \quad (2.32)$$

Note that this result basically coincides with Equation 2.19 obtained earlier in discussing coordinate transformations. The $\boldsymbol{\varepsilon}_r$, $\boldsymbol{\varepsilon}_\theta$ coordinate system of Figure XP2.6d replaces the X', Y' system of Figure 2.4. Applying the transformation to the components of $\boldsymbol{v}$ and $\boldsymbol{a}$ from Equations (ii) and (iii) gives

$$\begin{Bmatrix} v_X \\ v_Y \end{Bmatrix} = \begin{bmatrix} \cos 135° & -\sin 135° \\ \sin 135° & \cos 135° \end{bmatrix} \begin{Bmatrix} v_r \\ v_\theta \end{Bmatrix} = \begin{bmatrix} -0.707 & -0.707 \\ 0.707 & -0.707 \end{bmatrix} \begin{Bmatrix} 0.785 \\ 1.99 \end{Bmatrix} = \begin{Bmatrix} -1.96 \\ -0.852 \end{Bmatrix} \text{m/s},$$

and

$$\begin{Bmatrix} a_X \\ a_Y \end{Bmatrix} = \begin{bmatrix} \cos 135° & -\sin 135° \\ \sin 135° & \cos 135° \end{bmatrix} \begin{Bmatrix} a_r \\ a_\theta \end{Bmatrix} = \begin{bmatrix} -0.707 & -0.707 \\ 0.707 & -0.707 \end{bmatrix} \begin{Bmatrix} -3.31 \\ 8.23 \end{Bmatrix} = \begin{Bmatrix} -3.34 \\ -8.30 \end{Bmatrix} \text{m/s}^2.$$

Note that the same (numerical) coordinate transformation applies for converting $\boldsymbol{v}$ and $\boldsymbol{a}$.

In reviewing this example, note that Task a is fairly straightforward, consisting of direct substitution into Equations 2.31. Most of the effort is involved in transforming the components of $\boldsymbol{v}$ and $\boldsymbol{a}$ from polar coordinates to stationary coordinates. Generally speaking, the geometric developments of Figure XP2.6b and c are more tedious than developing and applying the coordinate transformation. Remember that the coordinate transformation of Equation 2.32 applies for any vector and any angle θ.

2.6 Particle Motion in a Plane: Normal-Tangential (Path) Coordinates

Figure 2.8 illustrates a particle P moving along a trajectory $y(x)$ in the X, Y coordinate system. At time $t = 0$, P is located at O'. At time t, P is located at the distance s along the path $y(x)$. At any location on the path, the velocity of the particle is tangent to the trajectory and

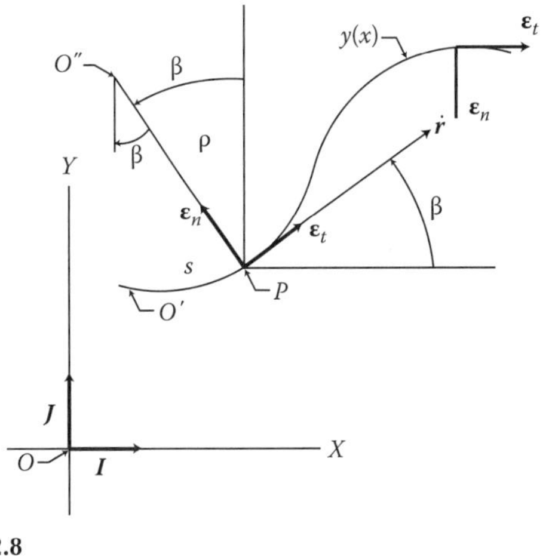

FIGURE 2.8
Path coordinates' unit vectors. Note that $\tan\beta = dy/dx$.

aligned with the unit vector $\boldsymbol{\varepsilon}_t$. The radius of curvature ρ extends from O'' to P. The unit vector $\boldsymbol{\varepsilon}_n$ is normal to $\boldsymbol{\varepsilon}_t$ and pointed opposite to $\boldsymbol{\rho}$. In comparing Figure 2.8 to Figure 2.3 (used to introduce velocity and acceleration in Cartesian coordinates), and Figure 2.5 (used to introduce these concepts for polar coordinates), note the absence of a displacement vector $\boldsymbol{r}$. The normal and tangential components are used to define velocity and acceleration components with regard to a trajectory or path, but do not lead directly to displacement components. The absence of a displacement vector causes an understandable confusion for students, given the importance of the $\boldsymbol{r}$ vector in prior developments; however, $\boldsymbol{r}$ is simply not a factor in customary path-coordinate developments or usage.

In Figure 2.8, note the $\boldsymbol{\varepsilon}_t$, $\boldsymbol{\varepsilon}_n$ coordinate orientation at the top right-hand position. For this position, the origin of the radius of curvature is below the curve, and $\boldsymbol{\varepsilon}_n$ has "flipped" directions. This feature of a changeable direction for $\boldsymbol{\varepsilon}_n$ is the second confusing issue with regards to the path-coordinate system. To "get it right," users of the path-coordinate system must be conscious of their location on a path and the corresponding direction for $\boldsymbol{\varepsilon}_n$.

Returning to Figure 2.8, the velocity vector $\dot{\boldsymbol{r}}$ continues to be the velocity of point P with respect to the X, Y coordinate system and can be stated

$$\dot{\boldsymbol{r}} = \dot{s}\boldsymbol{\varepsilon}_t = v\boldsymbol{\varepsilon}_t. \tag{2.33}$$

Again, s defines the distance traveled along the path $y(x)$ from the reference point O'. The radius of curvature of a path is defined formally as

$$\frac{1}{\rho} = \frac{|y''|}{\left[1 + (y')^2\right]^{3/2}},$$

where
$y' = dy/dx = \tan\beta$
$y'' = d^2y/dx^2$

In Figure 2.8, β defines the orientation of the vector $\dot{\boldsymbol{r}}$ with respect to the horizontal axis. The unit vector $\boldsymbol{\varepsilon}_n$ points from the path back toward O''.

When you look at this figure and try to connect it to the polar-coordinate Figure 2.5, there is some similarity between ρ, β and r, θ; however, in the present circumstances, the point O'' is moving constantly as the particle moves along the path, versus the polar-coordinate definition where one end of $r\boldsymbol{\varepsilon}_r$ is "anchored" at the origin. In parallel with the polar-coordinate velocity terms, $\dot{s}$ and $\dot{\beta}$ are related by

$$\dot{s} = v = \rho\dot{\beta}. \tag{2.34}$$

This is the (only) velocity definition for path coordinates.

Starting with Equation 2.33, an expression is needed for $\ddot{\boldsymbol{r}}$, the acceleration of point P with respect to the X, Y coordinate system, and we want the answer in terms of the $\boldsymbol{\varepsilon}_t$, $\boldsymbol{\varepsilon}_n$ components. To this end, we take the time derivative of $\dot{\boldsymbol{r}}$ with respect to the X, Y coordinate system, obtaining

$$\ddot{\boldsymbol{r}} = \left.\frac{d\dot{\boldsymbol{r}}}{dt}\right|_{X,Y} = \dot{v}\boldsymbol{\varepsilon}_t + v\left.\frac{d\boldsymbol{\varepsilon}_t}{dt}\right|_{X,Y} = \dot{v}\boldsymbol{\varepsilon}_t + v\dot{\boldsymbol{\varepsilon}}_t. \tag{2.35}$$

This result requires $\dot{\boldsymbol{\varepsilon}}_t$, the time derivative of $\boldsymbol{\varepsilon}_t$ with respect to the X, Y system. Hence, we have returned to the problem of differentiating a unit vector with respect to the X, Y coordinate system that arose earlier in the polar-coordinate developments. To achieve this goal, we will need the following expressions for $\boldsymbol{\varepsilon}_t$, $\boldsymbol{\varepsilon}_n$ in terms of their X, Y components:

$$\boldsymbol{\varepsilon}_t = \boldsymbol{I}\cos\beta + \boldsymbol{J}\sin\beta$$
$$\boldsymbol{\varepsilon}_n = -\boldsymbol{I}\sin\beta + \boldsymbol{J}\cos\beta.$$

Differentiating the first equation while holding $\boldsymbol{I}$ and $\boldsymbol{J}$ constant yields

$$\dot{\boldsymbol{\varepsilon}}_t = \left.\frac{d\boldsymbol{\varepsilon}_t}{dt}\right|_{X,Y} = \dot{\beta}(-\boldsymbol{I}\sin\beta + \boldsymbol{J}\cos\beta) = \dot{\beta}\boldsymbol{\varepsilon}_n.$$

Substituting this result into Equation 2.35 then gives

$$\ddot{\boldsymbol{r}} = \dot{v}\boldsymbol{\varepsilon}_t + v\dot{\beta}\boldsymbol{\varepsilon}_n. \tag{2.36}$$

Equation 2.34 provides the following alternative expressions for $\ddot{\boldsymbol{r}}$:

$$\ddot{\boldsymbol{r}} = \dot{v}\boldsymbol{\varepsilon}_t + \frac{v^2}{\rho}\boldsymbol{\varepsilon}_n, \tag{2.37}$$

or

$$\ddot{\boldsymbol{r}} = \dot{v}\boldsymbol{\varepsilon}_t + \rho\dot{\beta}^2\boldsymbol{\varepsilon}_n. \tag{2.38}$$

Since

$$\ddot{\boldsymbol{r}} = a_t\boldsymbol{\varepsilon}_t + a_n\boldsymbol{\varepsilon}_n,$$

Equations 2.36 through 2.39 provide the following component definitions for a_t and a_n:

$$a_t = \ddot{s} = \dot{v}, \quad a_n = v\dot{\beta} = \frac{v^2}{\rho} = \rho\dot{\beta}^2. \tag{2.39}$$

Equation 2.37 provides the more generally useful expression, incorporating $a_n = v^2/\rho$.

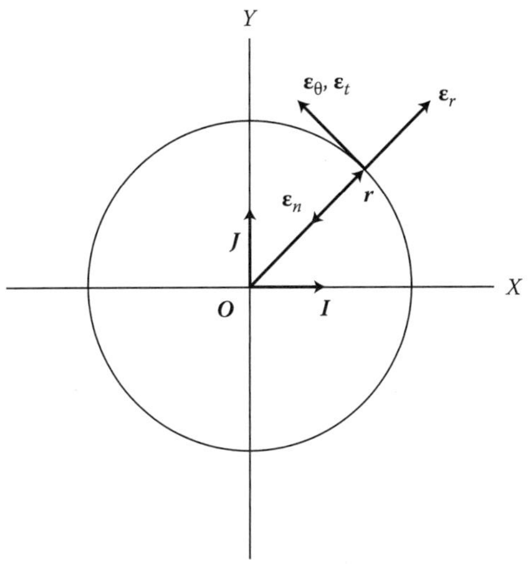

FIGURE 2.9
Constant-radius circular motion with the ε_r, ε_θ and ε_t, ε_n systems.

To get some idea how the polar and path coordinates are related, consider Figure 2.9 for strictly circular motion with $\dot{r}=0$. In this case $\rho=r$, $\dot{\theta}=\dot{\beta}$, $\dot{r}=0$, and ε_n and ε_r are oppositely directed. Hence the polar-coordinate model gives

$$v_r = 0,\quad v_\theta = r\dot{\theta},\quad a_r = -r\dot{\theta}^2,\quad a_\theta = r\ddot{\theta}.$$

For this reduced case, $v=r\dot{\theta}$; the normal-acceleration term v^2/r coincides with the centrifugal-acceleration term, $r\dot{\theta}^2$, and the path tangential-acceleration term $\dot{v}$ coincides with $r\ddot{\theta}$.

Example Problem 2.7

As illustrated in Figure XP2.7a, a track lies in the horizontal plane and is defined by $Y=kX^2$ with X and Y in meters and $k=1/400\ \text{m}^{-1}$. At $X=100$ m, the velocity and acceleration components of a vehicle traveling *along the path* are 20 m/s and −2 m/s², respectively.

Tasks:

a. Determine the normal and tangential components of $\boldsymbol{v}$ and $\boldsymbol{a}$.
b. Determine $\boldsymbol{v}$ and $\boldsymbol{a}$'s components in the X, Y system.

Solution

From the definitions of the ε_t, ε_n coordinate system, $\boldsymbol{v}$ and ε_t are co-linear, and both are directed along the tangent of the path. Hence, $v=\varepsilon_t\, 20\ \text{m/s}^2$. The velocity vector $\boldsymbol{v}$ has no component along ε_n. The problem statement gives $a_t=\dot{v}=-2\ \text{m/s}^2$. From Equation 2.39, the normal component is $a_n=v^2/\rho$. We are given $v=20$ m/s; however, we need to define the radius of curvature. Applying

$$\frac{1}{\rho}=\frac{|Y''|}{\left[1+(Y')^2\right]^{3/2}},$$

for $Y=kX^2$, we get $Y'=2kX$, and $Y''=2k$. Hence,

$$\left.\frac{1}{\rho}\right|_{X=100\ \text{m}} = \frac{2/400}{\left[1+(2\times 100/400)^2\right]^{3/2}} = 3.57\times 10^{-3}\ \text{m}^{-1} \Rightarrow \rho = 280\ \text{m},$$

and

$$a_n = \frac{20^2\ (\text{m/s})^2}{280\ (\text{m})} = 1.43\ \text{m/s}^2.$$

The answer for Task a is

$$\boldsymbol{v} = 20\varepsilon_t\ \text{m/s},\quad \boldsymbol{a} = -2\varepsilon_t + 1.43\varepsilon_n\ \text{m/s}^2.$$

Moving to Task b, the first question to answer in finding the components of $\boldsymbol{v}$ and $\boldsymbol{a}$ in the X, Y system is, "How are ε_t, ε_n oriented in the X, Y system?" Since ε_t is directed along the tangent of the path, we can find the orientation of ε_t with respect to the X-axis via

$$Y'|_{X=100\ \text{m}} = 2kX|_{X=100\ \text{m}} = 2\times\frac{1}{400}\times 100 = 0.5 \Rightarrow \beta = \tan^{-1}(0.5) = 26.6°.$$

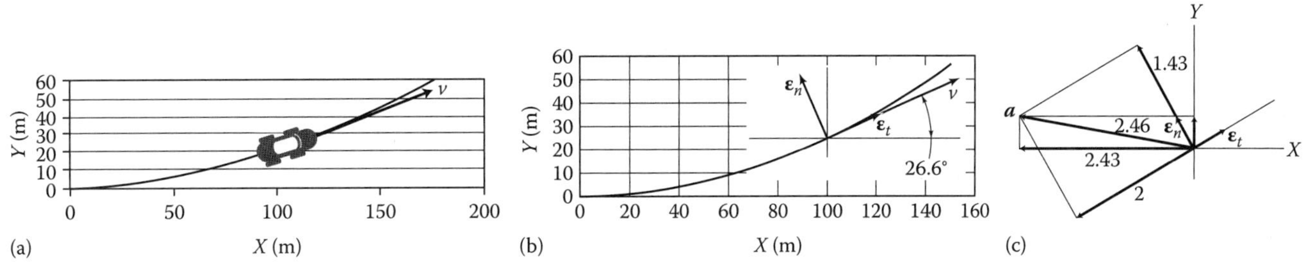

FIGURE XP2.7
(a) Track segment in the horizontal X, Y system. (b) ε_t, ε_n orientation at $X=100$ m. (c) Components of $\boldsymbol{a}$ in the path and X, Y system.

Figure XP2.7b shows this orientation of the $\boldsymbol{\varepsilon}_t$, $\boldsymbol{\varepsilon}_n$ coordinate system at $X = 100$ m.

From this figure, $\boldsymbol{v}$'s X and Y components are

$$v_X = v\cos 26.6^\circ = 20 \times 0.894 = 17.9 \text{ m/s},$$
$$v_Y = v\sin 26.6^\circ = 20 \times 0.448 = 8.96 \text{ m/s}.$$

Note that the X, Y and path components produce the same vector.

To find $\boldsymbol{a}$'s components in the X, Y system, consider the components of the arbitrary vector $\boldsymbol{B}$ in Figure 2.10. This figure is very similar to Figure XP2.6d that we developed to move from components in the $\boldsymbol{\varepsilon}_r$, $\boldsymbol{\varepsilon}_\theta$ system to components in the X, Y system. Summing components in the X- and Y-directions gives

$$B_X = B_t \cos\beta - B_n \sin\beta, \quad B_Y = B_t \sin\beta + B_n \cos\beta.$$

In matrix notation, these results become

$$\begin{Bmatrix} B_X \\ B_Y \end{Bmatrix} = \begin{bmatrix} \cos\beta & -\sin\beta \\ \sin\beta & \cos\beta \end{bmatrix} \begin{Bmatrix} B_t \\ B_n \end{Bmatrix}. \tag{2.40}$$

Substituting the acceleration components and $\beta = 26.6^\circ$ gives

$$\begin{Bmatrix} a_X \\ a_Y \end{Bmatrix} = \begin{bmatrix} \cos 26.6^\circ & -\sin 26.6^\circ \\ \sin 26.6^\circ & \cos 26.6^\circ \end{bmatrix} \begin{Bmatrix} a_t \\ a_n \end{Bmatrix}$$
$$= \begin{bmatrix} 0.894 & -0.447 \\ 0.447 & 0.894 \end{bmatrix} \begin{Bmatrix} -2 \\ 1.43 \end{Bmatrix} = \begin{Bmatrix} -2.43 \\ 0.384 \end{Bmatrix} \text{m/s}^2.$$

Note that $\boldsymbol{a}$'s magnitude is unchanged by the transformation.

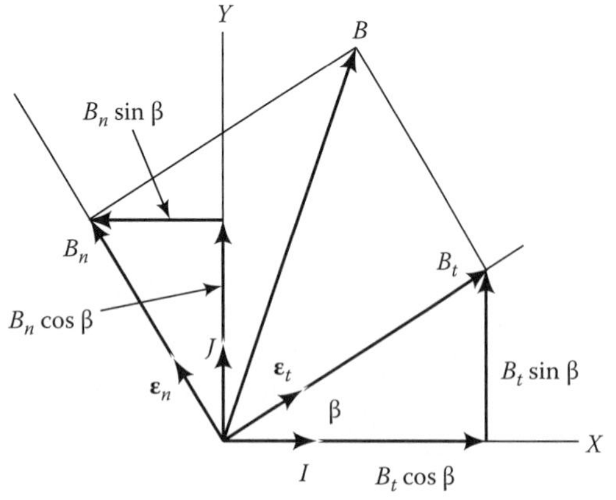

FIGURE 2.10
Coordinate transformation development in moving from components in the $\boldsymbol{\varepsilon}_t$, $\boldsymbol{\varepsilon}_n$ to the X, Y system.

Figure XP2.7c shows the component definitions for $\boldsymbol{a}$ in the two systems, demonstrating that the same vector is produced. This step concludes Task b.

In reviewing the steps involved in working out this example, applying the definitions to find the components of $\boldsymbol{v}$ and $\boldsymbol{a}$ in the path-coordinate system is relatively straightforward. The essential first step in finding $\boldsymbol{v}$'s and $\boldsymbol{a}$'s components in the X, Y system is in recognizing that $\boldsymbol{\varepsilon}_t$ lies along the path's tangent. Following this insight, projecting $\boldsymbol{v}$'s components into the X, Y system is simple, as is finding $\boldsymbol{a}$'s components via the coordinate transformation.

2.7 Moving between Cartesian, Polar, and Path-Coordinate Definitions for Velocity and Acceleration Components

We started the discussion of two-dimensional planar motion in Section 2.3 with Cartesian coordinates, then moving on to polar and path coordinates in Sections 2.5 and 2.6, respectively. On occasions, students fail to understand that the *same* velocity $\boldsymbol{v}$ and acceleration $\boldsymbol{a}$ vectors are being considered in these alternative coordinate systems. The three examples considered in this section are provided to emphasize that, while the *component* descriptions change for $\boldsymbol{v}$ and $\boldsymbol{a}$, the *vectors* do not. For each example, we will first define $\boldsymbol{v}$ and $\boldsymbol{a}$ in terms of their components in one coordinate type [(X, Y), (r, θ), or $(\boldsymbol{\varepsilon}_t, \boldsymbol{\varepsilon}_n)$] and then go on to obtain the components in the remaining systems.

2.7.1 An Example That Is Naturally Analyzed with Cartesian Components

Figure 2.11 illustrates a mechanism for controlling the position of point P via lag screws. Because guide A is

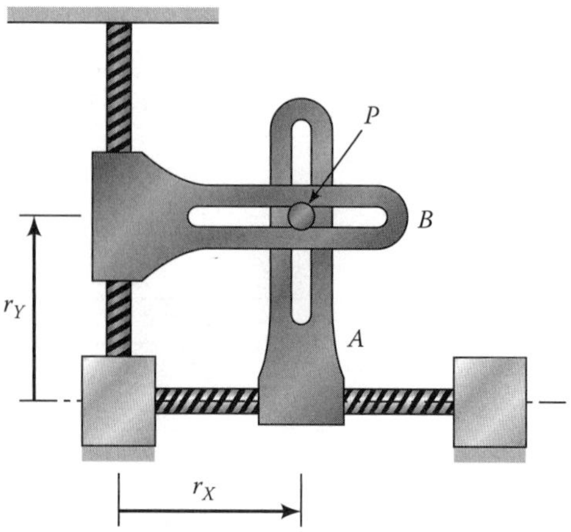

FIGURE 2.11
A lag-screw-driven mechanism.

threaded around the horizontal lag screw and is prevented from rotating out of the plane of the figure, turning this screw will cause guide A to move horizontally. Similarly, turning the vertical lag screw will cause guide B to move vertically. Hence, the mechanism can be used to directly control the Cartesian coordinates of P. Assume that control is applied to the screws such that

$$r_X(t) = A + a\cos(\omega t), \quad r_Y(t) = B + b\sin(2\omega t)$$
$$A = B = 1\text{ mm}, \quad a = 1.25\text{ mm}, \quad b = 2.5\text{ mm}, \quad \omega = 0.5\text{ Hz}. \tag{i}$$

Tasks: For $\omega t = 30°$, determine the components of P's velocity and acceleration vectors in the $[(X, Y), (r, \theta)$, and $(\boldsymbol{\varepsilon}_t, \boldsymbol{\varepsilon}_n)]$ coordinate systems. Draw pictures that demonstrate that the same $\boldsymbol{v}$ and $\boldsymbol{a}$ vectors are defined in all coordinate systems.

Figure 2.12 illustrates the trajectory. The path is called a Lissajous figure. (Jules Antoine Lissajous, March 1822–June 1880, was primarily noted for accomplishments in acoustics.) Moving to the engineering-analysis tasks, the obvious way to obtain components of P's velocity vector is to differentiate the components of the position vector obtaining

$$v_X = \dot{r}_X = -a\omega\sin(\omega t), \quad v_Y = \dot{r}_Y = 2b\omega\cos(2\omega t).$$

A second differentiation yields the acceleration components:

$$a_X = \ddot{r}_X = -a\omega^2\cos(\omega t), \quad a_Y = \ddot{r}_Y = -4b\omega^2\sin(2\omega t).$$

Plugging in the numbers for the components of $\boldsymbol{r}$, $\boldsymbol{v}$, and $\boldsymbol{a}$ from the data of Equation (i) at $\omega t = 30°$ nets

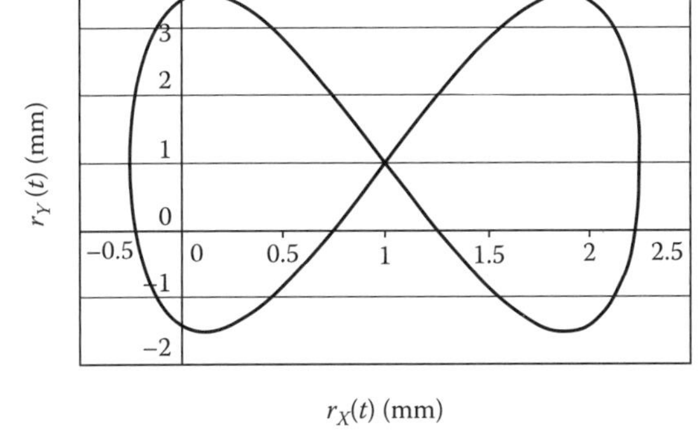

FIGURE 2.12
Lissajous figure for r_X and r_Y.

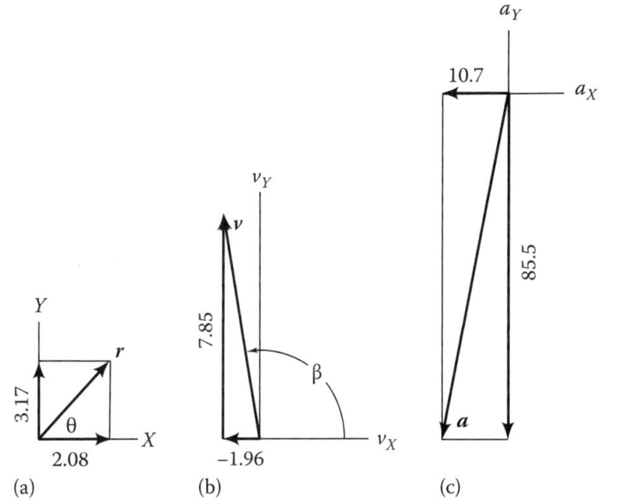

FIGURE 2.13
(a) Position, (b) velocity, and (c) acceleration vectors in the X, Y system at $\omega t = 30°$; mm/s units.

$$\begin{aligned}
r_X &= 1 + 1.25 \times 0.866 = 2.08\text{ mm} \\
r_Y &= 1 + 2.5 \times 0.86 = 3.17\text{ mm} \\
v_X &= -1.25 \times 3.1416 \times 0.5 = -1.96\text{ mm/s} \\
v_Y &= 2 \times 2.5 \times 3.1416 \times 0.5 = 7.85\text{ mm/s} \\
a_X &= -1.25 \times (3.1416)^2 \times 0.866 = -10.7\text{ mm/s}^2 \\
a_Y &= -4 \times 2.5(3.1416)^2 \times 0.866 = -85.5\text{ mm/s}^2.
\end{aligned} \tag{ii}$$

Figure 2.13 illustrates the vectors defined by these components.

With the results of Equation (ii), we are about one third of the way home, having obtained the requested vectors in terms of their components in the X, Y coordinate system. The coordinate-transformation results of Section 2.4 provide the quickest approach to obtain the components of $\boldsymbol{v}$ and $\boldsymbol{a}$ in terms of their (r, θ) components. Figure 2.14 shows the components of a vector $\boldsymbol{B}$ expressed initially in terms of its components in the X, Y coordinate system, B_X and B_Y. The components B_X and B_Y are then shown projected into components along the $\boldsymbol{\varepsilon}_r$, $\boldsymbol{\varepsilon}_\theta$ unit vectors. Summing components in the $\boldsymbol{\varepsilon}_r$-, $\boldsymbol{\varepsilon}_\theta$-directions gives

$$B_r = B_X\cos\theta + B_Y\sin\theta, \quad B_\theta = -B_X\sin\theta + B_Y\cos\theta,$$

or in matrix format

$$\begin{Bmatrix} B_r \\ B_\theta \end{Bmatrix} = \begin{bmatrix} \cos\theta & \sin\theta \\ -\sin\theta & \cos\theta \end{bmatrix} \begin{Bmatrix} B_X \\ B_Y \end{Bmatrix}. \tag{2.41}$$

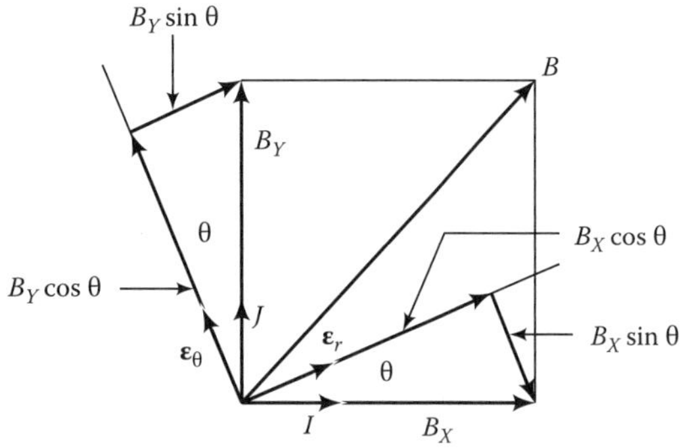

FIGURE 2.14
Components of $\boldsymbol{B}$ in the X, Y system projected along the $(\boldsymbol{\varepsilon}_r, \boldsymbol{\varepsilon}_\theta)$ unit vectors.

Appropriately, this is the same basic result provided earlier by Equation 2.15 and is the inverse transformation of that provided by Equation 2.32. From Figure 2.13a,

$$\theta = \tan^1\left(\frac{r_Y}{r_X}\right) = \tan^{-1}\left(\frac{3.17}{2.08}\right) = 56.65°. \qquad \text{(iii)}$$

Hence, from Equations (iii) and 2.41,

$$\begin{Bmatrix} v_r \\ v_o \end{Bmatrix} = \begin{bmatrix} \cos 56.65° & \sin 56.65° \\ -\sin 56.65° & \cos 56.65° \end{bmatrix} \begin{Bmatrix} v_x \\ v_y \end{Bmatrix}$$
$$= \begin{bmatrix} 0.550 & 0.835 \\ -0.835 & 0.550 \end{bmatrix} \begin{Bmatrix} -1.96 \\ 7.85 \end{Bmatrix}$$
$$= \begin{Bmatrix} 5.48 \\ 5.95 \end{Bmatrix} \text{mm/s}. \qquad \text{(iv)}$$

Continuing,

$$\begin{Bmatrix} a_r \\ a_o \end{Bmatrix} = \begin{bmatrix} 0.550 & 0.835 \\ -0.835 & 0.550 \end{bmatrix} \begin{Bmatrix} a_x \\ a_y \end{Bmatrix}$$
$$= \begin{bmatrix} 0.550 & 0.835 \\ -0.835 & 0.550 \end{bmatrix} \begin{Bmatrix} -10.68 \\ -85.47 \end{Bmatrix}$$
$$= \begin{Bmatrix} -77.2 \\ -38.2 \end{Bmatrix} \text{mm/s}^2. \qquad \text{(v)}$$

You can confirm that the transformation has not changed either $\boldsymbol{v}$ or $\boldsymbol{a}$'s magnitude. Equations (iv) and (v) conclude the second step in the engineering task.

When faced with the coordinate-transformation approach, students regularly ask (hopefully), Can I use another approach to get these results? The short answer is yes; however, the coordinate-transformation approach is generally quicker and more efficient than alternative approaches. The same coordinate transformation matrix is used to get both velocity and acceleration components, and the approach works with very little regard or requirement for physical insight. Equation 2.41 is valid for any θ value and any choice of B_X and B_Y. If you put in correct values for (θ, B_X, B_Y) and do the arithmetic correctly, you will get the correct B_r, B_θ components.

Facing the last requirement of obtaining the $(\boldsymbol{\varepsilon}_t, \boldsymbol{\varepsilon}_n)$ components for $\boldsymbol{v}$ and $\boldsymbol{a}$, the first question is as follows: Where are $\boldsymbol{\varepsilon}_t$ and $\boldsymbol{\varepsilon}_n$? Recall that $\boldsymbol{\varepsilon}_t$ is directed along the path of the trajectory. At any instant in time, $\boldsymbol{v}$ must also be pointed along the particle's trajectory; hence, ε_t must be co-linear with $\boldsymbol{v}$. From Figure 2.13b, $\boldsymbol{\varepsilon}_t$ is pointed at the angle β relative to the X-axis, where

$$\beta = \tan^{-1}\left(\frac{v_Y}{v_X}\right) = \tan^{-1}\left(\frac{7.85}{-1.96}\right) = 104°. \qquad \text{(vi)}$$

Also,

$$v_t = v = \left(v_X^2 + v_Y^2\right)^{1/2} = 8.09 \text{ mm/s}; \quad v_n = 0. \qquad \text{(vii)}$$

Finding the acceleration components a_t, a_n is our last task. Figure 2.15 illustrates a vector $\boldsymbol{B}$, initially defined by its components B_X and B_Y in the X, Y coordinate system. B_X and B_Y are then shown projected in terms of their own components along the $(\boldsymbol{\varepsilon}_t, \boldsymbol{\varepsilon}_n)$ unit vectors. Summing components in the $(\boldsymbol{\varepsilon}_t, \boldsymbol{\varepsilon}_n)$ directions gives

$$B_t = B_X \cos\beta + B_Y \sin\beta, \quad B_n = -B_X \sin\beta + B_Y \cos\beta,$$

or in matrix format,

$$\begin{Bmatrix} B_t \\ B_n \end{Bmatrix} = \begin{bmatrix} \cos\beta & \sin\beta \\ -\sin\beta & \cos\beta \end{bmatrix} \begin{Bmatrix} B_X \\ B_Y \end{Bmatrix}. \qquad (2.42)$$

Figure 2.15 largely coincides with Figure 2.14, and the development of Equation 2.42 closely parallels the

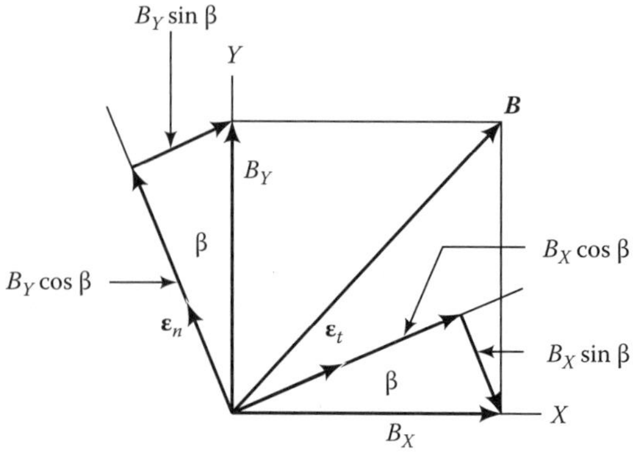

FIGURE 2.15
Components of the vector $\boldsymbol{B}$ in the X, Y system projected into the $(\boldsymbol{\varepsilon}_t, \boldsymbol{\varepsilon}_n)$ system.

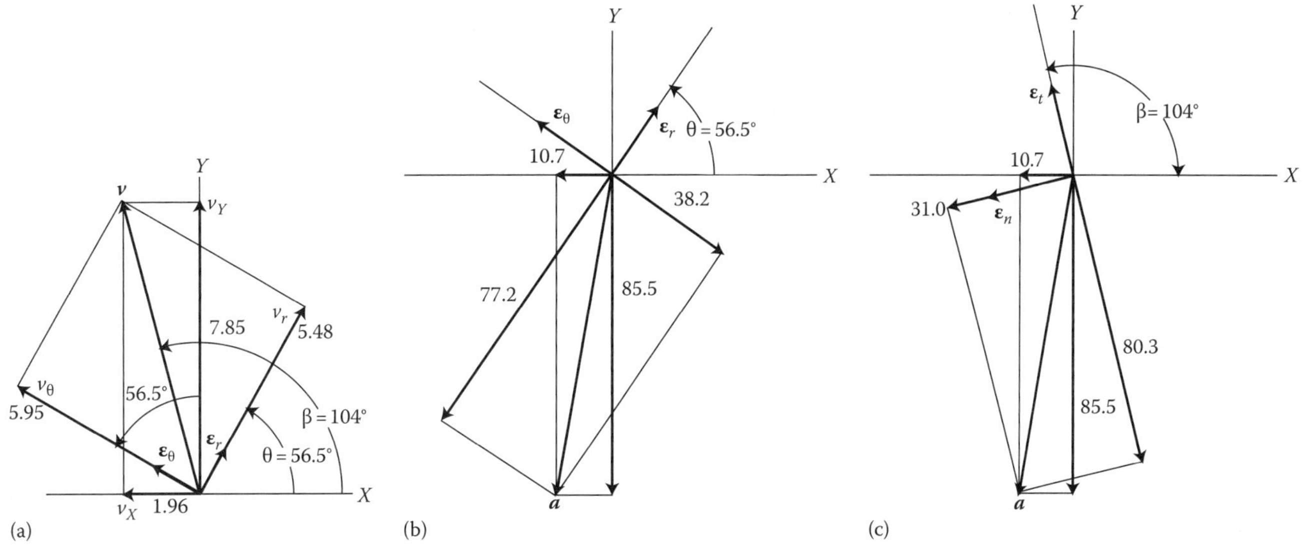

FIGURE 2.16
(a) Components of v in the three-coordinate systems. Acceleration components (b) X, Y and polar systems, and (c) X, Y and path systems; mm/s^2.

development of Equation 2.41. Applying Equation 2.42 to find a_t, a_n gives

$$\begin{Bmatrix} a_t \\ a_n \end{Bmatrix} = \begin{bmatrix} \cos 104^\circ & \sin 104^\circ \\ -\sin 104^\circ & \cos 104^\circ \end{bmatrix} \begin{Bmatrix} a_X \\ a_Y \end{Bmatrix} = \begin{bmatrix} -0.242 & 0.970 \\ -0.970 & -0.242 \end{bmatrix} \begin{Bmatrix} -10.68 \\ -85.47 \end{Bmatrix} = \begin{Bmatrix} -80.3 \\ 31.0 \end{Bmatrix} \text{mm/s}^2. \quad \text{(viii)}$$

You can confirm (again) that $\boldsymbol{a}$'s magnitude is not changed by the transformation. Also, despite the confident drawing of Figure 2.15, we really don't know where $\boldsymbol{\varepsilon}_n$ is directed until the calculation for a_n is completed. The positive sign for a_n calculated in Equation (viii) means that (at this instant) the direction drawn for $\boldsymbol{\varepsilon}_n$ in Figure 2.15 is correct. A negative outcome would imply that $\boldsymbol{\varepsilon}_n$ (at this instant) is actually directed negatively from the direction shown in Figure 2.15.

Figure 2.16a shows v's components in the three-coordinate systems. The Cartesian components of this figure coincide with those of Figure 2.13b. The polar components shown use θ from Equation (iii) and v_r, v_θ from Equation (iv). The path coordinates use β from Equation (vi) and v from Equation (vii). Figure 2.16b shows $\boldsymbol{a}$'s components in the Cartesian (Equation (ii)) and polar (Equation (v)) systems. Figure 2.16c shows them in the Cartesian and path (Equation (viii)) systems. The intended lesson of this example and section is that we are looking at the *same* $\boldsymbol{a}$ and $\boldsymbol{v}$ vectors in three different coordinate systems. The component definitions change, but the vectors do not.

2.7.2 An Example That Is Naturally Analyzed Using Polar Coordinates

Figure 2.17 illustrates a mechanism that includes a rotating hollow bar whose orientation is defined by θ. An internal lag-screw mechanism is used to drive point P radially. The motion is defined by

$$\theta = \frac{\pi}{4}\cos(\omega t), \quad r = 10 + 5\cos(2\omega t)\ \text{mm},$$

where $\omega = 60$ cycles/min.

Tasks: At $\omega t = \pi/6 = 30^\circ$, determine the components of the velocity and acceleration vector for point P, and state your results in the (X, Y), (r, θ), and $(\boldsymbol{\varepsilon}_t, \boldsymbol{\varepsilon}_n)$ coordinate systems. This example's nature strongly argues for polar-coordinate kinematics in initially defining $\boldsymbol{v}$ and $\boldsymbol{a}$. Moving forward in developing the necessary time derivatives,

$$\dot{\theta} = -\omega\frac{\pi}{4}\sin(\omega t)\ \text{rad/s}, \quad \dot{r} = -10\omega\sin(2\omega t)\ \text{mm/s}$$

$$\ddot{\theta} = -\omega^2\frac{\pi}{4}\cos(\omega t)\ \text{rad/s}^2, \quad \ddot{r} = -20\omega^2\cos(2\omega t)\ \text{mm/s}^2.$$

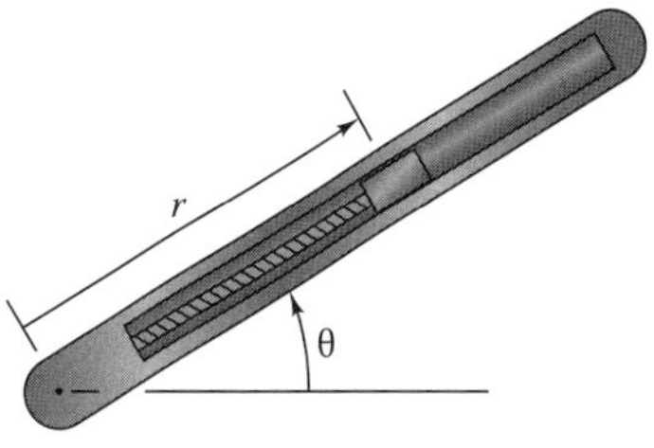

FIGURE 2.17
A rotating-bar/lag-screw mechanism.

Hence, for $\omega t = 30°$ and $\omega = 60$ cycles/min × 1 min/60 s × 2π rad/cycle = 2π rad/s,

$$\theta = 0.680 \text{ rad} = 39.0°, \quad r = 12.5 \text{ mm}$$
$$\dot{\theta} = -2.467 \text{ rad/s}, \quad \dot{r} = -54.41 \text{ mm/s}$$
$$\ddot{\theta} = -26.85 \text{ rad/s}^2, \quad \ddot{r} = -394.8 \text{ mm/s}^2.$$

From Equations 2.31,

$$v_\theta = r\dot{\theta} = 12.5 \times -2.467 = -30.83 \text{ mm/s}$$
$$v_r = \dot{r} = -54.41 \text{ mm/s}$$
$$a_\theta = r\ddot{\theta} + 2\dot{r}\dot{\theta} = 12.5 \times -26.85 + 2 \times -54.41 \times -2.467 = -67.17 \text{ mm/s}^2$$
$$a_r = \ddot{r} - r\dot{\theta}^2 = -394.8 - 12.5 \times (-2.467)^2 = -470.9 \text{ mm/s}.$$

This result concludes about one-third of the tasks. We still need to calculate the components of $\boldsymbol{v}$ and $\boldsymbol{a}$ in terms of Cartesian and path components.

In working Example Problem XP2.6, we drew Figure XP2.6d to develop Cartesian components B_X, B_Y, starting with components B_r, B_θ to obtain

$$\begin{Bmatrix} B_X \\ B_Y \end{Bmatrix} = \begin{bmatrix} \cos\theta & -\sin\theta \\ \sin\theta & \cos\theta \end{bmatrix} \begin{Bmatrix} B_r \\ B_\theta \end{Bmatrix}. \quad (2.32)$$

Recalling that the direction-cosine matrix is orthogonal, we could have stated this result directly from Equation 2.41. Substituting in the polar components for v gives

$$\begin{Bmatrix} v_X \\ v_Y \end{Bmatrix} = \begin{bmatrix} \cos\theta & -\sin\theta \\ \sin\theta & \cos\theta \end{bmatrix} \begin{Bmatrix} v_r \\ v_\theta \end{Bmatrix} = \begin{bmatrix} \cos 39.0° & -\sin 39.0° \\ \sin 39.0° & \cos 39.0° \end{bmatrix} \begin{Bmatrix} -54.41 \\ -30.83 \end{Bmatrix} = \begin{Bmatrix} -22.88 \\ -58.20 \end{Bmatrix} \text{ mm/s}. \quad \text{(i)}$$

Similarly for $\boldsymbol{a}$,

$$\begin{Bmatrix} a_X \\ a_Y \end{Bmatrix} = \begin{bmatrix} \cos\theta & -\sin\theta \\ \sin\theta & \cos\theta \end{bmatrix} \begin{Bmatrix} a_r \\ a_\theta \end{Bmatrix} = \begin{bmatrix} \cos 39.0° & -\sin 39.0° \\ \sin 39.0° & \cos 39.0° \end{bmatrix} \begin{Bmatrix} -470.9 \\ -67.17 \end{Bmatrix} = \begin{Bmatrix} -323.7 \\ -348.5 \end{Bmatrix} \text{ mm/s}^2. \quad \text{(ii)}$$

We can use the results of Equation (i) to define v's orientation in the X, Y system via

$$\beta = \tan^{-1}\left(\frac{v_Y}{v_X}\right) = \tan^{-1}\left(\frac{-58.20}{-22.8}\right) = 248.6°. \quad \text{(iii)}$$

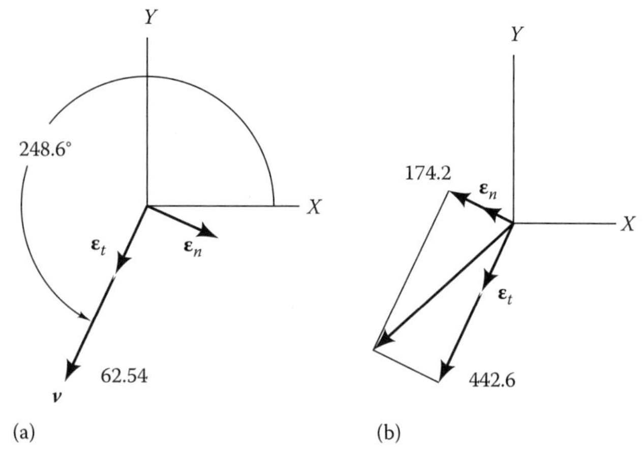

FIGURE 2.18
(a) Velocity v from Equation (i). (b) Acceleration $\boldsymbol{a}$ from Equation (v); mm/s units.

The components of v in the path system are

$$v = v_t = (58.2^2 + 22.9^2)^{1/2} = 62.54 \text{ mm/s}, \quad v_n = 0. \quad \text{(iv)}$$

We can obtain a_t, a_n from

$$\begin{Bmatrix} B_t \\ B_n \end{Bmatrix} = \begin{bmatrix} \cos\beta & \sin\beta \\ -\sin\beta & \cos\beta \end{bmatrix} \begin{Bmatrix} B_X \\ B_Y \end{Bmatrix}, \quad (2.41)$$

as

$$\begin{Bmatrix} a_t \\ a_n \end{Bmatrix} = \begin{bmatrix} \cos\beta & \sin\beta \\ -\sin\beta & \cos\beta \end{bmatrix} \begin{Bmatrix} a_X \\ a_Y \end{Bmatrix} = \begin{bmatrix} \cos 248.6° & \sin 248.6° \\ -\sin 248.6° & \cos 248.6° \end{bmatrix} \begin{Bmatrix} -323.7 \\ -348.5 \end{Bmatrix} = \begin{Bmatrix} 442.6 \\ -174.2 \end{Bmatrix} \text{ mm/s}^2. \quad \text{(v)}$$

Figure 2.18a and b illustrate, respectively, the solution for the path component description of v and $\boldsymbol{a}$. The negative sign for a_n in Equation (v) means that $\boldsymbol{\varepsilon}_n$ is directed in the opposite direction from the assumed direction in Figure 2.15. Note in Figure 2.18b that the direction of $\boldsymbol{\varepsilon}_n$ is reversed in accordance with the negative sign for a_n in Equation (v).

2.7.3 An Example That Is Naturally Analyzed with Path-Coordinate Components

Figure 2.19 illustrates a vehicle traveling along a path in a vertical plane defined by

$$Y = A\sin\left(\frac{2\pi X}{L}\right), \quad A = 100 \text{ m}, \quad L = 2000 \text{ m}. \quad \text{(i)}$$

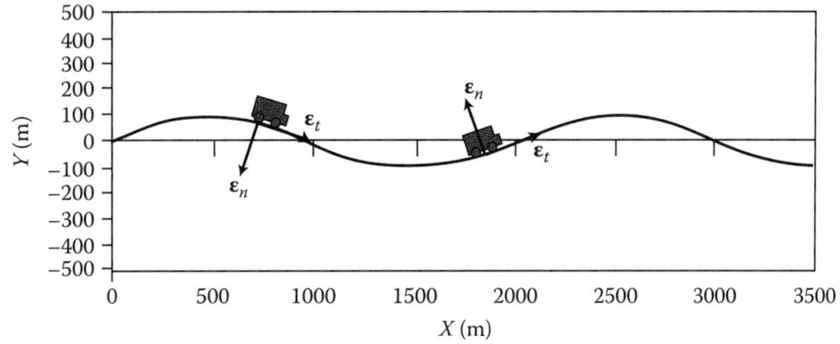

FIGURE 2.19
Vehicle following the curved path $Y = A\sin(2\pi X/L)$, $A = 100$ m, $L = 2000$ m.

The velocity and acceleration of the vehicle along the path are $v = 100$ km/h, and $a_t = 2$ m/s^2 at two locations defined by $X = 750$ m and $X = 1750$ m.

Tasks: Determine the components of $\boldsymbol{v}$ and $\boldsymbol{a}$ at $X = 750$ m and state the components in the $[X, Y, (r, \theta)$, and $(\boldsymbol{\varepsilon}_t\boldsymbol{\varepsilon}_n)]$ systems. Also, for $X = 1750$ m, determine the $(\boldsymbol{\varepsilon}_t\boldsymbol{\varepsilon}_n)$ and X, Y components of $\boldsymbol{v}$ and $\boldsymbol{a}$.

Starting with the $X = 750$ m location, the velocity direction in the X, Y system is obtained by differentiating Equation (i) with respect to X, obtaining

$$Y' = \frac{dY}{dX} = \frac{2A\pi}{L}\cos\left(\frac{2\pi X}{l}\right) = \frac{2\times 100\pi}{2000}\times\cos\left(\frac{2\pi 750}{2000}\right)$$
$$= -0.222 \Rightarrow \beta = \tan^{-1}(-0.222) = -12.52^\circ.$$

Hence, $\boldsymbol{\varepsilon}_t$ and $\boldsymbol{v}$ are pointed at 12.52° below the horizontal, and the velocity components in the X, Y system are

$$v_X = v\cos\beta = 27.78 \text{ m/s}\times\cos(-12.52^\circ) = 27.12 \text{ m/s}$$
$$v_Y = v\sin\beta = 27.78 \text{ m/s}\times\sin(-12.52^\circ) = -6.02 \text{ m/s},$$

where $v = 100$ km/h $\times$ 1000 m/h $\times$ 1 h/3600 s $= 27.78$ m/s. Figure 2.20a illustrates these components.

To find $a_n = v^2/\rho$, we need ρ. We obtained Y' above but still need Y'', defined as

$$Y'' = -A\left(\frac{2\pi}{L}\right)^2\sin\left(\frac{2\pi X}{L}\right)$$
$$= -100\left(\frac{2\pi}{2000}\right)^2\times\sin\left(\frac{2\pi 750}{2000}\right) = -6.98\times 10^{-4}\text{ m}^{-1}.$$

Hence,

$$\frac{1}{\rho} = \frac{|Y''|}{\left[1+(Y'')^2\right]^{3/2}} = \frac{6.98\times 10^{-4}}{\left[1+(-0.222)^2\right]^{3/2}}$$
$$= 6.49\times 10^{-4}\text{ m}^{-1} \Rightarrow \rho = 1540 \text{ m},$$

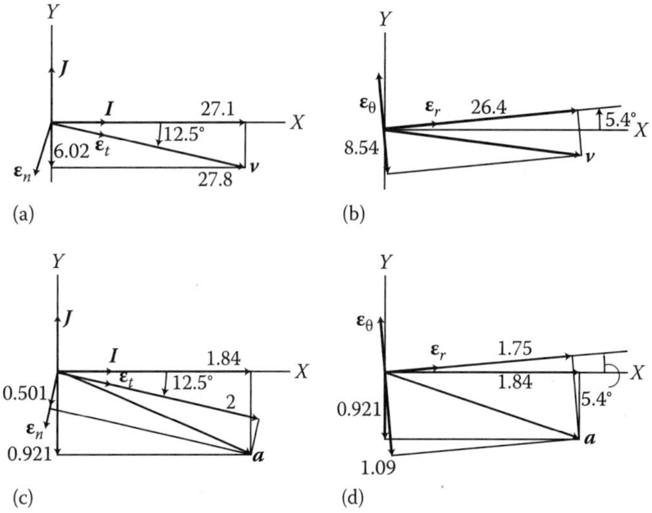

FIGURE 2.20
Velocity components (m/s) at $X = 750$ m: (a) path and Cartesian components, (b) polar components. Acceleration components (m/s^2) at $X = 750$ m: (c) path and Cartesian components, (d) polar components.

and

$$a_n = 27.78^2\ (\text{m/s})^2/1540\text{ m} = 0.501\text{ m/s}^2.$$

We were given $a_t = 2$ m/s^2, so we now have the $(\boldsymbol{\varepsilon}_t, \boldsymbol{\varepsilon}_n)$ components of $\boldsymbol{a}$ and need to find $\boldsymbol{a}$'s X and Y components. Compare the orientation of the $(\boldsymbol{\varepsilon}_t, \boldsymbol{\varepsilon}_n)$ vectors in Figure 2.19 at $X = 750$ m with Figure 2.8 that we used to derive the path velocity and acceleration components. The 180° "flip" in direction of $\boldsymbol{\varepsilon}_n$ from Figures 2.8 through 2.19 will regularly be experienced in using path unit vectors. This flip in direction for $\boldsymbol{\varepsilon}_n$ means that we cannot use the coordinate transformation results of Figure 2.10 and Equation 2.40. We could start over and redraw Figure 2.10 with $\boldsymbol{\varepsilon}_n$ flipped; however, it is easier to simply project the components of a_t, a_n into the X, Y system. Figure 2.20c illustrates $\boldsymbol{a}$'s path coordinates. Summing components in the X- and Y-directions gives

$$a_X = a_t \cos 12.5^\circ - a_n \sin 12.5^\circ$$
$$= 2 \times 0.976 - 0.501 \times 0.216 = 1.84 \text{ m/s}^2$$
$$a_Y = -a_t \sin 12.5^\circ - a_n \cos 12.5^\circ$$
$$= -2 \times 0.216 - 0.501 \times 0.976 = -0.921 \text{ m/s}^2.$$

These component results are shown in Figure 2.20c.

Proceeding to the polar-coordinate component definitions, we need to first find $r_Y = Y\,(X = 750 \text{ m}) = 100 \text{ m} \times \sin(2\times 750\ \pi/2000) = 70.7$ m to define θ as $\theta = \tan^{-1}(r_Y/r_X) = \tan^{-1}(70.7/750) = 5.39^\circ$.

Applying the coordinate transformation of Equation 2.41 to the Cartesian velocity and acceleration components gives

$$\begin{Bmatrix} v_r \\ v_\theta \end{Bmatrix} = \begin{bmatrix} \cos 5.39^\circ & \sin 5.39^\circ \\ -\sin 5.39^\circ & \cos 5.39^\circ \end{bmatrix} \begin{Bmatrix} v_X \\ v_Y \end{Bmatrix}$$
$$= \begin{bmatrix} 0.996 & 0.0939 \\ -0.0939 & 0.996 \end{bmatrix} \begin{Bmatrix} 27.12 \\ -6.02 \end{Bmatrix}$$
$$= \begin{Bmatrix} 26.4 \\ -8.54 \end{Bmatrix} \text{ m/s}, \qquad \text{(ii)}$$

and

$$\begin{Bmatrix} a_r \\ a_\theta \end{Bmatrix} = \begin{bmatrix} \cos 5.39^\circ & \sin 5.39^\circ \\ -\sin 5.39^\circ & \cos 5.39^\circ \end{bmatrix} \begin{Bmatrix} a_X \\ a_Y \end{Bmatrix}$$
$$= \begin{bmatrix} 0.996 & 0.0939 \\ -0.0939 & 0.996 \end{bmatrix} \begin{Bmatrix} -1.84 \\ -0.921 \end{Bmatrix}$$
$$= \begin{Bmatrix} 1.75 \\ -1.09 \end{Bmatrix} \text{ m/s}^2. \qquad \text{(iii)}$$

The velocity component results are illustrated in Figure 2.20b; the acceleration components are shown in Figure 2.20d and conclude the tasks for the $X = 750$ m location.

The $X = 1750$ m location results start by finding β from

$$Y' = \frac{dY}{dX} = \frac{2A\pi}{L}\cos\left(\frac{2\pi X}{l}\right) = \frac{2\times 100\pi}{2000} \times \cos\left(\frac{2\pi 750}{2000}\right)$$
$$= -.222 \Rightarrow \beta = \tan^{-1}(-.222) = -12.52^\circ.$$

Hence, v and ε_t are pointed 12.52° above the horizontal. Also, note in Figure 2.19 that ε_n has flipped directions by 180°, and that v, a_t, a_n, ρ are unchanged. The X, Y components of the velocity are

$$v_X = v\cos\beta = 27.78 \text{ m/s} \times \cos(12.52^\circ) = 27.12 \text{ m/s}$$
$$v_Y = v\sin\beta = 27.78 \text{ m/s} \times \sin(12.52^\circ) = 6.02 \text{ m/s}.$$

The orientation of ($\boldsymbol{\varepsilon}_t$, $\boldsymbol{\varepsilon}_n$) is the same in Figure 2.19 at $X = 1750$ m as in Figure 2.10; hence, Equation 2.40 applies and can be used to obtain $\boldsymbol{a}$'s components as

$$\begin{Bmatrix} a_X \\ a_Y \end{Bmatrix} = \begin{bmatrix} \cos\beta & -\sin\beta \\ \sin\beta & \cos\beta \end{bmatrix} \begin{Bmatrix} a_t \\ a_n \end{Bmatrix}$$
$$= \begin{bmatrix} \cos 12.52^\circ & -\sin 12.52^\circ \\ \sin 12.52^\circ & \cos 12.52^\circ \end{bmatrix} \begin{Bmatrix} 2 \\ 0.501 \end{Bmatrix}$$
$$= \begin{Bmatrix} 1.844 \\ 0.9226 \end{Bmatrix} \text{ m/s}^2.$$

This concludes the Cartesian-coordinate results. The procedure for obtaining the polar-coordinate components from the Cartesian components starts with

$$\theta = \tan^{-1}\left(\frac{r_Y}{r_X}\right) = \tan^{-1}\left(\frac{-70.7}{1750}\right) = -2.313^\circ$$

and then follows the same steps used earlier.

2.8 Time-Derivative Relationships in Two Coordinate Systems

In each of the preceding sections, two coordinate systems were used to develop useful relationships for the velocity and acceleration of a point P with respect to the X, Y coordinate system. In the polar-coordinate developments, the X, Y and $\boldsymbol{\varepsilon}_r$, $\boldsymbol{\varepsilon}_\theta$ references were used, while in the path-coordinate developments, the X, Y and $\boldsymbol{\varepsilon}_t$, $\boldsymbol{\varepsilon}_n$ references applied. In Section 2.9, velocity and acceleration relationships will be used to derive similar expressions for two Cartesian coordinate systems. However, before taking this step, a fundamental kinematic relationship is needed for the time derivative of a vector in two Cartesian-coordinate systems. Figure 2.21 illustrates our earlier X, Y system and a new x, y system. The x', y'

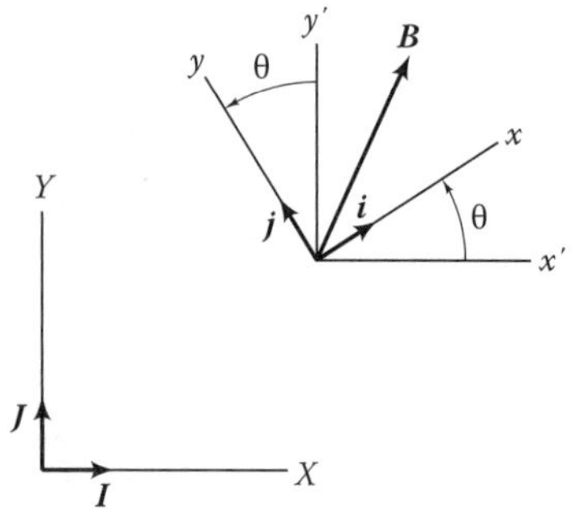

FIGURE 2.21
X, Y and x, y coordinate systems.

system has a common origin with the x, y system and the x'- and y'-axes are parallel, respectively, to the X- and Y-axes. The orientation of the x-, y-axes with respect to the X-, Y-axes is defined by the time-varying angle θ (t). The origin of the x-, y-axes can be moving around arbitrarily in the X, Y coordinate system. A vector $\boldsymbol{B}$ is shown, and following Equations 2.15 and 2.19, its component definitions in the x, y and X, Y coordinate systems are related by

$$\begin{Bmatrix} B_x \\ B_y \end{Bmatrix} = \begin{bmatrix} \cos\theta(t) & \sin\theta(t) \\ -\sin\theta(t) & \cos\theta(t) \end{bmatrix} \begin{Bmatrix} B_X \\ B_Y \end{Bmatrix}, \tag{2.43}$$

and

$$\begin{Bmatrix} B_X \\ B_Y \end{Bmatrix} = \begin{bmatrix} \cos\theta(t) & -\sin\theta(t) \\ \sin\theta(t) & \cos\theta(t) \end{bmatrix} \begin{Bmatrix} B_x \\ B_y \end{Bmatrix}. \tag{2.44}$$

The only "new" feature of the present situation, versus the results of Section 2.4, relating the components of a vector in two coordinate systems, is that θ is now a function of time. The time derivatives of $\boldsymbol{B}$ with respect to the X, Y and x, y coordinate system are

$$\dot{\boldsymbol{B}} = \left.\frac{d\boldsymbol{B}}{dt}\right|_{X,Y} = \boldsymbol{I}\dot{B}_X + \boldsymbol{J}\dot{B}_Y. \tag{2.45}$$

and

$$\widehat{\dot{\boldsymbol{B}}} = \left.\frac{d\boldsymbol{B}}{dt}\right|_{x,y} = \boldsymbol{i}\dot{B}_x + \boldsymbol{j}\dot{B}_y. \tag{2.46}$$

The "hat" convention denotes the derivative of $\boldsymbol{B}$ with respect to the x, y coordinate system and will be used throughout this chapter. The question to be addressed is as follows: How are $\dot{B}$ and $\widehat{\dot{\boldsymbol{B}}}$ related?

As a first step toward answering this question, we define the angular velocity vector of the x, y coordinate system relative to the X, Y system as

$$\boldsymbol{\omega} = \boldsymbol{k}\dot{\theta} = \boldsymbol{K}\dot{\theta}. \tag{2.47}$$

The positive sign for ω is defined by the right-hand screw convention illustrated in Figure 2.22. The screw illustrated in Figure 2.22 would advance in the $\boldsymbol{K}$ (or $\boldsymbol{k}$) direction due to increasing θ (counterclockwise rotation of the x-, y-axes relative to X-, Y-axes). To obtain the desired relationship, we start with

$$\boldsymbol{B} = \boldsymbol{i}B_x + \boldsymbol{j}B_y, \tag{2.48}$$

which has $\boldsymbol{B}$ defined in terms of its components in the x, y system (although we want the time derivative of $\boldsymbol{B}$ with respect to the X, Y system). Note that in the X, Y system, $\boldsymbol{i}$ and $\boldsymbol{j}$ are time varying. Differentiating Equation 2.48 with respect to time in the X, Y system yields

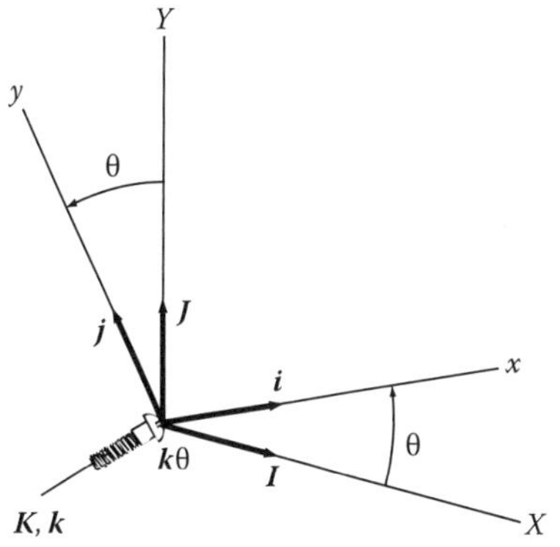

FIGURE 2.22
Right-hand screw convention to define a positive sign for $\boldsymbol{\omega}$.

$$\dot{\boldsymbol{B}} = \boldsymbol{i}\dot{B}_x + \boldsymbol{j}\dot{B}_y + \dot{\boldsymbol{i}}B_x + \dot{\boldsymbol{j}}B_y, \tag{2.49}$$

where

$$\dot{\boldsymbol{i}} = \left.\frac{d\boldsymbol{i}}{dt}\right|_{X,Y}, \quad \dot{\boldsymbol{j}} = \left.\frac{d\boldsymbol{i}}{dt}\right|_{X,Y}.$$

To find $\dot{\boldsymbol{i}}$ and $\dot{\boldsymbol{j}}$, we first state $\boldsymbol{i}$ and $\boldsymbol{j}$ in terms of their X, Y components:

$$\boldsymbol{i} = \boldsymbol{I}\cos\theta + \boldsymbol{J}\sin\theta, \quad \boldsymbol{j} = -\boldsymbol{I}\sin\theta + \boldsymbol{J}\cos\theta. \tag{2.50}$$

Then, differentiating with respect to time while holding $\boldsymbol{I}$ and $\boldsymbol{J}$ constant gives

$$\begin{aligned} \dot{\boldsymbol{i}} &= \dot{\boldsymbol{\theta}}(-\boldsymbol{I}\sin\theta + \boldsymbol{J}\cos\theta) = \boldsymbol{j}\dot{\theta} \\ \dot{\boldsymbol{j}} &= \dot{\theta}(-\boldsymbol{I}\cos\dot{\theta} - \boldsymbol{J}\sin\dot{\theta}) = -\boldsymbol{i}\dot{\theta}. \end{aligned}$$

Note that the same result is obtained by

$$\begin{aligned} \dot{\boldsymbol{i}} &= \boldsymbol{\omega} \times \boldsymbol{i} = \boldsymbol{K}\dot{\theta} \times (\boldsymbol{I}\cos\dot{\theta} + \boldsymbol{J}\sin\theta) = \boldsymbol{K}\dot{\theta} \times \boldsymbol{i} = \boldsymbol{j}\dot{\theta} \\ \dot{\boldsymbol{j}} &= \boldsymbol{\omega} \times \boldsymbol{j} = \boldsymbol{K}\dot{\theta} \times (-\boldsymbol{I}\sin\dot{\theta} + \boldsymbol{J}\cos\theta) = \boldsymbol{K}\dot{\theta} \times \boldsymbol{j} = -\boldsymbol{i}\dot{\theta}, \end{aligned}$$

with $\boldsymbol{i}$ and $\boldsymbol{j}$ defined by Equation 2.50. Hence, Equation 2.49 can be written

$$\begin{aligned} \dot{\boldsymbol{B}} &= \boldsymbol{i}\dot{B}_x + \boldsymbol{j}\dot{B}_y + \boldsymbol{\omega} \times \boldsymbol{i}B_x + \boldsymbol{\omega} \times \boldsymbol{j}B_y \\ &= \boldsymbol{i}\dot{B}_x + \boldsymbol{j}\dot{B}_y + \boldsymbol{\omega} \times (\boldsymbol{i}B_x + \boldsymbol{j}B_y). \end{aligned} \tag{2.51}$$

From Equation 2.46, the final result is

$$\dot{\boldsymbol{B}} = \overset{\frown}{\dot{\boldsymbol{B}}} + \boldsymbol{\omega} \times \boldsymbol{B}, \tag{2.52a}$$

or

$$\left.\frac{d\boldsymbol{B}}{dt}\right|_{X,Y} = \left.\frac{d\boldsymbol{B}}{dt}\right|_{x,y} + \boldsymbol{\omega} \times \boldsymbol{B}. \tag{2.52b}$$

This is a fundamental result that can be used to relate time derivatives of the same vector in two coordinate systems. Note particularly that $\boldsymbol{B}$ can be *any* kinematic vector, that is, position, velocity, etc. We will have a serious need for this relationship in Section 2.9.

This development used vector relationship to derive Equations 2.51, but simple differentiation can also be used to obtain the same relationship. This material assumes that you know some matrix algebra and is optional. Equation 2.43 can be written symbolically as

$$(B)_I = [A]^T(B)_i, \tag{2.53}$$

where, as you may recall, $[A]$ is the direction-cosine matrix, $[A]^T$ is its transpose, and

$$[A]^T[A] = [I], \tag{2.54}$$

with $[I]$ the identity matrix. Comparing Equation 2.53 and its expanded version in Equation 2.44 shows that $(B)_I$ is a column vector containing $\boldsymbol{B}$'s components in the X, Y systems, and $(B)_i$ is a column vector containing the components of $\boldsymbol{B}$ in the x, y systems.

Differentiating Equation 2.44 gives

$$\begin{Bmatrix} \dot{B}_X \\ \dot{B}_Y \end{Bmatrix} = \begin{bmatrix} \cos\theta & -\sin\theta \\ \sin\theta & \cos\theta \end{bmatrix} \begin{Bmatrix} \dot{B}_x \\ \dot{B}_y \end{Bmatrix} + \dot{\theta}\begin{bmatrix} -\sin\theta & -\cos\theta \\ \cos\theta & -\sin\theta \end{bmatrix} \begin{Bmatrix} B_x \\ B_y \end{Bmatrix}. \tag{2.55}$$

Observe that the column vectors given here from left to right contain components of $\dot{\boldsymbol{B}}$, $\overset{\frown}{\dot{\boldsymbol{B}}}$, and $\boldsymbol{B}$. Restating Equation 2.55 symbolically gives

$$(\dot{B})_I = [A]^T(\overset{\frown}{\dot{\boldsymbol{B}}})_i + [\dot{A}]^T(B)_i.$$

Inserting $[A]^T[A] = [I]$ into the second term gives

$$(\dot{B})_I = [A]^T\left\{(\overset{\frown}{\dot{B}})_i + [A][\dot{A}]^T(B)_i\right\}. \tag{2.56}$$

Substitution into the last term on the right for $[A]$ and $[\dot{A}]^T$ gives

$$\begin{aligned}[A][\dot{A}]^T(B)_i &= \dot{\theta}\begin{bmatrix} \cos\theta & \sin\theta \\ -\sin\theta & \cos\theta \end{bmatrix}\begin{bmatrix} -\sin\theta & -\cos\theta \\ \cos\theta & -\sin\theta \end{bmatrix}\begin{Bmatrix} B_x \\ B_y \end{Bmatrix} \\ &= \begin{bmatrix} 0 & -\dot{\theta} \\ \dot{\theta} & 0 \end{bmatrix}\begin{Bmatrix} B_x \\ B_y \end{Bmatrix} = \begin{Bmatrix} -\dot{\theta}\ B_y \\ \dot{\theta}\ B_x \end{Bmatrix} = \begin{Bmatrix} (\boldsymbol{\omega} \times B)_x \\ (\omega \times B)_y \end{Bmatrix}.\end{aligned} \tag{2.57}$$

The correctness of the last step in Equation 2.57 follows from

$$\boldsymbol{B} = \boldsymbol{i}B_x + \boldsymbol{j}B_y, \quad \boldsymbol{\omega} = \boldsymbol{k}\dot{\theta},$$

which gives

$$\boldsymbol{\omega} \times \boldsymbol{B} = \boldsymbol{i}(\omega \times B)_x + \boldsymbol{j}(\omega \times B)_y = -\boldsymbol{i}\dot{\theta}B_y + \boldsymbol{j}\dot{\theta}B_x.$$

Substitution from Equation 2.57 into Equation 2.56 yields

$$(\dot{B})_I = [A]^T(\overset{\frown}{\dot{B}} + \omega \times B)_i. \tag{2.58}$$

This result is a matrix statement of Equation 2.52 with the components of $\dot{B}$ given in the X, Y system, and the components of $\overset{\frown}{\dot{\boldsymbol{B}}}$ and $(\boldsymbol{\omega} \times \boldsymbol{B})$ given in the x, y system.

2.9 Velocity and Acceleration Relationships in Two Cartesian Coordinate Systems

We have been dealing with the planar kinematics problems involving two coordinate systems in the previous three sections, treating in sequence, polar coordinates, path coordinates, and, finally, time-derivative relationships in two Cartesian coordinate systems. This section extends the trend to velocity and acceleration relationships in two coordinate systems.

Contrary to your possible suspicions, the idea of using two coordinate systems is introduced in kinematics to make problem solutions *easier*, by making it possible to break a large task into two smaller (and more easily visualized) tasks. To demonstrate this point, Figure 2.23 illustrates a passenger moving along the aisle of an airplane. All of the motion occurs in the plane of the figure, and the "pitch" attitude of the airplane with respect to ground is defined by θ. A point on the plane (e.g., the plane's mass center) has displacement,

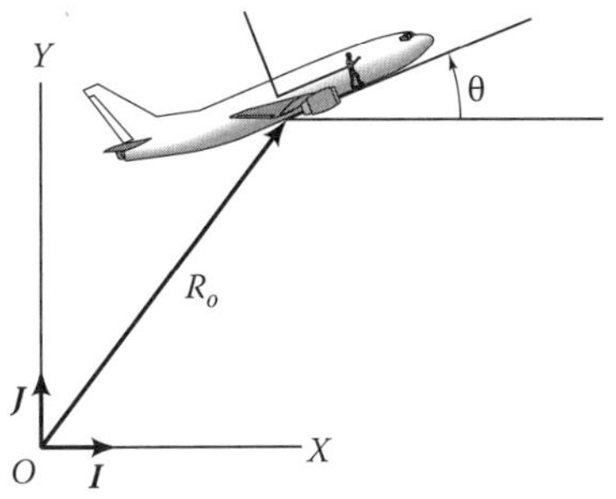

FIGURE 2.23
An airplane passenger moving down the aisle as the airplane moves with respect to ground and pitches upward.

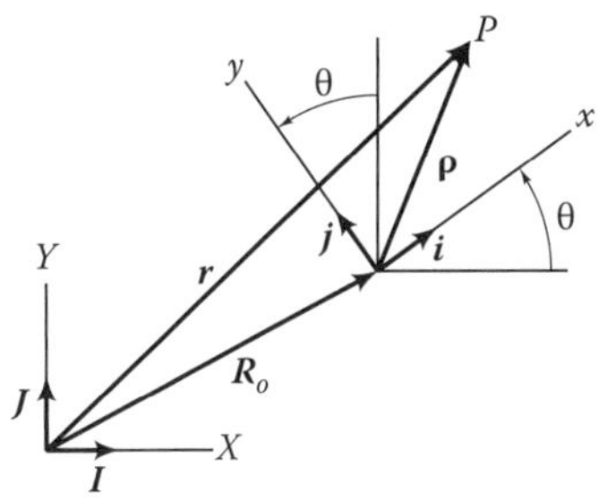

FIGURE 2.24
Two-coordinate model to locate passenger motion relative to ground.

velocity, and acceleration vectors with respect to ground. The passenger can have a position, velocity, and acceleration relative to the airplane, and the pitching rate $\dot{\theta}$ and pitch-acceleration rate $\ddot{\theta}$ can vary. Given this situation, someone wants to know the velocity and acceleration of the passenger with respect to ground.

This complicated problem can be simplified considerably by introducing a new coordinate system fixed to the plane, keeping track of the passenger with respect to this new coordinate system, and then accounting for motion of the origin of the new coordinate system with respect to ground. This idea is illustrated in Figure 2.24, with the X, Y system fixed to ground, and the x, y system fixed to the airplane. The vector $\boldsymbol{R}_o = \boldsymbol{I}R_{oX} + \boldsymbol{J}R_{oY}$ locates the origin of the x, y system in the X, Y system, and $\boldsymbol{\rho} = \boldsymbol{i}x + \boldsymbol{j}y$ locates the passenger (point P) in the x, y system. The angle θ defines the orientation of the x, y system (and the airplane) relative to the X, Y system.

The vector $\boldsymbol{r}$ locates point P in the X, Y system, and we want to develop *useful* relationships for $\dot{\boldsymbol{r}}$ (the velocity) and $\ddot{\boldsymbol{r}}$ (the acceleration) of P with respect to the X, Y system. Note from Figure 2.24 that

$$\boldsymbol{r} = \boldsymbol{R}_o + \boldsymbol{\rho}.$$

Taking the time derivative of this equation with respect to the X, Y system yields

$$\left.\frac{d\boldsymbol{r}}{dt}\right|_{X,Y} = \left.\frac{d\boldsymbol{R}_o}{dt}\right|_{X,Y} + \left.\frac{d\boldsymbol{\rho}}{dt}\right|_{X,Y}. \tag{2.59}$$

We return to $\dot{\boldsymbol{B}} = \hat{\boldsymbol{B}} + \boldsymbol{\omega} \times \boldsymbol{B}$ to obtain

$$\left.\frac{d\boldsymbol{\rho}}{dt}\right|_{X,Y} = \left.\frac{d\boldsymbol{\rho}}{dt}\right|_{x,y} + \boldsymbol{\omega} \times \boldsymbol{\rho}, \tag{2.60}$$

where

$$\boldsymbol{\omega} = \boldsymbol{k}\dot{\theta} = \boldsymbol{K}\dot{\theta} \tag{2.61}$$

is the angular velocity of the x, y system relative to the X, Y system. Substitution from Equation 2.60 into Equation 2.59 gives

$$\left.\frac{d\boldsymbol{r}}{dt}\right|_{X,Y} = \left.\frac{d\boldsymbol{R}_o}{dt}\right|_{X,Y} + \left.\frac{d\boldsymbol{\rho}}{dt}\right|_{x,y} + \boldsymbol{\omega} \times \boldsymbol{\rho},$$

or

$$\dot{\boldsymbol{r}} = \dot{\boldsymbol{R}}_o + \hat{\boldsymbol{\rho}} + \boldsymbol{\omega} \times \boldsymbol{\rho}. \tag{2.62}$$

Some authors denote $\hat{\boldsymbol{\rho}}$ as v_{rel} for velocity (relative). This notation can be confusing, since it suggests that "somewhere" in contrast to this "relative" velocity, there is an "absolute" velocity, which is simply not the case. All position, velocity, and acceleration vectors are defined "relative" to some reference system.

The acceleration of point P, with respect to the X, Y system, is obtained by differentiation of Equation 2.62, yielding

$$\ddot{\boldsymbol{r}} = \left.\frac{d^2\boldsymbol{r}}{dt^2}\right|_{X,Y} = \ddot{\boldsymbol{R}}_o + \left.\frac{d\hat{\boldsymbol{\rho}}}{dt}\right|_{X,Y} + \dot{\boldsymbol{\omega}} \times \boldsymbol{\rho} + \boldsymbol{\omega} \times \left.\frac{d\boldsymbol{\rho}}{dt}\right|_{X,Y}, \tag{2.63}$$

where

$$\dot{\boldsymbol{\omega}} = \boldsymbol{k}\ddot{\theta} = \boldsymbol{K}\ddot{\theta} = \frac{d\boldsymbol{\omega}}{dt}|_{X,Y} \tag{2.64}$$

is the angular-acceleration vector. The same "right-hand-rule" convention holds true here as for $\boldsymbol{\omega}$. The direct application of $\dot{\boldsymbol{B}} = \hat{\boldsymbol{B}} + \boldsymbol{\omega} \times \boldsymbol{B}$ to the second right-hand term in Equation 2.63 yields

$$\left.\frac{d\hat{\boldsymbol{\rho}}}{dt}\right|_{X,Y} = \left.\frac{d\hat{\boldsymbol{\rho}}}{dt}\right|_{x,y} + \boldsymbol{\omega} \times \hat{\boldsymbol{\rho}} = \hat{\hat{\boldsymbol{\rho}}} + \boldsymbol{\omega} \times \hat{\boldsymbol{\rho}}.$$

Substituting this result and our earlier definition for $\left.\frac{d\boldsymbol{\rho}}{dt}\right|_{x,y}$ from Equation 2.60 provides

$$\ddot{\boldsymbol{r}} = \ddot{\boldsymbol{R}}_o + \overset{\frown\frown}{\boldsymbol{\rho}} + 2\boldsymbol{\omega} \times \overset{\frown}{\boldsymbol{\rho}} + (\dot{\boldsymbol{\omega}} \times \boldsymbol{\rho}) + \boldsymbol{\omega} \times (\boldsymbol{\omega} \times \boldsymbol{\rho}), \quad (2.65)$$

which is the final desired result.

2.9.1 Comparisons to Polar-Coordinate Definitions

Equations 2.62 for $\dot{\boldsymbol{r}}$ and Equation 2.65 for $\ddot{\boldsymbol{r}}$ may strike you as less than the promised "useful" relationship. However, they are not as alien or difficult as they may first appear. The first term in Equation 2.62b, $\dot{\boldsymbol{R}}_o$, gives the velocity of point o with respect to the system, and the first term of Equation 2.65 gives o's acceleration, $\ddot{\boldsymbol{R}}_o$, with respect to the X, Y system. The remaining terms have a one-to-one correspondence to terms in our earlier polar-coordinate results.

Parallels between the present vector results and earlier polar-coordinate definitions for velocity and acceleration become more apparent if we require that $\boldsymbol{R}_o = 0$ and $\boldsymbol{\rho}$ lie along the x-axis; that is,

$$\boldsymbol{\rho} = \boldsymbol{i}x. \quad (2.66)$$

Hence,

$$\overset{\frown}{\boldsymbol{\rho}} = \left.\frac{d\boldsymbol{\rho}}{dt}\right|_{x,y} = \boldsymbol{i}\dot{x}, \quad \boldsymbol{\omega} \times \boldsymbol{\rho} = \boldsymbol{k}\dot{\theta} \times \boldsymbol{i}x = \boldsymbol{j}(x\dot{\theta})$$
$$\overset{\frown\frown}{\boldsymbol{\rho}} = \left.\frac{d^2\boldsymbol{\rho}}{dt^2}\right|_{x,y} = \boldsymbol{i}\ddot{x}, \quad \dot{\boldsymbol{\omega}} \times \boldsymbol{\rho} = \boldsymbol{k}\ddot{\theta} \times \boldsymbol{i}x = \boldsymbol{j}(x\ddot{\theta}) \quad (2.67)$$
$$\boldsymbol{\omega} \times (\boldsymbol{\omega} \times \boldsymbol{\rho}) = \boldsymbol{k}\dot{\theta} \times \boldsymbol{j}(x\dot{\theta}) = -\boldsymbol{i}(x\dot{\theta}^2),$$
$$\boldsymbol{\omega} \times \overset{\frown}{\boldsymbol{\rho}} = \boldsymbol{k}\dot{\theta} \times \boldsymbol{i}\dot{x} = \boldsymbol{j}(\dot{\theta}\dot{x}).$$

To get $\dot{\boldsymbol{r}}$, we substitute the first two terms into Equation 2.63 (with $\dot{\boldsymbol{R}}_o = 0$), obtaining

$$\dot{\boldsymbol{r}} = \boldsymbol{i}\dot{x} + \boldsymbol{j}\dot{\theta}x.$$

Comparing this result to Equation 2.27 shows the following parallels in physical terms:

$$\boldsymbol{\varepsilon}_r\dot{r} = \boldsymbol{i}\dot{x} = \overset{\frown}{\rho}$$
$$\boldsymbol{\varepsilon}_\theta(r\dot{\theta}) = \boldsymbol{j}(x\dot{\theta}) = \boldsymbol{\omega} \times \boldsymbol{\rho}.$$

For comparison of the acceleration terms, substituting the last four terms from Equation 2.67 into Equation 2.65 gives (with $\ddot{\boldsymbol{R}}_o = 0$)

$$\ddot{\boldsymbol{r}} = \boldsymbol{i}\ddot{x} + 2\boldsymbol{j}\dot{\theta}\dot{x} + \boldsymbol{j}\ddot{\theta}x - \boldsymbol{i}\dot{\theta}^2 x$$
$$= \boldsymbol{i}(\ddot{x} - x\dot{\theta}^2) + \boldsymbol{j}(x\ddot{\theta} + 2\dot{\theta}\dot{x}). \quad (2.68)$$

By comparison to the polar-coordinate definition of Equation 2.28, the following physical equivalence of terms is established:

$$\boldsymbol{\varepsilon}_r\ddot{r} = \boldsymbol{i}\ddot{x} = \hat{\hat{\boldsymbol{\rho}}}$$
$$2\boldsymbol{\varepsilon}_\theta\dot{r}\dot{\theta} = 2\boldsymbol{j}\dot{x}\dot{\theta} = 2\boldsymbol{\omega} \times \hat{\boldsymbol{\rho}} \text{ (Coriolis acceleration term)}$$
$$\boldsymbol{\varepsilon}_\theta r\ddot{\theta} = \boldsymbol{j}x\ddot{\theta} = \dot{\boldsymbol{\omega}} \times \boldsymbol{\rho} \text{ (Circumferential acceleration term)}$$
$$-\boldsymbol{\varepsilon}_r r\dot{\theta}^2 = -\boldsymbol{i}x\dot{\theta}^2$$
$$= \boldsymbol{\omega} \times (\boldsymbol{\omega} \times \boldsymbol{\rho}) \text{ (Centripetal acceleration term).}$$

Hence, Equation 2.65 merely presents old physical terms in a new vector format.

2.9.2 Coordinate-System Expressions for Kinematic Equations

The *vector* velocity and acceleration expressions derived above are

$$\dot{\boldsymbol{r}} = \dot{\boldsymbol{R}}_o + \hat{\boldsymbol{\rho}} + \boldsymbol{\omega} \times \boldsymbol{\rho} \quad (2.69a)$$
$$\ddot{\boldsymbol{r}} = \ddot{\boldsymbol{R}}_o + \hat{\hat{\boldsymbol{\rho}}} + 2\boldsymbol{\omega} \times \hat{\boldsymbol{\rho}} + \dot{\boldsymbol{\omega}} \times \boldsymbol{\rho} + \boldsymbol{\omega} \times (\boldsymbol{\omega} \times \boldsymbol{\rho}). \quad (2.69b)$$

These equations are valid for any coordinate system, but to apply them for a specific problem, a choice of coordinate systems must be made. To be useful, *every term* in a vector equation must be defined in terms of its components in the *same* coordinate system. There are "natural" choices for components descriptions of the terms cited in Equations 2.69; that is, by our definitions

$$\hat{\boldsymbol{\rho}} = \left.\frac{d\boldsymbol{\rho}}{dt}\right|_{x,y} = \boldsymbol{i}\dot{x} + \boldsymbol{j}\dot{y}$$
$$\dot{\boldsymbol{R}}_o = \left.\frac{d\boldsymbol{R}_o}{dt}\right|_{X,Y} = \boldsymbol{I}\dot{R}_{Xo} + \boldsymbol{J}\dot{R}_{Yo}.$$

In terms of "naturalness" and convenience, the vectors $\boldsymbol{\rho}$, $\hat{\boldsymbol{\rho}}$, and $\hat{\hat{\boldsymbol{\rho}}}$ are normally given in terms of their components in the x, y system, and $\dot{\boldsymbol{R}}_o$, $\ddot{\boldsymbol{R}}_o$, $\dot{\boldsymbol{r}}$, and $\ddot{\boldsymbol{r}}$ are normally given in the X, Y system. Given that $\boldsymbol{k} = \boldsymbol{K}$, the angular-velocity vector $\boldsymbol{\omega}$ can be expressed comfortably as either $\boldsymbol{\omega} = \boldsymbol{k}\dot{\theta}$ or $\boldsymbol{\omega} = \boldsymbol{K}\dot{\theta}$. Naturalness aside, *any* of these vectors can be expressed in terms of their components in *either* coordinate system.

Assuming that the components of $\dot{\boldsymbol{r}}$, and $\ddot{\boldsymbol{r}}$ are desired in the X, Y system, the natural component expression for Equations 2.69 are

$$(\dot{\boldsymbol{r}})_I = (\dot{\boldsymbol{R}}_o)_I + [A]^T\left\{\hat{\boldsymbol{\rho}} + (\boldsymbol{\omega} \times \boldsymbol{\rho})\right\}_i$$
$$(\ddot{\boldsymbol{r}})_I = (\ddot{\boldsymbol{R}}_o)_I + [A]^T\left\{\hat{\hat{\boldsymbol{\rho}}} + 2\boldsymbol{\omega} \times \hat{\boldsymbol{\rho}} + \dot{\boldsymbol{\omega}} \times \boldsymbol{\rho} + \boldsymbol{\omega} \times (\boldsymbol{\omega} \times \boldsymbol{\rho})\right\}_i. \quad (2.70)$$

Observe that $[A]^T$ is used to transform component expressions for vectors from the x, y to the X, Y system, following the developments of Section 2.4. If the components of $\dot{\boldsymbol{r}}$ and $\ddot{\boldsymbol{r}}$ are required in the x, y system, the following alternate expressions apply

$$(\dot{\boldsymbol{r}})_i = [A](\dot{\boldsymbol{R}}_o)_I + \left(\hat{\dot{\boldsymbol{\rho}}} + \boldsymbol{\omega} \times \boldsymbol{\rho}\right)_i$$
$$(\ddot{\boldsymbol{r}})_i = [A](\ddot{\boldsymbol{R}}_o)_I + \left\{\hat{\ddot{\boldsymbol{\rho}}} + (2\boldsymbol{\omega} \times \hat{\dot{\boldsymbol{\rho}}}) + (\dot{\boldsymbol{\omega}} \times \boldsymbol{\rho}) + \boldsymbol{\omega} \times (\boldsymbol{\omega} \times \boldsymbol{\rho})\right\}_i. \tag{2.71}$$

In this case, $[A]$ is used to transform the components of $\dot{\boldsymbol{R}}_o$ and $\ddot{\boldsymbol{R}}_o$ from the X, Y system to the x, y system.

In planar kinematics, one-coordinate kinematic definitions (using Cartesian, polar, and path-coordinate systems) are much more commonly used (and appropriate) than the two-coordinate kinematics results of Equations 2.69. However, the two coordinate system kinematic equations are frequently required to define the velocity and acceleration of a particle in three dimensions.

2.9.3 Coordinate System Observers

The preceding material of this section concentrated on obtaining definitions for vectors and the derivatives of vectors with respect to coordinate systems. The immediately preceding material concerns choices in stating vector equations in coordinate systems. One can easily get lost in terms of vectors, components, direction cosines, etc. and simply wonder, "What do all these vector expressions mean in terms of something approximating everyday experience?" The concept of "observers" fixed in coordinate systems can be useful in interpreting vector expressions, provides some answers to this question, and is the subject of this section.

Figure 2.25 largely repeats the results of Figure 2.24; however, two "observers" have been added to the figure. Observers O and o are riding in the X, Y and x, y coordinate systems, respectively. Some of the terms in the vector kinematic expressions,

$$\dot{\boldsymbol{r}} = \dot{\boldsymbol{R}}_o + \hat{\dot{\boldsymbol{\rho}}} + \boldsymbol{\omega} \times \boldsymbol{\rho}$$
$$\ddot{\boldsymbol{r}} = \ddot{\boldsymbol{R}}_o + \hat{\ddot{\boldsymbol{\rho}}} + 2\boldsymbol{\omega} \times \hat{\dot{\boldsymbol{\rho}}} + \dot{\boldsymbol{\omega}} \times \boldsymbol{\rho} + \boldsymbol{\omega} \times (\boldsymbol{\omega} \times \boldsymbol{\rho}),$$

can be explained in terms of the separate velocities and accelerations witnessed by these observers.

Observer O is only assigned to watch (1) point o (the origin of the x, y system located by $\boldsymbol{R}_o$), (2) point P (located by $\boldsymbol{r}$), and (3) θ, which defines the orientation of the x, y system with respect to the X, Y system. The vectors $\dot{\boldsymbol{r}}$ and $\dot{\boldsymbol{R}}_o$, can then be viewed, respectively, as the velocities of points P and o, "as seen by an observer in the X, Y system." Similarly, $\ddot{\boldsymbol{r}}$ and $\ddot{\boldsymbol{R}}_o$ are, respectively, the acceleration of points P and o as seen by an observer in the X, Y system.

Observer o is fixed in the x, y system and is assigned to watch point P and the angle θ. Hence, $\boldsymbol{\rho}$, $\hat{\dot{\boldsymbol{\rho}}}$, $\hat{\ddot{\boldsymbol{\rho}}}$, are, respectively, the position, velocity, and acceleration vectors of point P as seen by observer o fixed in the x, y system.

$\boldsymbol{\omega} = \boldsymbol{k}\dot{\theta} = \boldsymbol{K}\dot{\theta}$ is the angular velocity vector of the x, y system relative to the X, Y system and is the same vector whether seen by observer O or observer o. Moreover, the time derivative of $\boldsymbol{\omega}$ is the same whether performed with respect to the X, Y system (Observer O) or the x, y system (Observer o). To confirm this statement, Equation 2.52 shows

$$\left.\frac{d\boldsymbol{\omega}}{dt}\right|_{X,Y} = \left.\frac{d\boldsymbol{\omega}}{dt}\right|_{x,y} + \boldsymbol{\omega} \times \boldsymbol{\omega} = \left.\frac{d\boldsymbol{\omega}}{dt}\right|_{x,y} + 0,$$

or

$$\dot{\boldsymbol{\omega}} \times \hat{\dot{\boldsymbol{\omega}}}. \tag{2.72}$$

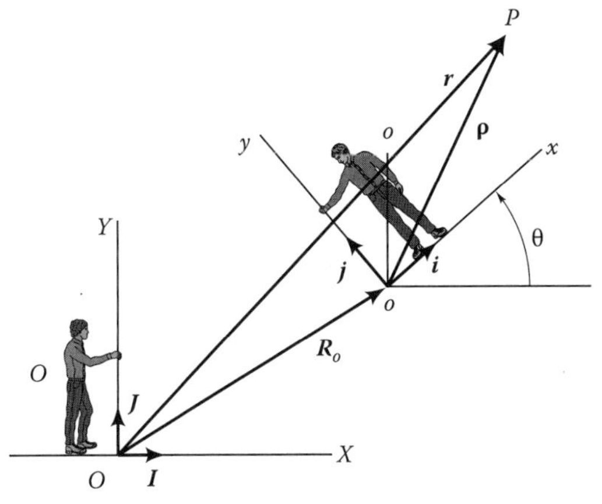

FIGURE 2.25
Observer O fixed in the X, Y system and observer o fixed in the x, y system.

Example Problem 2.8

Figure XP2.8 illustrates an airfreighter just before liftoff. The X, Y coordinate system is attached to ground; the x, y coordinate system is attached to the airplane with its origin at the airplane's mass center. The velocity and acceleration of the airplane's mass center are in the direction of flight and are 260 km/h and 6.5 m/s^2, respectively. The airplane has a pitch angle relative to the horizontal of 15° and has a constant positive pitch rate of 2° per second.

A washer is loose and sliding on the airplane's cargo-hold deck. The cargo-hold deck is parallel to the airplane's axis and 1.5 m below the freighter's mass center; that is, it is located at $y = -1.5$ m. At a given instant of time, the washer is located 6 m forward of the airplane's mass center. It is sliding toward the airplane's tail with velocity

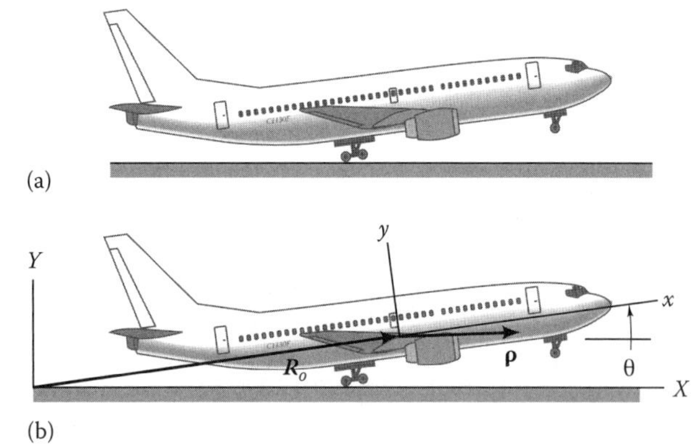

FIGURE XP2.8
(a) Air freighter at liftoff. (b) x, y and X, Y coordinate systems.

and acceleration (relative to the airplane) of 2.5 m/s and 1.0 m/s^2, respectively.

Tasks: Determine the velocity and acceleration of the washer relative to ground and state the answers in terms of components in the x, y and X, Y coordinate systems.

Solution

Developing the solution for this type of problem largely involves "mapping" the problem statement into the coordinate systems of Figure 2.24 and Equations 2.69

$$\dot{\boldsymbol{r}} = \dot{\boldsymbol{R}}_o + \hat{\dot{\boldsymbol{\rho}}} + \boldsymbol{\omega} \times \boldsymbol{\rho} \tag{2.69a}$$

$$\ddot{\boldsymbol{r}} = \ddot{\boldsymbol{R}}_o + \hat{\ddot{\boldsymbol{\rho}}} + 2\boldsymbol{\omega} \times \hat{\dot{\boldsymbol{\rho}}} + \dot{\boldsymbol{\omega}} \times \boldsymbol{\rho} + \boldsymbol{\omega} \times (\boldsymbol{\omega} \times \boldsymbol{\rho}). \tag{2.69b}$$

Figure XP2.8b shows the coordinate system and vector mapping corresponding to Figure 3.24. The X, Y coordinate system is attached to ground; the x, y coordinate system is attached to the airplane with its origin at the mass center. The vector $\boldsymbol{R}_o$ in Figure XP2.8b extends from the origin of the X, Y coordinate system to the airplane's mass center. The vector $\boldsymbol{\rho}$ extends from the airplane's mass center (origin of the x, y coordinate system) to the washer. Matching the problem-statement information with the variables of Equations 2.69 gives

$$\dot{\boldsymbol{R}}_o = \boldsymbol{i}260\ \frac{\text{km}}{\text{h}} \times \frac{1000\ \text{m}}{1\ \text{km}} \times \frac{1\ \text{h}}{3600\ \text{s}} = \boldsymbol{i}72.2\ \text{m/s}$$

$$\ddot{\boldsymbol{R}}_o = \boldsymbol{i}6.5\ \text{m/s}^2$$

$$\boldsymbol{\omega} = \boldsymbol{k}\ 2\ \frac{\text{degrees}}{\text{s}} \times \frac{2\pi\ \text{rad}}{360^\circ} = \boldsymbol{k}3.49 \times 10^{-2}\ \text{rad/s}$$

$$\dot{\boldsymbol{\omega}} = 0$$

$$\boldsymbol{\rho} = \boldsymbol{i}6 - \boldsymbol{j}1.5\ \text{m},\quad \hat{\dot{\boldsymbol{\rho}}} = -\boldsymbol{i}2.5\ \text{m/s},\quad \hat{\ddot{\boldsymbol{\rho}}} = -\boldsymbol{i}1\ \text{m/s}^2.$$

All of these vectors are defined in terms of their components in the x, y coordinate system, and we can proceed to determine the components of $\dot{\boldsymbol{r}}$ and $\ddot{\boldsymbol{r}}$ in terms of their components in the x, y system. Applying Equation 2.69a, the washer's velocity with respect to ground is

$$\begin{aligned}\dot{\boldsymbol{r}} &= \dot{\boldsymbol{R}}_o + \hat{\dot{\boldsymbol{\rho}}} + \boldsymbol{\omega} \times \boldsymbol{\rho}\\ &= \boldsymbol{i}72.2 - \boldsymbol{i}2.5 + \boldsymbol{k}3.49 \times 10^{-2} \times (\boldsymbol{i}6 - \boldsymbol{j}1.5)\ \text{m/s}\\ &= \boldsymbol{i}72.2 - \boldsymbol{i}2.5 + (\boldsymbol{j}0.209 + \boldsymbol{i}0.052) = \boldsymbol{i}69.7 + \boldsymbol{j}0.209\ \text{m/s}.\end{aligned}$$

Similarly, plugging into Equation 2.69b defines the washer's acceleration with respect to ground as

$$\begin{aligned}\ddot{\boldsymbol{r}} &= \ddot{\boldsymbol{R}}_o + \hat{\ddot{\boldsymbol{\rho}}} + 2\boldsymbol{\omega} \times \hat{\dot{\boldsymbol{\rho}}} + \dot{\boldsymbol{\omega}} \times \boldsymbol{\rho} + \boldsymbol{\omega} \times (\boldsymbol{\omega} \times \boldsymbol{\rho})\\ &= \boldsymbol{i}6.5 - \boldsymbol{i}1 + (2\boldsymbol{k}3.49 \times 10^{-2} \times -\boldsymbol{i}2.5) + 0\\ &\quad \times \boldsymbol{i}6 + \left[\boldsymbol{k}3.49\ 10^{-2} \times (\boldsymbol{j}0.209 + \boldsymbol{i}0.052)\right]\ \text{m/s}^2\\ &= \boldsymbol{i}5.5 - \boldsymbol{j}0.175 + (-\boldsymbol{i}7.3 \times 10^{-3} + \boldsymbol{j}1.8 \times 10^{-3})\\ &= \boldsymbol{i}5.49 - \boldsymbol{j}0.163\ \text{m/s}^2.\end{aligned}$$

Moving along to find the components of the velocity and acceleration in the X, Y coordinate system, Figure 2.4 and Equation 2.19 can be used to state the coordinate transformation, relating components of a vector in the x, y and X, Y coordinate systems. Applying Equation 2.19 for $\alpha = \theta = 15^\circ$,

$$\begin{aligned}\begin{Bmatrix} \dot{r}_X \\ \dot{r}_Y \end{Bmatrix} &= \begin{bmatrix} \cos\theta & -\sin\theta \\ \sin\theta & \cos\theta \end{bmatrix} \begin{Bmatrix} \dot{r}_X \\ \dot{r}_Y \end{Bmatrix}\\ &= \begin{bmatrix} 0.967 & -0.259 \\ 0.259 & 0.967 \end{bmatrix} \begin{Bmatrix} 69.7 \\ 0.209 \end{Bmatrix} = \begin{Bmatrix} 67.3 \\ 18.2 \end{Bmatrix}\ \text{m/s}.\end{aligned}$$

Similarly, the washer's acceleration components in the X, Y system are defined by

$$\begin{aligned}\begin{Bmatrix} \ddot{r}_X \\ \ddot{r}_Y \end{Bmatrix} &= \begin{bmatrix} 0.967 & -0.259 \\ 0.259 & 0.967 \end{bmatrix} \begin{Bmatrix} 5.49 \\ -0.163 \end{Bmatrix}\\ &= \begin{Bmatrix} 5.35 \\ 1.26 \end{Bmatrix}\ \text{m/s}^2.\end{aligned}$$

2.10 Relative Position, Velocity, and Acceleration Vectors between Two Points in the Same Coordinate System

The preceding sections dealt with the position, velocity, and acceleration of a point with respect to one or two coordinate systems. Our interest here concerns *relative* motion of two points in the *same* coordinate system. We will start with motion in a straight line before moving to planar motion. Figure 2.26 illustrates points A and B

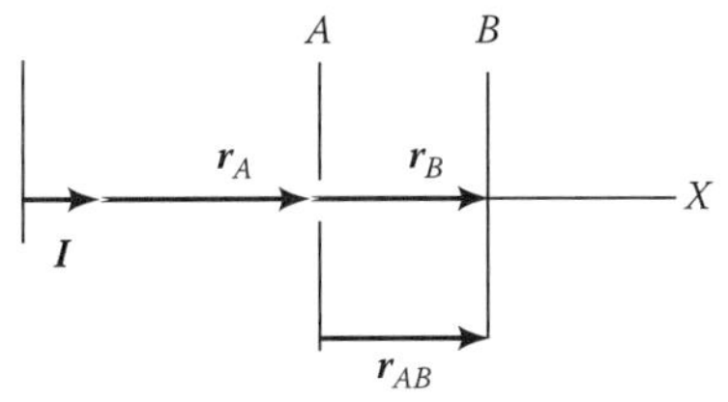

FIGURE 2.26
Relative positions of points A and B.

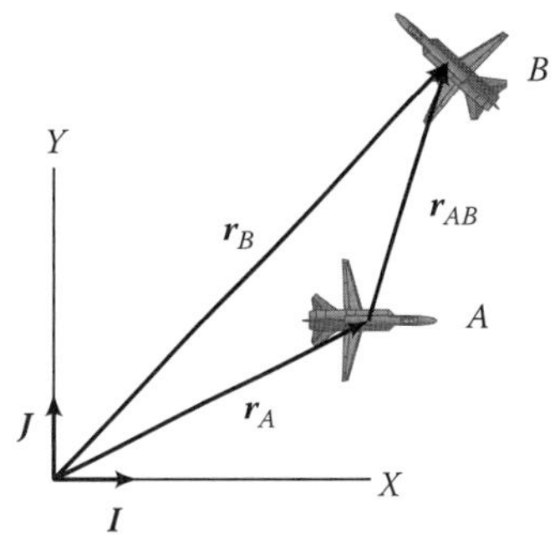

FIGURE 2.27
Points A and B located in the X, Y system by vectors $\boldsymbol{r}_A$ and $\boldsymbol{r}_B$.

located on the X-axis by vectors, $\boldsymbol{r}_A = \boldsymbol{I}r_A$, $\boldsymbol{r}_B = \boldsymbol{I}r_B$. The vectors are related by

$$\boldsymbol{I}r_B = \boldsymbol{I}(r_A + r_{AB}), \tag{2.73}$$

where $\boldsymbol{r}_{AB} = \boldsymbol{I}r_{AB}$ extends from A to B. Along the X-axis, the velocity of point B relative to point A is defined by

$$v_{B/A} = \dot{r}_{AB} = \dot{r}_B - \dot{r}_A. \tag{2.74}$$

Positive and negative signs of $v_{B/A}$ denote, respectively, that the relative velocity vector has the same or opposite direction as $\boldsymbol{I}$. The sign of $v_{B/A}$ does not necessarily tell us whether points B and A are moving toward or away from each other. To illustrate this point consider the following two cases:

a. $r_A = 5 + 10t \text{ m} \Rightarrow \dot{r}_A = 10 \text{ m/s},$
$r_B = 10 + 15t \text{ m} \Rightarrow \dot{r}_B = 15 \text{ m/s}$
$r_{AB} = r_B - r_A = 5 + 5t \text{ m}$
$v_{B/A} = \dot{r}_B - \dot{r}_A = 5 \text{ m/s}.$

b. $r_A = 5 + 10t \text{ m} \Rightarrow \dot{r}_A = 10 \text{ m/s},$
$r_B = -10 + 15t \text{ m} \Rightarrow \dot{r}_B = 15 \text{ m/s}$
$r_{AB} = r_B - r_A = -15 + 5t \text{ m}$
$v_{B/A} = \dot{r}_B - \dot{r}_A = 5 \text{ m/s}.$

In case (a), point B is initially ($t = 0$) to the right of point A, and both points have positive velocities (to the right). Because point B has a higher velocity, the distance between the two points increases progressively with time. In case (b), r_A is unchanged, but point B is now initially to the left of the origin. Hence, the distance between the two points will decrease for the first 3 s until point B overtakes point A. Note that the relative velocity is the same in both cases and conveys (only) the information that the velocity of point B exceeds that of point A by 5 m/s to the right. Returning to Figure 2.26, the acceleration of point B relative to point A along the X-axis is

$$a_{B/A} = \ddot{r}_{AB} = \ddot{r}_B - \ddot{r}_A. \tag{2.75}$$

Moving to two dimensions, Figure 2.27 illustrates two airplanes that are located in the X, Y plane by vectors $\boldsymbol{r}_A$ and $\boldsymbol{r}_B$. These vectors are related by

$$\boldsymbol{r}_B = \boldsymbol{r}_A + \boldsymbol{r}_{AB}, \tag{2.76}$$

and the velocity of point B relative to point A is defined by

$$\boldsymbol{v}_{B/A} = \dot{\boldsymbol{r}}_{AB} = \dot{\boldsymbol{r}}_B - \dot{\boldsymbol{r}}_A = \boldsymbol{v}_B - \boldsymbol{v}_A. \tag{2.77}$$

As with motion in a straight line, the relative velocity vector $\boldsymbol{v}_{B/A}$ does not tell us (by itself) whether airplanes A and B are moving toward or away from each other. The relative acceleration vector is defined as

$$\boldsymbol{a}_{B/A} = \ddot{\boldsymbol{r}}_{AB} = \ddot{\boldsymbol{r}}_B - \ddot{\boldsymbol{r}}_A = \boldsymbol{a}_B - \boldsymbol{a}_A. \tag{2.78}$$

Note that only one coordinate system is involved with this development, versus the two coordinates of the preceding section. Hence, $\boldsymbol{v}_{B/A}$ cannot generally be interpreted as, "the velocity of point B as seen by an observer in airplane A."

To illustrate this statement, consider Figure 2.28, which now includes the x, y coordinate system fixed to

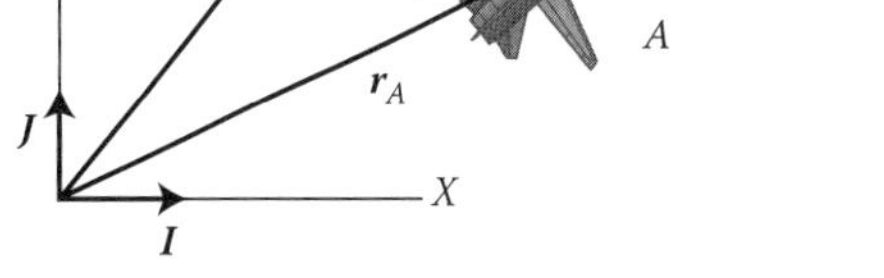

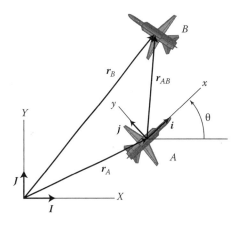

FIGURE 2.28
Two airplanes located in the X, Y system with the x, y system fixed to airplane A.

airplane A. The vector $\boldsymbol{r}_{AB}$ extends from point A to point B and (now) locates airplane B in the x, y system. Applying Equation 2.52 to $\boldsymbol{R}_B$ in Equation 2.76 yields

$$\left.\frac{d\boldsymbol{r}_B}{dt}\right|_{X,Y} = \left.\frac{d\boldsymbol{r}_A}{dt}\right|_{X,Y} + \left.\frac{d\boldsymbol{r}_{AB}}{dt}\right|_{x,y} + \boldsymbol{\omega} \times \boldsymbol{r}_{AB} \qquad (2.79)$$

or

$$\dot{\boldsymbol{r}}_B = \dot{\boldsymbol{r}}_A + \hat{\dot{\boldsymbol{r}}}_{AB} + \boldsymbol{\omega} \times \boldsymbol{r}_{AB} = \dot{\boldsymbol{r}}_A + \hat{\dot{\boldsymbol{r}}}_{AB} + \boldsymbol{k}\dot{\boldsymbol{\theta}} \times \boldsymbol{r}_{AB}, \qquad (2.80)$$

where $\boldsymbol{\omega} = \boldsymbol{k}\dot{\theta}$ is the angular velocity of the x, y coordinate system with respect to the X, Y system. $\hat{\dot{\boldsymbol{r}}}_{AB}$ is the velocity of airplane B as seen by an observer in airplane A (rotating with the airplane in the x, y coordinate system). By comparison to Equation 2.77,

$$\boldsymbol{v}_{B/A} = \boldsymbol{v}_B - \boldsymbol{v}_A = \dot{\boldsymbol{r}}_B - \dot{\boldsymbol{r}}_A = \hat{\dot{\boldsymbol{r}}}_{AB} + \boldsymbol{\omega} \times \boldsymbol{r}_{AB}. \qquad (2.81)$$

Hence, $\boldsymbol{v}_{B/A}$ (the velocity of plane B relative to plane A in the X, Y system) only coincides with $\hat{\dot{\boldsymbol{r}}}_{AB}$ (the velocity of point B as seen by an observer in plane A), if airplane A is flying in a straight line such that its angular velocity with respect to ground is zero; that is, $\boldsymbol{\omega} = \boldsymbol{k}\dot{\theta} = 0$. Restating Equation 2.81 as

$$\hat{\dot{\boldsymbol{r}}}_{AB} = \boldsymbol{v}_{B/A} - (\boldsymbol{\omega} \times \boldsymbol{r}_{AB}) \qquad (2.82)$$

provides a definition for in terms of $\boldsymbol{v}_{B/A}$ and $(\boldsymbol{\omega} \times \boldsymbol{r}_{AB})$.

Understanding the last term in this equation takes some thought. To simplify matters, think of points A and B as being fixed (motionless) in the X, Y plane; hence, $\boldsymbol{v}_B = \boldsymbol{v}_A = \boldsymbol{v}_{B/A} = 0$. However, in these simplified circumstances, Equation 2.82 states that $\hat{\dot{\boldsymbol{r}}}_{AB} = -(\boldsymbol{\omega} \times \boldsymbol{r}_{AB}) \neq 0$, provided that both $\boldsymbol{\omega}$ (the angular velocity of the x, y coordinates fixed in airplane A) and $\hat{\dot{\boldsymbol{r}}}_{AB}$ are not zero. Visualize a lighthouse situated at A and rotating with an angular velocity $\boldsymbol{\omega} = \boldsymbol{k}\dot{\theta}$. As the light beam sweeps past a target at point B (in the dark), an observer rotating with the light beam would see the target moving from left to right across his or her field of vision. Alternatively, visualize sitting in a chair that is stationary in the X, Y plane except for a counterclockwise rotation about a vertical Z-axis. While sitting in this chair, all of your surroundings would *appear* to be rotating in a clockwise direction. The vector $\hat{\dot{\boldsymbol{r}}}_{AB}$ defines the apparent or "sensed" velocity of an observer in the rotating x, y frame, and the term $(\boldsymbol{\omega} \times \boldsymbol{r}_{AB})$ is essential to capture this sensed velocity.

A similar outcome holds for relative accelerations as can be seen by differentiating Equation 2.80 with respect to the X, Y coordinate system to obtain

$$\begin{aligned}
\ddot{\boldsymbol{r}}_B &= \ddot{\boldsymbol{r}}_A + \left.\frac{d\hat{\dot{\boldsymbol{r}}}_{AB}}{dt}\right|_{X,Y} + \dot{\boldsymbol{\omega}} \times \boldsymbol{r}_{AB} + \boldsymbol{\omega} \times \left.\frac{d\boldsymbol{r}_{AB}}{dt}\right|_{X,Y} \\
&= \ddot{\boldsymbol{r}}_A + (\hat{\ddot{\boldsymbol{r}}}_{AB} + \boldsymbol{\omega} \times \hat{\dot{\boldsymbol{r}}}_{AB}) + \dot{\boldsymbol{\omega}} \times \boldsymbol{r}_{AB} + \boldsymbol{\omega} \times (\hat{\dot{\boldsymbol{r}}}_{AB} + \boldsymbol{\omega} \times \boldsymbol{r}_{AB}) \\
&= \ddot{\boldsymbol{r}}_A + \hat{\ddot{\boldsymbol{r}}}_{AB} + 2\boldsymbol{\omega} \times \hat{\dot{\boldsymbol{r}}}_{AB} + \dot{\boldsymbol{\omega}} \times \boldsymbol{r}_{AB} + \boldsymbol{\omega} \times (\boldsymbol{\omega} \times \boldsymbol{r}_{AB}) \\
&= \ddot{\boldsymbol{r}}_A + \hat{\ddot{\boldsymbol{r}}}_{AB} + 2\boldsymbol{k}\dot{\theta} \times \hat{\dot{\boldsymbol{r}}}_{AB} + \boldsymbol{k}\ddot{\theta} \times \boldsymbol{r}_{AB} + \boldsymbol{k}\dot{\theta} \times (\boldsymbol{k}\dot{\theta} \times \boldsymbol{r}_{AB}).
\end{aligned}$$

Hence,

$$\begin{aligned}
\boldsymbol{a}_{B/A} &= \boldsymbol{a}_B - \boldsymbol{a}_A = \ddot{\boldsymbol{r}}_B - \ddot{\boldsymbol{r}}_A \\
&= \hat{\ddot{\boldsymbol{r}}}_{AB} + 2\boldsymbol{k}\dot{\theta} \times \hat{\dot{\boldsymbol{r}}}_{AB} + \boldsymbol{k}\ddot{\theta} \times \boldsymbol{r}_{AB} + \boldsymbol{k}\dot{\theta} \times (\boldsymbol{k}\dot{\theta} \times \boldsymbol{r}_{AB}),
\end{aligned} \qquad (2.83)$$

and $\boldsymbol{a}_{B/A}$ (the acceleration of plane B relative to plane A in the X, Y system) only coincides with $\hat{\ddot{\boldsymbol{r}}}_{AB}$ (the acceleration of point B as seen by an observer in plane A) if airplane A is flying in a straight line such that $\boldsymbol{\omega} = \boldsymbol{k}\dot{\theta} = 0$ and $\boldsymbol{\omega} = \boldsymbol{k}\ddot{\theta} = 0$.

The relative velocity and acceleration relationships of this section are of minor value compared to the two-coordinate results of the preceding section. They are used to answer specific kinematic questions concerning motion of two points in a plane. The two-coordinate equations are used more frequently in kinematics analysis and will be used subsequently to develop governing equations of motion for particles and rigid bodies.

Example Problem 2.9

Figure XP2.9 illustrates two airplanes. Airplane A is flying in a circle with diameter $D = 800$ m at the constant speed $v_A = 185$ km/h $= 51.8$ m/s. Airplane B is flying to the east at a constant speed $v_B = 205$ km/h $= 56.9$ m/s. At the time of interest, airplane B is 800 m north of the center of airplane A's circular path.

Tasks:

a. Determine airplane B's velocity relative to airplane A, for airplane A in positions 1 through 4.
b. Determine the velocity of airplane B as seen by an observer in airplane A in positions 1 through 4, assuming that airplane B's position is unchanged.
c. Determine airplane B's acceleration relative to airplane A for airplane A at position 1.
d. Determine the acceleration of airplane B as seen by an observer in airplane A at position 1.

Solution

The angular velocity of airplane A and the attached x, y coordinate system relative to the X, Y system is $\boldsymbol{K}\dot{\theta} = \boldsymbol{k}\dot{\theta}$.

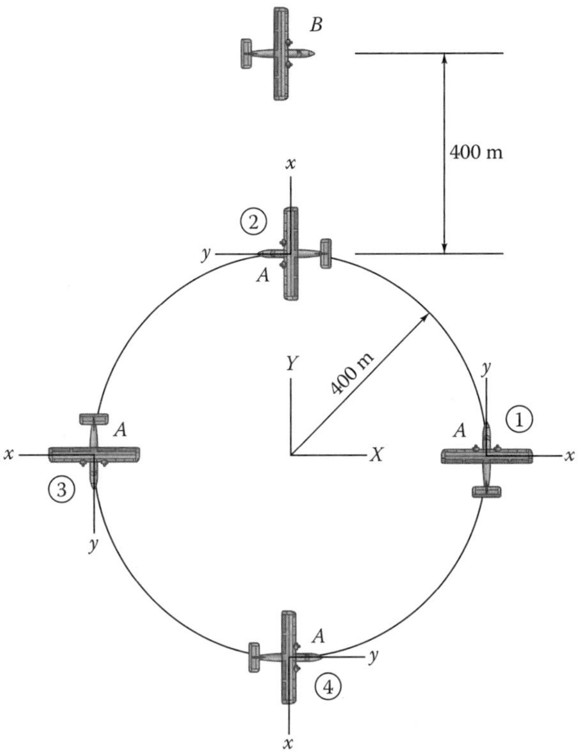

FIGURE XP2.9
Two airplanes in a horizontal plane.

The angular velocity $\dot{\theta}$ is determined from the circumferential-velocity component definition,

$$v_\theta = r\dot{\theta} \Rightarrow \dot{\theta} = 51.8\ (\text{m/s})/400\ \text{m} = 0.1295\ \text{rad/s}$$
$$\Rightarrow \boldsymbol{\omega} = \boldsymbol{K}0.1295\ \text{rad/s}.$$

Applying Equation 2.77 for airplane A's four designated locations gives

$$\begin{aligned}
&1.\ \boldsymbol{v}_{B/A} = \boldsymbol{v}_B - \boldsymbol{v}_A = 56.9\boldsymbol{I} - 51.8\boldsymbol{J}\ \text{m/s}\\
&2.\ \boldsymbol{v}_{B/A} = \boldsymbol{v}_B - \boldsymbol{v}_A = 56.9\boldsymbol{I} - (-51.8\boldsymbol{J})\\
&\qquad = 136.7\boldsymbol{I}\ \text{m/s}\\
&3.\ \boldsymbol{v}_{B/A} = \boldsymbol{v}_B - \boldsymbol{v}_A = 56.9\boldsymbol{I} - (-51.8\boldsymbol{J}) \qquad \text{(i)}\\
&\qquad = 56.9\boldsymbol{I} + 51.8\boldsymbol{J}\ \text{m/s}\\
&4.\ \boldsymbol{v}_{B/A} = \boldsymbol{v}_B - \boldsymbol{v}_A = 56.9\boldsymbol{I} - 51.8\boldsymbol{I} = 5.14\boldsymbol{I}\ \text{m/s}.
\end{aligned}$$

Note that these vector results are all stated in terms of their components in the X, Y coordinate system. Think of these results in terms of the planes' motion (as points) that would be observed on a radar screen. At position 1, airplane B is moving in a south easterly direction toward airplane A. In position 2, airplane B's velocity relative to A is to the east at the combined closure velocity of the two airplanes. At position 3, airplane B's velocity relative to A is in a north-easterly direction. In position 4, airplane B's velocity relative to A is toward the east at a slow velocity. Note that the relative locations of the airplanes do not enter or modify these results. Airplane B could be located several kilometers away from airplane A in any direction, and the same results would hold.

Moving to Task b, we will employ Equation 2.82 and the results of Equation (i) to obtain the specified relative velocities of airplane B as seen by an observer in airplane A as

$$\begin{aligned}
1.\ \hat{\dot{\boldsymbol{r}}}_{AB} &= \boldsymbol{v}_{B/A} - \boldsymbol{\omega}\times\boldsymbol{r}_{AB} = (56.9\boldsymbol{I} - 51.8\boldsymbol{J})\ (\text{m/s})\\
&\quad - 0.1295\ (\text{rad/s})\boldsymbol{K}\times(-400\boldsymbol{I} + 800\boldsymbol{K})\ (\text{m})\\
&= [-56.9\boldsymbol{I} - 51.8\boldsymbol{J} + 51.8\boldsymbol{J} + 103.6\boldsymbol{I}]\ (\text{m/s})\\
&= (160.5\boldsymbol{I} - 0\boldsymbol{J})\ \text{m/s}\\
2.\ \hat{\dot{\boldsymbol{r}}}_{AB} &= \boldsymbol{v}_{B/A} - \boldsymbol{\omega}\times\boldsymbol{r}_{AB} = 136.7\ (\text{m/s})\boldsymbol{I}\\
&\quad - 0.1295\ (\text{rad/s})\boldsymbol{K}\times 400\ (\text{m})\boldsymbol{J}\\
&= (136.7\boldsymbol{I} + 51.8\boldsymbol{I})\ (\text{m/s})\\
&= (188.5\boldsymbol{I})\ \text{m/s} \qquad \text{(ii)}\\
3.\ \hat{\dot{\boldsymbol{r}}}_{AB} &= \boldsymbol{v}_{B/A} - \boldsymbol{\omega}\times\boldsymbol{r}_{AB} = (56.9\boldsymbol{I} + 51.8\boldsymbol{J})\ (\text{m/s})\\
&\quad - 0.1295\ (\text{rad/s})\boldsymbol{K}\times(400\ \text{m}\boldsymbol{I} + 800\boldsymbol{J})\\
&= (56.9\boldsymbol{I} + 51.8\boldsymbol{J} - 51.8\boldsymbol{J} + 103.6\boldsymbol{I})\ (\text{m/s})\\
&= (160.5\boldsymbol{I} + 0\boldsymbol{J})\ \text{m/s}\\
4.\ \hat{\dot{\boldsymbol{r}}}_{AB} &= \boldsymbol{v}_{B/A} - \boldsymbol{\omega}\times\boldsymbol{r}_{AB} = 5.14\ (\text{m/s})\boldsymbol{I}\\
&\quad - 0.1295\ (\text{rad/s})\boldsymbol{K}\times 1200\ (\text{m})\boldsymbol{J}\\
&= (5.14\boldsymbol{I} + 155.4\boldsymbol{I})\ (\text{m/s}) = 160.5\boldsymbol{I}\ \text{m/s}.
\end{aligned}$$

For convenience we have stated vector components in the fixed X, Y coordinate system. Note that quite different *component* descriptions would result if the x, y coordinate system (fixed in airplane A) had been used. Comparing the results of Equations (i) and (ii) emphasizes the difference between the definitions for $\hat{\dot{\boldsymbol{r}}}_{AB}$ (the velocity of point B as seen by an observer in airplane A) and $\boldsymbol{v}_{B/A}$ (the velocity of point B relative to the velocity of point A in the X, Y system). Unlike the relative-velocity results of Equation (i), the results for $\hat{\dot{\boldsymbol{r}}}_{AB}$ depend heavily on the relative locations of the airplanes as defined by the vector $\boldsymbol{r}_{AB}$.

Proceeding with Task c, the acceleration of airplane B with respect to the X, Y coordinate system is $\boldsymbol{a}_B = 0$; the acceleration of airplane A at position 1 is

$$\begin{aligned}
\boldsymbol{a}_A &= -\boldsymbol{I}r\dot{\theta}^2 = -\boldsymbol{I}400\ (\text{m})\times 0.1295^2\ (\text{rad/s})^2 = -\boldsymbol{I}6.71\ \text{m/s}^2\\
&= -\boldsymbol{I}v^2/\boldsymbol{r} = -\boldsymbol{I}51.8^2\ (\text{m/s})^2/400\ \text{m} = -\boldsymbol{I}6.71\ \text{m/s}^2.
\end{aligned}$$

Hence, the acceleration of airplane B relative to A is

$$\boldsymbol{a}_{B/A} = \boldsymbol{a}_B - \boldsymbol{a}_A = 0 - (-\boldsymbol{I}6.71) = \boldsymbol{I}6.71\ \text{m/s}^2, \qquad \text{(iii)}$$

which concludes Task c.

Finally, applying Equation 2.83 for Task d gives

$$\begin{aligned}
\hat{\ddot{\boldsymbol{r}}}_{AB} &= \boldsymbol{a}_{B/A} - 2\boldsymbol{k}\dot{\theta}\times\hat{\dot{\boldsymbol{r}}}_{AB} - \boldsymbol{k}\ddot{\theta}\times\boldsymbol{r}_{AB} - \boldsymbol{k}\dot{\theta}\times(\boldsymbol{k}\dot{\theta}\times\boldsymbol{r}_{AB})\\
&= \boldsymbol{I}6.71\left(\text{m/s}^2\right) - 2\boldsymbol{K}0.1295\ (\text{rad/s})\times 160.5\ \text{m}\boldsymbol{I}\\
&\quad - \boldsymbol{K}0\left(\text{rad/s}^2\right)\times(-400\boldsymbol{I} + 800\boldsymbol{J})\ \text{m}\\
&\quad - 0.1295\ (\text{rad/s})\boldsymbol{K}\times[-0.1295\ (\text{rad/s})\boldsymbol{K}\\
&\quad \times(-400\boldsymbol{I} + 800\boldsymbol{J})\ \text{m}]\\
&= (\boldsymbol{I}6.71 - \boldsymbol{J}41.6 + \boldsymbol{I}6.71 - 13.4\boldsymbol{J})\ \text{m/s}^2\\
&= 13.4\boldsymbol{I} - 55.1\boldsymbol{J}\ \text{m/s}^2. \qquad \text{(iv)}
\end{aligned}$$

Comparing the results of Equations (iii) and (iv) confirms the differences between $\boldsymbol{a}_{B/A}$ (the acceleration of point B relative to point A in the X, Y system) and $\hat{\ddot{\boldsymbol{r}}}_{AB}$ (the acceleration of point B as seen by an observer in airplane A).

2.11 Summary and Discussion

The material introduced in this section has the (misleadingly) simple objective of locating a point P in a plane and defining the velocity and acceleration of the point with respect to a coordinate system that is fixed in the plane. A Cartesian (X, Y) coordinate system was initially used to define the position, velocity, and acceleration of point P, relative to the X, Y system, by the vectors

$$\boldsymbol{r} = \boldsymbol{I}r_X + \boldsymbol{J}r_Y, \quad \dot{\boldsymbol{r}} = \left.\frac{d\boldsymbol{r}}{dt}\right|_{X,Y} = \boldsymbol{I}\dot{r}_X + \boldsymbol{J}\dot{r}_Y,$$

$$\ddot{\boldsymbol{r}} = \left.\frac{d\dot{\boldsymbol{r}}}{dt}\right|_{X,Y} = \boldsymbol{I}\ddot{r}_X + \boldsymbol{J}\ddot{r}_Y.$$

These results demonstrate the procedure for obtaining the derivative of a vector with respect to a coordinate system. Specifically, the time derivative of a vector $\boldsymbol{B}$ with respect to a specific coordinate system is obtained by stating $\boldsymbol{B}$ in terms of its components in the system and then differentiating $\boldsymbol{B}$ while holding the unit vectors constant.

The components of the same vector in two different coordinate systems can be conveniently related via the direction-cosine matrix. Coordinate transformations based on the direction-cosine matrix are introduced in Figure 2.4 and Section 2.4. The relationships

$$\begin{Bmatrix} B_{X'} \\ B_{Y'} \end{Bmatrix} = \begin{bmatrix} \cos\alpha & \sin\alpha \\ -\sin\alpha & \cos\alpha \end{bmatrix} \begin{Bmatrix} B_X \\ B_Y \end{Bmatrix} \tag{2.15}$$

and

$$\begin{Bmatrix} B_X \\ B_Y \end{Bmatrix} = \begin{bmatrix} \cos\alpha & -\sin\alpha \\ \sin\alpha & \cos\alpha \end{bmatrix} \begin{Bmatrix} B_{X'} \\ B_{Y'} \end{Bmatrix}, \tag{2.19}$$

and their shorthand notation $(B)_{I'} = [A](B)_I$ and $(B)_I = [A]^T(B)_{I'}$ are of exceptional value in transforming vector components. Three by three versions of the direction-cosine matrix are needed for 3D motion.

A substantial amount of work is needed to appreciate the differences between the following distinct concepts:

a. The time derivatives of a vector $\boldsymbol{B}$ with respect to different coordinate system will generally yield *different vectors* if the coordinate systems are moving with respect to each other, and
b. The vector $\boldsymbol{B}$ can be stated correctly in terms of its components in any coordinate system, and *different components* will result in different coordinate systems.

Polar-component definitions were obtained for $\dot{\boldsymbol{r}}$ and $\ddot{\boldsymbol{r}}$ as

$$v_r = \dot{r}, \quad v_\theta = r\dot{\theta}$$
$$a_r = \ddot{r} - r\dot{\theta}^2, \quad a_\theta = r\ddot{\theta} + 2\dot{r}\dot{\theta}. \tag{2.31}$$

These are very useful definitions and will be used regularly in the balance of this book.

The path-component definitions for acceleration are

$$a_t = \dot{v}, \quad a_n = v\dot{\beta} = \frac{v^2}{\rho} = \rho\dot{\beta}^2, \tag{2.39}$$

where ρ is the radius of curvature. These expressions are less useful than the Cartesian or polar definitions. They are mainly used to determine the acceleration of a particle that is moving along a given path defined by $Y = Y(X)$. The radius of curvature D can be defined in terms of the function $Y = Y(X)$ and its derivatives.

Developing an understanding that (1) the same vectors $\dot{\boldsymbol{r}}$ and $\ddot{\boldsymbol{r}}$ can be defined in terms of Cartesian, polar, or path components, and (2) coordinate transformations can be used to "jump" back and forth between the coordinate systems to get different component definitions is essential to mastering particle kinematics.

The two-coordinate time-derivative relationship

$$\dot{\boldsymbol{B}} = \hat{\dot{\boldsymbol{B}}} + \boldsymbol{\omega} \times \boldsymbol{B}, \tag{2.52a}$$

or

$$\left.\frac{d\boldsymbol{B}}{dt}\right|_{X,Y} = \left.\frac{d\boldsymbol{B}}{dt}\right|_{x,y} + \boldsymbol{\omega} \times \boldsymbol{B}. \tag{2.52b}$$

is of fundamental value in developing particle and rigid-body kinematic equations. Equations of exactly this form arise in 3D kinematics. Equations 2.52 were used to develop the two-coordinate velocity and acceleration equations:

$$\dot{\boldsymbol{r}} = \dot{\boldsymbol{R}}_o + \hat{\dot{\boldsymbol{\rho}}} + \boldsymbol{\omega} \times \boldsymbol{\rho} \tag{2.69a}$$

$$\ddot{\boldsymbol{r}} = \ddot{\boldsymbol{R}}_o + \hat{\ddot{\boldsymbol{\rho}}} + 2\boldsymbol{\omega} \times \hat{\dot{\boldsymbol{\rho}}} + \dot{\boldsymbol{\omega}} \times \boldsymbol{\rho} + \boldsymbol{\omega} \times (\dot{\boldsymbol{\omega}} \times \boldsymbol{\rho}). \tag{2.69b}$$

These definitions are required much less frequently in planar particle kinematics than the one-coordinate, Cartesian and polar relationships. However, they will be used in Chapter 4 on planar kinematics of rigid bodies and in Chapter 5 for planar kinetics of rigid bodies. Identical vector equations are required for 3D vector kinematics.

The relative velocity and relative acceleration relationships of Section 2.10 are used much less frequently than the remaining kinematic definitions in this chapter. They are needed to answer very specific questions about relative motion and are not revisited in this book.

As noted in this chapter's introduction, kinematics is the essential skeleton for particle and rigid-body dynamics. Competency in dynamics requires a mastery of kinematics. Particle kinetics equations based on Newton's second law of motion, $\Sigma f = m\ddot{r}$, require that the acceleration $\ddot{r}$ be stated correctly. Similarly, applying the work-energy equation involving particle kinetic energy $T = mv^2/2$ requires a correct velocity definition. On the bright side, the essentials of kinematics have been introduced in this chapter.

Problems

2.1 Figure P2.1 illustrates a point P located in the X, Y coordinate system by polar coordinates r and θ.

Task: Derive the polar-coordinate expressions for the velocity and acceleration components of the point P with respect to the X, Y coordinate system; that is, derive expressions for v_r, v_θ, a_r, a_θ.

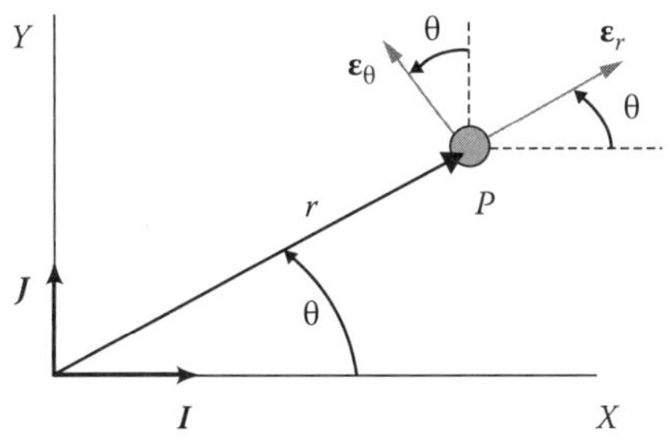

FIGURE P2.1

2.2 A particle is following the path shown in Figure P2.2. By definition, its velocity vector $v = v\boldsymbol{\varepsilon}_t$ is directed tangent to the path. The normal vector, $\boldsymbol{\varepsilon}_n$, is perpendicular to $\boldsymbol{\varepsilon}_t$. The radius of curvature is ρ.

Task: Show that the acceleration vector can be expressed as

$$a = \dot{v}\boldsymbol{\varepsilon}_t + \frac{v^2}{\rho}\boldsymbol{\varepsilon}_n = \ddot{s}\boldsymbol{\varepsilon}_t + \frac{v^2}{\rho}\boldsymbol{\varepsilon}_n.$$

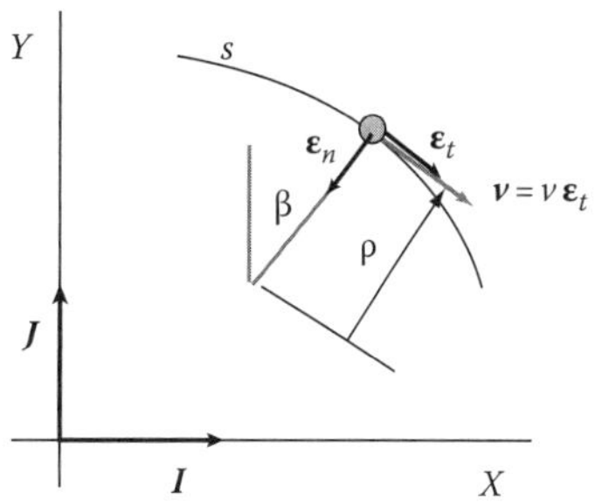

FIGURE P2.2

2.3 Use the information provided in the Figure P2.3 to derive expressions for the normal and tangential components of acceleration: that is, find a_t and a_n.

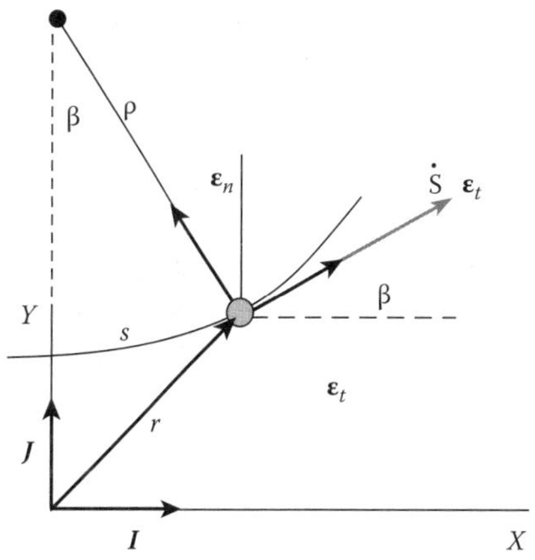

FIGURE P2.3

2.4 In Figure P2.4, X and Y motions of the guides A and B control the curvilinear motion of connecting pin P, which slides in both slots. For a short interval, the motion is governed by $X_p = 20 + \frac{1}{4}t^2$ and $Y_p = 15 - \frac{1}{6}t^3$, where X and Y are in millimeters and t is in seconds.

Task: At $t = 2$ s, perform the following:

(a) Determine the velocity and acceleration components in Cartesian coordinates.

(b) Determine the velocity and acceleration components in path coordinates.

(c) Determine the velocity and acceleration components in polar coordinates.

(d) Draw vector diagrams to demonstrate that the same velocity and acceleration vectors are obtained for all three-coordinate systems. Use figures showing labeled components, unit vectors, and units.

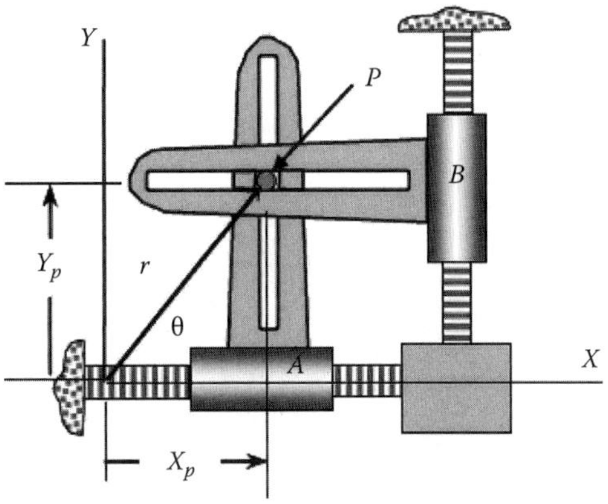

FIGURE P2.4

2.5 In Figure P2.4, P's position is controlled by the guided slots such that $X_p = 0.2 + 0.1\cos(\omega t)$ m and $Y_p = 0.6 + 0.3\sin(2\omega t)$ m where $\omega t = 22.5°$, $\omega = 10$ rad/s.

Tasks:
(a) Determine the velocity and acceleration components in Cartesian coordinates.
(b) Determine the velocity and acceleration components in path coordinates.
(c) Determine the velocity and acceleration components in polar coordinates.
(d) Draw vector diagrams to demonstrate that the same velocity and acceleration vectors are obtained for all three-coordinate systems. Use figures showing labeled components, unit vectors, and units.

2.6 In Figure P2.4, point P's coordinates are defined by $X_p = D\cos(\omega t)$, $Y_p = 2D\sin(\omega t)$, where $D = 62.1$ mm, and $\omega = 12.56$ rad/s.

Tasks:
(a) For $\omega t = \pi/4$ rad, determine P's velocity and acceleration vectors (with respect to the X, Y system) in terms of components in the Cartesian, polar, and path-coordinate systems.
(b) Describe (sketch) the path that P makes.
(c) Draw vector diagrams to demonstrate that the same velocity and acceleration vectors are obtained for all three-coordinate systems. Use figures showing labeled components, unit vectors, and units.

2.7 The mechanism of Figure P2.4 is used to control the position of point P such that the components of P's position vectors are

$$X_p = 10 + 5t + \frac{t^2}{5}\ \text{mm}, \quad Y_p = 10 - 5t - \frac{t^2}{8}\ \text{mm}$$

Tasks: At $t = 1$ s, for point P, perform the following:
(a) Determine the velocity and acceleration components in Cartesian coordinates.
(b) Determine the velocity and acceleration components in polar coordinates.
(c) Determine the velocity and acceleration components in path coordinates.
(d) Draw vector diagrams to demonstrate that the same velocity and acceleration vectors are obtained for all three-coordinate systems. Use figures showing labeled components, unit vectors, and units.

2.8 In Figure P2.8, the point P moves in the fixed parabolic slot whose shape is given by $Y = X^2/4$, with X and Y in inches. The guide with the vertical slot is given a horizontal oscillatory motion according to $X = 4\sin 2t$, where X is in inches and t is in seconds.

Tasks: At $t = \pi/12$ s, perform the following:
(a) Determine the velocity and acceleration components in Cartesian coordinates.
(b) Determine the velocity and acceleration components in polar coordinates.
(c) Determine the velocity and acceleration components in path coordinates.
(d) Draw vector diagrams to demonstrate that the same velocity and acceleration vectors are obtained for all three-coordinate systems. Use figures showing labeled components, unit vectors, and units.

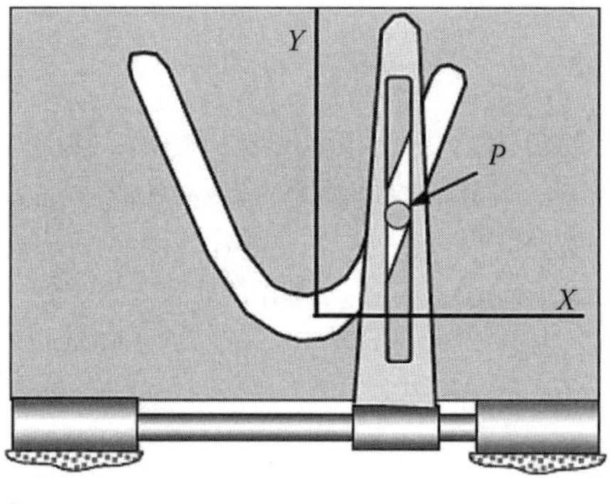

FIGURE P2.8

2.9 In Figure P2.9, a projectile is fired with an initial velocity of 150 mph at an angle of 35° above the horizontal plane. It impacts an inclined plane that is at an angle of 30° from the vertical after 7.6 s.

Tasks: Neglecting drag due to air, perform the following:
(a) Determine the X and Y components of the velocity of the projectile at the point of impact.
(b) Transform the X and Y velocity components into components parallel and perpendicular to the inclined plane.
(c) Draw a diagram showing the velocity components at the time of impact for both reference systems.

(d) Draw a diagram with acceleration components at the time of impact for both coordinate systems.

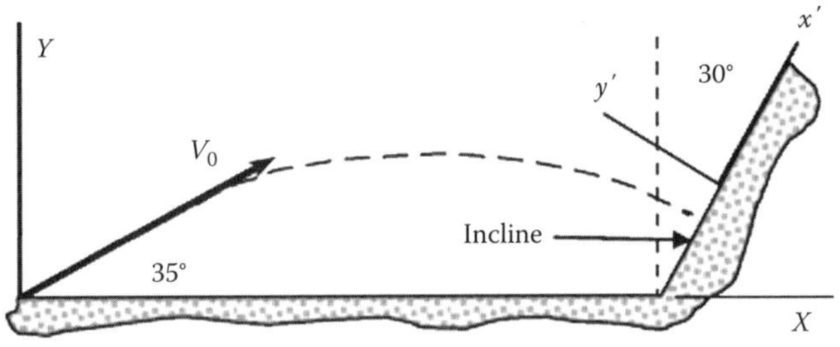

FIGURE P2.9

2.10 In Figure P2.10, r and θ are given by

$$r = r_0 + \frac{r_0}{2}\cos(\omega t), \quad \theta = \frac{\pi}{6}\sin(\omega t),$$

where r_0 and ω are constant.

Task: For $\omega t = \pi/2$, for point P, perform the following:

(a) Determine the velocity and acceleration components in polar coordinates.
(b) Determine the velocity and acceleration components in Cartesian coordinates.
(c) Determine the velocity and acceleration components in path coordinates.
(d) Draw vector diagrams to demonstrate that the same velocity and acceleration vectors are obtained for all three-coordinate systems. Use figures showing labeled components, unit vectors, and units.

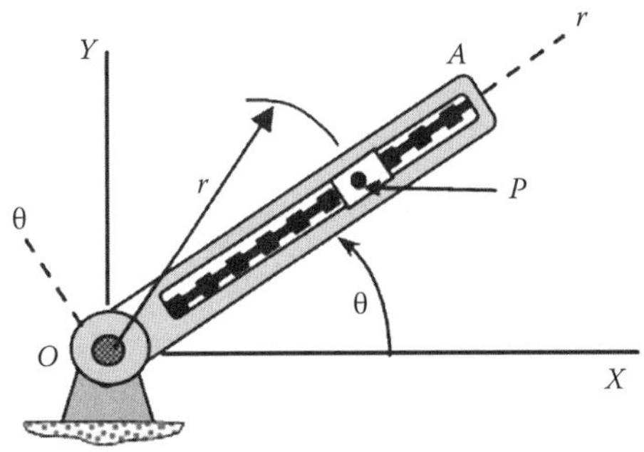

FIGURE P2.10

2.11 In Figure P2.10, $\dot{\theta} = 8$ rad/s and $\ddot{\theta} = -20$ rad/s^2, $r = 200$ mm, $\dot{r} = -300$ mm/s, and $\ddot{r} = 0$.

Tasks: For $\theta = 30°$, for point P, perform the following:

(a) Find the velocity and acceleration components in polar coordinates.
(b) Find the velocity and acceleration components in Cartesian coordinates.
(c) Find the velocity and acceleration components in path coordinates.
(d) Draw vector diagrams to demonstrate that the same velocity and acceleration vectors are obtained for all three-coordinate systems. Use figures showing labeled components, unit vectors, and units.

2.12 In Figure P2.12, rod OA's rotation about O is defined by $\theta = 2t^2$ where θ is in radians, and t is in seconds. The distance from collar B to O is defined by $r = 60t^2 - 0.9t^3$, with r in meters and t in seconds.

Tasks: At $t = 1$ s, for the collar B, perform the following:

(a) Find the velocity and acceleration components in polar coordinates.
(b) Find the velocity and acceleration components in Cartesian coordinates.
(c) Find the velocity and acceleration components in path coordinates.
(d) Draw vector diagrams to demonstrate that the same velocity and acceleration vectors are obtained for all three-coordinate systems. Use figures showing labeled components, unit vectors, and units.

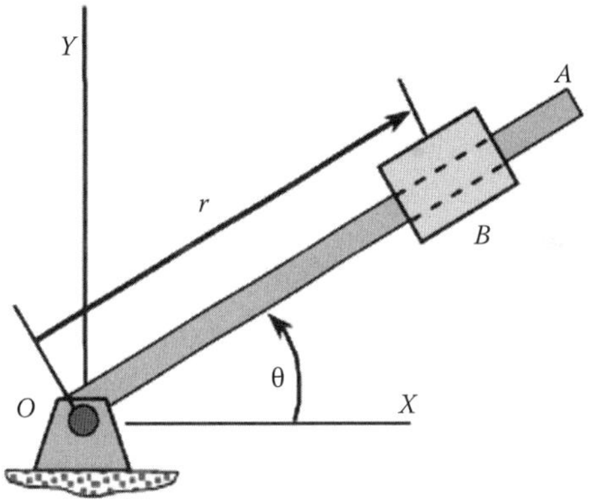

FIGURE P2.12

2.13 For the mechanism of Problem 2.12, the motion is defined by

$$\theta = \frac{\pi}{4}\cos(\omega t) \text{ rad}, \quad r = 10 + 5\cos(2\omega t) \text{ mm},$$

where $\omega = 60$ cycles/s (*constant*), and $\omega t = 30°$.

Tasks: For point B, perform the following:

(a) Find the velocity and acceleration components in polar coordinates.
(b) Find the velocity and acceleration components in Cartesian coordinates.
(c) Find the velocity and acceleration components in path coordinates.

(d) Draw vector diagrams to demonstrate that the same velocity and acceleration vectors are obtained for all three-coordinate systems. Use figures showing labeled components, unit vectors, and units.

2.14 For the rocket system illustrated in Figure P2.14,

$$r = 2200 \text{ m} \quad \dot{r} = 500 \text{ m/s} \quad \ddot{r} = 5 \text{ m/s}^2$$
$$\theta = 45^\circ \quad \dot{\theta} = 0.1 \text{ rad/s} \quad \ddot{\theta} = -0.01 \text{ rad/s}^2.$$

Tasks: Perform the following for the position illustrated:

(a) Find the velocity and acceleration components in polar coordinates.
(b) Find the velocity and acceleration components in Cartesian coordinates.
(c) Find the velocity and acceleration components in path coordinates.
(d) Draw vector diagrams to demonstrate that same velocity and acceleration vectors are obtained for all three-coordinate systems. Use figures showing labeled components, unit vectors, and units.

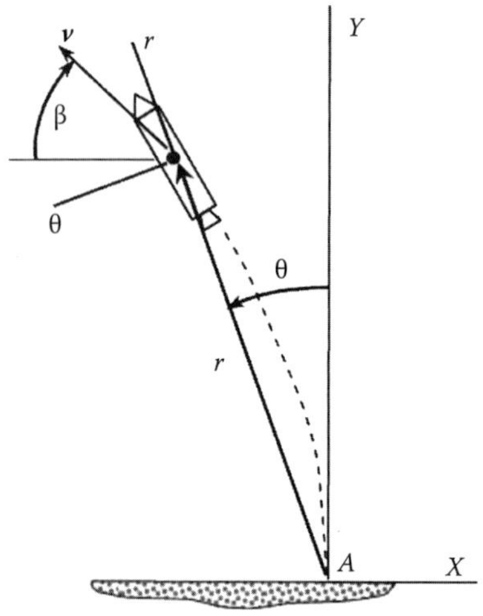

FIGURE P2.14

2.15 The cam in Figure P2.15 is designed so that the center of the roller at A follows the contour and moves on a limacon defined by $r = b - c\cos\theta$.

Tasks: For $b = 1$ m, $c = 0.248$ m, $\dot{\theta} = 1$ cycles/s, $\theta = 135^\circ$, and $\ddot{\theta} = 0$, perform the following:

(a) Find the velocity and acceleration components in polar coordinates.
(b) Find the velocity and acceleration components in Cartesian coordinates.
(c) Find the velocity and acceleration components in path coordinates.
(d) Draw vector diagrams to demonstrate that same velocity and acceleration vectors are obtained for all three-coordinate systems. Use figures showing labeled components, unit vectors, and units.

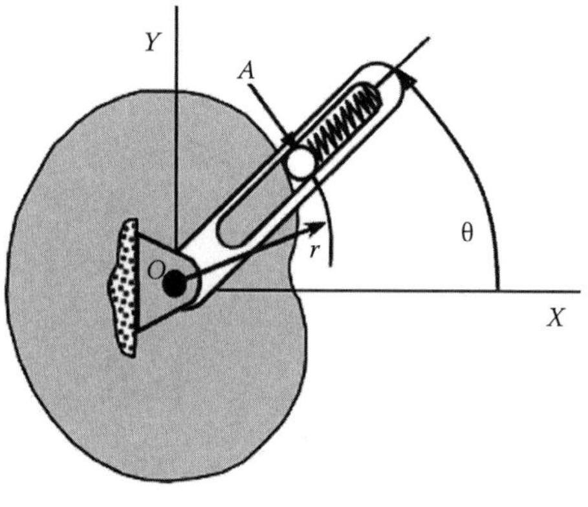

FIGURE P2.15

2.16 Illustrated in Figure P2.16 is a bead sliding down a wire due to gravity. The shape of the wire for a segment is defined by $Y = -X + \frac{1}{4}X^2$ ft. Neglecting damping, the bead's velocity along the wire is defined by $v = (-2gY)^{1/2}$ ft/s, and its acceleration along the path is defined by $a_t = g \sin\beta$.

Tasks:

(a) For $X = 1$ ft, what are the components of velocity and acceleration for path, Cartesian, polar coordinates? Solve using $g = 32.17$ ft/s^2.
(b) Draw vector diagrams to demonstrate that same velocity and acceleration vectors are obtained for all three-coordinate systems. Use figures showing labeled components, unit vectors, and units.

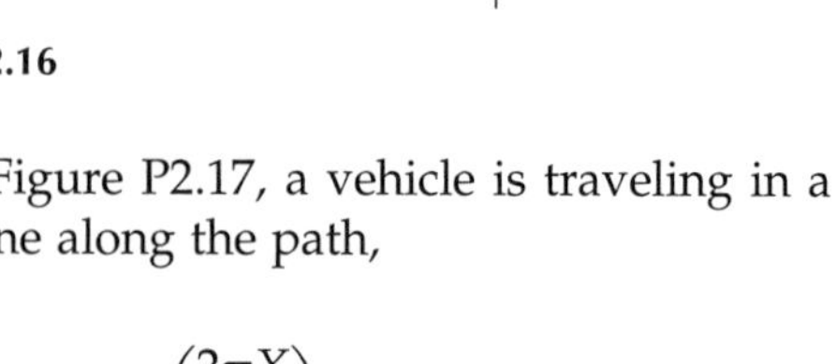

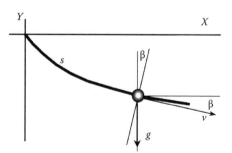

FIGURE P2.16

2.17 In Figure P2.17, a vehicle is traveling in a vertical plane along the path,

$$Y = A\sin\left(\frac{2\pi X}{L}\right), \quad A = 100 \text{ m}, \quad L = 2000 \text{ m}.$$

At $X = 1500$ m, the vehicle's velocity and acceleration along the path are $v = 100$ km/h, and $a_t = 2$ m/s^2.

Tasks: For the specified car position, perform the following:

(a) Determine the velocity and acceleration components in path coordinates.
(b) Determine the velocity and acceleration components in Cartesian coordinates.
(c) Determine the velocity and acceleration components in polar coordinates.
(d) Draw vector diagrams to demonstrate that same velocity and acceleration vectors are obtained for all three-coordinate systems. Use figures showing labeled components, unit vectors, and units.

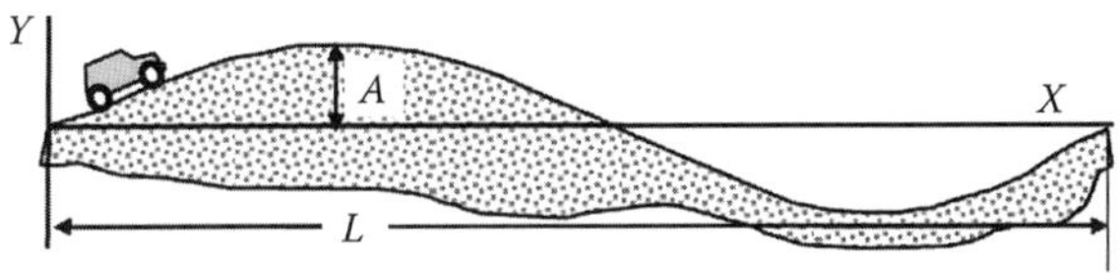

FIGURE P2.17

2.18 In Figure P2.18, the airplane travels along the vertical parabolic path, $Y = 0.4X^2$ (note that X and Y are in km, not m). At point A (5, 10) km, the airplane's speed along the path is 200 m/s, and it is decreasing at 1 m/s^2.

Tasks: At the instant illustrated, perform the following:

(a) Determine the velocity and acceleration components in path coordinates.
(b) Determine the velocity and acceleration components in Cartesian coordinates.
(c) Determine the velocity and acceleration components in polar coordinates.
(d) Draw vector diagrams to demonstrate that same velocity and acceleration vectors are obtained for all three-coordinate systems. Use figures showing labeled components, unit vectors, and units.

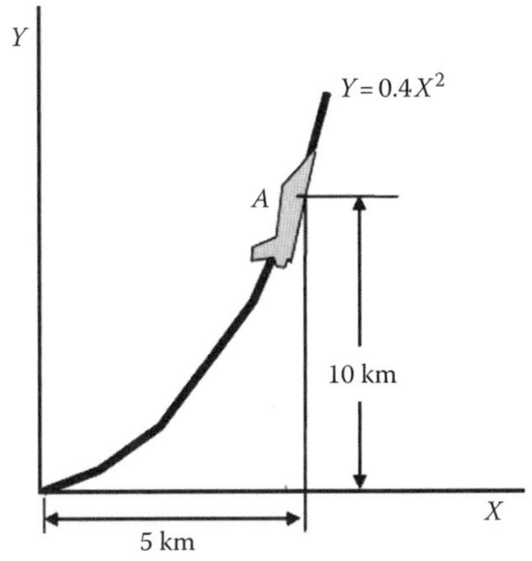

FIGURE P2.18

2.19 The track illustrated in Figure P2.19 lies in a horizontal plane and is defined by $Y = kX^2$ for X and Y in meters with $k = 1/400\ \text{m}^{-1}$. At $X = 100$ m, the velocity and acceleration components along the path are 20 m/s and 2 m/s^2, respectively.

Tasks: At the given instant, perform the following:

(a) Determine the velocity and acceleration components in path coordinates.
(b) Determine the velocity and acceleration components in Cartesian coordinates.
(c) Determine the velocity and acceleration components in polar coordinates.
(d) Draw vector diagrams to demonstrate that same velocity and acceleration vectors are obtained for all three-coordinate systems. Use figures showing labeled components, unit vectors, and units.

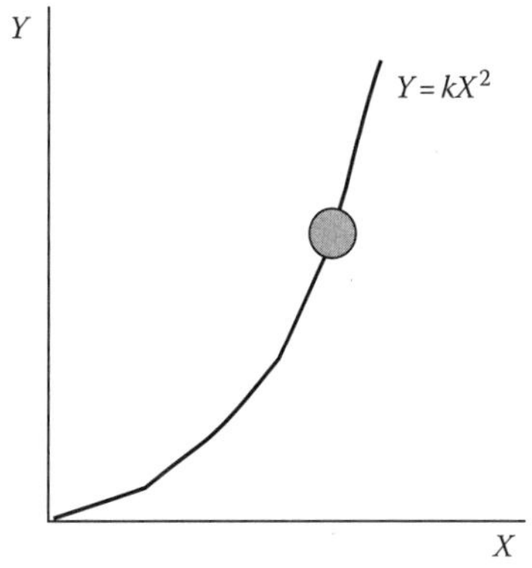

FIGURE P2.19

2.20 As shown in Figure P2.20, a portion of a roller coaster's path is defined by the parabola $Y = 0.05(X - X_O)^2 + 0.5(X - X_O) + 17$ with X and Y in feet.

When $X = 30$ ft and $X_O = 50$ ft, the lead car is moving at 40 ft/s relative to the track and accelerating at 0.32 ft/s^2 in the direction of motion.

Tasks: At the given instant perform the following:

(a) Determine the velocity and acceleration components in path coordinates.
(b) Determine the velocity and acceleration components in Cartesian coordinates.
(c) Determine the velocity and acceleration components in polar coordinates.
(d) Draw vector diagrams to demonstrate that the same velocity and acceleration vectors are obtained for all three-coordinate systems. Use figures showing labeled components, unit vectors, and units.

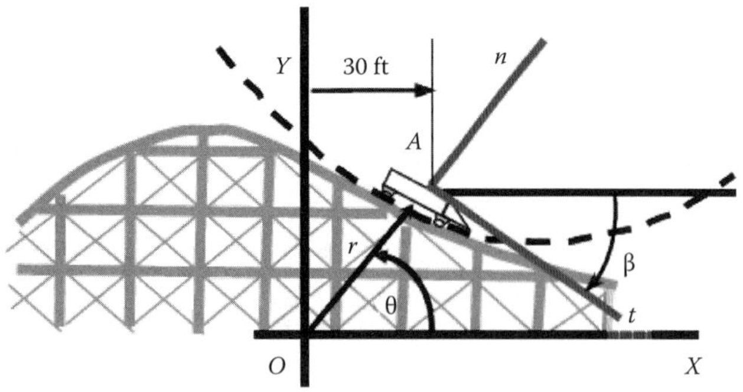

FIGURE P2.20

2.21 A particle travels along a path in a vertical plane defined by

$$Y = -0.3X^3 - 0.7X^2 + 5X,$$

where Y and X are in meters. At $X = 2.25$ m, the velocity and acceleration of the particle along the path are $v = 5$ m/s, and $a_t = -1.5$ m/s^2. Assume the particle is moving from the left to the right.

Hint: Y'' defines the curvature of the path.

Tasks: At the given instant, perform the following:

(a) Determine the velocity and acceleration components in path coordinates.

(b) Determine the velocity and acceleration components in Cartesian coordinates.

(c) Determine the velocity and acceleration components in polar coordinates.

(d) Draw vector diagrams to demonstrate that the same velocity and acceleration vectors are obtained for all three-coordinate systems. Use figures showing labeled components, unit vectors, and units.

3

Planar Kinetics of Particles

3.1 Introduction

As noted in Chapter 2, kinematics is the study, using geometric methods, of motion (position, velocity, acceleration) without regard for the forces that either cause or result from motion. By contrast, kinetics concerns the prediction of motion of particles or bodies that are acted on by known forces. Kinetics can also involve solutions for reaction forces acting on a particle or body when the motion is specified. This chapter considers the motion of particles in one dimension (motion in a straight line) or two dimensions (motion in a plane). Mastery of the kinematics material of the preceding chapter is essential for understanding the concepts and mechanics of this chapter. There are many new variables introduced in this chapter, and the following nomenclature is provided to help you keep track of them.

Nomenclature

$(a)_i$	ith eigenvector for a multi-degree-of-freedom (MDOF) system	
$[A]$	Matrix of un-normalized eigenvectors	
$[A^*]$	Matrix of normalized eigenvectors	
c	Damping coefficient	[F s/L]
$[C]$	Damping matrix for an MDOF system	
e	Coefficient of restitution, Equation 3.193	[—]
f	Force	[F]
$f_n = \omega_n/2\pi$	Undamped natural frequency in cycles per second or hertz	$[s^{-1}]$
$f_d = \omega_d/2\pi$	Damped natural frequency in cycles per second or hertz	$[s^{-1}]$
g	Acceleration of gravity	$[L/s^2]$
$G(r)$	Amplification factor for base excitation, Equation 3.48, Figure 3.17	[—]
$H(r)$	Amplification factor for forced excitation, Equation 3.40, Figure 3.15	[—]
$\boldsymbol{H}_o$	Moment-of-momentum vector for a particle about point O, Equation 3.198	$[ML^2/s]$
$[I]$	Identity matrix	
$J(r)$	Amplification factor for either (1) relative motion due to base excitation, Equation 3.51 or (2) response due to rotating imbalance, Equation 3.55, Figure 3.19	[—]
k	Stiffness coefficient	[F/L]
m	Mass	[M]
$[K]$	Stiffness matrix for an MDOF system	
q_i	ith modal coordinate for an MDOF system	
Q_i	ith modal force for an MDOF system, Equation 3.153	
q factor $= 1/2\zeta$	A measure of viscous damping, Table 3.1, Equation 3.44	[—]
$r = \omega/\omega_n$	Frequency ratio	[—]
sgn	The sign function defined in Equation 3.56b	[—]
T	Kinetic energy	[FL]
V	Potential energy	[FL]
w	Weight	[F]
δ	Logarithmic decrement, Table 3.1	[—]
$\zeta = c/2m\omega_n$	Damping factor	[—]
ζ_i	ith modal damping factor for an MDOF system	[—]

$[\Lambda]$	Diagonal matrix of natural frequencies for an MDOF system, Equation 3.152	$[\text{rad}^2/\text{s}^2]$
μ_d	Dynamic Coulomb-friction factor, Section 3.2.6	[—]
μ_s	Static Coulomb-friction factor, Section 3.2.6	[—]
$\tau_n = 2\pi/\omega_n$	Undamped period of motion	[s]
$\tau_d = 2\pi/\omega_d$	Damped period of motion	[s]
$\psi(r)$	Phase for forced harmonic excitation Equation 3.41, Figure 3.16	[rad]
ω	Excitation frequency	[rad/s]
$\omega_n = \sqrt{k/m}$	Natural frequency	[rad/s]
$\omega_d = \omega_n\sqrt{1-\zeta^2}$	Damped natural frequency	[rad/s]
ω_{ni}	ith undamped natural frequency for an MDOF system	[rad/s]

Subscripts

h	Homogeneous solution
i	Denotes ith mode for an MDOF system
p	Particular solution

Abbreviations

1DOF	One degree of freedom
2DOF	Two degrees of freedom
EOM	Equation of motion
MDOF	Multi degrees of freedom

Newton's laws of motion for a particle are based on experimental observations and can be summarized as follows:

***Law* 1**. Unless a force is applied to a particle it will either remain at rest or continue to move in a straight line at constant velocity.

***Law* 2**. The acceleration of a particle in an inertial reference frame is proportional to the force acting on the particle.

***Law* 3**. For every action (force), there is an equal and opposite reaction (force).

The second law of motion is normally written

$$\sum f = m\ddot{\mathbf{r}}, \tag{3.1}$$

where

- $\sum f$ is the resultant force acting on the particle
- m is the particle's mass
- $\ddot{\mathbf{r}}$ is the particle's acceleration *with respect to an inertial reference frame*

The idea of an ''inertial reference frame'' leads to a circular definition, since an inertial coordinate system is defined to be a coordinate system for which Equation 3.1 is valid. For the purpose of this book and most engineering practice, the earth's surface provides a valid inertial reference frame, even though the earth is rotating around its polar axis every 24 hours and precessing around the sun about every 365 days. Using the earth's surface as an inertial coordinate system means that any reference system that has a measurable acceleration with respect to the earth's surface is not an inertial reference frame.

Figure 3.1 illustrates an inertial X, Y reference frame fixed to ''ground'' and a second x, y coordinate system fixed to an automobile. The automobile is located in the X, Y system by $\mathbf{I}X$, and a particle in the car is located in the x, y system by $\mathbf{i}x$. Hence the particle is located with respect to an inertial frame by $\mathbf{I}(X+x)$, and a correct version of Newton's second law of motion for the particle is

$$f = m(\ddot{X} + \ddot{x}), \tag{3.2}$$

where

- f is the net force acting on the particle
- m is the mass of the particle

An incorrect result would be obtained by trying to apply Newton's law in terms of $\ddot{x}$, the acceleration of the particle with respect to the automobile; that is, $f \neq m\ddot{x}$.

This result should be consistent with your own experience; that is, when you accelerate or decelerate a car with respect to ground, loose objects in the car can move around (with respect to the car) apparently without any forces acting on them. This observation is perfectly well-explained by Equation 3.2, particularly if we rewrite it as

$$f - m\ddot{X} = m\ddot{x},$$

FIGURE 3.1
A car located in the X, Y (inertial system) by $\mathbf{I}X$ and a particle located in the x, y system by $\mathbf{i}x$.

which shows that the acceleration of the particle, with respect to the car, is caused by both the external force and the acceleration of the car with respect to ground. However, the object's motion due to the car's acceleration cannot be explained by the *incorrect* relation $f = m\ddot{x}$, since there is no real physical force (push or pull) acting on the particle and present on the left-hand side of the equation.

The correct application of Newton's second law is critical to correct dynamic formulation and is generally straightforward. Just remember to make sure that the acceleration $\ddot{r}$ used in $\sum f = m\ddot{r}$ is with respect to the earth's surface. Exceptions to this rule can occur in motion of satellites or space probes for which the acceleration of the earth with respect to the sun becomes a factor.

Figure 3.2 illustrates an inertial X, Y coordinate system. A particle of mass m is moving along the X-axis and is acted upon by a force f. From Newton's second law of motion, the *differential equation of motion* is

$$f = m\ddot{X}. \tag{3.3}$$

Note that the force and acceleration have the same positive sign convention; that is, the force f and acceleration $\ddot{X}$ are both positive when acting in the $\mathbf{I}$-direction. This convention also applies to displacement and velocity.

The kinetic energy of the particle in Figure 3.2 is defined as

$$T = K.E. = \frac{m\dot{X}^2}{2}$$

where $\dot{X}$ is (again) the velocity of the particle *with respect to an inertial coordinate system.*

Kinetics largely involves the following sequential steps: (1) correctly deriving equations of motion, and (2) correctly (or appropriately) integrating the differential equations of motion to obtain solutions for the velocity and displacement as functions of time. The contents of the next section are devoted to derivation of governing equations of motion for a particle or particles moving in a straight line and various analytical options for their solution.

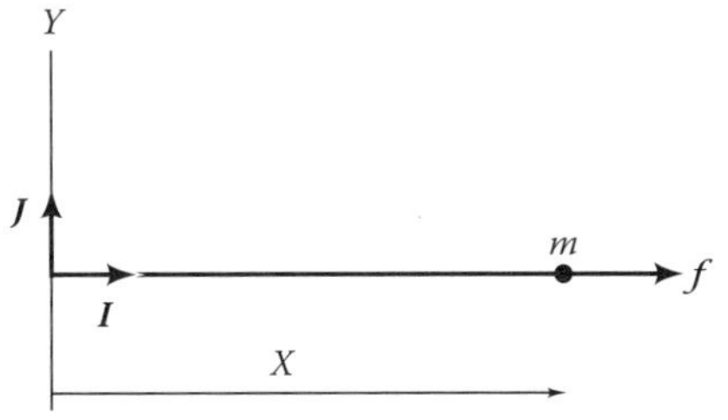

FIGURE 3.2
A particle located in the X, Y (inertial) system by IX.

3.2 Differential Equations of Motion for a Particle Moving in a Straight Line: An Introduction to Physical Modeling

As applied to a particle moving in a straight line, Newton's second law of motion yields a single second-order differential equation of motion. The equation varies depending on the physics of the problem at hand, and this section is primarily concerned with the derivation of correct equations of motion based on commonly-recurring physical situations in engineering practice. A limited amount of discussion is also provided concerning solution (integration) of the equations of motion.

3.2.1 Constant Acceleration: Free Fall of a Particle without Drag

The simplest differential equations of motion for a particle arise for constant acceleration. Figure 3.3 illustrates a particle of mass m falling under gravity without a drag force. (The last part of this section shows the influence of aerodynamic drag on a falling body.) Neglecting drag, the only force acting on the body is weight, $w = mg$. In Figure 3.3a, the particle is located with respect to ground by Y. The positive sign for Y identifies the positive sense for velocity $\dot{Y}$, acceleration $\ddot{Y}$, and the forces acting on the body. With this in mind, Newton's second law of motion gives

$$m\ddot{Y} = \sum f_Y = -w = -mg \Rightarrow \ddot{Y} = -g. \tag{3.4}$$

The negative sign holds for the weight because it acts in the $-Y$ direction. The acceleration is constant, equals the acceleration of gravity, and is in the $-Y$ direction.

Figure 3.3b illustrates the same physical problem; however, we now choose to keep track of the particle by using Y', which locates the particle with respect to a point that is a fixed distance above "ground." Hence, we can use $\ddot{Y}'$ as our acceleration in applying Newton's second law of motion. As you can see, Y' is positive downward; hence,

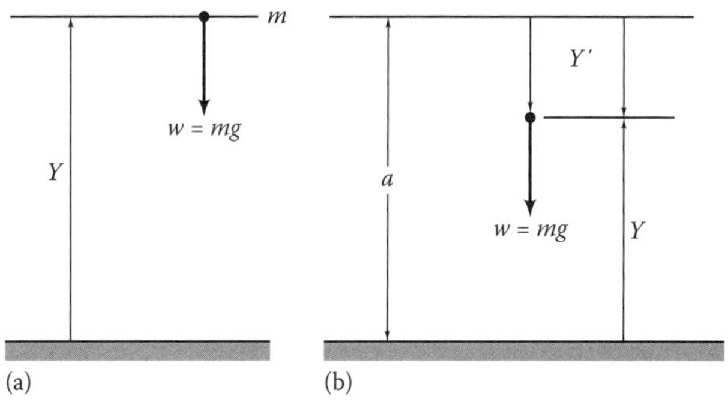

FIGURE 3.3
Particle acted on by its weight; (a) Y coordinate, (b) Y' coordinate.

in applying Newton's second law, $\ddot{Y}'$ and w are also positive downward, and the correct differential equation of motion with Y' used to locate the mass is

$$m\ddot{Y}' = \sum f_{Y'} = w = mg \Rightarrow \ddot{Y}' = g. \tag{3.5}$$

This equation conveys precisely the same physical information as Equation 3.4; namely, the acceleration is constant and vertically downward with magnitude g.

Now, suppose we think about solving the differential equation presented by Equation 3.4

$$\ddot{Y} = \frac{d^2Y}{dt^2} = -g. \tag{3.6}$$

First, observe that this is a second-order, *linear* differential equation with Y the dependent variable and t the independent variable. In dynamics, we are normally concerned with "initial-value" problems such that, at an initial time (normally $t=0$), we know the initial velocity $\dot{Y}(0) = \dot{Y}_0$ and initial position $Y(0) = Y_0$. By contrast, a "boundary-value" problem would arise if the position is specified at two different times; for example, $Y(0) = Y_0$, $Y(t_1) = Y_1$. A solution can still generally be developed for boundary-value problems, but can be much more difficult. We will consider some simple boundary-value problems in Section 3.4.1 covering trajectory motion in a plane. The solution consists of $\dot{Y}(t)$, and $Y(t)$, for $t \geq 0$. Equation 3.6 can be readily integrated with respect to time yielding

$$\frac{dY}{dt} = \dot{Y}(t) = \dot{Y}_0 - gt, \tag{3.7}$$

where the initial velocity condition $\dot{Y}(t=0) = \dot{Y}_0$ has been introduced. A second integration yields

$$Y(t) = Y_0 + \dot{Y}_0 t - \frac{gt^2}{2}, \tag{3.8}$$

and introduces the position initial condition $Y(t=0) = Y_0$. Equations 3.7 and 3.8 are the desired solution for the velocity and displacement as a function of time.

The first integral of the equation of motion with respect to time reflects a useful physical principle. The first integral to Equation 3.4 provided by Equation 3.7 can be multiplied by m to obtain

$$m\dot{Y}(t) - m\dot{Y}_0 = \int_0^t (-w)d\tau.$$

The product of mass times velocity on the left is the "linear momentum" of the particle. The integral on the right is called "linear impulse," and the equation embodies the following physical principle:

$$\text{Change in linear momentum} = \text{Linear impulse}. \tag{3.9}$$

We will return to linear impulse–momentum problems in Section 3.7.

Returning to Equation 3.4 and Figure 3.3, consider the following question: If the particle is released from rest at a height H above the ground, how fast will it be going when it strikes the ground; that is, "What is $\dot{Y}$ when $Y=0$?" We can answer the question by directly applying the initial conditions $\dot{Y} = 0$, $Y_0 = H$ to Equations 3.7 and 3.8, obtaining

$$\dot{Y}(t) = \dot{Y}_0 - gt = -gt$$

$$Y(t) = Y_0 - \frac{gt^2}{2} = H - \frac{gt^2}{2}.$$

The unknown time $\bar{t}$ is the time when the particle strikes the ground, $Y(\bar{t}) = 0$, and is defined from the second equation as

$$0 = H - \frac{g\bar{t}^2}{2} \Rightarrow \bar{t} = \sqrt{\frac{2H}{g}}.$$

Substitution for $\bar{t}$ into the first equation then gives

$$\dot{Y}(\bar{t}) = -g\sqrt{\frac{2H}{g}} = -\sqrt{2gH}.$$

A more direct approach to answer a question concerning the particle's velocity at a specified position follows from the chain rule for differentiation. Note that

$$\ddot{Y} = \frac{d\dot{Y}}{dt} = \frac{d\dot{Y}}{dY}\frac{dY}{dt} = \dot{Y}\frac{d\dot{Y}}{dY} = \frac{d}{dY}\left(\frac{\dot{Y}^2}{2}\right), \tag{3.10}$$

called the "energy-integral substitution." Substituting this result into Equation 3.4 yields

$$m\frac{d}{dY}\left(\frac{\dot{Y}^2}{2}\right) = -w = -mg.$$

Multiplying through by dY and integrating from H to Y gives

$$\frac{m\dot{Y}^2}{2} - \frac{m\dot{Y}_0^2}{2} = -mg\int_H^Y dy = mg(H - Y). \tag{3.11}$$

Since $\dot{Y}_0 = 0$, Equation 3.11 provides the answer for the velocity when the particle hits ground at $Y=0$ as

$$\dot{Y}(Y=0) = \sqrt{2gH}.$$

Returning to Equation 3.11, $m\dot{Y}^2/2$ is the kinetic energy of the particle, and the integral on the right is the work done by w acting through the distance $(H - Y)$. Equation 3.11 embodies the following general physical principle:

$$\text{Work} = \text{Change in kinetic energy.} \qquad (3.12)$$

We will spend a good deal of time working problems in Section 3.6 using work–energy principles. The natural evolution of the equation of motion into a work–energy expression from $\ddot{Y} = d(\dot{Y}^2/2)/dY$ leads to its designation as the "energy-integral substitution."

Now, a strong word of caution: Equations 3.7, 3.8, and 3.11 apply *only* for *constant* acceleration. In moments of desperation occasioned by exams, students have been known to try to make these equations work for all sorts of inappropriate conditions. Please consciously remember that the equations only apply for problems involving constant acceleration and believe that problems involving constant acceleration rarely occur outside of undergraduate dynamics courses.

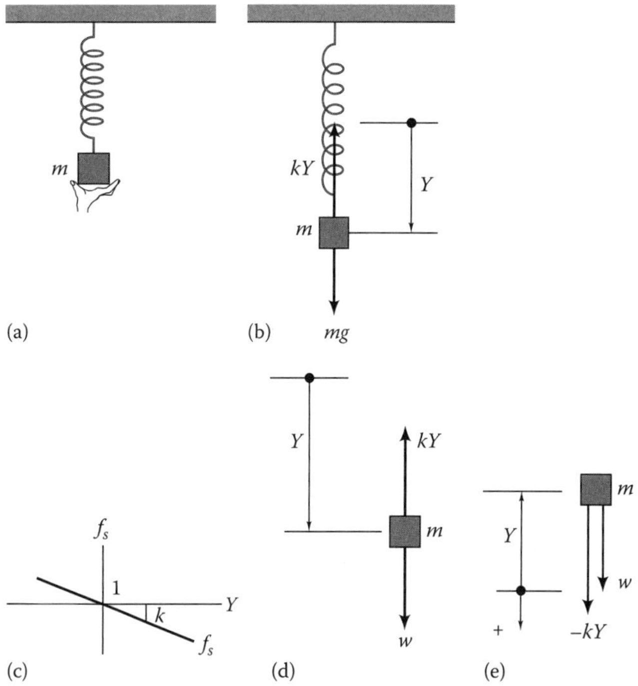

FIGURE 3.4
(a) Particle being held prior to release, (b) Y coordinate defining m's position below the release point, (c) spring reaction force $f_s = -kY$, (d) free-body diagram for $+Y$ displacement, (e) free-body diagram for $-Y$ displacement.

3.2.2 Acceleration as a Function of Displacement: Spring Forces

Dynamics involves both developing models and producing solutions from the model. The model to be developed in this section applies for a one-degree-of-freedom (1DOF) mass-spring system.

3.2.2.1 Deriving the Equation of Motion Starting with the Spring Undeflected

Figure 3.4a illustrates a particle of mass m that is being supported prior to release. It is attached by a linear spring to a fixed support. The spring in Figure 3.4a is undeflected prior to m's release. We want to derive the equation of motion for m following its release. Our first step is choosing the coordinate that will locate m. Figure 3.4b shows the choice Y that defines the distance that m moves vertically downward and away from its starting point. From Figure 3.4b, a negative value for Y implies that m is above its release point. *The conscious choice of coordinates is the single most important step in dynamics.* The choice of $+Y$ as vertically downward requires that the force components acting on m from gravity and the spring, the velocity $\dot{Y}$, and the acceleration $\ddot{Y}$ are also positive in this direction.

Note in Figure 3.4b that m is displaced in the $+Y$ direction. As we will see below, the same equation is obtained if we choose to draw a free-body diagram for m displaced upward. However, developing the equation of motion is generally easier if it is drawn with the body displaced in the positive Y-direction.

For our choice of Y the spring force f_s is zero when Y is zero. It is a linear function of Y as illustrated in Figure 3.4c. Figure 3.4d provides a free-body diagram of the body with w acting downward and the spring force opposing the downward displacement of the mass with the force $-kY$. The minus shows that the spring force is acting in the $-Y$ direction.

Applying Newton's second law of motion to m in Figure 3.4b gives

$$m\ddot{Y} = \sum f_Y = w - kY \Rightarrow m\ddot{Y} + kY = w. \qquad (3.13)$$

In writing Σf_Y, the signs of w and $-kY$ are positive and negative, respectively, depending on whether they are in the same direction (positive) or the opposite direction (negative) as Y. We derived this equation of motion assuming that m was released from rest with the spring undeflected. Equation 3.13 defines Y as the position of m below the release point; however, the equation is correct for any position or velocity of m irrespective of its prior position or history. The *solution* to the differential equation depends on initial conditions of position and velocity; the equation does not.

Continuing with Y directed vertically downward and $f_s = 0$ for $Y = 0$, Figure 3.4e provides a free-body diagram for m displaced upward in the $-Y$ direction. The weight w continues to act in the $+Y$ direction. For this position, the spring is in compression, the spring force has changed direction and now acts in the $+Y$ direction.

As shown, Y is negative; that is, $Y = -|Y|$. Hence, $f_s = -kY = -k(-|Y|) = +|kY|$, and the spring-force definition $f_s = -kY$ continues to be correct for $Y < 0$. The equation of motion development,

$$m\ddot{Y} = \sum f_Y = w - kY \Rightarrow m\ddot{Y} + kY = w,$$

based on $f_s = -kY$ continues to be correct for $0 \le Y \le 0$. Accepting that the equation is valid for any value of Y, this development shows that deriving the equation of motion is generally easier if you draw a free-body diagram with the mass displaced in a positive coordinate direction.

A large part of dynamics involves deriving equations of motion. As with this example, the first step in applying $\sum f = m\ddot{r}$, is the choice of coordinates. Without a choice for a coordinate and its attendant sign definitions for $f, Y, \dot{Y}, \ddot{Y}$, a free-body diagram cannot be converted into an equation of motion. With a coordinate choice, moving from a correct free-body diagram is straightforward.

3.2.2.2 Deriving the Equation of Motion for Motion about Equilibrium

The circumstances of Figure 3.4 where the mass is initially supported, with the spring in an unstretched position, is less common than motion about an equilibrium position as shown in Figure 3.5a. The free-body diagram for equilibrium is shown in Figure 3.5b and yields $k\bar{Y} = w$. The free-body diagram of Figure 3.5c applies for m displaced the distance y below the equilibrium point. The additional spring displacement generates the spring reaction force, $f_s = k(\bar{Y} + y)$, and yields the following equation of motion:

$$m\ddot{y} = \sum f_Y = w - k(\bar{Y} + y) = w - k\bar{Y} - ky = -ky$$
$$\therefore\ m\ddot{y} + ky = 0. \tag{3.14}$$

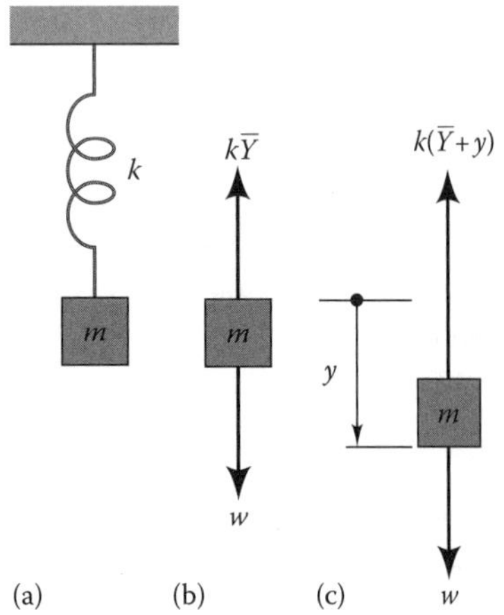

FIGURE 3.5
(a) Mass m in equilibrium, (b) equilibrium free-body diagram, (c) general-position free-body diagram.

This result holds for a *linear* spring and shows that w is eliminated, leaving only the perturbed spring force ky.

A different outcome holds for a nonlinear-spring model such as $f_s = -kY(1 + \alpha|Y|)$. This model defines a "hardening" spring, since its local stiffness increases with increasing Y For $Y \ge 0$, the local stiffness defined by the slope $df_s/dY = k(1 + 2\alpha Y)$ increases with increasing Y. A similar result holds for $Y \le 0$. Using this model equilibrium due to weight is defined by $k\bar{Y}(1 + \alpha|\bar{Y}|) = w$. This quadratic equation can be solved for the positive equilibrium position $\bar{Y}$. For motion about the equilibrium position $\bar{Y}$, with $Y \ge 0, Y = \bar{Y} + y \Rightarrow \ddot{Y} = \ddot{y}$. Summing forces gives the equation of motion

$$m\ddot{y} = \sum f_Y = w - k(\bar{Y} + y)[1 + \alpha(\bar{Y} + y)]$$
$$\therefore\ m\ddot{y} + ky(1 + 2\alpha\bar{Y} + \alpha y) = w - k\bar{Y}(1 + \alpha\bar{Y}) = 0,$$

and w is eliminated. For small motion about the equilibrium position, the quadratic term y^2 can be neglected, and the linearized equation is

$$m\ddot{y} + [k(1 + 2\alpha\bar{Y})]y = 0.$$

This local stiffness $k(1 + 2\alpha\bar{Y})$ depends on the equilibrium position $\bar{Y}$. For the linear model, the stiffness is constant and independent of the equilibrium position.

Figure 3.6a shows a mass m, supported by two springs that is in equilibrium. In terms of the spring deflections, equilibrium can be arrived at in several ways. The equilibrium in the left frame of Figure 3.6b might correspond to starting with the springs undeflected and slowly lowering m until equilibrium is achieved, causing the top spring to be in tension with force $f_1 = k_1\bar{Y}$ and the bottom spring to be in compression with force $f_2 = k_2\bar{Y}$. Equilibrium requires: $w = f_1 + f_2 = (k_1 + k_2)\bar{Y}$. The right frame of Figure 3.6b provides a free-body diagram for m displaced downward from the equilibrium position a distance y. This displacement increases the stretch in the top spring to $\bar{Y} + y$ producing the force $k_1(\bar{Y} + y)$. The initial displacement in the bottom spring is reduced to $\bar{Y} - y$ to produce the reaction force $k_2(\bar{Y} - y)$. Summing forces produces the equation of motion

$$m\ddot{y} = \sum f_y = w - k_1(\bar{Y} + y) - k_2(\bar{Y} + y)$$
$$= w - (k_1 + k_2)\bar{Y} - (k_1 + k_2)y = 0 - (k_1 + k_2)y$$
$$\therefore\ m\ddot{y} + (k_1 + k_2)y = 0.$$

As with the prior example, for motion about the equilibrium position defined by y, the weight does not enter the equations of motion. In looking at this equation, you should be thinking that the effective stiffness $k_{eff} = k_1 + k_2$ follows from the rules for springs in parallel.

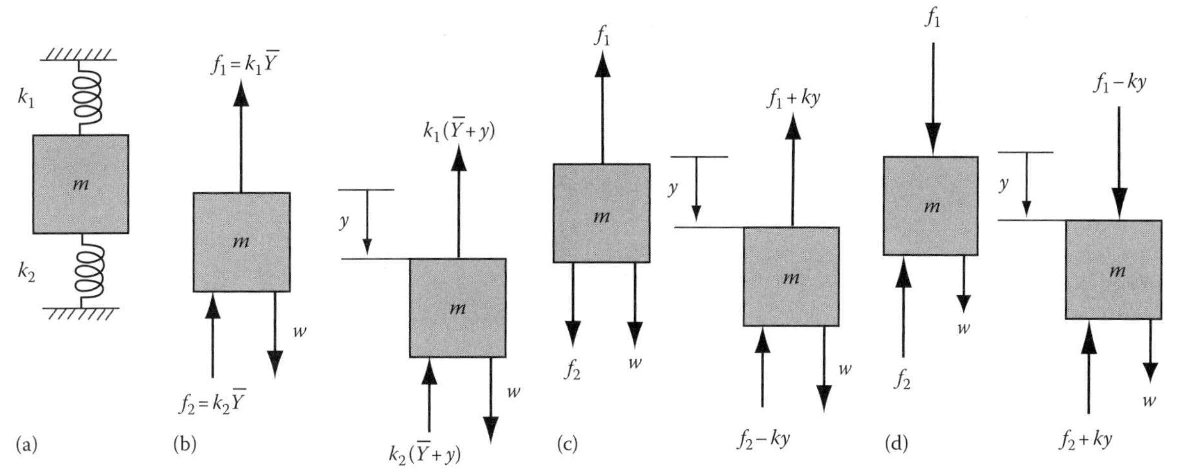

FIGURE 3.6
(a) Equilibrium, (b) equilibrium with spring 1 in tension and spring 2 in compression, (c) equilibrium with both springs in tension, (d) equilibrium with both springs in compression.

The left frame of Figure 3.6c shows another equilibrium condition with both springs in tension. The equilibrium requirement is $w = f_1 - f_2$. As shown in the right frame of Figure 3.6c, displacing the mass downward from the equilibrium position through y increases the top spring tension to $f_1 + k_1 y$, while decreasing the bottom spring tension to $f_2 - k_2 y$. Summing forces gives the equation of motion

$$\begin{aligned} m\ddot{y} &= \sum f_y = w - (f_1 + k_1 y) + (f_2 - k_2 y) \\ &= (w - f_1 + f_2) - (k_1 + k_2)y = 0 - (k_1 + k_2)y \\ \therefore\ & m\ddot{y} + (k_1 + k_2)y = 0. \end{aligned}$$

Note that this outcome does not depend on the magnitudes of f_1, f_2.

Figure 3.6d illustrates yet another equilibrium condition with both springs in compression, netting the equilibrium requirement $w = f_2 - f_1$. Displacing m downward through y decreases the compression in the top spring to $f_1 - k_1 y$, while increasing the compression in the bottom spring to $f_2 + k_2 y$. Summing forces nets the same equation of motion.

$$\begin{aligned} m\ddot{y} &= \sum f_y = w + (f_1 - k_1 y) - (f_2 + k_2 y) \\ &= (w + f_1 - f_2) - (k_1 + k_2)y = 0 - (k_1 + k_2)y \\ \therefore\ & m\ddot{y} + (k_1 + k_2)y = 0. \end{aligned}$$

In these developments, the springs are assumed to be attached to the mass. Hence, for the circumstances of Figure 3.6d, the compression in the bottom spring $f_2 - k_2 y$ could change sign and shift to tension without changing the equation of motion. Without attaching the spring to m, the spring could disengage rather than shift to tension. *In summary, with linear springs and motion about equilibrium, the same equation applies, independently of the actual condition of equilibrium as defined by reaction forces, static displacements, etc.*

Example Problem 3.1

Figure XP3.1 illustrates a mass m that is in equilibrium and supported by two spring–pulley combinations. The mass weighs 100 lbs, and the spring constant is $k = 200$ lb/in.

Tasks: (a) For motion about equilibrium, draw a free-body diagram and derive the equations of motion, and (b) determine the natural frequency.

Solution

Figure XP3.1b shows the coordinate choice y directed downward and the free-body diagram. Note that y is the displacement away from equilibrium. Because m starts in equilibrium, $w = 4\overline{T}_c$. The weight is included in the free-body diagram but we only need to define δT_c, the change in the spring forces due to the displacement y away from equilibrium. Applying $\Sigma f_y = m\ddot{y}$ to the free-body diagram of Figure XP3.1b gives

$$m\ddot{y} = \sum f_y = w - 4(\overline{T}_c + \delta T_c) = -4\delta T_c. \quad \text{(i)}$$

The reaction forces are defined as $\delta T_c = k y_s$ where y_s is the change in length of the spring due to the displacement y. We need a relationship between y and y_s. The cord wrapped around a pulley is inextensible with length l_c. Point A in Figure XP3.1a locates the end of the cord. From this figure, $l_c = a + (a - d) = 2a - d$. Figure XP3.1c shows the displaced position and provides the following relationship:

$$\begin{aligned} l_c &= (a + y) + [(a + y) - (d + y_s)] = (2a - d) + 2y - y_s \\ &= l_c + 2y - y_s \quad \text{(ii)} \\ \therefore\ & y_s = 2y. \end{aligned}$$

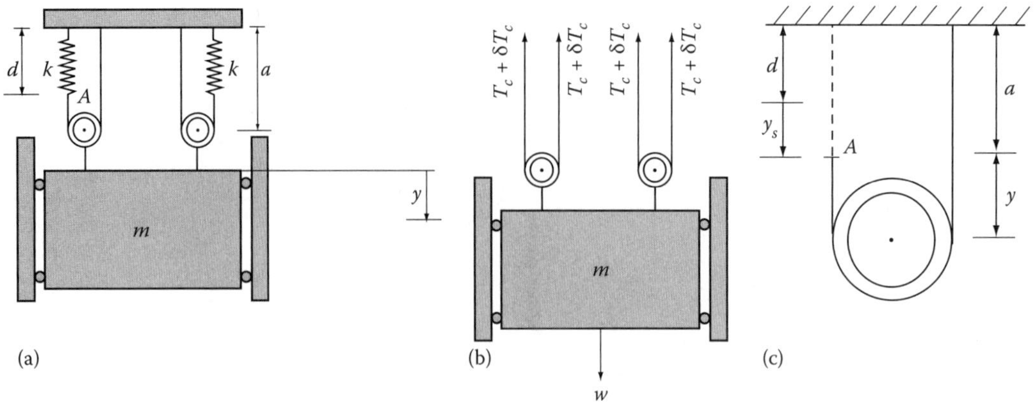

FIGURE XP3.1
(a) Mass in equilibrium, (b) free-body diagram, (c) kinematic-constraint relation.

Applying these results to Equation (i) gives

$$\begin{aligned} m\ddot{y} &= -4\delta T_c = -4ky_s == -4k_s(2y) \\ &= -8ky \Rightarrow m\ddot{y} + 8ky = 0. \end{aligned} \quad \text{(iii)}$$

This step concludes Task a.

Equation (i) has the form $m\ddot{y} + k_{eq}y = 0$ where $k_{eq} = 8k$ is the "equivalent stiffness." Dividing through by m puts the equation into the form $\ddot{y} + \omega_n^2 y = 0$; hence,

$$\begin{aligned} \omega_n^2 &= \frac{k_{eq}}{m} = \frac{8 \times 200 \text{ lb/in.}}{[200 \text{ lb}/(386 \text{ in./s}^2)]} \\ &= 3088\left(\frac{\text{rad}}{\text{s}}\right)^2 \Rightarrow \omega_n = 55.6 \text{ rad/s} \end{aligned}$$

Note the conversion from weight to mass via $m = w/g$ where for the inch-pound-second system, $g = 386.$ in./s^2.

3.2.2.3 Developing a Time Solution for the Equation of Motion

The time solution for $m\ddot{Y} + kY = w$ is normally obtained by first dividing through by m to obtain

$$\ddot{Y} + \omega_n^2 Y = g, \quad (3.15)$$

where

$$\omega_n = \sqrt{\frac{k}{m}}. \quad (3.16)$$

The parameter ω_n is called the "natural frequency." Using the kg-meter-second system for length, mass, and time, the dimensions for ω_n^2 follow as

$$\omega_n^2 = \frac{k\ (\text{N/m})}{m\ (\text{kg})} = kg\left(\frac{\text{kg m}}{\text{s}^2} \times \frac{1}{\text{m}}\right) \times \frac{1}{m\ (\text{kg})} = \frac{k}{m}\ (\text{s}^{-2}). \quad (3.17)$$

Hence, the dimension for ω_n is rad/s, or s^{-1} (since the radian is dimensionless). The natural frequency can also be stated in terms of Hertz (Hz), which is cycles per second, defined as

$$\begin{aligned} f_n\left(\frac{\text{cycles}}{\text{s}}\right) &= \omega_n\left(\frac{\text{rad}}{\text{s}}\right) \times \left(\frac{1 \text{ cycle}}{2\pi \text{ rad}}\right) = \frac{\omega_n}{2\pi}\frac{\text{cycles}}{\text{s}} \\ &= \frac{\omega_n}{2\pi} \text{ Hz}. \end{aligned}$$

Note that Equation 3.15 requires ω_n, not f_n. The "period" of the oscillation is defined by

$$\tau_n = \frac{1}{f_n} = \frac{2\pi}{\omega_n}\frac{\text{s}}{\text{cycle}}. \quad (3.18)$$

We will show that the solution to Equation 3.15 is periodic. Specifically, if it contains the point, $Y(\bar{t})$, it will return to this point at $t = \bar{t} + \tau_n$; that is, $Y(\bar{t} + \tau_n) = Y(\bar{t})$. (Heinrich Rudolph Hertz (1857–1894) was a German physicist who first produced, detected, and identified radio waves. Appropriately radio frequency transmissions are specified in Hertz.)

The homogeneous version of $\ddot{Y} + \omega_n^2 Y = g$ is

$$\ddot{Y}_h + \omega_n^2 Y_h = 0.$$

You would expect that pulling down and releasing a mass supported by a spring would result in harmonic motion. Guessing a harmonic solution of the form $Y_h = A\cos\omega t \Rightarrow \ddot{Y}_h = -\omega^2 A\cos\omega t$ and substituting gives

$$(-\omega^2 + \omega_n^2)A\cos\omega t = 0$$

Since neither A nor $\cos\omega t$ is zero, this equation is satisfied by $\omega = \omega_n = \sqrt{k/m}$; that is, the motion is harmonic at a frequency equal to the undamped natural frequency. We could have obtained a similar result by guessing the harmonic solution $Y_h = A\sin\omega t \Rightarrow \ddot{Y}_h = -A\omega^2\sin\omega t$; hence, the complete homogeneous solution is

$$Y_h = A\cos\omega_n t + B\sin\omega_n t.$$

You can confirm that the particular solution,

$$Y_p = \frac{g}{\omega_n^2} = \frac{w}{k},$$

satisfies the right-hand side of $\ddot{Y} + \omega_n^2 Y = g$. This is also the static solution, defining the steady-state deflection of the spring due to the weight w. The complete solution is

$$Y = Y_h + Y_p = A\cos\omega_n t + B\sin\omega_n t + \frac{w}{k}. \quad (3.19)$$

The two constants A and B are used *with the complete solution* to satisfy the problem's initial conditions. For this example, the particle was released from rest at $Y=0$. From Equation 3.19, the displacement initial condition gives

$$Y(t=0) = Y_0 = 0 = A + \frac{w}{k} \Rightarrow A = -\frac{w}{k}.$$

The velocity initial condition is implemented into the solution by differentiating Equation 3.19 to obtain

$$\dot{Y} = -A\omega_n\sin\omega_n t + B\omega_n\cos\omega_n t.$$

Evaluating this expression at $t=0$ gives

$$\dot{Y}(0) = \dot{Y}_0 = 0 = B\omega_n \Rightarrow B = 0$$

Back substitution for A and B into Equation 3.19 yields the complete solution

$$Y = \frac{w}{k}(1 - \cos\omega_n t). \quad (3.20)$$

Had we slowly lowered the weight onto the spring in Figure 3.4a, the static deflection would be $\overline{Y} = w/k$. The solution of Equation 3.20 has a minimum of zero and a peak amplitude, $Y_{max} = 2\overline{Y} = 2w/k$. This solution applies for the specific set of initial conditions $(Y(0) = \dot{Y}(0) = 0)$ with a constant applied force due to weight. Appendix B provides a refresher on differential equations and includes the development of general solution for arbitrary initial conditions and a range of forcing functions, for example, a force that is linearly increasing with time, quadratically increasing with time, etc.

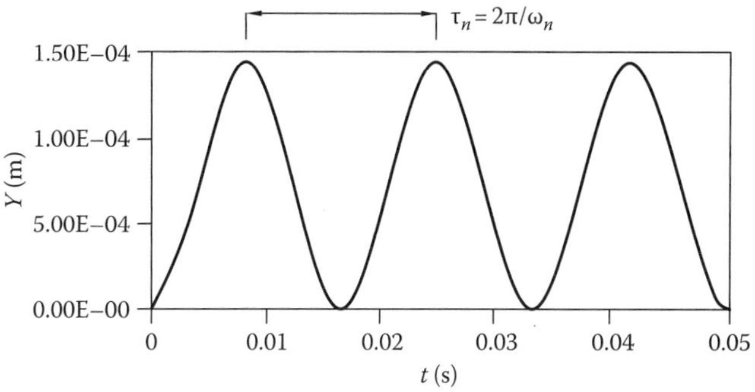

FIGURE 3.7
$Y(t)$ from Equation 3.20 with $k = 5.685 \times 10^6$ N/m and $m = 40$ kg, yielding $\omega_n = 377$ Hz, $f_n = \omega_n/(2\pi) = 60$ Hz and $\tau_n = 1/f_n = .0167$ cycles/s.

Equation 3.20's solution is illustrated in Figure 3.7 for $k = 5.685 \times 10^6$ N/m and $m = 40$ kg and consists of a periodic oscillation at the natural frequency $\omega_n = \sqrt{5.865 \times 10^6/40} = 377$ rad/s. In Hz, the frequency is $f_n = 377/(2\pi) = 60$ Hz with the corresponding period $\tau_n = 1/f_n = 0.0167$ s. Figure 3.7 shows a steady oscillation about the static equilibrium position. This result should be generally consistent with your own experiences and expectation; that is, if you visualize holding the mass with the spring undeflected and then releasing it, a periodic oscillation would be expected. The solution differs from first-hand experience to the extent that it continues oscillating at the same amplitude indefinitely; whereas, free vibration amplitudes generally decrease as time increases. The next section will add energy dissipation, yielding a reduction in amplitude with continuing time.

3.2.2.4 Developing a Solution for $\dot{Y}$ as a Function of Y

Returning to $m\ddot{Y} + kY = w$, the EOM can be restated as

$$m\ddot{Y} = w - kY = f(Y).$$

The right-hand side is a function of Y only (no $\dot{Y}$ or t). A solution for $\dot{Y}$ as a function of Y is possible when $\ddot{Y} = \ddot{Y}(Y)$ using the energy-integral substitution,

$$\ddot{Y} = \frac{d\dot{Y}}{dt} = \frac{d\dot{Y}}{dY}\frac{dY}{dt} = \dot{Y}\frac{d\dot{Y}}{dY} = \frac{d}{dY}\left(\frac{\dot{Y}^2}{2}\right), \quad (3.10)$$

to obtain

$$m\frac{d}{dY}\left(\frac{\dot{Y}^2}{2}\right) = w - kY.$$

Multiplying through by dY and integrating both side of the equation yields

$$\frac{m\dot{Y}^2}{2} - \frac{m\dot{Y}_0^2}{2} = w(Y - Y_0) - \left(\frac{kY^2}{2} - \frac{kY_0^2}{2}\right).$$

This equation states that the system's mechanical (kinetic plus potential) energy is conserved. Solving for $\dot{Y}(Y)$ gives

$$\dot{Y}(Y) = \sqrt{\dot{Y}_0^2 + g(Y - Y_0) - \omega_n^2\left(\frac{Y^2}{2} - \frac{Y_0^2}{2}\right)},$$

For some systems that are modeled by nonlinear equations, an analytical time solution cannot be developed, and getting $\dot{Y}(Y)$ is the best that we can do.

3.2.2.5 Negative Sign for the Stiffness Coefficient

Students have been known to get the sign wrong on the spring force yielding

$$m\ddot{Y} - kY = w.$$

The solution to this equation is

$$Y = A\cosh\omega_n t + B\sinh\omega_n t + \frac{w}{k}.$$

The terms "$\cosh\omega_n t$" and "$\sinh\omega_n t$" are the hyperbolic cosine and sine functions, respectively,

$$\cosh\omega_n t = \frac{(e^{\omega_n t} + e^{-\omega_n t})}{2}, \quad \sinh\omega_n t = \frac{(e^{\omega_n t} - e^{-\omega_n t})}{2},$$

and they grow exponentially with increasing time. A solution that grows exponentially with increasing time requires a continuous input of energy and is not consistent with our present expectations.

The model $m\ddot{Y} - kY = 0$ is said to be statically unstable, and this type of model can occur in real-world experiences. Figure 3.8 shows an inverted compound pendulum, whose motion (for small θ) is governed by

$$\ddot{\theta} - \frac{3g}{2l}\theta = 0.$$

The governing differential equation for a compound pendulum is derived in Section 5.6.1. The negative coefficient for θ is appropriate for Figure 3.8, because the prediction of increasing amplitude with increasing time is consistent with physical reality; namely, the motion is unstable. Any small perturbation of the pendulum away from the vertical will yield a rapidly increasing value for θ as time increases. Check the sign for the stiffness coefficient in your differential equations of motion. Unless you expect the system to be statically unstable (as with an inverted pendulum), a negative-stiffness coefficient in the equation of motion is invariably wrong.

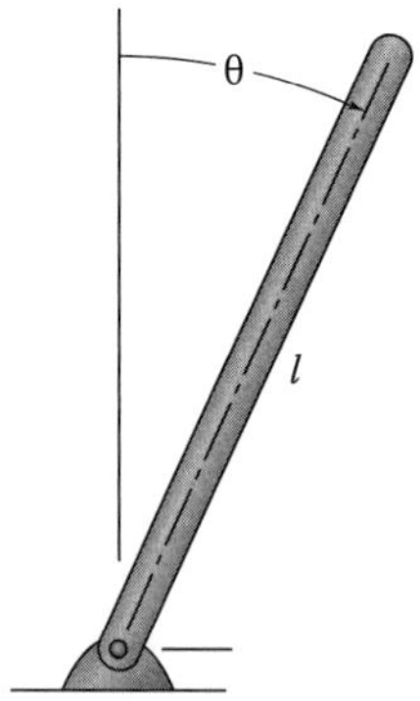

FIGURE 3.8
Inverted compound pendulum.

3.2.3 Energy Dissipation: Viscous Damping

3.2.3.1 Viscous Damper

Figure 3.9 illustrates a linear "dashpot" damper consisting of a piston that can move inside a cylinder. A small radial clearance between the cylinder and piston contains a lubricant. If the top end of the piston is pulled upward, fluid shearing forces around the cylinder will create a resisting force that is proportional to the piston's velocity relative to the cylinder. The proportionality factor c in Figure 3.9 is called a linear damping coefficient. Similar devices are used in automatic door closers and automobile shock absorbers to limit the amplitudes of response.

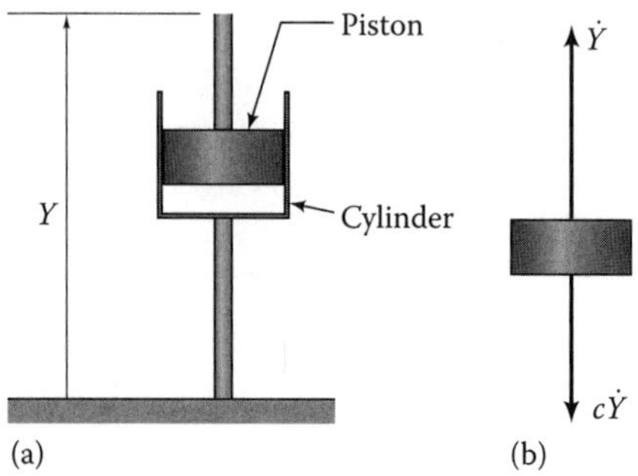

FIGURE 3.9
(a) Linear dashpot. (b) Viscous-damping force.

3.2.3.2 Deriving the Equation of Motion for a Mass-Spring-Damper System

Figure 3.10a illustrates a mass m that is connected to a stationary support by two linear springs with stiffness coefficients $k/2$ and a viscous damper with coefficient c. As shown, the springs are undeflected. The position of the mass is defined by Y, and the spring is undeflected when $Y = 0$. The free-body diagram in Figure 3.10b corresponds to a positive displacement Y and positive velocity $\dot{Y}$. Summing forces in the $+Y$ direction yields from $\sum f = m\ddot{r}$

$$\begin{aligned} m\ddot{Y} &= \sum f_Y = w - \frac{kY}{2} - c\dot{Y} - \frac{kY}{2}, \quad \text{or} \\ m\ddot{Y} &+ c\dot{Y} + kY = w. \end{aligned} \tag{3.21}$$

Since the damping force is defined by $f_d = -c\dot{Y}$, the damping coefficient has dimensions defined by

$$c = -\frac{f_d(\text{N})}{\dot{Y}\,(\text{m/s})} = -\frac{f_d}{\dot{Y}}\left(\frac{\text{N-s}}{\text{m}}\right).$$

In the U.S. conventional ft-lb-s unit system, c's dimensions are lb-s/ft. In the in.-lb-s system, the dimensions are lb-s/in.

Figure 3.10c shows another free-body diagram for the model, corresponding to an upward velocity of m with $\dot{Y} < 0$; hence, $\dot{Y} = -|\dot{Y}|$. The reaction force opposes the velocity direction and points downward in the $+Y$-direction. The damping force model $f_d = -c\dot{Y}$ gives this result, via $f_d = -c\dot{Y} = -c(-|\dot{Y}|) = |c\dot{Y}|$. As expected, the model of Equation 3.22 is correct for any value of $\dot{Y}$. A comparison of our present development proceeding from Figure 3.10c and the development from Figure 3.10b supports the following general conclusion: Assuming a positive velocity direction when drawing a free-body diagram makes the derivation of equations of motion easier.

3.2.3.3 Motion about the Equilibrium Position

Deriving equations of motion for motion about equilibrium was discussed earlier in regard to mass-spring systems. For the present circumstances, restating Equation 3.21 as

$$m\ddot{Y} = \sum f_Y = w - c\dot{Y} - kY$$

provides the equilibrium definition $\sum f_Y = 0$ that requires (1) $\dot{Y} = 0$, and (2) $Y_{\text{equilibrium}} = \overline{Y} = w/k$. Motion about the equilibrium position is defined by $Y = \overline{Y} + y \Rightarrow \dot{Y} = \dot{y} \Rightarrow \ddot{Y} = \ddot{y}$.

Substituting these results gives

$$\begin{aligned} &m\ddot{y} + c\dot{y} + k(\overline{Y} + y) = w, \quad \text{or since } k\overline{Y} = w \\ &m\ddot{y} + c\dot{y} + ky = 0. \end{aligned}$$

Comparing this equation to the results from Equation 3.14 without the damper, $m\ddot{y} + ky = 0$, the viscous damper only adds the $c\dot{y}$ term. Motion about equilibrium impacts the spring and weight terms but not the damper term.

3.2.3.4 Developing a Time Solution for the Equation of Motion

Proceeding to a time solution for $m\ddot{Y} + c\dot{Y} + kY = w$ starts by dividing through by m and yields

$$\ddot{Y} + 2\zeta\omega_n\dot{Y} + \omega_n^2 Y = g, \tag{3.22}$$

where $\omega_n = \sqrt{k/m}$, and

$$2\zeta\omega_n = \frac{c}{m} \Rightarrow \zeta = \frac{c}{2m\omega_n} = \frac{c}{2\sqrt{km}}. \tag{3.23}$$

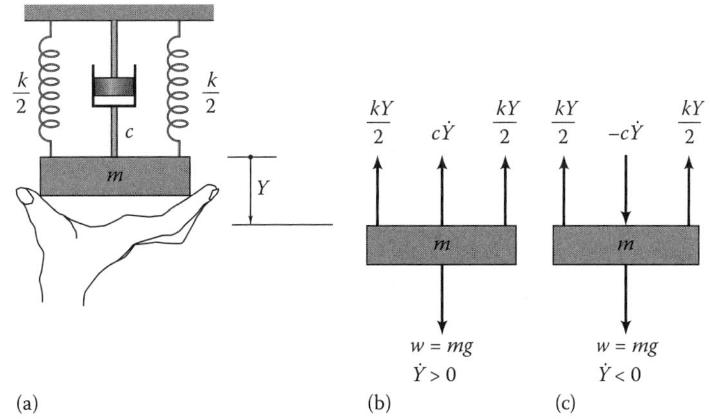

FIGURE 3.10
(a) Particle acted on by its weight and supported by two (undeflected) springs and a viscous damper, (b) free-body diagram for $Y > 0$, $\dot{Y} > 0$, and (c) free-body diagram for $Y > 0$, $\dot{Y} < 0$.

For reasons that will become apparent, the coefficient ζ is called the viscous-damping factor. Note that ζ is dimensionless. The solution to the homogeneous equation,

$$\ddot{Y}_h + 2\zeta\omega_n\dot{Y}_h + \omega_n^2 Y_h = 0,$$

is obtained by assuming a solution of the form $Y_h = Ae^{st}$, which yields

$$(s^2 + 2\zeta\omega_n s + \omega_n^2)Ae^{st} = 0.$$

Since neither A nor e^{st} are generally equal to zero, s must satisfy

$$s^2 + 2\zeta\omega_n s + \omega_n^2 = 0.$$

This equation is called the "characteristic equation." The condition $\zeta = 1$ corresponds to "critical damping." For $\zeta \geq 1$ the solution decays exponentially without oscillating. Appendix B provides solutions for $\zeta \geq 1$; however, we are normally more concerned with oscillatory motion corresponding to $\zeta < 1$. For this condition, the equation's two roots are

$$s = -\zeta\omega_n \pm j\omega_n\sqrt{1-\zeta^2} = -\zeta\omega_n + j\omega_d,\ -\zeta\omega_n - j\omega_d$$
$$\omega_d = \omega_n\sqrt{1-\zeta^2}. \tag{3.24}$$

Again, for reasons that will become apparent, $\omega_d = \omega_n\sqrt{1-\zeta^2}$ is called the "damped natural frequency." There are two roots corresponding to the $\pm$ signs in the square root of Equation 3.24; hence, the homogeneous solution looks like

$$Y_h = \boldsymbol{A}_1 e^{(-\zeta\omega_n + j\omega_d)t} + \boldsymbol{A}_2 e^{(-\zeta\omega_n - j\omega_d)t},$$

where $\boldsymbol{A}_1$, $\boldsymbol{A}_2$ are *complex* coefficients. Substituting the identities,

$$e^{j\omega_d t} = \cos\omega_d t + j\sin\omega_d t, \quad e^{-j\omega_d t} = \cos\omega_d t - j\sin\omega_d t,$$

yields a final homogeneous solution of the form

$$Y_h = e^{-\zeta\omega_n t}(A\cos\omega_d t + B\sin\omega_d t). \tag{3.25}$$

where A and B are *real* constants. (This solution is developed in more detail in Appendix B.) With damping present, the frequency of oscillation is seen to be ω_d. You may verify that the particular solution from Equation 3.19, $Y_p = w/k$, remains valid for the present differential equation. Hence, the complete solution is

$$Y = e^{-\zeta\omega_n t}(A\cos\omega_d t + B\sin\omega_d t) + \frac{w}{k}. \tag{3.26}$$

We repeat our prior initial conditions for motion without damping; namely, $Y(0) = \dot{Y}(0) = 0$. Hence, Equation 3.26 yields

$$Y(0) = 0 = A + \frac{w}{k} \Rightarrow A = -\frac{w}{k}.$$

To evaluate B, we differentiate Equation 3.26, obtaining

$$\dot{Y} = e^{-\zeta\omega_n t}\omega_d\,(-A\sin\omega_d t + B\cos\omega_d t)$$
$$- \zeta\omega_n e^{-\zeta\omega_n t}(-A\cos\omega_d t + B\sin\omega_d t).$$

Evaluating this expression at $t = 0$ gives

$$\dot{Y}(0) = 0 = B\omega_d - \zeta\omega_d A.$$

Hence,

$$B = \frac{\zeta\omega_n A}{\omega_d} = \frac{\zeta A}{\sqrt{1-\zeta^2}} = -\frac{\zeta}{\sqrt{1-\zeta^2}}\frac{w}{k}$$

The complete solution is

$$Y = \frac{w}{k}\left[1 - e^{-\zeta\omega_n t}\left(\cos\omega_d t + \frac{\zeta}{\sqrt{1-\zeta^2}}\sin\omega_d t\right)\right]. \tag{3.27}$$

Figure 3.11 illustrates this solution for the data $m = 1$ kg, $k = 3948.$ N/m, $c = 12.57$ N s/m, which yield

$$\omega_n = \sqrt{\frac{k}{m}} = 62.8\ \text{rad/s}; \quad f_n = \frac{\omega_n}{2\pi} = 10\ \text{cycles/s} = 10\ \text{Hz}$$
$$\zeta = \frac{c}{2m\omega_n} = .10, \quad \omega_d = \omega_n\sqrt{1-\zeta^2} = 62.2\ \text{rad/s}. \tag{3.28}$$

For damped motion, the period is

$$\tau_d = \frac{2\pi}{\omega_d}. \tag{3.29}$$

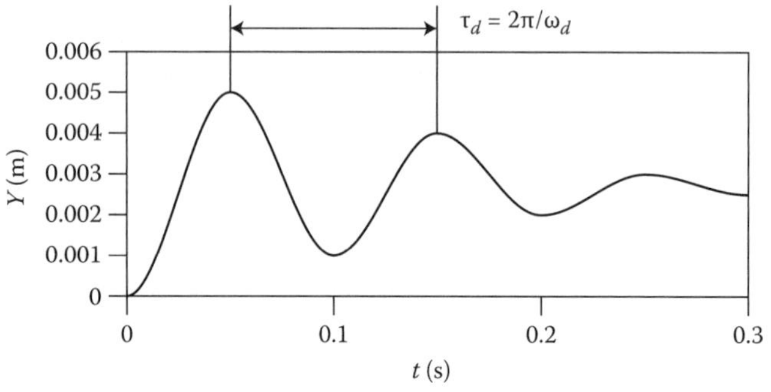

FIGURE 3.11
Time solution of Equation 3.27 with the data of Equation 3.28.

Observe in Figure 3.11 that an oscillation about the static equilibrium still occurs but decays exponentially with increasing time, which should be consistent with your every-day experiences with vibrations. For example, pushing down on the front end of an automobile and then releasing it is a standard check to see if the shock absorbers are worn out. With new shock absorbers, there is virtually no vibration (large ζ); with worn shocks, vibration persists following the initial disturbance.

The next two example problems illustrate the development of time solutions for 1DOF mass-spring-damper systems. They are provided as both examples for deriving equations of motion and a review for the solution of differential equations.

Example Problem 3.2

Figure XP3.2a illustrates a cart that is rolling along at the speed v_0 with negligible rolling resistance. It has a collision-absorption unit consisting of a spring and damper connected to a rigid plate of negligible mass. At $t=0$, it collides with a vertical wall.

Tasks:

a. Select a coordinate and derive the equation of motion.
b. For $m=100$ kg, $k=9.87E+04$ N/m, $c=3141.6$ N s/m, and $v_0=10$ m/s determine the solution for motion following collision while the cart remains in contact with the wall.
c. For the given k and m values, illustrate how a range of damping constants c producing $0 \leq \zeta < 1$ will reduce the stopping time and peak amplitude.

Solution

Figure XP3.2b shows the coordinate choice x. Contact occurs for $x \geq 0$. The free-body diagram in Figure XP3.2b corresponds to $x > 0$ and $\dot{x} > 0$, requiring compression in the spring and damper. Applying Newton's laws gives

$$m\ddot{x} = \sum f = -kx - c\dot{x} \Rightarrow m\ddot{x} + c\dot{x} + kx = 0.$$

The spring and damper forces are negative in the force equation because they are acting in the $-x$ direction. The initial conditions: $x(0) = 0, \dot{x}(0) = v_0$ apply. There is no

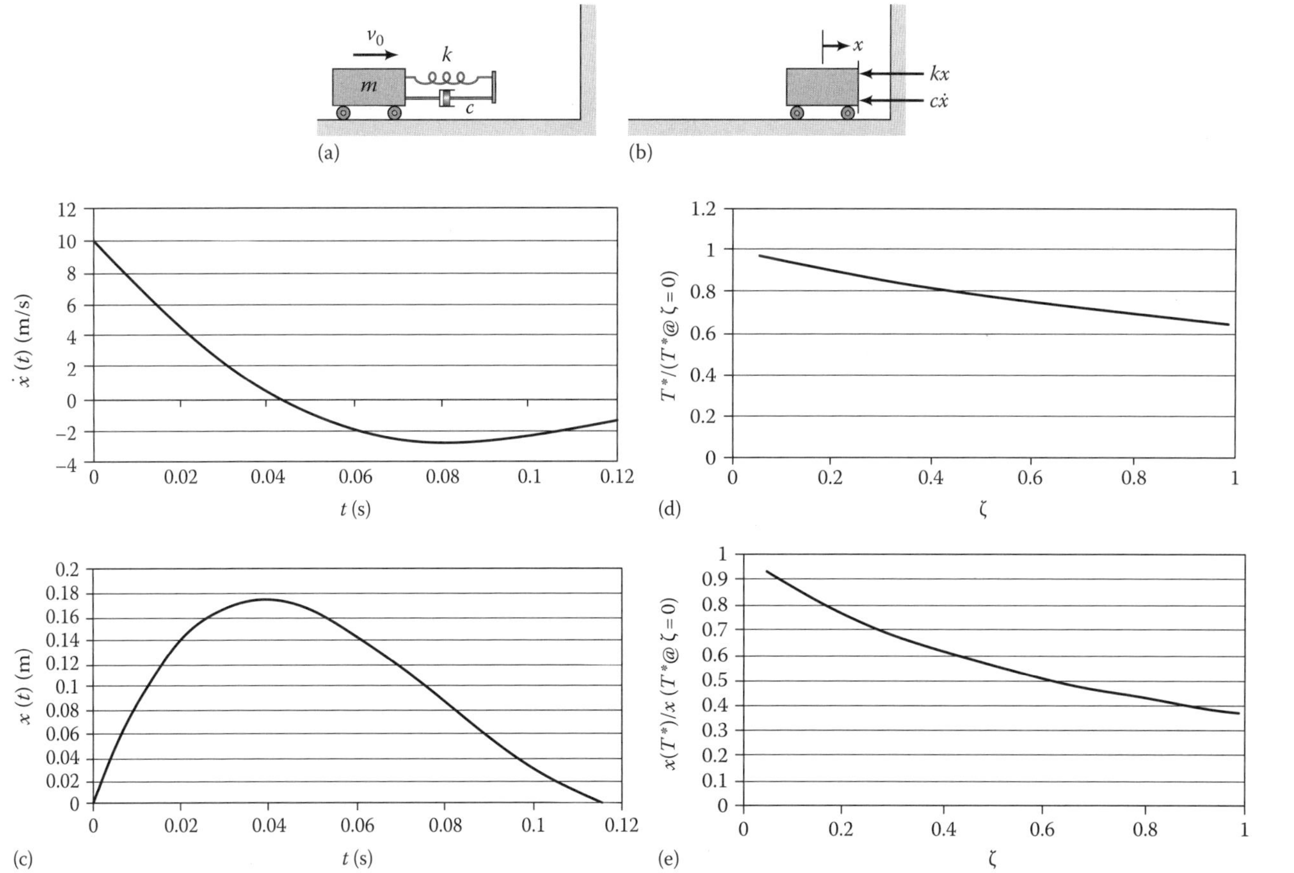

FIGURE XP3.2
(a) Cart approaching collision, (b) free-body diagram during contact, (c) $x(t)$ and $\dot{x}(t)$ following impact, (d) ratio of stopping times for $0 \leq \zeta < 1$, (e) ratio of peak amplitudes for $0 \leq \zeta < 1$.

forcing function on the right-hand side of the equation, implying $x_p = 0$ for the particular solution; hence, the homogeneous solution of Equation 3.25,

$$x = x_h = e^{-\zeta\omega_n t}(A\cos\omega_d t - B\sin\omega_d t), \quad \text{(i)}$$

is the complete solution. Differentiation with respect to time gives

$$\begin{aligned}\dot{x} = &-\zeta\omega_n e^{-\zeta\omega_n t}(A\cos\omega_d t + B\sin\omega_d t)\\ &+ e^{-\zeta\omega_n t}\omega_d\,(-A\sin\omega_d t + B\cos\omega_d t). \quad \text{(ii)}\end{aligned}$$

Applying the initial conditions gives

$$x(0) = 0 = A, \quad \dot{x}(0) = v_0 = -\zeta\omega_n A + B\omega_d \Rightarrow B = \frac{v_0}{\omega_d}.$$

Substituting back into (i) and (ii) gives

$$\begin{aligned}x &= \frac{v_0}{\omega_d}e^{-\zeta\omega_n t}\sin\omega_d t\\ \dot{x} &= -\frac{\zeta v_0}{\sqrt{1-\zeta^2}}e^{-\zeta\omega_n t}\sin\omega_d t + v_0 e^{-\zeta\omega_n t}\cos\omega_d t,\end{aligned} \quad \text{(iii)}$$

where $\omega_d = \omega_n\sqrt{1-\zeta^2}$. The cart loses contact with the wall when $x(T_f) = 0$. T_f is defined from the first of Equation (iii) by $\sin(\omega_d T_f) = 0 \Rightarrow \omega_d T_f = \pi$. (The equation $\sin(\omega_d T_f) = 0$ has an infinite number of solutions defined by $\omega_d T_f = n\pi$; $n = 0, 1, 2, \ldots$.) The problem parameters give

$$\begin{aligned}\omega_n &= \sqrt{\frac{k}{m}} = \sqrt{\frac{9.87E+4\ \text{N/m}}{100\ \text{kg}}} = 31.416\ \text{rad/s}\\ &\Rightarrow f_n = \frac{\omega_n}{2\pi} = 5\ \frac{\text{cy}}{\text{s}} = 5\ \text{Hz}\\ 2\zeta\omega_n &= \frac{c}{m} \Rightarrow \zeta = \frac{3141.6\ \text{Ns/m}}{2\times 31.416\ \text{s}^{-1}\times 100\ \text{kg}} = 0.5\\ \omega_d &= \omega_n\sqrt{1-\zeta^2} = 31.416\sqrt{1-0.25} = 26.97\ \text{rad/s}\\ f_d &= \frac{\omega_d}{2\pi} = 4.29\ \text{Hz} \Rightarrow \tau_d = \frac{1}{f_d} = 0.233\ \text{s}.\end{aligned}$$

Plots for $x(t)$ and $\dot{x}(t)$ are provided in Figure XP3.2c for $0 \le t \le T_f$.

Moving to Task c, the cart's forward motion stops, and the peak deflection occurs at the time T^* defined by $\dot{x}(T^*) = 0$. From the second of Equation (iii), this "stopping time" T^* is defined by

$$\begin{aligned}\dot{x}(T^*) = 0 = &-\frac{\zeta v_0}{\sqrt{1-\zeta^2}}e^{-\zeta\omega_n T^*}\sin\omega_d T^*\\ &+ v_0 e^{-\zeta\omega_n T^*}\cos\omega_d T^*\\ \therefore \tan\omega_d T^* = &\frac{\sqrt{1-\zeta^2}}{\zeta}, \quad T^* = \frac{1}{\omega_d}\tan^{-1}\left(\frac{\sqrt{1-\zeta^2}}{\zeta}\right). \quad \text{(iv)}\end{aligned}$$

The maximum value for T^* occurs for $\zeta = 0$ and is defined by

$$T^*_{max} = \frac{1}{\omega_n}\tan^{-1}\left(\frac{1}{0}\right) = \frac{1}{\omega}\frac{\pi}{2},$$

that is, one fourth of the natural period $\tau_n = 2\pi/\omega_n$. Figure XP3.2d illustrates the stopping-time ratio

$$\frac{T^*}{T^*_{max}} = \frac{2}{\pi\sqrt{1-\zeta^2}}\tan^{-1}\left(\frac{\sqrt{1-\zeta^2}}{\zeta}\right)$$

versus ζ. Increasing ζ from 0 to 1 reduces the stopping time by 32%.

To determine the influence of damping on the peak amplitude, from Equation (iv), $\sin(\omega_d T^*) = \sqrt{1-\zeta^2}/\sqrt{(1-\zeta^2)+\zeta^2} = \sqrt{1-\zeta^2}$. The peak amplitude is defined by substituting this result into the first of Equation (iii), netting

$$x(T^*) = x_{max} = \frac{v_0}{\omega_d}e^{-\zeta\omega_n T^*}\sqrt{1-\zeta^2} = \frac{v_0}{\omega_n}e^{-\zeta\omega_n T^*} \quad \text{(v)}$$

For $\zeta = 0$, the peak deflection is $x(T^*)|_{\zeta=0} = v_0/\omega_n$. Figure XP3.2e illustrates the peak deflection ratio

$$\frac{x_{max}}{x_{max}\ (\zeta = 0)} = e^{-\zeta\omega_n T^*}$$

versus ζ. Increasing the damping ratio to $\zeta \cong 1$ decreases the peak amplitude by 56%. We could increase the damping constant c such that $\zeta \ge 1$; however, the solution provided by Equation (i) is no longer valid. For $\zeta > 1$, and the over-damped solution of Equation B.21 applies. For $\zeta = 1$, the critically damped solution of Equation B.22 applies.

Example Problem 3.3

Figure XP3.3a illustrates a spring-mass-damper system characterized by the following parameters: ($m = 100$ kg, $k = 9.87E+04$ N/m, $c = 3141.6$ N s/m), with the forcing function illustrated. The mass starts from rest with the spring undeflected. The force increases linearly until $t = t_1 = \pi/\omega_d = \tau_d/4$ when it reaches a magnitude equal to $w = mg$. For $t \ge t_1$, $f(t) = 0$.

Tasks: Determine the mass's position and velocity at $t = t_2 = t_1 + 2\pi/\omega_d$. Plot $Y(t)$ for $0 \le t \le t_1$ and $t_1 \le t \le t_2$.

Solution

This example problem has the same k and m values as Example Problem 3.2; hence,

$$\omega_n = 31.416\ \frac{\text{rad}}{\text{s}} \Rightarrow f_n = 5\ \text{Hz}, \quad \zeta = 0.5,$$
$$\omega_d = 27.21\ \frac{\text{rad}}{\text{s}}$$

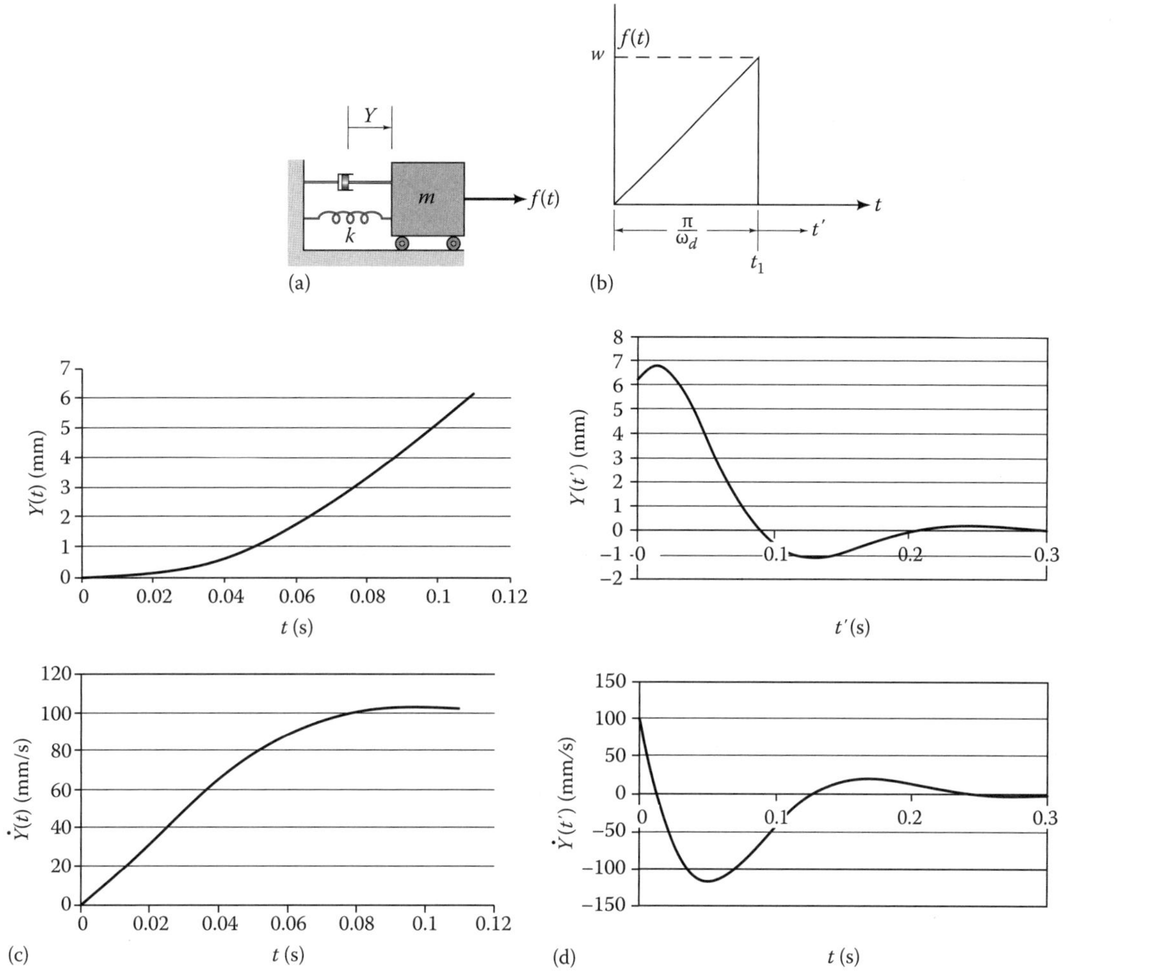

FIGURE XP3.3
(a) Spring-mass system with an applied force, (b) applied force definition, (c) $Y(t)$ and $\dot{Y}(t)$ for $0 \le t \le t_1$, (d) $Y(t')$ and $\dot{Y}(t')$ for $t_1 \le$ or $t' > 0$.

The equation of motion for $0 \le t < \pi/\omega_d = 0.016$ s is

$$m\ddot{Y} + c\dot{Y} + kY = ht$$

$$h = \frac{w}{\pi/\omega_d}$$

$$= \left(100\,\text{kg} \times 9.81\frac{\text{m}}{\text{s}^2} \times 27.21s^{-1}\right)\Big/\pi = 8497 \cdot \frac{N}{s} \quad \text{(i)}$$

For $t > \pi/\omega_d$, the equation of motion is $m\ddot{Y} + c\dot{Y} + kY = 0$. We will use the solution for $m\ddot{Y} + c\dot{Y} + kY = ht$ at $t = t_1 = \pi/\omega_d$ to determine $Y(t_1)$ and $\dot{Y}(t_1)$ that we will then use as initial conditions in stating the solution to $m\ddot{Y} + c\dot{Y} + kY = 0$ for $t > t_1$.

From Equation 3.25, the homogeneous solution is

$$Y_h = e^{-\zeta\omega_n t}(A\cos\omega_d t + B\sin\omega_d t). \quad \text{(ii)}$$

From Table B.2, the particular solution for $\ddot{Y} + 2\zeta\omega_n\dot{Y} + \omega_n^2 Y = at$, is

$$Y_p = \frac{a}{\omega_n^2}\left(t - \frac{2\zeta}{\omega_n}\right)$$

Hence, the particular solution for $\ddot{Y} + 2\zeta\omega_n\dot{Y} + \omega_n^2 = (h/m)t$ is

$$Y_p = \frac{h}{m\omega_n^2}\left(t - \frac{2\zeta}{\omega_n}\right) = \frac{h}{k}\left(t - \frac{2\zeta}{\omega_n}\right),$$

and the complete solution is

$$Y = Y_h + Y_p$$

$$= e^{-\zeta\omega_n t}(A\cos\omega_d t + B\sin\omega_d t) + \frac{h}{k}\left(t - \frac{2\zeta}{\omega_n}\right). \quad \text{(iii)}$$

The velocity is

$$\dot{Y} = -\zeta\omega_n e^{-\zeta\omega_n t}(A\cos\omega_d t + B\sin\omega_d t) + \omega_d e^{-\zeta\omega_n t}(-A\sin\omega_d t + B\cos\omega_d t) + \frac{h}{k}. \quad \text{(iv)}$$

Solving for A and B from the initial conditions gives

$$Y(0) = 0 = A - \frac{2\zeta h}{k\omega_n} \Rightarrow A = \frac{2\zeta h}{k\omega_n}$$
$$\dot{Y}(0) = 0 = -\zeta\omega_n A + \zeta\omega_d B + \frac{h}{k} \Rightarrow B = \frac{h(2\zeta^2 - 1)}{k\omega_d}.$$

The complete solution satisfying the initial conditions is

$$\dot{Y}(t) = \frac{h}{k}(1 - e^{-\zeta\omega_n t}\cos\omega_d t)$$
$$Y(t) = \frac{h}{k}\left\{e^{-\zeta\omega_n t}\left[\frac{2\zeta}{\omega_n}\cos\omega_d t + \frac{(2\zeta^2 - 1)}{\omega_d}\sin\omega_d t\right] + \left(t - \frac{2\zeta}{\omega_n}\right)\right\}; \quad 0 \le t < t_1 \quad \text{(v)}$$

and is illustrated in Figure XP3.3c, showing a growth in the displacement and velocity due to the linearly increasing applied force.

The final conditions for Figure XP3.3c are $\dot{Y}(t_1) = 9.92E\text{–}02$ m/s, and $Y(t_1) = 6.766E\text{–}3$ m at $t_1 = \pi/\omega_d$ $\pi/(27.207 \text{ rad/s}) = 0.1155$ s. For $t > \pi/\omega_d = t_1$, the equation of motion becomes $m\ddot{Y} + c\dot{Y} + kY = 0$ with the solution

$$Y = Y_h = e^{-\zeta\omega_n t}(A\cos\omega_d t + B\sin\omega_d t)$$

Differentiating with respect to time gives

$$\dot{Y} = -\zeta\omega_n e^{-\zeta\omega_n t}(A\cos\omega_d t + B\sin\omega_d t) + \omega_d e^{-\zeta\omega_n t}(-A\sin\omega_d t + B\cos\omega_d t).$$

We could solve for the unknown constants A and B via $Y(t_1) = 6.766E\text{–}3$ m and $\dot{Y}(t_1) = 9.92E\text{–}02$ m/s and produce a solution for $t > t_1$. We are going to take an easier task by restarting time via the definition $t' = t - t_1$. Since,

$$\frac{DY}{dt'} = \frac{DY}{dt}\frac{dt}{dt'} = \frac{DY}{dt},$$

changing the independent variable from t to t' does not change the equation of motion. The solution in terms of t' is

$$Y = e^{-\zeta\omega_n t'}(A\cos\omega_d t' + B\sin\omega_d t'),$$

with the derivative

$$\dot{Y} = \zeta\omega_n e^{-\zeta\omega_n t'}(A\cos\omega_d t' + B\sin\omega_d t') + \omega_d e^{-\zeta\omega_n t'}(-A\sin\omega_d t' + B\cos\omega_d t'),$$

where $t' = t - t_1$, $Y(t' = 0) = Y_0 = Y(t = t_1) = 6.766E\text{–}3$ m, and $\dot{Y}(t' = 0) = \dot{Y}_0 = \dot{Y}(t = t_1) = 9.92E - 02$ m/s. Again, we are basically restarting time to simplify this solution. Solving for the constants in terms of the initial conditions gives

$$Y_0 = A, \quad \dot{Y}_0 - \zeta\omega_n A + \omega_d B \Rightarrow B = \frac{(\dot{Y}_0 + \zeta\omega_n Y_0)}{\omega_d}$$

and the complete solution in terms in terms of the initial conditions is

$$\dot{Y}(t') = e^{-\zeta\omega_n t'}\left[\dot{Y}_0\cos\omega_d t' - \frac{(\zeta\dot{Y}_0 + \omega_n Y_0)}{\sqrt{1-\zeta^2}}\sin\omega_d t'\right]$$
$$Y(t') = e^{-\zeta\omega_n t'}\left[Y_0\cos\omega_d t' + \frac{(\dot{Y}_0 + \zeta\omega_n Y_0)}{\omega_d}\sin\omega_d t'\right] \quad t' \ge 0$$

This solution is plotted in Figure XP3.3d, showing a generally steady decay after the force drops to zero.

3.2.3.5 Characterizing Damping

Engineers frequently use the "logarithmic decrement" or simply "log dec" δ instead of the damping factor ζ as a relative measure of the damping that is present in a vibrating system. The log dec can be determined directly from an experimentally-measured transient response. From Equation 3.26, the motion about the equilibrium position can be stated

$$y = e^{-\zeta\omega_n t}(A\cos\omega_d t + B\sin\omega_d t) = e^{-\zeta\omega_n t}D\cos(\omega_d t - \phi),$$

where $D = (A^2 + B^2)^{-1/2}$, and $\phi = \tan^{-1}(-B/A)$. Peaks in the response curves occur when $\cos(\omega_d t - \phi) = 1$, at time intervals equal to the damped period $\tau_d = 2\pi/\omega_d$. Hence, the ratio of two successive peaks would be

$$\frac{Y_1}{Y_2} = \frac{e^{-\zeta\omega_n t_1}}{e^{-\zeta\omega_n(t_1+\tau_d)}} = e^{\zeta\omega_n\tau_d},$$

and the log dec δ is defined as

$$\delta = \ln\left(\frac{Y_1}{Y_2}\right) = \zeta\omega_n\tau_d = \zeta\omega_n\frac{2\pi}{\omega_n\sqrt{1-\zeta^2}} = \frac{2\pi\zeta}{\sqrt{1-\zeta^2}}. \quad (3.30)$$

The log dec can also be defined in terms of the ratio of the first peak to the nth successive peak as

$$\delta = \frac{1}{(n-1)}\ln\left(\frac{Y_1}{Y_2}\right) = \frac{1}{(n-1)}\ln\left[\frac{e^{-\zeta\omega_n t_1}}{e^{-\zeta\omega_n(t_1+(n-1\tau_d)}}\right] = \zeta\omega_n\tau_d = \frac{2\pi\zeta\omega_n}{\omega_n\sqrt{1-\zeta^2}} = \frac{2\pi\zeta}{\sqrt{1-\zeta^2}}. \quad (3.31)$$

TABLE 3.1

Relationship between Damping Characterizations

Output Column	ζ	δ	% of Critical Damping	q Factor
ζ	1	$\delta/[(2\pi)^2+\delta^2]^{1/2}$	% of critical damping/100	$1/(2q)$
δ	$2\pi\zeta/(1-\zeta^2)^{1/2}$	1	Find ζ first	Find ζ first
% of critical damping	$100\times\zeta$	Find ζ first	1	$50/q$
q factor	$1/(2\zeta)$	Find ζ first	50/% of critical damping	1

Note in Equation 3.30 that δ becomes unbounded as $\zeta \to 1$. In many dynamic systems, the damping factor is small, $\zeta \ll 1$, and $\delta \cong 2\pi\zeta$. In many engineering applications, the log dec is measured, calculated, and quoted exclusively instead of the damping factor ζ. In stability calculations for systems with unstable eigenvalues, negative log dec's are regularly stated. Solving for the damping factor ζ in terms of δ from Equation 3.30 gives

$$\zeta = \frac{\delta}{[(2\pi)^2+\delta^2]^{1/2}}.$$

The effect of damping in the model $m\ddot{x}+c\dot{x}+kx=0$ can be characterized in terms of the damping factor, ζ, and the log-dec, δ. Recalling that $\zeta=1$ corresponds to critical damping, a "percent of critical damping" is also used to specify the amount of available damping. For example, $\zeta=0.1$ implies 10% of critical damping. Table 3.1 demonstrates how to proceed from one characterization to another. The q factor characterization is introduced in Section 3.2.5 related to steady-state response due to harmonic excitation, and is defined in Equation 3.44.

Example Problem 3.4

Figure XP3.4 illustrates a transient response result for a mass that has been disturbed from its equilibrium position. The first peak occurs at $t=1.6$ s with an amplitude of 0.018 m. The fourth peak occurs at $t=10.6$ s with an amplitude of 0.003 m.

Tasks: Determine the log dec and the damping factor. Also, what is the damped natural frequency?

SOLUTION

Applying Equation 3.31 gives

$$\delta = \frac{1}{n-1}\ln\left(\frac{Y_1}{Y_n}\right) = \frac{1}{3}\ln\left(\frac{Y_1}{Y_4}\right) = \frac{1}{3}\ln\left(\frac{.018}{.003}\right) = 0.597$$

for the log dec. From Table 3.1, the damping factor is

$$\zeta = \frac{\delta}{[(2\pi^2+\delta^2)]^{1/2}} = \frac{0.597}{\sqrt{4\pi^2+.597^2}} = 0.095.$$

3.2.3.6 *Solution for $\dot{Y}$ as a Function of Y Including Damping?*

Without damping, we were able to use the energy-integral substitution $\ddot{Y}=d(\dot{Y}^2/2)/dY$ to obtain a solution to the equation of motion in the form $\dot{Y}(Y)$. We can try the same path here to get a solution, obtaining

$$m\frac{d}{dY}\left(\frac{\dot{Y}^2}{2}\right) = -kY + w - c\dot{Y}.$$

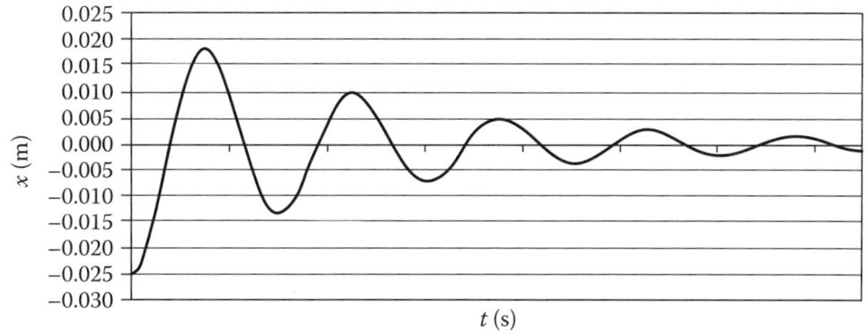

FIGURE XP3.4
Transient response.

We can still multiply through by dY, but we cannot complete the integral,

$$\int_{Y_0}^{Y} c\dot{Y}dY = \int_{t_0}^{t} c\dot{Y}\frac{dY}{dt}dt = \int_{Y_0}^{Y} c\dot{Y}^2 dt$$

since $\dot{Y}$ is generally a function of t, not Y. Hence, with viscous damping, developing the solution $\dot{Y}(Y)$ is not feasible. The energy dissipated by a viscous damper can be calculated from this integral *if you already have a solution* $Y(t)$, but it is of no value in developing a solution.

Integrating the equation of motion with respect to Y yields,

$$\left[\frac{m\dot{Y}^2}{2} - wY + \frac{kY^2}{2}\right] = \left[\frac{m\dot{Y}_0^2}{2} - wY_0 + \frac{kY_0^2}{2}\right] - \int_{t_0}^{t} c\dot{Y}^2\, dt.$$

This result translates physically into the statement that the current mechanical (kinetic plus potential) energy equals the initial mechanical energy minus the energy dissipated by damping.

3.2.3.7 Negative Damping and Dynamic Instability

Earlier, we considered the possibility of erroneously obtaining a negative-stiffness coefficient for a spring, concluding that it would yield a statically unstable differential-equation model. Suppose that the wrong sign had been selected for the damping coefficient model of Equation 3.22 yielding

$$\ddot{Y} - 2\zeta\omega_n\dot{Y} + \omega_n^2 Y = g.$$

In place of Equation 3.26, the general solution is now

$$Y = e^{\zeta\omega_n t}(A\cos\omega_d t + B\sin\omega_d t) + \frac{w}{k}.$$

Instead of decaying exponentially with time, this solution grows exponentially with time. This outcome is not consistent with a linear damper; however, many physical systems can become *dynamically* unstable leading to serious failures. In developing the equations of motion for a nominally stable dynamic system, arriving at a negative damping coefficient is a certain indication that you have made a mistake.

3.2.4 Base Excitation for a Spring-Mass-Damper System

The examples in the preceding subsection showed different external forces $f(t)$ arising for a spring-mass-damper system for the equation of motion $m\ddot{y} + c\dot{y} + ky = f(t)$. Figure 3.12a shows a mass supported on a base by two identical springs with spring coefficients $k/2$ and a linear damper with damping coefficient c. The forces in the springs and the damper can be either in compression or tension, depending on the relative positions of the base and the mass m. We want the equation of motion for the supported mass m. We will assume two distinct forms for the relative motion and then draw free-body diagrams in accordance with the assumed motion to separately arrive at the same equation of motion.

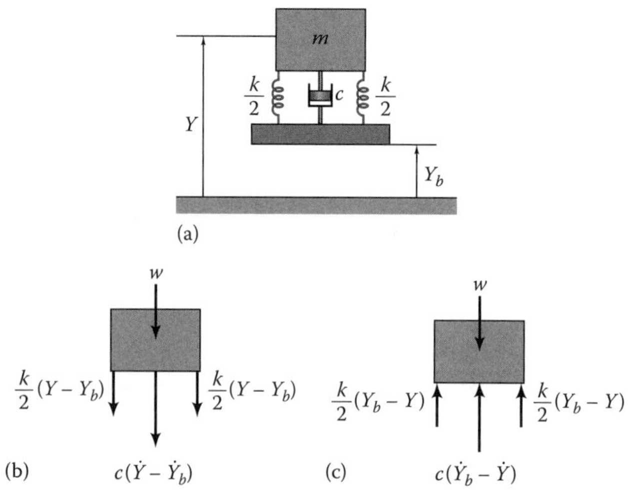

FIGURE 3.12
(a) Suspended mass and movable base, (b) free-body diagram for tension in the springs and damper, (c) free-body diagram for compression in the springs and damper.

3.2.4.1 Deriving the Equation of Motion

For both assumed-motion cases, we will suppose that the base and the mass m have positive displacements ($Y > 0$, $Y_b > 0$) and velocities ($\dot{Y} > 0, \dot{Y}_b > 0$). For zero base motion ($Y_b = 0$), the springs are undeflected when $Y = 0$. Assuming that m's displacement is greater than the base displacement ($Y > Y_b$) means that the springs are in tension, and the spring forces are defined by $f_s = k(Y - Y_b)/2$. Similarly, assuming that m's velocity is greater than the base velocity ($\dot{Y} > \dot{Y}_b$), the damper is also in tension, and the damper force is defined by $f_b = c(\dot{Y} - \dot{Y}_b)$. The free-body diagram of Figure 3.12b is consistent with this assumed motion. Applying $\Sigma f = m\ddot{r}$ to the free-body diagram of Figure 3.12b gives the differential equation of motion as

$$m\ddot{Y} = \Sigma f_Y = -w - c(\dot{Y} - \dot{Y}_b) + k(Y - Y_b)$$
$$\therefore\; m\ddot{Y} + c\dot{Y} + kY = -w + kY_b + c\dot{Y}_b.$$

The negative signs to the right-hand side of the first of these equations apply because the forces are pointing in the $-Y$ direction. Note the positive signs for the stiffness and damping coefficients on the left of the resulting equation. A negative sign for the stiffness coefficient would indicate a *statically* unstable system. A negative sign for the damping coefficient would indicate a *dynamically* unstable system. For the present stable system, negative signs would indicate that a mistake has been made in deriving the equation of motion.

For the second assumed form of relative motion, the base displacement is greater than the mass displacement ($Y_b > Y$). With this assumed relative motion, the springs are in compression, and the spring forces are defined by $f_s = k(Y_b - Y)/2$. Similarly, assuming that the base velocity is greater than m's ($\dot{Y}_b > \dot{Y}$) causes the damper to also be in compression, and the damper force is defined by $f_b = c(\dot{Y}_b - \dot{Y})$. The free-body diagram of Figure 3.12c is consistent with this assumed motion and leads to the same differential equation of motion

$$m\ddot{Y} = \sum f_Y = -w + c(\dot{Y}_b - \dot{Y}) + k(Y_b - Y)$$
$$\therefore\ m\ddot{Y} + c\dot{Y} + kY = -w + kY_b + c\dot{Y}_b.$$

The spring and damper forces in the first equation are positive because they are pointed in the $+Y$ direction. The short lesson from these developments is as follows: The same governing equation should result for any assumed motion, since the governing equation applies for any combination of position and velocity. Note that the following procedural steps were taken in arriving at the equation of motion:

a. Coordinates were selected.
b. The nature of the motion was assumed, for example, ($Y_b > Y$).
c. The spring or damper force was then stated in a manner consistent with the assumed motion, for example, $f_s = k(Y_b - Y)/2$.
d. The free-body diagram was drawn in a manner consistent with the assumed motion and its resultant spring and damper forces, that is, in tension or compression.
e. Newton's second law of motion $\sum \boldsymbol{f} = m\ddot{\boldsymbol{r}}$ was applied to the free-body diagram to obtain the equation of motion.

After applying these steps to obtain the equation of motion, you should always check to make sure that the stiffness and damping coefficients have positive signs. A negative sign means that a mistake has been made. The model of Figure 3.12 is only one of several possible base-excitation possibilities for one-degree-of-freedom vibration models. These same steps will be followed in Section 3.5, dealing with problems having more than one degree of freedom.

The earlier discussion of motion about an equilibrium position applies here. Specifically, if m is initially in equilibrium with Y defining the displacement away from equilibrium, the base-excitation equation of motion reduces to

$$m\ddot{Y} + c\dot{Y} + kY = kY_b + c\dot{Y}_b, \tag{3.32}$$

with w eliminated.

Example Problem 3.5

Figure XP3.5a illustrates a cart that is connected to a movable base via a spring and a damper. The cart's change of position is defined by x. The movable base's location is defined by x_b. The spring is undeflected for $x = x_b = 0$. First, we need to derive the cart's equation of motion, starting with an assumption for the relative positions and velocities. The free-body diagram of Figure XP3.5b is drawn assuming that $x_b > x > 0$ (placing the spring in tension) and $\dot{x}_b > \dot{x} > 0$ (placing the damper in tension). Applying $\sum \boldsymbol{f} = m\ddot{\boldsymbol{r}}$ nets

$$m\ddot{x} = \sum f_x = k(x_b - x) + c(\dot{x}_b - \dot{x})$$
$$\therefore\ m\ddot{x} + c\dot{x} + kx = c\dot{x}_b + kx_b \tag{i}$$

In Equation (i), the spring and damper forces are positive because they are in the $+x$ direction.

The model of Equation (i) can be used to *approximately* predict the motion of a small trailer being towed behind a much larger vehicle with the spring and damper used to model the hitch connection between the trailer and the towing vehicle. To *correctly* model two vehicles connected by a spring and damper, we would need to write $\sum \boldsymbol{f} = m\ddot{\boldsymbol{r}}$ for both bodies and account for the influence of the trailer's mass on the towing vehicle's motion. We will consider motion of connected bodies in Section 3.5.

Assume that the cart and the base are initially motionless with the spring undeflected. The base is given the constant acceleration $\ddot{x}_b = g/5$.

Tasks:

a. State the equation of motion.
b. State the homogeneous and particular solutions, and state the complete solution satisfying the initial conditions.
c. For $w = 300$ lbs, $k = 776.$ lb/in., $c = 24.4$ lb s/in., plot the cart's motion for two periods of damped oscillations.

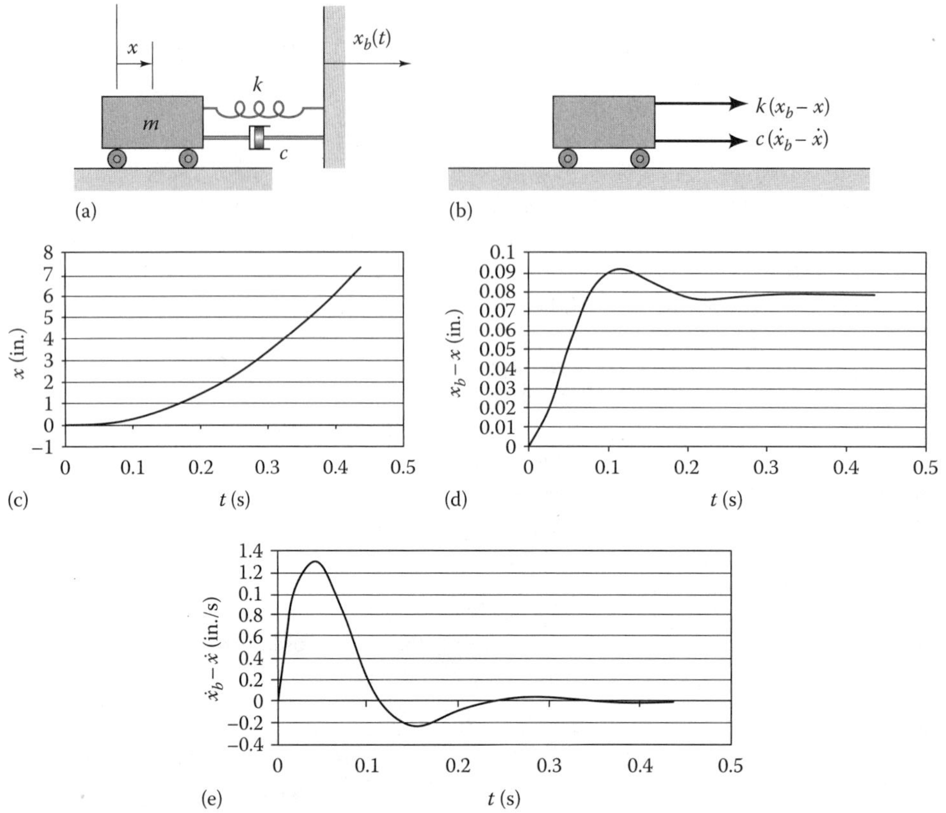

FIGURE XP3.5
Movable cart with base excitation. (a) Coordinates, (b) free-body diagram for $x_b > x > 0$; $\dot{x}_b > \dot{x} > 0$, (c) $x(t)$, (d) $x_b(t) - x(t)$, and (e) $\dot{x}_b(t) - \dot{x}(t)$.

Stating the equation of motion simply involves proceeding from $\ddot{x}_b = g/5$ to find $\dot{x}_b, x_b$ and plugging them into Equation (i), via

$$\dot{x}_b = \dot{x}_b(0) + \int_0^t \ddot{x}_b \, d\tau = 0 + \int_0^t \frac{g}{5} \, d\tau = \frac{gt}{5}$$

$$x_b = x_b(0) + \int_0^t \dot{x}_b \, d\tau = 0 + \frac{gt^2}{10}.$$

Substituting these results into Equation (i) produces

$$m\ddot{x} + c\dot{x} + kx = c\dot{x}_b + kx_b = \frac{g}{5}\left(ct + k\frac{t^2}{2}\right),$$

and concludes Task a. From Equation 3.25, the homogeneous solution is

$$x_h = e^{-\zeta\omega_n t}(A\cos\omega_d t + B\sin\omega_d t).$$

We can pick the particular solutions from Table B.2 as follows

$$\frac{gc}{5m}t \Rightarrow x_{p1} = \frac{gc}{5m}\frac{1}{\omega_n^2}\left(t - \frac{2\zeta}{\omega_n}\right) = \frac{2\zeta g}{5\omega_n}\left(t - \frac{2\zeta}{\omega_n}\right)$$

$$\frac{gk}{10m}t^2 \Rightarrow x_{p2} = \frac{gk}{10m}\frac{1}{\omega_n^2}\left[t^2 - \frac{4\zeta t}{\omega_n} - \frac{2}{\omega_n^2}(1 - 4\zeta^2)\right]$$
$$= \frac{g}{10}\left[t^2 - \frac{4\zeta t}{\omega_n} - \frac{2}{\omega_n^2}(1 - 4\zeta^2)\right]$$

The complete solution is

$$x = x_h + x_{p1} + x_{p2}$$
$$= e^{-\zeta\omega_n t}(A\cos\omega_d t + B\sin\omega_d t) + \frac{2\zeta g}{5\omega_n}\left(t - \frac{2\zeta}{\omega_n}\right)$$
$$+ \frac{g}{10}\left[t^2 - \frac{4\zeta t}{\omega_n} - \frac{2}{\omega_n^2}(1 - 4\zeta^2)\right].$$

The constant A is obtained via,

$$x(0) = 0 = A - \frac{4\zeta^2 g}{5\omega_n^2} - \frac{g}{5\omega_n^2}(1 - 4\zeta^2) \Rightarrow A = \frac{g}{5\omega_n^2}.$$

To obtain B, first

$$\dot{x} = -\zeta\omega_n e^{-\zeta\omega_n t}(A\cos\omega_d t + B\sin\omega_d t) + \zeta\omega_d e^{-\zeta\omega_n t}(-A\sin\omega_d t + B\cos\omega_d t) + \frac{gt}{5}.$$

Hence,

$$\dot{x}(0) = 0 = -\zeta\omega_n A + \omega_d B + \frac{2\zeta g}{5\omega_n} - \frac{2\zeta g}{5\omega_n} \Rightarrow B = \frac{g\zeta}{5\omega_n^2\sqrt{1-\zeta^2}}.$$

The complete solution satisfying the initial conditions is

$$\dot{x} = \frac{gt}{5} - \frac{g}{5\omega_d t}\sin\omega_d t e^{-\zeta\omega_n t}$$

$$x = e^{-\zeta\omega_n t}\frac{g}{5\omega_d t}\left(\cos\omega_d t + \frac{\zeta}{\sqrt{1-\zeta^2}}\sin\omega_d t\right) - \frac{g}{5\omega^2} + \frac{gt^2}{10}. \quad \text{(ii)}$$

Equation (ii) completes Task b.

Moving toward Task c, $m = w/_g = 300$ lbs/(386.4 in./s^2) = 0.766 lb/(s^2 in.).

$$\omega_n = \sqrt{\frac{k}{m}} = \sqrt{\frac{776.\ \text{lb/in.}}{.776\ \text{lb s}^2/\text{in.}}} = 31.62\ \text{rad/s}$$

$$\therefore f_n = 31.62\ \frac{\text{rad}}{\text{s}} \times \frac{1\ \text{cycle}}{2\pi\ \text{rad}} = 5.03\ \frac{\text{cycles}}{\text{s}} = 5.03\ \text{Hz}$$

$$\zeta = \frac{c}{2m\omega_n} = 24.4\ \frac{\text{lbs}}{\text{in.}} \times \frac{1}{2\times .776\ \text{snails} \times 31.62\ \text{s}^{-1}} = 0.497$$

$$\omega_d = \omega_n\sqrt{1-\zeta^2} = 31.62\sqrt{1-0.497^2} = 27.4\ \text{rad/s}$$

$$\therefore f_d = 27.4\ \frac{\text{rad}}{\text{s}} \times \frac{1\ \text{cycle}}{2\pi\ \text{rad}} = 4.36\ \frac{\text{cycles}}{\text{s}} = 4.36\ \text{Hz}.$$

Two cycles of damped oscillations will be completed in $2\times\tau_d$ s, where $\tau_d \doteq 1/f_d = 1/4.36$ (cycles/s) = 0.229 s/cycle. The solutions are shown in Figure XP3.5c through e. After an initial transient, the relative displacement $(x_b - x)$ approaches a constant value $(x_b - x) \cong 0.078$ in., while the relative velocity approaches zero; that is, $(\dot{x}_b - \dot{x}) \cong 0$. The scaling for $x(t)$ hides the initial transient, but the plot shows a quadratic increase with time, which is consistent with its constant acceleration of $g/5$. The spring force $k(x - x_b) \cong$ (776. lbs/in.) $\times$.078 in. $\cong$ 60.1 lbs, while the force required to accelerate the mass at $g/5$ is

$$f = m\ddot{x} = 0.776\left(\frac{\text{lbs}}{\text{in.}^2}\right) \times \frac{386.4}{5} = 60\ \text{lbs}.$$

Hence, the asymptotic value for $(x_b - x)$ creates the spring force required to accelerate the mass at $g/5$.

Example Problem 3.6

Think about an initially motionless cart being "snagged" by a moving vehicle with velocity v_0 through a connection consisting of a parallel spring-damper assembly. The connecting spring is undeflected prior to contact. That circumstance can be modeled by giving the base end of the model in Figure XP3.5a the velocity v_0; that is, $\dot{x}_b = v_0 \Rightarrow x_b = x_b(0) + v_0(t) = v_0 t$. Hence, the governing equation of motion is

$$m\ddot{x} + c\dot{x} + kx + cv_0 + kv_0 t, \quad \text{(iii)}$$

with initial conditions $x(0) = \dot{x}(0) = 0$.

Tasks:

a. Determine the complete solution that satisfies the initial conditions.
b. For the data set, $m = 10$ kg, $k = 1590$ N/m, and $c = 1257$ N/m, determine ω_n, ζ, ω_d.
c. For $v_0 = 20$ km/h produce plots for the mass displacement $x(t)$, relative displacement $x_b(t) - x(t)$, and the mass velocity $\dot{x}(t)$ for two cycles of motion.

Solution

From Table B.2, the particular solutions corresponding to the right-hand terms are

$$\frac{cv_0}{m} \Rightarrow x_{p1} = \frac{cv_0}{m}\frac{1}{\omega_n^2} = \frac{2\zeta v_0}{\omega_n}$$

$$\frac{kv_0}{m}t \Rightarrow x_{p2} = \frac{kv_0}{m}\frac{1}{\omega_n^2} = \left(t - \frac{2\zeta}{\omega_n}\right) = v_0\left(t - \frac{2\zeta}{\omega_n}\right).$$

The complete solution is

$$x = x_h + x_{p1} + x_{p2} = e^{-\zeta\omega_n t}(A\cos\omega_d t + B\sin\omega_d t) + \frac{2\zeta v_0}{\omega_n} + v_0\left(t - \frac{2\zeta}{\omega_n}\right)$$
$$= e^{-\zeta\omega_n t}(A\cos\omega_d t + B\sin\omega_d t) + v_0 t$$

Solving for A,

$$x(0) = 0 = A \Rightarrow A = 0.$$

Solving for B from

$$\dot{x} = -\zeta\omega_n e^{-\zeta\omega_n t}(A\cos\omega_d t + B\sin\omega_d t) + \omega_d e^{-\zeta\omega_n t}(-A\sin\omega_d t + B\cos\omega_d t) + v_0,$$

gives

$$\dot{x}(0) = 0 = -\zeta\omega_n A + \omega_d B + v_0 \Rightarrow B = -\frac{v_0}{\omega_d}.$$

The complete solution satisfying the initial conditions is

$$\dot{x} = v_0 + v_0 e^{-\zeta\omega_n t}\left(\frac{\zeta}{\sqrt{1-\zeta^2}}\sin\omega_d t - \cos\omega_d t\right) \quad \text{(iv)}$$

$$x = v_0 t - \frac{v_0}{\omega_d}e^{-\zeta\omega_n t}\sin\omega_d t,$$

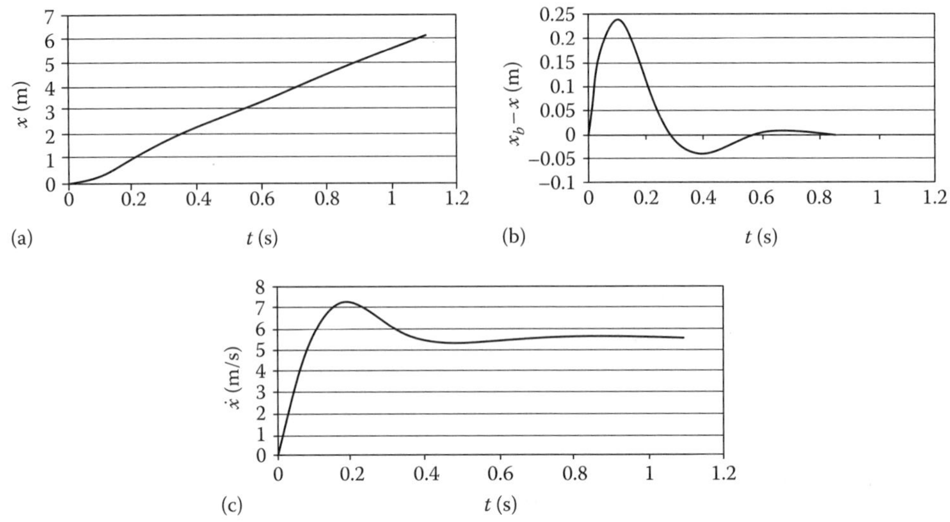

FIGURE XP3.6
(a) $x(t)$, (b) $x_b(t)$, (c) $\dot{x}(t)$.

which completes Task a. For Task b,

$$\omega_n = \sqrt{\frac{k}{m}} = \sqrt{\frac{1580}{10}} = 12.57 \text{ rad/s}$$

$$\therefore\ f_n = 12.57\ \frac{\text{rad}}{\text{s}} \times \frac{1\text{ cycle}}{2\pi\text{ rad}} = 2\ \frac{\text{cycles}}{\text{s}} = 2\text{ Hz}$$

$$\zeta = \frac{c}{2m\omega_n} = 1257\ \frac{\text{Ns}}{\text{m}} \times \frac{1}{2 \times 10\text{ kg} \times 12.57\text{ s}^{-1}} = 0.5$$

$$\omega_d = \omega_n\sqrt{1-\zeta^2} = 12.57\sqrt{1-0.5^2} = 10.89 \text{ rad/s}$$

$$\therefore\ f_d = 10.89\ \frac{\text{rad}}{\text{s}} \times \frac{1\text{ cycle}}{2\pi\text{ rad}} = 1.733\ \frac{\text{cycle}}{\text{s}} = 1.733\text{ Hz.}$$

From the last of these results, the period for a damped oscillation is $\tau_d = 1/f_d = 0.577$ s. For Task c, $v_0 = 20$ km/h $\times$ 1 h/3600 s $\times$ 100 m/km $= 5.55$ m/s. Figure XP3.6 illustrates the solution for two cycles of motion.

Figure XP3.6a and c show the mass rapidly moving toward the towing velocity $v_0 = 5.55$ m/s. Figure XP3.6b shows the relative position approaching zero. Note that at about 0.3 s, the towed cart appears to pass the towing vehicle when $(x_b - x)$ becomes negative. While a collision of the two bodies is possible, for this case, if the initial spring separation between the cart and the towing vehicle is greater than roughly 0.4 m, no collision would occur. However, for some range of parameters, the towed vehicle could "crash" into the back of the towing vehicle.

3.2.4.2 Relative Motion due to Base Excitation

On occasions, we are interested in the "relative displacement" between the moving base defined by Y_b and the mass motion defined by Y. For example, fatigue failure of the support springs would arise because of the relative displacement $\delta y = Y - Y_B$. Consider the definition

$$\delta y = Y - Y_b \Rightarrow \delta\dot{y} = \dot{Y} - \dot{Y}_b \Rightarrow \delta\ddot{y} = \ddot{Y} - \ddot{Y}_b.$$

Restating the results from Equation 3.32,

$$m\ddot{Y} + c\dot{Y} + kY = K\dot{Y}_b + c\dot{Y}_b, \tag{3.32b}$$

and substituting for δy and its derivatives gives

$$m(\ddot{Y} - \ddot{Y}_b) + c(\dot{Y} - \dot{Y}_b) + k(Y - Y_b)$$
$$= -m\ddot{Y}_b \Rightarrow m\delta\ddot{y} + c\delta\dot{y} + k\delta\ddot{y} = -m\ddot{Y}_b \tag{3.33}$$

This result shows that the dynamic relative deflection δy arises due to base acceleration.

Example Problem 3.7

The tasks here are to restate the equations of motion and initial conditions of Example Problems XP3.3 and XP3.6 in terms of relative deflections. The starting equation of motion for both these examples is $m\ddot{x} + c\dot{x} + kx = c\dot{x}_b + kx_b$, where x locates m with respect to ground, and x_b defines the base motion. The spring is undeflected when $x = x_b = 0$. Restated in terms of the relative deflection $\delta x = x - x_b$, the governing equation is $m\delta\ddot{x} + c\delta\dot{x} + k\delta x = -m\ddot{x}_b$. For Example Problem 3.3, $\ddot{x}_b = g/5$, so the governing equation is

$$m\delta\ddot{x} + c\delta\dot{x} + k\delta x = -m\ddot{x}_b = -\frac{mg}{5}. \tag{i}$$

Both the mass and base start from rest, so the initial conditions are $\delta x(0) = x(0) - x_b(0) = 0 - 0 = 0$, and $\delta\dot{x}(0) = \dot{x}(0) - \dot{x}_b(0) = 0 - 0 = 0$. Solving Equation (i) for $\delta x(t)$ follows exactly the same steps used in moving from Equations 3.22 through 3.29.

3.2.5 Harmonic Excitation for a 1DOF, Spring-Mass-Damper System: Solution for Motion in the Frequency Domain

In the preceding subsection, we considered the homogeneous and particular solutions for

$$m\ddot{Y} + c\dot{Y} + kY = w, \tag{3.21}$$

finding that the mass oscillates about its equilibrium position at its damped natural frequency, and that the oscillations decay at an exponential rate due to viscous damping. Many problems of interest to engineering involve harmonic excitation, yielding the equation of motion

$$m\ddot{Y} + c\dot{Y} + kY = f_o \sin \omega t. \tag{3.34}$$

You have probably experienced harmonic excitation in an automobile; for example, an unbalanced front tire causes a harmonic excitation at the rotational speed of the tire. The pistons and drive shaft of an automobile engine have been "balanced" to minimize harmonic excitation at the engine running speed. Many conveyor systems use oscillatory motion to assist in transferring material.

Dividing through by m in Equation 3.34 yields

$$\ddot{Y} + 2\zeta\omega_n\dot{Y} + \omega_n^2 Y = \left(\frac{f_o}{m}\right)\sin \omega t. \tag{3.35}$$

Equation 3.25 provides the homogeneous solution to this differential equation. We need a particular solution for the right-hand term to complete the solution. Looking at Equation 3.35, one would reasonably expect a particular solution of the form

$$\begin{aligned} Y_p &= C \sin \omega t + D \cos \omega t \\ \dot{Y}_p &= \omega C \cos \omega t - \omega D \sin \omega t \\ \ddot{Y}_p &= -\omega^2 C \sin \omega t - \omega^2 D \cos \omega t. \end{aligned} \tag{3.36}$$

The "reasonableness" argument for this solution is that the system is being excited harmonically at the frequency ω and would be expected to respond harmonically at the same frequency. Substituting the assumed solution into Equation 3.35 yields

$$\begin{aligned} &(-\omega^2 C \sin \omega t - \omega^2 D \cos \omega t) \\ &+ 2\zeta\omega_n(C\omega \cos \omega t - \omega D \sin \omega t) \\ &+ \omega_n^2(C \sin \omega t + D \cos \omega t) = \left(\frac{f_o}{m}\right)\sin \omega t. \end{aligned}$$

The only general solution to the equation $A\cos\omega t + B\sin\omega t = 0$ is $A = B = 0$; hence, the coefficients of $\sin\omega t$ and $\cos\omega t$ must independently be zero to satisfy this equation. Gathering these coefficients gives

$$\begin{aligned} \sin \omega t &: -\omega^2 C - 2\zeta\omega_n\omega D + \omega_n^2 C = \frac{f_o}{m} \\ \cos \omega t &: -\omega^2 D + 2\zeta\omega_n\omega C + \omega_n^2 D = 0. \end{aligned}$$

The matrix equation for the unknowns C and D is

$$\begin{bmatrix} (\omega_n^2 - \omega^2) & -2\zeta\omega_n\omega \\ 2\zeta\omega_n\omega & (\omega_n^2 - \omega^2) \end{bmatrix} \begin{Bmatrix} C \\ D \end{Bmatrix} = \begin{Bmatrix} f_o/m \\ 0 \end{Bmatrix}.$$

Using Cramer's rule (see Appendix A) for their solution gives

$$\begin{aligned} C &= \frac{1}{\Delta}\begin{vmatrix} f_o/m & -2\zeta\omega_n\omega \\ 0 & (\omega_n^2 - \omega^2) \end{vmatrix} = \frac{f_o(\omega_n^2 - \omega^2)}{m\Delta} \\ D &= \frac{1}{\Delta}\begin{vmatrix} (\omega_n^2 - \omega^2) & f_o/m \\ 2\zeta\omega_n\omega & 0 \end{vmatrix} = \frac{-f_o 2\,\zeta\omega_n\omega}{m\Delta}, \end{aligned} \tag{3.37}$$

where Δ is the determinant of the coefficient matrix,

$$\Delta = \left(\omega_n^2 - \omega^2\right)^2 + 4\zeta^2\omega_n^2\omega^2.$$

The solution defined by Equation 3.36 can be restated as

$$\begin{aligned} Y_p(t) &= C \sin \omega t + D \cos \omega t = Y_{op} \sin(\omega t + \psi) \\ &= Y_{op}(\sin \omega t \cos \psi + \cos \omega t \sin \psi). \end{aligned} \tag{3.38}$$

where

Y_{op} is the amplitude of the solution

ψ is the phase between the solution $Y_p(t)$ and the input excitation force $f(t) = f_o \sin \omega t$

Figure 3.13 illustrates the phase relation between $f(t)$ and $Y_p(t)$, showing $Y_p(t)$ "leading" $f(t)$ by the phase angle ψ.

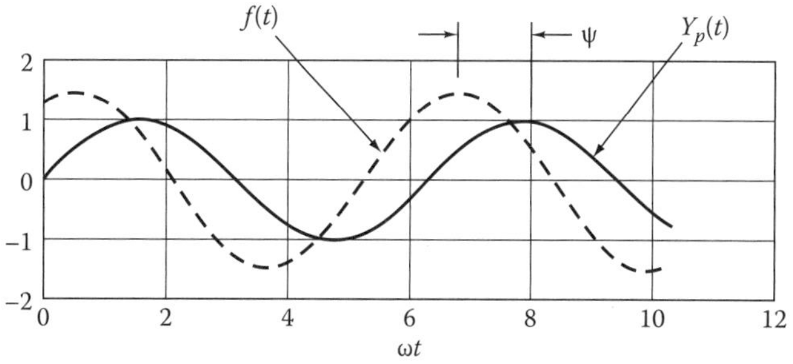

FIGURE 3.13
Phase relation between the response $Y_p(t) = Y_{op}(\omega)\sin(\omega t + \psi)$ and the input harmonic-excitation force $f(t) = f_o \sin \omega t$.

From the solutions for C and D of Equation 3.37, the steady-state solution amplitude is

$$
\begin{aligned}
Y_{op} = \sqrt{C^2 + D^2} &= \frac{f_o}{m} \cdot \frac{1}{\Delta} \left[\left(\omega_n^2 - \omega^2 \right)^2 + 4\zeta^2 \omega_n^2 \omega^2 \right]^{1/2} \\
&= \frac{f_o}{m} \cdot \frac{1}{\left[\left(\omega_n^2 - \omega^2 \right)^2 + 4\zeta^2 \omega_n^2 \omega^2 \right]^{1/2}} \\
&= \frac{f_o}{m(k/m)} \cdot \frac{1}{\left\{ \left[1 - (\omega/\omega_n)^2 \right]^2 + 4\zeta^2 (\omega/\omega_n)^2 \right\}^{1/2}}
\end{aligned} \tag{3.39}
$$

In the last of these expressions, f_o/k is the static deflection due to a constant force f_o. The ratio of the dynamic amplitude Y_{op} to the static amplitude f_o/k is called the "amplification factor" and is defined from Equation 3.39 as

$$
\frac{Y_{op}}{f_o/k} = \frac{1}{\left[(1 - r^2)^2 + 4\zeta^2 r^2 \right]^{1/2}} = H(r); \quad r = \frac{\omega}{\omega_n}, \tag{3.40}
$$

where r is the "frequency ratio."

The phase ψ is defined from Equations 3.37 and 3.38 as

$$
\psi(r) = \tan^{-1}\left(\frac{D}{C} \right) = \tan^{-1}\left[\frac{-2\zeta\omega_n\omega}{\left(\omega_n^2 - \omega^2 \right)} \right] = -\tan^{-1}\left(\frac{2\zeta r}{1 - r^2} \right), \tag{3.41}
$$

and is also a function of the frequency ratio. Note that $\psi(r)$ is negative, with an amplitude lying between zero and π. Hence, the response $Y_p(t)$ "lags" (is behind) the excitation force $f_0 \sin \omega t$.

The differential equation, $m\ddot{Y} + c\dot{Y} + kY = f_o \sin \omega t$, had the assumed solution $Y_p = C \sin \omega t + D \cos \omega t$, which we restated as $Y_p = Y_{op} \sin(\omega t + \psi)$. Think about changing the differential equation to $m\ddot{Y} + c\dot{Y} + kY = f_o \cos \omega t$. Fortunately, the complete solution procedure for the amplification factor and phase does not need to be repeated. Note that $\cos(\omega t) = \sin(\omega t + \pi/2)$; hence, the steady-state solution for the forcing function $f_o \cos(\omega t) = f_o \sin(\omega t + \pi/2)$ is $Y_p = Y_{op} \sin(\omega t + \pi/2 + \psi) = Y_{op} \cos(\omega t + \psi)$, with the amplitude and phase definition provided by Equations 3.40 and 3.41, respectively. Similarly, the general harmonic forcing function, $f(t) = f_{os} \sin \omega t + f_{oc} \cos \omega t = f_o \sin(\omega t + \varphi)$, has the solution, $Y_p(t) = Y_{op} \sin(\omega t + \varphi + \psi)$, with Y_{op} and ψ defined by Equations 3.40 and 3.41, respectively. In words, for any harmonic-excitation forcing function the steady-state solution's amplitude is defined by the amplification factor Equation 3.40. The phase of the steady-state response (relative to the input force phase) is defined by Equation 3.41.

Before discussing the physical implications of the amplification factor and phase, recall that the complete solution combines the homogeneous solution of Equation 3.25 and the particular solution of Equation 3.40 to obtain

$$
Y = Y_h + Y_p = e^{-\zeta\omega_n t}(A \cos \omega_d t + B \sin \omega_d t) + Y_{op} \sin(\omega t + \psi).
$$

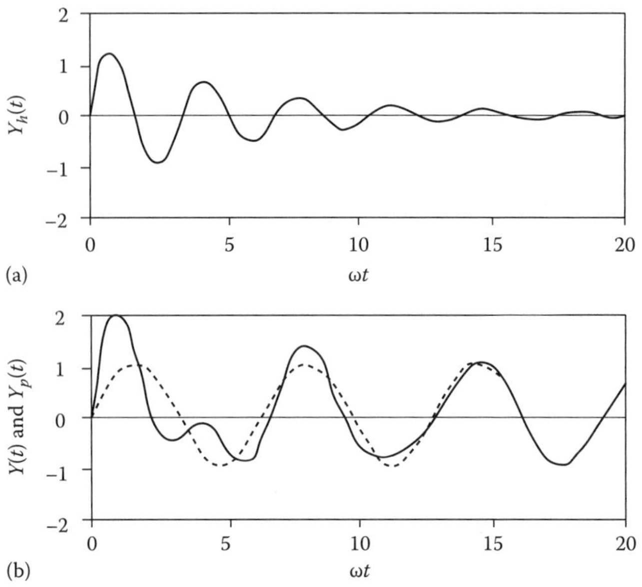

FIGURE 3.14
(a) Homogeneous solution for $\zeta < 1$, (b) complete $[Y(t) = Y_h(t) + Y_p(t)]$ and steady-state particular solution $Y_p(t)$ for $f(t) = f_o \sin \omega t$.

Figure 3.14 illustrates the nature of this solution, with the "transient" homogeneous solution decaying with increasing time, while the particular solution persists. The particular solution is called the "steady-state" solution because of this "persisting" nature. It continues indefinitely after the homogeneous solution has disappeared.

Figures 3.15 and 3.16 illustrate $H(r)$ and $\psi(r)$, respectively. Starting with the amplification factor $H(r)$, Figure 3.15 shows that the dynamic and static response coincide when the excitation frequency ω is much lower than the natural frequency ω_n. For low damping factors ($\zeta \ll 1$), as $r = \omega/\omega_n \Rightarrow 1$, (the excitation frequency approaches the natural frequency) the steady-state amplitude grows rapidly. The coincidence of excitation frequency ω with the natural frequency ω_n is called a "resonance" condition. From Equation 3.40 at $\omega/\omega_n = 1$, the amplification factor is $H(r=1) = 1/2\zeta$, and becomes unbounded for $\zeta = 0$. Recall that the correct particular solution to the differential equation $\ddot{y} + \omega_n^2 y = (f_o/m) \sin \omega_n t$ has the form $y_p = At \sin \omega_n t$ at resonance instead of the solution provided by Equation 3.40. For this physically unlikely condition, the solution grows *linearly* with time.

For large values of the damping factor ($\zeta \geq 0.5$), the dynamic amplitude Y_{op} remains smaller than the static

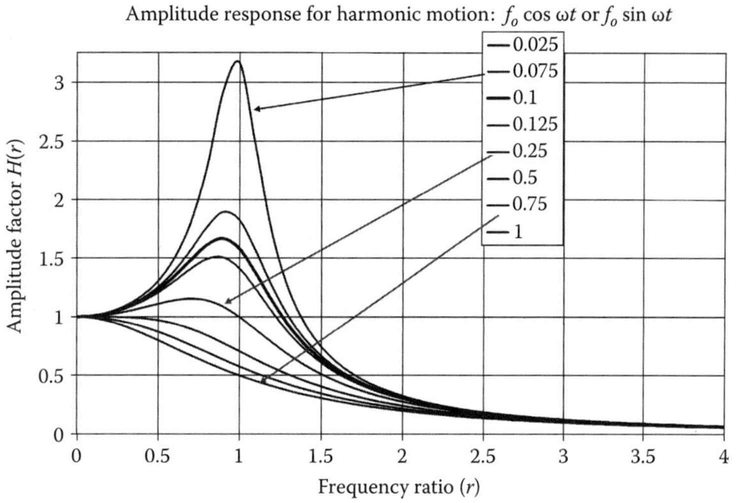

FIGURE 3.15
Amplification factor $H(r)$ for harmonic motion.

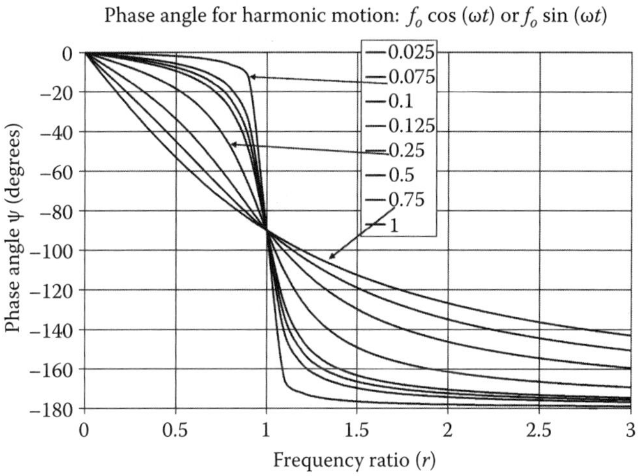

FIGURE 3.16
Phase angle for harmonic excitation.

deflection f_o/k. For all damping values, $H(r)$ approaches zero as $\omega/\omega_n \Rightarrow \infty$. From Equation 3.40, the peak value for the amplification factor $H(r)$ is found from $dH(r)/dr = 0$ as

$$r_{\max} = \sqrt{1 - 2\zeta^2} \tag{3.42}$$

Substituting $r = r_{\max} = \sqrt{1 - 2\zeta^2}$ into Equation 3.40 gives

$$H_{\max} = \frac{1}{2\zeta\sqrt{1 - 2\zeta^2}}. \tag{3.43}$$

For low values of ζ, $H_{\max} \cong 1/2\zeta$.

Since H is nominally the ratio of a response at a frequency ω to the response amplitude at a low-frequency, steady-state, experimental results that includes the peak and low-frequency response amplitudes can be used to directly determine $H_{\max}$ as

$$H_{\max} \cong \frac{|Y(\omega \cong \omega_n)|}{|Y(\omega \cong 0)|}.$$

The difficulty with this approach is that the amplitudes at or near resonance may be too large and potentially destructive to make measurements. Occasionally, engineers characterize the damping in a structure by its quality factor or "q factor" defined in terms of the damping factor by

$$q \text{ factor} = \frac{1}{2\zeta}. \tag{3.44}$$

The customary "real" engineering problems associated with harmonic excitation are excessive amplitudes that arise for lightly damped systems when the excitation frequency is near the natural frequency. Excessive vibration amplitudes due to harmonic excitation near resonance are normally reduced by one of the following approaches:

a. "Detune" the system; that is, change the natural frequency by changing k or m
b. Increase damping
c. Reduce the force excitation levels

For $\zeta = 0$ the phase plot in Figure 3.16 shows that the response is (1) in phase ($\psi = 0$) with the input excitation for $r<1 \Rightarrow \omega<\omega_n$, (2) 180° behind the excitation for $r>1 \Rightarrow \omega>\omega_n$, and (3) 90° behind the excitation at $r=1 \Rightarrow \omega=\omega_n$. At low frequencies ($r<1$), increasing damping causes a more rapid phase shift with increasing frequency. At higher frequencies ($r>1$), increasing damping reduces the phase shift for increasing frequency. For all values of ζ, the response is 90° behind the excitation for $\omega=\omega_n$.

The phase angle does not have the same strong physical impact in engineering as the large amplitudes associated with excitation near resonance. It can be important in vibration diagnostics in determining the locations of natural frequencies. Real structures have multiple degrees of freedom and multiple natural frequencies. (Two-degree-of-freedom problems are introduced in Section 3.5.) Experimental structural-dynamics work frequently involves tests to determine the multiple natural frequencies. In structures with high levels of damping, resonance conditions may be difficult to determine from the amplitude-versus-excitation frequency results but may be more easily identified from the phase-versus-excitation-frequency plot. Specifically, resonance occurs when the measured phase lag is 90°.

Example Problem 3.8

The spring-mass-damper system of Figure XP3.8 is acted on by the external harmonic force, $f=f_o\cos\omega t$, netting the differential equation of motion

$$m\ddot{Y}+c\dot{Y}+kY=f_o\cos\omega t. \qquad \text{(i)}$$

Tasks: For the data, $m=30$ kg, $k=12{,}000$ N/m, $f_o=200$ N, do the following:

a. For $c=0$, determine the range of excitation frequencies for which the amplitudes will be less than 76 mm.
b. Determine the damping value that will keep the steady-state response below 76 mm for all excitation frequencies.

Solution

First, note that Equation (i) has $f_o\cos\omega t$ as the excitation term, versus $f_o\sin\omega t$ in Equation 3.34. This change means

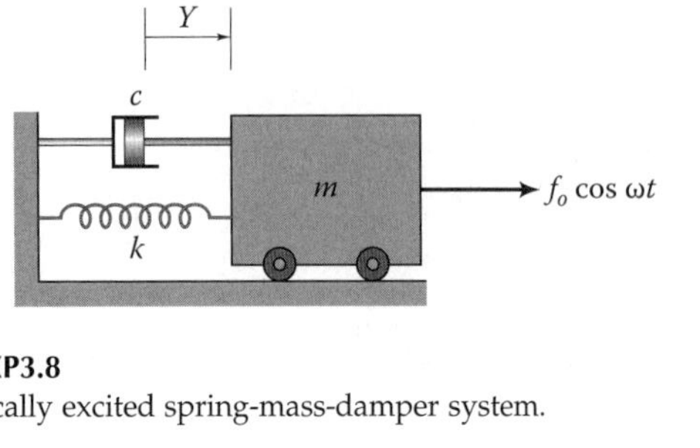

FIGURE XP3.8
Harmonically excited spring-mass-damper system.

that the steady-state response is now $Y_p(t)=Y_{op}\cos(\omega t+\psi)$ instead of $Y_p(t)=Y_{op}\sin(\omega t+\psi)$ of Equation 3.38. The steady-state amplification factor and phase continue to be defined by Equations 3.40 and 3.41, respectively. From the data provided, we can calculate

$$\omega_n=\sqrt{\frac{k}{m}}=\sqrt{\frac{12{,}000\,\text{N/m}}{30\,\text{kg}}}=\sqrt{\frac{12{,}000\,\text{kg-m/s}^2/\text{m}}{30\,\text{kg}}}$$
$$=20\,\text{rad/s}$$

$$\delta_{\text{static}}=\frac{f_o}{k}=200\,\text{N}/(1200\,\text{N/m})=.01667\,\text{m}=16.67\,\text{mm}.$$

Note that we used the derived units for the Newton, kg m/s^2, to solve for ω_n. The amplification factor corresponding to the amplitude $Y_{op}=76$ mm $=.076$ m is $H(r)=.076$ m/(.0167 m) $=4.56$. Looking back at Figure 3.15, this amplification factor would be associated with frequency ratios that are fairly close to $r=1$. For zero damping, we can use Equation 3.40 to solve for the two frequency ratios, via

$$\frac{Y_{op}}{f_o/k}=\frac{1}{[(1-r^2)^2+4\zeta^2r^2]^{1/2}}\Rightarrow 4.56=\frac{1}{[(1-r^2)^2]^{1/2}}.$$

Restating the last equation gives

$$1-2r^2+r^4=\frac{1}{4.56^2}=.0481\Rightarrow r^4-2r^2+.952=0.$$

Solving this quadratic equation in r^2 defines the two frequencies as

$$r_1^2=.781\Rightarrow r_1=.883,$$
$$\omega_1=r_1\times\omega_n=.883\times 20\text{ rad/s}=17.7\text{ rad/s}$$
$$r_2^2=1.22\Rightarrow r_2=1.10,$$
$$\omega_2=r_2\times\omega_n=1.10\times 20\text{ rad/s}=22.1\text{ rad/s}.$$

As expected, ω_1 and ω_2 are close to $\omega_n=20$ rad/s. The steady-state amplitudes will be less than the specified 76 mm, for the frequency ranges, $0<\omega<\omega_1=17.7$ rad/s and $\omega>\omega_2=22.1$ rad/s, and we have completed Task a.

Moving to Task b, From Equations 3.42 and 3.43, the amplification factor is a maximum at $r=\sqrt{1-2\zeta^2}$, and its maximum value is $H_{\max}=1/2\zeta\sqrt{1-2\zeta^2}$. Hence,

$$H_{\max}=\frac{1}{2\zeta\sqrt{1-2\zeta^2}}=4.56\Rightarrow\zeta=.0122,$$

and the required damping to achieve this ζ value is, from Equation 3.22,

$$c=2\zeta\omega_n m=2(.0122)\left(20.\ \frac{\text{rad}}{\text{s}}\right)(30.\ \text{kg})=14.6\ \text{N s/m}.$$

Again, we have used the derived units for the Newton (kg m/s^2) to obtain this result. Damping coefficients of this value or higher will keep the peak response amplitudes at less than the specified 76 mm for all excitation frequencies, and Task b is completed.

3.2.5.1 Base Excitation

The harmonic excitation provided by the force $f_o \sin \omega t$ in Equation 3.34 is one of several ways that forced harmonic excitation can be provided to a spring-mass-damper system. The earlier discussion of base excitation for motion defined by Y about an equilibrium produced the equation of motion

$$m\ddot{Y} + c\dot{Y} + kY = kY_b + c\dot{Y}_b, \tag{3.32}$$

where $Y_b(t)$ defines the base motion. Harmonic base motion can be defined by $Y_b = A \sin \omega t$, $\dot{Y}_b = A\omega \cos \omega t$, and the equation of motion becomes

$$m\ddot{Y} + c\dot{Y} + kY = kA \sin \omega t + cA\omega \cos \omega t.$$

Dividing through by m gives

$$\begin{aligned}\ddot{Y} + 2\zeta\omega_n\dot{Y} + \omega_n^2 Y &= A\left(\omega_n^2 \sin \omega t + 2\zeta\omega\omega_n \cos \omega t\right)\\ &= C \sin(\omega t + \phi)\\ &= C(\sin \omega t \cos \phi + \cos \omega t \sin \phi).\end{aligned} \tag{3.45}$$

Noting that $C \cos \phi = A\omega_n^2$ and $C \sin\phi = 2\zeta A\omega\omega_n$, C and ϕ can be determined as

$$C = A\omega_n^2\left[1 + 4\zeta^2\left(\frac{\omega}{\omega_n}\right)^2\right]^{1/2}, \quad \phi = \tan^{-1}\left[2\zeta\left(\frac{\omega}{\omega_n}\right)\right]. \tag{3.46}$$

Looking at Equation 3.45, our present interest is the steady-state solution due to the harmonic-excitation term $B \sin(\omega t + \phi)$, arising from base motion, not the homogeneous solution due to initial conditions. Based on the earlier results of Equation 3.38, the expected steady-state solution to Equation 3.45 is

$$Y_p = Y_{op} \sin(\omega t + \phi + \psi), \tag{3.47}$$

with ψ defined by Equation 3.41. This solution is sinusoidal at the input frequency ω, having the same phase lag ψ with respect to the input force excitation $C \sin(\omega t + \phi)$ as determined earlier for the harmonic force excitation $f(t) = f_o \sin(\omega t)$. To find the steady-state-solution amplitude for Equation 3.45, a comparison of Equations 3.35 and 3.45, shows C replacing f_o/m Substituting C into Equation 3.40 gives

$$\begin{aligned}Y_{op} &= \frac{C}{\omega_n^2}\frac{1}{\{[1-(\omega/\omega_n)^2]^2 + 4\zeta^2(\omega/\omega_n)^2\}^{1/2}}\\ &= A\frac{\left[1 + 4\zeta^2(\omega/\omega_n)^2\right]^{1/2}}{\left\{\left[1-(\omega/\omega_n)^2\right]^2 + 4\zeta^2(\omega/\omega_n)^2\right\}^{1/2}}.\end{aligned}$$

Hence, the ratio of the steady-state-response amplitude Y_{op} to the base-excitation amplitude A is

$$\frac{Y_{op}}{A} = \frac{\left[1 + 4\zeta^2 r^2\right]^{1/2}}{\left[(1-r^2)^2 + 4\zeta^2 r^2\right]^{1/2}} = G(r). \tag{3.48}$$

Figure 3.17 illustrates $G(r)$, showing a strong similarity to $H(r)$ of Figure 3.15 with the peak amplitudes occurring near $r = \omega/\omega_n = 1$ for small damping ratios. Note that, as $r \to \infty$, $G(r) \to 0$. For the limiting value, $\zeta = 0$, the two transfer functions coincide. The frequency that produces a peak value for $G(r)$ is obtained via $dG(r)/dr = 0$, yielding

$$2\zeta^2 r^4 + r^2 - 1 = 0 \Rightarrow r_{\text{max-amplitude}} = \frac{1}{2\zeta}\left[(1 + 8\zeta^2)^{1/2} - 1\right]^{1/2} \tag{3.49}$$

Note that $\zeta = 0 \Rightarrow r_{\text{max-amplitude}} = 1$.

From the trigonometric identity, $\tan(\alpha + \beta) = (\tan \alpha + \tan \beta)/(1 - \tan \alpha \tan \beta)$, we can use Equations 3.41 and 3.46 to define the phase between the steady-state solution $Y_p = Y_{op} \sin(\omega t + \phi + \psi)$ and the base-motion excitation $Y_b = A \sin \omega t$ as

$$\phi + \psi = \tan^{-1}\left[\frac{-2\zeta r^3}{1 - r^2(1 - 4\zeta^2)}\right]. \tag{3.50}$$

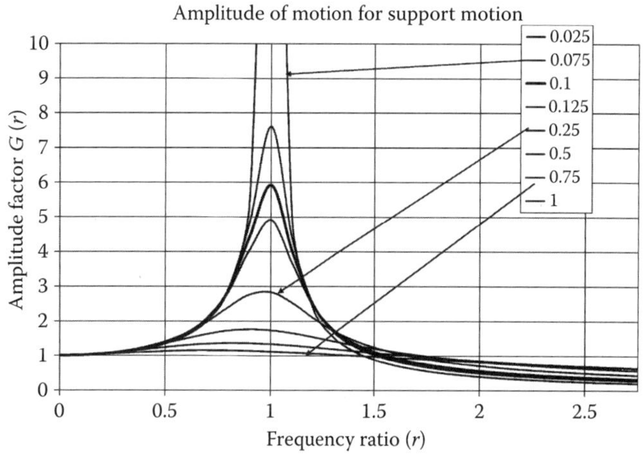

FIGURE 3.17
Amplification factor $G(r)$ for base excitation.

Example Problem 3.9

Figure XP3.9 shows a vehicle model moving horizontally to the right at velocity v along a path that has the harmonic undulation defined by

$$Y_b = A\sin\left(\frac{2\pi x}{L}\right),\quad A = 0.03\text{ m},\quad L = 0.6\text{ m}. \qquad \text{(i)}$$

The surface defined by Equation (i) could correspond to a rough "washboard" gravel road. The model would be a reasonable representation for a single-axle trailer. The mass of the vehicle is m, and the springs and damper are used to model the vehicle's suspension, including the tires. The vehicle has a constant horizontal velocity $v = \dot{x} = 100$ km/h. For convenience, we will assume that it starts at $x = 0$; hence,

$$x = x_o + \dot{x}t = 0 + 100\left(\frac{\text{km}}{\text{h}}\right) \times \left(\frac{1\text{ h}}{3600\text{ s}}\right) \times \frac{1000\text{ m}}{1\text{ km}}t$$
$$= 27.8t\text{ m}.$$

Substituting into Equation (i), $Y_b = A\sin(2\pi x/l) = A\sin(2\pi v/l)t = A\sin\omega t$

$$Y_b(t) = 0.03(m)\sin\left(\frac{2\pi \times 27.8t}{0.6}\right)$$
$$= 0.03\sin(291t)\text{ m}. \qquad \text{(ii)}$$

From Equation (ii), as long as the vehicle's tires remain in contact with the road surface, the vehicle's steady velocity to the right will generate base excitation at the frequency $\omega = 291$ rad/s $= 46.3$ Hz and amplitude 0.03 m. Driving faster will increase the excitation frequency; driving slower will decrease it.

Tasks: Tests show that the vehicle's damped natural frequency is 2.62 Hz, and the damping factor is $\zeta = 0.3$. Do the following:

a. Determine the amplitude of vehicle motion.
b. Determine the speed for which the response is a maximum and determine the response amplitude at this speed.

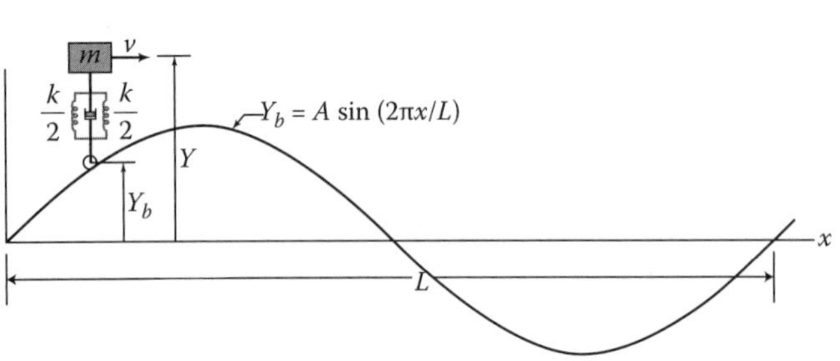

FIGURE XP3.9
Vehicle moving horizontally along a path defined by $Y_b = A\sin(2\pi x/L)$.

Solution

First, the undamped natural frequency is defined by $f_n = f_d/\sqrt{1-\zeta^2} = 2.62/\sqrt{1-0.3^2} = 2.5$ Hz. The frequency ratio is $r = \omega/\omega_n = f/f_n = 46.3/2.5 = 18.5$. Hence, from Equation 3.48,

$$\frac{Y_{op}}{A} = \frac{\left[1+4(0.3^2)(18.5)^2\right]^{1/2}}{\left[(1-18.5^2)^2+4(0.3^2)(18.5)^2\right]^{1/2}]^{1/2}} = 0.0326$$
$$\therefore Y_{op} = 0.03 \times 0.0326 = 0.977\times 10^{-3}\text{ m} = 0.977\text{ mm}. \qquad \text{(iii)}$$

The response amplitude will be a maximum when $r = r_{\text{max-amplitude}}$ as defined in Equation 3.49; that is,

$$r_{\max} = \frac{1}{2\zeta}\left[(1+8\zeta^2)^{1/2}-1\right]^{1/2}$$
$$= \frac{1}{2\times 0.3}\left\{\left[1+8(0.3)^2\right]^{1/2}-1\right\}^{1/2} = 0.930.$$

Hence,

$$\omega = \frac{2\pi \times v\text{ (m/s)}}{0.6\text{ m}} = 0.930 \times \omega_n$$
$$= 0.930 \times 2.5\,\frac{\text{cycle}}{\text{s}} \times \frac{2\pi\text{ rad}}{1\text{ cycle}} = 14.6\text{ rad/s}$$
$$\therefore\ v = \frac{L\omega}{2\pi} = \frac{0.6(14.6)}{2\pi} = 1.40\ \frac{\text{m}}{\text{s}}$$

Hence, the peak-amplitude speed is $v = 1.40 \times [(100\text{ km/h})/(27.8\text{ m/s})] = 5.04$ km/h. From Equation 3.48, at this speed, the steady-state amplification factor is

$$\frac{Y_{op}}{A} = \frac{\left[1+4(0.3^2)0.930^2\right]^{1/2}}{\left[(1-0.930^2)^2+4(0.3^2)0.930^2\right]^{1/2}} = 1.99,$$
$$\therefore Y_{op} = 0.03(1.99) = 0.060\text{ m}.$$

This example reflects a common experience of driving on a rough road; namely, the perceived vibration level grows worse with increasing speed until, at some limiting speed, the vibration levels drop sharply and then remain at a reduced level as the speed increases. "Rumble strips" reflect another driving experience where closely spaced grooves are cut into a smooth road surface. The high-frequency oscillation induced in the vehicle by the grooves is designed to get the driver's attention concerning an approaching hazard or drifting out of the driving lane.

Example Problem 3.10

An instrument package is to be attached to the housing of a rotating machine. Measurements on the casing show a vibration at 3600 rpm with an acceleration level of 0.25 g. The instrument package has a mass of 0.5 kg. Tests show that the support bracket to be used in attaching the package to the vibrating structure has a stiffness of 10^5 N/m.

Task: Determine how much damping is needed to keep the instrument package vibration levels below 0.5 g.

SOLUTION

The frequency of excitation at 3600 rpm converts to $\omega = 3600 \text{ rev/min} \times (1 \text{ min}/60 \text{ s}) \times (2\pi \text{ rad/rev}) = 377$ rad/s. With harmonic motion, the housing's amplitude of motion is related to its acceleration by $a = 25 \text{ g} = -\omega^2 A$; hence, the amplitude corresponding to the housing-acceleration levels of 0.25 g is

$$A = \frac{0.25(9.81 \text{ m/s}^2)}{3.77^2} = 1.726 \times 10^{-5} \text{ m} = 0.017 \text{ mm}.$$

Similarly, a 0.5 g acceleration level for the instrument package means its steady-state amplitude is

$$Y_{op} = \frac{0.5(9.81 \text{ m/s}^2)}{377^2} = 0.034 \text{ mm}.$$

Hence, the target amplification factor is $G = 0.034/0.017 = 2$.

The natural frequency of the instrument package is $\omega_n = \sqrt{k/m} = \sqrt{10^5 \text{ N}/0.5 \text{ kg}} = 447.2$ rad/s. Hence, the frequency ratio is $r = \omega/\omega_n = 377/447 = 0.75$. Plugging into Equation 3.48 gives

$$\frac{Y_{op}}{A} = G(.75) = 2 = \frac{\left[1 + 4\zeta^2 0.75^2\right]^{1/2}}{\left[(1 - 0.75^2)^2 + 4\zeta^2 0.75^2\right]^{1/2}}$$
$$\Rightarrow 4 = \frac{1 + 2.25\zeta^2}{0.1914 + 2.25\zeta^2}.$$

The solution to this equation is $\zeta^2 = 0.0347 \Rightarrow \zeta = 0.18628$. Hence, the required damping is

$$c = 2\zeta\omega_n m = 2(0.18628)\left(447.2 \frac{\text{rad}}{\text{s}}\right)(0.5 \text{ kg})$$
$$= 83.3 \text{ N s/m},$$

which concludes the engineering-analysis task. Since only one excitation frequency is involved, we could have simply taken the ratio of the acceleration levels directly to get $G = 2$.

3.2.5.2 *Steady-State Relative Motion due to Base Excitation*

In Section 3.2.4.2, we were interested in the "relative displacement" between the moving base defined by Y_b and the mass motion defined by Y. The definition for relative motion, defined by, $\delta y = Y - Y_B \Rightarrow \delta\dot{y} = \dot{Y} - \dot{Y}_B \Rightarrow \delta\ddot{y} = \ddot{Y} - \ddot{Y}_B$ converted the customary base-excitation equation, $m\ddot{Y} + c\dot{Y} + kY = kY_b + c\dot{Y}_b$ into

$$m\delta\ddot{y} + c\delta\dot{y} + k\delta y = -m\ddot{Y}_b. \qquad (3.33)$$

This result shows that the dynamic relative deflection δy arises due to base acceleration. For harmonic base excitation defined by $Y_b = A \sin \omega t$, the equation of motion is

$$m\delta\ddot{y} + c\delta\dot{y} + k\delta y = mA\omega^2 \sin \omega t.$$

We want a steady-state solution to this equation of the form, $\delta y = \Delta \sin(\omega t + \beta)$. This equation has the same form as Equation 3.34, except $mA\omega^2$ has replaced f_o. Hence, by comparison to Equation 3.40, the steady-state relative amplitude due to harmonic base excitation is

$$\frac{\Delta}{mA\omega^2/k} = \frac{\Delta}{A\omega^2(m/k)} = \frac{\Delta}{mA(\omega/\omega_n)^2}$$
$$= \frac{\Delta}{Ar^2} = H(r); \quad r = \frac{\omega}{\omega_n} \qquad (3.51)$$
$$\therefore \frac{\Delta}{A} = r^2 H(r) = \frac{r^2}{\left[(1 - r^2)^2 + 4\zeta^2 r^2\right]^{1/2}} = J(r).$$

As shown in the next section, this is also the amplification factor for response due to rotating imbalance. The function $J(r)$ is illustrated in Figure 3.19. Note that $J(r) \to 1$ as $r \to \infty$. From Equation 3.41, the phase angle β between δ and $Y_B = A \sin \omega t$ is

$$\beta(r) = \pi + \psi(r) = \pi - \tan^{-1}\left(\frac{2\zeta r}{1 - r^2}\right). \qquad (3.52)$$

The π term in this phase is required to account for the 180° phase shift between base *position* defined by $Y_B = A \sin \omega t$ and base *acceleration* defined by $\ddot{Y}_B = -A\omega^2 \sin \omega t$.

Returning to Example Problem 3.9, from Equation 3.51, the steady-state *relative* amplitude at $v = 100$ km/h is

$$\frac{\Delta}{A} = \frac{18.5^2}{\left[(1 - 18.5^2)^2 + 4(0.3^2)(18.5)^2\right]^{1/2}} = 1.002$$
$$\therefore \Delta = 0.03 \times 1.002 = 0.030 \text{ m} = 30 \text{ mm}.$$

This value is much greater than the absolute amplitude of motion $Y_{op} = 0.977$ mm that we calculated earlier. This result shows that, while the vehicle has small absolute vibration amplitudes, its base is following the ground contour; hence, the relative deflection (across the spring and damper) is approximately equal to the amplitude of the base oscillation.

3.2.5.3 *Rotating-Imbalance Excitation*

Figure 3.18a provides an end view of an imbalanced electric motor. The motor has mass m_u and is rotating at the constant angular velocity ω. The bearing axis of

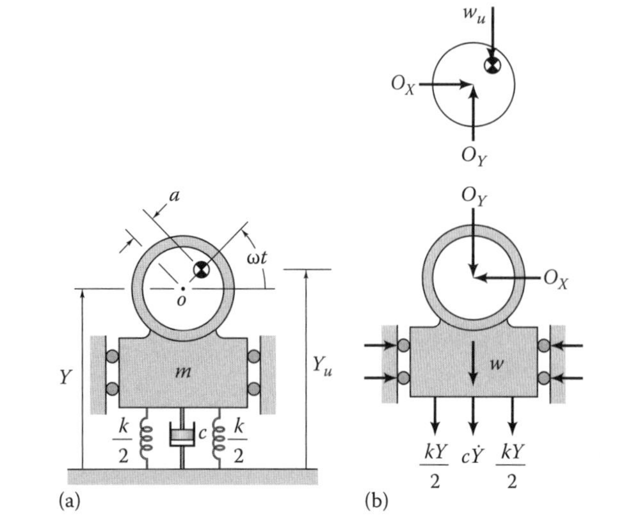

FIGURE 3.18
(a) Imbalanced motor with mass m_u supported by a housing mass m, (b) free-body diagram for $Y > 0$, $\dot{Y} > 0$.

the rotor is denoted by O. The rotor is assumed to be a rigid body that is attached to the housing via rigid bearings. The motor's mass center is displaced the distance a from the axis of rotation defined by the motor's bearings. Note that a's magnitude is greatly exaggerated in Figure 3.18a. The product $m_u a$ is called the "imbalance vector." Considerable care is exercised in balancing electric motors to minimize this vector. The motor support structure has mass m and can only move vertically. Y defines m_u's vertical position with respect to ground, and Figure 3.18a gives

$$Y_u = Y + a \sin \omega t \Rightarrow \ddot{Y}_u = \ddot{Y} - a\omega^2 \sin \omega t. \tag{3.53}$$

Figure 3.18b provides free-body diagrams for the rotor and support masses. The free-body diagram for the support mass applies for upward motion of the support housing that causes tension in the support $(Y > 0)$ springs and the support damper $(\dot{Y} > 0)$. The internal reaction force (acting at the motor's bearings) between the motor and the support mass is defined by the components O_X, O_Y.

Applying $\sum f = m\ddot{r}$ to the individual masses of Figure 3.18b gives

$$\text{Rotor: } m_u\ddot{Y}_u = m_u(\ddot{Y} - a\omega^2 \sin \omega t) = \sum f_Y = -w_u + O_Y$$
$$\text{Stator: } m\ddot{Y} = \sum f_Y = -w - O_Y - kY - c\dot{Y},$$

where Equation 3.53 has been used to eliminate $\ddot{Y}_u$. Adding these equations eliminates the vertical reaction force, O_Y, and gives the single equation of motion

$$(m + m_u)\ddot{Y} + c\dot{Y} + kY = -(w + w_u) + m_u a\omega^2 \sin \omega t.$$

This equation resembles Equation 3.34, except $(m + m_u)$ and $m_u a\omega^2$ have replaced m and f_o, respectively. Defining Y as the displacement from the equilibrium position eliminates $(w + w_u)$, and dividing by $(m + m_u)$ gives

$$\ddot{Y} + 2\zeta\omega_n\dot{Y} + \omega_n^2 Y = \frac{m_u a\omega^2}{M} \sin \omega t, \tag{3.54a}$$

where,

$$M = (m + m_u), \quad 2\zeta\omega_n = \frac{c}{M}, \quad \omega_n^2 = \frac{k}{M}. \tag{3.54b}$$

We want the steady-state solution to Equation 3.54a due to the rotating-imbalance excitation term $[(m_u a\omega^2)/M] \sin \omega t$ and are not interested in the homogeneous solution due to initial conditions. Equation 3.54a has the same form as Equation 3.34 except $m_u a\omega^2 lM$ has replaced f_o/m. Hence, by comparison to Equation 3.40, the steady-state response amplitude is

$$Y_{op} = \frac{m_u a\omega^2}{M} \cdot \frac{1}{\left[\left(\omega_n^2 - \omega^2\right)^2 + 4\zeta^2\omega_n^2\omega^2\right]^{1/2}}$$
$$= \frac{m_u a\omega^2}{M\omega_n^2} \cdot \frac{1}{\left\{\left[1 - (\omega/\omega_n)^2\right]^2 + 4\zeta^2(\omega/\omega_n)^2\right\}^{1/2}},$$

and the amplification factor due to the rotating imbalance is

$$\frac{Y_{op}}{a(m_u/M)} = \frac{r^2}{\left[(1 - r^2)^2 + 4\zeta^2 r^2\right]^{1/2}} = J(r). \tag{3.55}$$

Figure 3.19 illustrates $J(r)$ for a range of damping-ratio values. The major differences between $J(r)$ and $H(r)$ are

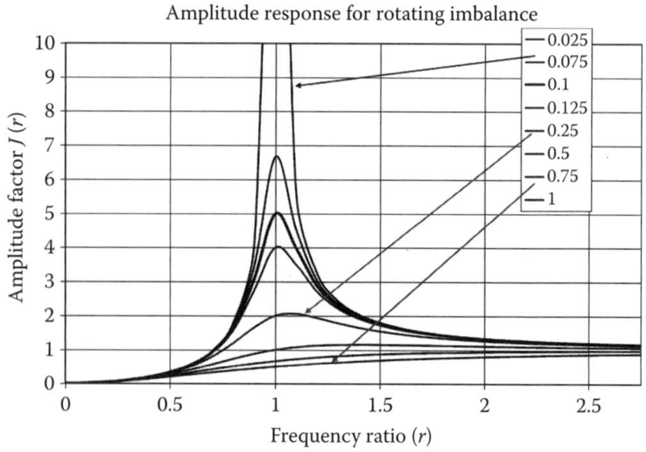

FIGURE 3.19
$J(r)$ versus frequency ratio r from Equation 3.55 for a range of damping-ratio values.

(1) $J(r)=0$ for $r=\omega/\omega_n=0$, while $G(r)=1$, and (2) The asymptotic value is $J(r\Rightarrow\infty)=1$, while $H(r\Rightarrow\infty)=0$. Near resonance, with $r\cong 1$, both $H(r)$ and $J(r)$ can have very large values for small damping ratios. Hence, the excitation frequency is higher than the undamped natural frequency, which is higher than the damped natural frequency. This result contrasts with the peak-excitation amplitude frequency for the model $m\ddot{Y}+2\zeta\omega_n\dot{Y}+\omega_n^2 Y=(f_o/m)\sin\omega t$ of Equations 3.42 and 3.43 where the peak-amplitude frequency is below ω_n.

The frequency ratio at the peak $J(r)$ amplitude is obtained from $dJ(r)/dr=0$ as

$$r_{\text{max-amplitude}}=\frac{1}{\sqrt{1-2\zeta^2}}\Rightarrow\omega_{\text{max-amplitude}}$$
$$=\frac{\omega_n}{1-2\zeta^2}>\omega_n>\omega_d=\omega_n\sqrt{1-\zeta^2}.$$

Comparing the earlier forced-response equation of motion $\ddot{Y}+2\zeta\omega_n\dot{Y}+\omega_n^2 Y=(f_o/m)\sin\omega t$ to the present rotating-imbalance equation $\ddot{Y}+2\zeta\omega_n\dot{Y}+\omega_n^2 Y=(m_u a\omega^2/M)\sin\omega t$ shows that the forced-response phase definition of Equation 3.41 and Figure 3.16 apply for rotating-imbalance excitation. These results are amenable to a reasonable "physical" interpretation for the rotating-imbalance case. At low frequencies ($r\ll 1$), the base mass m is moving up and down, respectively, when the rotating imbalance is moving up and down, that is, m's motion is in phase with the rotating imbalance. At high frequencies ($r\gg 1$), the base motion is moving, respectively, down and up when the rotating imbalance is moving up and down; that is, m's motion is 180° out of phase with the rotating imbalance.

At resonance, when the imbalance is at zero degrees ($\omega t=0$), pointed directly to the right, the support mass m is at its lowest position with zero velocity and a positive upward acceleration. When the imbalance is vertical ($\omega t=\pi/2$ rad $=90°$), m is at the zero location, but moving upward at maximum velocity. When the imbalance is pointed directly to the left ($\omega t=\pi$ rad $=180°$), m is at its highest position with zero velocity and a downward acceleration. When the imbalance is pointing vertically downward ($\omega t=3\pi/2$ rad $=270°$), m is at the zero location with a maximum downward velocity. In all these cases, m's motion is 90° behind the phase of the rotating imbalance.

Example Problem 3.11

Figure XP3.11 illustrates an industrial blower mounted on a framed support structure. The support structure is much stiffer in the vertical direction than in the horizontal; hence, the model of Equation 3.54a holds for horizontal motion. This unit runs at 500 rpm, and has been running at "high" vibration levels. A "rap" test has been performed by mounting an accelerometer to the fan base, hitting the support structure below the fan with a large hammer, and recording the output of the accelerometer. This test shows a damped natural frequency of 10.4 Hz = 625 rpm with very little damping ($\zeta\cong 0.005$). The rotating mass of the fan is 114 kg. The total weight of the fan (including the rotor) and its base plate was stated to be 1784 N by the manufacturer.

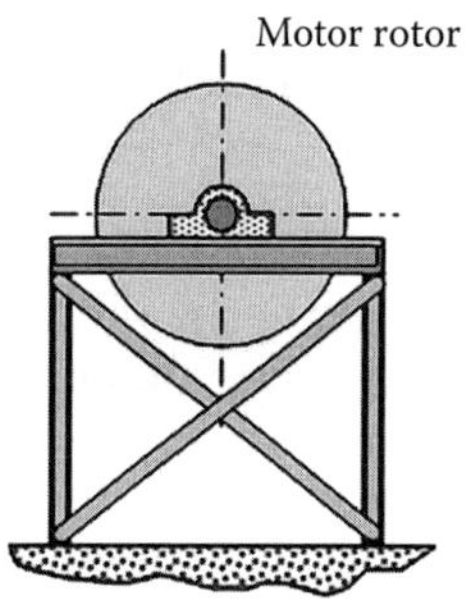

FIGURE XP3.11
Industrial blower supported by a welded-frame structure.

Tasks:

a. Assuming that the vibration level on the fan is to be less than 0.1 g, how well should the fan be balanced; that is, to what value should a be reduced?
b. Assuming that the support structure could be stiffened laterally by approximately 50%, how well should the fan be balanced?

Solution

In terms of the steady-state amplitude of the housing, the acceleration magnitude is $|a_p|=\omega^2 Y_{op}=0.1$ g, where g = 9.81 m²/s. Hence, at

$$\omega=500\left(\frac{\text{rev}}{\text{min}}\right)\times\left(\frac{1\text{ min}}{60\text{ s}}\right)\times\left(\frac{2\pi\text{ rad}}{1\text{ rev}}\right)=52.36\text{ rad/s},$$

the housing-amplitude specification is $Y_{op}\leq 0.1\times 9.81/52.36^2=3.578\times 10^{-4}$ m $=0.358$ mm. Applying the notation of Equation 3.54b gives $M=W/g=1784.$ N/(9.81 m/s²) = 181.9 kg, and $m_u=114$ kg.

Because the damping is quite low, $\omega_n=\omega_d/\sqrt{1-\zeta^2}=625/\sqrt{1-0.005^2}=625$ rpm $=68.07$ rad/s, and $r=\omega/\omega_n=500/625=0.8$. Substituting into Equation 3.55 gives

$$\frac{Y_{op}}{a(m_u/M)}=\frac{0.64}{[(1-0.64)^2+4.(0.005)^2(0.64)]^{1/2}}=1.77.$$

Hence, the imbalance-vector magnitude a should be no more than

$$a=\frac{[Y_{op}/(m_u/M)]}{1.77}=0.358\text{ mm}/(114/182)/1.77$$
$$=0.323\text{ mm}=0.013(\text{in.})$$

which concludes Task a. An American Petroleum Institute (API) balancing requirement specifies the imbalance per bearing at $m_u a$ (ounce–inches) $= 4W$ (lb)/rpm. This formula is not dimensionally consistent; however, for the present example, the required imbalance vector is $a = 0.00254$ mm $= 0.001$ in. Hence, balancing the rotor such that $a \leq 0.323$ mm (0.013 in.) for a 114 kg fan rotor is feasible. However, many fans accumulate debris that can cause the imbalance to grow with operating time.

Moving to Task b, stiffening the housing has not changed the damping coefficient, which is

$$c = 2\zeta\omega_n(\text{old})M = 2 \times 0.005 \times 68.07\ \frac{\text{rad}}{\text{r s}} \times 181.9\ \text{kg}$$
$$= 123.8\ \frac{\text{N s}}{\text{m}}.$$

Assuming that the structural stiffening does not appreciably increase the mass of the fan assembly, increasing the lateral stiffness by 50% would change the natural frequency to

$$\omega_n(\text{new}) = \sqrt{\frac{1.5 \times k}{M}} = \sqrt{1.5} \times \omega_n(\text{old})$$
$$= 1.225 \times 68.07\ \frac{\text{rad}}{\text{s}} = 83.39\ \text{rad/s}.$$

With a stiffened housing, the new damping factor is

$$\zeta(\text{new}) = \frac{c}{2\omega_n(\text{new})M} = \frac{123.8\ \text{N s/m}}{[2 \times 83.39\ (\text{rad/s}) \times 181.91\ \text{kg}]}$$
$$= 0.004$$

Also, $r_{\text{new}} = \omega/\omega_n = 500/765 = 0.653$, $r^2_{\text{new}} = 0.427$. Hence,

$$\frac{Y_{op}}{a(m_u/M)} = \frac{0.427}{[(1 - 0.427)^2 + 4(0.004)^2(0.427)]^{1/2}} = 0.745,$$

and $a = Y_{op}/(m_u/M)/0.745 = 0.358$ mm/(114/182)/0.745 = 0.767 mm (0.030 in.). By elevating the system natural frequency, the imbalance-vector magnitude can be $767/323 = 2.4$ times greater without exceeding the vibration limit. Stated differently, the fan can tolerate a much higher imbalance when its operating speed ω is further away from resonance.

Typically, appreciable damping is very difficult to introduce into this type of system; moreover, increasing the damping factor to $\zeta = 0.05$ in Task a reduces only slightly the required value for a to meet the housing-acceleration level specification. Damping would be more effective for $r = \omega/\omega_n \cong 1$. Of course, the vibration amplitudes would also be much higher.

3.2.5.4 Summary and Extensions

In this section, we have considered the same basic differential equation with the following three forms of harmonic excitation:

a. Excitation due to a harmonic force
b. Excitation due to base excitation
c. Excitation due to a rotating imbalance

The same basic steady-state response solution is obtained for all three forms of excitation. In practice, harmonic force excitation can arise due to many factors. For turbomachinery units such as steam or gas turbines, unsteady aerodynamic forces can provide harmonic excitation to turbine blades. Aerodynamic forces in rotating machinery normally have many frequency components, and the turbine blades being excited have many natural frequencies. A great deal of engineering-vibration analysis involves predicting and measuring natural frequencies, since the coincidence of a harmonic-excitation frequency with a lightly damped natural frequency will normally lead to large amplitudes and structural failures. Most engineering-vibration problems involving harmonic excitation are more complicated than $m\ddot{Y} + c\dot{Y} + kY = f_o \sin \omega t$ in that (1) more degrees of freedom are involved, and (2) more excitation frequencies are present. However, the same basic 1DOF ideas hold in analyzing these more complicated systems.

The base-excitation model and results apply for many practical vibration problems. The vehicle example of Problem XP3.9 provides a common application of base excitation; however, structural failures due to base excitation arise in many technical circumstances. Various devices are attached to the casings of rotating machinery or to the decks or structures of power boats, offshore structures, etc. Coincidence or nearness to resonance conditions of these "add-on" structures can cause large amplitudes and structural failure due to fatigue.

The rotating-imbalance excitation of Equation 3.49 is always present when dealing with rotating machinery, for example, electric motors, pumps, compressors, or turbines. However, the 1DOF model provided by Figure 3.18a is misleading in that horizontal motion is not allowed. Looking at this figure should suggest to you that a real electric motor would vibrate in both the vertical and *horizontal* directions, since the horizontal reaction force, $O_X(t)$, of Figure 3.18b would also be harmonic. While Equation 3.49 provides a manageable 1DOF model and a reasonable idea of the expected response, a real electric-motor/base assembly would have at least two degrees of freedom (one vertical and one horizontal) and two natural frequencies. The horizontal-motion natural frequency is typically lower than the vertical

TABLE 3.2

Forced-Excitation Results where $r = \omega/\omega_n$

Application	Transfer Function	Amplitude Ratio	r for Maximum Response
Forced response	$\frac{Y_{op}}{f_o/k} = H(r)$	$\frac{1}{[(1-r^2)^2 + 4\zeta^2 r^2]^{1/2}}$	$r_{max} = \sqrt{1-2\zeta^2}$
Base excitation	$\frac{Y_{op}}{A} = G(r)$	$\frac{[1+4\zeta^2 r^2]^{1/2}}{[(1-r^2)^2 + 4\zeta^2 r^2]^{1/2}}$	$r_{max} = \frac{1}{2\zeta}\sqrt{(1+8\zeta^2)^{1/2} - 1}$
Relative deflection with base excitation	$\frac{\Delta}{A} = J(r)$	$\frac{r^2}{[(1-r^2)^2 + 4\zeta^2 r^2]^{1/2}}$	$r_{max} = \frac{1}{\sqrt{1-2\zeta^2}}$
Rotating imbalance	$\frac{Y_{op}}{a(m_u/M)} = J(r)$	$\frac{r^2}{[(1-r^2)^2 + 4\zeta^2 r^2]^{1/2}}$	$r_{max} = \frac{1}{\sqrt{1-2\zeta^2}}$

natural frequency because the lateral stiffness is lower. A two-degree-of-freedom (2DOF) model for the motion of an axisymmetric rotor is introduced in Section 5.7.4. Childs (1993) provides a review of analysis techniques for lateral vibrations of rotating machinery.

1DOF systems incorporating springs, viscous dampers, and a mass comprise a large part of the practice of engineering vibrations. Understanding the 1DOF solutions for free and forced motion will help you considerably in engineering practice and in the subsequent analysis of multi-degree-of-freedom systems. The introduction provided in this section and book are no substitute for a real vibrations text, and Dimarogonas (1995) text is a good choice. Table 3.2 provides a summary of the steady-state, forced-response amplitude ratios.

3.2.6 Energy Dissipation: Coulomb Damping

We began this section by discussing viscous (linear) dampers. Viscous damping is commonly provided in mechanical systems to deliberately dissipate energy by developing a reaction force that is proportional to velocity, and it causes an exponential decay of the free (unforced) motion of a mass-damper-spring system. It can be used to reduce the amplitudes of forced harmonic motion near resonance conditions. Many other energy-dissipation devices arise in nature and are normally required in models of mechanical systems to reflect reality. Coulomb friction is a commonly-observed energy-dissipation mechanism and is the subject of the present section. [Charles Coulomb (1736–1806) was a French physicist who worked on electricity, magnetism, and friction. He is most widely known for Coulomb's inverse-square law of electricity.]

Figure 3.20 illustrates a mass m that is sliding on a dry horizontal surface. Since there is no vertical acceleration, the weight w of the mass pushing down on the plane is opposed by a normal reaction force, N, pushing up on the bottom of the mass. The friction force is proportional to N and opposes motion of the mass; hence, in Figure 3.20a

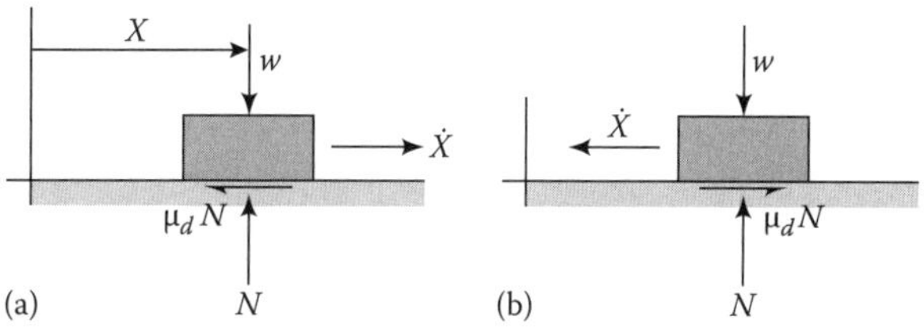

FIGURE 3.20
Sliding mass with Coulomb damping. (a) Free-body for sliding motion to the right, (b) free-body for sliding motion to the left.

with velocity to the right, the friction force is to the left, while in the Figure 3.20b the velocity is to the left, and the friction force is to the right. Applying $\sum f = m\ddot{r}_g$ yields

$$m\ddot{X} = \sum f_X = -\mu_d N \, sgn(\dot{X}), \qquad (3.56a)$$

where

$$sgn(\dot{X}) = +1 \quad \text{for } \dot{X} > 0$$
$$sgn(\dot{X}) = -1 \quad \text{for } \dot{X} < 0. \qquad (3.56b)$$

The differential equation is linear, providing that the sign of $\dot{X}$ is constant and accordingly is said to be "piece-wise" linear in $\dot{X}$.

Equation 3.56 does not define the Coulomb-friction force when $\dot{X} = 0$, a deficiency that we will now try to remedy. Figure 3.21 considers a *static* situation where an external force is applied to the mass. The external force increases linearly and, at some time $\bar{t}$, the mass begins to move. Figure 3.22 shows the parallel change in the Coulomb-friction force over this time interval. For $t \le \bar{t}$, the friction force is equal and opposite to the applied force; that is, $f_f = -f(t)$. As the mass begins to slip, the friction force drops to $-\mu_d N$ as defined by Equation 3.56. The friction force at the time slipping begins is characterized by the "static" Coulomb-friction coefficient, $f_f(\max) = -\mu_s N$. The static Coulomb-friction coefficient μ_s is seen to be greater than the dynamic Coulomb-friction coefficient μ_d.

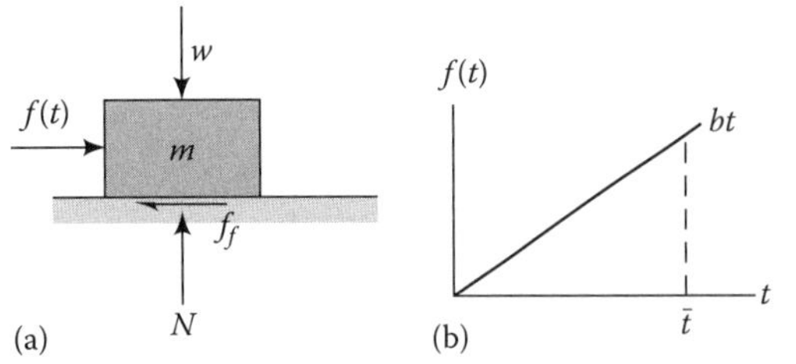

FIGURE 3.21
(a) Mass with linearly increasing applied force and Coulomb-friction reaction force, (b) increasing applied force with motion initiating at $t = \bar{t}$.

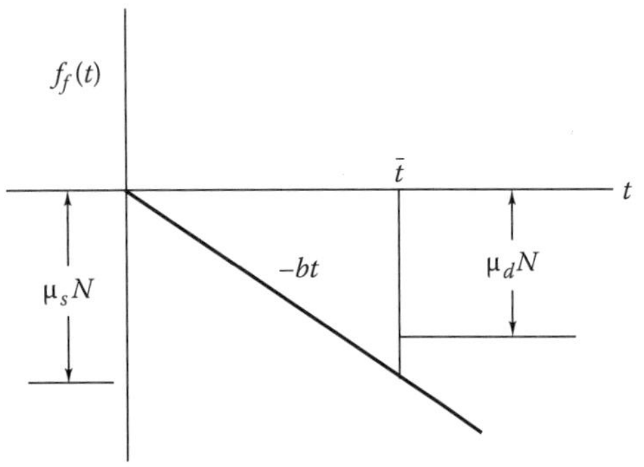

FIGURE 3.22
Reaction force due to Coulomb damping for the applied force of Figure 3.21.

Decelerating a car by applying brakes provides an example of the difference between static and dynamic Coulomb friction. The maximum deceleration is achieved when the tires are just starting to slip relative to the roadway. Once slipping initiates, the braking force drops markedly. Automatic braking systems (ABS) on vehicles are designed specifically to sense and avoid slipping between tires and the roadway.

The Coulomb-friction model states that the friction force is proportional to the normal force, *independent of the velocity magnitude*. This is a reasonable engineering approximation for dry-friction sliding at low velocities. However, sliding friction is a quite complicated phenomenon, and circumstances arise for which the simple Coulomb model stops being appropriate. The following two example problems are presented to clarify the meaning of Coulomb damping and the distinction between static and dynamic Coulomb-friction factors.

Example Problem 3.12

For Figure XP3.12, the initial conditions are

$$X(0) = 0, \quad \dot{X}(0) = \dot{X}_0 > 0 \text{ m/s},$$

and the dynamic Coulomb-friction factor is μ_d.

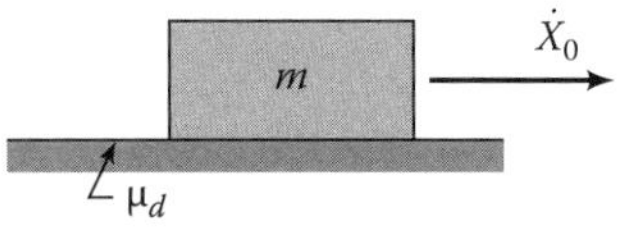

FIGURE XP3.12
Mass m sliding to the right on a surface with μ_d, the dynamic Coulomb coefficient of friction.

Tasks: How far will the mass slide before it comes to rest? How long will motion last?

Solution

From Equation 3.56, the differential equation of motion is

$$m\ddot{X} = \sum f_X = -\mu_d N = -\mu_d mg \Rightarrow \ddot{X} = -\mu_d g.$$

This equation has the same constant-acceleration form as Equation 3.4 for a particle falling under the influence of gravity. The time solution can be obtained by direct integration, yielding

$$\dot{X} = \dot{X}_0 - \mu_d g t, \quad X = 0 + \dot{X}_0 t - \frac{\mu_d g t^2}{2}.$$

Motion stops when the velocity is zero; hence, $\bar{t}$, the time at which motion stops, is $\dot{X}(\bar{t}) = 0 \Rightarrow \bar{t} = \dot{X}_0/(\mu_d g)$. The distance traveled before stopping is

$$X(\bar{t}) = d = \dot{X}_0\left(\frac{\dot{X}_0}{\mu_d g}\right) - \frac{\mu_d g}{2}\left(\frac{\dot{X}_0}{\mu_d g}\right)^2 = \frac{\dot{X}_0^2}{2\mu_d g}.$$

This result can also be obtained directly from the energy-integral substitution, $\ddot{X} = d(\dot{X}^2/2)/dX$, to get

$$m\frac{d}{dX}\left(\frac{\dot{X}^2}{2}\right) = -\mu_d mg,$$

Multiplying by dX reduces both sides of the equation to exact differentials. Integration yields

$$\frac{m\dot{X}^2}{2} - \frac{m\dot{X}_0^2}{2} = -\mu_d mg \int_0^X dX = -\mu_d mgX. \qquad \text{(i)}$$

Requiring zero velocity when $X = d$ yields our earlier result

$$d = \frac{\dot{X}_0^2}{2\mu_d g}.$$

Equation (i) states that the (negative) work done on the body by friction equals the body's change (loss) in kinetic energy.

Example Problem 3.13

Figure XP3.13 shows a mass resting on an inclined plane at an angle θ to the horizontal. The angle θ is increased slowly from zero until slipping is initiated and then held constant.

Tasks:

a. If $\mu_s = 0.15$, for what value of θ does slipping initiate?
b. For $\mu_d = 0.1$, once slipping initiates, how long does it take for the mass to move 10 m, and how fast will it be going when it has moved 10 m?

Solution

There is no acceleration normal to the plane; hence, equilibrium yields

$$\sum f_y = 0 = N - w\cos\theta = 0. \quad \text{(i)}$$

Just before slipping initiates, the acceleration of the mass down the plane is zero, yielding

$$\sum f_x = 0 = w\sin\bar{\theta} - \mu_s N. \quad \text{(ii)}$$

Substituting for N from (i) into (ii) yields

$$w\sin\bar{\theta} - \mu_s(w\cos\bar{\theta}) = 0$$
$$\therefore \tan\bar{\theta} = \mu_s = .15 \Rightarrow \bar{\theta} = 8.53°.$$

Just after slipping occurs, the dynamic Coulomb coefficient $\mu_d = 0.1$ applies, and from $\sum \boldsymbol{f} = m\ddot{\boldsymbol{r}}_g$ the equation of motion is defined from the free-body diagram as

$$m\ddot{x} = \sum f_x = w\sin\bar{\theta} - \mu_d(w\cos\bar{\theta});$$
$$\therefore \ \ddot{x} = g\cos\bar{\theta}(\tan\bar{\theta} - \mu_d) = .05\,g\cos\bar{\theta} = .0494\,g\ \text{m/s}^2, \quad \text{(iii)}$$

where $g = 9.81$ m/s^2 is the acceleration due to gravity. The initial conditions for Equation (iii) are $x(0) = \dot{x}(0) = 0$; hence, integration with respect to time yields

$$\dot{x}(t) = .0494gt \text{ m/s}, \quad x(t) = .0494\frac{gt^2}{2} \text{ m.} \quad \text{(iv)}$$

Substituting $x(\bar{t}) = 10$ m yields

$$10 = \frac{.0494(9.81)t^{-2}}{2} \Rightarrow \bar{t} = 6.42 \text{ s.}$$

Back substituting $\bar{t}$ into Equation (iv) gives

$$\dot{x}(\bar{t}) = .0494 \times 9.81\,(\text{m/s}^2) \times 6.42\ (\text{s}) = 3.11\text{m/s.} \quad \text{(v)}$$

Since the acceleration is constant, substituting $\ddot{x} = d(\dot{x}^2/2)/dx$ into Equation (iii) and integrating with respect to x yields $m\dot{x}^2/2 = .0494\,mgx$. Substituting $x = 10$ m gives

$$\dot{x}(x = 10\text{ m}) = \left[2 \times .0494 \times 9.81\,(\text{m/s}^2) \times 10\text{ m}\right]^{1/2} = 3.11\text{m/s},$$

which repeats the result of Equation (v).

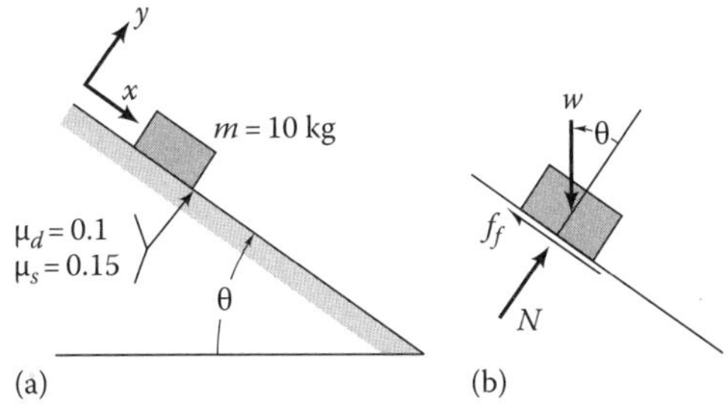

FIGURE XP3.13
Mass m on an inclined plane. (a) Coordinate system and geometry and (b) free-body diagram.

3.2.7 Quadratic Damping: Aerodynamic Drag

A projectile moving through a fluid normally has a drag force proportional to its velocity squared. Figure 3.23 illustrates a sphere falling vertically through air, acted on by its weight and a drag force that opposes motion and has a magnitude of $C_d\dot{Y}^2$. Applying $\sum \boldsymbol{f} = m\ddot{\boldsymbol{r}}_g$ gives

$$m\ddot{Y} = \sum f_Y = w - C_d\dot{Y}^2 sgn(\dot{Y}),$$

where

$$sgn(\dot{Y}) = 1 \quad \text{for } \dot{Y} > 0, \quad sgn(\dot{Y}) = -1 \quad \text{for } \dot{Y} < 0.$$

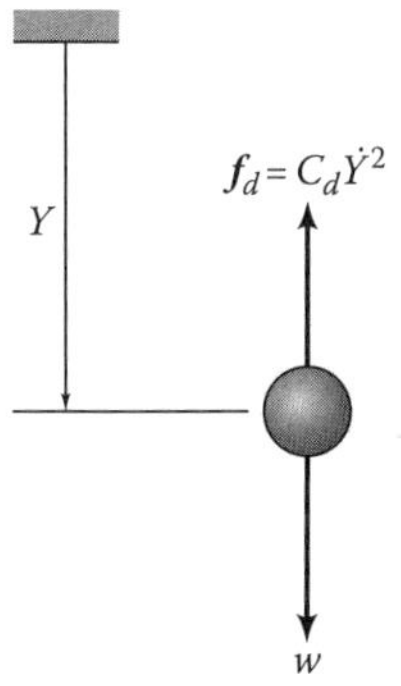

FIGURE 3.23
Free-body diagram for a mass falling with a quadratic drag force.

For a falling body, $\dot{Y} > 0$; hence, $sgn(\dot{Y}) = 1$, and the equation of motion is simply

$$m\ddot{Y} = w - C_d\dot{Y}^2. \tag{3.57}$$

3.2.7.1 Terminal Velocity Calculation

As the velocity increases, at some point the drag force will equal the body's weight. This limiting velocity is referred to as the terminal velocity, v_t, and is determined from Equation 3.57 as

$$\sum f_y = 0 = w - C_d v_t^2 \Rightarrow v_t = \sqrt{\frac{w}{C_d}}. \tag{3.58}$$

Equation 3.57 can be rearranged as

$$m\frac{dv}{dt} = C_d\left[\frac{w}{C_d} - v^2\right] = C_d[v_t^2 - v^2] \Rightarrow \int_0^v \frac{dv}{v_t^2 - v^2} = \int_0^t \frac{C_d}{m}\,d\tau.$$

with $\dot{Y} = v$, and v_t defined by Equation 3.58. This approach always works if the acceleration is a function of the velocity (only); however, these circumstances arise rarely. Integrating both sides of the equation yields

$$\frac{1}{v_t}\tanh^{-1}\left(\frac{v}{v_t}\right) = \frac{C_d t}{m} = \frac{gC_d t}{w} = \frac{gt}{v_t^2}.$$

Rearranging and taking the tanh of both sides gives

$$v = v_t \tanh\left(\frac{gt}{v_t}\right). \tag{3.59}$$

Since $\tanh x = (e^x - e^{-x})/(e^x + e^{-x})$, as $x \Rightarrow \infty$, $\tanh x \Rightarrow 1$. Hence, from Equation 3.59, $v(t \to \infty) = v_t$, giving the earlier asymptotic velocity result of Equation 3.58. A second integral follows from Equation 3.59 as

$$\frac{dY}{dt} = v \Rightarrow \int_0^Y dy = v_t \int_0^t \tanh\left(\frac{g\tau}{v_t}\right) d\tau,$$

yielding (for $Y(0) = 0$),

$$Y = \frac{v_t^2}{g}\ln\left[\cosh\left(\frac{gt}{v_t}\right)\right]. \tag{3.60}$$

Figure 3.24 illustrates the solution of Equations 3.59 and 3.60 for $m = 1$ kg, $C_d = 8.48 \times 10^{-4}$ N s^2/m^2. The results are consistent with our expectations in the approach to the terminal velocity of 107.6 m/s (200 mi/h), and the continuing linear increase of Y, at the terminal velocity, with increasing time.

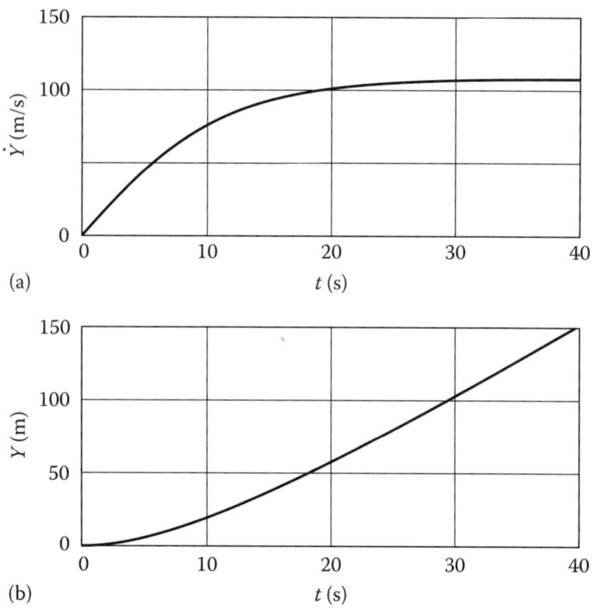

FIGURE 3.24
(a) $v(t) = \dot{Y}(t)$, (b) $Y(t)$ from Equations 3.59 and 3.60, respectively.

Example Problem 3.14

This problem is inspired by a February 2005 letter to *Mechanical Engineering* by Joseph Prusa concerning the loss of the Columbia Space Shuttle. As you may know, the leading edge of Columbia's wing was damaged by a collision with foam that came off the external tank, leading to loss of the vehicle and crew. Prusa was responding to a September 2004 article that found the impact velocity to be "surprising." He set out to demonstrate that a simple analysis could readily demonstrate that very high differential velocities could be predicted. The analysis given here retains Prusa's model, but uses an alternative analysis view point.

Prusa stated that Columbia was at an altitude of 30 km when the incident occurred yielding a reduction in the air density to about 2% of sea level; hence, $\rho = 0.0241$ kg/m^3. He proposed an aerodynamic-drag model of the form

$$f_d = -\rho C_d A\frac{v^2}{2}, \tag{i}$$

where

A is the foam's cross-sectional area
C_d is the drag coefficient

He suggested the following data $C_d = 1$, $A = 0.2$ m^2, and $m = 1$ kg, where m is the foam's mass. At the time the foam broke loose, Prusa suggested that the shuttle was traveling at 800 m/s with an acceleration of 1 g. Also, the foam was located approximately 30 m ahead of the wing surfaces.

Tasks: Determine the relative velocity of collision for the foam with the shuttle assuming (1) the foam's deceleration rate remains constant at its initial value (Prusa made this assumption), and (2) the foam decelerates according to the aerodynamic-drag model of Equation (i).

Solution

The equations of motion and initial conditions for the foam and the shuttle are

$$m\ddot{x}_f = -\rho C_d A \frac{\dot{x}_f^2}{2}, \quad \dot{x}_f(0) = 800\ \text{m/s}; \quad x_f(0) = 30\ \text{m}$$
$$\ddot{x}_s = g; \quad \dot{x}_s(0) = 800\ \text{m/s};\ x_s(0) = 0,$$

where x_f and x_s are the position of the foam and the shuttle, respectively, with respect to a local inertial coordinate system. Both bodies are initially traveling at 800 m/s. The foam is 30 m forward of the shuttle when it separates from the external tank. The initial acceleration of the foam is

$$\ddot{x}_f(0) = -0.0241 \times 1 \times 0.2 \times \frac{800^2}{2} = -1542\ \text{m/s}^2 = -157g.$$

Assuming that the foam acceleration is constant gives

$$\ddot{x}_f = -157g, \quad \dot{x}_f = 800 - 157t, \quad x_f = 30 + 800t - 157\frac{gt^2}{2}$$
$$\ddot{x}_s = g, \quad \dot{x}_s = 800 + gt, \quad x_s = 0 + 800t + \frac{gt^2}{2}$$

The foam collides with the shuttle when $x_s(\bar{t}) = x_f(\bar{t})$, netting

$$30 + 800\bar{t} - \frac{157g\bar{t}^2}{2} = 800\bar{t} + \frac{g\bar{t}^2}{2} \Rightarrow 158g\bar{t}^2 = 30$$
$$\Rightarrow \bar{t} = 0.197\ \text{s}. \qquad \text{(ii)}$$

The relative velocity at collision is

$$\begin{aligned}\delta\dot{x}(\bar{t}) &= \dot{x}_s(\bar{t}) - \dot{x}_f(\bar{t}) = (800 + g\bar{t}) - (800 - 157g\bar{t})\\ &= 802 - 496\\ &= 305\ \text{m/s} = 1098\ \text{km/h} = 682\ \text{mi/h},\end{aligned}$$

which (as Prusa noted) is clearly alarming. Note that almost all of the relative velocity arises from the foam's deceleration from its initial 800 m/s to its collision velocity of 496 m/s. The constant-acceleration assumption gives us a worst-case solution, since Equation (i) shows that the actual deceleration drops steadily.

Keeping the foam acceleration variable gives the following solution for $\dot{x}_f(t)$

$$\ddot{x}_f = \frac{d\dot{x}f}{dt} = -b\dot{x}_f^2 \Rightarrow \dot{x}_f^{-2}d\dot{x}_f = -bdt \Rightarrow -\frac{1}{\dot{x}_f} + \frac{1}{\dot{x}_{f_0}} = -bt$$
$$\Rightarrow \dot{x}_f(t) = \frac{\dot{x}_{f_0}}{1 + b\dot{x}_{f_0}t}.$$

where $b = .00241$. We can integrate the velocity $\dot{x}_f(t)$ to get $x_f(t)$ via

$$\dot{x}_f = \frac{dx_f}{dt} = \frac{\dot{x}_{f_0}}{1 + b\dot{x}_{f_0}t} \Rightarrow x_f(t) - 30 = \frac{1}{b}\int_0^t \frac{b\dot{x}_{f_0}}{1 + b\dot{x}_{f_0}t}dt$$
$$= \frac{1}{b}\Big|_0^t \ln(1 + b\dot{x}_{f_0}t). = \frac{1}{b}\ln(1 + b\dot{x}_{f_0}t)$$
$$\therefore\ x_f(t) = 30 + \frac{1}{b}\ln(1 + b\dot{x}_{f_0}t)$$

The collision time continues to be defined by $x_f(\bar{t}) = x_s(\bar{t})$, which now gives us

$$30 + \frac{1}{0.00241}\ln(1 + 0.00241 \times 800\bar{t})$$
$$= 30 + 415.\ln(1 + 1.92\bar{t}) = 800\bar{t} + 4.91\bar{t}^2.$$

Getting an analytical solution for this nonlinear equation is not feasible, but Figure XP3.14 for $x_f(t)$, $x_s(t)$ shows collision occurring at $\bar{t} \cong 0.22$ s, versus the earlier approximation of $\bar{t} \cong 0.197$ s. For $\bar{t} = 0.22$ s, the differential velocity is

$$\delta\dot{x}(\bar{t}) = \dot{x}_s(\bar{t}) - \dot{x}_f(\bar{t}) = (800 + g\bar{t}) - \frac{800}{1 + 1.92\bar{t}} = 802 - 562$$
$$= 240\ \text{m/s} = 864\ \text{km/h} = 537\ \text{mi/h}.$$

Although reduced from the worst-case assumption, this remains an alarming differential velocity for collision.

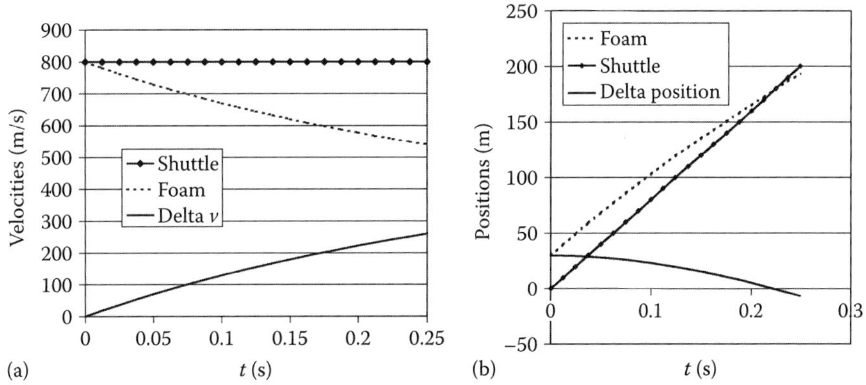

FIGURE XP3.14
(a) Absolute and relative velocities, (b) positions of the foam and the shuttle.

Even with these velocities, one could question whether the "soft" foam would cause damage on the carbon–carbon composite wing surfaces. However, the following is taken from *Popular Mechanics*, May 30, 2003: "Evidence of the devastating effect polyurethane foam can have when hitting the leading edge of a space shuttle wing became clear during tests ordered by the Columbia Accident Investigation Board (CAIB). Engineers at Southwest Research Institute in San Antonio, Texas, obtained the stunning result on May 29 as they performed the second round of such impact tests. Using a high-pressure nitrogen gun, the researchers shot a 1.7-pound chunk of foam at a replica of a shuttle wing. As the projectile hit the wing's leading edge at a 20° angle and 530 mph, one of the T-seals sitting between adjacent U-shape reinforced carbon–carbon (RCC) panels was dislodged, leaving a 22 in.-long gap, CAIB spokesman Woody Woodyard says." The Shuttle's first flight was on April 12, 1981. The question is *Why weren't these tests done years ago, when they could have done some good*?

3.2.8 Closure and Review

For a particle moving in a straight line, this section has looked at various commonly-recurring force elements that arise in deriving governing equations of motion for particles using Newton's second law of motion. The contents of this section were provided to teach the following lessons:

a. Newton's second law, $\sum f = m\ddot{r}$, results in a second-order differential equation. Starting with initial conditions, this differential equation completely defines a particle's motion.

b. Consciously selecting coordinates to locate particles is the first and most important step to developing a correct model.

c. In applying $\sum f = m\ddot{r}$, the choice of coordinate defines the positive direction that applies to *position*, *velocity*, *acceleration*, and *forces*.

d. The first integral of Newton's second law with respect to time,

$$\int_0^t f(\tau)\,d\tau = mv(t) - mv_0,$$

yields the following physical principle:

Linear impulse = Change in linear momentum

This first integral defines the velocity as a function of time and can be accomplished if the acceleration is an explicit function of time; that is, $a = a(t)$.

e. If the acceleration is an explicit function of displacement Y only; that is, $md^2Y/dt^2 = f(Y)$, a first integral can be accomplished with respect to displacement Y via the energy-integral substitution,

$$\frac{d^2Y}{dt^2} = \frac{d}{dY}\left(\frac{\dot{Y}^2}{2}\right), \tag{3.61}$$

yielding

$$md\left(\frac{\dot{Y}^2}{2}\right) = f(Y)\,dY.$$

Both sides of this equation are exact differentials, and integration gives

$$\frac{m\dot{Y}^2}{2} - \frac{m\dot{Y}_0^2}{2} = \int_{Y_0}^{Y} f(y)\,dy.$$

This result defines velocity as a function of displacement and reflects the physical principle

Change in kinetic energy = Work

The substitution provided by Equation 3.61 can always be used (productively) if the acceleration is an explicit function of Y alone (not $\dot{Y}$ or t). It is not helpful (does not lead to an integrable form) if the force acting on the body is a function of t or $\dot{Y}$.

f. If the acceleration is constant; that is, $\ddot{Y} = c$, integration with respect to time yields

$$\begin{aligned}\dot{Y} &= \dot{Y}_0 + ct;\\ Y &= Y_0 + \dot{Y}_0 t + \frac{ct^2}{2}.\end{aligned} \tag{3.62}$$

Integrating $\ddot{Y} = c$ with respect to displacement Y, using the energy-integral substitution gives

$$\dot{Y}^2 = \dot{Y}_0^2 + 2c(Y - Y_0). \tag{3.63}$$

These results only apply for constant acceleration. Equations 3.62 are used to find $Y(t)$ and $\dot{Y}(t)$; Equation 3.63 is used to find $\dot{Y}(Y)$.

g. Only a small percentage of dynamics problems yield differential-equation models that can be solved analytically. Most require numerical solutions, and numerical integration packages are widely available for the solution of initial-value problems. This text has an accompanying (free) set of examples that are solved using MATLAB®.

3.3 More Motion in a Straight Line: Degrees of Freedom and Equations of Kinematic Constraints

The concept of "degrees of freedom" is extremely important in dynamics and concerns the idea of the minimum number of coordinates that are required to define the position or orientation of a dynamic system. To get some idea concerning the meaning of the preceding (somewhat vague) sentence, consider the simple pendulum shown in Figure 3.25. This type of pendulum could be made by tying a small weight on the end of an inextensible string. Suppose that we wanted to keep track of the position of the mass m in the x, y plane. Our choices would be (1) use θ, or (2) use x and y. For simplicity, we would probably use the single coordinate θ rather than the two coordinates, x and y. This example is said to have "one degree of freedom" because only one coordinate is required to define the orientation of the pendulum. Note, that a decision to use two coordinates (x and y), where only one coordinate is required, means that x and y are not independent, since they are related by

$$x^2 + y^2 = l^2. \tag{3.64}$$

Equation 3.64 is a "kinematic-constraint" equation relating x and y. Using more than the minimum number of coordinates to define the orientation of a dynamic system will always lead to kinematic equations of constraint. Situations regularly arise in dynamics, where it makes sense to use more than the minimum number of coordinates and deal with kinematic-constraint equations. Systems for which a minimum number of coordinates have been used are said to be described by "generalized coordinates," and are used in "Lagrangian" dynamics.

Figure 3.26 illustrates a "double pendulum," made from two strings and two weights. This system has two degrees of freedom. Note that θ_1 and θ_2 completely define the locations of the two masses, m_1 and m_2. Choosing Cartesian coordinates (x_1, y_1, x_2, y_2) to define the positions of the two masses will result in the two kinematic-constraint equations:

$$\begin{aligned} x_1^2 + y_1^2 &= l_1^2, \\ (x_2 - x_1)^2 + (y_2 - y_1)^2 &= l_2^2, \end{aligned} \tag{3.65}$$

relating the four coordinates. This case reflects the following general outcome:

Number of constraint equations = (number of coordinates) − (number of degrees of freedom)

The following definition follows from these examples:

The number of degrees of freedom $\triangleq$ *the least number of coordinates which are required to completely define the position and orientation of a dynamic system.*

The ideas of "degrees of freedom" and "equations of constraint" will be used throughout the remainder of this book. Their application to the rectilinear motion of particles is considered in the next subsection.

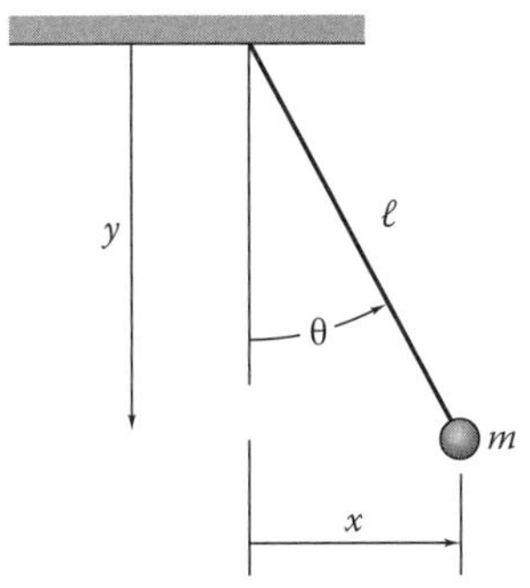

FIGURE 3.25
Simple pendulum.

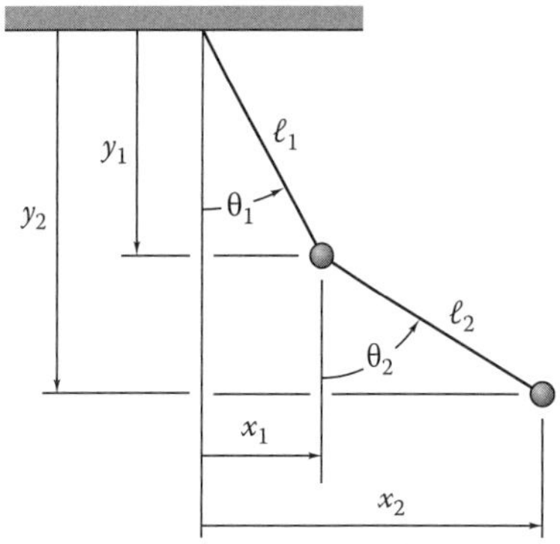

FIGURE 3.26
Double pendulum.

3.3.1 Pulleys: Equations of Motion and Equations of Constraint

Figure 3.27a illustrates two masses connected by an inextensible cord of length l_c extending over a pulley. As shown, the bottom mass is supported. Figure 3.27b illustrates the coordinate choices with x_1 locating m_1, and x_2 locating m_2 with respect to ground. Drawing Figure 3.27b with the coordinates is the first crucial step in deriving the equation of motion. Figure 3.27c provides a free-body diagram for the two bodies after they have been released. The weight w_2 is pulling m_2 down and pulling m_1 toward the pulley. Suppose we delay thinking about degrees of freedom for a short time

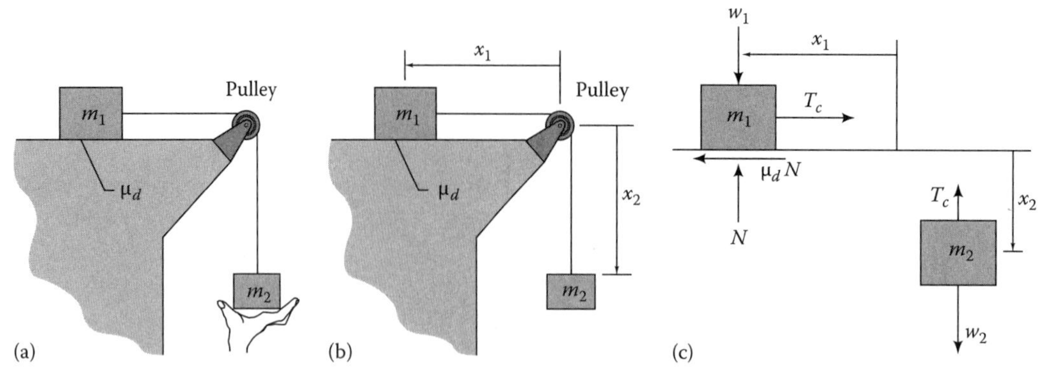

FIGURE 3.27
(a) Two masses initially supported, (b) coordinate selection, (c) free-body diagram.

and simply derive equations of motion for the two masses from the free-body diagram. Note that the same cord tension force T_c is shown acting on both bodies. This is an approximate but reasonable outcome, based on the following assumptions: (1) the inertia of the pulley is negligible, (2) the friction in the pulley bearing is negligible, and (3) the mass of the cord is negligible.

In the vertical direction, the free-body diagram for body 1 shows that m_1 is acted on by both the weight w_1 and the normal reaction force $N = w_1$. The Coulomb-friction force $\mu_d N = \mu_d w_1$ opposes the sliding motion of m_1 along the horizontal plane. For body 1, summing forces in the $+x_1$ direction yields

$$m_1\ddot{x}_1 = \sum f_{x_1} = \mu_d N - T_c, \tag{3.66}$$

where T_c is the tension in the cord. Summing forces in the $+x_2$ direction for body 2 yields

$$m_2\ddot{x}_2 = \sum f_{x_2} = w_2 - T_c. \tag{3.67}$$

We now have two equations but three unknowns ($\ddot{x}_1$, $\ddot{x}_2$, and T_c). Students regularly find themselves in this situation; specifically, Newton's second law of motion has been applied (correctly), but there still aren't enough equations to solve for all the unknowns. This outcome basically means that more coordinates have been used than the number of degrees of freedom, and one (or more) kinematic-constraint equations are needed. In the present case, the relationship between x_1 and x_2 is simply

$$x_1 + x_2 = l_c. \tag{3.68}$$

Differentiating this equation twice with respect to time gives

$$\ddot{x}_1 = -\ddot{x}_2, \tag{3.69}$$

which is the required kinematic-constraint equation. Subtracting Equation 3.67 from Equation 3.66 eliminates T_c, yielding

$$m_2\ddot{x}_2 - m_1\ddot{x}_1 = w_2 - \mu_d w_1. \tag{3.70}$$

Substitution from Equation 3.69 for $\ddot{x}_1 = -\ddot{x}_2$ yields the single EOM

$$(m_1 + m_2)\ddot{x}_2 = w_2 - \mu_d w_1. \tag{3.71}$$

Equation 3.71 states that the resultant force $(w_2 - \mu_d w_1)$ acting on the two particles, causes the resultant mass $(m_1 + m_2)$, to have the acceleration $\ddot{x}_2$. This equation defines constant acceleration for $\ddot{x}_2$ and can be integrated with respect to time or, using the energy-integral substitution $\ddot{x}_2 = d(\dot{x}_2^2/2)/dx_2$, integrated with respect to x_2 to obtain $\dot{x}_2$ as a function of x_2.

Note particularly, the acceleration coefficient $(m_1 + m_2)$ *in* Equation 3.71. An error in separately applying $\sum f = m\ddot{r}$ to the masses, in drawing the free-body diagrams, or stating the kinematic-constraint equation(s) can lead to *erroneous* negative mass entries in the final equation of motion. Negative mass entries do not exist in the physical world. If in developing an equation of motion, you arrive at a negative mass entry, be assured that your results are wrong, and start looking for your error(s).

The coordinate system of Figure 3.27b was selected to yield the simple kinematic-constraint result of Equation 3.68. Figure 3.28 shows an alternative choice using x_1' to locate m_1. The coordinate x_1' defines the displacement of m_1 to the right of a point *o fixed to ground*. Point *o* is at a fixed distance l_1 from the center of the pulley; hence, the constraint equation becomes

$$l_c = (l_1 - x_1') + x_2. \tag{3.72}$$

Differentiating this equation twice gives

$$\ddot{x}_1' = \ddot{x}_2. \tag{3.73}$$

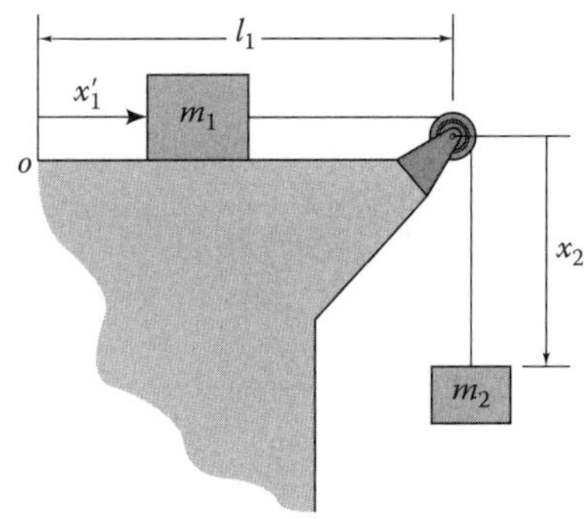

FIGURE 3.28
Alternative kinematics for the two-mass system of Figure 3.27b.

The free-body diagram of Figure 3.27b continues to apply for body 2. For m_1, summing forces in the $+x_1'$ direction yields

$$m_1\ddot{x}_1' = T_c - \mu_d w_1. \tag{3.74}$$

Adding Equations 3.67 and 3.74 eliminates T_c and gives

$$m_1\ddot{x}_1' + m_2\ddot{x}_2 = w_2 - \mu_d w_1. \tag{3.75}$$

Substituting for $\ddot{x}_1'$ from Equation 3.73 gives (again)

$$(m_1 + m_2)\ddot{x}_2 = w_2 - \mu_d w_1. \tag{3.76}$$

The message from this outcome is the following: You can select the kinematics coordinates in a somewhat arbitrary fashion, provided that you follow your selected positive coordinate directions in stating the equations of motion.

Figure 3.29a introduces a new "wrinkle" to our problem by adding a pulley connected to mass m_1. In terms of x_1 and x_2, the kinematic constraint is now

$$l_c = 2x_1 + x_2 \Rightarrow \ddot{x}_1 = -\frac{\ddot{x}_2}{2}. \tag{3.77}$$

Alternatively, using x_1', the kinematic-constraint equation is

$$l_c = 2(l_1 - x_1') + x_2 \Rightarrow \ddot{x}_1' = \frac{\ddot{x}_2}{2}. \tag{3.78}$$

The free-body diagrams for the double-pulley problem is given in Figure 3.29c. Note that our free-body-diagram "cut" has been made across both cord segments running over the pulley attached to mass m_1, exposing two contributions of T_c, the tensile force carried uniformly through the cord. Summing forces in the $+x_1'$ direction gives

$$\Sigma f_{x_1'} = 2T_c - \mu_d w_1 = m_1\ddot{x}_1'. \tag{3.79}$$

Equation 3.67, the differential equation of motion for body 2, is unchanged. Solving for T_c in Equation 3.79 and substituting into Equation 3.67 gives

$$m_2\ddot{x}_2 = w_2 - \frac{(m_1\ddot{x}_1' + \mu_d w_1)}{2}, \tag{3.80}$$

or

$$m_2\ddot{x}_2 + m_1\frac{\ddot{x}_1'}{2} = w_2 - \frac{\mu_d w_1}{2}. \tag{3.81}$$

Substituting for $\ddot{x}_1'$ from Equation 3.78 gives the final EOM for $\ddot{x}_2$:

$$\left(m_2 + \frac{m_1}{4}\right)\ddot{x}_2 = w_2 - \frac{\mu_d w_1}{2}. \tag{3.82}$$

Note that the m_1 and m_2 entries in the inertia coefficient $((m_2 + m_1)/4)$ have positive signs. Negative entries would be wrong.

Many "pulley" problems are possible, involving multiple arrangements of pulleys. The central point to recognize is that one or more kinematic-constraint equations are needed to account for the constant length of the cord(s) that connect the bodies.

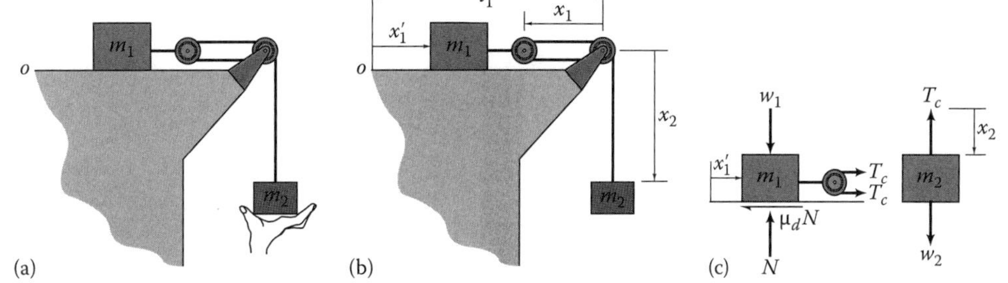

FIGURE 3.29
(a) Two masses connected by two pulleys, initially supported, (b) coordinate choices, and (c) free-body diagrams.

Example Problem 3.15

Figure XP3.15a illustrates two masses connected to each other by an inextensible cord of length l_c, running over two (light) pulleys. The left mass m_1 is connected to ground by a linear spring with spring coefficient k and a viscous damper with damping coefficient c. The right mass m_2 is supported by rollers on a plane that is inclined to the horizontal by the angle θ. The spring is initially undeflected.

Tasks:

a. Select coordinates to define the bodies' positions.
b. Neglecting friction, draw free-body diagrams for the two bodies.
c. Using $\sum \boldsymbol{f} = m\ddot{\boldsymbol{r}}$, state the equations of motion for each body.
d. State the kinematic-constraint equation between the variables, x_1, x_2 and use it to obtain one equation of motion for x_1.
e. Determine the natural frequency for the two-body system.
f. Determine the damping factor for the two-body system.

Solution

Figure XP3.15b illustrates the coordinates used to define the positions of the two bodies and satisfies Task a. Selecting the coordinates is the essential first step in deriving the equation of motion. The arrangement illustrated was chosen to simplify the development of the kinematic-constraint equation. We will also look at using x_1'.

Figure XP3.15c provides the free-body diagrams for the two bodies. Note particularly the free body for mass m_1. The spring is undeflected in the initial position. A change in length of the spring δx_1, corresponding to a positive displacement in the $+x_1$ direction, places the spring in compression. The spring is undeflected when $\delta x_1 = 0$. Note that the spring force is proportional to the change in position away from the unstretched position, i.e., $f_s = k\delta x_1$. The alternate choice $f_s = kx_1$ is wrong, since x_1 locates m_1 but does not define the change in length of the spring. Frequently, students have a sense that the spring force in Figure XP3.15c should be in tension because of the weight w_2, reasoning (correctly) that if the system is released from rest, the subsequent motion will produce tension. This is a reasonable position; however, the real tests of validity for the system differential equation are the following: (1) When the governing equation is completed, is the stiffness coefficient positive (implying that the system is stable)?, and (2) Does m_1 accelerate to the right when released from rest? We will return to these questions after the differential equation has been completed.

Note also that the damping force in the free-body diagram is directed to the right (in the negative x_1-direction) corresponding to a positive $\delta\dot{x}_1 = \dot{x}_1$. The check on the correctness of this figure occurs when we get the differential equation; specifically, is the damping coefficient in the differential equation positive (correct)?

Applying $\sum \boldsymbol{f} = m\ddot{\boldsymbol{r}}$ to the free-body diagrams for two bodies gives

$$\begin{aligned} m\delta\ddot{x}_1 &= \sum f_{x_1} = -2T_c - c\delta\dot{x}_1 - k\delta x_1, \\ m_2\delta\ddot{x}_2 &= \sum f_{x_2} = w_2 \sin\theta - T_c, \end{aligned} \quad \text{(i)}$$

which concludes Task b. We have two equations in the three unknowns, $\delta\ddot{x}_1, \delta\ddot{x}_2, T_c$. Eliminating T_c gives

$$\begin{aligned} m_1\delta\ddot{x}_1 &= -2(w_2 \sin\theta - m_2\delta\ddot{x}_2) - c\delta\dot{x}_1 - k\delta x_1 \\ &\Rightarrow m_1\delta\ddot{x}_1 - 2m_2\delta\ddot{x}_2 + c\delta\dot{x}_1 + k\delta x_1 \\ &= -2w_2 \sin\theta. \end{aligned} \quad \text{(ii)}$$

Looking back at Figure XP3.15b, the coordinates x_1, x_2 are related by

$$2x_1 + x_2 + \text{constant} = l_c, \quad \text{(iii)}$$

where the "constant" accounts for the small amount of cord that is wrapped around the left-hand pulley. Proceeding from Equation (iii) gives

$$\delta x_2 = -2\delta x_1, \quad \delta\dot{x}_2 = -2\delta\dot{x}_1, \quad \delta\ddot{x}_2 = -2\delta\ddot{x}_1. \quad \text{(iv)}$$

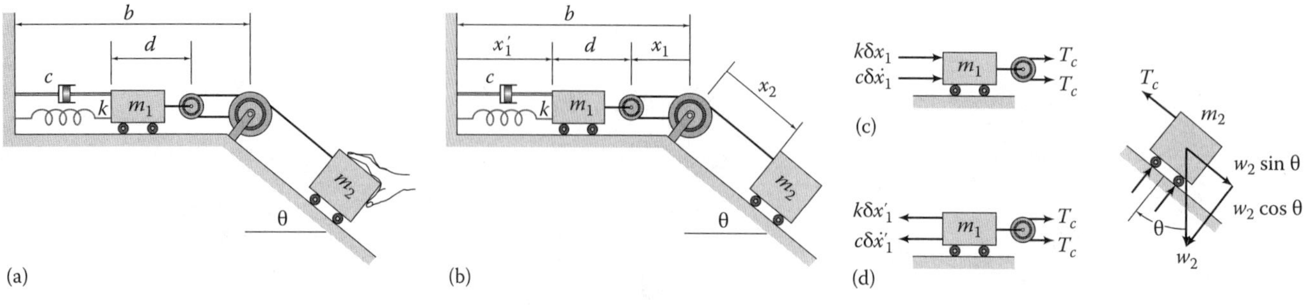

FIGURE XP3.15
(a) Spring-mass-damper system prior to release, (b) coordinate choices: (1) x_1, x_2 or (2) x_1', x_2, (c) free-body diagram for x_1, x_2 coordinates, (d) free-body diagram for x_1', x_2, coordinates.

Substituting for $\delta\ddot{x}_2$ from Equation (iv) into Equation (ii) gives

$$\begin{aligned} &m_1\delta\ddot{x}_1 - 2m_2(-2\delta\ddot{x}_1) + c\delta\dot{x}_1 + k_1\delta x_1 = -2w_2\sin\theta \\ &\therefore\ (m_1 + 4m_2)\delta\ddot{x}_1 + c\delta\dot{x}_1 + k\delta x_1 = -2w_2\sin\theta, \end{aligned} \quad \text{(v)}$$

and concludes Task d. Note that for $\delta x_1(0) = \delta\dot{x}_1(0) = 0$, the bodies' initial acceleration is negative (in the negative x_1-direction), which implies, from Figure XP3.15a, that m_i's acceleration is (as expected) to the right. Moreover, the spring and damping coefficients are positive, implying (as expected) that the system is stable. Note also that the resultant mass coefficient $(m_1 + 4m_2)$ has *positive* contributions from the individual masses. Mistakes in stating either the individual force equations or the kinematic-constraint equations tend to show up as negative contributions to the resultant mass coefficient. The physical world does not admit "negative mass" terms, and a negative entry in the final mass coefficient is invariably wrong.

From Equation (v), the natural frequency for the system is $\omega_n = [k/(m_1 + 4m_2)]^{1/2}$ which concludes Task e. The damping factor is defined as

$$2\zeta\omega_n = \frac{c}{(m_1 + 4m_2)} \Rightarrow \zeta = \frac{c}{2\sqrt{k(m_1 + 4m_2)}},$$

concluding Task f.

Figure XP3.15b shows a different coordinate choice, with x_1' now locating m_1 with respect to the fixed left-hand wall. The spring is initially undeflected. A displacement $\delta x_1'$ in the $+x_1'$ direction creates tension in the spring as reflected by the free-body diagram of Figure XP3.15d. Similarly, a positive velocity $\delta\dot{x}_1' = \dot{x}_1'$ causes tension in the damper. Hence, m_1's EOM is

$$m_1\delta\ddot{x}_1' = \sum f_{x_1'} = 2T_c - c\delta\dot{x}_1' - k\delta x_1' \quad \text{(vi)}$$

The equation of motion for m_2 is unchanged. Solving for T_c (again) from m_2's EOM and substituting into Equation (vi) gives

$$\begin{aligned} &m_1\delta\ddot{x}_1' = 2(w_2\sin\theta - m_2\delta\ddot{x}_2) - c\delta\dot{x}_1' - k\delta x_1' \\ &\Rightarrow m_1\delta\ddot{x}_1' + 2m_2\delta\ddot{x}_2 + c\delta\dot{x}_1' + k\delta x_1' \\ &= 2w_2\sin\theta. \end{aligned} \quad \text{(vii)}$$

From Figure XP3.15b, the kinematic relationship between x_1' and x_2 is

$$\begin{aligned} &2(b - x_1' - d) + x_2 + \text{constant} = l_c \Rightarrow \delta x_2 = 2\delta x_1', \\ &\delta\dot{x}_2 = 2\delta\dot{x}_1', \quad \delta\ddot{x}_2 = 2\ddot{x}_1'. \end{aligned} \quad \text{(viii)}$$

Substituting these results into Equation (vii) gives

$$\begin{aligned} &m_1\delta\ddot{x}_1' + 2m_2(2\delta\ddot{x}_1') + c\,\delta\dot{x}_1' + k\delta x_1' = 2w_2\sin\theta \\ &\therefore\ (m_1 + 4m_2)\delta\ddot{x}_1' + c\delta\dot{x}_1' + k\delta x_1' = 2w_2\sin\theta. \end{aligned} \quad \text{(ix)}$$

This equation is very similar to Equation (iv) with $\delta x_1'$ and its derivatives replacing δx_1 and its derivatives plus a change in sign for the right-hand term. However, for $\dot{x}_1'(0) = x_1'(0) = 0$, the positive right-hand term still implies an initial acceleration to the right ($+x_1'$ direction). Hence, Equations (v) and (ix) provide the same physical model.

We have worked this example with two different coordinate systems, obtaining the same physical answer. Note that the following sequential steps were used in developing the equations of motion:

a. The coordinates were selected and drawn on the system illustration.
b. Free-body diagrams were drawn, and the spring and damper forces were drawn in accordance with the assumed positive direction of the spring coordinate and velocity. For $+x_1$, the spring was compressed and is drawn as such in Figure XP3.15c; for $+x_1'$, the spring is stretched and is drawn in tension in Figure XP3.15d. For $+\dot{x}_1$, the damper is compressed and is drawn as such in Figure XP3.15c; for $+\dot{x}_1'$, the damper is drawn in tension in Figure XP3.15d.
c. The equations of motion $\sum \boldsymbol{f} = m\ddot{\boldsymbol{r}}$ are stated for the coordinates selected. In stating Equation (i), $+x_1$ is to the left, and forces are positive in this direction. In stating Equation (vi), $+x_1'$ is to the right, and forces are positive to the right.
d. The kinematic-constraint equation is written from a direct inspection of the system illustration with the coordinates drawn in. We arrived at Equation (iii) involving x_1 and Equation (viii) involving $+x_1'$ by an inspection of Figure XP3.15b.

Mistakes are easily made in developing equations of motion. However, they typically show up in fairly obvious and predictable ways. A negative sign in the stiffness, damping, or inertia coefficients is a clear sign of a mistake. Similarly, you can look at the equation of motion to see whether the predicted initial acceleration agrees with your expectations.

3.3.2 Linkage Problems: More Equations of Constraint

Figure 3.30a illustrates two blocks, having the same mass m and connected by an inextensible (and light) link of length l. Friction between the masses and their guides is assumed to be negligible. The system has one degree of freedom; that is, if we know the position of one mass, we can immediately calculate the position of the second mass. We want to derive the governing differential equation of motion for this two-mass system.

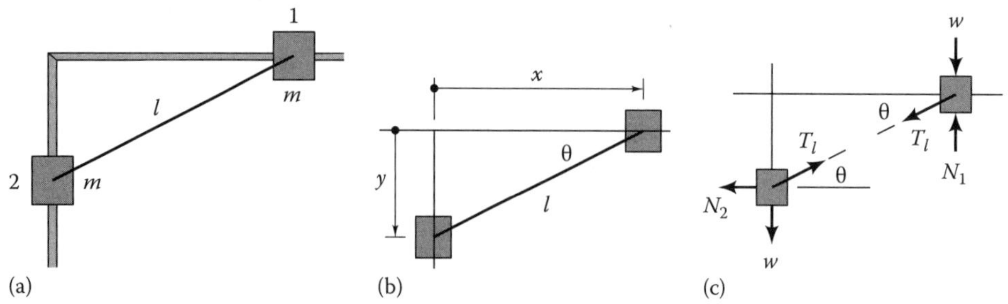

FIGURE 3.30
(a) Two masses connected by a link, (b) coordinates, (c) free-body diagram.

Figure 3.30b shows the x and y coordinates selected to locate the bodies and to be used in applying Newton's second law of motion. The free-body diagram of Figure 3.30c follows from neglecting Coulomb friction between the masses and their guides and assuming that the mass of the link is negligible. As drawn, the link is assumed to be in tension, with magnitude T_l.

Applying $\sum f = m\ddot{r}$ to the bodies yields

$$\text{Body 1:}\quad m\ddot{x} = \sum f_x = -T_1 \cos\theta. \tag{3.83}$$

$$\text{Body 2:}\quad m\ddot{y} = \sum f_y = w - T_1 \sin\theta. \tag{3.84}$$

We now have two equations for the three unknowns $\ddot{x}, \ddot{y}, T_1$. The tension T_l can be eliminated by the operation Equation 3.84 × cos θ − Equation 3.84 × sin θ, yielding

$$m\ddot{y}\cos\theta - m\ddot{x}\sin\theta = w\cos\theta. \tag{3.85}$$

The "natural" coordinate for this problem is θ; hence, instead of trying to eliminate x in favor of y or y in favor of x (from $x^2 + y^2 = l^2$), we will eliminate both x and y in terms of θ. Starting with

$$x = l\cos\theta, \quad y = l\sin\theta,$$

and differentiating with respect to time gives

$$\dot{x} = -l\sin\theta\dot{\theta}, \quad \dot{y} = l\cos\theta\dot{\theta}. \tag{3.86a}$$

The second derivative in time gives

$$\ddot{x} = -l\cos\theta\dot{\theta}^2 - l\sin\theta\ddot{\theta}, \quad \ddot{y} = -l\sin\theta\dot{\theta}^2 + l\cos\theta\ddot{\theta}. \tag{3.86b}$$

Substituting for $\ddot{x}$ and $\ddot{y}$ into Equation 3.85 yields

$$\begin{aligned} w\cos\theta &= m\cos\theta(-l\sin\theta\dot{\theta}^2 + l\cos\theta\ddot{\theta}) \\ &\quad - m\sin\theta(-l\cos\theta\dot{\theta}^2 - l\sin\theta\ddot{\theta}) \\ &= ml\ddot{\theta}, \end{aligned}$$

or

$$\ddot{\theta} = \frac{g\cos\theta}{l}. \tag{3.87}$$

Time solutions to this nonlinear governing differential equation for $\theta(t)$ and $\dot{\theta}(t)$ are easily obtained numerically and will be presented at the end of this subsection.

A partial analytical solution as $\dot{\theta}(\theta)$ can be obtained from the energy-integral substitution $\ddot{\theta} = d(\dot{\theta}^2/2)/d\theta$, yielding

$$\begin{aligned} \frac{\dot{\theta}^2}{2} &= \frac{\dot{\theta}_0^2}{2} + \int_{\theta_0}^{\theta} \frac{g\cos x}{l} dx = \frac{\dot{\theta}_0^2}{2} + \frac{g}{l}(\sin\theta - \sin\theta_0) \\ \therefore\ \dot{\theta} &= \sqrt{\dot{\theta}_0^2 + 2\frac{g}{l}(\sin\theta - \sin\theta_0)}. \end{aligned} \tag{3.88}$$

We can use this result to answer the question: If the assembly is released from rest at $\theta = 0$, how fast will the bodies be going when $\theta = \pi/2$? The answer is obtained from Equations 3.88 and 3.86a as

$$\dot{\theta}\left(\theta = \frac{\pi}{2}\right) = \sqrt{\frac{2g}{l}}$$

$$\dot{x}\left(\theta = \frac{\pi}{2}\right) = -l\sqrt{\frac{2g}{l}},$$

$$\dot{y}\left(\theta = \frac{\pi}{2}\right) = 0.$$

The force T_l can also be stated as a function of θ (only). Performing the operation Equation 3.83 × cos θ + Equation 3.84 × sin θ gives

$$T_1 = -m\ddot{x}\cos\theta + (w - m\ddot{y})\sin\theta$$

Substituting for $\ddot{x}$ and $\ddot{y}$ from Equation 3.86b then yields

$$T_1 = ml\dot{\theta}^2 + w\sin\theta,$$

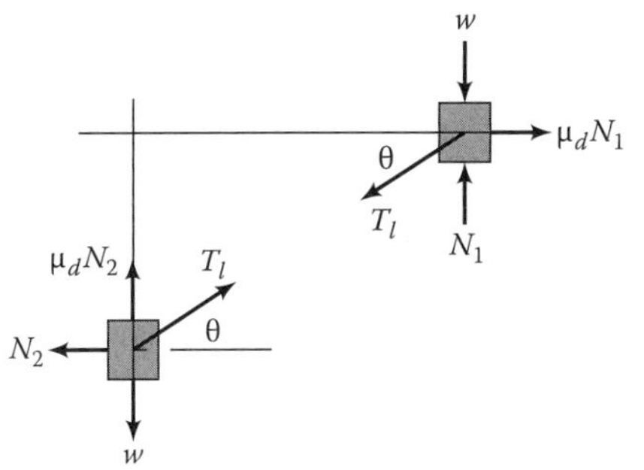

FIGURE 3.31
Free-body diagram for Figure 3.30 with Coulomb friction between the bodies and their guides.

which defines T_l as a function of $\dot\theta$ and θ. To find T_l as a function of θ (alone), we substitute for $\dot\theta^2$ from Equation 3.88 to obtain

$$T_1 = ml\left[\dot\theta_0^2 - \frac{2g}{l}(\cos\theta - \cos\theta_0)\right] + w\sin\theta.$$

For the initial conditions $\dot\theta_0 = \theta_0 = 0$, this equation defines $T_l(\theta = \pi/2) = 3w$; hence, at the bottom position, the tension in the link is three times the weight.

Figure 3.31 illustrates the addition of Coulomb-friction forces to the example of Figure 3.30. The Coulomb-friction forces shown in Figure 3.31 oppose motion that results when body 1 is falling, and body 2 is moving to the left; that is, the free-body diagram is correct for $\dot\theta > 0$. Reversing the direction of motion would reverse the Coulomb-friction-force directions. Application of Newton's second law to the two bodies yields

$$\text{Body 1:}\quad m\ddot x = \sum f_x = \mu_d N_1 sgn(\dot\theta) - T_l\cos\theta$$
$$\sum f_y = 0 = w - N_1 + T_l\sin\theta$$
$$\text{Body 2:}\quad m\ddot y = \sum f_y = w - \mu_d N_2 sgn(\dot\theta) - T_l\sin\theta$$
$$\sum f_x = 0 = T_l\cos\theta - N_2$$

Eliminating N_1 and N_2 and substituting for $\ddot x$ and $\ddot y$ from Equations 3.86b gives:

$$\begin{aligned} m\ddot x &= m(-l\cos\theta\dot\theta^2 - l\sin\theta\ddot\theta)\\ &= \mu_d sgn(\dot\theta)(w + T_l\sin\theta) - T_l\cos\theta\\ m\ddot y &= m(-l\sin\theta\dot\theta^2 + l\cos\theta\ddot\theta)\\ &= w - \mu_d T_l\cos\theta\, sgn(\dot\theta) - T_l\sin\theta. \end{aligned} \quad (3.89)$$

In matrix notation, the equations for $\ddot\theta$ and T_l become

$$\begin{bmatrix} ml\cos\theta & [\sin\theta + \mu_d\cos\theta\, sgn(\dot\theta)]\\ -ml\sin\theta & [\cos\theta - \mu_d\sin\theta\, sgn(\dot\theta)] \end{bmatrix}\begin{Bmatrix}\ddot\theta\\ T_l\end{Bmatrix} = \begin{Bmatrix} w + ml\sin\theta\dot\theta^2\\ \mu_d w\, sgn(\dot\theta) + ml\cos\theta\dot\theta^2\end{Bmatrix}. \quad (3.90)$$

Solving Equation 3.90 via Cramer's rule (see Appendix A) gives

$$\ddot\theta = \frac{[w\cos\theta(1-\mu_d^2) - \mu_d sgn(\dot\theta)(2w\sin\theta + ml\dot\theta^2)]}{ml} \quad (3.91a)$$

$$T_l = ml\dot\theta^2 + w\sin\theta + \mu_d w\cos\theta\, sgn(\dot\theta). \quad (3.91b)$$

From Equation 3.91a, the θ equation of motion is nonlinear and can be written as

$$\ddot\theta + \mu_d\dot\theta^2 sgn(\dot\theta) - \frac{g}{l}(1-\mu_d^2)\cos\theta + \frac{2g}{l}\mu_d sgn(\dot\theta)\sin\theta = 0. \quad (3.92)$$

Note that Equation 3.92 reduces to Equation 3.87 for $\mu_d = 0$. The nonlinear character eliminates the possibility of obtaining an analytical solution for $\dot\theta(t)$ and $\theta(t)$, but numerical integration in terms of initial conditions is straightforward.

Numerical integration algorithms are normally set up to solve first-order *vector* differential equations of the form

$$\dot x_i = f_i(x_i, t); \quad i = 1, n.$$

We can replace the second-order Equation 3.92 with two first-order equations via the substitution $\theta = x_1$, $\dot\theta = x_2$, obtaining

$$\dot x_2 = \frac{g}{l}\cos x_1(1-\mu_d^2) - \mu_d sgn(x_2)\left[\frac{2g}{l}\sin x_1 + x_2^2\right]$$
$$\dot x_1 = x_2.$$

This set of first-order equations is entirely equivalent to the second-order model of Equation 3.92 and is said to be the "state-variable" model. The state-variable form is commonly used in developing control models for mechanical systems. Figure 3.32 presents numerical solutions for the state-variable model with $\theta(0) = \dot\theta(0) = 0$ for $\mu_d = 0.0$, 0.1, 0.2. Obviously, from the model definition in Figure 3.30, the solution is over when $\theta = \pi/2$. (The upper mass collides with the vertical wall.) The addition of Coulomb damping means that the bodies take longer to reach this final value and have a slower velocity when they reach it.

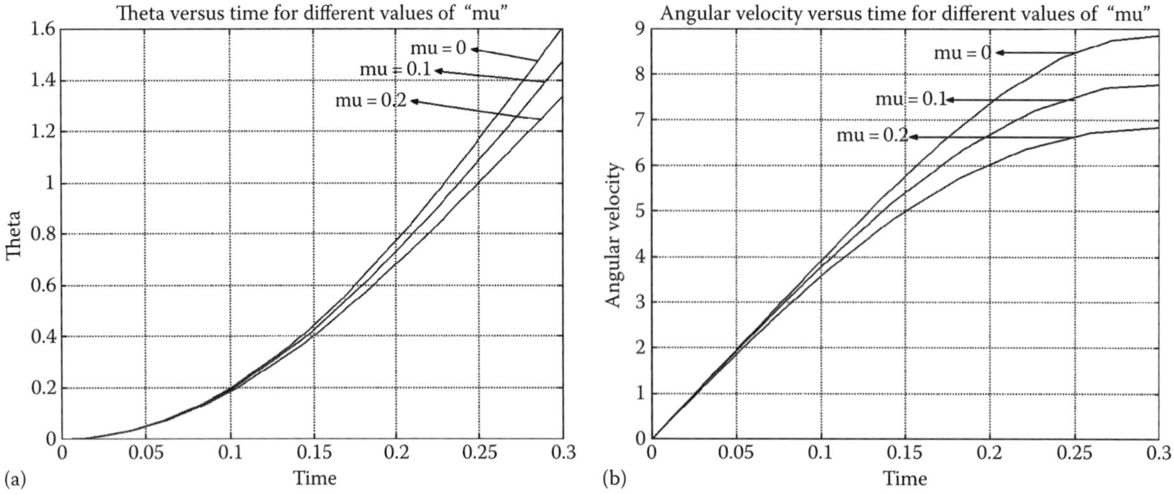

FIGURE 3.32
Numerical solutions for (a) $\dot{\theta}(t)$ and (b) $\theta(t)$ with $\mu_d = 0.0, 0.1, 0.2$.

Remarkably, the energy-integral substitution $\ddot{\theta} = d(\dot{\theta}^2/2)/d\theta$ is effective with Equation 3.92 yielding, for $\dot{\theta} > 0$,

$$\frac{dX}{d\theta} + 2\mu_d X = \frac{g}{l}(1 - \mu_d^2)\cos\theta - \frac{2\mu_d}{l}\sin\theta, \quad X = \frac{\dot{\theta}^2}{2}.$$

This is a first-order, linear, constant-coefficient differential equation, where θ is the independent variable and $X = \dot{\theta}^2/2$ is the dependent variable. From Appendix B, the homogeneous solution is $X_h = Ae^{-2\mu_d\theta}$. By inspection, the particular solution has the form

$$X_p = A\cos\theta + B\sin\theta.$$

Substituting this solution and equating coefficients for $\cos\theta$ and $\sin\theta$ gives two equations in the two unknowns, A and B. A solution is provided in Section 3.4.2 for a similar differential equation.

3.4 Motion in a Plane: Equations of Motion and Forces of Constraint

In the preceding sections of this chapter, we were concerned with the application of Newton's second law of motion for a particle or connected particles moving in straight lines. We have considered examples involving the development and integration of equations of motion, including equations of constraint. In this section, we extend these developments to two-dimensional motion in a plane.

The application of $\sum f = m\ddot{r}$ in a plane requires the kinematic results of Chapter 2. Figure 3.33a illustrates the Cartesian and polar-coordinate systems; Figure 3.33b illustrates the path-coordinate representation. By specifying the X, Y coordinate system of Figure 3.33 to be an inertial coordinate system, the application of $\sum f = m\ddot{r}$ for a particle at point P yields

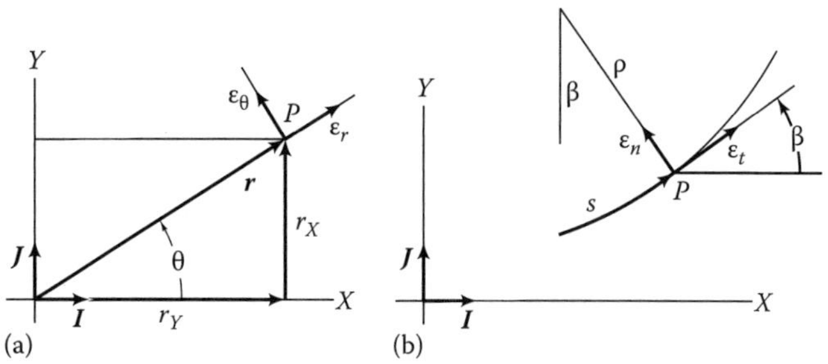

FIGURE 3.33
Planar kinematic representations. (a) Cartesian and polar, and (b) path coordinates.

Cartesian

$$m\ddot{r}_X = \sum f_X \quad m\ddot{r}_Y = \sum f_Y \tag{3.93a}$$

Polar

$$\begin{aligned} ma_r &= m(\ddot{r} - r\dot{\theta}^2) = \sum f_r \\ ma_\theta &= m(r\ddot{\theta} + 2\dot{r}\dot{\theta}) = \sum f_\theta, \end{aligned} \tag{3.93b}$$

Path

$$\begin{aligned} ma_t &= m\dot{v} = \sum f_t \\ ma_n &= m\dot{\beta}v = m\sigma\dot{\beta}^2 = \frac{mv^2}{\rho} = \sum f_n. \end{aligned} \tag{3.93c}$$

The definitions provided by Equations 3.93a through c are alternative component definitions for the *same* vector equation $\sum f = m\ddot{r}$. Equations 3.93 will be used in this section to (1) derive governing equations of motion and (2) solve for reaction forces in situations where the motion is known.

3.4.1 Cartesian-Coordinate Applications: Trajectory Motion in a Vertical Plane

3.4.1.1 Drag-Free Motion

Section 3.2 considered the vertical motion of a particle falling due to gravity without drag. Here, we are still interested in the drag-free motion of a particle but expand the problem by considering motion in the horizontal direction. Using X and Y as coordinates and applying $\sum f = m\ddot{r}$ to the particle of Figure 3.34 yields

$$\begin{aligned} m\ddot{X} &= \sum f_X = 0 \Rightarrow \ddot{X} = 0 \\ m\ddot{Y} &= \sum f_Y = -w \Rightarrow \ddot{Y} = -g. \end{aligned} \tag{3.94}$$

Motion in the two directions is seen to be uncoupled; that is, the differential equation for X does not include Y or its derivatives, and vice versa. Equations 3.94 can be integrated immediately in term of time yielding

$$\begin{aligned} \dot{X} &= \dot{X}_0, \quad X = X_0 + \dot{X}_0 t \\ \dot{Y} &= \dot{Y}_0 - gt, \quad Y = Y_0 + \dot{Y}_0 t - g\frac{t^2}{2}. \end{aligned} \tag{3.95}$$

Also, the energy-integral substitution, $\ddot{Y} = d(\dot{Y}^2/2)/dY$, converts the last of Equation 3.94 to

$$\frac{d}{dY}\left(\frac{\dot{Y}^2}{2}\right) = -g.$$

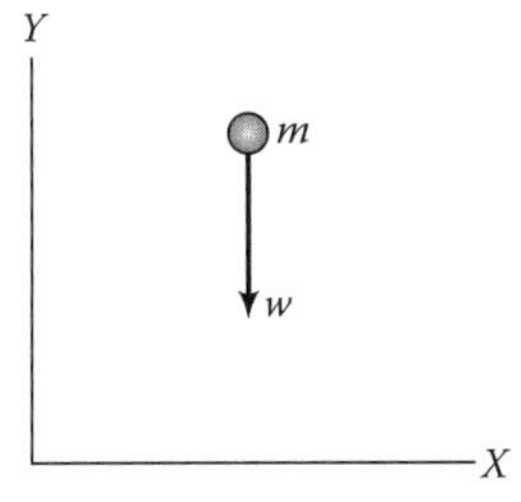

FIGURE 3.34
Free-body diagram for drag-free motion of a particle that is falling in a vertical X, Y plane.

Multiplying through by dY and integrating yields

$$\frac{\dot{Y}^2}{2} = \frac{\dot{Y}_0^2}{2} - g(Y - Y_0). \tag{3.96}$$

Application of these equations for the solution of simple trajectories are provided by the following problems.

Example Problem 3.16

As illustrated in Figure XP3.16, a projectile is fired with initial velocity v_0 m/s at an angle α to the horizontal.

Tasks: Answer the following questions.

a. How long will the particle take to hit the ground at point B?
b. How far down the range will the projectile go (What is D?).
c. What angle α gives a maximum value for D?
d. How high will the projectile go (What is H?).

SOLUTION

From Figure XP3.14, the initial conditions for position and velocity are

$$X_o = Y_o = 0; \quad \dot{X}_o = v_o \cos\alpha; \quad \dot{Y}_o = v_o \sin\alpha.$$

Designating the time at which the particle reaches point B as $\bar{t}$, a direct evaluation of Equation 3.95 gives

$$\begin{aligned} X(\bar{t}) &= D = v_0\bar{t}\cos\alpha \\ Y(\bar{t}) &= 0 = v_0\bar{t}\sin\alpha - \frac{g\bar{t}^2}{2}. \end{aligned} \tag{i}$$

Solving the second of these equations for $\bar{t}$ gives $\bar{t} = 0$, and

$$\bar{t} = \frac{2v_0 \sin\alpha}{g}\,\text{s}. \tag{ii}$$

Evaluating the first of (i) with $\bar{t}$ from (ii) yields

$$D = v_0 \cos\alpha \cdot \frac{2v_0 \sin\alpha}{g} = \frac{v_0^2}{g}\sin 2\alpha.$$

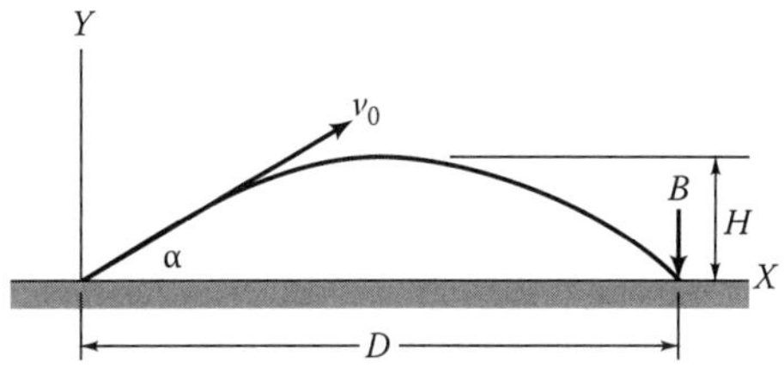

FIGURE XP3.16
Projectile with an initial velocity v_0 at an initial angle α to the horizontal.

The angle α to maximize D is obtained from

$$\frac{dD}{d\alpha} = 0 = \frac{2v_0^2}{g}\cos 2\alpha \Rightarrow \alpha_{opt} = \frac{\pi}{4}\text{ rad} = 45°.$$

Y's maximum value occurs when $\dot{Y} = 0$. Designating the time for which $Y = H$ as $\hat{t}$ and substituting into Equation 3.95 gives

$$0 = \dot{Y}_0 - g\hat{t} \Rightarrow \hat{t} = \frac{v_0 \sin\alpha}{g}.$$

Substituting $\hat{t}$ into $Y(t)$ gives

$$H = v_0 \sin\alpha\left(\frac{v_0 \sin\alpha}{g}\right) - \frac{g}{2}\left(\frac{v_0 \sin\alpha}{2}\right)^2,$$

or

$$H = \frac{v_0^2 \sin^2\alpha}{2g}.$$

The same result can be obtained directly from Equation 3.96 by setting $\dot{Y}(Y = H) = 0$, to obtain

$$0 = \frac{(v_0 \sin\alpha)^2}{2} - gH \Rightarrow H = \frac{v_0^2 \sin^2\alpha}{2g}.$$

The trajectory of the path followed in this problem can also be obtained from Equation 3.95 by eliminating t to obtain $Y(X)$. Specifically, $t = X/(v_0 \cos\alpha)$, and

$$\begin{aligned} Y &= v_0 \sin\alpha\left(\frac{X}{v_0 \cos\alpha}\right) - \frac{g}{2}\left(\frac{X}{v_0 \cos\alpha}\right)^2 \\ &= X\tan\alpha - \frac{gX^2}{2v_0^2\cos^2\alpha} \\ &= X\tan\alpha\left(1 - \frac{X}{D}\right). \end{aligned}$$

Example Problem 3.17

Figure XP3.17 shows a rifle at position A, which is pointed directly at a target that is located 250 m horizontally to the right and 50 m vertically above the rifle. The muzzle velocity of the bullet leaving the rifle is 400 m/s.

Task: Neglecting air-friction drag, how far below the target will the bullet strike?

Solution

Initially, $X(0) = Y(0) = 0$, and

$$\begin{aligned} \dot{X}(0) &= \dot{X}_0 = v_0 \cos\alpha = 400 \cos(11.3) = 392.2 \text{ m/s} \\ \dot{Y}(0) &= \dot{Y}_0 = v_0 \sin\alpha = 400 \sin(11.3) = 78.45 \text{ m/s}. \end{aligned}$$

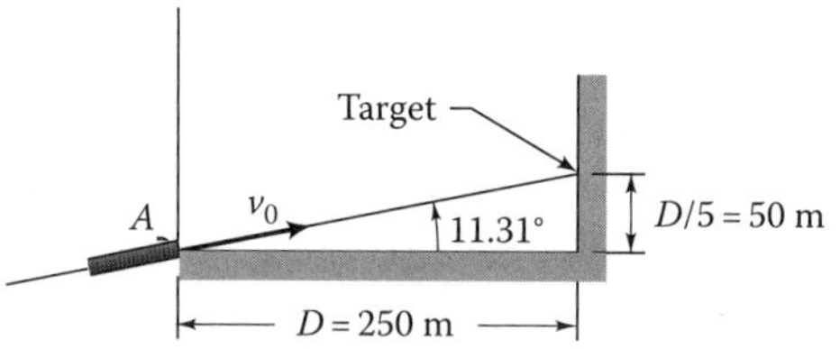

FIGURE XP3.17
Marksman aiming at a target.

From Equation 3.95, the time solution for $X(t)$ and $Y(t)$ are

$$X(t) = \dot{X}_0 t, \quad Y(t) = \dot{Y}_0 t - \frac{gt^2}{2}.$$

Denoting $\bar{t}$ as the time at which the bullet strikes the vertical wall yields from $X(\bar{t}) = D$,

$$\bar{t} = \frac{D}{\dot{X}_0} = \frac{250}{392.2} = 0.637 \text{ s}.$$

Substituting into $Y(t)$ gives

$$\begin{aligned} Y(\bar{t}) &= \dot{Y}_0\bar{t} - \frac{g\bar{t}^2}{2} = 78.45(0.637) - \frac{9.81}{1}(0.637)^2 \\ &= 48.01 \text{ m}. \end{aligned}$$

Since the target is 50 m vertical, the bullet will strike approximately 2 m "low."

Example Problem 3.18

In Example Problem 3.17 what two angles α_1 and α_2 could the marksman choose to hit the target? *Note*: This is a "boundary-value" problem versus previous initial-value problems, since the position of the body is specified at both the initial and final times.

Solution

From Figure XP3.14, the initial conditions are (again)

$$\begin{aligned} &X_0 = X(0) = 0, \quad Y_0 = Y(0) = 0 \\ &\dot{X}(0) = \dot{X}_0 = v_0 \cos\alpha, \quad \dot{Y}(0) = \dot{Y}(0) = \dot{Y}_0 = v_0 \sin\alpha_0, \end{aligned}$$

and the time solution is

$$X(t) = v_0 t \cos\alpha, \quad Y(t) = v_0 t \sin\alpha - \frac{gt^2}{2}.$$

Evaluating these solutions at the final time, $\bar{t}$ gives

$$X(\bar{t}) = 250 = 400\bar{t}\cos\alpha \text{ m} \qquad \text{(i)}$$

$$Y(\bar{t}) = 50 = 400\bar{t}\sin\alpha - \frac{9.81}{2}\bar{t}^2 \text{ m}. \qquad \text{(ii)}$$

We now have two nonlinear equations for the two unknowns α and $\bar{t}$. Equation (i) defines $\bar{t}$ as

$$\bar{t} = \frac{0.625}{\cos\alpha}.$$

Substituting this into (ii) gives the single nonlinear equation in α:

$$50 = 400(0.625)\frac{\sin\alpha}{\cos\alpha} - \frac{9.81(0.625)^2}{2\cos^2\alpha}.$$

Multiplying through by $\cos^2\alpha$ gives

$$50\cos^2\alpha = 250\sin\alpha\cos\alpha - 1.916. \qquad \text{(iii)}$$

By using the trigonometric identities

$$\cos^2\alpha = \frac{1+\cos 2\alpha}{2}, \qquad \sin\alpha\cos\alpha = \frac{\sin 2\alpha}{2},$$

Equation (iii) becomes

$$\cos 2\alpha = 5\sin 2\alpha - 1.077,$$

which can be restated as

$$\left[1 - (\sin 2\alpha)^2\right]^{1/2} = 5\sin 2\alpha - 1.077.$$

Squaring this equation, plus some algebra gives the following quadratic equation

$$26z^2 - 10.77z + 160 = 0, \quad z = \sin 2\alpha.$$

The two roots are

$$\begin{aligned} z_1 &= 0.3987 \Rightarrow \alpha = 11.75° \Rightarrow \bar{t} = 0.638\,\text{s} \\ z_2 &= 0.0155 \Rightarrow \alpha = 89.56° \Rightarrow \bar{t} = 80.6\,\text{s} \end{aligned} \qquad \text{(iv)}$$

The answer α = 11.75° in Equation (iv) is intuitively reasonable. If we return to the preceding example problem and correct the elevation after the first shot by aiming a distance 1.99 m higher, the elevation angle would be

$$\tan\alpha = \left(\frac{51.99}{250}\right) \Rightarrow \alpha = 11.75°.$$

The second answer is less intuitive and requires that the rifle be aimed very near the vertical to "lob" the bullet on the target. Air resistance might be negligible in the first case, but would be a much greater factor for the second solution since the time of flight is over 125 times greater.

Concerning the effort involved in solving the nonlinear Equation (iii) producing the answers of Equation (iv), nonlinear *algebraic* equations (like nonlinear differential equations) cannot be generally solved analytically. However, most are readily solved numerically.

3.4.1.2 Trajectory Motion with Aerodynamic Drag

Figure 3.35 provides a free-body diagram for a particle moving in a vertical plane and acted on by its weight and a drag force $f_d = C_d v^2 sgn(v)$. This is the same drag force introduced in Section 3.2.7 where the velocity was straight down and the drag force was directed straight up in an exactly opposite direction. For the present situation, the drag force f_d is a two-component vector but still must be pointed in a direction opposite to the velocity v and be proportional to v^2, where $v = (\dot{x}^2 + \dot{y}^2)^{1/2}$. For the trajectory motion of interest, v will not change sign; hence, we will set $sgn(v) = 1$. Applying $\sum f = m\ddot{r}$ yields

$$m\ddot{X} = \sum f_x = -f_d\cos\alpha = -C_d v^2\left(\frac{\dot{X}}{v}\right)$$

$$m\ddot{Y} = \sum f_Y = -w - f_d\sin\alpha = -w - C_d v^2\left(\frac{\dot{Y}}{v}\right),$$

and the governing equations become

$$\begin{aligned} m\ddot{X} + C_d\dot{X}\,(\dot{X}^2 + \dot{Y}^2)^{-1/2} &= 0 \\ m\ddot{Y} + C_d\dot{Y}\,(\dot{X}^2 + \dot{Y}^2)^{-1/2} &= -w. \end{aligned}$$

These coupled nonlinear equations cannot be solved analytically and must generally be solved numerically. The equations of motion must be put into state-variable form for numerical integration, and this development begins by defining $X = x_1, \dot{X} = \dot{x}_1 = x_2, Y = \dot{x}_3, \dot{Y} = \dot{x}_3 = x_4$, with the equations of motion

$$\begin{Bmatrix} \dot{x}_1 \\ \dot{x}_2 \\ \dot{x}_3 \\ \dot{x}_4 \end{Bmatrix} = \begin{Bmatrix} x_2 \\ -C_d x_2(x_2^2 + x_4^2)^{-1/2}/\text{m} \\ x_4 \\ -C_d x_4(x_2^2 + x_4^2)^{-1/2}/\text{m-g} \end{Bmatrix}.$$

This set of four first-order differential equations is entirely equivalent to the initial set of two second-order

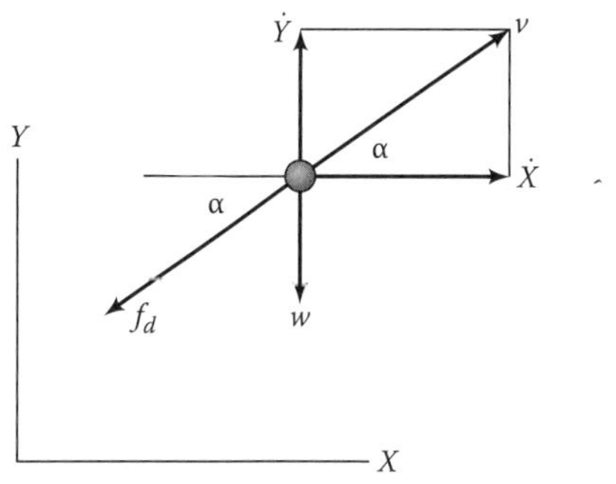

FIGURE 3.35
Free-body diagram for aerodynamic drag for motion in a vertical X, Y plane.

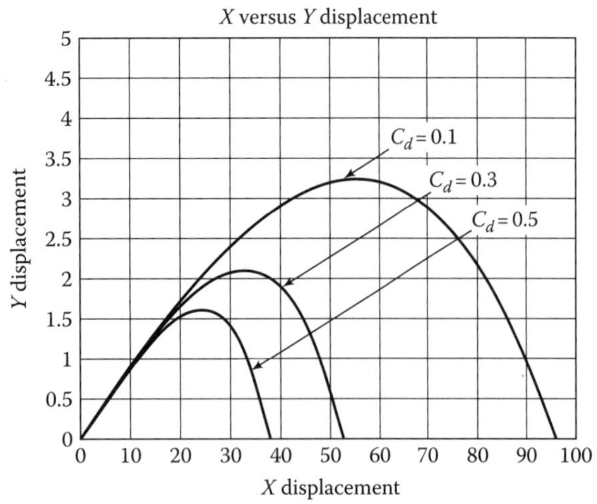

FIGURE 3.36
Calculated trajectories for $v_o = 400$ m/s, $\alpha = 45°$ with a range of values for C_d (N s^2/m^2).

differential equations. The state-variable form is not unique. We could have introduced any number of variables to obtain a state-variable set of equations.

Figure 3.36 compares computed trajectories with various levels of aerodynamic drag. As expected, increasing the drag reduces the peak height and the distance traveled. The drag coefficients used for this figure are much larger than would occur in practice.

3.4.1.3 *Trajectory Motion and Coulomb Drag*

Figure 3.37a gives a vertical-view free-body diagram for a sliding hockey puck. The weight is acting straight down on the plane of the figure and is opposed by the normal force. The drag force $\mu_d w$ acts opposite to the direction of the velocity. Summing forces in the X- and Y-directions yields the following equations of motion:

$$m\ddot{X} = \sum f_X = -\mu_d w \cos\alpha = -\mu_d w \frac{\dot{X}}{\sqrt{\dot{X}^2 + \dot{Y}^2}}$$
$$m\ddot{Y} = \sum f_Y = -\mu_d w \sin\alpha = -\mu_d w \frac{\dot{Y}}{\sqrt{\dot{X}^2 + \dot{Y}^2}}.$$

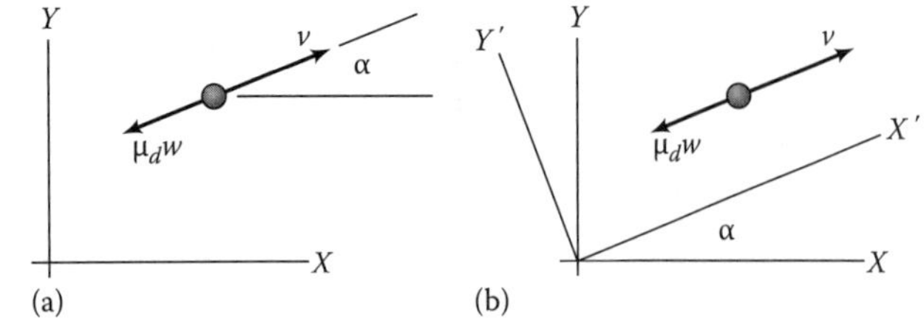

FIGURE 3.37
(a) Free-body diagram for motion of a sliding body on a horizontal plane with Coulomb friction, (b) alternative kinematics with X' directed along the linear path of motion.

These unpleasant looking equations can be "collapsed" by multiplying the first and second equations by $\cos\alpha$ and $\sin\alpha$, respectively, and adding the results to obtain

$$m(\ddot{X}\cos\alpha + \ddot{Y}\sin\alpha) = m\frac{dv}{dt} = -\mu_d w.$$

Note that $(\ddot{X}\cos\alpha + \ddot{Y}\sin\alpha)$ is the acceleration component in the direction of the velocity. This result could have been obtained directly by writing the equation of motion in the X'-direction of Figure 3.37b, since the velocity direction does not change; that is,

$$m\frac{d^2X'}{dt^2} = m\frac{dv'}{dt} = \sum f_{X'} = -\mu_d w.$$

This equation can be integrated directly just like the related one-degree-of-freedom problem of Section 3.2.6.

3.4.2 Polar-Coordinate Applications

The polar-coordinate version of Newton's second law,

$$\begin{aligned} ma_r &= m(\ddot{r} - r\dot{\theta}^2) = \sum f_r \\ ma_\theta &= m(r\ddot{\theta} + 2\dot{r}\dot{\theta}) = \sum f_\theta, \end{aligned} \tag{3.93b}$$

is particularly useful because many applications involve circular motion and are naturally described using polar-coordinate kinematics.

3.4.2.1 *Particle Sliding on the Inside of a Horizontal Cylinder without Friction*

Figure 3.38a provides an application for Equations 3.93b. A particle of mass m is sliding on the inside of a horizontal cylindrical surface, and θ has been selected as the coordinate to locate m. Neglecting friction (for now) between the particle and the surface, we want to first derive the governing equations of motion. Figure 3.38a shows a local coordinate system defined by the $\boldsymbol{\varepsilon}_r$, $\boldsymbol{\varepsilon}_\theta$ unit vectors in which the governing equations are to be developed. Figure 3.38b provides the free-body diagram. Application of the governing equations requires that (1) the forces N and w in Figure 3.38b be resolved into their $\boldsymbol{\varepsilon}_r$ and $\boldsymbol{\varepsilon}_\theta$ components to obtain Σf_r and Σf_θ on the right side of the equations and (2) that the kinematic conditions $r = r_0, \dot{r} = \ddot{r} = 0$ can be substituted into the acceleration terms. The resulting equations are

$$ma_r = m(-r_0\dot{\theta}^2) = \sum f_r = -N + w\sin\theta \tag{3.97a}$$
$$ma_\theta = mr_0\ddot{\theta} = \sum f_\theta = w\cos\theta. \tag{3.97b}$$

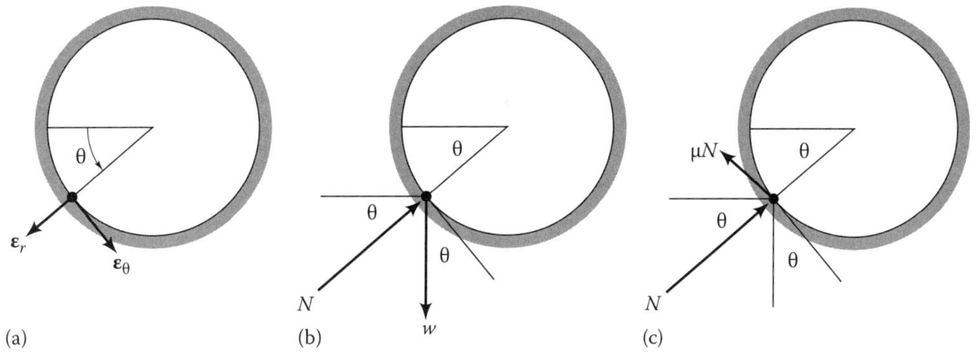

FIGURE 3.38
(a) Particle moving on the surface of a horizontal cylinder, (b) free-body diagram without Coulomb friction, (c) free-body diagram including Coulomb friction with the mass sliding in a counterclockwise direction.

The signs for the force components on the right arise because of the positive directions for $\boldsymbol{\varepsilon}_r$, $\boldsymbol{\varepsilon}_\theta$.

Having arrived at the governing equations of motion, consider the following tasks:

a. Assuming that m starts from rest at $\theta = 0$, find its angular velocity $\dot{\theta}$ as a function of θ.
b. Find the reaction force, N, as a function of θ (only).

Equation 3.97b can be readily integrated to obtain $\dot{\theta}$ as a function of θ via the energy-integral substitution, obtaining

$$\ddot{\theta} = \frac{d}{d\theta}\left(\frac{\dot{\theta}^2}{2}\right) = \frac{g}{r_0}\cos\theta. \tag{3.98}$$

Multiplying by $d\theta$ and integrating both sides yields

$$\frac{\dot{\theta}^2}{1} - \frac{\dot{\theta}_0^2}{2} = \frac{g}{r_0}\int_{\theta_0}^{\theta}\cos u\,du = \frac{g}{r_0}(\sin\theta - \sin\theta_0), \tag{3.99}$$

Applying the initial condition, $\dot{\theta} = 0$ at $\theta = 0$, gives

$$\dot{\theta}^2 = \frac{2g}{r_0}\sin\theta,$$

which completes Task a. Returning to Equation 3.97a, one can now specify the normal force N as a function of θ (alone) by eliminating $\dot{\theta}^2$ to obtain

$$N = mr_0\left(\frac{2}{r_0}g\sin\theta\right) + w\sin\theta = 3w\sin\theta.$$

Hence, in the lowest position ($\theta = \pi/2$), the normal force is three times the weight of the particle.

Note that Equation 3.99 can be rearranged as

$$\frac{m}{2}(r_0\dot{\theta})^2 - \frac{m}{2}(r_0\dot{\theta}_0)^2 = w(r_0\sin\theta - r_0\sin\theta_0),$$

which expresses the physical principle that the change in kinetic energy of the particle equals the work done by the weight as the angle increases from θ_0 to θ.

3.4.2.2 Particle Sliding on the Inside of a Horizontal Cylinder with Coulomb Friction

Figure 3.38c shows the modification of the free-body diagram that is required when the example is expanded to account for Coulomb friction on the sliding surface. Specifically, the friction force component $\mu_d N$ is now applied in the $-\boldsymbol{\varepsilon}_\theta$ direction, which is consistent with the particle sliding down the surface; that is, $\dot{\theta} > 0$. If the particle were sliding up the surface, $\mu_d N$ would be directed oppositely. The governing equations are now

$$\begin{aligned} ma_r &= -mr_0\dot{\theta}^2 = \sum f_r = -N + w\sin\theta \\ ma_\theta &= mr_0\ddot{\theta} = \sum f_\theta = w\cos\theta - \mu_d N\,sgn(\dot{\theta}), \end{aligned} \tag{3.100}$$

with the sgn function defined in Equation 3.56. Using the first equation to eliminate N gives

$$mr_0\ddot{\theta} = w\cos\theta - \mu_d\,(mr_0\dot{\theta}^2 + w\sin\theta)sgn(\dot{\theta}),$$

or

$$\ddot{\theta} + \mu_d\dot{\theta}^2 sgn(\dot{\theta}) = \frac{g}{r_0}[\cos\theta - \mu_d\sin\theta\,sgn(\dot{\theta})]. \tag{3.101}$$

We will solve this equation for $\dot{\theta}(\theta)$ to answer the following question: For the same initial condition

$\dot{\theta}(\theta = 0) = 0$ and $\mu_d = 0.1$, what is m's velocity when $\theta = \pi/2$? In addressing this question, note that $\dot{\theta} > 0$ for $0 \leq \theta \leq \pi/2$; hence, for the period of interest, $sgn(\dot{\theta}) = 1$, somewhat simplifying Equation 3.101. Even so, it remains nonlinear and requires numerical integration from initial conditions to obtain time solutions $\theta(t), \dot{\theta}(t)$. Numerical solutions are presented at the end of this subsection.

The solution for $\dot{\theta}(\theta)$ begins by introducing the energy-integral substitution $\ddot{\theta} = d(\dot{\theta}^2/2)/d\theta$, obtaining, (for $sgn(\dot{\theta}) = 1$),

$$\frac{dX}{d\theta} + 2\mu_d X = \frac{g}{r_0}(\cos\theta - \mu_d \sin\theta); \quad X = \frac{\dot{\theta}^2}{2}. \tag{3.102}$$

This constant-coefficient, first-order equation is easily solved by summing the particular and homogeneous solutions. The homogeneous equation, $dX_h/d\theta + 2\mu_d X_h = 0$, has the solution

$$X_h = Ae^{-2\mu_d \theta}.$$

The particular solution is obtained by substituting

$$X_p = B\cos\theta + C\sin\theta \Rightarrow \frac{dX_p}{d\theta} = -B\sin\theta + C\cos\theta,$$

into Equation 3.102, obtaining

$$-B\sin\theta + C\cos\theta + 2\mu_d\,(B\cos\theta + C\sin\theta) = \frac{g}{r_0}(\cos\theta - \mu_d\sin\theta).$$

Gathering $\sin\theta$ and $\cos\theta$ coefficients on both sides of the equation yields

$$\sin\theta : -B + 2\mu_d C = \frac{g\mu}{r_0}$$
$$\cos\theta : 2\mu_d B + C = \frac{g}{r_0}.$$

Solving for B and C and substituting gives

$$X_p = \frac{3g\mu_d}{r_0(1+4\mu_d^2)}\cos\theta + \frac{g(1-2\mu_d^2)}{r_0(1+4\mu_d^2)}\sin\theta.$$

The complete solution, $X = \dot{\theta}^2/2 = X_p + X_h$, is now stated

$$\frac{\dot{\theta}^2}{2} = Ae^{-2\mu_d\theta} + \frac{3g\mu_d}{r_0(1+4\mu_d^2)}\cos\theta + \frac{g(1-2\mu_d^2)}{r_0(1+4\mu_d^2)}\sin\theta.$$

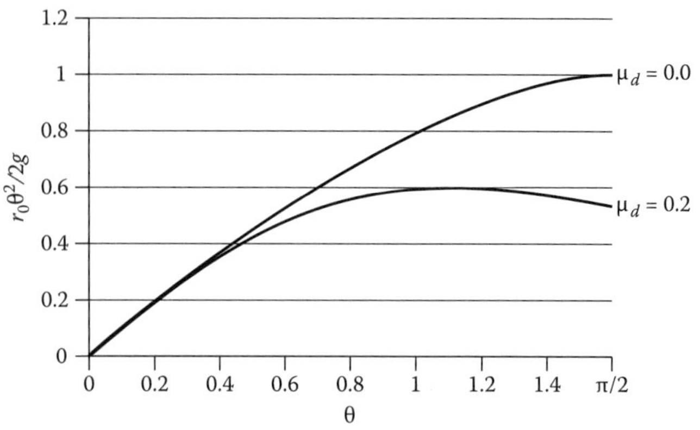

FIGURE 3.39
Solution from Equation 3.103 for $r_0\dot{\theta}^2/2g$ versus θ for $\mu_d = 0.0, 0.2$.

Imposing the initial condition $\dot{\theta}(\theta = 0) = 0$ defines A and yields the final solution

$$\frac{\dot{\theta}^2}{2} = \frac{3g\mu_d\,(\cos\theta - e^{2\mu_d\theta})}{r_0(1+4\mu_d^2)} + \frac{g(1-2\mu_d^2)\sin\theta}{r_0(1+4\mu_d^2)}. \tag{3.103}$$

The significance of this result does not "leap to the eye." However, note that setting $\mu_d = 0$ gives our original solution, $\theta^2 = 2g\sin\theta/r_0$. Figure 3.39 presents plots of this solution for the nondimensional variable $x = r_0\dot{\theta}^2/2g$ for two values of μ_d. Observe that $x(\theta)$ decreases as μ_d increases. Also, the velocity at $\theta = \pi/2$ is lower, and the peak velocity occurs sooner. These results should be consistent with your expectations.

At the lowest point defined by $\theta = \pi/2$, the angular velocity is

$$\dot{\theta}^2\left(\theta = \frac{\pi}{2}\right) = \frac{2g(1-2\mu_d^2)}{r_0(1+4\mu_d^2)} - \frac{6g\mu_d e^{-\mu_d x}}{r_0(1+4\mu_d^2)}$$

By comparison to the previous solution, Coulomb friction is seen to reduce the angular velocity. Evaluating this equation for $\mu_d = 0.1$ gives

$$\dot{\theta}^2\left(\theta = \frac{\pi}{2}\right) = 0.732\left(\frac{2g}{r_0}\right) \Rightarrow \dot{\theta}\left(\theta = \frac{\pi}{2}\right) = 0.855\sqrt{\frac{2g}{r_0}},$$

showing about a 15% reduction in angular velocity for $\mu_d = 0.1$.

The present analytical solution is incomplete in that we have not solved for $\dot{\theta}$ or θ as functions of time, but have instead solved for $\dot{\theta}^2$ as a function of θ. Equation 3.92 is very similar to the present Equation 3.101, and its state-variable model and time-transient solution is given in Figure 3.32.

3.4.2.3 The Simple Pendulum

The simple pendulum of Figure 3.25 consists of a point mass connected to a pivot point via a massless string. The designation "simple" is used to distinguish between the present pendulum consisting of a particle at the end of a string and the "compound" pendulum of Section 5.6. This idealized model could be used to model the motion of a child on a swing, with the child's mass concentrated at the end of the swing. Figure 3.40 provides a free-body diagram for the simple pendulum of Figure 3.25. Neglecting the drag forces from the air through which the pendulum is swinging and assuming that the mass of the string is negligible, we need to complete the following tasks:

a. Derive the differential equation of motion for the pendulum.
b. Derive a linearized equation of motion that would apply for "small" motion of the pendulum. From the linearized differential equation of motion, what is the natural frequency of motion?
c. Assume that the pendulum is released from rest at $\theta = 90° = \pi/2$ rad, and determine the tension in the cord as a function of θ (alone).

We start on Task a by applying Equation 3.93b to the free-body diagram to obtain

$$\begin{aligned} ma_r &= -ml\dot{\theta}^2 = \sum f_r = -T_c + w\cos\theta \\ ma_\theta &= ml\ddot{\theta} = \sum f_\theta = -w\sin\theta \end{aligned} \tag{3.104}$$

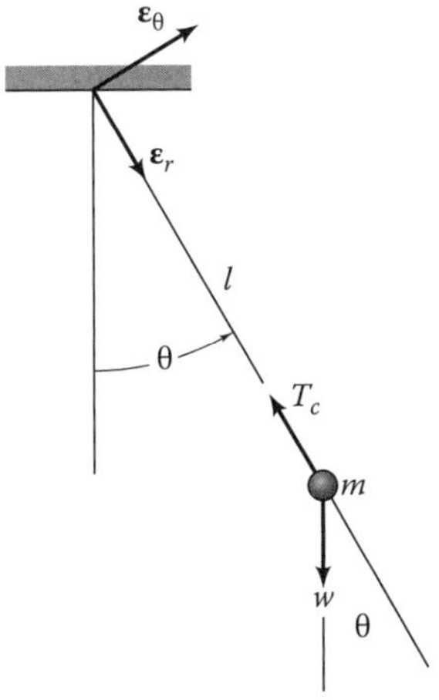

FIGURE 3.40
Free-body diagram for the simple pendulum of Figure 3.25 without drag forces.

The second of these equations defines the governing, nonlinear differential equation of motion

$$\ddot{\theta} + \frac{g}{l}\sin\theta = 0 \tag{3.105}$$

and completes Task a.

The $\sin\theta$ term provides the nonlinearity in Equation 3.105, and it can be linearized for small θ ($\theta \leq$ about $15° = 0.262$ rad), by discarding higher-order terms in the Taylor-series expansion:

$$\sin\theta = \theta - \frac{\theta^3}{6} + \frac{\theta^5}{120} + \cdots \tag{3.106}$$

Sin $\theta \cong \theta$, and the linearized equation is

$$\ddot{\theta} + \frac{g}{l}\theta = 0 \Rightarrow \ddot{\theta} + \omega_n^2\theta = 0. \tag{3.107}$$

This equation has the same form as $\ddot{x} + \omega_n^2 x = 0$ for the translational motion of a mass supported by a spring. From Equation 3.107, the natural frequency is

$$\omega_n = \sqrt{\frac{g}{l}}, \tag{3.108}$$

and completes Task b.

Returning to the approximation, $\sin\theta \cong \theta$, at $\theta = 15° = 0.2618$ rad, note the following magnitude of terms in the Taylor series:

$$\begin{aligned} \sin 15° &= \sin(0.2618 \text{ rad}) = 0.2588 \\ \theta &= 0.2618 \\ \frac{\theta^3}{6} &= 0.002990 \\ \frac{\theta^5}{120} &= 0.000010. \end{aligned} \tag{3.109}$$

Looking at these numbers, one can more readily appreciate the validity (and limitations) of the approximation.

The model provided by the second-order differential-equation model $\ddot{\theta} + (g/l)\sin\theta = 0$ must be put into state variable form for numerical solution. This task is started via the substitution $\theta = x_1, \dot{\theta} = x_2 = \dot{x}_1$, yielding

$$\dot{x}_1 = x_2, \quad \dot{x}_2 = \frac{g}{l}\sin x_1.$$

This is an entirely equivalent model to $\ddot{\theta} + (g/l)$ $\sin\theta = \theta$. Figure 3.41 provides comparisons of the linearized and nonlinear numerical solution for $l = 5$ cm with the initial conditions $\dot{\theta}(0) = 0$, $\theta(0) = \pi/2$ rad $= 90°$. Note that the period for the nonlinear solution is appreciably longer than for the linear solution.

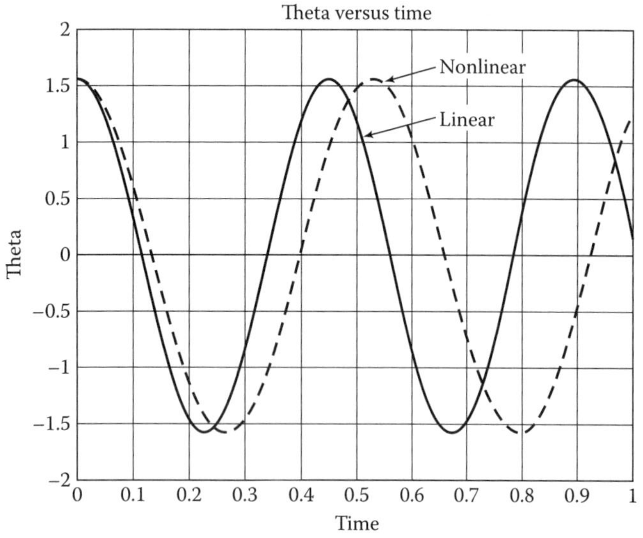

FIGURE 3.41
Linearized solution versus numerical nonlinear solutions for $\dot{\theta}(0)=0$ with $\theta(0)=\pi/2$ rad $=90°$, and $l=5$ cm.

Looking to Task c, the second of Equations 3.104 defines the tension T_c; however, T_c is defined in terms of both θ and $\dot{\theta}^2$ rather than θ alone. We can obtain $\dot{\theta}^2$ as a function of θ by introducing the energy-integral substitution into Equation 3.105 to obtain

$$\ddot{\theta}=\frac{d}{d\theta}\left(\frac{\dot{\theta}^2}{2}\right)=-\frac{g}{l}\sin\theta. \quad (3.110)$$

We have used the nonlinear equation of motion here (instead of the linearized Equation 3.107) because θ starts at 90°, which is considerably larger than the upper limit of 15° for which the linearization applies. Multiplying through by $d\theta$ and integrating both sides of Equation 3.110 gives

$$\frac{\dot{\theta}^2}{2}-\frac{\dot{\theta}_o^2}{2}=-\int_{\frac{\pi}{2}}^{\theta}\left(\frac{g}{l}\sin x\right)dx\Rightarrow\frac{\dot{\theta}^2}{2}=\frac{g}{l}\Big|_{\frac{\pi}{2}}^{\theta}\cos x=\frac{g}{l}\cos\theta.$$

Substituting for $\dot{\theta}^2$ from this result into the first of Equation 3.110 gives

$$T_c=w\cos\theta+ml\times 2\frac{g}{l}\cos\theta=3w\cos\theta.$$

Hence, when the pendulum reaches the bottom of its trajectory ($\theta=0$), the tension in the string will be three times the weight of the particle. We have completed Task c.

The simple pendulum presents the first demonstration of linearization of a nonlinear differential equation to obtain a solution for small motion about an equilibrium position. This is a vital and continuing feature of dynamics and many other fields of engineering analysis, and examples of problems that require this type of analysis will continue to "pop up" throughout the book.

3.4.2.4 The Simple Pendulum with Damping

In Section 3.4.2.3 we considered the motion of a particle of mass m supported by a spring of stiffness k that lead to a differential equation of the form $m\ddot{x}+kx=f(t)$. The pendulum example considered above yields a similar differential equation after linearization. Various damping possibilities were also introduced in Section 3.2, related to translational motion, including viscous (linear), aerodynamic (quadratic), and Coulomb damping. In this subsection, we will consider the consequence of adding viscous and aerodynamic damping to the motion of a simple pendulum. We have already considered a child on a swing as an example of a simple pendulum. Before starting with analysis of a damped pendulum, think back to your experiences in pushing a child on a swing. Basically, you have to periodically push the swing to maintain motion. Without your energy input, the motion diminishes, because of energy dissipation. In dynamics, engineers generally use some form of damping to account for energy dissipation.

Consider the following engineering-analysis tasks for the simple pendulum:

a. Neglecting any drag forces developed by the pendulum's string, assume that the pendulum's end mass is acted on by a viscous-damping force, and derive the governing differential equation of motion.
b. For small motion about the equilibrium position defined by $\theta=0$, determine the damping factor.
c. Continuing to neglect any drag forces on the string, assume that the pendulum's end mass is acted on by an aerodynamic-drag force, and derive the governing differential equation of motion.

Figure 3.42 provides the appropriate free-body diagram for Task a. The pendulum is assumed to be rotating in the $+\dot{\theta}$ direction, yielding the viscous drag force $c_d(l\dot{\theta})$ in the $-\boldsymbol{\varepsilon}_\theta$ direction. Hence, the second of Equation 3.104 becomes

$$ml\ddot{\theta}=\Sigma f_\theta=-w\sin\theta-c_d l\dot{\theta},$$

with the resultant governing differential equation of motion

$$ml\ddot{\theta}+c_d l\dot{\theta}+w\sin\theta=0.$$

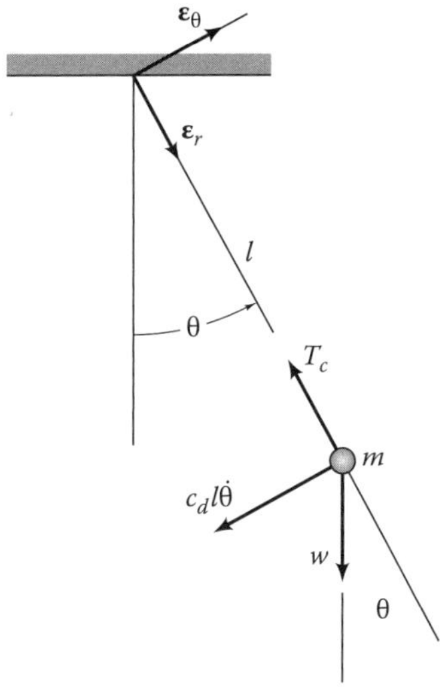

FIGURE 3.42
Free-body diagram for the simple pendulum of Figure 3.25 with viscous damping at the end mass for $\dot{\theta} > 0$.

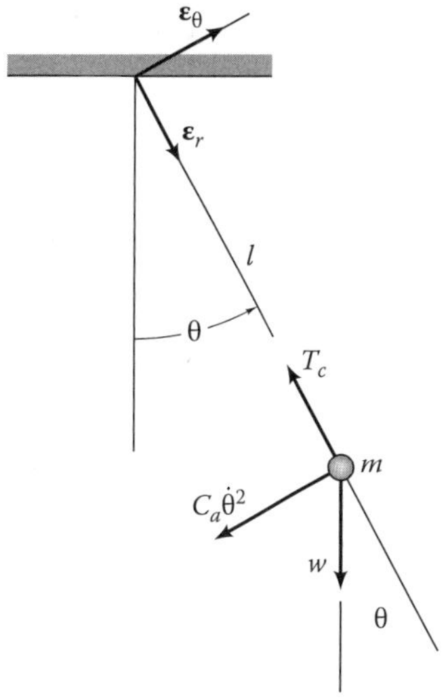

FIGURE 3.43
Free-body diagram for the simple pendulum of Figure 3.25 with aerodynamic damping at the end mass for $\dot{\theta} > 0$.

For general (large angle) motion, this nonlinear equation would need to be integrated numerically, starting from prescribed initial conditions. The nonlinear $\sin\theta$ term can be linearized, following the Taylor-series development of Equation 3.106, to yield

$$\ddot{\theta} = \frac{c_d}{m}\dot{\theta} + \frac{g}{l}\theta \Rightarrow \ddot{\theta} + 2\varsigma\omega_n\dot{\theta} + \omega_n^2\theta = 0, \qquad (3.111)$$

which completes Task a. The damping factor is defined by,

$$2\varsigma\omega_n = \frac{c_d}{m} \Rightarrow \varsigma = \frac{c_d}{2\sqrt{km}},$$

with ω_n continuing to be defined by $\omega_n = \sqrt{g/l}$. Equation 3.111 resembles $\ddot{Y} + 2\varsigma\omega_n\dot{Y} + \omega_n^2 Y = 0$ for translational motion of a mass, about equilibrium, supported by a linear spring and damper. We have completed Task b.

Figure 3.43 provides the appropriate free-body diagram for the simple pendulum with aerodynamic drag acting at the mass m. As illustrated, the pendulum is swinging in the $+\dot{\theta}$ direction. The aerodynamic-drag force is proportional to the square of the velocity magnitude and is opposed to the direction of the velocity; that is, $f_a = -C_d(l\dot{\theta})^2 sgn(\dot{\theta}) = -C_a\dot{\theta}^2 sgn(\dot{\theta})$, where $C_a = C_d l^2$. Applying the polar version of $\Sigma f = m\ddot{r}$ provided by Equation 3.7b to the pendulum gives

$$-ml\dot{\theta}^2 = \sum f_r = -T_c + w\cos\theta$$
$$ml\ddot{\theta} = \sum f_\theta = -w\sin\theta - C_a\dot{\theta}^2 sgn(\dot{\theta}),$$

with the EOM,

$$\ddot{\theta} + \frac{g}{l}\sin\theta + \frac{C_a}{ml}\dot{\theta}^2 sgn(\dot{\theta}) = 0. \qquad (3.112)$$

This equation is nonlinear due to both the $\sin\theta$ and $sgn(\dot{\theta})$ terms. Numerical integration is the only practical approach for obtaining a transient solution for $\dot{\theta}(t), \theta(t)$. Putting Equation 3.112 into state-variable form proceeds via $\theta = x_1, \dot{\theta} = x_2 = \dot{x}_1$ and gives

$$\begin{Bmatrix} \dot{x}_1 \\ \dot{x}_2 \end{Bmatrix} = \begin{Bmatrix} x_2 \\ -\sin x_1 - x_2^2 sgn x_2 \end{Bmatrix}. \qquad (3.113)$$

Figure 3.44 provides a comparison between viscous and quadratic damping for the data

$$m = 10\text{ kg}, \quad g = 9.81\text{ m/s}^2, \quad l = 5\text{ cm}, \quad C_a = 14\text{ N s}^2,$$

with the initial conditions $\theta(0) = \pi/2$, $\dot{\theta}(0) = 0$. Viscous and quadratic damping are seen to remove energy from the pendulum, progressively slowing its motion. Quadratic damping is more effective (reduces the amplitude more rapidly) at large amplitudes than viscous damping. However, at small amplitudes, viscous damping is more effective. Assuming that the viscous $C_d l\dot{\theta}$ and quadratic $C_a\dot{\theta}^2$ damping forces are equal at $\dot{\theta} = 1$ rad/s, then the quadratic damping force will be greater than the viscous force for $\dot{\theta} > 1$ and less for $\dot{\theta} < 1$. For $\dot{\theta} \ll 1$, the quadratic damping force will be exceptionally small. We will continue to encounter and analyze problems involving various forms of damping, and you should expect to encounter damping in virtually all dynamic applications in engineering.

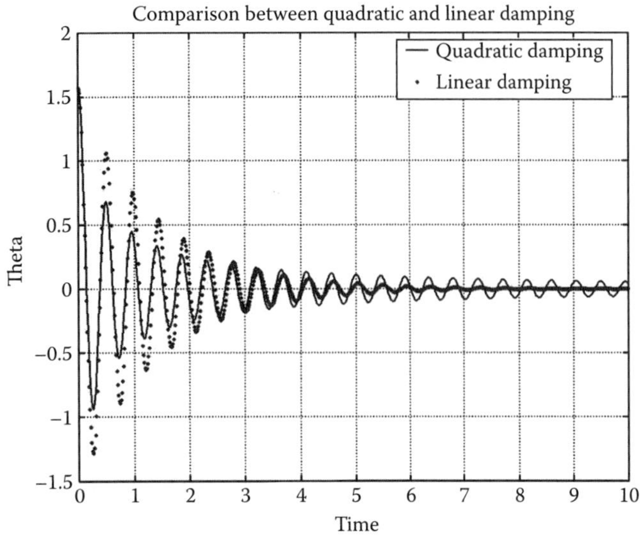

FIGURE 3.44
Pendulum response with viscous and quadratic damping.

3.4.3 Path-Coordinate Applications

The path-coordinate version of Newton's second law is

$$\begin{aligned} ma_t &= m\dot{v} = \sum f_t \\ ma_n &= m\dot{\beta}v = m\rho\dot{\beta}^2 = m\frac{v^2}{\rho} = \sum f_n, \end{aligned} \tag{3.93c}$$

where ρ is the radius of curvature defined by

$$\rho^{-1} = \frac{|y''|}{\left[1 + (y')^2\right]^{3/2}}. \tag{3.114}$$

These equations are used most frequently in solving prescribed-motion problems where a particle follows a specified path $y = y(x)$ in the x, y plane. For a particular position and specified velocity v, the second of Equations 3.93c can then be used to determine the normal force acting on the mass. Situations that call for the direct application of the path-coordinate version of Equation 3.93c *for derivation of equations of motion* are uncommon as compared to the Cartesian or polar-coordinate forms, since integration of the path-coordinate Equation 3.93c defines the velocity *but will not generally define the position of a particle.*

Figure 3.45 illustrates a bead sliding (without friction) on a wire whose shape is defined by $y = -2x + x^2$. The bead is released from rest of $x = y = 0$. Note that the distance traveled by the bead along the path is s, and the path velocity of the bead is $v = \dot{s}$. The local slope of the path defines the angle β, via $\tan\beta = -y' = -dy/dx$. Note in the free-body diagram of Figure 3.45b that $\boldsymbol{\varepsilon}_n$ is directed inward along the radius of curvature toward the instantaneous center of rotation. Proper attention to the directions of $\boldsymbol{\varepsilon}_n$, $\boldsymbol{\varepsilon}_t$ is crucial to the correct application of Newton's law with path coordinates. Applying Equation 3.93c to the free-body diagram of Figure 3.45b gives the following equations of motion

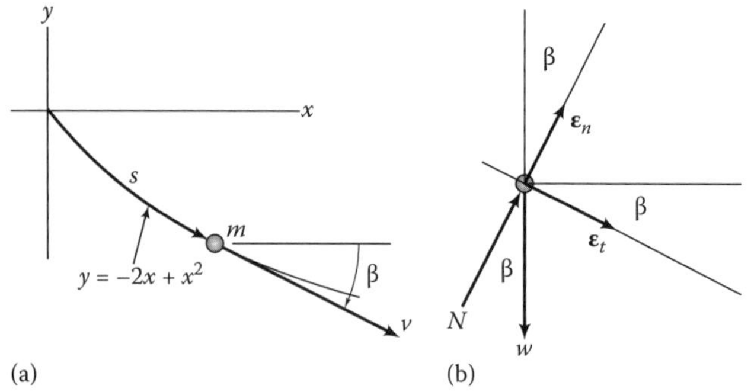

FIGURE 3.45
(a) Bead sliding down a wire (without friction), (b) free-body diagram.

$$\begin{aligned} ma_t &= m\dot{v} = \sum f_t = w\sin\beta \\ ma_n &= \frac{mv^2}{\rho} = \sum f_n = N - w\cos\beta. \end{aligned} \tag{3.115}$$

The most likely use for these equations would be the solution for the normal force N at a specified value of x, and (as will be shown below) the selection $a_n = v^2/\rho$ from the three available definitions for a_n simplifies this task. Solving for N is straightforward if v^2 can be determined, since Equation 3.114 can be used to solve for ρ at any specified position x. In fact, v^2 can be easily deduced from conservation of energy, that is,

$$T + V = T_0 + V_0 = \text{constant},$$

where T and V are, respectively, the kinetic and potential energy of the particle. (You should remember this from physics; if you need a refresher, check Section 3.6.) If $y = 0$ is used as the potential-energy datum, this equation gives

$$\frac{mv^2}{2} + mgy = 0 \Rightarrow v^2 = -2gy. \tag{3.116}$$

Applying specific numbers, at $x = 0.5$ m gives

$$y = -2\left(\frac{1}{2}\right) + \left(\frac{1}{2}\right)^2 = -0.75 \text{ m},$$

and from Equation 3.116

$$v^2 = -2(9.81 \times -0.75) = 19.71 \text{ (m/s)}^2 \Rightarrow v = 3.84 \text{ m/s}.$$

The radius of curvature is defined via Equation 3.114 from ($y' = -2 + 2x$, $y'' = 2$) to be

$$\rho^{-1}|_{x=0.5} = \frac{2}{[1+(-1)^2]^{2/3}} \Rightarrow \rho = 1.41 \text{ m}.$$

The angle β is obtained from

$$\tan\beta = \frac{dy}{dx} \Rightarrow \beta|_{x=1/2} = -\tan^{-1}(-1) = 45^\circ.$$

Substituting into the second of Equation 3.115 then gives

$$N|_{x=1/2} = \frac{w}{\sqrt{2}} + w\left[\frac{3.84^2 \text{ (m/s)}^2}{1.41 \text{ (m)} \times 9.81 \text{ (m/s}^2)}\right]$$
$$= w(0.707 + 1.06) = 1.77w.$$

Hence, the normal acceleration component has significantly increased N.

The path acceleration at $x = 0.5$ m is defined from the first of Equation 3.115 as

$$a_t = \dot{v}|_{x=1/2} = g \sin(45^\circ) = 0.707g \text{ (m/s}^2).$$

The normal acceleration at $x = 0.5$ m is

$$a_n = \frac{v^2}{\rho} = \frac{3.84^2 \text{ (m/s)}^2}{1.41 \text{ (m)}} = 10.5 \text{ m/s}^2 = 1.06 \text{ g}.$$

These results for N, a_t, and a_n conclude a "customary" application of the normal and tangential version of Newton's second law. The solution steps leading to these results start with the correct development of free-body diagrams in Figure 3.45 leading to the equations of motion. The second essential step in applying the equations is the choice $a_n = v^2/\rho$. The other two choices for a_n provided in Equation 3.93c lead nowhere. The final steps in moving toward a solution are (1) solving for v^2 from conservation of energy, and (2) evaluating ρ from its definition. Note (in contrast to the earlier application of Cartesian and polar coordinates) that we did not solve for the position of the particle. The position is given, and we are solving for the reaction force and acceleration components.

Equation 3.116 can be obtained directly by integration of Equation 3.115 instead of using conservation of energy. First, substituting

$$\dot{v} = \frac{dv}{dt} = \frac{dv}{ds}\frac{ds}{dt} = v\frac{dv}{ds} = \frac{d}{ds}\left(\frac{v^2}{2}\right) \tag{3.117}$$

gives

$$d\left(\frac{v^2}{2}\right) = g\sin\beta\, ds. \tag{3.118}$$

Since $\tan\beta = \dfrac{\sin\beta}{\cos\beta} = -y' = \dfrac{-y'}{1}$,

$$\sin\beta = \frac{-y'}{[1+(y')^2]^{1/2}}. \tag{3.119}$$

Also,

$$ds = \sqrt{dx^2 + dy^2} = dx\left[1 + \left(\frac{dy}{dx}\right)^2\right]^{1/2}$$
$$= [1+(y')^2]^{1/2}dx. \tag{3.120}$$

Substituting Equations 3.118 through 3.120 into Equation 3.118 gives

$$d\left(\frac{v^2}{2}\right) = g\frac{-dy/dx}{[1+(y')^2]^{1/2}}[1+(y')^2]^{1/2}dx = -g\,dy$$
$$\Rightarrow v^2 = -2gy. \tag{3.121}$$

Obviously, direct application of conservation of energy is a better choice than integration in this case.

Example Problem 3.19

Figure XP3.19 illustrates a particle sliding on a surface defined by the curve $y = -x^3$. The bead is released from rest at $x = y = 0$.

Task: Determine the position at which the particle leave the surface.

Solution

Compare the free-body diagrams of Figures 3.45 and XP3.19b, and note that the direction of $\boldsymbol{\varepsilon}_n$ is different, although, in both figures, $\boldsymbol{\varepsilon}_n$ is pointing toward

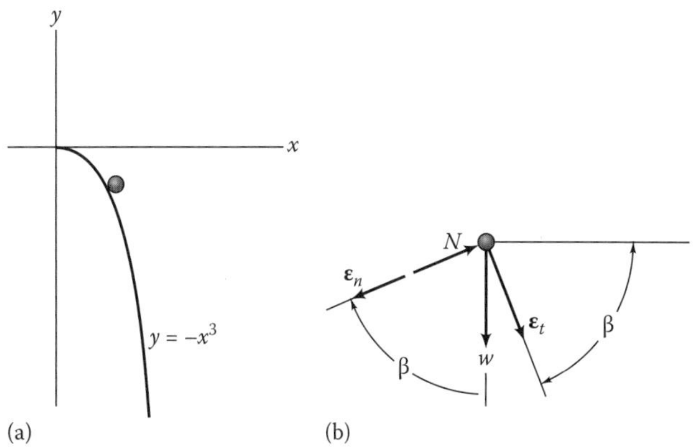

FIGURE XP3.19
(a) Particle sliding down a smooth surface, (b) free-body diagram.

the instantaneous center of rotation. Applying the path version of Newton's second law provided by Equation 3.93c to the free-body diagram of Figure XP3.17b gives

$$ma_t = m\dot{v} = \sum f_i = w \sin\beta$$
$$ma_n = \frac{mv^2}{\rho} = \sum f_n = -N + w\cos\beta. \qquad \text{(i)}$$

Note that the signs are different for N and $w\cos\beta$ in the second of Equations 3.115 and (i), because of the reversal in direction of $\boldsymbol{\varepsilon}_n$. Returning to the stated task, the particle will leave the surface when N is zero. Solving for N in the second of Equation (i) gives

$$N = w\cos\beta - \frac{mv^2}{\rho}. \qquad \text{(ii)}$$

We need to solve for $\cos\beta$, v^2, and ρ in terms of x and substitute their values back into Equation (ii) to determine N as a function of x. Setting $N = 0$ in the resultant equation will allow us to solve for $\bar{x}$, the value of x that the particle leaves the surface.

As with the preceding example, conservation of energy allows us to (easily) solve for v^2; specifically, setting the reference datum for potential energy at zero for $y = 0$ gives

$$T_0 + V_0 = T + V \Rightarrow 0 = \frac{mv^2}{2} + mgy$$
$$\therefore\ mv^2 = -2mgy = 2mgx^3. \qquad \text{(iii)}$$

The solution for $\cos\beta$ required for Equation (ii) proceeds from

$$\tan\beta = -\frac{dy}{dx} = 3x^2 \Rightarrow \cos\beta = \frac{1}{[1+(y')^2]^{1/2}} = \frac{1}{[1+9x^4]^{1/2}}. \qquad \text{(iv)}$$

The radius of curvature ρ is defined via Equation 3.118 from $y' = -3x^2$, $y'' = -6x$ to be

$$\frac{1}{\rho} = \frac{6x}{[1+9x^4]^{3/2}}. \qquad \text{(v)}$$

Substituting the results of Equations (iii) through (v) into Equation (ii) yields

$$N = \frac{w}{[1+9x^4]^{1/2}} - \frac{12mgx^4}{[1+9x^4]^{3/2}}.$$

The second term on the right arises from the normal acceleration term. The equation is physically valid for positive N. (Negative values for N would imply tension between the particle and the surface). Setting $N = 0$, and solving for $\bar{x}$ gives

$$1 + 9\bar{x}^4 = 12\bar{x}^4 \Rightarrow \bar{x} = \left(\frac{1}{3}\right)^{1/4} \text{ m.}$$

The most likely mistake to be made in working out this problem is choosing the wrong direction for $\boldsymbol{\varepsilon}_n$ Remember, $\boldsymbol{\varepsilon}_n$ is directed inward along the radius of curvature toward the instantaneous center of rotation You might want to confirm that a particle would never leave the curve $y = -x^2$.

3.4.4 Summary and Overview

The central task of dynamics is the statement of correct governing equations of motion from Newton's second law, $\sum \boldsymbol{f} = m\ddot{\boldsymbol{r}}$. The selection of an appropriate kinematics set to use in stating the governing equation is the first major step toward the successful execution of this task. The same *vector* equation is always used; however, the component equations assume different forms depending on the choice of kinematic variables.

Sections 3.4.1 through 3.4.3 have provided typical applications of $\sum \boldsymbol{f} = m\ddot{\boldsymbol{r}}$ in a plane. Applying this vector equation in a plane requires a decision on the appropriate choice of coordinate system. More specifically, the first question to be answered in trying to apply these equations is the following: *Shall I use Cartesian, polar, or path coordinates*? An inspection of the problem's geometry will normally determine which coordinate system naturally fits. Situations that require keeping track of the X and Y coordinates of a particle will normally be resolved in favor of Cartesian coordinates. Polar coordinates apply naturally to situations that involve circular motion.

Path coordinates are used less frequently and normally apply when the path of motion is prescribed in the X, Y plane by an equation of the form $y = y(x)$. Equations of motion that are formulated in Cartesian and polar coordinates can be integrated to determine a particle's position as a function of time. Problems that are formulated using path coordinates are generally used to determine reaction force or acceleration components *at a position* and are not profitably integrated to define the position as a function of time. Three choices are provided for the normal acceleration in path coordinates, specifically, $a_n = \dot{\beta}v = \rho\dot{\beta}^2 = v^2/\rho$. The first two choices are rarely employed. Most applications for the path coordinates use $a_n = v^2/\rho$ to determine the normal reaction force component and require integration (or conservation of energy) to determine v^2.

Having selected the appropriate coordinate system, the next crucial step toward applying $\sum \boldsymbol{f} = m\ddot{\boldsymbol{r}}$ involves the development of a (correct) free-body diagram. The free-body diagram should include the unit vectors of the selected coordinate system (i.e., I and J, $\boldsymbol{\varepsilon}_r$ and $\boldsymbol{\varepsilon}_\theta$, or $\boldsymbol{\varepsilon}_t$, and $\boldsymbol{\varepsilon}_n$). In proceeding from the free-body diagram to the component equations make sure that you observe signs for the acceleration and force components that are consistent with the unit vectors. Also, recall that a unit vector defines a positive sense for *all* vector-component values (force, acceleration, velocity, and displacement). In polar coordinates, $\boldsymbol{\varepsilon}_\theta$ also must have a direction consistent with the positive direction of rotation for θ.

Dynamics problems follow patterns, and learning dynamics involves working enough problems to recognize the patterns. At this point, you should have worked enough problems to begin seeing some patterns. All problems involve deriving or applying component forms of $\sum f = m\ddot{r}$, and this task requires a formal choice of the kinematic coordinate system. Assuming that you have accomplished these steps correctly, *the information requested by the problem provides a "code" that can be used to guide your solution efforts*. The table below contains this code to help you translate the information requested by a problem into a strategy for its solution.

Information Requested	Appropriate Action with the Equations of Motion
Position or velocity at a given time	Integrate the equations (or one component) with respect to time. The first integral of $\sum f = m\ddot{r}$ with respect to time yields the linear-momentum equations for a particle which are covered in Section 3.7.
Velocity at a position	Use the energy-integral substitution, $\frac{dv}{dt} = \frac{dv}{ds}\frac{ds}{dt} = \frac{d}{ds}\left(\frac{v^2}{2}\right)$ and integrate with respect to displacement s, or (if energy is conserved) use the conservation-of-energy equation directly. The integration of $\sum f = m\ddot{r}$ terms of displacement yields the work–energy equation that is covered in Section 3.6.
Reaction-force component	The equations of motion will need to be used directly to solve for a force component. If the force component is required *at a specified time*, you probably need to integrate the equations of motion with respect to time. If the force component is required *at a specified position*, you probably need to integrate the equations of motion with respect to displacement (using the energy-integral substitution) or use conservation of energy.

3.5 Particle Kinetics Examples with More Than One Degree of Freedom

To this point, all examples have had one degree of freedom. Not surprisingly, the real world of dynamics contains problems with multiple degrees of freedom. Structural-dynamics models for buildings or space structures might contain thousands or hundreds of thousands of degrees of freedom. Fortunately, most of the features of multi-mass structural-dynamics problems can be demonstrated by simple two-degree-of-freedom examples, and we are going to consider two such examples in this section, including the double pendulum of Figure 3.26 and a two-mass vibration problem. These examples lead to 2DOF linear vibration problems. The two-mass vibration problem starts as a linear vibration problem; the double-pendulum example starts as a nonlinear vibration problem that we will linearize following the procedure used with the simple pendulum of the preceding section.

Particle kinetics involves the following two steps: (1) derive the correct differential equations of motion, and (2) develop a solution to the governing differential equations. The material of Section 3.5.1 concentrates on deriving and developing the equations of motion versus solution procedures. Section 3.5.2 introduces solution procedures for linear multi-degree-of-freedom vibration problems using two-degree-of-freedom examples. This section relies more heavily on matrix algebra than the preceding material, and you may need look back into Appendix A.

3.5.1 Developing Equations of Motion for Problems Having More Than One Degree of Freedom

3.5.1.1 Developing Equations of Motion for a Two-Mass Vibration Example

Figure 3.46a illustrates a two-mass vibration example. The two masses move on a horizontal plane supported by small frictionless rollers. Mass m_1 is connected to "ground" via a linear spring with stiffness coefficient

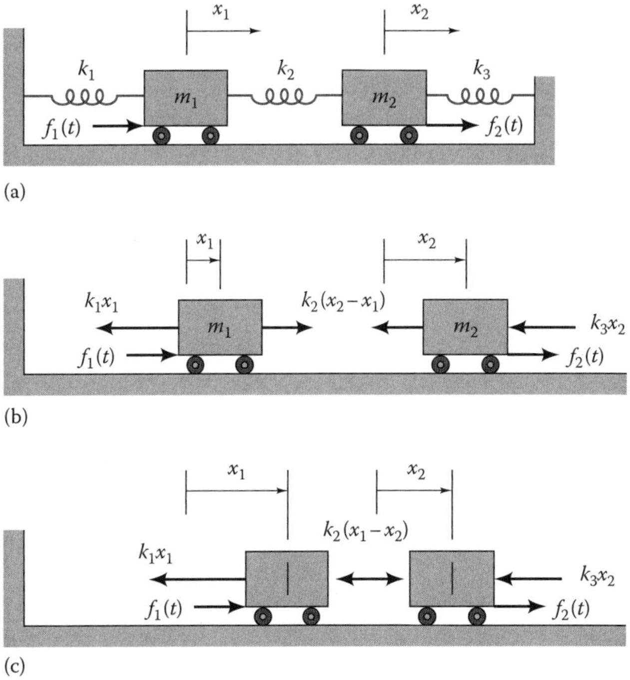

FIGURE 3.46
(a) Two-mass, linear vibration system with spring connections, (b) free-body diagrams, (c) alternative free-body diagram.

k_1; mass m_2 is connected to mass m_1 by a linear spring with the stiffness coefficient k_2 and connected to the right-hand wall with a spring having stiffness k_3. The displacements x_1 and x_2 define, respectively, the changes in position of m_1 and m_2 from equilibrium positions. Spring forces are developed on m_1 and m_2 due to changes in their positions. The masses m_1 and m_2 are also acted on by horizontal forces $f_1(t), f_2(t)$, respectively.

Our first task concerns deriving the governing equations of motion. The main new feature introduced by the 2DOF concerns the reaction force developed in the spring that connects the two masses. Moving both masses the same distance to the right or left will create no *relative* deflection between the two bodies, and accordingly, there will be no reaction force in spring 2. You should be able to think of various ways in which the two bodies could be moved to create either compression or tension in the connecting spring. Uncertainties in modeling the connecting spring, frequently leads to mistakes in the equations of motion. We will go through the development of governing equations for this system using two assumed relative positions for m_1 and m_2, showing that the same governing equations result. This development follows closely the base-excitation development for equations of motion (Section 3.2.4).

First, assume that both masses have been displaced to the right with mass m_2 moved further than m_1. Hence, $x_2 > x_1$, the connecting spring is in tension, and the connecting-spring-force magnitude is $f_{12} = k_2(x_2 - x_1)$. Figure 3.46b provides the free-body-diagram for the two masses corresponding to this assumed relative position. Applying $\sum f = m\ddot{r}$ to these free-body diagrams yields

$$\begin{aligned} \text{mass } m_1:\quad m_1\ddot{x}_1 &= \sum f_{x_1} = f_1(t) - k_1x_1 + k_2(x_2 - x_1) \\ \text{mass } m_2:\quad m_2\ddot{x}_2 &= \sum f_{x_2} = f_2(t) - k_2(x_2 - x_1) - k_3x_2, \end{aligned} \tag{3.122}$$

with the resultant differential equations

$$\begin{aligned} m_1\ddot{x}_1 + (k_1 + k_2)x_1 - k_2x_2 &= f_1(t) \\ m_2\ddot{x}_2 - k_2x_1 + (k_2 + k_3)x_2 &= f_2(t). \end{aligned} \tag{3.123}$$

Next, assume that both masses have moved to the right; however, this time, mass m_1 has moved further than m_2, the spring is in compression, and the connecting-spring-force magnitude is $f_{12} = k_2(x_1 - x_2)$. Figure 3.46c provides free-body-diagrams for the two masses corresponding to this assumed relative position. Applying $\sum f = m\ddot{r}$ to these free-body diagrams yields:

$$\begin{aligned} \text{mass } m_1:\quad m_1\ddot{x}_1 &= \sum f_{x_1} = f_1(t) - k_1x_1 - k_2(x_1 - x_2) \\ \text{mass } m_2:\quad m_2\ddot{x}_2 &= \sum f_{x_2} = f_2(t) + k_2(x_1 - x_2) - k_3x_2. \end{aligned}$$

Rearranging these differential equations gives (as it should) Equations 3.123. The lesson from this short exercise is that you can obtain the correct differential equations of motion for this type of problem by the following steps:

a. Assume a displaced position for the bodies and decide whether the connecting-spring forces are in tension or compression.
b. Draw free-body diagrams that conform to the assumed displacement positions and their resultant reaction forces (i.e., tension or compression).
c. Apply $\sum f = m\ddot{r}$ to the free-body diagrams to obtain the governing equations of motion.

The matrix statement of Equations 3.123 is

$$\begin{bmatrix} m_1 & 0 \\ 0 & m_2 \end{bmatrix} \begin{Bmatrix} \ddot{x}_1 \\ \ddot{x}_2 \end{Bmatrix} + \begin{bmatrix} (k_1 + k_2) & -k_2 \\ -k_2 & (k_2 + k_3) \end{bmatrix} \begin{Bmatrix} x_1 \\ x_2 \end{Bmatrix} = \begin{Bmatrix} f_1(t) \\ f_2(t) \end{Bmatrix}. \tag{3.124}$$

Note that the mass matrix is diagonal, and the stiffness matrix is symmetric. A theorem due to Castigliano states that the stiffness matrix for a neutrally stable structure must be symmetric. The present system is neutrally stable, since the springs can store energy but cannot dissipate or create energy. A stiffness matrix that is not symmetric and cannot be made symmetric by multiplying one or more of its rows by constants indicates a system that is or can be dynamically unstable. You have made a mistake, if in working through the example problems in this chapter, you arrive at a nonsymmetric stiffness matrix.

Note in Equation 3.124, that the off-diagonal terms in the stiffness matrix due to the connecting-spring-coefficient k_2 "couple" x_1 and x_2. For $k_2 = 0$, we have two *uncoupled* spring-mass systems. In the next section, we will be dealing with means for uncoupling multi-degree-of-freedom vibration equations that are coupled.

For a neutrally stable system, the diagonal entries must be greater than zero. Also, for a diagonal-stiffness-matrix entry such as $(k_1 + k_2)$, the individual contributions from separate springs must be positive. Specifically, $(-k_1 + k_2)$ or $(k_1 - k_2)$ *would be wrong*. Real springs make a positive increase to the reaction forces acting on a mass, and they accordingly make a positive contribution to a diagonal-stiffness-matrix entry. If, in developing equations of motion for a system composed of springs and masses, you arrive at a diagonal-stiffness-matrix entry with negative spring coefficients, you can confidently start looking for your error(s).

As with the 1DOF vibration example in Section 3.2.4, MDOF spring-mass systems can also involve base

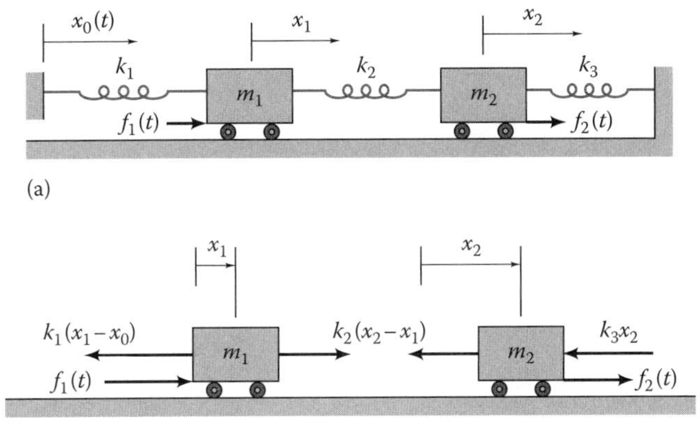

FIGURE 3.47
(a) Two-mass, linear vibration system with motion of the left-hand support, (b) free-body diagram for assumed motion $x_2 > x_1 > x_0 > 0$.

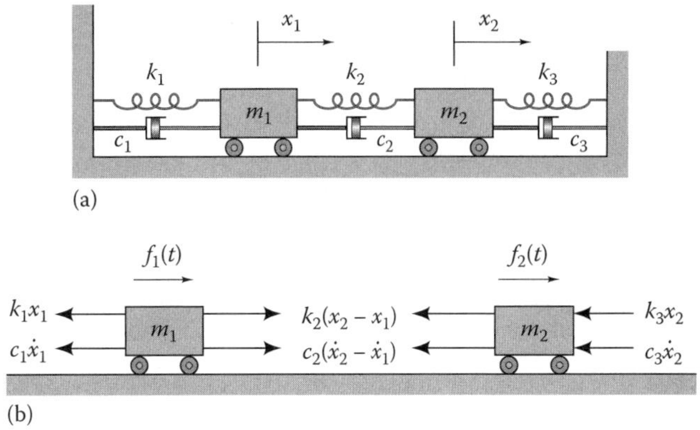

FIGURE 3.48
(a) Two-mass, linear vibration system with spring and damper connections; external forces $f_1(t)$ and $f_2(t)$ are not shown, (b) free-body diagram; $x_2 > x_1 > 0, \dot{x}_2 > \dot{x}_1 > 0$.

excitation. Figure 3.47a shows a variation to the two-body model of Figure 3.46, with the left-hand wall in motion defined by $x_0(t)$. We follow the same logical process in developing the equations of motion for this system, starting by assuming the nature of the displaced motion. For example, assuming $x_2 > x_1 > x_0 > 0$ means that the left spring and the center connecting spring are in tension, and the right spring is in compression. For $x_1 > x_0$ the left-hand spring force is $k_1(x_1 - x_0)$. For $x_2 > x_1$, the center-spring-force magnitude is $k_2(x_2 - x_1)$. Applying $\sum f = m\ddot{r}$ to the free-body diagram of Figure 3.47b gives

$$\begin{aligned} \text{mass } m_1\text{:} \quad & m_1\ddot{x}_1 = \sum f_{x_1} = f_1(t) - k_1(x_1 - x_0) + k_2(x_2 - x_1) \\ \text{mass } m_2\text{:} \quad & m_2\ddot{x}_2 = \sum f_{x_2} = f_2(t) - k_2(x_2 - x_1) - k_3x_2. \end{aligned} \tag{3.125}$$

Rearranging these equations and stating them in matrix format gives

$$\begin{bmatrix} m_1 & 0 \\ 0 & m_2 \end{bmatrix} \begin{Bmatrix} \ddot{x}_1 \\ \ddot{x}_2 \end{Bmatrix} + \begin{bmatrix} (k_1 + k_2) & -k_2 \\ -k_2 & (k_2 + k_3) \end{bmatrix} \begin{Bmatrix} x_1 \\ x_2 \end{Bmatrix} = \begin{Bmatrix} f_1(t) + k_1x_0(t) \\ f_2(t) \end{Bmatrix}. \tag{3.126}$$

The term $k_1x_0(t)$ in the right-hand vector accounts for base excitation of the left-hand wall, and is the only new term introduced in Equation 3.124. Base excitation is frequently harmonic; that is, $x_0 = A\cos\omega t$ would constitute typical motion. Earthquake excitation is a typical structural-dynamics issue in seismically active areas and is a classical base-excitation example.

Now that you are feeling modestly confident in modeling systems comprised of springs and masses, suppose that we try additional connections of the bodies with linear dampers. Figure 3.48a illustrates a modified version of the two-mass system of Figure 3.46a with the additions of connecting linear dampers. The damper connecting mass m_1 to ground has the linear damping coefficient c_1, the damper connecting masses m_1 and m_2 has the linear damping coefficient c_2, and the damper connecting m_2 to ground has damping coefficient c_3. Figure 3.48b provides free-body diagrams for the two masses and represents the following assumed-motion conditions:

a. Both m_1 and m_2 are moving to the right ($\dot{x}_1 > 0$; $\dot{x}_2 > 0$).
b. The velocity of m_2 is greater than the velocity of m_1 ($\dot{x}_2 > \dot{x}_1 > 0$).

Based on this assumed motion, tension is developed in left and center dampers, but compression is developed in the right damper. The tension in damper 1 is $c_1\dot{x}_1$, the tension in damper 2 is $c_2(\dot{x}_2 - \dot{x}_1)$, and the compression in damper 3 is $c_3\dot{x}_2$. Applying $\sum f = m\ddot{r}$ to the free-body diagrams of Figure 3.48b gives

$$\begin{aligned} \text{mass } m_1\text{:} \quad & m_1\ddot{x}_1 = \sum f_{x_1} = f_1(t) - k_1x_1 \\ & \qquad + k_2(x_2 - x_1) - c_1\dot{x}_1 + c_2(\dot{x}_2 - \dot{x}_1) \\ \text{mass } m_2\text{:} \quad & m_2\ddot{x}_2 = \sum f_{x_2} = f_2(t) - k_2(x_2 - x_1) \\ & \qquad - k_3x_2 - c_2(\dot{x}_2 - \dot{x}_1) - c_3\dot{x}_2. \end{aligned} \tag{3.127}$$

These equations can be rearranged as

$$\begin{aligned} & m_1\ddot{x}_1 + (c_1 + c_2)\dot{x}_1 - c_2\dot{x}_2 + (k_1 + k_2)x_1 - k_2x_2 = f_1(t) \\ & m_2\ddot{x}_2 - c_2\dot{x}_1 + (c_2 + c_3)\dot{x}_2 - k_2x_1 + (k_2 + k_3)x_2 = f_2(t). \end{aligned} \tag{3.128}$$

Note that the new damping terms "mirror" the old stiffness terms. Changing the stiffness coefficients to corresponding damping coefficients ($k_1 \Rightarrow c_1; k_2 \Rightarrow c_2$) and changing displacements to velocities ($x_1 \Rightarrow \dot{x}_1$; $x_2 \Rightarrow \dot{x}_2$) gives the corresponding damping terms. This observation is not a suggestion that you obtain the damping terms in this fashion, but it does provide a good check at the end of the development.

As with the spring terms, the same damping terms would be obtained in the equations if different assumptions had been made at the outset concerning the directions and amplitudes of the velocities. You can assume any sign and relative amplitudes for $\dot{x}_1$, $\dot{x}_2$. You then must decide whether the assumed motion will create tension or compression in the damper, calculate the resulting damper forces, and draw your free-body diagrams in accordance with your assumptions. Remember, the governing equations of motion are valid for any possible combination of positions, velocities, and accelerations.

The matrix statement of Equation 3.128 is

$$\begin{bmatrix} m_1 & 0 \\ 0 & m_2 \end{bmatrix}\begin{Bmatrix} \ddot{x}_1 \\ \ddot{x}_1 \end{Bmatrix} + \begin{bmatrix} (c_1+c_2) & -c_2 \\ -c_2 & (c_2+c_3) \end{bmatrix}\begin{Bmatrix} \dot{x}_1 \\ \dot{x}_2 \end{Bmatrix} + \begin{bmatrix} (k_1+k_2) & -k_2 \\ -k_2 & (k_2+k_3) \end{bmatrix}\begin{Bmatrix} x_1 \\ x_2 \end{Bmatrix} = \begin{Bmatrix} f_1(t) \\ f_2(t) \end{Bmatrix}. \tag{3.129}$$

The damping matrix is also seen to be symmetric. Nonsymmetric damping matrices do not imply system instability; however, for the problems of this chapter, all of the damping matrices should be symmetric. A *diagonal* damping-matrix term, c_{ii}, that is negative would imply a dynamic instability and should not result from your developments.

Note the positive signs for c_1 and c_2 in the diagonal damping-matrix entry (c_1+c_2). Negative signs for these terms would be wrong, since dampers act additively to dissipate energy for motion of the mass to which they are attached. Negative contributions to a diagonal damping-matrix entry from a damper are a certain indication of error(s) in developing the equations of motion. If you arrive at this type of outcome, start looking for mistakes in your development.

Forces can also be developed by base-motion excitation through a damper. Figure 3.49a shows the right-hand-side support in motion as defined by $x_3(t)$. Developing the equations of motion for the model of Figure 3.49a starts by assuming the nature of the motion. Assuming $x_3 > x_2 > x_1 > 0$ for displacements causes all of the springs to be in tension. Similarly, assuming $\dot{x}_3 > \dot{x}_2 > \dot{x}_1 > 0$ for velocities causes all of the dampers to be in tension, and the free-body diagram of Figure 3.49b results. Applying $\sum f = m\ddot{r}$ to the free-body diagram nets

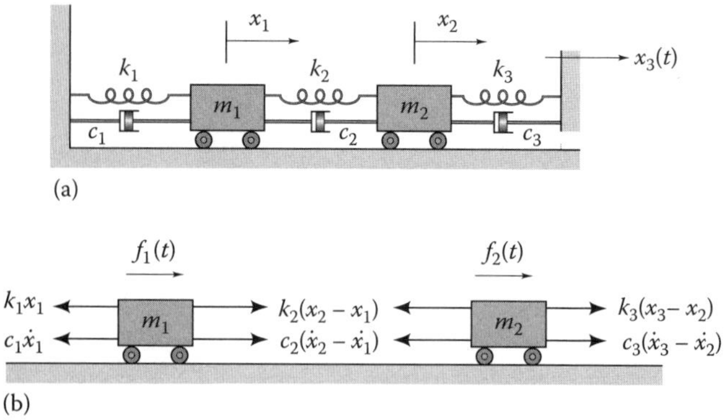

FIGURE 3.49
(a) Coupled two-mass system with motion of the right-hand support defined by $x_3(t)$; external forces $f_1(t)$ and $f_2(t)$ are not shown, (b) free-body diagram corresponding to assumed motion defined by $x_3 > x_2 > x_1 > 0$ and $\dot{x}_3 > \dot{x}_2 > \dot{x}_1 > 0$.

$$\begin{aligned} \text{mass } m_1:\ m_1\ddot{x}_1 &= \sum f_{x_1} = f_1(t) - k_1x_1 \\ &\quad + k_2(x_2 - x_1) - c_1\dot{x}_1 + c_2(\dot{x}_2 - \dot{x}_1) \\ \text{mass } m_2:\ m_2\ddot{x}_2 &= \sum f_{x_2} = f_2(t) - k_2(x_2 - x_1) \\ &\quad + k_3(x_3 - x_2) - c_2(\dot{x}_2 - \dot{x}_1) + c_3(\dot{x}_3 - \dot{x}_2). \end{aligned} \tag{3.130}$$

Restating these equations in matrix format gives

$$\begin{bmatrix} m_1 & 0 \\ 0 & m_2 \end{bmatrix}\begin{Bmatrix} \ddot{x}_1 \\ \ddot{x}_2 \end{Bmatrix} + \begin{bmatrix} (c_1+c_2) & -c_2 \\ -c_2 & (c_2+c_3) \end{bmatrix}\begin{Bmatrix} \dot{x}_1 \\ \dot{x}_2 \end{Bmatrix} + \begin{bmatrix} (k_1+k_2) & -k_2 \\ -k_2 & (k_2+k_3) \end{bmatrix}\begin{Bmatrix} x_1 \\ x_2 \end{Bmatrix} = \begin{Bmatrix} f_1(t) \\ f_2(t) + k_3x_3(t) + c_3\dot{x}_3(t) \end{Bmatrix}.$$

The new term, $[k_3x_3(t) + c_3\dot{x}_3(t)]$, provides base excitation from motion of the right-hand support. You might want to compare this result to the 1DOF model for base excitation provided by Equation 3.32.

We could proceed with this example to include Coulomb and aerodynamic (quadratic damping). However, the damping available in multi-dimensional vibration problems is, as a rule, both poorly understood and poorly predicted, arising from material hysteresis, rubbing contacts, aerodynamic damping, etc. Circumstances in which an engineer has a reasonably good qualitative and quantitative grasp of damping are generally restricted to damping that has been deliberately introduced to a system, such as shock-absorbers in automobiles.

3.5.1.2 Developing Equations of Motion for a Double Pendulum

Figure 3.50 provides free-body diagrams for the two particles in the double pendulum of Figure 3.26. You

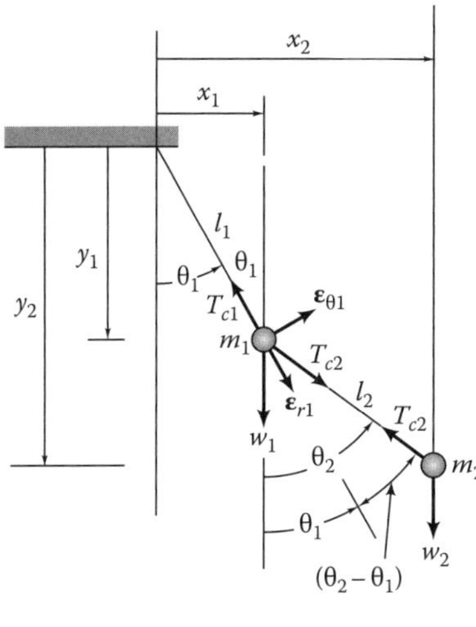

FIGURE 3.50
Free-body diagram for the double pendulum of Figure 3.26.

might want to return to Figure 3.40 for a quick review of the comparable free-body diagram for the simple pendulum. We can use polar coordinates to state the following equations of motion for mass m_1:

$$\begin{aligned} m_1a_r &= m_1(\ddot{r} - r\dot{\theta}_1^2) = -m_1l_1\dot{\theta}_1^2 = \sum f_{r1} \\ &= w_1\cos\theta_1 + T_{c2}\cos(\theta_2-\theta_1) - T_{c1} \\ m_1a_\theta &= m_1(r\ddot{\theta}_1 + 2\dot{r}\dot{\theta}_1) = m_1l_1\ddot{\theta}_1 = \sum f_{\theta 1} \\ &= T_{c2}\sin(\theta_2-\theta_1) - w_1\sin\theta_1. \end{aligned} \quad (3.131)$$

By comparison to Equation 3.104, for the simple pendulum, new terms are introduced in these equations by the tension T_{c2} acting along the string connecting the two masses. The second of Equations 3.131 provides one equation in the two unknowns $(T_{c2}, \ddot{\theta}_1)$.

The equations of motion for mass m_2 are not as easily stated in terms of polar coordinates; hence, we will use the Cartesian coordinates of Figure 3.50. Applying $\sum f = m\ddot{r}$ for mass m_2 along the x_2 and y_2 (vertically downward in Figure 3.50) axes gives

$$\begin{aligned} m_2\ddot{x}_2 &= \sum f_{x2} = -T_{c2}\sin\theta_2 \\ m_2\ddot{y}_2 &= \sum f_{y2} = w_2 - T_{c2}\cos\theta_2. \end{aligned} \quad (3.132)$$

We now have two additional unknowns $(\ddot{y}_2, \ddot{x}_2)$. These components of the acceleration of mass m_2 are developed using the following kinematics from Figure 3.50:

$$\begin{aligned} x_2 &= l_1\sin\theta_1 + l_2\sin\theta_2, \\ y_2 &= l_1\cos\theta_1 + l_2\cos\theta_2. \end{aligned} \quad (3.133a)$$

Differentiating these equations once with respect to time gives the velocity components:

$$\begin{aligned} \dot{x}_2 &= l_1\cos\theta_1\dot{\theta}_1 + l_2\cos\theta_2\dot{\theta}_2 \\ \dot{y}_2 &= -l_1\sin\theta_1\dot{\theta}_1 - l_2\sin\theta_2\dot{\theta}_2. \end{aligned} \quad (3.133b)$$

Differentiating again with respect to time gives the acceleration components:

$$\begin{aligned} \ddot{x}_2 &= l_1\cos\theta_1\ddot{\theta}_1 - l_1\sin\theta_1\dot{\theta}_1^2 + l_2\cos\theta_2\ddot{\theta}_2 - l_2\sin\theta_1\dot{\theta}_2^2 \\ \ddot{y}_2 &= -l_1\sin\theta_1\ddot{\theta}_1 - l_1\cos\theta_1\dot{\theta}_1^2 - l_2\sin\theta_2\ddot{\theta}_2 - l_2\cos\theta_1\dot{\theta}_2^2. \end{aligned} \quad (3.133c)$$

Substituting these results into Equation 3.132 gives

$$\begin{aligned} & m_2(l_1\cos\theta_1\ddot{\theta}_1 - l_1\sin\theta_1\dot{\theta}_1^2 + l_2\cos\theta_2\ddot{\theta}_2 - l_2\sin\theta_1\dot{\theta}_2^2) \\ &\quad = -T_{c2}\sin\theta_2 \\ & m_2(-l_1\sin\theta_1\ddot{\theta}_1 - l_1\cos\theta_1\dot{\theta}_1^2 - l_2\sin\theta_2\ddot{\theta}_2 - l_2\cos\theta_1\dot{\theta}_2^2) \\ &\quad = w_2 - T_{c2}\cos\theta_2. \end{aligned} \quad (3.134)$$

The second of Equations 3.131 and Equations 3.134 now provide three equations for the unknowns $(T_{c2}, \ddot{\theta}_1, \ddot{\theta}_2)$. The tension T_{c2} can be eliminated from Equations 3.134 by the following steps: (i) multiply the first equation by $-\cos\theta_2$, (ii) multiply the second equation by $\sin\theta_2$, and (iii) add
the results to obtain

$$\begin{aligned} & m_2[l_2\ddot{\theta}_2 + l_1\ddot{\theta}_1(\cos\theta_1\cos\theta_2 + \sin\theta_2\sin\theta_1) \\ &\quad + l_1\dot{\theta}_1^2(\sin\theta_2\cos\theta_1 - \cos\theta_2\sin\theta_1)] \\ &\quad = m_2[l_2\ddot{\theta}_2 + l_1\ddot{\theta}_1\cos(\theta_2-\theta_1) \\ &\quad\quad + l_1\dot{\theta}_1^2\sin(\theta_2-\theta_1)] = -w_2\sin\theta_2 \end{aligned} \quad (3.135)$$

This is the first governing differential equation.

Looking back at the second of Equation 3.131, we would like to use Equations 3.134 to solve for T_{c2} sin $(\theta_2-\theta_1)$ to eliminate this term and obtain the final equation of motion. Multiplying the first of Equation 3.134 by $-\cos\theta_1$, multiplying the second by $\sin\theta_1$, and then adding the results gives

$$\begin{aligned} & T_{c2}(\sin\theta_2\cos\theta_1 - \cos\theta_2\sin\theta_1) = T_{c2}\sin(\theta_2-\theta_1) \\ &\quad = -w_2\sin\theta_1 + m_2[-l_1\ddot{\theta}_1 - l_2\ddot{\theta}_2\cos(\theta_2-\theta_1) \\ &\quad\quad + l_2\dot{\theta}_2^2\sin(\theta_2-\theta_1)] \end{aligned}$$

Substituting this result into the second of Equation 3.131 gives

$$\begin{aligned} m_1l_1\ddot{\theta}_1 = &-w_2\sin\theta_1 + m_2[-l_1\ddot{\theta}_1 - l_2\ddot{\theta}_2\cos(\theta_2-\theta_1) \\ & + l_2\dot{\theta}_2^2\sin(\theta_2-\theta_1)] - w_1\sin\theta_1, \end{aligned} \quad (3.136)$$

which is the second (and last) required differential equation.

Restating Equations 3.135 and 3.136 in matrix format gives

$$\begin{bmatrix} l_1(m_1+m_2) & m_2 l_2 \cos(\theta_2-\theta_1) \\ m_2 l_1 \cos(\theta_2-\theta_1) & m_2 l_2 \end{bmatrix} \begin{Bmatrix} \ddot{\theta}_1 \\ \ddot{\theta}_2 \end{Bmatrix} = \begin{Bmatrix} -(w_1+w_2)\sin\theta_1 + m_2 l_2 \dot{\theta}_2^2 \sin(\theta_2-\theta_1) \\ -w_2 \sin\theta_2 - m_1 l_1 \dot{\theta}_1^2 \sin(\theta_2-\theta_1) \end{Bmatrix}. \quad (3.137)$$

Note that this inertia matrix is neither diagonal nor symmetric, but it can be made symmetric; for example, multiply the first equation by l_1 and the second equation by l_2. As with the stiffness matrix, the inertia matrix should be either symmetric, or capable of being made symmetric.

The linearized version of this equation is obtained by assuming that both θ_1 and θ_2 are small (i.e., $\sin\theta_1, = \theta_1$; $\cos\theta_1 = 1$) and can be stated

$$\begin{bmatrix} (m_1+m_2)l_1^2 & m_2 l_2 l_1 \\ m_2 l_1 l_2 & m_2 l_2^2 \end{bmatrix} \begin{Bmatrix} \ddot{\theta}_1 \\ \ddot{\theta}_2 \end{Bmatrix} + \begin{bmatrix} (w_1+w_2)l_1 & 0 \\ 0 & w_2 l_2 \end{bmatrix} \begin{Bmatrix} \theta_1 \\ \theta_2 \end{Bmatrix} = 0. \quad (3.138)$$

(The linearization procedure is discussed at length in developing the equations of motion for a simple pendulum; see Equations 3.106 and 3.109.) We have "symmetrized" the inertia matrix, and now have a diagonal stiffness matrix. The inertia coupling of the coordinates in this example presents an interesting contrast to the model of Equation 3.124 for which the stiffness matrix causes coordinate coupling.

Note the positive contributions from m_1 and m_2 to the diagonal inertia matrix entry $(m_1+m_2)l_1^2$. A negative contribution in this term would be a certain sign of an error in developing the equations of motion. Individual mass terms always act to increase the inertia of a system and consequently act to increase the inertia in a diagonal inertia matrix entry.

Deriving equations of motion for this example is complicated, first by kinematics (finding $(\ddot{x}_1, \ddot{x}_2)$) and then by algebra to eliminate T_{c2} from the equations of motion. Drawing the free-body diagrams of Figure 3.50 and applying $\Sigma f = m\ddot{r}$ to obtain the equations of motion is comparatively straightforward, versus the first example where getting the correct equations of motion (assuming the nature of the motion to arrive at the correct free-body diagrams) is the central difficulty. The two examples presented here are reasonably representative of multi-degree-of-freedom systems.

3.5.2 Analyzing Multi-Degree-of-Freedom Vibration Problems

This section is a digression in the central thrust of the book, namely, developing governing equations of motion and is to some extent "optional." It speaks to a reasonable question following Section 3.5.1; namely, Now that I have these coupled differential equations, how do I *analyze* them? Note that this question considers analysis versus solution. Time solutions can be developed for the differential-equation models of Section 3.5.1 via numerical integration starting from initial conditions. The interest here concerns uncoupling an n-degree-of-freedom vibration problem to obtain n single-degree-of vibration problems that can be analyzed in a manner similar to the single-degree-of-freedom, spring-mass-damper problems of Section 3.2.

3.5.2.1 Analyzing Undamped Two-Degree-of-Freedom Vibration Problems

Linear MDOF vibration problems have mass, stiffness, and damping matrices. Our 2DOF examples will have 2×2 mass, stiffness, and (sometimes) damping matrices that couple the motion of the individual degrees of freedom. A linear vibration problem with 100 degrees of freedom will have matrices that are 100×100, etc. Fortunately, these vibration problems can be "uncoupled," so that a problem with 100 degrees of freedom described by 100 coupled second-order differential equations can be treated as 100 separate, uncoupled, second-order differential equations. The uncoupling steps are referred to as "modal analysis" and will entail several new terms such as eigenvalues, eigenvectors, and mode shapes. You may be thinking that this modal analysis doesn't sound like such a good deal, since it does not appear to reduce the number of degrees of freedom but only transforms them to an uncoupled state. One hundred uncoupled equations may not sound any better to you than one hundred coupled equations. In actual vibration analysis, modal analysis can be used to reduce problem dimensionality (dramatically). It also provides a mental picture that allows one to interpret the motion of large vibration problems in terms of known characteristics of 1DOF vibration problems.

Prior material in this section has been aimed at correctly deriving governing equations of motion without any particular regard for their solutions. Suppose that we start thinking about solving coupled linear differential equations by considering the problem of developing a solution to the following homogeneous version of Equation 3.124

$$\begin{bmatrix} m_1 & 0 \\ 0 & m_2 \end{bmatrix} \begin{Bmatrix} \ddot{x}_1 \\ \ddot{x}_2 \end{Bmatrix} + \begin{bmatrix} (k_1+k_2) & -k_2 \\ -k_2 & (k_2+k_3) \end{bmatrix} \begin{Bmatrix} x_1 \\ x_2 \end{Bmatrix} = 0. \quad (3.139)$$

Earlier in Section 3.2 we faced the simpler problem of finding a solution to the comparable 1DOF problem $m\ddot{x} + kx = 0$. Recall that we "guessed" a solution of the form $x = A\cos\omega t \Rightarrow \ddot{x} = -\omega^2 A\cos\omega t$. Substituting this result into the governing equation netted

$$\left(-\omega^2 + \frac{k}{m}\right)A\cos\omega t = 0.$$

The term $\cos\omega t$ is not zero, and a nontrivial solution $(A \neq 0)$ requires that $\omega = \omega_n = \sqrt{k/m}$. We are going to follow a very similar procedure to obtain a solution to Equation 3.139 by assuming a solution of the form

$$\begin{Bmatrix} x_1 \\ x_2 \end{Bmatrix} = \begin{Bmatrix} a_1 \\ a_2 \end{Bmatrix}\cos\omega t \Rightarrow \begin{Bmatrix} \ddot{x}_1 \\ \ddot{x}_2 \end{Bmatrix} = -\omega^2\begin{Bmatrix} a_1 \\ a_2 \end{Bmatrix}\cos\omega t. \tag{3.140}$$

Substituting this result into Equation 3.139 gives

$$\left[-\omega^2\begin{bmatrix} m_1 & 0 \\ 0 & m_2 \end{bmatrix} + \begin{bmatrix} (k_1+k_2) & -k_2 \\ -k_2 & (k_2+k_3) \end{bmatrix}\right] \times \begin{Bmatrix} a_1 \\ a_2 \end{Bmatrix}\cos\omega t = \begin{Bmatrix} 0 \\ 0 \end{Bmatrix}.$$

Since, $\cos\omega t \neq 0$, it can be eliminated. Gathering terms yields

$$\begin{bmatrix} [-m_1\omega^2 + (k_1+k_2)] & -k_2 \\ -k_2 & [-m_2\omega^2 + (k_2+k_3)] \end{bmatrix} \times \begin{Bmatrix} a_1 \\ a_2 \end{Bmatrix} = \begin{Bmatrix} 0 \\ 0 \end{Bmatrix}. \tag{3.141}$$

Solving this equation for the unknowns a_1,a_2 via Cramer's rule gives

$$a_1 = \frac{1}{\Delta}\begin{vmatrix} 0 & -k_2 \\ 0 & -m_2\omega^2 + (k_2+k_3) \end{vmatrix} = \frac{0}{\Delta}$$
$$a_2 = \frac{1}{\Delta}\begin{vmatrix} -m_1\omega^2 + (k_1+k_2) & 0 \\ -k_2 & 0 \end{vmatrix} = \frac{0}{\Delta},$$

where Δ is the determinant of the coefficient matrix. The only way that these equations can have a nontrivial solution (a_1, $a_2 \neq 0$) is for $\Delta = 0$; that is, the coefficient matrix must be singular. From Equation 3.141, this requirement gives

$$\begin{aligned}\Delta &= [-(k_1+k_2) - m_1\omega^2][(k_2+k_3) - m_2\omega^2] - k_2^2 \\ &= m_1m_2\omega^4 - [m_1(k_2+k_3) + m_2(k_1+k_2)]\omega^2 \\ &\quad + (k_1+k_2)(k_2+k_3) - k_2^2 = 0.\end{aligned} \tag{3.142}$$

This is the "characteristic equation" for the model of Equation 3.139, and it defines two natural frequencies ω_{n1}^2, ω_{n2}^2 versus the single natural frequency for the 1DOF vibration examples.

We need "numbers" to proceed with this development and will use the following assumed mass and stiffness data:

$$m_1 = 1\ \text{kg}, \quad m_2 = 2\ \text{kg}, k_1 = k_2 = k_3 = 1\ \text{N/m}. \tag{3.143}$$

For these data, the differential Equation 3.139 becomes

$$\begin{bmatrix} 1 & 0 \\ 0 & 2 \end{bmatrix}\begin{Bmatrix} \ddot{x}_1 \\ \ddot{x}_2 \end{Bmatrix} + \begin{bmatrix} 2 & -1 \\ -1 & 2 \end{bmatrix}\begin{Bmatrix} x_1 \\ x_2 \end{Bmatrix} = 0, \tag{3.144}$$

and the frequency Equation 3.142 becomes

$$\omega^4 - 3\omega^2 + \frac{3}{2} = 0 \Rightarrow \omega^2 = \frac{3}{2} \pm \frac{\sqrt{3}}{2},$$

with the solutions

$$\begin{aligned}\omega_{n1}^2 &= .6340\ \text{s}^{-2} \Rightarrow \omega_{n1} = .7962\ \text{rad/s} \\ \omega_{n2}^2 &= 2.366\ \text{s}^{-2} \Rightarrow \omega_{n2} = 1.538\ \text{rad/s}.\end{aligned} \tag{3.145}$$

The first (lowest) root $\omega_{n1}^2 = .634\ \text{s}^{-2}$ is the first "eigenvalue" and defines the first natural frequency $\omega_{n1} = .7962$ rad/s. The next root $\omega_{n2}^2 = 2.366\ \text{s}^{-2}$ is the second eigenvalue and defines the second natural frequency $\omega_{n2} = 1.538$ rad/s.

Having obtained the natural frequencies in the assumed solution of Equation 3.141, we need to solve (in some sense) for the a_1, a_2 coefficients. Substituting the data of Equation 3.143 into Equation 3.141 gives

$$\begin{bmatrix} -\omega^2 + 2 & -1 \\ -1 & -2\omega^2 + 2 \end{bmatrix}\begin{Bmatrix} a_1 \\ a_2 \end{Bmatrix} = 0.$$

Now substituting $\omega^2 = \omega_{n1}^2 = .633975\ \text{s}^{-2}$ gives

$$\begin{bmatrix} -.633975 + 2 & -1 \\ -1 & -1.26796 + 2 \end{bmatrix}\begin{Bmatrix} a_{11} \\ a_{21} \end{Bmatrix} = 0.$$

You can confirm that the determinant of the coefficient matrix is zero which implies that there is only one independent equation for the two unknowns. Hence, we can use either equation to solve for the ratios of the two unknowns. Setting $a_{11} = 1$ gives

$$\begin{aligned}&1.366063(1) - a_{21} = 0 \Rightarrow a_{21} = 1.36603 \\ &-1(1) + .732050a_{21} = 0 \Rightarrow a_{21} = 1.36603.\end{aligned}$$

Hence, the first "eigenvector" is

$$(a_{i1}) = \begin{Bmatrix} a_{11} \\ a_{21} \end{Bmatrix} = \begin{Bmatrix} 1 \\ 1.36603 \end{Bmatrix}. \tag{3.146}$$

Note that multiplying this vector by any finite constant (positive or negative) will yield an equally valid first eigenvector, since the vector is defined only in terms of the ratio of its components. In vibration problems, an eigenvector is also called a "mode shape."

The second eigenvector is obtained in exactly the same fashion by substituting $\omega^2 = \omega_{n2}^2 = 2.3360\ \text{s}^{-2}$ into Equation 3.141, obtaining

$$(a_{i2}) = \begin{Bmatrix} a_{21} \\ a_{22} \end{Bmatrix} = \begin{Bmatrix} 1 \\ -0.36603 \end{Bmatrix}.$$

Figure 3.51 illustrates the two eigenvectors. Note that the two masses move in the same direction (in phase) in the first eigenvector and in opposite directions (out of phase) in the second eigenvector or mode shape. The matrix of eigenvectors is

$$[A] = \begin{bmatrix} 1.0 & 1.0 \\ 1.36603 & -0.36603 \end{bmatrix}. \tag{3.147}$$

At this point, you could reasonably be thinking, "That's nice, but why are we doing this, and when are we going to solve a dynamics problem?" The answer is that we are going to use the matrix of eigenvectors in Equation 3.147 to *uncouple* Equations 3.144, so that we can solve problems associated with our 2DOF vibration example in terms of 1DOF results.

Consider the following coordinate transformation with $[A]$ defined from Equation 3.147:

$$(x_i) = [A](q_i) \Rightarrow \begin{Bmatrix} x_1 \\ x_2 \end{Bmatrix} = \begin{bmatrix} 1.0 & 1.0 \\ 1.36603 & -0.36603 \end{bmatrix} \begin{Bmatrix} q_1 \\ q_2 \end{Bmatrix}, \tag{3.148a}$$

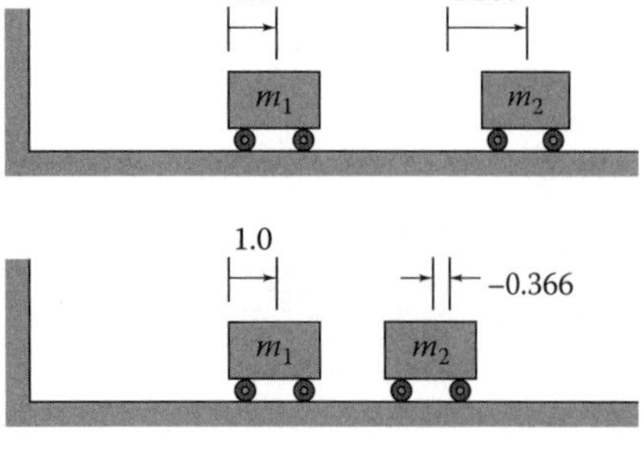

FIGURE 3.51
Eigenvectors for the two-mass system of Figure 3.46, with the numerical values of Equation 3.143.

where the right vector (q_i) is the vector of *modal* coordinates. Also, since $[A]$ is constant,

$$(\ddot{x}_i) = [A](\ddot{q}_i). \tag{3.148b}$$

Substituting for $(x)_i$ and $(\ddot{x})_i$ into Equation 3.139 gives

$$[M][A](\ddot{q}_i) + [K][A](q_i) = (f_i).$$

Premultiplying by the transpose of $[A]$ gives

$$[A]^T[M][A](\ddot{q}_i) + [A]^T[K][A](q_i) = [A]^T(f_i). \tag{3.149}$$

We can now show by substitution (for this example) that

$$[A]^T[M][A] = [M_q], \quad [A]^T[K][A] = [K_q],$$

where the "modal-mass matrix" $[M_q]$ and "modal-stiffness matrix" $[K_q]$ are diagonal. Note

$$\begin{aligned}[A]^T[M][A] &= \begin{bmatrix} 1.0 & 1.36603 \\ 1.0 & -0.36603 \end{bmatrix} \begin{bmatrix} 1 & 0 \\ 0 & 2 \end{bmatrix} \begin{bmatrix} 1.0 & 1.0 \\ 1.36603 & -0.36603 \end{bmatrix} \\ &= \begin{bmatrix} 1.0 & 1.36603 \\ 1.0 & -0.36603 \end{bmatrix} \begin{bmatrix} 1.0 & 1.0 \\ 2.73206 & -0.73206 \end{bmatrix} \\ &= \begin{bmatrix} 4.732 & 0.0000 \\ 0.0000 & 1.268 \end{bmatrix} = [M_q].\end{aligned}$$

The modal-mass matrix $[M_q]$ is diagonal, with the first and second "modal masses" defined by $m_{q1} = 4.732$; $m_{q2} = 1.268$.

For a 1DOF vibration problem with the equation of motion $m\ddot{x} + kx = f(t)$, we found it useful to divide through the equation by m to obtain $\ddot{x} + \omega_n^2 x = f(t)/m$. We want to do the same thing with the present 2DOF example, by *normalizing the eigenvectors with respect to the mass matrix* such that the modal-mass matrix $[M_q]$ reduces to the identity matrix $[I]$. The diagonal modal-mass coefficient for the jth mode is defined by $(a_{ji})^T\ [M](a_{ji}) = M_{qj}$. Dividing the jth eigenvector (a_{ji}) by $M_{qj}^{1/2}$ will yield an eigenvector with a modal mass equal to 1, yielding

$$[A^*]^T[M][A^*] = [I]. \tag{3.150}$$

Normalizing the current eigenvector set means dividing the first and second eigenvectors by $M_{q1}^{1/2} = \sqrt{4.732} = 2.175$ and $M_{q2}^{1/2} = \sqrt{1.268} = 1.126$, respectively, obtaining

$$[A^*] = \begin{bmatrix} 0.45970 & 0.88807 \\ 0.62796 & -0.32506 \end{bmatrix}. \tag{3.151}$$

You may want to repeat calculations for this set of eigenvectors to confirm that the modal-mass matrix is now the identity matrix. Proceeding with this normalized version of the eigenvector matrix to verify that the modal-stiffness matrix is diagonal yields

$$[A^*]^T[K][A^*] = [A^*]^T\begin{bmatrix} 2 & -1 \\ -1 & 2 \end{bmatrix}\begin{bmatrix} 0.45970 & 0.88807 \\ 0.62796 & -0.32506 \end{bmatrix}$$
$$= \begin{bmatrix} 0.45970 & 0.62796 \\ 0.88807 & -0.32506 \end{bmatrix}\begin{bmatrix} 0.29144 & 2.10120 \\ 0.79622 & -1.53819 \end{bmatrix}$$
$$= \begin{bmatrix} 0.63345 & -0.00000 \\ -0.00000 & 2.3660 \end{bmatrix} = [K_q].$$

The normalized matrix of eigenvectors yields a diagonalized modal-stiffness matrix $[K_q]$; moreover, the diagonal entries are the eigenvalues defined in Equation 3.145; that is,

$$[A^*]^T[K][A^*] = [\Lambda], \tag{3.152}$$

where $[\Lambda]$ is the diagonal matrix of eigenvalues.

With normalized eigenvectors, the matrix version of the modal Equation 3.149 is

$$(\ddot{q})_i[\Lambda](q_i) = (Q_i) = [A^*]^T(f_i) = \begin{bmatrix} \overset{*}{A}_{11} & \overset{*}{A}_{21} \\ \overset{*}{A}_{12} & \overset{*}{A}_{22} \end{bmatrix}\begin{Bmatrix} f_1 \\ f_2 \end{Bmatrix}$$
$$= \begin{Bmatrix} \overset{*}{A}_{11} f_1 & + & \overset{*}{A}_{21} f_2 \\ \overset{*}{A}_{12} f_1 & + & \overset{*}{A}_{22} f_2 \end{Bmatrix}. \tag{3.153}$$

The vector on the right (Q_i) is called the modal force vector, and the individual (uncoupled) modal differential equations look like the following:

$$\begin{aligned} \ddot{q}_1 + \omega_{n1}^2 q_1 &= Q_1 = \overset{*}{A}_{11} f_1 + \overset{*}{A}_{21} f_2 \\ \ddot{q}_2 + \omega_{n2}^2 q_2 &= Q_2 = \overset{*}{A}_{12} f_1 + \overset{*}{A}_{22} f_2. \end{aligned} \tag{3.154}$$

Obviously, a demonstration for one example does not constitute a proof; however, Meirovitch (1997) and numerous other authors provide a general proof that a matrix of normalized eigenvectors always diagonalizes *symmetric* mass and stiffness matrices in precisely the manner demonstrated here. Hence, we will proceed, agreed that eigenvalues and an associated matrix of eigenvectors can be developed satisfying $[A^*]^T[M][A^*] = [I]$ and $[A^*]^T\,[K]\,[A^*] = [\Lambda]$.

Given that $[A^*]^T\,[M]\,[A^*] = [I]$, and $[A^*]^T\,[K]\,[A^*] = [\Lambda]$, the units for an entry in normalized eigenvector matrix $[A^*]$ is $\text{mass}^{-1/2}$. Hence, for the SI system, $[A^*]$'s units are $\text{kg}^{-1/2}$; for the U.S. standard unit system, the units are $\text{slug}^{-1/2}$. From the coordinate transformation $(x)_i = [A^*]\,(q)_i$, the units for a modal coordinate is $\text{mass}^{1/2} \times$ length. For the SI and U.S. standard systems, the appropriate units are, respectively, m $\text{kg}^{1/2}$ and ft $\text{slug}^{1/2}$. Looking at the first of Equation 3.154, a dimensional analysis yields

$$\begin{aligned} &\ddot{q}_1(Lm^{1/2}T^{-2}) + \omega_{n1}^2(T^{-2})q_1(Lm^{1/2}) \\ &\quad = a_{11}(m^{-1/2})f_1(F) + a_{21}(m^{-1/2})f_2(F) \\ &\therefore\ mLT^{-2} = F, \end{aligned}$$

confirming the correctness of the dimensions. These dimensional considerations hold for the present model. Obviously, a different result would hold for the double-pendulum example.

Calculation procedures are particularly efficient for undamped eigenvalues and eigenvectors defined by symmetric stiffness and inertia matrices. Hence, the considerable effort involved in manually working through the example of this section is not representative of the requirements of "real" engineering problems. Example Problem 3.20 demonstrates a similar eigenanalysis for the double-pendulum model for small motion.

Example Problem 3.20

Calculate the eigenvalues and normalized eigenvectors for the double pendulum of Figure 3.50 for small motion, as modeled by Equation 3.138 with the data

$$l_1 = l_2 = 10\text{ cm},\quad m_1 = 0.1\text{ kg},\quad m_2 = 0.2\text{ kg}. \tag{i}$$

Solution

Substituting the data of Equation (i) into Equation 3.138

$$\begin{bmatrix} (m_1+m_2)l_1^2 & m_2 l_2 l_1 \\ m_2 l_1 l_2 & m_2 l_2^2 \end{bmatrix}\begin{Bmatrix} \ddot{\theta}_1 \\ \ddot{\theta}_1 \end{Bmatrix} + \begin{bmatrix} (w_1+w_2)l_1 & 0 \\ 0 & w_2 l_2 \end{bmatrix}\begin{Bmatrix} \theta_1 \\ \theta_2 \end{Bmatrix} = 0 \tag{3.138}$$

yields

$$\begin{bmatrix} (0.1+0.2)(0.1)^2 & 0.2(0.1)(0.1) \\ 0.2(0.1)(0.1) & 0.2(0.1)^2 \end{bmatrix}\begin{Bmatrix} \ddot{\theta}_1 \\ \ddot{\theta}_2 \end{Bmatrix} + \begin{bmatrix} 9.81(0.1+0.2)(0.1) & 0 \\ 0 & 9.81(0.2)(0.1) \end{bmatrix}\begin{Bmatrix} \theta_1 \\ \theta_2 \end{Bmatrix} = 0, \tag{ii}$$

where $g = 9.81\text{ m/s}^2$ is the acceleration of gravity. Completing the arithmetic in Equation (ii) gives

$$\begin{bmatrix} 0.003 & 0.002 \\ 0.002 & 0.002 \end{bmatrix}\begin{Bmatrix} \ddot{\theta}_1 \\ \ddot{\theta}_2 \end{Bmatrix} + \begin{bmatrix} 0.2943 & 0 \\ 0 & 0.1962 \end{bmatrix}\begin{Bmatrix} \theta_1 \\ \theta_2 \end{Bmatrix} = 0.$$

Assuming a solution of the form

$$(\theta_i) = (a_i)\cos\omega t \Rightarrow (\ddot{\theta}_i) = -\omega^2(a_i)\cos\omega t$$

yields

$$\begin{bmatrix} (-0.003\omega^2+0.2943) & -0.002\omega^2 \\ -0.002\omega^2 & (-0.002\omega^2+0.1962) \end{bmatrix} \begin{Bmatrix} a_1 \\ a_2 \end{Bmatrix} = 0. \quad \text{(iii)}$$

Setting the determinant of the coefficient matrix equal to zero defines the characteristic equation

$$2.0\times10^{-6}\omega^4 - 1.1722\times10^{-3}\omega^2 + 5.774\times10^{-2} = 0$$
$$\therefore\ \omega^4 - 588.60\omega^2 + 28870.8 = 0.$$

The roots of this equation define the eigenvalues and natural frequencies:

$$\begin{aligned} \omega_{n1}^2 &= 54.005\ \text{s}^{-2} \Rightarrow \omega_{n1} = 7.348\ \text{rad/s} \\ \omega_{n2}^2 &= 534.6\ \text{s}^{-2} \Rightarrow \omega_{n2} = 23.12\ \text{rad/s}. \end{aligned} \quad \text{(iv)}$$

The first eigenvector is obtained by substituting the first eigenvalue $\omega_{n1}^2 = 54.005$ back into Equation (iii), obtaining

$$\begin{bmatrix} (-0.003\times54.005+0.2943) & -0.002\times54.005 \\ -0.002\times54.005 & (-0.002\times54.005+0.1962) \end{bmatrix} \times \begin{Bmatrix} a_{11} \\ a_{21} \end{Bmatrix} = 0, \text{ or}$$

$$\begin{bmatrix} 0.132285 & -0.10801 \\ -0.10801 & 0.08819 \end{bmatrix} \begin{Bmatrix} a_{11} \\ a_{21} \end{Bmatrix} = 0.$$

You can confirm that the determinant of this coefficient matrix is very nearly zero, yielding the two dependent equations

$$0.132285a_{11} - 0.10801a_{21} = 0 \Rightarrow a_{11} = 1,\ a_{21} = -1.2247$$
$$-0.10801a_{11} + 0.08819a_{21} = 0 \Rightarrow a_{11} = 1,\ a_{21} = 1.2247.$$

Hence, the first (nonnormalized) eigenvector is

$$(a_{i1}) = \begin{Bmatrix} 1 \\ 1.2247 \end{Bmatrix}.$$

Substituting the second eigenvalue $\omega_{n2}^2 = 534.6$ into Equation (iii) yields

$$-1.3095a_{12} - 1.0692a_{22} = 0 \Rightarrow a_{12} = 1,\ a_{22} = -1.2247$$
$$-1.0692a_{12} - 0.8730a_{22} = 0 \Rightarrow a_{12} = 1,\ a_{22} = -1.2247,$$

and the second (nonnormalized) eigenvalue is

$$(a_{i2}) = \begin{Bmatrix} 1 \\ -1.2247 \end{Bmatrix}.$$

Figure XP3.20 illustrates the eigenvectors. Note that the two pendulum elements are moving in the same direction for the first mode (in phase) and moving in opposite directions in the second mode (out of phase).

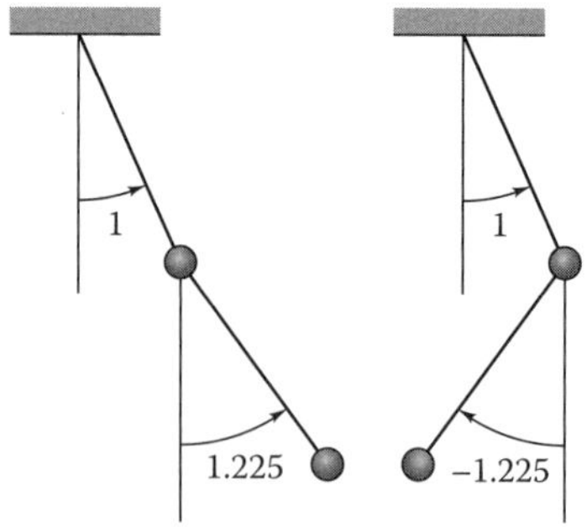

FIGURE XP3.20
Mode shapes for the double pendulum of Figure 3.50 with the data of Equation (i).

The matrix of nonnormalized matrix of eigenvectors is

$$[A] = \begin{bmatrix} 1.0 & 1.0 \\ 1.2247 & -1.2247 \end{bmatrix}$$

yielding the modal-mass matrix,

$$\begin{aligned} [M_q] &= [A]^T[M][A] \\ &= [A]^T \begin{bmatrix} 0.003 & 0.002 \\ 0.002 & 0.002 \end{bmatrix} \begin{bmatrix} 1.0 & 1.0 \\ 1.2247 & -1.2247 \end{bmatrix} \\ &= \begin{bmatrix} 1.0 & 1.2247 \\ 1.0 & -1.2247 \end{bmatrix} \begin{bmatrix} 5.4494\times10^{-3} & 5.5060\times10^{-4} \\ 4.4494\times10^{-3} & -4.4494\times10^{-4} \end{bmatrix} \\ &= \begin{bmatrix} m_{q1} & 0.0 \\ 0.0 & m_{q2} \end{bmatrix} = \begin{bmatrix} 1.0899\times10^{-2} & 0.000 \\ 0.000 & 1.10955\times10^{-3} \end{bmatrix} (\text{kg m}^2) \end{aligned}$$

The eigenvectors are normalized with respect to the mass matrix by dividing the ith vector by $m_{qi}^{-1/2}$, yielding

$$[A^*] = \begin{bmatrix} 9.5789 & 30.213 \\ 11.7313 & -37.002 \end{bmatrix} (\text{kg}^{-1/2}/\text{m})$$

Verifying that this matrix can be used to satisfactorily diagonalize our inertia and stiffness matrices proceeds as follows:

$$\begin{aligned} [A^*]^T[M][A^*] &= [A^*]^T \begin{bmatrix} 0.003 & 0.002 \\ 0.002 & 0.002 \end{bmatrix} \begin{bmatrix} 9.5789 & 30.213 \\ 11.7313 & -37.0016 \end{bmatrix} \\ &= \begin{bmatrix} 9.5789 & 11.7313 \\ 30.213 & -37.0016 \end{bmatrix} \times \begin{bmatrix} 5.21193\times10^{-2} & 1.6635\times10^{-2} \\ 4.26204\times10^{-2} & -1.35778\times10^{-2} \end{bmatrix} \\ &= \begin{bmatrix} 1.0000 & 0.0001 \\ 0.0001 & 1.0050 \end{bmatrix}, \end{aligned}$$

and

$$\begin{aligned} [A^*]^T[K][A^*] &= [A^*]^T \begin{bmatrix} 0.2943 & 0.0 \\ 0.0 & 0.1962 \end{bmatrix} \begin{bmatrix} 9.5789 & 30.213 \\ 11.7313 & -37.0016 \end{bmatrix} \\ &= \begin{bmatrix} 9.5789 & 11.7313 \\ 30.213 & -37.0016 \end{bmatrix} \begin{bmatrix} 2.8191 & 8.8916 \\ 2.30168 & -7.2597 \end{bmatrix} \\ &= \begin{bmatrix} 54.00 & 0.00 \\ 0.00 & 5373 \end{bmatrix} \text{s}^{-2}. \end{aligned}$$

You can verify by comparison to Equation (iv) that the diagonal elements of this matrix approximately equal the eigenvalues.

3.5.2.2 Free Motion from Initial Conditions (the Homogeneous Solution)

Following the same pattern that we used with a single-degree-of-freedom vibration problem, we will start our solution process by developing a homogeneous solution with a disturbance to the model provided by initial conditions. The initial conditions for the modal differential equations are obtained from the physical-coordinate initial conditions via the coordinate transformation relationships $(x) = [A^*](q)$. Premultiplying this equation by $[A^*]^T[M]$ gives

$$[A^*]^T[M](x) = [A^*]^T[M][A^*](q) = [I](q)$$
$$\therefore\ (q) = [A^*]^T[M](x), \quad [A^*]^{-1} = [A^*]^T[M].$$

Hence, the modal-coordinate initial conditions are defined by

$$(q_0) = [A^*]^{-1}(x_0) = [A^*]^T[M](x_0).$$

Similarly, the modal-velocity initial conditions are defined by

$$(\dot{q}_0) = [A^*]^T[M](\dot{x}_0).$$

For the prior example, the modal-coordinate initial conditions are defined by

$$\begin{Bmatrix} q_{10} \\ q_{10} \end{Bmatrix} = [A^*]^T[M]\begin{Bmatrix} x_{10} \\ x_{20} \end{Bmatrix} = \begin{bmatrix} 0.45970 & 0.62796 \\ 0.88807 & -0.32506 \end{bmatrix}\begin{bmatrix} 1 & 0 \\ 0 & 2 \end{bmatrix}\begin{Bmatrix} x_{10} \\ x_{10} \end{Bmatrix} = \begin{bmatrix} 0.45970 & 1.2559 \\ 0.88807 & -0.65012 \end{bmatrix}\begin{Bmatrix} x_{10} \\ x_{20} \end{Bmatrix} = \begin{Bmatrix} 0.45970x_{10} & +1.2559x_{20} \\ 0.88807x_{10} & -0.65012x_{20} \end{Bmatrix},$$

The initial conditions for modal velocities are obtained similarly as

$$\begin{Bmatrix} \dot{q}_{10} \\ \dot{q}_{20} \end{Bmatrix} = [A^*]^T[M]\begin{Bmatrix} \dot{x}_{10} \\ \dot{x}_{20} \end{Bmatrix} = \begin{Bmatrix} 0.45970\dot{x}_{10} & +1.2559\dot{x}_{20} \\ 0.88807\dot{x}_{10} & -0.65012\dot{x}_{20} \end{Bmatrix}.$$

With these initial conditions, we can solve Equations 3.154 for any specified modal force terms $Q_1(t)$; $Q_2(t)$, obtaining a modal-coordinate solution defined by $q_1(t)$, $q_2(t)$. From $(x)_i = [A^*](q)_i$, the solution for the physical coordinates can be stated

$$\begin{Bmatrix} x_1(t) \\ x_2(t) \end{Bmatrix} = \begin{bmatrix} A_{11}^* & A_{12}^* \\ A_{21}^* & A_{22}^* \end{bmatrix}\begin{Bmatrix} q_1(t) \\ q_2(t) \end{Bmatrix} = q_1(t)\begin{Bmatrix} A_{11}^* \\ A_{22}^* \end{Bmatrix} + q_2(t)\begin{Bmatrix} A_{12}^* \\ A_{22}^* \end{Bmatrix}. \tag{3.155}$$

Hence, the physical-coordinate response vector is a linear sum of the modal solutions $q_1(t)$, $q_2(t)$ times their respective mode shapes (or eigenvectors).

As a demonstration of this outcome, free motion for the present example with $\dot{q}_{10} = \dot{q}_{20} = 0$ would be defined from the homogeneous version of Equation 3.154, $\ddot{q}_t + \omega_{ni}^2 q_i = 0$, as

$$q_1 = q_{10}\cos\omega_{n1}t = q_{10}\cos .7962t,$$
$$q_2 = q_{20}\cos\omega_{n2}t = q_{20}\cos 1.538t,$$

with the natural frequencies defined by Equation 3.145. The corresponding physical-coordinate solution is

$$\begin{Bmatrix} x_1 \\ x_2 \end{Bmatrix} = q_{10}\cos(0.7962t)\begin{Bmatrix} 0.4597 \\ 0.6280 \end{Bmatrix} + q_{20}\cos(1.538t)\begin{Bmatrix} 0.8881 \\ -0.3251 \end{Bmatrix}, \tag{3.156}$$

with the eigenvectors defined by Equation 3.151. The lesson from Equation 3.155 is as follows: Any disturbance to the system from initial conditions (or external forces) will result in combined motion at both natural frequencies. Figure 3.52 illustrates a solution from Equation 3.156 with $x_{10} = 1$ cm, $x_{20} = 0 \Rightarrow q_{10} = 0.4597$ $\text{kg}^{1/2}$ cm, $q_{20} = 0.8881$ $\text{kg}^{1/2}$ cm. Note the presence of both modal frequencies in the transient solution.

Example Problem 3.21

The 2-mass model illustrated in Figure XP3.21a is moving along at a constant velocity v_0 without drag forces and is just about to collide with a wall. Both springs are undeflected prior to impact. The right-hand spring will cushion the shock of the collision. Before collision, the system model is

$$\begin{bmatrix} m_1 & 0 \\ 0 & m_2 \end{bmatrix}\begin{Bmatrix} \ddot{x}_1 \\ \ddot{x}_2 \end{Bmatrix} + \begin{bmatrix} k_1 & -k_1 \\ -k_1 & k_1 \end{bmatrix}\begin{Bmatrix} x_1 \\ x_2 \end{Bmatrix} = 0. \tag{i}$$

Once contact is established, the model of Figure XP3.21b applies and the EOM is

$$\begin{bmatrix} m_1 & 0 \\ 0 & m_2 \end{bmatrix}\begin{Bmatrix} \ddot{x}_1 \\ \ddot{x}_2 \end{Bmatrix} + \begin{bmatrix} k_1 & -k_1 \\ -k_1 & (k_1+k_2) \end{bmatrix}\begin{Bmatrix} x_1 \\ x_2 \end{Bmatrix} = 0. \tag{ii}$$

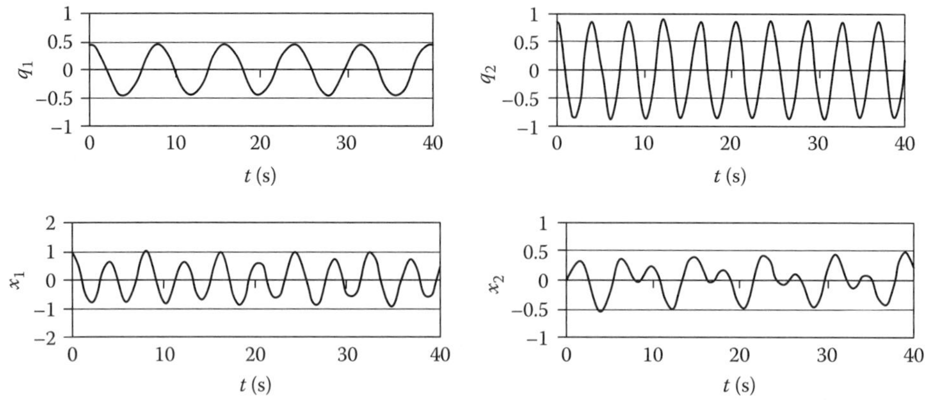

FIGURE 3.52
Solution from Equation 3.156 for q_1, q_2, x_1, and x_2 for $x_{10} = 1$ cm, $x_{20} = 0$ cm $\Rightarrow q_{10} = 0.4597$ kg$^{1/2}$ cm, $q_{20} = 0.8881$ kg$^{1/2}$ cm.

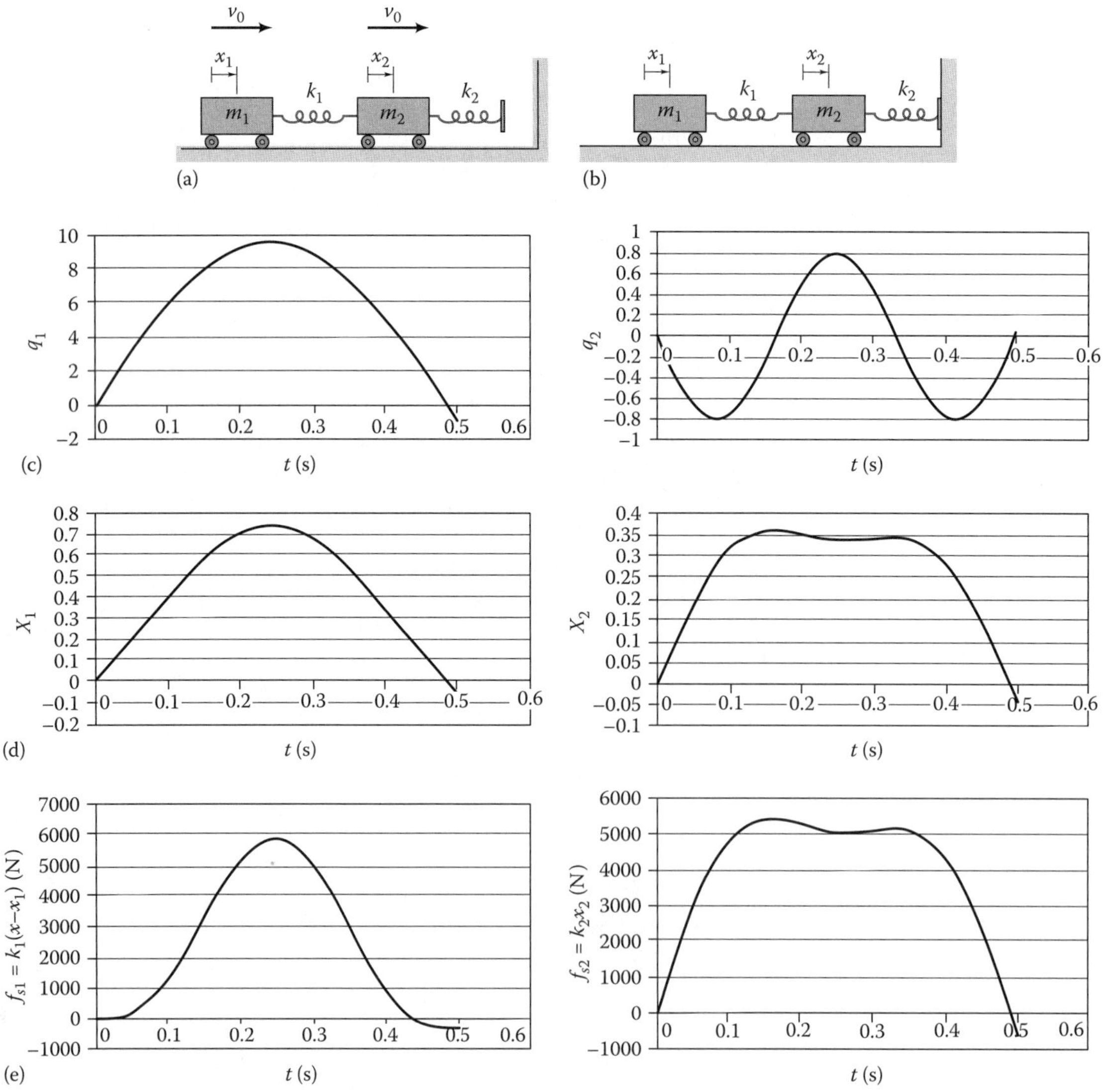

FIGURE XP3.21
(a) Coupled cars prior to collision, (b) coupled cars after collision, (c) modal coordinates, (d) physical-coordinate solutions, (e) reaction-force solutions.

The physical parameters, $m_1 = 150$ kg, $m_2 = 100$ kg, $k_1 = k_2 = 1.5 \times 10^4$ N/m, yield

$$\begin{bmatrix} 150 & 0 \\ 0 & 100 \end{bmatrix} \begin{Bmatrix} \ddot{x}_1 \\ \ddot{x}_2 \end{Bmatrix} + \begin{bmatrix} 1.5 \times 10^4 & -1.5 \times 10^4 \\ -1.5 \times 10^4 & 3.0 \times 10^4 \end{bmatrix} \begin{Bmatrix} x_1 \\ x_2 \end{Bmatrix} = 0. \quad \text{(iii)}$$

At the moment of contact, the physical initial conditions are

$$x_1(0) = x_2(0) = 0, \quad \dot{x}_1(0) = \dot{x}_2(0) = v_0 = 4 \text{ m/s}.$$

Tasks: Solve for $x_1(t)$, $x_2(t)$ and the reaction forces $k_1[x_1(t) - x_2(t)]$, $k_2 x_2(t)$.

Solution

Following the procedures of the preceding examples, the eigenvalues and natural frequencies are

$$\begin{aligned} \omega_{nl}^2 &= 41.886 \text{ s}^{-2} \Rightarrow \omega_{nl} = 6.4720 \text{ rad/s} \\ \omega_{nl}^2 &= 358.11 \text{ s}^{-2} \Rightarrow \omega_{n2} = 18.924 \text{ rad/s}. \end{aligned} \quad \text{(iv)}$$

The matrix of un-normalized eigenvectors is

$$[A] = [(a)_1, (a)_2] = \begin{bmatrix} A_{11} & A_{12} \\ A_{21} & A_{22} \end{bmatrix} = \begin{bmatrix} 1 & 1 \\ .5811 & -2.5811 \end{bmatrix}.$$

The matrix of normalized eigenvectors follows from $[A^*]^T [M][A^*] = [I]$ as

$$[A^*] = \begin{bmatrix} 0.073767 & 0.035002 \\ 0.042866 & -0.090344 \end{bmatrix}.$$

Hence, the model is

$$\ddot{q}_1 + 41.886 q_1 = 0, \quad \ddot{q}_2 + 358.11 q_2 = 0,$$

$$\begin{Bmatrix} x_1 \\ x_2 \end{Bmatrix} = \begin{bmatrix} 0.073767 & 0.035002 \\ 0.042866 & -0.090344 \end{bmatrix} \begin{Bmatrix} q_1 \\ q_2 \end{Bmatrix}.$$

From $(q_0) = [A^*]^T [M](x_0)$, the modal-coordinate initial conditions are zero. From $(\dot{q}_0) = [A^*]^T [M](\dot{x}_0)$ the modal-velocity initial conditions are

$$\begin{aligned} \begin{Bmatrix} \dot{q}_{10} \\ \dot{q}_{20} \end{Bmatrix} &= \begin{bmatrix} 0.073767 & 0.042866 \\ 0.035002 & -0.090344 \end{bmatrix} \begin{bmatrix} 150 & 0 \\ 0 & 100 \end{bmatrix} \begin{Bmatrix} 4 \\ 4 \end{Bmatrix} \\ &= \begin{bmatrix} 11.065 & 4.2866 \\ 5.2503 & -9.0344 \end{bmatrix} \begin{Bmatrix} 4 \\ 4 \end{Bmatrix} \\ &= \begin{Bmatrix} 61.4 \\ -15.136 \end{Bmatrix} \text{m kg}^{1/2}/\text{s} \end{aligned} \quad \text{(v)}$$

In terms of initial conditions, the solution to $\ddot{q}_i + \omega_{ni}^2 q_i = 0$ is $q_i = q_{i0} \cos \omega_{ni} t + (\dot{q}_{i0}/\omega_{ni}) \sin \omega_{ni} t$.

There is no forcing functions; hence, there is no particular solution, and the complete modal solutions are

$$\begin{aligned} q_1(t) &= q_{1h}(t) + 0 = \frac{61.406}{6.472} \sin 6.472t \\ &= 9.488 \sin 6.472t \text{ (m kg}^{1/2}) \\ q_2(t) &= q_{2h}(t) + 0 = \frac{-15.136}{18.924} \sin 18.92t \\ &= -0.800 \sin 18.924t \text{ (m kg}^{1/2}). \end{aligned} \quad \text{(vi)}$$

Figures XP3.21c illustrate these solutions.

From $(x) = [A^*](q)$, the solution for the physical coordinates is

$$\begin{Bmatrix} x_1(t) \\ x_2(t) \end{Bmatrix} = q_1(t) \begin{Bmatrix} 0.0738 \\ 0.0429 \end{Bmatrix} + q_2(t) \begin{Bmatrix} 0.0350 \\ -0.0903 \end{Bmatrix} \text{(m)}. \quad \text{(vii)}$$

Figures XP3.21d illustrate the physical solutions, showing that $q_1(t)$ provides most of the solution for $x_1(t)$. This outcome is explained from Equations (vi) that show $q_1(t)$'s magnitude to be much larger that $q_2(t)$'s. Further, from Equation (vii), $q_1(t)$'s contribution to $x_1(t)$ is approximately twice as large as $q_2(t)$'s. Figure XP3.21d's solution for $x_2(t)$ shows comparable contributions from both $q_1(t)$ and $q_2(t)$.

Figure XP3.21e illustrates the two spring reaction forces. Note that the spring-mass system loses contact with the wall when $f_{s2} = k_2 x_2(t)$ and changes sign at about $t = 0.044$ s. After this time, the right spring is disengaged, Equation (ii) stops being valid and

$$\begin{bmatrix} m_1 & 0 \\ 0 & m_2 \end{bmatrix} \begin{Bmatrix} \ddot{x}_1 \\ \ddot{x}_2 \end{Bmatrix} + \begin{bmatrix} k_1 & -k_1 \\ -k_1 & k_1 \end{bmatrix} \begin{Bmatrix} x_1 \\ x_2 \end{Bmatrix} = 0 \quad \text{(i)}$$

becomes the valid model.

The solutions of Figures XP3.21 were generated using a very simple spreadsheet program. However, solutions are frequently generated from large programs. A "reality check" is regularly helpful to answer the following question: Do the numbers seem right? In this case, we can do a quick calculation to see if the peak force and deflection seem to be reasonable. Suppose, for example, that both bodies are combined into one body having mass $m = m_1 + m_2 = 100 + 150 = 250$ kg. Then, a conservation-of-energy equation to find the peak deflection is

$$T_0 + V_0 = T_f + V_f$$

$$\therefore \frac{m}{2} v_0^2 = \frac{k}{2} \Delta^2 \Rightarrow \frac{250}{2} 4^2 = \frac{1.5 \times 10^4}{2} \Delta^2 \Rightarrow \Delta = 0.516 \text{ m}.$$

The predicted peak deflection for $x_2(t)$ in Figure XP3.2c is approximately 0.35 m, which is the same order of magnitude for this estimate, but lower. We would expect the correct number to be lower, because two masses with a spring between them will produce a lower collision force than a single rigid body with an equivalent mass.

3.5.2.3 Modal Damping Models

As with the 1DOF vibration problems, the addition of viscous damping to the model will cause an exponential decay in the transient (homogeneous) solution. The analysis of free motion with a general damping matrix, as presented by Equation 3.129, and stated

$$[M](\ddot{x}_i) + [C](\dot{x}_i) + [K](x_i) = (f_i)$$

involves complex eigenvalues and eigenvectors. The matrix of real eigenvectors, based on a symmetric stiffness $[K]$ and mass $[M]$ matrices, will only diagonalize a damping matrix that can be stated as a linear summation of these matrices; that is,

$$[C] = \alpha[M] + \beta[K].$$

For a damping matrix of this particular (and very unlikely) form, the modal damping matrix is defined by

$$\begin{aligned}[C_q] &= [A]^T[C][A] = \alpha[A]^T[M][A] + \beta[A]^T[K][A] \\ &= \alpha[I] + \beta[\Lambda].\end{aligned}$$

With this damping-matrix format, an m-degree-of-freedom vibration problem would have modal differential equations of the form

$$\begin{aligned}&\ddot{q}_1 + (\alpha + \beta\omega_{n1}^2)\dot{q}_1 + \omega_{n1}^2 q_1 = Q_1 \\ &\ddot{q}_2 + (\alpha + \beta\omega_{n2}^2)\dot{q}_2 + \omega_{n2}^2 q_2 = Q_2 \\ &\cdots \\ &\ddot{q}_m + (\alpha + \beta\omega_{mn}^2)\dot{q}_m + \omega_{mn}^2 q_m = Q_m.\end{aligned}$$

This is not a very useful or generally applicable result. For lightly damped systems, damping is more often introduced directly in the undamped modal equations via

$$\begin{aligned}&\ddot{q}_1 + 2\zeta_1\omega_{n1}\dot{q}_1 + \omega_{n1}^2 q_1 = Q_1 \\ &\ddot{q}_2 + 2\zeta_2\omega_{n1}\dot{q}_2 + \omega_{n2}^2 q_2 = Q_2 \\ &\cdots \\ &\ddot{q}_m + 2\zeta_m\omega_{mn}\dot{q}_m + \omega_{mn}^2 q_m = Q_m.\end{aligned} \tag{3.157}$$

The damping factors $\zeta_1, \zeta_2, \ldots, \zeta_m$ are specified for each modal differential equation, based on measurements or experience. This approach is approximate, based on the assumption that the system damping does not appreciably change the undamped mode shapes. An approximation based on specifying modal damping factors is normally consistent with the usual (poor) level of knowledge on damping in an actual structure.

Procedures for performing eigenanalysis for vibration problems with general damping matrices, as represented by the model of Equation 3.129, are beyond the scope of this book. Interested readers can consult Meirovich (1997) or other references in mechanical vibrations.

Example Problem 3.22

Figure XP3.22 represents the same physical problem as Example Problem XP3.21, except that now viscous dampers have been added between the cars and in the shock absorbing unit in front of car 2. The vibration model is now

$$\begin{bmatrix} m_1 & 0 \\ 0 & m_2 \end{bmatrix}\begin{Bmatrix} \ddot{x}_1 \\ \ddot{x}_2 \end{Bmatrix} + \begin{bmatrix} c_1 & -c_1 \\ -c_1 & (c_1 + c_2) \end{bmatrix}\begin{Bmatrix} \dot{x}_1 \\ \dot{x}_2 \end{Bmatrix} + \begin{bmatrix} k_1 & -k_1 \\ -k_1 & (k_1 + k_2) \end{bmatrix}\begin{Bmatrix} x_1 \\ x_2 \end{Bmatrix} = 0. \tag{3.129}$$

Tasks: Rework Example Problem XP3.21 using assumed modal damping of 10% for each mode; that is, $\zeta_1 = \zeta_2 = 0.10$.

Solution

For Example Problem XP3.21, $\omega_{n1} = 6.472$ rad/s, and $\omega_{n2} = 18.92$ rad/s; hence,

$$\begin{aligned}2\zeta_1\omega_{n1} &= 2 \times 0.1 \times 6.472 = 1.294\,\text{s}^{-1}, \\ 2\zeta_2\omega_{n2} &= 2 \times 0.1 \times 18.92 = 3.785\,\text{s}^{-1}\end{aligned}$$

From Equation 3.157, the differential-equation model including modal damping is

$$\begin{aligned}&\ddot{q}_1 + 1.294\dot{q}_1 + 41.886q_1 = 0, \\ &\ddot{q}_2 + 3.785\dot{q}_2 + 358.11q_2 = 0\end{aligned}$$

$$\begin{Bmatrix} x_1 \\ x_2 \end{Bmatrix} = \begin{bmatrix} 0.07377 & 0.03500 \\ 0.04287 & -0.09034 \end{bmatrix}\begin{Bmatrix} q_1 \\ q_2 \end{Bmatrix}. \tag{i}$$

The modal differential equations are homogeneous. The physical initial conditions are $x_1(0) = x_2(0) = 0$, and $\dot{x}_1(0) = \dot{x}_2(0) = v_0 = 4$ m/s. From Example Problem XP3.21, the corresponding modal initial conditions are $q_1(0) = q_2(0) = 0$, and $\dot{q}_1(0) = 61.41$ m kg$^{1/2}$/s, $\dot{q}_2(0) = -15.54$ m kg$^{1/2}$/s. We need to find the modal-coordinate solutions $q_1(t)$, $q_2(t)$ and then solve for the physical-coordinate solution $x_1(t)$, $x_2(t)$. The complete solution for the equation of motion $\ddot{q}_i + 2\zeta_i\omega_{ni}\dot{q}_i + \omega_{ni}^2 q_i = 0$ with initial conditions $q_1(0) = 0$, $q_1(0) = q_{10}$ is $q_i = (\dot{q}_{i0}/\omega_{di})e^{-\zeta\omega_{ni}t} \sin \omega_{di}t$. Starting with

$$\begin{aligned}\omega_{d1} &= 6.472\sqrt{1 - 0.1^2} = 6.440 \text{ rad/s}, \\ \omega_{d2} &= 18.92\sqrt{1 - 0.1^2} = 18.82 \text{ rad/s},\end{aligned}$$

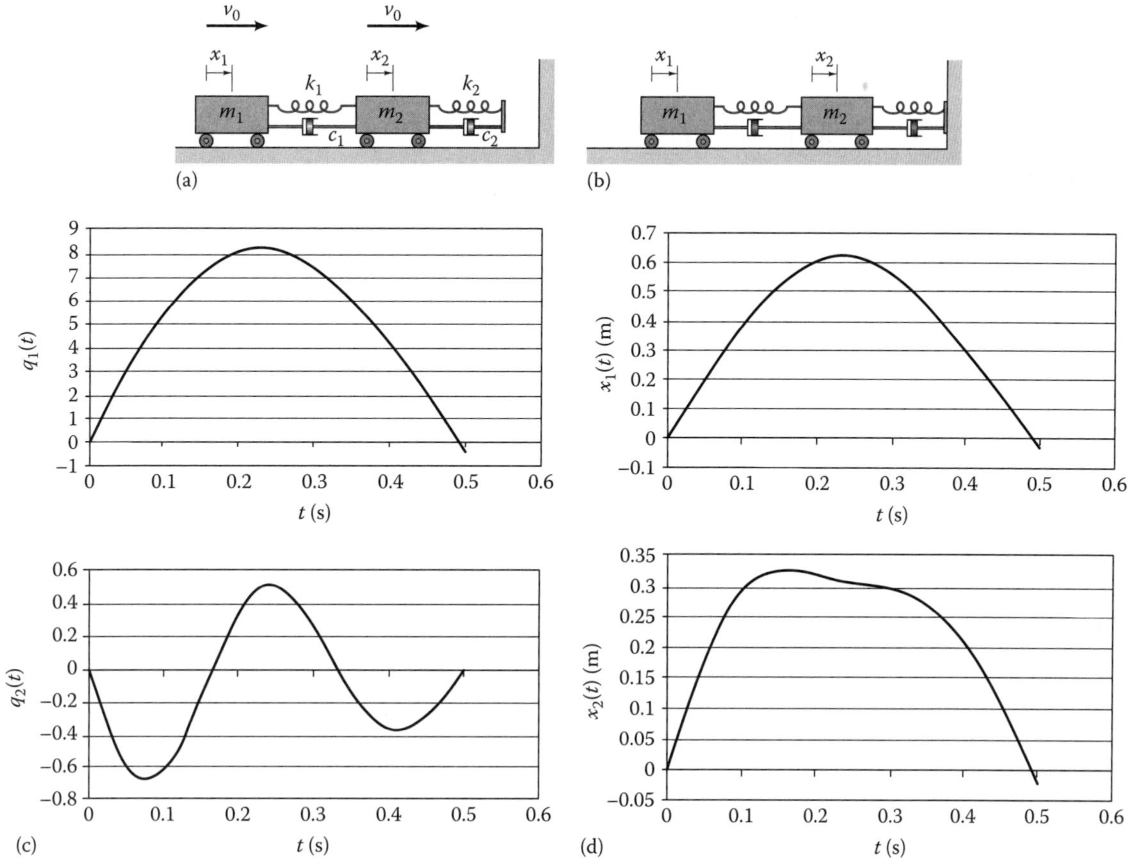

FIGURE XP3.22
(a) Two-car assembly in motion prior to impact, (b) motion following impact, (c) modal-coordinate solution, (d) physical-coordinate solution.

and applying these results to the present example gives the modal-coordinate solutions

$$q_1 = \frac{61.41}{6.440} e^{-0.6472t} \sin 6.440t$$

$$= 9.535e^{-0.6472t} \sin 6.404t \ (\text{m kg}^{1/2})$$

$$q_2 = \frac{-15.14}{18.82} e^{-1.892t} \sin 18.82t$$

$$= -.804e^{-1.892t} \sin 18.82t \ (\text{m kg}^{1/2}).$$

The transformation to obtain the physical coordinates gives

$$\begin{Bmatrix} x_1(t) \\ x_2(t) \end{Bmatrix} = q_1(t) \begin{Bmatrix} 0.07377 \\ 0.04287 \end{Bmatrix} + q_2(t) \begin{Bmatrix} 0.03500 \\ -0.09034 \end{Bmatrix} \ (\text{m}).$$

Figure XP3.22c illustrates the modal-coordinate solution. Comparing this result to Figure XP3.21c shows a slight reduction in the peak value for $q_1(t)$ and a brisk reduction in $q_2(t)$. This comparatively greater reduction in $q_2(t)$ arises because $q_2(t)$ experiences 1.5 cycles of motion, while $q_1(t)$ is experiencing about one half cycle. Comparing the physical coordinates with damping (Figure XP3.22d) and without (Figure XP3.21d) shows that damping produces a reduction in amplitudes.

3.5.2.4 Steady-State Solutions due to Harmonic Excitation

We developed steady-state solutions for a single-degree-of-freedom, damped harmonic oscillator in Section 3.2.5 and will consider the development of a similar solution for a 2DOF example here. Consider the following version of Equation 3.124:

$$\begin{bmatrix} m_1 & 0 \\ 0 & m_2 \end{bmatrix} \begin{Bmatrix} \ddot{x}_1 \\ \ddot{x}_1 \end{Bmatrix} + \begin{bmatrix} (k_1+k_2) & -k_2 \\ -k_2 & (k_2+k_3) \end{bmatrix} \begin{Bmatrix} x_1 \\ x_2 \end{Bmatrix} = \begin{Bmatrix} f_{10} \\ f_{20} \end{Bmatrix} \sin \omega t, \quad (3.158)$$

with a harmonic-excitation force vector on the right-hand side.

Continuing with the model defined by Equations 3.124 and 3.153 we can write the modal differential Equations 3.154 as

$$\begin{Bmatrix} \ddot{q}_1 + 0.6340q_1 \\ \ddot{q}_2 + 2.336q_2 \end{Bmatrix} = \begin{bmatrix} 0.4597 & 0.6280 \\ 0.8881 & -0.3251 \end{bmatrix} \begin{Bmatrix} f_{10} \\ f_{20} \end{Bmatrix} \sin \omega t$$
$$= \begin{Bmatrix} 0.4597f_{10} + 0.6280f_{20} \\ 0.8881f_{20} - 0.3251f_{20} \end{Bmatrix} \sin \omega t.$$

To complete a numerical solution, we arbitrarily set $f_{10} = f_{20} = 1$ N to obtain the component modal differential equations as follows:

$$\ddot{q}_1 + 0.6340q_1 = 1.087 \sin \omega t$$
$$\ddot{q}_2 + 2.366q_2 = 0.563 \sin \omega t.$$

Assuming steady-state solutions of the form $q_{1ss} = b_1 \sin \omega t$; $q_{2ss} = b_2 \sin \omega t$, and solving for the unknowns yields

$$\begin{aligned} q_{1ss}(\omega, t) &= \left(\frac{1.087}{\omega_{n_1}^2}\right) \frac{\sin \omega t}{1 - (\omega/\omega_{n1})^2}, \\ \omega_{n1} &= \sqrt{0.6340} = 0.7962 \text{ rad/s} \\ q_{2ss}(\omega, t) &= \left(\frac{0.563}{\omega_{n_2}^2}\right) \frac{\sin \omega t}{1 - (\omega/\omega_{n2})^2}, \\ \omega_{n2} &= \sqrt{2.3660} = 1.538 \text{ rad/s}. \end{aligned} \quad (3.159)$$

The steady-state, modal-coordinate amplitudes are functions of ω/ω_{ni}, the ratio of excitation frequency to the natural frequency. These results are comparable to the (damped) solution amplitude defined in Equation 3.38 with $\zeta = 0$. Obviously, a modal coordinate q_i becomes very large when the excitation frequency ω is near its natural frequency ω_{ni}.

The physical coordinates are obtained from the coordinate transformation $(x_i) = [A^*](q_i)$ as

$$\begin{Bmatrix} x_1(\omega,t) \\ x_2(\omega,t) \end{Bmatrix}_{ss} = q_{1ss}(\omega,t) \begin{Bmatrix} 0.4597 \\ 0.6280 \end{Bmatrix} + q_{2ss}(\omega,t) \begin{Bmatrix} 0.8881 \\ 0.3251 \end{Bmatrix} \text{ (m)}. \quad (3.160)$$

Looking at the solution provided by Equations 3.159 and 3.160, think about slowly increasing the excitation frequency from nearly zero to a frequency well above the second natural frequency. For $\omega \cong \omega_{nI}$, very large amplitudes result for $q_{1ss}(\omega, t)$. In this frequency range, the physical solution will be dominated by the first modal coordinate and, at any specified value of time, the solution vector (x_1, x_2), will "look like" the first mode shape. Similarly, for $\omega \cong \omega_{n2}$ the physical solution will be dominated by the second modal coordinate and will "look like" the second mode shape.

Figure 3.53a illustrates the amplitudes and phase for the modal coordinates of Equation 3.159. These solutions have the form $q_{iss} = b_i(\omega) \sin(\omega t + \phi_i)$ and basically are a repeat of the 1DOF amplitude and phase results of Figures 3.15 and 3.16. The amplitudes for $q_{1ss}(\omega)$, $q_{2ss}(\omega)$ peak when ω is near their respective natural frequencies ω_{n1}, ω_{n2}. The phase ϕ_i is zero for $\omega < \omega_{ni}$; it is 180° for $\omega > \omega_{ni}$, and 90° for $\omega = \omega_{ni}$.

The physical coordinates of Equation 3.160 can be stated

$$\begin{aligned} x_1 &= A_1(\omega) \sin[\omega t + \phi_1(\omega)] + A_2(\omega) \sin[\omega t + \phi_2(\omega)] \\ &= A_1[\sin \omega t \cos \phi_1 + \cos \omega t \sin \phi_1] \\ &\quad + A_2[\sin \omega t \cos \phi_2 + \cos \omega t \sin \phi_2] \\ &= x_{1c} \cos \omega t + x_{1s} \sin \omega t \\ &= X_i \sin(\omega t + \Phi_i), \end{aligned}$$

where

$$x_{1c} = A_1 \sin \phi_1 + A_2 \sin \phi_2, \quad x_{1s} = A_1 \cos \phi_1 + A_2 \cos \phi_2$$
$$X_i = (x_{ic}^2 + x_{is}^2)^{1/2}, \quad \Phi_i = \tan^{-1}\left(\frac{x_{ic}}{x_{is}}\right).$$

The angle Φ_i defines the phase of the x_i coordinate with respect to the input excitation vector that is driving the system at $\sin\omega t$. Figure 3.53b provides the amplitude and phase solutions for the physical coordinates. Note that $x_1(\omega)$, $x_2(\omega)$ are in phase when $\omega \cong \omega_{nI}$ and out of phase when $\omega \cong \omega_{n2}$, which is consistent with the mode shapes stated in Equation 3.147 and illustrated in Figure 3.51. Also, the ratios of the physical amplitudes near the peaks are the same as predicted by the eigenvectors. Specifically, when $\omega \cong \omega_{n2}$, the response amplitudes (x_1, x_2) have the same signs and relative amplitudes as the first eigenvector; when $\omega \cong \omega_{n2}$ the response amplitudes (x_1, x_2) have the same signs and relative amplitudes as the second eigenvector.

Look back at the modal solution provided by Equation 3.159. Note that the modal steady-state amplitude for the ith modal coordinate is inversely proportional to the eigenvalue ω_{nt}^2; that is, $q_{1ss} \approx 1/\omega_{n1}^2$ and $q_{2ss} \approx 1/\omega_{n2}^2$. Think of a problem with many degrees of freedom with harmonic excitation over a restricted frequency range $[0, \overline{\omega}]$. The solution format of Equation 3.159 suggests that a mode with a natural frequency ω_{ni} that is much larger than the top excitation frequency $\overline{\omega}$ can be "dropped" from the model, because (1) its contribution is primarily static; that is, $\omega_{nt}/\overline{\omega} \ll 1$ and $1/[1 - (\omega/\omega_{nt})^2] \simeq 1$, (2) the static contribution (being proportional to $1/\omega_{nt}^2$) is itself small. This "modal truncation" approach is regularly used to reduce the dimensionality of large vibration and structural-dynamics problems. A model with several hundred degrees of

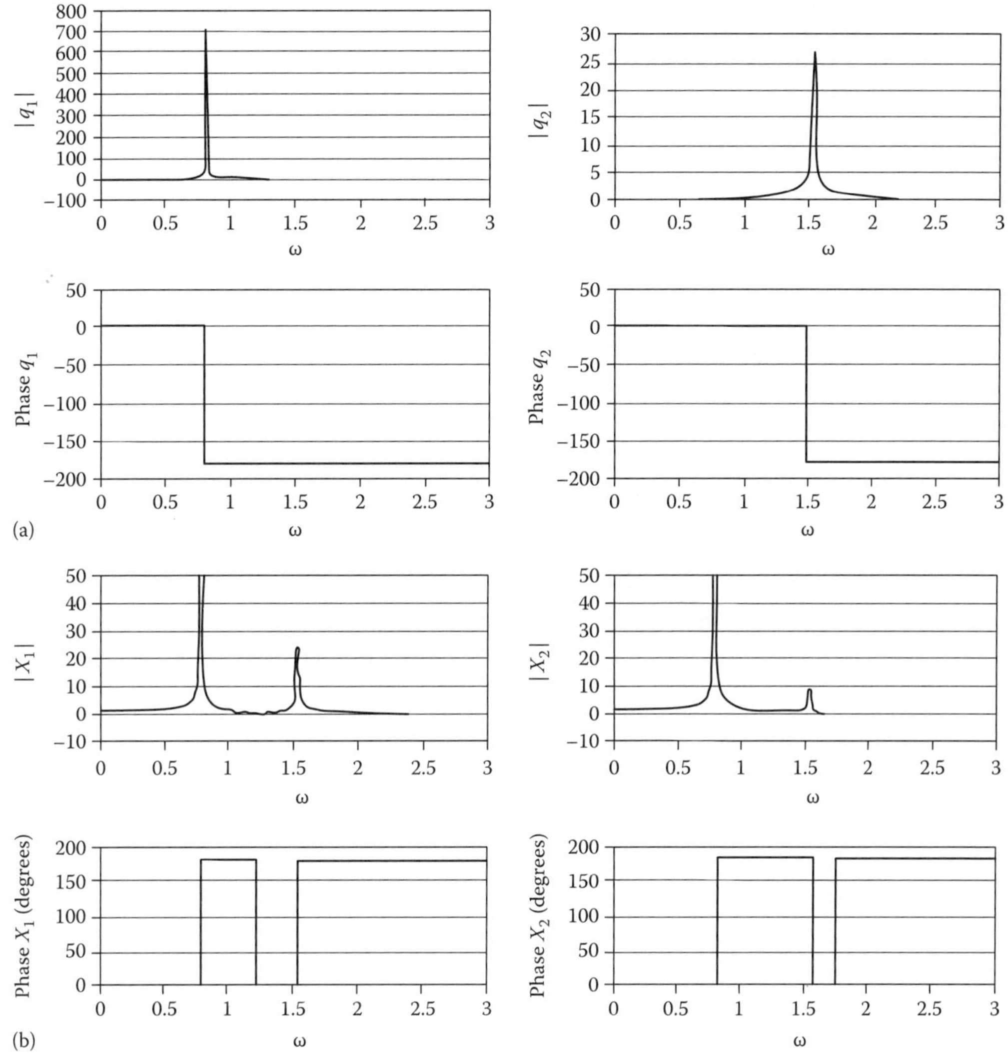

FIGURE 3.53
(a) Steady-state amplitude and phase for modal-coordinate solution of Equation 3.159 for $0 \le \omega \le 3$ rad/s, (b) steady-state amplitudes and phase for the physical-coordinates of Equation 3.160 for $0 \le \omega \le 3$ rad/s.

freedom can frequently be modeled adequately with a greatly-reduced-dimension model after dropping higher-order modes.

The modal-solution development has been covered to give some insight to the *nature* of a harmonic solution in terms of modal response. In fact, a harmonic solution can be developed more directly in terms of the physical coordinates (alone). To demonstrate the approach, we substitute the assumed steady-state solution, $x_1 = X_1 \sin\omega t$, $x_2 = X_2 \sin\omega t$, into Equation 3.158 to obtain

$$\begin{bmatrix} -m_1\omega^2 + (k_1 + k_2) & -k_2 \\ -k_2 & -m_2\omega^2 + (k_2 + k_3) \end{bmatrix} \begin{Bmatrix} X_1 \\ X_2 \end{Bmatrix} = \begin{Bmatrix} f_{10} \\ f_{20} \end{Bmatrix}. \tag{3.161}$$

Solving for the unknowns using Cramer's rule gives

$$X_1 = \frac{f_{10}[-m_2\omega^2 + (k_2 + k_3)] + f_{20}k_{10}}{m_1m_2\omega^4 - [m_1(k_2 + k_3) + m_2(k_1 + k_2)]\omega^2 + (k_1 + k_2)(k_2 + k_3) - k_2^2}$$
$$X_2 = \frac{f_{10}k_2 + f_{20}[(k_1 + k_2) - m_1\omega^2]}{m_1m_2\omega^4 - [m_1(k_2 + k_3) + m_2(k_1 + k_2)]\omega^2 + (k_1 + k_2)(k_2 + k_3) - k_2^2}. \tag{3.162}$$

Go back and look at the frequency Equation 3.142 and note that it is obtained by setting the denominator in Equations 3.162 equal to zero. Hence, when the excitation frequency ω coincides with either of the natural frequencies ω_{n1}, ω_{n2}, the denominator is zero, and the response amplitudes become unbounded. For the same numerical input data, Equations 3.162 provide precisely the same answer as illustrated in Figure 3.53b.

3.5.2.5 Harmonic Response with Damping

Calculating the steady-state response due to harmonic excitation is generally straightforward. Consider the following general n-dimensional matrix version of the 2DOF model of Equation 3.129:

$$[M](\ddot{x}_i) + [C](\dot{x}_t) + [K](x_i) = (f_{i0}) \sin \omega t, \tag{3.163}$$

where (x_i) is an n-dimensional vector of physical coordinates, and (f_{i0}) is an n-dimensional vector of physical forces. In an exact parallel to the assumed solution for the steady-state solution for the 1DOF damped harmonic oscillator of Equation 3.34, we assume a solution of the form

$$\begin{aligned}
(x_{iss}) &= (x_{is}) \sin \omega t + (x_{ic}) \cos \omega t \\
(\dot{x}_{iss}) &= \omega(x_{is}) \cos \omega t - \omega(x_{ic}) \sin \omega t \\
(\ddot{x}_{iss}) &= -\omega^2(x_{is}) \sin \omega t - \omega^2(x_{ic}) \cos \omega t.
\end{aligned} \tag{3.164}$$

Substituting these results into Equation 3.163 gives

$$\begin{aligned}
&[M]\{-\omega^2(x_{is}) \sin \omega t - \omega^2(x_{ic}) \cos \omega t\} \\
&\quad + [C]\{\omega(x_{is}) \cos \omega t - (x_{ic}) \sin \omega t\} \\
&\quad + [K]\{(x_{is}) \sin \omega t + (x_{ic}) \cos \omega t\} = \{f_{10}\} \sin \omega t.
\end{aligned}$$

From arguments of linear independence, the coefficients of $\sin\omega t$, $\cos\omega t$ on both sides of this equation must be equal and give the following:

$$\begin{aligned}
\cos \omega t: &\quad -\omega^2[M](x_{ic}) + \omega[C](x_{is}) + [K](x_{ic}) = (0) \\
\sin \omega t: &\quad -\omega^2[M](x_{is}) - \omega[C](x_{ic}) + [K](x_{is}) = (f_{i0}).
\end{aligned}$$

Combining these equations gives

$$\begin{bmatrix} -\omega^2[M] + [K] & \omega[C] \\ -\omega[C] & -\omega^2[M] + [K] \end{bmatrix} \begin{Bmatrix} (x_{ic}) \\ (x_{is}) \end{Bmatrix} = \begin{Bmatrix} (0) \\ (f_{i0}) \end{Bmatrix}, \tag{3.165}$$

a single matrix equation in the 2-n unknowns (x_{ic}), (x_{is}).

The solutions for the steady-state response components is normally given in terms of amplitude and phase as defined by

$$x_i = x_{is} \sin \omega t + x_{ic} \cos \omega t = X_i \sin(\omega t + \phi_i),$$

where

$$X_i = (x_{ic}^2 + x_{is}^2)^{1/2}, \quad \phi_i = \tan^{-1}\left(\frac{x_{ic}}{x_{is}}\right). \tag{3.166}$$

Note that the phase angle defines the phase of the response with respect to the input excitation vector that is driving the system at $\sin\omega t$.

For a numerical example, we will use the following mass and stiffness matrices from Equation 3.144:

$$[M] = \begin{bmatrix} 1 & 0 \\ 0 & 2 \end{bmatrix} \text{kg} \quad [K] = \begin{bmatrix} 2 & -1 \\ -1 & 2 \end{bmatrix} \text{N/m}. \tag{3.167a}$$

These matrices define the undamped natural frequencies $\omega_{n1} = .7962$ rad/s $\Rightarrow f_{n1} = .1267$ Hz, and $\omega_{n2} = 1.538$ rad/s $\Rightarrow f_{n2} = .2448$ Hz. The damping matrix is defined (arbitrarily) by the coefficients $c_1 = .001$ N s/m; $c_2 = .001$ N s/m as

$$[C] = \begin{bmatrix} .002 & -.001 \\ -.001 & .001 \end{bmatrix} \text{N s/m}. \tag{3.167b}$$

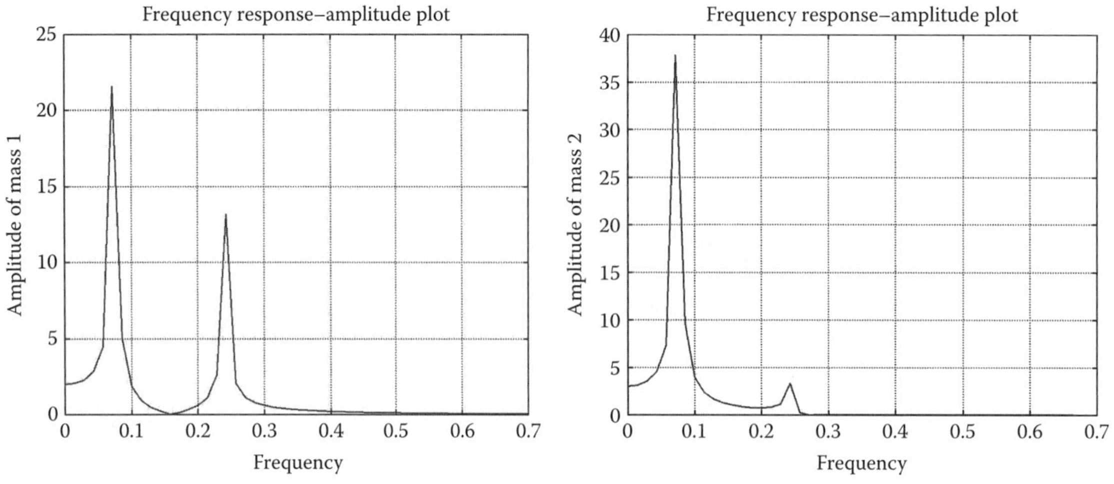

FIGURE 3.54
Steady-state amplitudes for the 2DOF example defined by Equations 3.163 and 3.167 for $0 \leq \omega \leq 0.7$ Hz.

We will continue with the excitation force vector used in the prior undamped excitation example, specifically,

$$\begin{Bmatrix} f_1(t) \\ f_2(t) \end{Bmatrix} = \begin{Bmatrix} f_{10} \\ f_{20} \end{Bmatrix} \sin \omega t, \quad f_{10} = f_{20} = 1\text{ N}. \qquad (3.167c)$$

Figure 3.54 illustrates the steady-state amplitudes for the x_1 and x_2 coordinates of the model defined by Equations 3.163, 3.165, and 3.167. By comparison to Figures 3.53, damping is seen (as expected) to limit the peak responses associated with excitation at or near the natural frequencies.

3.6 Work–Energy Applications for One-Degree-of-Freedom Problems in Plane Motion

The preceding sections dealt most directly with the derivation of equations of motion for one or two particles, with a secondary interest in solving the equations. In solving the differential equations of motion for 1DOF examples, we regularly used the "energy-integral substitution" to integrate equations with respect to displacement. This subsection considers the energy integral more formally and directly and considers classes of problems that can be modeled conveniently from an energy perspective.

We will also begin to consider a reversal of the thought process in which a free-body diagram is used as the starting point to derive differential equations of motion that are then integrated via the energy-integral substitution:

Integrate (Differential equation of motion)
$\Rightarrow$ (Work–energy expression),

to a thought process of starting with a work–energy equation and differentiating to obtain the equation of motion; that is,

Differentiate (Work–energy expression)
$\Rightarrow$ (Differential equation of motion).

Stated differently, you will find that (sometimes) the equation of motion can be more easily derived from a work–energy expression than from $\sum f = m\ddot{r}$ and free-body diagrams. In fact, the work–energy equation provides an alternative basis for the analysis and interpretation of dynamics.

Note that the procedures that are employed here only apply for single-degree-of-freedom examples. The 2DOF examples of Section 3.5 cannot be developed by the procedures of this section. Equations of motion for multi-degree-of-freedom systems can be developed from more sophisticated work–energy formulations or "variational" principles, using Lagrange's equations of motion.

3.6.1 The Work–Energy Equation and Its Application

Our objective in this subsection is the derivation of a general work–energy expression. The derivation is as easily done in three dimensions as in two and will be developed in a Cartesian-coordinate system. Figure 3.55 illustrates a particle of mass m in the X, Y, Z inertial coordinate system. The resultant force vector $\boldsymbol{f}$ acts on the particle causing the acceleration $\ddot{\boldsymbol{r}}$. The component differential equations of motion are

$$f_X = m\ddot{r}_X = m\ddot{X} = m\frac{d}{dX}\left(\frac{\dot{X}^2}{2}\right)$$

$$f_Y = m\ddot{r}_Y = m\ddot{Y} = m\frac{d}{dY}\left(\frac{\dot{Y}^2}{2}\right)$$

$$f_Z = m\ddot{r}_Z = m\ddot{Z} = m\frac{d}{dZ}\left(\frac{\dot{Z}^2}{2}\right),$$

where the energy-integral substitutions

$$\ddot{X} = \frac{d\dot{X}}{dt} = \frac{d\dot{X}}{dX}\frac{dX}{dt} = \frac{d}{dX}\left(\frac{\dot{X}^2}{2}\right)$$

etc. have been employed. Multiplying the first, second, and third component equations by dX, dY, and dZ, respectively, and adding the resultant expressions gives

$$\boldsymbol{f}\cdot d\boldsymbol{r} = f_X dX + f_Y dY + f_Z dZ = \frac{m}{2}d[(\dot{X}^2) + (\dot{Y}^2) + (\dot{Z}^2)].$$

Note that the term on the left-hand side is the differential work done by the force vector $\boldsymbol{f} = \boldsymbol{I}f_X + \boldsymbol{J}f_Y + \boldsymbol{K}f_Z$

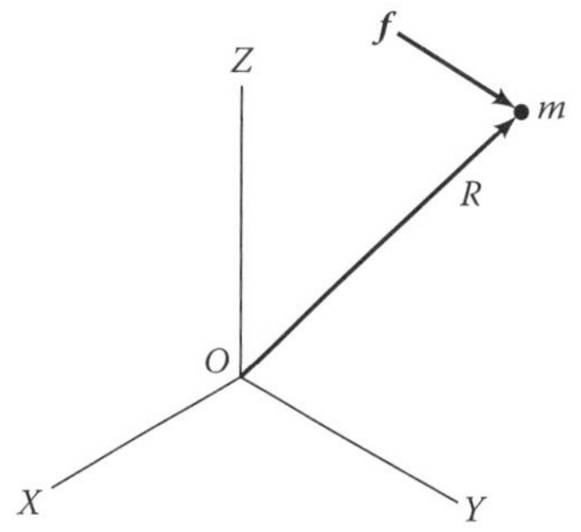

FIGURE 3.55
A particle of mass m acted on by an external force f.

while acting through the differential-displacement vector $\boldsymbol{dr} = \boldsymbol{I}dX + \boldsymbol{J}dY + \boldsymbol{K}dZ$, and defined formally by $\boldsymbol{f} \cdot \boldsymbol{dr}$. The term on the right-hand side is

$$dT = d\left(\frac{mv^2}{2}\right),$$

the differential kinetic energy of the particle, where

$$v^2 = \dot{X}^2 + \dot{Y}^2 + \dot{Z}^2 = v_X^2 + v_Y^2 + v_Z^2.$$

Integrating the differential scalar work–energy expression gives

$$\int \boldsymbol{f} \cdot \boldsymbol{dr} = \int dT$$

$$\int d\ (\text{Work}) = \int d\ (\text{Kinetic energy}).$$

Integrating the last expression from point 1 ($\boldsymbol{I}X_1 + \boldsymbol{J}Y_1 + \boldsymbol{K}Z_1$) to point 2 ($\boldsymbol{I}X_2 + \boldsymbol{J}Y_2 + \boldsymbol{K}Z_2$) with respect to the path coordinate s gives the final work–energy expression

$$\int_1^2 \boldsymbol{f} \cdot \boldsymbol{dr} = \text{Work}_{1\to 2} = T_2 - T_1. \quad (3.168)$$

Note that the three components of the *vector* equation $\sum \boldsymbol{f} = m\ddot{\boldsymbol{r}}$ have been reduced to a single scalar expression.

As we shall see in the balance of this section, Equation 3.168 is extremely useful in analyzing and formulating dynamics problems. It is easy to apply in terms of the kinetic-energy expression on the right-hand side; that is, $T = mv^2/2$ is easily stated. Understanding and developing the work expression on the left-hand side can be more difficult.

To begin thinking about the work term, consider the example provided in Figure 3.56a. The mass is supported by the springs and a linear damper and is acted on by the time-varying force $f(t)$. The spring is undeflected at $Y = 0$. The free-body diagram is shown in Figure 3.56b.

Summing forces in the $+Y$ direction yields

$$m\ddot{Y} = \Sigma f_Y = f(t) - kY - c\dot{Y} - w.$$

Introducing the energy-integral substitution for $\ddot{Y}$ into this equation gives

$$m\frac{d}{dY}\left(\frac{\dot{Y}^2}{2}\right) = (-kY - w) + f(t) - c\dot{Y}.$$

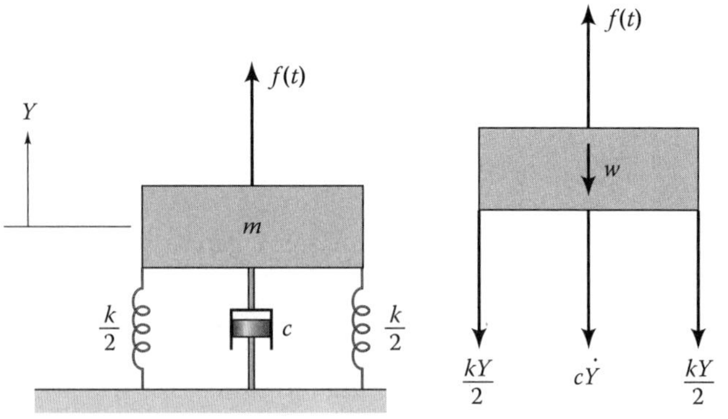

(a) (b)

FIGURE 3.56
(a) Mass supported by linear springs and a damper and acted on by $f(t)$, (b) free-body diagram for the springs initially undeflected and $Y > 0$, $\dot{Y} > 0$.

Multiplying through by dY and integrating both sides yields

$$m\left(\frac{\dot{Y}_2^2}{2}\right) - m\left(\frac{\dot{Y}_1^2}{2}\right) = \Bigg|_1^2 \left[-k\left(\frac{Y^2}{2}\right) - wY\right] + \int_1^2 [f(t) - c\dot{Y}]\,dY. \quad (3.169)$$

The first two force terms on the right were integrated directly with respect to Y and are called *conservative*. The second two force terms cannot be integrated directly and are called *nonconservative*. A conservative force is defined formally as a force which can be expressed as the negative gradient of a *potential-energy function*, and the present example shows

$$f_{\text{spring}} = -kY = -\frac{d}{dY}\left[k\left(\frac{Y^2}{2}\right)\right] = -\frac{d}{dY}(V_{\text{spring}})$$

$$f_{\text{gravity}} = -w = -\frac{d}{dY}(wY) = -\frac{d}{dY}(V_{\text{gravity}}).$$

Hence, the potential-energy functions for a linear spring and gravity are, respectively,

$$V_{\text{spring}} = \text{spring potential energy function} = k\frac{Y^2}{2}$$

$$V_{\text{gravity}} = \text{gravity potential energy function} = wY.$$

The potential-energy function for a linear spring is $kY^2/2$, where Y is the change in length of the spring *from its undeflected position*. The potential function due to gravity is the weight of a body times the distance *above* an arbitrary elevation datum. For this problem, $Y = 0$ is the specified zero potential-energy-function datum for gravity. We now combine the separate potential-energy functions for the spring and gravity to obtain the *system potential-energy* function

$$V = V_{\text{spring}} + V_{\text{gravity}} = k\left(\frac{Y^2}{2}\right) + wY. \qquad (3.170)$$

Substituting from Equation 3.170 into Equation 3.169 gives

$$(T_2 + V_2) - (T_1 + V_1) = \int_1^2 [f(t) - c\dot{Y}]\,dY = \text{Work}_{n.c},$$

$$(3.171)$$

where

$$V_1 = k\frac{Y_1^2}{2} + wY_1, \quad T_1 = m\frac{\dot{Y}^2}{2},$$

etc., and $\text{Work}_{n.c.}$ is the nonconservative work. The intelligent application of Equation 3.171 greatly simplifies the development of equations of motion for many 1DOF dynamics examples. Kinematics is simplified because only velocities are needed, versus the requirement for accelerations in applying $\sum f = m\ddot{r}$. If all the forces acting on the body are conservative, the right-hand side of Equation 3.171 is zero, and it reduces to the particularly simple form

$$T_2 + V_2 = T_1 + V_1 = T + V, \qquad (3.172)$$

which states that the *mechanical* energy of the system is conserved (is constant). Equation 3.172 is normally called the *Conservation of Mechanical Energy Equation.* When it applies, even the free-body diagram may be optional. We will be using conservation-of-energy principles to analyze several examples in this section.

Conservative and nonconservative forces differ on both a physical and a mathematical basis. On a physical basis, nonconservative forces change the mechanical energy $(T + V)$ of the system in an irreversible fashion. The linear damper of the current example (plus Coulomb and aerodynamic damping) dissipates mechanical energy as heat. The applied force $f(t)$ adds or subtracts mechanical energy to the system either from some external energy source or to an external sink.

A conservative force is defined formally as a force that can be expressed as the negative gradient of a scalar potential (energy) function; that is,

$$f_{conservative} = -\nabla V,$$

where ∇ (nabla) is the vector operator

$$\nabla = \boldsymbol{I}\frac{\partial}{\partial \mathbf{X}} + \boldsymbol{J}\frac{\partial}{\partial \mathbf{Y}} + \boldsymbol{K}\frac{\partial}{\partial \mathbf{Z}}.$$

This definition obscures the really important feature of conservative forces; namely, for a conservative force, the integrand can be integrated as an exact differential, that is,

$$\text{Work}_{conservative} = \int_1^2 f_{conservative} \cdot \boldsymbol{dr} = -\int_1^2 \nabla V \cdot \boldsymbol{dr}$$

$$= -\int_1^2 dV = -\Big|_1^2 V.$$

Because dV is an exact differential, the value of the integral is independent of the path that is taken in its evaluation. This is not the case with nonconservative forces. In the present example, the nonconservative work done by the external force and the damper are defined by the integral

$$\text{Work}_{n.c.} = \int_1^2 [f(t) - c\dot{Y}]dY.$$

The value of the integral depends upon the path followed between points 1 and 2; hence, it could only be evaluated (numerically) in the form

$$\text{Work}_{n.c.} = \int_{Y_1}^{Y_2} [f(t) - c\dot{Y}]dY = \int_{t_1}^{t_2} [f(t) - c\dot{Y}(t)]\dot{Y}(t)\,dt,$$

after the problem's solution in terms of $Y(t)$ and $\dot{Y}(t)$ are known. From this perspective, Equation 3.171 is rarely used *directly* in developing a solution (i.e., solving the differential equation of motion) for dynamics problems. Problems do arise, in which the mechanical energy $(T + V)$ of a system is known at two different times, and Equation 3.171 can be used to evaluate the nonconservative work due to a change in $T + V$, that is,

$$\text{Work}_{n.c.} = (T_2 + V_2) - (T_1 + V_1) \qquad (3.173)$$

Equation 3.173 can sometimes be used to efficiently *derive* equations of motion that include nonconservative forces, and examples of this nature will be considered.

3.6.1.1 More on Spring Forces and Spring Potential-Energy Functions

The definition of potential energy for a linear spring with spring constant k is $V = k\delta_s^2/2$, where δ_s *is the change in length of the spring from its undeflected length.* The definition of the potential energy due to a body's

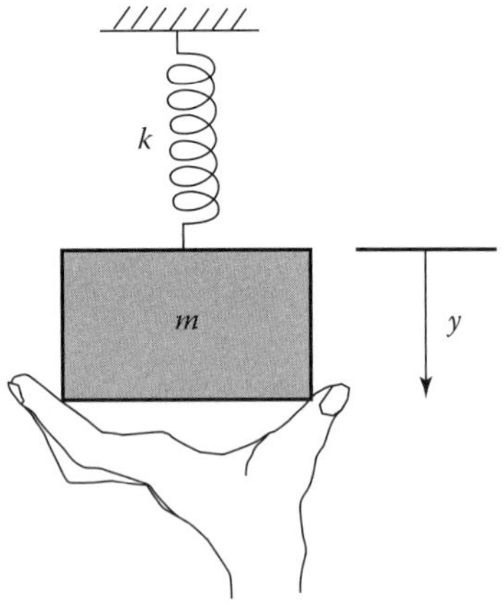

FIGURE 3.57
Supported system with spring undeflected.

weight is the weight times the distance above a datum plane. Figure 3.57 repeats Figure 3.4a and shows a mass m connected to a support point by a linear spring with stiffness coefficient k. We derived the equation of motion for this model as Equation 3.13 using $\sum f = m\ddot{r}$ and are now going to get the same result starting from conservation of energy.

The mass is initially supported with the spring undeflected. Y defines the downward displacement of the body and is initially zero. With this coordinate choice, $V_g = -wY$, and $V_s = kY^2/2$. Since energy is preserved, we can state

$$T + V = T_0 + V_0 \Rightarrow m\frac{\dot{Y}^2}{2} - wY + k\frac{Y^2}{2} = 0$$

differentiating with respect to Y gives

$$m\frac{d(\dot{Y}^2/2)}{dY} - w + kY = 0 \Rightarrow m\ddot{Y} + kY = w \qquad (3.174)$$

This outcome repeats Equation 3.13 and was obtained without recourse to a free-body diagram.

Equation 3.174 defines the equilibrium position $\overline{Y}$ from $m\ddot{Y} = \sum f_Y = 0 = w - k\overline{Y}$ as $\overline{Y} = w/k$. We can define the motion about equilibrium by $Y = \overline{Y} + y \Rightarrow \dot{Y} = \dot{y}$. Substituting this result into Equation 3.174 gives

$$m\frac{\dot{y}^2}{2} - w(\overline{Y} + y) + k\frac{(\overline{Y} + y)^2}{2} = 0.$$

Differentiating with respect to y gives

$$m\frac{d(\dot{y}^2/2)}{dy} - w + k(\overline{Y} + y) = 0 \Rightarrow m\ddot{y} + ky = w - K\overline{Y} = 0. \qquad (3.175)$$

This equation repeats Equation 3.14 demonstrating that w is eliminated if y defines motion about the equilibrium position. As developed in Equation 3.175 and in the earlier Newtonian development, this result can be viewed as an algebraic outcome. However, in terms of potential energy, eliminating w from the equation of motion has a distinct physical meaning. Potential energy can be viewed as a measure of the "capacity to do work." For the initially supported mass, release of m causes w to convert its potential energy $-wY$ into kinetic energy $m\dot{Y}^2/2$. However, at equilibrium, $w = k\overline{Y}$, and w can no longer do work on the body. Think about starting with the spring undeflected as shown in Figure 3.57 and slowly easing the body downward until equilibrium is reached. The action of slowly lowering the body actually removes w's potential energy. Hence, for motion about equilibrium,

$$T + V = T_0 + V_0 \Rightarrow m\frac{\dot{y}^2}{2} + k\frac{y^2}{2} = 0, \qquad (3.176a)$$

applies, and w's potential-energy contribution does not enter the equation.

Spring potential-energy functions can be developed for nonlinear as well as linear springs, and the balance of this subsection deals with nonlinear springs and their associated potential-energy functions. This topic is included for completeness. Most classroom (and real) dynamics problems involve linear springs. Figure 3.58 shows a "hardening" nonlinear spring which leads to the reaction-force definition

$$f_{\text{spring}} = -k(Y + \alpha Y|Y|),$$

where Y is the change in position of the spring from its undeflected position. This force model can be stated,

$$Y \geq 0;\quad f_{\text{spring}} = -k(Y + \alpha Y^2),$$
$$Y \leq 0;\quad f_{\text{spring}} = -k(Y - \alpha Y^2),$$

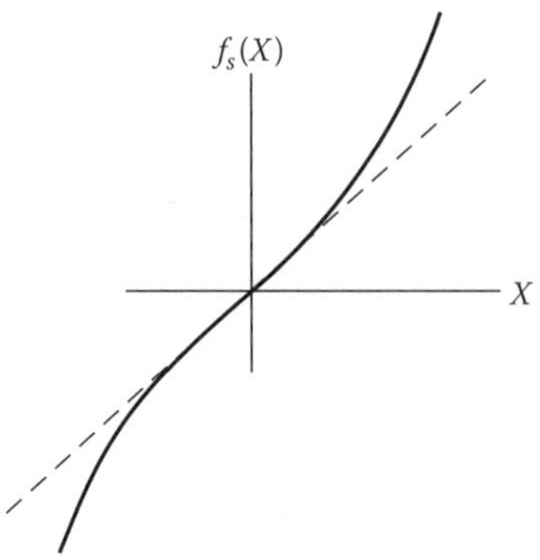

FIGURE 3.58
Nonlinear spring with a hardening characteristic.

which yields the "two-sided" potential-energy function

$$Y \geq 0; \quad V_{\text{spring}} = \frac{k}{2}Y^2 + \frac{k\alpha}{3}Y^3,$$

$$Y \leq 0; \quad V_{\text{spring}} = \frac{k}{2}Y^2 - \frac{k\alpha}{3}Y^3.$$

Replacing the linear spring in Figure 3.57 with this nonlinear-spring model and assuming that $Y \geq 0$ gives

$$T + V = T_0 + V_0 \Rightarrow m\frac{\dot{Y}^2}{2} - wY + \frac{k}{2}Y^2 + \frac{k\alpha}{3}Y^3 = 0. \tag{3.176b}$$

Differentiating with respect to Y gives

$$m\frac{d(\dot{Y}^2/2)}{dY} - w + kY(1 + \alpha Y) = 0$$
$$\Rightarrow m\ddot{Y} + kY(1 + \alpha Y) = w; \quad Y \geq 0.$$

The equilibrium position is defined from $m\ddot{Y} = \Sigma f_Y = 0 = w - k\overline{Y}(1 + \alpha\overline{Y})$. The positive root for this quadratic equation defines the equilibrium position $\overline{Y}$. For motion about equilibrium defined by $Y = \overline{Y} + y \Rightarrow \ddot{Y} = \ddot{y}$, the equation of motion becomes

$$\begin{aligned} m\ddot{y} &= \Sigma f_Y = w - k(\overline{Y} + y)[1 + \alpha(\overline{Y} + y)] \\ &= [w - k\overline{Y}(1 + \alpha\overline{Y})] - ky(1 + 2\alpha\overline{Y} + \alpha y) \\ &= 0 - ky(1 + 2\alpha\overline{Y} + \alpha y) \end{aligned} \tag{3.177}$$
$$\therefore \; m\ddot{y} + ky(1 + 2\alpha\overline{Y} + \alpha y) = 0$$

As with the linear-spring model, w is eliminated. However, if you wanted to eliminate w's contribution to the potential-energy function in Equation 3.176b, you would need the potential-energy function $\overline{V} = ky^2(1 + 2\alpha\overline{Y})/2 + k\alpha y^3/3$ to produce,

$$f_{sy} = -\frac{\partial\overline{V}}{\partial y} = -ky(1 + 2\alpha\overline{Y} + \alpha y^2),$$

on the right-hand side of Equation 3.177. Hence, unlike the linear-spring result of Equation 3.176a, you cannot simply drop w's contribution in developing the potential-energy function.

3.6.1.2 More on the Force of Gravity and the Potential-Energy Function for Gravity

The preceding development showed that the potential-energy function for a linear spring is difficult to "mess up" in terms of the *sign* of the displacement function because of the quadratic nature of the function $V_s = K\delta^2/2$, where δ is the change in length of the spring from its undeflected state. Figure 3.59 illustrates a simple problem with opportunities for confusion in defining the potential-energy function for gravity.

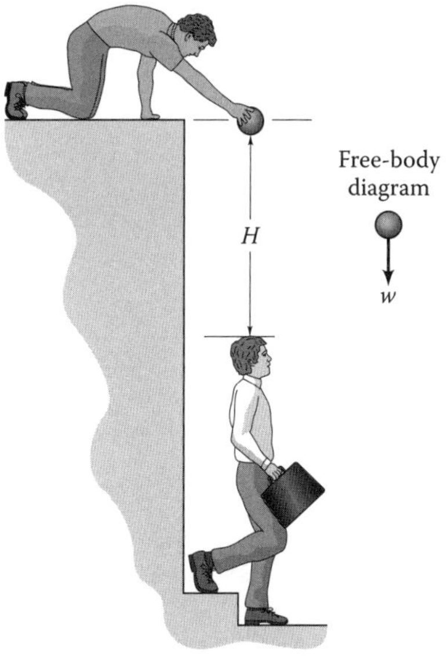

FIGURE 3.59
Falling-mass example.

As illustrated, an engineering student is conducting an experiment involving dropping a water balloon on a class mate from a height H. The question to be asked is the following: How fast will the balloon be going when (and if) it hits the target? Neglecting drag forces on the balloon, there are no nonconservative forces, and energy is conserved, hence,

$$T_1 + V_1 = T + V = \text{Constant}.$$

The potential energy due to gravity is the weight of the body w times the vertical distance *above* an arbitrary zero potential-energy datum. We choose the point of release as the zero potential-energy datum. In Figure 3.60a, Y is chosen as the coordinate to locate the mass m, and $Y = 0$ is the zero datum for gravity potential energy. Hence, the potential-energy function due to gravity is $V_g = wY = mgY$. The force due to gravity is defined from the potential-energy function to be $f_g = -dV_g/dY = -w$; that is, the weight is acting downward in the $-Y$ direction. Applying the conservation-of-energy equation to answer the velocity question at $Y = -H$ gives

$$T_1 + V_1 = T_2 + V_2$$
$$0 + 0 = \frac{m}{2}v_2^2 + w(-H) \Rightarrow v_2 = \sqrt{2gH}.$$

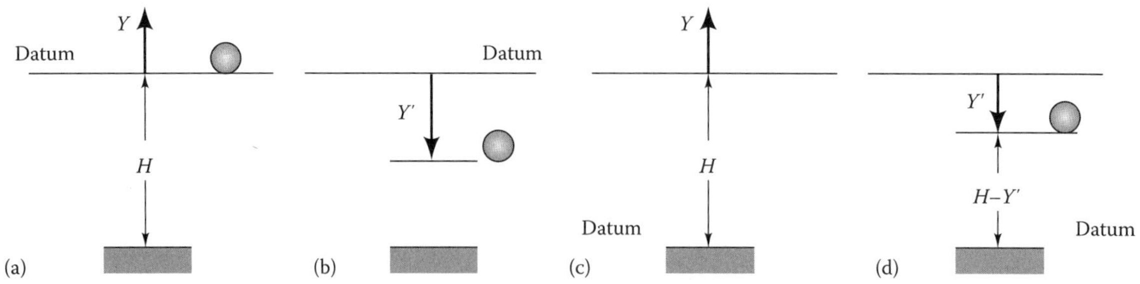

FIGURE 3.60
(a) Y locates m with the point of release as datum, (b) Y' locates m with the point of release as datum, (c) Y locates m with the point of contact as datum, (d) Y' locates m with the point of contact as datum.

In this equation, T_1 is zero because the mass starts from rest ($v_1 = 0$), and V_1 is zero because Y is zero at the initial position.

Suppose, as illustrated in Figure 3.60b, we keep the same datum but choose Y' as the coordinate to keep track of the m position. Recalling that the potential-energy function due to gravity is the weight of the mass times the distance *above* the datum, the potential-energy function is now $V_g = w(-Y') = -mgY'$, defining the force due to gravity as $f_g = -dV_g/dY' = w$. With Y' as the coordinate, the correct sign for w is now positive; that is, downward in the $+Y'$ direction. In terms of Y', the conservation-of-energy equation becomes

$$T_1 + V_1 = T + V$$

$$0 + 0 = \frac{m}{2}v^2 + w(-Y').$$

Setting $Y' = +H$ gives our earlier solution for v_2.

So far, this example shows that you can choose the coordinate locating the mass arbitrarily. A subsequent definition of V_g that is consistent with the definition (weight times vertical distance *above* the datum) will give both the correct sign for the weight force in the equation of motion and a correct implementation of the energy equation. In working problems using the energy equation, arriving at the end with a solution for a negative v^2 is a good clue that the wrong sign was used in defining V_g.

In Figure 3.60c, the problem has been redrawn from the "target's" point of view, with the datum now being the point of contact. Retaining Y as the coordinate means that the potential-energy function is now $V_g = w(Y + H)$; hence, $V_{g1} = w(0 + H) = wH$, and $V_{g2} = w(-H + H) = 0$.

Conservation of energy gives

$$T_1 + V_1 = T_2 + V_2 \Rightarrow 0 + wH = \frac{m}{2}v_2^2 + 0 \Rightarrow v_2 = \sqrt{2gH}. \quad (3.178)$$

Figure 3.60d illustrates Y' as the coordinate with the point of contact as datum. The potential-energy function becomes $V_g = w(H - Y')$; hence, $V_{g1} = w(H - 0) = wH$, and $V_{g2} = w(H - H) = 0$. The results of Equation 3.178 continue to apply.

The following lesson should be learned by working through the possibilities of Figure 3.60. The datum and coordinate choices are largely arbitrary, depending on your views in setting up the problem. However, to get a correct and meaningful formulation, the potential-energy-function definition must be consistent with your choice for the coordinate and the potential-energy datum.

Example Problem 3.23

Figure XP3.23a illustrates a particle that is released from rest on a (smooth) cylindrical surface. It slides to the bottom of the surface, and then zips along a (rough) horizontal surface until it hits a linear spring (spring constant k). The roughness of the horizontal surface is characterized by the Coulomb-friction coefficient μ_d.

Task: Find the maximum spring deflection.

Solution

We observe that energy is conserved in moving from point 0 to point 1 (no friction), but that the Coulomb friction will

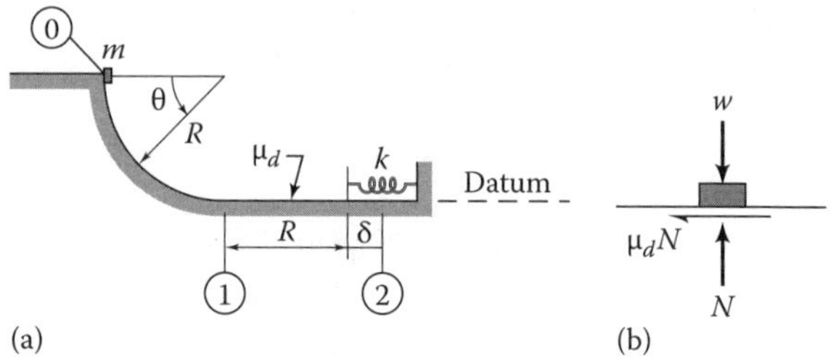

FIGURE XP3.23
(a) Particle sliding on a combined circular and horizontal surface, (b) free-body diagram for m sliding on the horizontal surface.

dissipate energy in moving from point 1 to point 2. Hence, we will break the problem up into $[0 \Rightarrow 1]$ and $[1 \Rightarrow 2]$ stages.

Starting with the $[0 \Rightarrow 1]$ stage, polar coordinates are appropriate; that is, in defining the velocity, we would logically use $v = v_\theta = r\dot{\theta}$. To this point, in applying the work–energy expression, we have used Cartesian coordinates to define kinetic energy. However, the kinetic-energy definition is independent of the choice of coordinate systems and applies equally for Cartesian, polar, or path coordinates. The definition only requires that the velocity components define the particle's velocity with respect to an *inertial* coordinate system.

As illustrated in Figure XP3.23a, the horizontal plane has been chosen as the zero gravity, potential-energy datum. Applying conservation of energy to find the velocity of the particle at an arbitrary value for θ yields

$$T_0 + V_0 = T + V \Rightarrow wR + 0 = \frac{m}{2}v^2 + wR(1 - \sin\theta)$$
$$\Rightarrow v = \sqrt{2gR\sin\theta}. \quad \text{(i)}$$

The velocity at point 1 is $v_1 = v_\theta(\theta = \pi/2) = \sqrt{2gR}$.

Proceeding to the $[1 \Rightarrow 2]$ stage, $\text{Work}_{n.c.} = \Delta(T+V)$ *is* appropriate. The free-body diagram in Figure XP3.23b applies for the mass sliding on the horizontal plane and can be used to determine the work done by Coulomb friction. The friction force is $f_f = \mu_d N = \mu_d w$. Hence, the magnitude of the work done in moving from point 1 to point 2 is $\mu_d w(R+\delta)$. The correct sign is negative, since Coulomb friction *removes* energy from the system. The correctness of this choice is seen from $(T_1 + V_1) - \mu_d w(R+\delta) = (T_2 + V_2)$, which states that the initial mechanical energy is *reduced* by the work due to friction to yield the mechanical energy in state 2. Completion of the stage 2 solution follows from

$$(T_1 + V_1) - \mu_d w(R+\delta) = (T_2 + V_2)$$
$$\Rightarrow \frac{m}{2}v_1^2 + 0 - \mu_d w(R+\delta) = 0 + \frac{k}{2}\delta^2. \quad \text{(ii)}$$

T_2 was set equal to zero in the last of Equation (ii) to determine the maximum deflection of the spring, since the peak deflection occurs (momentarily) when m's velocity is zero. Substituting for $v_1^2 = 2gR$ into the last of (ii) gives the following quadratic equation in δ:

$$\frac{m}{2}(2gR) - \mu_d w(R+\delta) = \frac{k}{2}\delta^2$$
$$\Rightarrow k\delta^2 + 2\mu_d w\delta - 2wR(1 - \mu_d) = 0.$$

The positive root to this equation is the answer to our problem and the only physically correct root. A negative value for δ would imply tension in the spring, which is not physically consistent with our problem definition. In working this example, we have neglected Coulomb friction for m's motion on the cylindrical surface. This aspect of the example is discussed in detail in Section 3.4.2.

3.6.2 Deriving Equations of Motion from Work–Energy Relations

We have plowed along through this book, regularly using the energy-integral substitution to integrate equations of motion. In the prior subsection, we looked at the work–energy equation as a useful integrated form of Newton's differential equation of motion. In this section, we will concentrate on the possibility of using the work–energy equation as a starting point to derive the equation of motion.

Figure 3.61a illustrates two particles, each of mass m, connected by a bar of length l with negligible mass. The position of the mass on the left is defined by Y; the mass on the right is located by X. Both masses move in frictionless guides. Friction between the masses and the guide walls is neglected. This system has one degree of freedom, and we will (eventually) use θ as the coordinate. The problem at hand is as follows: Derive the differential equation of motion for θ. The intended lesson of this exercise is that the work–energy equation is frequently a better (easier) starting point for deriving the equation of motion than $\sum f = m\ddot{r}$. To illustrate this point, we will first derive the equations of motion using $\sum f = m\ddot{r}$. Then, for comparison, we will rework the example from an energy-equation starting point.

Figure 3.61b provides the free-body diagram. Both bodies are acted on by the reaction force f_b in the connecting bar, which is shown acting in compression. The horizontal reaction force $\mathbf{N}_A$ acts on the left body as does the weight w. The weight of the right body is reacted by $\mathbf{N}_B$. Applying $\sum f = m\ddot{r}$ separately to the bodies gives

$$\begin{aligned} \text{Body } A:\ & m\ddot{Y} = \sum f_Y = f_b\sin\theta - w \\ \text{Body } B:\ & m\ddot{X} = \sum f_X = f_b\cos\theta. \end{aligned} \quad (3.179)$$

We have two equations and three unknowns (f_b, $\ddot{X}$, and $\ddot{Y}$), a familiar situation indicating the need for kinematic relations between the coordinates. For

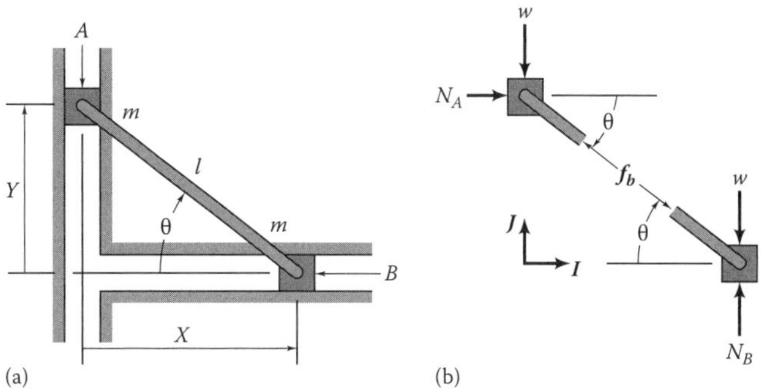

FIGURE 3.61
(a) Two bodies of mass m connected to a massless bar of length l, (b) free-body diagram.

the present problem, the requisite kinematics is provided by

$$Y = l\sin\theta \Rightarrow \dot{Y} = l\dot{\theta}\cos\theta \Rightarrow \ddot{Y} = -l\dot{\theta}^2\sin\theta + l\ddot{\theta}\cos\theta$$
$$X = l\cos\theta \Rightarrow \dot{X} = -l\dot{\theta}\sin\theta \Rightarrow \ddot{X} = -l\dot{\theta}^2\cos\theta - l\ddot{\theta}\sin\theta. \tag{3.180}$$

We have taken the trigonometric definitions for X and Y and used direct differentiation with respect to time to obtain $\dot{X}$, $\dot{Y}$ and $\ddot{X}$, $\ddot{Y}$. Substituting for $\ddot{X}$ and $\ddot{Y}$ into Equation 3.179 gives

$$m(-l\,\dot{\theta}^2\sin\theta + l\,\ddot{\theta}\cos\theta) = f_b\sin\theta - w$$
$$m(-l\,\dot{\theta}^2\cos\theta - l\,\ddot{\theta}\sin\theta) = f_b\cos\theta.$$

We have two equations in the two unknowns f_b, $\ddot{\theta}$, and we could use Cramer's rule to solve for these unknowns formally. Alternatively, we eliminate f_b, via (first × cos θ − second × sin θ) to obtain

$$m[l\dot{\theta}^2(-\sin\theta\cos\theta + \sin\theta\cos\theta)$$
$$+ l\ddot{\theta}(\cos^2\theta + \sin^2\theta) = -w\cos\theta$$
$$\therefore\ ml\ddot{\theta} + w\cos\theta = 0.$$

This is a considerable effort to obtain a *really* simple governing EOM.

Now let us try the same thing with energy. Noting that energy is conserved and using $Y = 0$ as a datum for gravity potential energy in Figure 3.61a gives

$$T_0 + V_0 = T + V \Rightarrow 0 + mgl\sin\theta_0$$
$$= \frac{m}{2}(\dot{X}^2 + \dot{Y}^2) + mgl\sin\theta.$$

We have assumed (arbitrarily) that the assembly starts from rest at the angle θ_0 to determine values for T_0 and V_0. Substituting for $\dot{X}$ and $\dot{Y}$ from Equation 3.180 gives

$$mgl(\sin\theta_0 - \sin\theta) = ml^2\frac{\dot{\theta}^2}{2}(\sin^2\theta + \cos^2\theta) = ml^2\frac{\dot{\theta}^2}{2}.$$

Differentiating this equation with respect to θ gives the same governing differential equation of motion

$$-w\cos\theta = ml\frac{d}{d\theta}\left(\frac{\dot{\theta}^2}{2}\right) = ml\ddot{\theta} \Rightarrow ml\ddot{\theta} + wl\cos\theta = 0. \tag{3.181}$$

The energy approach is much easier for this problem. A free-body diagram is not required. The kinematics is simpler, since only velocity terms are required (versus acceleration terms when using $\sum f = m\ddot{r}$). Finally, the algebra is generally simpler since we are starting with a scalar work–energy equation to obtain a scalar differential equation.

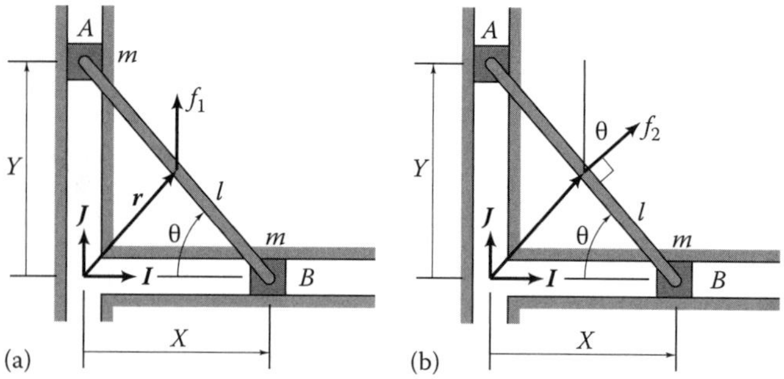

FIGURE 3.62
Variations of the system of Figure 3.61 with external forces. (a) Vertical force, (b) force perpendicular to the bar.

The frames of Figure 3.62 illustrate possible complications to our problem by the addition of external forces. Derivation of the equations of motion including the external forces illustrated in Figure 3.62 is the task at hand. We will proceed from the governing equation $\text{Work}_{n.c.} = \Delta(T + V)$. The "new" job presented by this equation is the calculation of the nonconservative work. This task is easily accomplished by proceeding from the definition

$$\text{Work} = \int \boldsymbol{f}\cdot d\boldsymbol{r}.$$

In this equation, $d\boldsymbol{r}$ is the change in the position vector $\boldsymbol{r}$ locating the point of application for force $\boldsymbol{f}$ in an inertial coordinate system. In Figure 3.62a, $\boldsymbol{r}$ is defined by

$$\boldsymbol{r} = \frac{l}{2}(\boldsymbol{I}\cos\theta + \boldsymbol{J}\sin\theta) \Rightarrow d\boldsymbol{r} = \frac{l}{2}(-\boldsymbol{I}\sin\theta + \boldsymbol{J}\cos\theta)\,d\theta.$$

Note that $d\boldsymbol{r}$ is the differential change in $\boldsymbol{r}$ due to a change in θ, not $\dot{\boldsymbol{r}}$ the derivative of $\boldsymbol{r}$ with respect to time. The differential work for Figure 3.62a is now

$$d\text{Work} = \boldsymbol{f}\cdot d\boldsymbol{r} = \boldsymbol{J}f_1\cdot\frac{l}{2}(-\boldsymbol{I}\sin\theta + \boldsymbol{J}\cos\theta)d\theta$$
$$= f_1\frac{l}{2}\cos\theta\,d\theta = Q_\theta\,d\theta.$$

The quantity $f_1 l\cos\theta/2$ is Q_θ, the *generalized force* component associated with θ. This designation will (hopefully) make more sense to you at the end of this development than at present.

Proceeding with the integral for work gives

$$\text{Work} = \int_\theta^\theta f_1\frac{l}{2}\cos\theta\,d\theta = \Big|_{\theta_0}^{\theta} f_1\frac{l}{2}\sin\theta = f_1\frac{l}{2}(\sin\theta - \sin\theta_0).$$

The work–energy expression is now

$$f_1\frac{l}{2}(\sin\theta - \sin\theta_0) = ml^2\frac{\dot{\theta}^2}{2} + mgl(\sin\theta - \sin\theta_0).$$

Differentiating with respect to θ gives the desired equation of motion,

$$ml^2\ddot{\theta} + wl\cos\theta = Q_\theta = f_1\frac{l}{2}\cos\theta. \tag{3.182}$$

Note the Q_θ term on the right-hand side of this equation. Reviewing the development for the *Work* expression, this outcome would be expected, since Q_θ is the integrand in the work integral. The subsequent differentiation of the work integral simply recovers this term. As we will demonstrate shortly, deriving this equation from $\sum \boldsymbol{f} = m\ddot{\boldsymbol{r}}$ is much more difficult than derivation from work–energy.

Figure 3.62b shows the same system, but now the applied force is perpendicular to the bar, $\boldsymbol{f} = f_2(\boldsymbol{I}\sin\theta + \boldsymbol{J}\cos\theta)$. The generalized force term associated with θ is obtained as

$$\begin{aligned}\boldsymbol{f}\cdot d\boldsymbol{r} &= f_2(\boldsymbol{I}\sin\theta + \boldsymbol{J}\cos\theta)\cdot\frac{l}{2}(-\boldsymbol{I}\sin\theta + \boldsymbol{J}\cos\theta)d\theta\\ &= \frac{f_2 l}{2}(-\sin^2\theta + \cos^2\theta)d\theta = \frac{f_2 l}{2}\cos 2\theta\, d\theta = Q_\theta\, d\theta\end{aligned}$$

where $Q_\theta = f_2(l/2)\cos 2\theta$. We could complete the integral for work in terms of θ and then differentiate with respect to θ to recover Q_θ; however, we can also simply place it on the right-hand side of Equation 3.181, obtaining

$$ml^2\ddot{\theta} + wl\cos\theta = Q_\theta = \frac{f_2 l}{2}\cos 2\theta.$$

The name "generalized force" in the preceding paragraphs makes sense from a work viewpoint. In one dimension, the differential-work expression in Cartesian coordinates is $d\text{Work} = f_X dX$. In the examples above, we obtained $d\text{Work} = Q_\theta d\theta$. Hence, in non-Cartesian coordinates, the generalized force component "looks like" the Cartesian force component. The developments of this subsection have crept into an alternative (non-Newtonian) viewpoint of dynamics, which is generically termed "variational" dynamics. As applied here, the approach can only be used to derive equations of motion for 1DOF problems. However, this is not a fundamental restriction on the variational approach. Lagrange's equations apply for deriving equations of motion for MDOF problems.

To appreciate the advantages of a derivation based on the work–energy principle versus $\sum \boldsymbol{f} = m\ddot{\boldsymbol{r}}$, consider the free-body diagrams for the problem of Figure 3.62a, as illustrated in Figure 3.63.

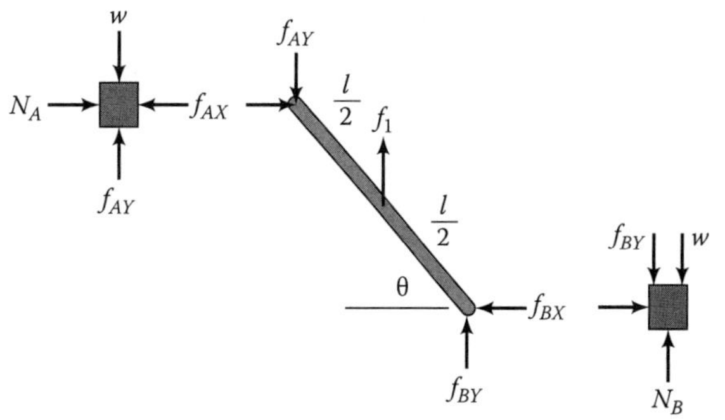

FIGURE 3.63
Free-body diagram for the problem of Figure 3.62a.

Because of the applied force f_1, the reaction force in the bar is not collinear with the bar's axis. Hence, the force components acting on masses A and B are unknown. Applying statics to the bar free-body diagram gives

$$\begin{aligned}&\sum f_X = 0 = f_{AX} - f_{BX}, \quad \sum f_Y = 0 = f_1 + f_{BY} - f_{AY}\\ &\sum M_0 = 0 = (f_{AX} + f_{BX})\frac{l}{2}\sin\theta - (f_{AY} + f_{BY})\frac{l}{2}\cos\theta.\end{aligned} \tag{3.183}$$

Clockwise moments are taken about the center of the rod. Stating $\sum \boldsymbol{f} = m\ddot{\boldsymbol{r}}$ for the two masses gives

$$\begin{aligned}\text{Body } A:\quad m\ddot{Y} &= m(-l\dot{\theta}^2\sin\theta + l\ddot{\theta}\cos\theta)\\ &= \sum f_Y = f_{AY} - w\\ \text{Body } B:\quad m\ddot{X} &= m(-l\dot{\theta}^2\cos\theta - l\ddot{\theta}\sin\theta) = \sum f_X = f_{BX}.\end{aligned} \tag{3.184}$$

The substitutions for $\ddot{X}$ and $\ddot{Y}$ are taken from Equation 3.180. Equations 3.183 and 3.184 provide five equations for the five unknowns: $f_{AX}, f_{AY}, f_{BX}, f_{BY}$, and $\ddot{\theta}$. We can use the first two of Equations 3.183 to solve for f_{AX} and f_{BY} in terms of f_{BX} and f_{BY} as ($f_{AX} = f_{BX}$; $f_{BY} = f_{AY} - f_1$), and then substitute into the last of Equations 3.183, obtaining

$$\begin{aligned}&(2f_{BX})\sin\theta - (2f_{AY} - f_1)\cos\theta = 0\\ &\Rightarrow f_{AY}\cos\theta - f_{BX}\sin\theta = \frac{f_1}{2}\cos\theta.\end{aligned} \tag{3.185}$$

Equations 3.184 and 3.185 now comprise three equations in the three unknowns f_{BX}, f_{AY}, and $\ddot{\theta}$. Returning to Equations 3.184 and forming (first × cos θ − second × sin θ) yields

$$ml\ddot{\theta} + w\cos\theta = f_{AY}\cos\theta - f_{BX}\sin\theta.$$

Substitution from Equation 3.185 gives

$$ml\,\ddot{\theta} + w\cos\theta = \frac{f_1}{2}\cos\theta,$$

which coincides with Equation 3.184 that we obtained (with much less effort) using the work–energy equation.

Looking back at Equations 3.184 and 3.185 and the statement that they comprise five equations for the five unknowns $f_{AX}, f_{AY}, f_{BX}, f_{BY}$, and $\ddot{\theta}$, you might ask What about $\dot{\theta}^2$, isn't it also an unknown? The short answer is *no*. In solving differential equations, initial conditions are required for $\dot{\theta}(0)$ and $\theta(0)$. Moreover, integration of the differential equations from the initial conditions will generate $\dot{\theta}(t)$ and $\theta(t)$ as continuous functions of time. Hence, for an adequately defined dynamic problem, $\dot{\theta}$ is always known.

At this point you might be thinking: The work–energy equation is wonderful, and I am never going to use $\sum \boldsymbol{f} = m\ddot{\boldsymbol{r}}$ again. Sorry, the work–energy equation is wonderful *when it works*, but it is not always advantageous as indicated by the example illustrated in Figure 3.64. We have returned to the example of Figure 3.61; however, we now have Coulomb friction in the guides for the sliding masses. The free-body diagram of Figure 3.64b includes Coulomb-friction reaction forces that are appropriate for ($\dot{\theta} > 0$; $\dot{Y} > 0, \dot{X} < 0$). For $\dot{\theta} < 0$, the signs of the Coulomb-friction forces would reverse. From the free-body diagram the equations of motion and equilibrium are

$$\begin{aligned}
\text{Body } A:\quad & \sum f_Y = f_b\sin\theta - w - \mu_d N_A sgn(\dot{\theta}) = m\ddot{Y} \\
& \sum f_X = 0 \Rightarrow N_A = f_b\cos\theta \\
\text{Body } B:\quad & \sum f_X = f_b\cos\theta + \mu_d N_A sgn(\dot{\theta}) = m\ddot{X} \\
& \sum f_Y = 0 \Rightarrow w + f_b\sin\theta = N_B.
\end{aligned} \tag{3.186a}$$

Eliminating N_A and N_B from the equations of motion and substituting from Equation 3.180 for $\ddot{X}$ and $\ddot{Y}$ gives

$$\begin{aligned}
\sum f_Y &= f_b\sin\theta - w - \mu_d f_b\cos\theta\, sgn(\dot{\theta}) \\
&= m(-l\,\dot{\theta}^2\sin\theta + l\,\ddot{\theta}\cos\theta) \\
\sum f_X &= f_b\cos\theta + \mu_d\,(w + f_b\sin\theta) sgn(\dot{\theta}) \\
&= m(-l\,\dot{\theta}^2\cos\theta - l\,\ddot{\theta}\sin\theta).
\end{aligned} \tag{3.186b}$$

We now have two equations for the two unknowns: f_b and $\ddot{\theta}$. Solving for $\ddot{\theta}$ via Cramer's rule gives the equation of motion

$$ml\,\ddot{\theta} = -w\cos\theta(1 - \mu_d^2) + (2\mu_d w\sin\theta - \mu_d ml\,\dot{\theta}^2)sgn(\dot{\theta}).$$

Note that the original governing equation, Equation 3.181, is recovered if μ_d is set equal to zero. Obtaining solutions for $\dot{\theta}(t)$ and $0(t)$ from this (notably unpleasant) second-order, nonlinear differential equation would be straightforward via numerical solutions from the initial conditions $\dot{\theta}(0) = \dot{\theta}_0$; $\theta(0) = \theta_0$.

Now let us try developing the equation of motion using the work–energy equation. From Figure 3.64b, the pairs of Coulomb-friction forces and their vector locations at bodies A and B are, respectively, $[-\boldsymbol{J}\mu_d N_A sgn(\dot{\theta}),\ \boldsymbol{r}_1 = \boldsymbol{J}l\sin\theta]$ and $[\boldsymbol{I}\mu_d N_B sgn(\dot{\theta}),\ \boldsymbol{r}_2 = \boldsymbol{I}l\cos\theta]$. Recall in looking at these Coulomb-friction forces that the free-body diagram of Figure 3.64b applies for $\dot{\theta} > 0$. The differential work done by the Coulomb-friction forces is

$$\begin{aligned}
d\text{Work}_{n.c.} &= -\boldsymbol{J}\mu_d N_A sgn(\dot{\theta})\cdot d\boldsymbol{r}_1 - \boldsymbol{I}\mu_d N_B sgn(\dot{\theta})\cdot d\boldsymbol{r}_2 \\
&= -\boldsymbol{J}\mu_d N_A sgn(\dot{\theta})\cdot \boldsymbol{J}l\cos\theta\, d\theta \\
&\quad + \boldsymbol{I}\mu_d N_B sgn(\dot{\theta})\cdot -\boldsymbol{I}l\sin\theta\, d\theta \\
&= -\mu_d l\, sgn(\dot{\theta})(N_A\cos\theta + N_B\sin\theta)d\theta.
\end{aligned}$$

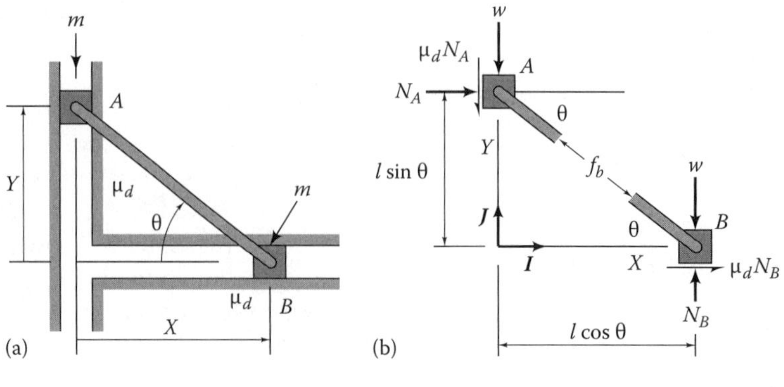

FIGURE 3.64
(a) Coulomb damping added to the example of Figure 3.61, (b) free-body diagram for $\theta > 0$ and $\dot{\theta} < 0$.

Note before proceeding further that we required the free-body diagram of Figure 3.64b *to arrive at this preliminary step*. To proceed with the differential-work expression, we can substitute from Equation 3.186 for N_A and N_B (obtained from the bar free-body diagram), to get

$$\begin{aligned} d\text{Work}_{n.c.} &= -\mu_d l\, sgn(\dot{\theta})[f_b(\cos^2\theta + \sin^2\theta) + w\sin\theta]d\theta \\ &= -\mu_d l\, sgn(\dot{\theta})(f_b + w\sin\theta)d\theta. \end{aligned}$$

Now the real troubles become apparent: We need to eliminate f_b. Equations 3.186b can be used to solve for f_b, but they are the component equation of $\sum f = m\ddot{r}$ for bodies *A* and *B*. The central idea of applying the work–energy equation was to avoid using the components of $\sum f = m\ddot{r}$, but in this example, we are forced to use them to develop the differential-work expression. Hence, deriving the equation of motion is *possible* using the work–energy equation, but it is a bad idea for this example.

Figure 3.65a illustrates the change in the example of Figure 3.61, caused by the addition of viscous damping at the guide-way slots. The same damping coefficient is assumed for contacts at both masses. The free-body diagram of Figure 3.65b applies for $\dot{X} > 0, \dot{Y} > 0$, the equations of motion are

$$\begin{aligned} \text{Body } A\text{:}\quad m\ddot{Y} &= \Sigma f_Y = f_b\sin\theta - w - c\dot{Y} \\ \text{Body } B\text{:}\quad m\ddot{X} &= \Sigma f_X = f_b\cos\theta - c\dot{X} \end{aligned}$$

Substituting for $\dot{X}, \ddot{X}, \dot{Y}, \ddot{Y}$ from Equation 3.180 gives

$$\begin{aligned} m(-l\,\dot{\theta}^2\sin\theta + l\,\ddot{\theta}\cos\theta) &= f_b\sin\theta - w - c(l\,\dot{\theta}\cos\theta) \\ m(-l\,\dot{\theta}^2\cos\theta - l\,\ddot{\theta}\sin\theta) &= f_b\cos\theta - c(l\,\dot{\theta}\sin\theta). \end{aligned}$$

We now have two equations in the two unknowns $\ddot{\theta}, f_b$. The governing EOM for $\ddot{\theta}$ is obtained by the operation, (first $\times \cos\theta$ − second $\times \sin\theta$), which eliminates f_b and $\dot{\theta}^2$, giving

$$ml\,\ddot{\theta} + cl\,\dot{\theta} + w\cos\theta = 0.$$

Note (again) that the original differential-equation model of Equation 3.181 is recovered for zero damping; that is, $c = 0$. Numerical solutions approaches would be appropriate for this second-order, nonlinear differential equation, proceeding from the initial conditions $\dot{\theta}(0) = \dot{\theta}_0;\ \ \theta(0) = \theta_0$.

Deriving this governing equation from the work–energy equation is possible and more reasonable than the preceding example that had Coulomb damping. The pair of viscous-damping force terms and their vector locations are defined by the free-body diagram of Figure 3.65b by $[-\boldsymbol{J}c\dot{Y}, \boldsymbol{r}_1 = \boldsymbol{J}l\sin\theta]$ and $[-\boldsymbol{I}c\dot{X}, \boldsymbol{r}_2 = \boldsymbol{I}l\cos\theta]$. The differential-work expression can be stated

$$\begin{aligned} d\text{Work}_{n.c.} &= -\boldsymbol{J}c\dot{Y}\cdot d\boldsymbol{r}_1 - \boldsymbol{I}c\dot{X}\cdot d\boldsymbol{r}_2 \\ &= -\boldsymbol{J}c\dot{Y}\cdot\boldsymbol{J}l\cos\theta\, d\theta - \boldsymbol{I}c\dot{X}\cdot -\boldsymbol{I}l\sin\theta\, d\theta \\ &= -cl(\dot{Y}\cos\theta - \dot{X}\sin\theta)\, d\theta. \end{aligned}$$

Substituting $\dot{Y} = l\cos\theta\dot{\theta}$ and $\dot{X} = -l\sin\theta\dot{\theta}$ from Equation 3.180 gives

$$\begin{aligned} d\text{Work}_{n.c.} &= -cl[(l\cos\theta\dot{\theta})\cos\theta - (-l\sin\theta\dot{\theta})\sin\theta]d\theta \\ &= -cl^2\dot{\theta}\, d\theta = Q_\theta\, d\theta \end{aligned}$$

Substituting the generalized force $Q_\theta = -cl^2\dot{\theta}$ into the right-hand side of Equation 3.181 gives

$$\begin{aligned} &ml^2\ddot{\theta} + wl\cos\theta = Q_\theta = -cl^2\dot{\theta} \\ &\Longrightarrow ml^2\ddot{\theta} + wl\cos\theta + cl^2\dot{\theta} = 0, \end{aligned}$$

which completes the development. Viscous damping reduces the advantage of the work–energy equation in developing the equation of motion in that (1) a free-body diagram is required to define the reaction forces due to viscous damping, and (2) about the same amount of algebra is required.

The lesson from the examples of Figure 3.64 and 3.65 is that *The work–energy equation is not always advantageous*

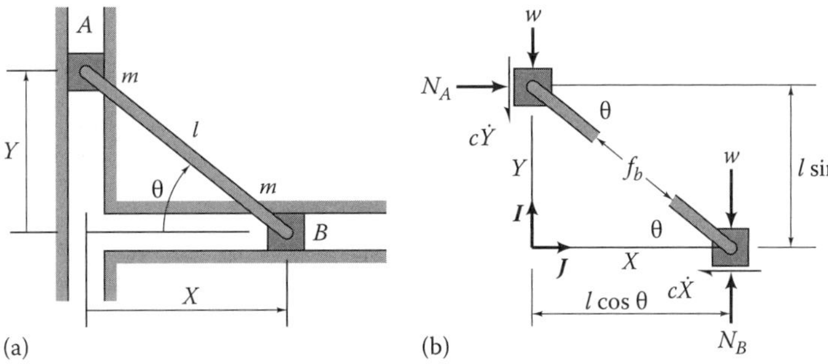

FIGURE 3.65
(a) Viscous damping added to the example of Figure 3.61, (b) free-body diagram for $\dot{X} > 0, \dot{Y} > 0$.

in deriving the equation of motion. Coulomb-friction forces can be particularly unpleasant in their ability to discourage application of the work–energy equation. However, even in cases where the work–energy equation is not appropriate, working through a problem (initially) without the troublesome energy-dissipation elements can be helpful to see the form of the governing differential equation. In both of these examples, recall that we checked to see that the governing differential equation of motion reduced to our original Equation 3.181 once the friction forces were eliminated.

Repeating the theme of this book, the central tasks of dynamics is the development of models to predict motion. Developing these models involves derivation of general kinematics equations and general equations of motion. The examples from this subsection provide the following lessons:

a. Equations of motion can frequently be obtained more easily from the work–energy equations than from $\sum f = m\ddot{r}$.
b. External forces can easily be incorporated into the work–energy equation through the work equation, Work $= \int f \cdot dr$.
c. The work–energy equation is not always advantageous for deriving governing equations of motion when energy dissipation is present in the form of Coulomb or viscous damping.

Deriving equations of motion in dynamics is not a choice between $\sum f = m\ddot{r}$ or the work–energy equation. Newton's second law of motion, $\sum f = m\ddot{r}$, always applies, but is not always the easiest way to derive the equation of motion. Conversely, the work–energy equation is not always appropriate, but (when it works) generally provides an easier approach to the derivation of equations of motion. Both of these complementary tools are needed in the dynamics tool box of a practicing engineer, and in some problems, a combination of both approaches is advantageous.

3.7 Linear-Momentum Applications in Plane Motion

We considered several problems in Section 3.2 for which Newton's differential equation of motion had the form

$$f(t) = m\frac{dv}{dt}.$$

This differential-equation form is easily integrated (once) by multiplying through by dt and integrating both side of the equation to obtain

$$\int_{t_0}^{t_1} f(t)dt = mv(t_0) - mv(t_1)$$

$$\text{Linear impulse} = \Delta(\text{linear momentem})$$

The integral on the left is called *linear impulse*. On the right-hand side, we have the change in *linear momentum*. The linear impulse–momentum equation is not particularly noteworthy or useful in comparison to the work–energy equation of the preceding section. It is simply a first integral of the differential equation of motion with respect to time. Linear momentum becomes important in dynamics problems when it is constant (conserved). This section considers circumstances for which linear momentum is conserved for more than one particle during a collision, beginning with motion in one dimension, and then proceeding to motion in a plane. The *coefficient of restitution* will be introduced to assist in solving collision problems between particles.

3.7.1 Collision Problems in One Dimension

Figure 3.66 provides an example of a useful application of *conservation of linear momentum*. As illustrated, two bodies are shown before and after a collision. The initial velocities of the two bodies $v_1(t_1)$ and $v_2(t_1)$ are known. We want to determine $v_1(t_2)$ and $v_2(t_2)$, the velocities following the impact. During impact, the equations of motion for the two bodies are

$$\text{Body 1:} \quad -f(t) = m_1\dot{v}_1$$
$$\text{Body 2:} \quad f(t) = m_2\dot{v}_2$$

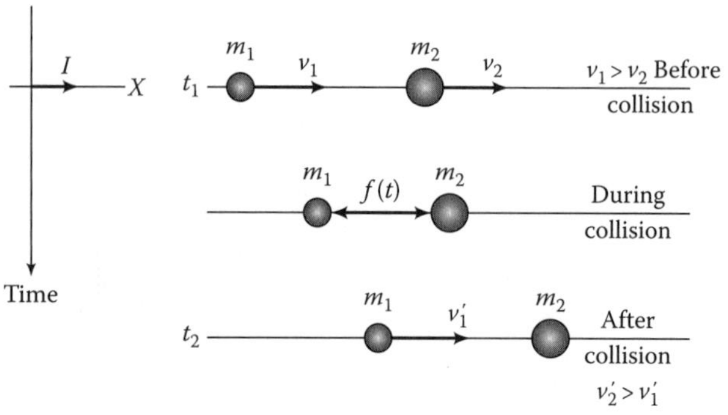

FIGURE 3.66
A collision between two bodies.

Note that integrating over the time interval of the collision $[t_1, t_2]$ gives

$$\text{Body 1:} \quad -\int_{t_1}^{t_2} f(t)dt = m_1 v_1(t_2) - m_1 v_1(t_1)$$

$$\text{Body 2:} \quad \int_{t_1}^{t_2} f(t)dt = m_2 v_2(t_2) - m_2 v_2(t_1).$$

Adding these two equations gives the useful result

$$m_1 v_1(t_1) + m_2 v_2(t_1) = m_1 v_1(t_2) + m_2 v_2(t_2), \quad \text{or}$$

Initial linear system momentum

= Final linear system momentum. (3.187)

This equation is quite useful and is a mathematical statement that linear momentum is conserved for the *system* of the two bodies. This outcome is explained by noting that *If* is the only force acting on the two bodies of Figure 3.66 and is *internal* to the system of the two bodies.

The good news from Equation 3.187 is that we have obtained an equation for our unknowns without any knowledge concerning the nature of the interaction force $f(t)$. The bad news is we have only one equation for the two unknowns $v_1(t_2)$ and $v_2(t_2)$. You might think of the following two possibilities to gain a second equation and resolve this deficit

a. Energy is conserved; that is,

$$\frac{m_1}{2}v_1^2(t_1) + \frac{m_2}{2}v_2^2(t_1) = \frac{m_1}{2}v_1^2(t_2) + \frac{m_2}{2}v_2^2(t_2). \tag{3.188}$$

b. The two bodies "stick" together after collision; hence, they have the common velocity

$$v_1(t_2) = v_2(t_2). \tag{3.189}$$

In regard to Equation 3.188, energy is *never conserved* during a collision (except in dynamics books). The mildest and least offensive collision between two rigid bodies announces itself to our ears via acoustic energy created during the collision. This energy is removed from the initial kinetic energy of the system, dissipated to the surroundings, and not returned to the bodies following the collision. More violent collisions, for example wrecks involving automobiles dissipate a huge amount of energy in the inelastic tearing and crushing of materials, plus a sizable amount of heat. While Equation 3.188 reflects one extreme of energy conservation, we shall see below that the kinematic constraint of Equation 3.189, requiring that the bodies have a common velocity following a collision, reflects a maximum dissipation of energy.

Example Problem 3.24

Use Equations 3.187 through 3.189 with the following data,

$$\begin{aligned} m_1 &= 20\text{ kg}, \quad m_2 = 10\text{ kg} \\ v_1(t_1) &= 10\text{ m/s}, \quad v_2(t_1) = -5\text{ m/s}, \end{aligned} \tag{i}$$

and determine the final velocities, assuming: (1) conservation of energy, and (2) the two bodies stick together.

Solution

The conservation-of-linear-momentum result provided by Equation 3.187 gives

$$\begin{aligned} &20\text{ kg} \times 10\text{ m/s} + 10\text{ kg} \times (-5\text{ m/s}) = 150\text{ kg m/s} \\ &= 20v_1(t_2) + 10v_2(t_2). \end{aligned} \tag{ii}$$

Initially assuming conservation of energy, Equation 3.188 give us the following equation in the two unknowns:

$$\begin{aligned} &\frac{20\text{ kg}}{2} \times 10^2 \frac{\text{m}^2}{\text{s}^2} + \frac{10\text{ kg}}{2} \times (-5)^2 \frac{\text{m}^2}{\text{s}^2} = T_1 = 1125\text{ N m} \\ &= \frac{20}{2}v_1^2(t_2) + \frac{10}{2}v_2^2(t_2). \end{aligned} \tag{iii}$$

Note that we have used the derived units for the Newton, kg m/s^2 (from $w = mg$) to arrive at the dimensions for T_1. Solving Equations (ii) and (iii) gives the following two solutions:

$$\begin{aligned} v_1(t_2) &= 0\text{ m/s}, \quad v_2(t_2) = 15\text{ m/s} \\ v_1(t_2) &= 10\text{ m/s}, \quad v_2(t_2) = -5\text{ m/s}. \end{aligned}$$

Both answers are mathematically correct; however, the second equation is not realizable, since it would require that body 2 somehow get past or through body 1 following the collision (to continue moving in the $-X$ direction).

Equation 3.189 applies when we assume that the bodies have a common velocity following the collision. Together with the conservation-of-linear-momentum result of Equation (ii), this assumption gives the solution

$$\begin{aligned} &v_1(t_2) = v_2(t_2) = 5\text{ m/s} \\ &\therefore\ T(t_2) = \frac{20\text{ kg}}{2} \times 5^2 \frac{\text{m}^2}{\text{s}^2} + \frac{10\text{ kg}}{2} \times 5^2 \frac{\text{m}^2}{\text{s}^2} = 375\text{ N m}. \end{aligned}$$

With this result, the kinetic energy following collision is one third the initial kinetic energy given in Equation (iii). We will show below that the kinematic requirement of

Equation 3.189 (the bodies have the same velocity following a collision) maximizes the dissipation of energy.

Clearly, some reasonable way to characterize a collision that lies somewhere between the extremes of energy conservation and maximum energy dissipation would be useful. The *coefficient of restitution*, which is introduced below, meets this goal.

3.7.2 The Coefficient of Restitution

The collision process between two bodies can be broken down into the following two stages:

a. A deformation process that starts with contact and continues until both bodies have reached their maximum state of deformation.
b. A restoration process in which the deformation is reversed and the two bodies proceed to rebound from each other.

At the instant between the deformation and restoration processes, the two bodies have the common velocity, v_0. When the restoration process is completed, the internal restoration force reaches zero, and the bodies can have either (1) a relative velocity of separation, or (2) move together with a common velocity and zero velocity of separation.

Figure 3.67 illustrates this process. Note that we have simplified our notation in this figure and are using a prime to denote the velocity of the two bodies following the collision. With this notation, the conservation-of-linear-momentum equation becomes

$$m_1v_1 + m_2v_2 = m_1v_1' + m_2v_2' = (m_1 + m_2)v_0 \quad (3.190)$$

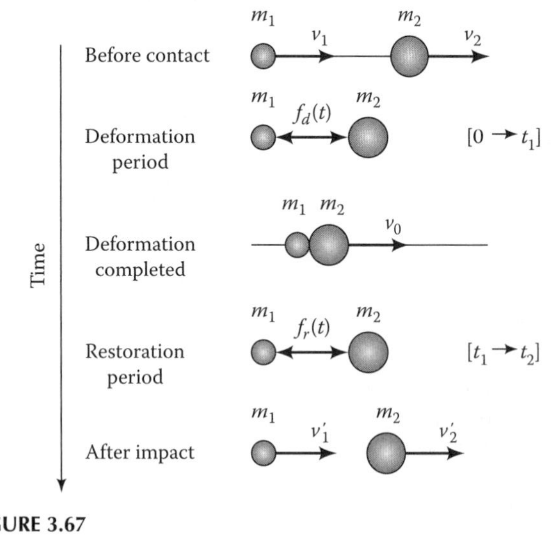

FIGURE 3.67
The collision process broken down into a deformation and restoration process.

Figure 3.67 illustrates the internal reaction deformation $f_d(t)$ and restoration $f_r(t)$ forces. The coefficient of restitution, e, is *formally* defined as the impulse during restoration divided by the impulse during deformation. To be honest, this is not a particularly motivating definition, but as we shall see, it yields a very useful outcome. To evaluate the integrals we assume that the impact starts at $t = 0$, that the deformation occurs over the interval $[0, t_1]$, and that the restoration takes place during the interval $[t_1, t_2]$. With this notation, applying the definition for e to body 1 gives

$$e = \frac{\int_{t_1}^{t_2} -f_r(t)\,dt}{\int_0^{t_1} -f_d(t)\,dt} = \frac{m_1v_1' - m_1v_0}{m_1v_0 - m_1v_1} = \frac{v_1' - v_0}{v_0 - v_1}. \quad (3.191a)$$

Applying the definition to body 2 gives

$$e = \frac{\int_{t_1}^{t_2} f_r(t)\,dt}{\int_0^{t_1} f_d(t)\,dt} = \frac{m_2v_2' - m_2v_0}{m_2v_0 - m_2v_2} = \frac{v_2' - v_0}{v_0 - v_2}. \quad (3.191b)$$

Either Equation 3.191a or b can be used to complete the definition for e. We will use Equation 3.191a, and substitute for v_0 from Equation 3.190 to obtain

$$e = \frac{v_1' - (m_1v_1' + m_2v_2')/(m_1 + m_2)}{(m_1v_1 + m_2v_2)/(m_1 + m_2) - v_1} = \frac{v_2' - v_1'}{v_1 - v_2}. \quad (3.192)$$

In words, the definition becomes

$$e = \frac{\text{Relative velocity of separation}}{\text{Relative velocity of closure}}. \quad (3.193)$$

As we will demonstrate shortly, e can range from 0 to 1. As defined by Equation 3.193, the condition $e = 0$ corresponds to zero separation velocity (common terminal velocity).

The upper limit for e is 1, as we will now demonstrate. Taken together, Equations 3.190 and 3.192 give us the following simultaneous equations for v_1' and v_2':

$$m_1v_1' + m_2v_2' = m_1v + m_2v_2$$
$$-v_1' + v_2' = ev_1 + ev_2$$

Solving for v_1' and v_2' defines

$$\begin{aligned} v_1' &= [(m_1 - em_2)v_1 + m_2(1 + e)v_2]/(m_1 + m_2) \\ v_2' &= [(m_2 - em_1)v_1 + m_1(1 + e)v_1]/(m_1 + m_2). \end{aligned} \quad (3.194)$$

Using these results to solve for the final kinetic energy, T_f, gives (after a great deal of algebra)

$$T_f = \frac{m_1}{2}v_1'^2 + \frac{m_2}{2}v_2'^2 = T_i - \frac{m_1m_2(1 - e^2)(v_1 - v_2^2)}{2(m_1 + m_2)},$$

where T_i is the initial kinetic energy. Note that $e=1$ implies conservation of energy during the collision; that is, $T_f = T_i$, which provides an upper limit for e. Also, note that $e=0$ defines the maximum energy dissipation.

To illustrate the e definition, we will rework Example Problem XP3.24 with $e=0.5$. The relative velocity of approach is $10-(-5)=15$ m/s; hence, Equation 3.193 gives the relative velocity of separation,

$$0.5 \times 15 \text{ m/s} = 7.5 \text{ m/s} = v_2' - v_1'.$$

The conservation-of-momentum equation result of Equation (ii) continues to be valid, contributing the second equation

$$15 \text{ m/s} = 2v_1' + v_2'.$$

Solving these equations gives: $v_1' = 2.5$ m/s; $v_2' = 10$ m/s. The final kinetic energy is

$$T_f = \frac{m_1}{2}v_1'^2 + \frac{m_2}{2}v_2'^2 = \frac{20 \text{ kg}}{2} \times (2.5)^2 + \frac{10 \text{ kg}}{2} \times (10)^2 \frac{\text{m}^2}{\text{s}^2}$$
$$= 562.5 \text{ N m},$$

which is 50% of the initial kinetic energy.

3.7.3 Collision Problems in Two Dimensions

If applied to cars, our discussions of collisions between particles could be characterized as "head-on" or "rear-end." By contrast, Figure 3.68 illustrates an *oblique* collision between two particles. Conditions are known before the collision; that is, m_1, m_2, v_1, v_2, θ_1, and θ_2 are known. The conditions following the collision are unknown, meaning that we need to develop four equations to solve for v_1', v_2', θ_1', and θ_2'. Note that the reaction force during impact is internal to the *system* of the two bodies; hence, system momentum is conserved. At the point of contact, the spheres' surfaces define a plane. The normal to the plane of contact is denoted in Figure 3.68 by $\boldsymbol{\varepsilon}_n$. In the $\boldsymbol{\varepsilon}_n$-direction, linear momentum is conserved for the system of two bodies and gives

$$\boldsymbol{\varepsilon}_n\text{:} \quad m_2v_2\sin\theta_2 - m_1v_1\sin\theta_1 = m_1v_1'\sin\theta_1' - m_2v_2'\sin\theta_2'. \tag{3.195}$$

The unit vector $\boldsymbol{\varepsilon}_t$ denotes the direction parallel to the plane of contact. There is no reaction force in the $\boldsymbol{\varepsilon}_t$-direction; hence, the linear momentum of the *individual* bodies is conserved in this direction, giving

$$\boldsymbol{\varepsilon}_t\text{:} \quad \begin{array}{ll} \text{body 1,} & m_1v_1\cos\theta_1 = m_1v_1'\cos\theta_1' \\ \text{body 2,} & m_2v_2\cos\theta_2 = m_2v_2'\cos\theta_2'. \end{array} \tag{3.196}$$

The coefficient-of-restitution definition of Equation 3.192 can be applied to the system along the $\boldsymbol{\varepsilon}_n$-direction as

$$e = \frac{\text{Relative velocity of separation}}{\text{Relative velocity of closure}} = \frac{v_1'\sin\theta_1' + v_2'\sin\theta_2'}{v_1\sin\theta_1 + v_2\sin\theta_2} \tag{3.197}$$

Reviewing, we have broken the problem into two parts, by working alternately along the $\boldsymbol{\varepsilon}_n$- and $\boldsymbol{\varepsilon}_t$-directions. Our prior developments for collisions in one dimension continue to work along the $\boldsymbol{\varepsilon}_n$-direction, using the conservation-of-momentum (Equation 3.187), and the coefficient-of-restitution (Equation 3.193). There are no forces along the $\boldsymbol{\varepsilon}_t$-direction; hence, in this direction, momentum of the *individual* bodies is conserved during the collision, yielding Equation 3.196. Equations 3.195 through 3.197 define the unknowns $(v_1', v_2', \theta_1', \text{and } \theta_2')$ following the collision, in terms of the known parameters and velocity conditions (m_1, m_2, v_1, v_2, θ_1, and θ_2) before the collision.

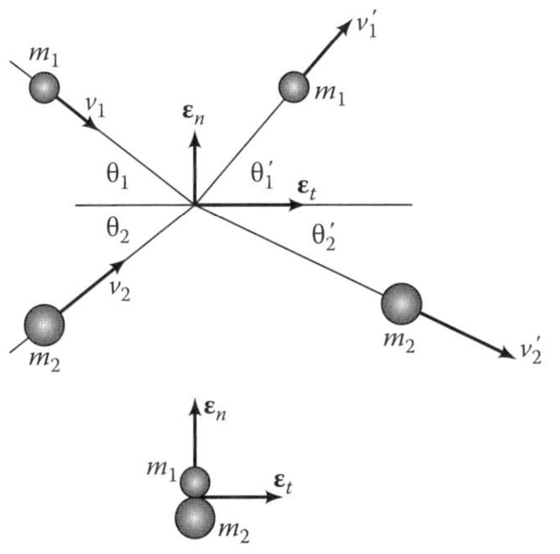

FIGURE 3.68
An oblique collision.

Example Problem 3.25

Use the following data,

$$m_1 = 20 \text{ kg}, \quad v_1 = 10 \text{ m/s}, \quad \theta_1 = 30°,$$
$$m_2 = 10 \text{ kg}, \quad v_2 = 5 \text{ m/s}, \quad \theta_2 = 10°, \quad e = 0.5,$$

for the situation illustrated in Figure 3.68 to determine the velocities immediately following impact. Also, determine the fraction of energy dissipated during the collision.

Solution

We need four equations for the four unknowns $(v_1', v_2', \theta_1', \text{and } \theta_2')$. Equation 3.195 gives the system conservation-of-linear-momentum result

$$10\text{ kg} \times 5\ \frac{\text{m}}{\text{s}} \sin 10° - 20\text{ kg} \times 10\ \frac{\text{m}}{\text{s}} \times \sin 30°$$
$$= 20\text{ kg} \times v_1'\ \frac{\text{m}}{\text{s}} \sin\theta_1' - 10\text{ kg} \times v_2'\ \frac{\text{m}}{\text{s}} \times \sin\theta_2'. \quad \text{(i)}$$

Equation 3.196 yields

$$\text{Body 1,}\quad 20\text{ kg} \times 10\ \frac{\text{m}}{\text{s}} \times \cos 30° = 173.2\text{ kg m/s}$$
$$= 20\text{ kg} \times v_1' \cos\theta_1' \Rightarrow v_1' \cos\theta_1 = 8.66\text{ m/s} \quad \text{(ii)}$$
$$\text{Body 2,}\quad 10\text{ kg} \times 5\ \frac{\text{m}}{\text{s}} \times \cos 10° = 49.24\text{ kg m/s}$$
$$= 10\text{ kg} \times v_2' \cos\theta_2' \Rightarrow v_2' \cos\theta_2 = 4.92\text{ m/s}$$

Finally, Equation 3.197 provides the coefficient-of-restitution relation

$$e = 0.5 = \frac{v_1' \sin\theta_1' + v_2' \sin\theta_2'}{10\ \frac{\text{m}}{\text{s}} \times \sin 30° + 5\ \frac{\text{m}}{\text{s}} \times \sin 10°}. \quad \text{(iii)}$$

Equations (i) and (iii) provide the following two linear equations in the unknowns $v_1' \sin\theta_1'$ and $v_2' \sin\theta_2'$:

$$2v_1' \sin\theta_1' - v_1' \sin\theta_1' = -4.983$$
$$v_1' \sin\theta_1' + v_2' \sin\theta_2' = 2.934.$$

Their solutions are

$$v_1' \sin\theta_1' = -.683\text{ m/s},\quad v_2' \sin\theta_2' = 3.617\text{ m/s}. \quad \text{(iv)}$$

Combining these results with Equation (ii) gives

$$\tan\theta_1' = -\frac{.683}{8.66} = -.0789 \Rightarrow \theta_1' = -4.51°$$
$$\tan\theta_2' = \frac{3.62}{4.92} = .735 \Rightarrow \theta_2' = 36.3°.$$

From either Equation (ii) or Equation (iv), the missing final velocities are

$$v_1' = 8.69\text{ m/s},\quad v_2' = 6.11\text{ m/s}.$$

Figure XP3.25 illustrates the solution, showing the velocities of the two particles before and after the collision. The illustrated solution reflects the expected outcome of a larger mass and velocity for body 1 before the collision.

The initial kinetic energy (before the collision) is

$$T_{\text{initial}} = \frac{m_1 v_1^2}{2} + \frac{m_2 v_2^2}{2} = \frac{20\text{ kg}}{2} \times 10^2\ \frac{\text{m}^2}{\text{s}^2} + \frac{10\text{ kg}}{2} \times 5^2\ \frac{\text{m}^2}{\text{s}^2}$$
$$= 1125\text{ N m}.$$

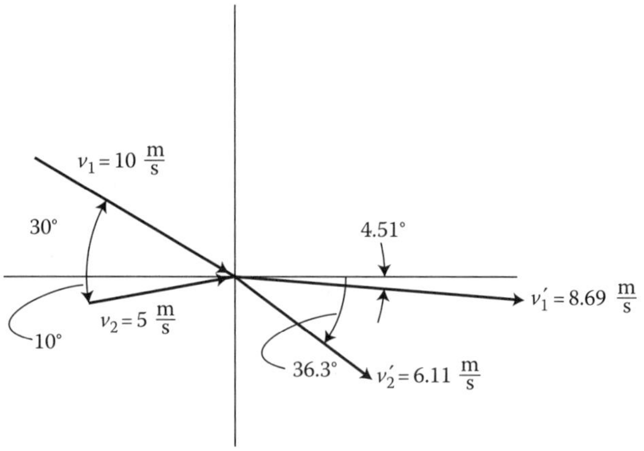

FIGURE XP3.25
Velocities of particles 1 and 2 "before and after" the collision.

The post-collision kinetic energy is

$$T_{\text{final}} = \frac{m_1 v_1'^2}{2} + \frac{m_2 v_2'^2}{2}$$
$$= \frac{20\text{ kg}}{2} \times 8.69^2\ \frac{\text{m}^2}{\text{s}^2} + \frac{10\text{ kg}}{2} \times 6.11^2\ \frac{\text{m}^2}{\text{s}^2}$$
$$= 942\text{ N m}.$$

Hence, the fractional loss of kinetic energy during the collision is $(T_{\text{initial}} - T_{\text{final}})/T_{\text{initial}} = (1125 - 942)/1125 = .163$, and the collision has dissipated about 16% of the initial kinetic energy. Again, real collisions always dissipate energy.

Example Problem 3.26

Figure XP3.26 illustrates mass $m_1 = 5$ kg moving to the right with velocity $v_0 = 5$ m/s. It will subsequently collide with mass $m_2 = 10$ kg. The coefficient of restitution for the collision is $e = 0.8$. Mass m_2 is supported by a linear spring with stiffness coefficient $k = 4.0 \times 10^4$ N/m and a linear damper with coefficient $c = 125$ N s/m. Prior to the collision, mass m_2 is at rest, and the spring is undeflected.

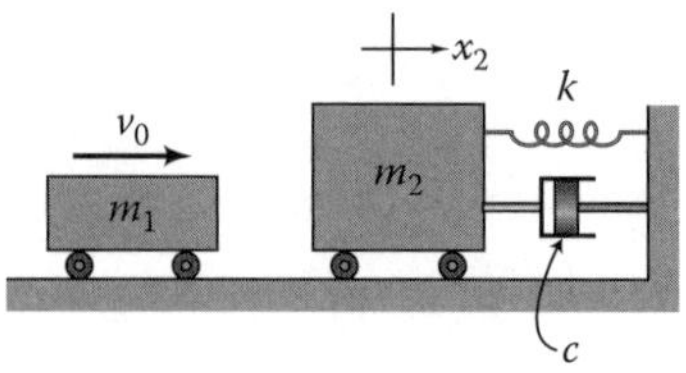

FIGURE XP3.26
Two bodies before impact.

Tasks:

a. Determine the velocity of mass m_2 and mass m_1 immediately after the collision.
b. What fraction of the kinetic energy is dissipated during the collision?
c. What is the undamped natural frequency, the damped natural frequency, and the damping factor for mass m_2?
d. What is m_2's maximum deflection following the collision?

Solution

The collision is assumed to occur "instantaneously" such that there is a change in velocities for both bodies but no change in position. Hence, during the collision, the spring supporting mass m_2 remains undeflected. In addition, the damper provides no reaction force until after the collision when mass m_2 has acquired velocity v_2'. The problem solution has the following sequential steps: (1) a collision problem defined by Task a that determines the initial velocity of m_2 following the collision, and (2) a one-degree-of-freedom vibration problem for mass m_2 following the collision.

Conservation for momentum during the course of the collision gives

$$m_1 v_0 = m_1 v_1' + m_2 v_2'. \tag{i}$$

The coefficient-of-restitution definition gives

$$e = \frac{\text{relative velocity of separation}}{\text{relative velocity of closure}} = \frac{v_2' - v_1'}{v_0}. \tag{ii}$$

We have two equations in the two unknowns v_1' and v_2' with the solutions

$$\begin{aligned} v_2' &= \frac{m_1(1+e)v_0}{m_1+m_2} = \frac{5\text{ kg} \times 1.8 \times 5\text{ m/s}}{(5+10)\text{ kg}} = 3\text{ m/s} \\ v_1' &= \frac{(m_1 - e m_2)v_0}{m_1+m_2} = \frac{(5-0.6\times 10)\text{ kg}\times 5\text{ m/s}}{15\text{ kg}} = 0.333\text{ m/s}, \end{aligned} \tag{iii}$$

which concludes Task a.

The kinetic energy immediately following the collision is

$$\begin{aligned} T' &= \frac{m_1 v_1'^2 + m_2 v_2'^2}{2} \\ &= \frac{5\text{ kg} \times 0.333^2\ (\text{m/s})^2 + 10\text{ kg}\times 3^2\ (\text{m/s})^2}{2} \\ &= 45.3\text{ N m/s}. \end{aligned}$$

Note that the unit conversion, Newton = kg m/s^2, was used. The initial kinetic energy is $T_0 = m_1 v_0^2/2 = 5\text{ kg} \times 5^2\ (\text{m/s})^2/2 = 62.5$ N m/s. The fractional change in kinetic energy is

$$\frac{\Delta T}{T_0} = \frac{45.3 - 62.5}{62.5} = -0.276.$$

Hence, 27.6% of the initial kinetic energy is dissipated in the collision, and we have concluded Task b.

From Equation 3.21, the differential equation of motion for mass m_2, following the collision is

$$m_2 \ddot{x}_2 + c\dot{x}_2 + k_2 x_2 = 0. \tag{iv}$$

For the present horizontal motion, w has been discarded. The undamped natural frequency ω_n, the damping factor ζ, and the damped natural frequency ω_d are

$$\begin{aligned} \omega_n &= \sqrt{\frac{k_2}{m_2}} = \sqrt{\frac{4.0\times 10^4\text{ N/m}}{10\text{ kg}}} = 63.24\text{ rad/s}, \\ \zeta &= \frac{c}{2m\omega_n} = \frac{125}{(2\times 10\times 63.24)} = 0.0989 \\ \omega_d &= \omega_n\sqrt{1-\zeta^2} = 63.24\sqrt{1-0.0988^2} = 62.94\text{ rad/s}. \end{aligned}$$

From Equation 3.25, the solution to Equation (iv) is

$$x_2 = e^{-\zeta\omega_n t}(A\cos\omega_d t + B\sin\omega_d t). \tag{v}$$

The initial conditions are $x_2(0) = 0$, and $\dot{x}_2(0) = v_2' = 3$ m/s. Using the first initial condition gives $A = 0$. Setting $A = 0$, and differentiating Equation (v) gives

$$\dot{x}_2 = -\zeta\omega_n e^{-\zeta\omega_n t} B\sin\omega_d t + e^{-\zeta\omega_n t} B\omega_d\cos\omega_d t. \tag{vi}$$

Hence,

$$\dot{x}_2(0) = v_2' = B\omega_d \Rightarrow B = \frac{v_2'}{\omega_d}. \tag{vii}$$

Substituting for A and B into Equation (v) gives

$$\begin{aligned} x_2 &= e^{-\zeta\omega_n t}\left(\frac{v_2'}{\omega_d}\right)\sin\omega_d t, \\ \dot{x}_2 &= -\zeta\omega_n e^{-\zeta\omega_n t}\left(\frac{v_2'}{\omega_d}\right)\sin\omega_d t + e^{-\zeta\omega_n t} v_2'\cos\omega_d t. \end{aligned} \tag{viii}$$

The maximum deflection occurs when

$$\begin{aligned} \dot{x}_2(\bar{t}) = 0 &= -\frac{\zeta\omega n}{\omega_d}\sin\omega_d\bar{t} + \cos\omega_d\bar{t} \\ \Rightarrow \tan\omega_d\bar{t} &= \frac{\omega_d}{\zeta\omega_n} = \sqrt{\frac{1-\zeta^2}{\zeta}}. \end{aligned}$$

Plugging in the numbers gives

$$\begin{aligned} \tan\omega_d\bar{t} &= \frac{\sqrt{1-\zeta^2}}{\zeta} = \frac{\sqrt{1-.0989^2}}{.0989} = 10.06 \\ &\Rightarrow \omega_d\bar{t} = 1.417\text{ rad} \\ \bar{t} &= \frac{1.417\text{ rad}}{62.94\text{ rad/s}} = .0233\text{ s}. \end{aligned}$$

This elapsed time is little less than one fourth of the damped period, $\tau_d = 2\pi/\omega_d = 2 \times 3.1416/62.94$ (rad/s) = .0998 s, which is an expected result. Specifically, in about one fourth of the period, the mass displacement will first grow to its peak displacement,

$$x_{2\,max} = x_2(\bar{t}) = e^{-\zeta\omega_n\bar{t}}\left(\frac{v_2'}{\omega_d}\right)\sin\omega_d\bar{t}$$
$$= e^{(-.0989\times63.24\times.0233)}\left(\frac{3\text{ m/s}}{62.94\text{ rad/s}}\right)$$
$$\times\sin\left(62.94\,\frac{\text{rad}}{\text{s}}\times.0233\text{ s}\right) = 4.09E-2\text{ m}$$

3.8 Moment of Momentum

In the preceding section, we found that the linear-momentum vector can frequently be useful in solving problems when it is conserved (constant). This observation is also true for the *moment-of-momentum* vector; however, before worrying about situations in which moment of momentum is conserved, we first need to define this vector. Section 3.8.1 is used to develop the vector, and applications are presented in Section 3.8.2.

3.8.1 Developing the Moment-of-Momentum Equation for a Particle

Figure 3.69 illustrates a particle of mass m, acted on by the *resultant* force f. The position vector $\boldsymbol{R} = \boldsymbol{I}R_X + \boldsymbol{J}R_Y + \boldsymbol{K}R_Z$ locates m in the inertial X, Y, Z coordinate system. The moment-of-momentum vector is defined with respect to O, the origin of the X, Y, Z system, by

$$\boldsymbol{H}_O = \boldsymbol{R}\times m\dot{\boldsymbol{R}} \qquad (3.198)$$

The designation, moment of momentum, is seen to be appropriate. (A large number of dynamics authors refer to this term as "angular momentum.") The moment about point O due to the force f is $M_0 = R\times f$. By comparison, $\boldsymbol{H}_O$ in Equation 3.198 is the moment about O of the linear momentum, $m\dot{\boldsymbol{R}}$. We would now like to demonstrate that $\boldsymbol{M}_O = \dot{\boldsymbol{H}}_O$, where $\dot{\boldsymbol{H}}_O$ is $d\boldsymbol{H}_O/dt|_{X,Y,Z}$, the time derivative of $\boldsymbol{H}_O$ with respect to the X, Y, Z system.

We can start with the following vector development

$$\dot{\boldsymbol{H}}_O = \left.\frac{d(\boldsymbol{R}\times m\dot{\boldsymbol{R}})}{dt}\right|_{X,Y,Z} = m(\dot{\boldsymbol{R}}\times\dot{\boldsymbol{R}} + \boldsymbol{R}\times\ddot{\boldsymbol{R}}) = \boldsymbol{R}\times m\ddot{\boldsymbol{R}}.$$

Returning to Equation 3.164, Newton's second law of motion is

$$\boldsymbol{f} = m\ddot{\boldsymbol{R}},$$

where $\ddot{\boldsymbol{R}}$ is the acceleration of m with respect *to* X, Y, Z. Taking moments about O of both sides of this equation gives

$$\boldsymbol{M}_O = \boldsymbol{R}\times\boldsymbol{f} = \boldsymbol{R}\times m\ddot{\boldsymbol{R}} = \dot{\boldsymbol{H}}_O. \qquad (3.199)$$

For particle dynamics, the usual application of this equation involves finding circumstances for which the moment is (always) zero about a *single* axis. For example, from Equations 3.198 and 3.199, a constant zero moment about the Z-axis would yield the following component equation:

$$M_{OZ} = 0 \Rightarrow H_{Oz} = \text{constant} = (\boldsymbol{R}\times m\dot{\boldsymbol{R}})_Z = [(\boldsymbol{I}R_X + \boldsymbol{J}R_Y + \boldsymbol{K}R_Z)\times m(\dot{\boldsymbol{R}}_X + \boldsymbol{J}\dot{R}_Y + \boldsymbol{K}\dot{R}_Z)]_Z$$
$$= m(R_X\dot{R}_Y - R_Y\dot{R}_X)$$

This Cartesian-coordinate version of $\boldsymbol{H}_O$ is much less frequently used than the polar-coordinate version. As illustrated in Figure 3.70, the polar-coordinate version follows from $\boldsymbol{R} = \boldsymbol{K}Z + \boldsymbol{\varepsilon}_r r$ and $\dot{\boldsymbol{R}} = \boldsymbol{K}Z + \boldsymbol{\varepsilon}_r\dot{r} + \boldsymbol{\varepsilon}_\theta r\dot{\theta}$ as

$$\boldsymbol{H}_O = (\boldsymbol{K}Z + \boldsymbol{\varepsilon}_r r)\times m(\boldsymbol{K}\dot{Z} + \boldsymbol{\varepsilon}_r\dot{r} + \boldsymbol{\varepsilon}_\theta r\dot{\theta})$$
$$= \boldsymbol{\varepsilon}_r mZr\dot{\theta} + \boldsymbol{\varepsilon}_\theta mr\dot{Z} + \boldsymbol{K}mr^2\dot{\theta}.$$

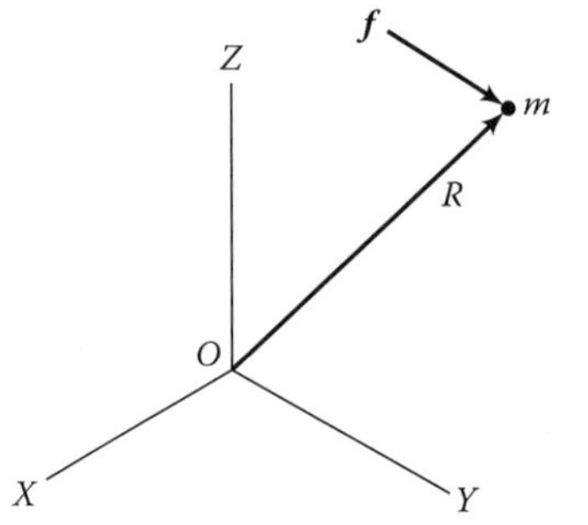

FIGURE 3.69
A particle acted on by a *resultant* force f in the X, Y, Z inertial system.

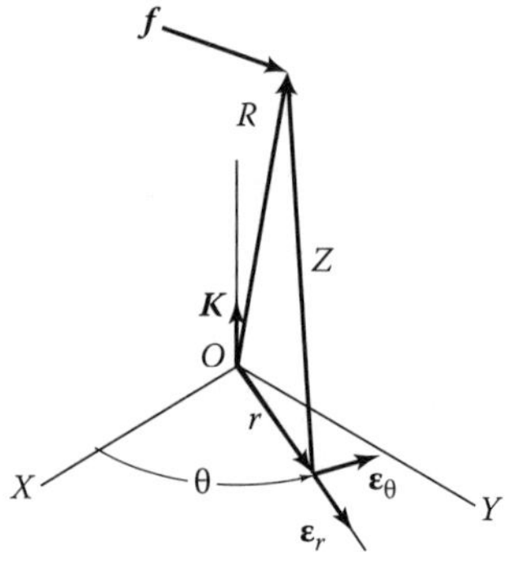

FIGURE 3.70
Particle acted on by force f in the inertial $\boldsymbol{\varepsilon}_r$, $\boldsymbol{\varepsilon}_\theta$, $\boldsymbol{K}$ system.

Hence, in this coordinate system,

$$M_{OZ} = 0 \Rightarrow H_{OZ} = mr(r\dot{\theta}) = \text{constant}. \quad (3.200)$$

This is the most useful form of the conservation-of-momentum equation.

3.8.2 Applying Conservation of Moment of Momentum for a Particle

3.8.2.1 Two Particles Connected by an Inextensible Cord

Figure 3.71a illustrates two particles, each of mass m, connected to each other by an inextensible cord of length L. Particle A moves without friction on a smooth horizontal plane. The connecting cord goes vertically down through the horizontal plane at o. Particle A is located on the plane by r and θ. Particle B is located at a distance $d = L - r$ below the hole in the plane. Note that this problem has the two degrees of freedom, r and θ. The following engineering-analysis tasks apply:

a. Draw free-body diagrams and derive the equations of motion for this system.
b. Assuming the initial conditions $\dot{r}(0) = 0$; $\dot{\theta}(0) = \omega_0$, and $\mathrm{r}(0) = \mathrm{L}$, use conservation of moment of momentum to uncouple the r differential equation of motion; that is, eliminate $\dot{\theta}$ from the r differential equation.

Figure 3.71b provides the free-body diagram for the two particles, where T_c is the tension in the cord. Using polar coordinates, particle A's equations of motion are

$$\begin{aligned} ma_r &= m(\ddot{r} - r\dot{\theta}^2) = \sum f_r = -T_c \\ ma_\theta &= m(r\ddot{\theta} - 2\dot{r}\dot{\theta}) = \sum f_\theta = 0 \end{aligned} \quad (3.201)$$

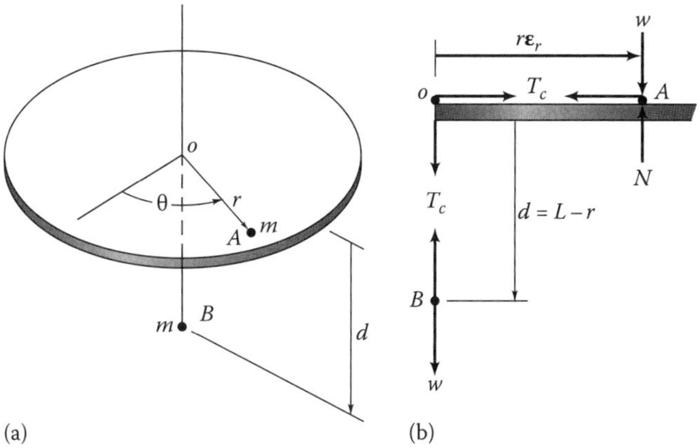

FIGURE 3.71
(a) Two particles, each of mass m, connected to each other by an inextensible cord of length L, (b) free-body diagram.

For particle B, we obtain

$$m\ddot{d} = m(-\ddot{r}) = \sum f_d = w - T_c \Rightarrow T_c = w + m\ddot{r}. \quad (3.202)$$

Substituting T_c back into the first of Equation 3.201 gives

$$2m\ddot{r} - mr\dot{\theta}^2 = -w. \quad (3.203)$$

Equation 3.203 and the last of Equation 3.201 provide two equations of motion in the two variables r and θ. As we will show below, conservation of momentum can be used to eliminate θ as a variable, netting a single differential equation in r.

Thinking about moment of momentum, no moment acts on a vertical axis through o; hence, from Equation 3.200, the moment of momentum about this axis, $H_{oZ} = r \times mv_\theta = r \times mr\dot{\theta}$, is conserved. In terms of the initial conditions, this condition gives

$$H_{oZ} = mr^2\dot{\theta} = mL^2\omega_0 \Rightarrow \dot{\theta} = \left(\frac{L}{r}\right)^2 \omega_0 \quad (3.204)$$

Note that differentiating $mr^2\dot{\theta} = \text{constant}$ with respect to time gives $r\ddot{\theta} + 2\dot{r}\dot{\theta} = 0$, which coincides with the second of Equation 3.201. The two results coincide because they arise from the same physical fact; namely, since there is no circumferential force on particle A, there is zero moment about a vertical axis through o. Equation 3.204 states that $\dot{\theta}$ is proportional *to* r^{-2}, so an increase in r will decrease $\dot{\theta}$, while a decrease in r will increase $\dot{\theta}$.

Substituting $\dot{\theta}$ from Equation 3.204 into Equation 3.203 gives

$$2m\ddot{r} - m\frac{L^4}{r^3}\omega_0^2 = -w \Rightarrow \ddot{r} - \frac{L^4\omega_0^2}{2r^3} = -g, \quad (3.205)$$

a single differential equation in r, and concludes Task a. Generally, time solutions for $\dot{r}(t)$, $r(t)$ to this nonlinear differential equations would require numerical integration from initial conditions.

3.8.2.2 Closing Comments

Conservation of momentum for particles represents a "side step" from the thrust of this book, namely, deriving equations of motion. However, as demonstrated, it is an important concept in explaining observed dynamics phenomena such as the acceleration and deceleration of a particle's circumferential velocities when decreasing or increasing (respectively) the radius. The phenomena are observable in vortex flow of fluid mechanics. For example, the surface circumferential velocity of fluid approaching the drain in a tub of water is observed to accelerate substantially as it moves radially inward.

We will return to the concept of moment of momentum in Section 5.8 for planar motion of rigid bodies. Eliminating a coordinate or coordinates by application of conservation of momentum is used in more advanced dynamic applications.

3.9 Summary and Discussion

A great deal of material is covered in this chapter, and separate summaries have been provided at the end of most sections to review and emphasize the intended lessons and viewpoints. Taking a broad view, this chapter has considered the following topics:

a. Applying Newton's second law of motion for a particle, $\sum f = m\ddot{r}$
b. Developing and applying the work–energy integral of $\sum f = m\ddot{r}$

$$\text{Work}_{n.c.} = (T_2 + V_2) - (T_1 + V_1). \tag{3.173}$$

c. Developing and applying the time integral of $\sum f = m\ddot{r}$ for conservation of linear momentum for a particle or a system of particles
d. Developing and applying the moment-of-momentum integrals for a particle from $\sum f = m\ddot{r}$

One-degree-of-freedom (1DOF) models of systems including a mass, viscous dampers, and springs were introduced in Section 3.2. Models incorporating these elements form the basis for engineering-vibration analysis, and examples were presented demonstrating the derivation of governing vibration differential equations (and their solutions) plus analysis techniques for 1DOF vibration models. Various important ideas and terminology were introduced including damped and undamped natural frequencies, damping factors, resonance, steady-state harmonic solutions, etc.

The equation for vertical motion of a 1DOF vibration system is

$$m\ddot{Y} + c\dot{Y} + kY = w,$$

where
Y is the directed downward
w is the weight

This differential equation requires a positive stiffness coefficient k for "static" stability. In nontechnical terms, a statically stable structure will support itself and the loads applied to it. By contrast, a statically unstable body (e.g., a two-legged stool) will fall over unless supported by an outside force. A positive damping coefficient c is required in the equation of motion to ensure "dynamically" stable motion. The vibratory motion of a dynamically stable system consists of *decaying* oscillations, versus *growing* or *diverging* oscillations for a dynamically unstable system. The principal lesson from this discussion is as follows: Unless you have reason to believe that a model is either statically or dynamically unstable, when deriving the equation of motion for a system, a negative result for either k or c means that you have made a mistake. Dynamic systems can be statically and dynamically unstable. Statically unstable examples are introduced in Chapter 5.

Newton's laws were used to derive the equation of motion for the simple pendulum. The governing equation was nonlinear and represented our first nonlinear vibration equation. This example introduced the very important concept of "linearization for small motion about an equilibrium position." Equilibrium positions and small motion about equilibrium positions are an essential part of engineering dynamic analysis and will be encountered throughout the book. Small motion about an equilibrium position is an essential concept in deciding whether the perturbed motion of a system will be stable or unstable.

Systems involving bodies connected by cords via pulleys were used to introduce the idea of a kinematic-constraint equation. These examples involved two bodies but only one degree of freedom, since the coordinates of the bodies were related via a separate equation. Kinematic-constraint equations occur regularly in dynamics and are required to completely define a system model. When Newton's laws have been applied correctly, and there are more unknowns than equations, kinematic constraints are missing and must be developed from the system's geometrical constraints.

In dynamic analysis, most mechanical systems have more than one degree of freedom, and 2DOF examples were introduced in Section 3.5. The two-mass/spring model of Figures 3.47 through 3.49 and the double-pendulum example of Figure 3.50 provide common 2DOF examples. The spring-mass system of Figure 3.47 is coupled through the stiffness matrix; the double-pendulum example of Figure 3.50 is coupled through angular-acceleration/inertia terms. Equations of motion for MDOF linear vibration systems can be coupled by either the stiffness, damping, or mass matrices.

Undamped, linearized vibration equations of motion for MDOF systems can be stated

$$[M](\ddot{x}_i) + [K](x_i) = (f_i),$$

where $[M]$ and $[K]$ are, respectively, the mass (inertia) and stiffness matrices. Diagonal $[M]$ and $[K]$ matrices would imply that the coordinates are uncoupled. However, most MDOF systems are coupled. To facilitate understanding and solution development, they can be uncoupled via modal analysis. Section 3.5.2 introduced modal analysis and provided examples for eigenanalysis, leading to the development of eigenvalues and eigenvectors (mode shapes). Solutions are presented for free and forced motion of 2DOF examples. Procedures for developing steady-state solutions due to harmonic excitation are also presented. The material presented in Section 3.5 provides a "crash" course in the essentials of mechanical vibration analysis. As noted previously however, it is no substitute for a real course or dedicated book on mechanical vibrations.

Section 3.6 developed the work–energy equation and presented its application for 1DOF examples. The examples presented showed the direct application of $\Delta(T+V) = \text{Work}_{n.c.}$ to determine an unknown velocity at a specific position. Also, the work–energy equation was presented as the basis for an alternative approach (to $\sum f = m\ddot{r}$ and free-body diagrams) in developing the equation of motion. Specifically, the equation of motion can be obtained by differentiating a general-position form of the work–energy equation with respect to the coordinate displacement (or rotation angle). Except for cases involving Coulomb damping, this approach will generally yield the equation of motion as quickly and easily as the Newtonian approach. In many cases, the avoidance of free-body diagrams and simpler kinematics favors the work–energy formulation. The lesson from Section 3.6 is the following: Equations of motion can be correctly developed via either Newtonian or energy approaches. The particular application (and experience) determines which approach is better. Also, the two approaches provide an independent check on the correctness of a derived differential equation of motion.

The work–energy approach of Section 3.6 does not apply for systems with more than one degree of freedom. However, the "variational" approach represented by the work–energy approach can be readily extended to MDOF problems via Lagrange's equations.

The conservation-of-linear-momentum results of Section 3.7 should be a review of earlier materials from your college-physics course. The coefficient-of-restitution definition of Section 3.7.2 was probably new to you and provides a practical engineering approach to model collisions between particles. However, in the author's experience, most practical applications of this coefficient involve the sporting pages of newspapers and magazines. Specifications of baseballs and golf balls are normally made in terms of the coefficient of restitution.

The conservation of moment-of-momentum results of Section 3.8 are of limited direct value; however, the concept is quite important, being used regularly to explain weather phenomena and simpler matters such as the acceleration of fluid in the whirlpool developed in the drain of a basin. With respect to this book, the moment-of-momentum results are a step toward the rigid-body equations of Chapter 5.

All of the essential features of dynamics have been introduced in Chapters 2 and 3. The Newtonian particle kinetics features of this chapter are extended in Chapter 5 to planar kinetics of rigid-bodies. Although not considered in this text, deriving the equation of motion for a 1DOF model from the work–energy equation can be extended to MDOF examples via Lagrange's equations.

Problems

Spring-Mass-Damper Applications

3.1 The buoy shown in Figure P3.1 has a circular cross section with diameter d, length L, and weight w. Most of the buoy's weight is concentrated in the base. The buoy is released from rest with one end in contact with the surface of the fluid with specific weight of γ_f. *Hint*: Apply Archimedes's theorem for buoyancy.

Tasks:

(a) Draw a free-body diagram and obtain the equation of motion for the system using Y as your coordinate.
(b) Determine the system's natural frequency.
(c) Determine the buoy's equilibrium position.
(d) Determine the equation of motion for motion about the equilibrium position.

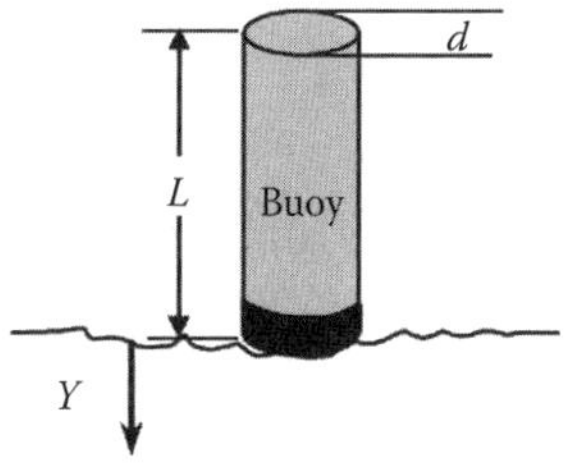

FIGURE P3.1

3.2 A body of mass m is supported at the end of a cantilever beam of length L and section modulus EI, Figure P3.2. Recall that the deflection δ at the end of the beam is defined by $\delta = (PL^3/3EI)$ where P is the load.

Tasks:

(a) Draw a free-body diagram (neglecting the mass of the beam) and obtain the equation of motion for Y.

(b) Determine the natural frequency.

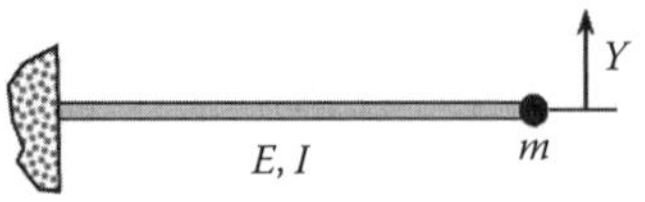

FIGURE P3.2

3.3 A body of mass m is supported at the end of uniform bar of length L, Figure P3.3. Recall that the deflection at the end of a uniform bar due to an axial load P is $\delta = (PL/AE)$ where A is the cross-sectional area, and E is the modulus of elasticity.

Tasks:

(a) Neglect the mass of the bar and determine the equation of motion using Y as your coordinate.

(b) Determine the natural frequency.

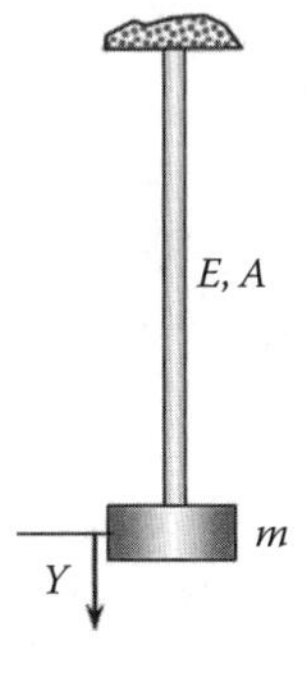

FIGURE P3.3

3.4 In Figure P3.4a through c, the springs are initially unstretched.

Tasks:

(a) Select a coordinate, draw the free-body diagram, and obtain the equation of motion.

(b) Apply work–energy principles and obtain the equation of motion.

(c) Calculate the natural frequency using $m_1 = 10$ kg, $k_1 = 1000$ N/m, $k_2 = 1000$ N/m, $k_3 = 1000$ N/m.

(d) Suppose the springs are initially stretched (or compressed), and consider motion about equilibrium; does the differential equation change?

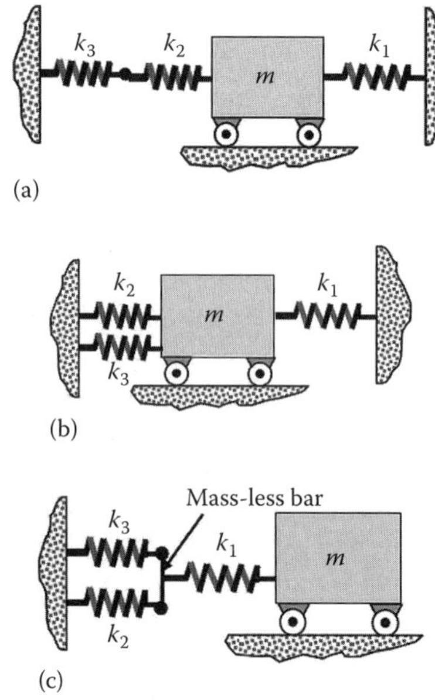

FIGURE P3.4

3.5 The mass shown in Figure P3.5 is attached to the two springs, which are in turn attached to the fixed walls. The springs are in compression at equilibrium. Assume that m is displaced from the equilibrium position and then released from rest.

Tasks:

(a) Select a coordinate, draw the free-body diagram, and obtain the equation of motion.

(b) Apply work–energy principles and obtain the equation of motion.

(c) Obtain the natural frequency using: $k_1 = 10$ kN/m, $k_2 = 15$ kN/m, and $m = 50$ kg.

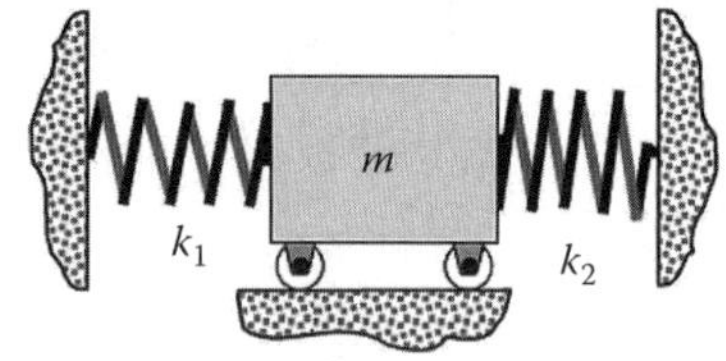

FIGURE P3.5

3.6 The mass shown in Figure P3.6 is held in place by the two springs. The springs are in tension when the mass is at equilibrium and are attached firmly to the walls and the mass. Friction at the contact between the mass and the inclined plane is negligible.

Tasks:

(a) Select a coordinate, draw the free-body diagram, and obtain the equation of motion if the mass is displaced from the equilibrium position and released.

(b) Apply work–energy principles and obtain the equation of motion.

(c) Obtain the natural frequency of the system using $k_1 = 100$ lb/in., $k_2 = 250$ lb/in., $w = 50$ lb, and $\alpha = 35°$.

(d) Repeat part (a), but assume that the springs are initially undeflected when the body is released from rest.

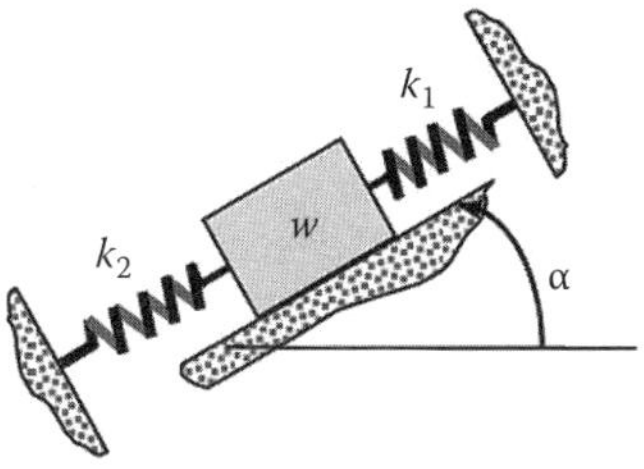

FIGURE P3.6

3.7 For the system illustrated in Figure P3.7, the springs are undeflected.

Tasks:

(a) Select a coordinate, draw a free-body diagram, and derive the equation of motion.

(b) For $k = 70$ kN/m, $m = 100$ kg, $c = 802$ N s/m, calculate the following:

i. Undamped natural frequency ω_n

ii. Damping factor ζ

iii. Damped natural frequency ω_d

(c) Repeat part (a) assuming that the body is displaced from equilibrium.

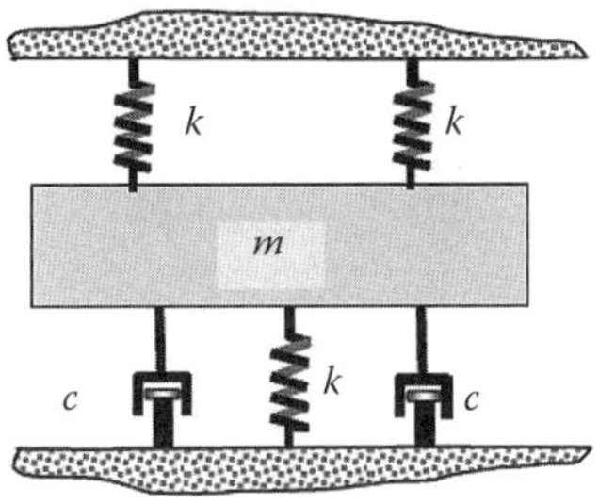

FIGURE P3.7

3.8 The spring shown in Figure P3.8 is undeflected when the mass is released from rest.

Tasks:

(a) Select a coordinate, draw the free-body diagram, and state the equation of motion.

(b) Using $k = 20$ lb/in., $W = 50$ lb, $\alpha = 25°$ determine the undamped natural frequency.

(c) For $c = 3.74$ lbs/ft determine the damping factor ζ and the damped natural frequency.

(d) Drop the damping element and obtain the equation of motion by applying work–energy principles.

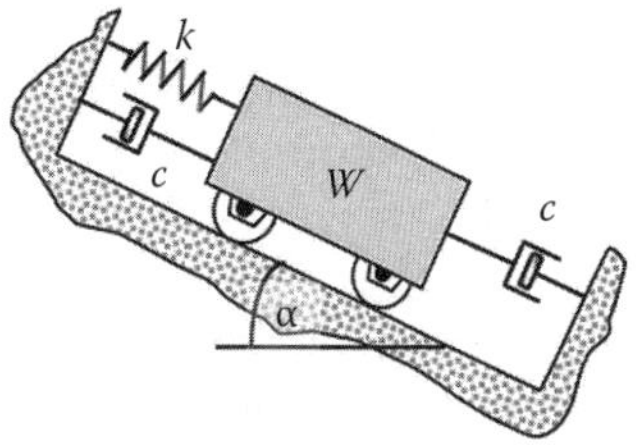

FIGURE P3.8

Spring-Mass-Damper Systems with Transient Excitation

Undamped Systems

3.9 A constant force of 2.5 kN acts on the cart shown in Figure P3.9 for $0.15\tau_n$ seconds, where τ_n is the period of the natural frequency of the system. The cart has a mass of 1000 kg, and the spring, k_a has a stiffness of 500 kN/m.

Tasks:

(a) Select coordinates, derive the equation of motion, and put it into a form that is consistent with Table B.1 to find the particular solution.

(b) Applying the initial conditions $X(0) = 0$; $\dot{X}(0) = 0$, determine the complete solution for motion of the system for the time interval, $0 < t < t_f = 2\tau_n$.

(c) Determine the peak response amplitude.

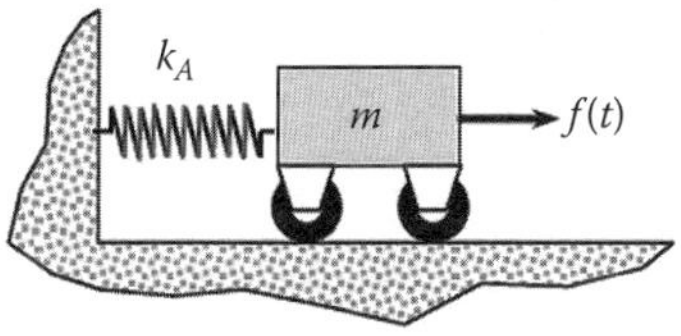

FIGURE P3.9

3.10 The cart shown in Figure P3.9 weighs 200 lb and is attached to ground by a spring with a stiffness of 2 kip/in. The force $f(t) = -25t$ lb acts on the cart for 0.5 s. For $t > 0.5$ s, $f(t) = 0$. Note, 1 kip = 1000 lb.

Tasks:

(a) Select a coordinate, derive the equation of motion for the system, and put into a form so that Table B.1 can be applied to obtain the particular solution.

(b) Develop the complete solution, satisfying the initial conditions $X(0) = 0$; $\dot{X}(0) = 0$, and determine the peak response for a total time period of 1 s.

3.11 The tower shown in Figure P3.11 has the same equation model as the cart system shown in Figure P3.9 where $k = k_A$. The force acting on the tower can be modeled by $f(t) = 75t - 30.5t^2$ kN with the force acting for 1 s. For $t > 1$, $f(t) = 0$.

Tasks:

(a) Select a coordinate, draw a free-body diagram, and develop the equation of motion in a form that is consistent with Table B.1 to obtain the particular solution.

(b) Assuming that the tower is initially undeflected and at rest, develop the complete solution for $0 < t < 5$ s. *Note*: you will need to restate the equation of motion and restart the problem for $t > 1$ s.

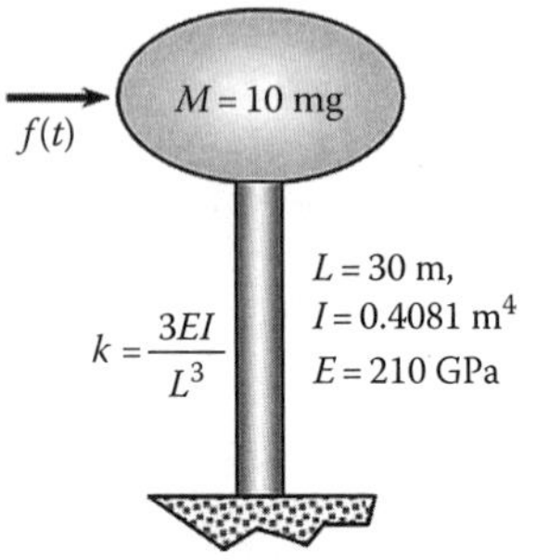

FIGURE P3.11

3.12 The cart shown in Figure P3.12 is connected to ground by a spring-damper system. A force with a constant amplitude of 350 N acts on the cart for $t = 2.5\tau_d$ s. The cart has a mass of 50 kg, the spring has a stiffness of 15,000 N/m, and the system has a damping ratio of 0.15. The spring is initially undeflected, and the cart is at rest.

Tasks:

(a) Select a coordinate, draw a free-body diagram, and derive the equation of motion.

(b) Solve for motion for the time interval, $0 < t < 1.5$ s.

(c) Determine when the peak response occurs. *Note*: You will need to break this solution into two periods covering (1) the time while the force is applied, and (2) the time after the force is turned off.

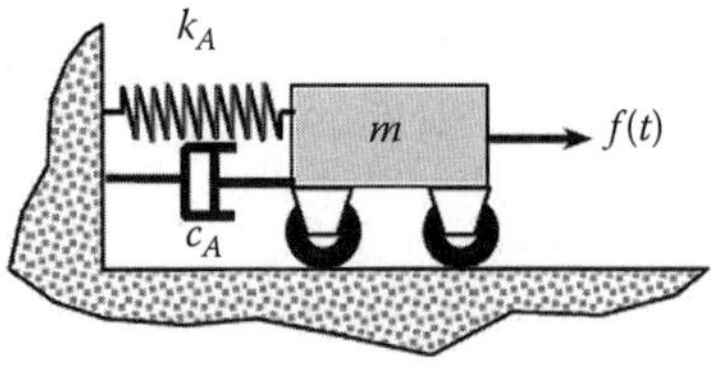

FIGURE P3.12

3.13 The cart shown in Figure P3.12 weighs 150 lbs, $k_A = 2000$ lb/in., and $c_A = 6.0$ in.-s/lb. The force $f(t) = 2t$ lb is applied for 0.25 s. For $t > 0.25$ s, $f(t) = 0$. The spring is initially undeflected, and the mass is at rest.

Tasks:

(a) Select a coordinate, draw a free-body diagram, and derive the equation of motion for (1) the forced and (2) the free motion.

(b) Solve for the physical response of the system for the time interval $0 < t < 0.4$ s.

(c) Determine when the peak amplitude of the system occurs. *Note*: You will need to break this solution into two periods covering (1) the time while the force is applied, and (2) the time after the force is turned off.

3.14 The cart shown in Figure P3.12 has a damped natural frequency of 5.02 Hz, a mass of 150 kg, and is connected to ground by a spring with a stiffness of 150 kN/m. The cart is acted upon by the force $f(t) = -5t^2 - 15t$ N for a period of 1.5 s. After this time, $f(t) = 0$. The spring is initially undeflected and the cart is at rest.

Tasks:

(a) Select a coordinate, draw a free-body diagram, and derive the equation of motion.

(b) Solve for the displacement over the time interval $0 < t < 3$ s using the results of Table B.2.

(c) Determine the peak amplitude. *Note*: You will need to break this solution into two periods covering: (1) the time while the force is applied, and (2) the time after the force is turned off.

Base Excitation

3.15 The equation of motion for the system shown in Figure P3.15 can be stated $m\ddot{X} + c\dot{X} + kX = kX_B + c\dot{X}_B$, or as $m\ddot{X} = k(X_b - X) + c(\dot{X}_b - \dot{X})$. Subtracting $m\ddot{X}_b$ from both sides of the previous expression results in $m\ddot{\delta} + c\dot{\delta} + k\delta = -m\ddot{X}_b$ where $\delta = X_b - X$ is the relative displacement of the mass with respect to the base.

Task: Assume the system has the initial conditions, $\delta(0) = \dot{\delta}(0) = 0$, and determine the relative response of the system for $\ddot{X}_b = D$.

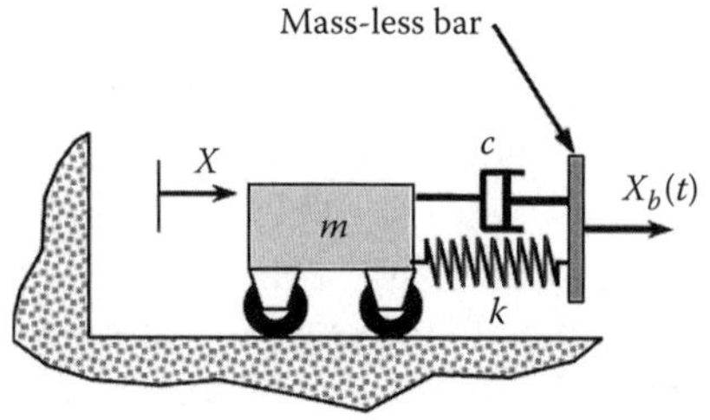

FIGURE P3.15

3.16 The cart shown in Figure P3.16 is connected to ground by spring k_A and damper c_A. It weighs 250 lb and has a damped natural frequency of

29.96 Hz; $k_A = 15$ kip/in., and $c_A = 5$ lbs/in. The damping ratio of the system is 0.05. The springs k_A and k_B are initially unstretched, and the mass m is at rest. For $0 < t < 0.25$ s, the plate at the end of the spring-damper combination has motion defined by $x(t) = 1.5 + 2.5t$ in. After 0.25 s, the connector that is represented by k_B and c_B fails, releasing the connection to m. *Note*: 1 kip = 1000 lb.

Tasks:

(a) Select a coordinate, draw a free-body diagram, and state the equation of motion.

(b) Solve for the cart's displacement as a function of time.

(c) Determine the cart's maximum velocity after $t = 0.25$ s.

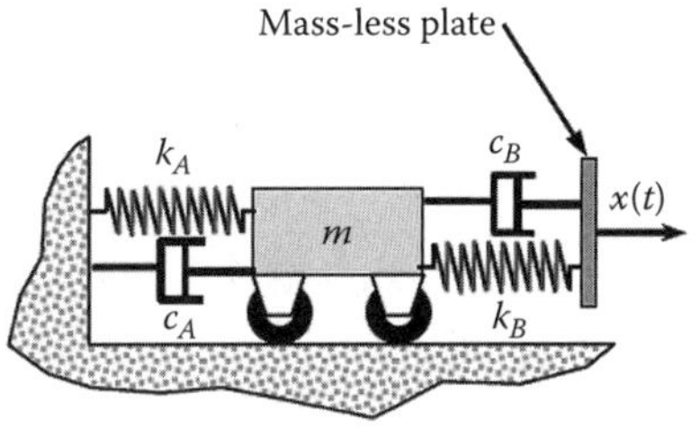

FIGURE P3.16

3.17 For Figure P3.16, motion for the plate on the right is defined by $x(t) = 20t$ mm for $0 < t_a = 2(2\pi/\omega_n)$. The springs k_A and k_B are initially unstretched, and the mass m is at rest. For $t > t_a$, the plate is held motionless at $x(t) = x(t_a)$ with zero velocity. The component properties are $m = 100$ kg, $k_A = 1000$ N/m, $k_B = 500$ N/m, $c_A = 2.5$ N s/m, and $c_B = 1.25$ N s/m.

Tasks:

(a) Select a coordinate, draw a free-body diagram, and determine the equations of motion for both conditions of the plate.

(b) Determine the cart's motion up to $t = 6$ s. Compare the force of the spring and damper between the cart and the plate at time t_a and adjust the excitation accordingly.

(c) Determine the cart's peak response amplitude.

3.18 The cart shown in Figure P3.16 weighs 500 lbs. Further, $k_A = 2000$ lb/in., $c_A = 50$ lb-s/in., $k_B = 200$ lb/in., and $c_B = 15$ lbs/in. The plate's motion is defined by $x(t) = 0.25t^2$ ft, where t is in seconds. For $t_a > 0.25$ s, the connector defined by k_B and c_B breaks.

Tasks:

(a) Select a coordinate, draw a free-body diagram, and determine the equation of motion.

(b) Solve for the cart's motion over the time interval $0 < t < 0.5$ s.

(c) Determine the peak amplitude of the system and the time at which it occurs.

Spring-Mass-Damper Systems with Harmonic Excitation

3.19 The system shown in Figure P3.19 is acted upon by the forcing function shown. The system parameters are $m = 15$ kg, $k = 75$ kN/m, $f_0 = 750$ N, $\omega = 15.13$ Hz.

Tasks: For motion about equilibrium, determine the steady-state amplitude and phase for

(a) Free vibration tests result in a *log dec*, δ, of 0.523.

(b) $c = 0$.

(c) Using the damping from part (a), determine the range of excitation frequencies such that the amplitude is 20 mm or less.

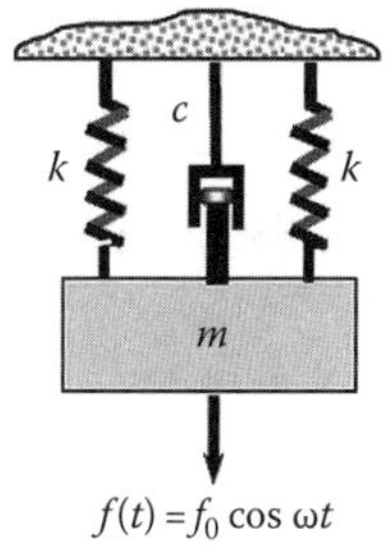

FIGURE P3.19

3.20 Refer to Figure P3.19 and use the following parameters: $W = 175$ lb, $k = 1.05(10^5)$ lb/ft, $f_0 = 500$ lb, $\omega = 24$ Hz. Figure P3.20 displays test results from a free vibration response. Use these results to determine the *log-dec* and the damping factor.

Tasks:

(a) Determine the steady-state amplitude and phase.

(b) Determine the operating frequency range to avoid so that the amplitude does not exceed 0.125 in.

(c) Determine the required damping value at resonance such that the peak amplitude does not exceed 0.3 in.

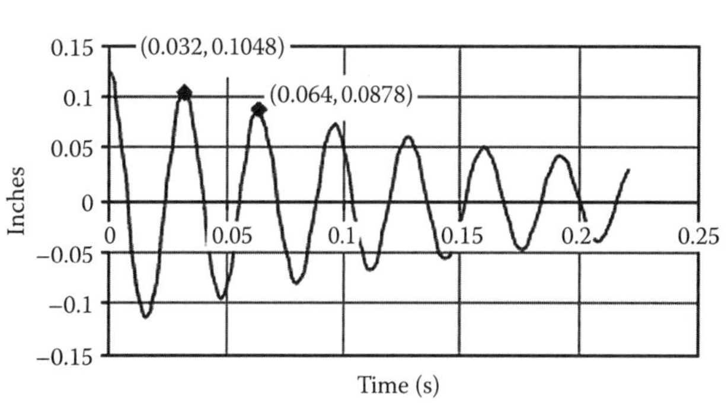

FIGURE P3.20

3.21 Given: The differential equation of motion $m\ddot{X} + c\dot{X} + kX = f_0 \sin \omega t$ with the parameters, $w = mg = 10$ lb, $k = 102$ lb/in., and $f_0 = 10$ lb.

Task:
For an excitation frequency above 11 Hz (cycles/s), select the lowest possible value for c such that the amplitude of forced motion $X(t)$ will be no more than 0.3 in. Also, determine the peak response amplitude at resonance for a damping value of 0.1 lb-s/in.

3.22 The system shown in Figure P3.22 is a configuration for seismic motion measurements. The system is mounted on a structure that has a vertical vibration at 15 Hz and a peak-to-peak amplitude of 2 mm. The sensing element has a mass $m = 2.0$ kg, and the spring has a stiffness $k = 1.75$ kN/m. The motion of the mass relative to the instrument base is recorded on a revolving drum and shows a peak-to-peak amplitude of 2.2 mm during the steady-state condition.

Task: Calculate the viscous-damping constant c.

Hint: Note that the pen's motion measures *relative* motion between the mass and the base.

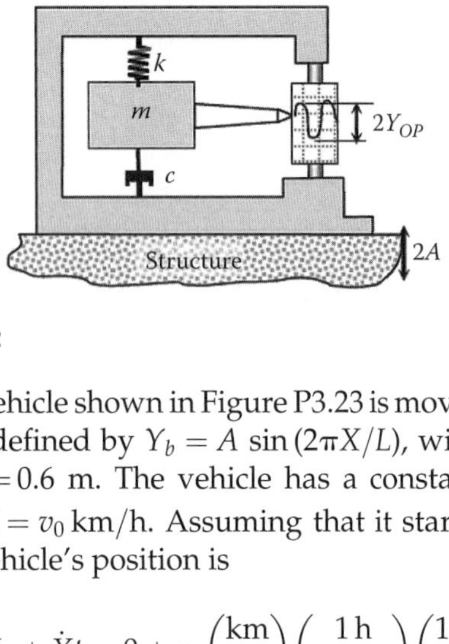

FIGURE P3.22

3.23 The vehicle shown in Figure P3.23 is moving along a path defined by $Y_b = A \sin(2\pi X/L)$, with $A = 0.03$ m, $L = 0.6$ m. The vehicle has a constant velocity $v = \dot{X} = v_0$ km/h. Assuming that it starts at $X = 0$, the vehicle's position is

$$X = X_0 + \dot{X}t = 0 + v_0 \left(\frac{\text{km}}{\text{h}}\right)\left(\frac{1\,\text{h}}{3600\,\text{s}}\right)\left(\frac{1000\,\text{m}}{1\,\text{km}}\right)t$$

$$\Rightarrow X(t) = 0.278 v_0 t \ \text{m}.$$

Tests show that the vehicle has a damped natural frequency of 4.84 Hz (cycles/s) with a damping factor of 0.251. The mass of the vehicle is 750 kg.

Tasks:
(a) Determine the speed of the vehicle that will cause the highest steady-state amplitude for the vehicle.
(b) Select the damping value c for the peak running speed in question (a) such that the vehicle's steady-state amplitude will be less than 0.12 m.

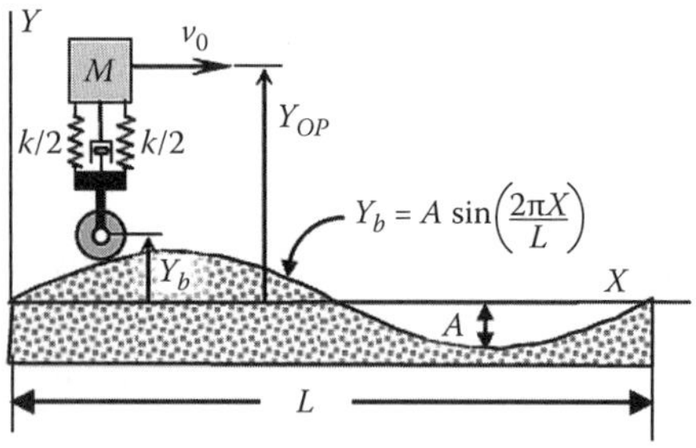

FIGURE P3.23

3.24 Refer to Problem 3.23 and Figure P3.23 and apply the following information: Tests show that the vehicle has a damped natural frequency of 4.84 Hz (cycles/s) with a damping factor of 0.251. The weight of the vehicle is 1750 lb, $L = 1.5$ ft, $A = 0.25$ ft.

Tasks:
(a) Determine the speed at which the peak amplitude will occur.
(b) Determine the required damping value if the peak amplitude is to be 3.75 in.

3.25 For the system shown in Figure P3.25 perform the following tasks:
(a) Select a coordinate, draw a free-body diagram, and obtain the equation of motion.
(b) Determine the steady-state response for the given parameters: $k_1 = 250$ N/m, $k_2 = 400$ N/m, $m = 50$ g, $c = 0.75$ N s/m, $A = 15$ mm, $\omega = 20$ Hz.
(c) Determine an excitation-frequency range so that a steady-state dynamic amplitude of 30 mm is not exceeded.

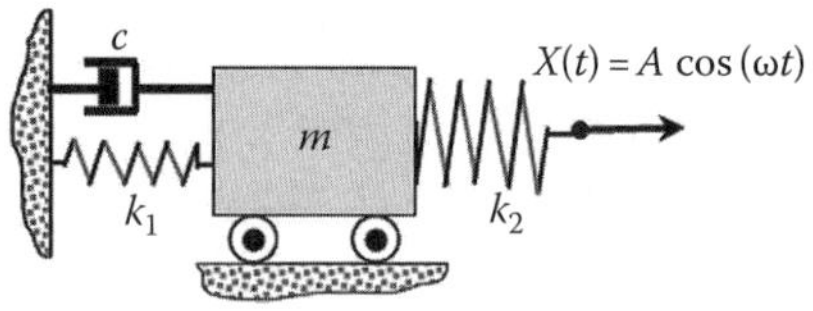

FIGURE P3.25

3.26 The rotor of an electric generator weighs 750 lbs and is attached to a platform weighing 7750 lbs. The motor has an eccentricity of 2.5 *mils*. The motor and platform can be modeled by the figure shown in Figure P3.26. The equivalent stiffness of the spring connecting the platform to ground is $7.5(10^5)$ lb/in., and the equivalent damping connection of the platform to ground is 80 lbs/in. The operating speed of the generator is 1800 rpm. Neglect the friction between the platform and ground.

Tasks:

(a) Determine the platform's peak steady-state displacement.

(b) Determine the peak response at resonance ($r = 1$).

(c) Determine the allowable amount of imbalance if the stiffness of the platform is increased by 25%, and the allowable peak response is to be under 4.9 mils. Assume that the mass of the platform and the amount of available damping does not change by an appreciable amount when the stiffness is increased.

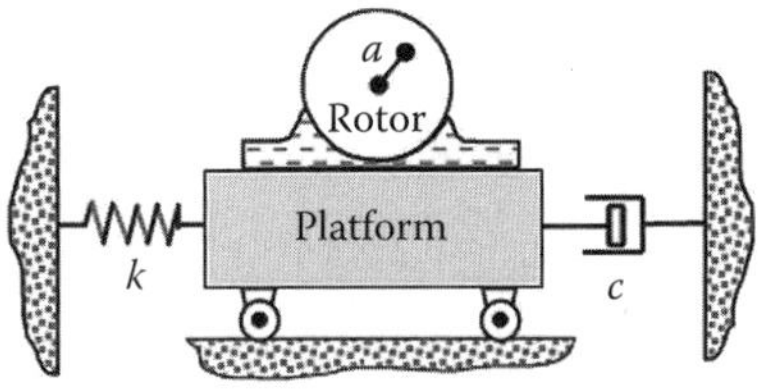

FIGURE P3.26

3.27 The model of Figure P3.26 has the following parameters: mass of platform, m_P, is 2000 kg, mass of rotor, m_R, is 700 kg, damped natural frequency of system, $\omega_d = 8.975$ Hz, and $\zeta \cong 0.075$.

Tasks:

(a) Determine the allowable amount of imbalance so that the steady-state amplitude does not exceed 2.5 mm at an operating speed of 600 rpm.

(b) Determine the allowable imbalance if the stiffness is increased by 35%. Assume that there is a negligible change in mass and damping when the stiffness is changed. Apply the allowable steady-state amplitude from part (a).

(c) Let the imbalance of the rotor be 0.75 mm. Determine the operating-speed range to avoid so that the response amplitude does not exceed 3 mm. Assume the system now has a damping factor of 0.02.

3.28 The industrial fan shown in Figure P3.28 has an electric-motor rotor that weighs 300 lb and a base support with an estimated weight of 200 lb. It has a steady operating speed of 3600 rpm. Rap tests show that the fan assembly has a damped natural frequency of 44.94 Hz and a log dec $\delta = 0.314$ for motion in the horizontal plane. The assembly is much stiffer for vertical motion.

Tasks:

(a) Determine the maximum imbalance displacement a so that the steady-state response of the assembly is below 50 mils when the running speed coincides with the damped natural frequency.

(b) Determine the response for the imbalance found in (a) at the steady-state running speed?

(c) The response resulting due to the imbalance in (b) is to be reduced by 25%. Determine the required imbalance to achieve this. Also, determine the required change in stiffness to achieve the same effect if $r > 1$.

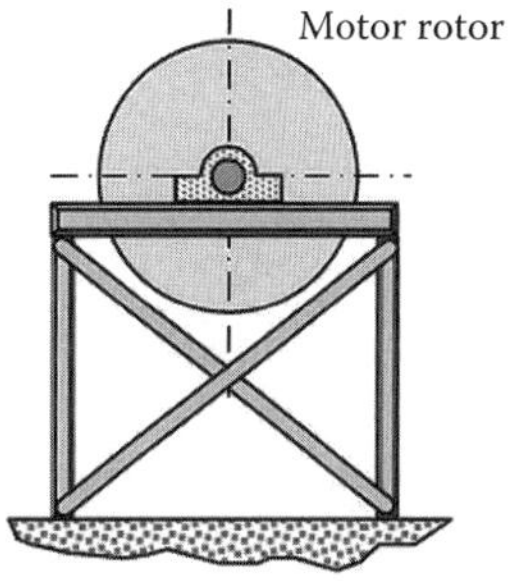

FIGURE P3.28

Motion with Constraints

Applications with Pulleys

3.29 For the system illustrated in Figure P3.29, the truck is moving to the right with a constant velocity of v_0 at $X = h$.

Task: For the general position shown, what is the tension in the cable?

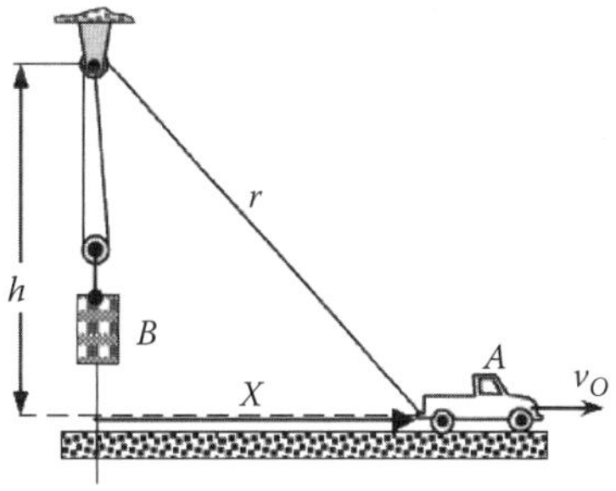

FIGURE P3.29

3.30 The bodies shown in Figure P3.30 are released from rest from the position shown. Neglect the pulley inertias.

Tasks:

(a) Select coordinates, draw the free-body diagrams, state the kinematic-constraint equation (s), and derive the equation of motion.

(b) Derive the equation of motion by applying work–energy principles.

(c) After 1 s, what are the velocities of the bodies?

(d) After body C has moved 1 m, what are the velocities of the two bodies?

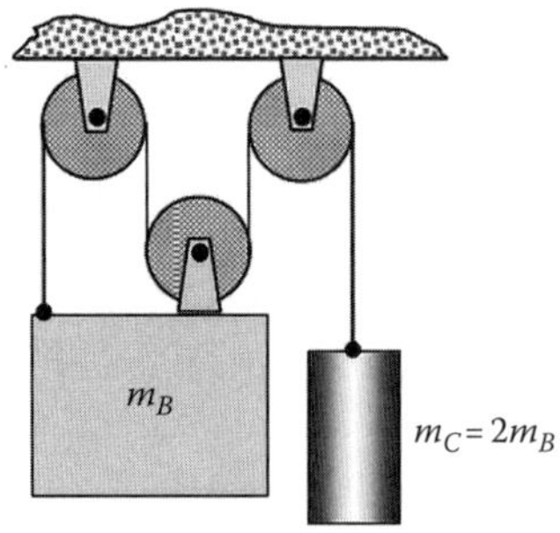

FIGURE P3.30

3.31 The system illustrated in Figure P3.31 is released from rest in the position shown. Neglect the pulley inertias.

Tasks:
(a) Select coordinates, draw the free-body diagrams, state the kinematic-constraint equation(s), and derive the equation of motion.
(b) What are the initial accelerations of the bodies?
(c) What are the velocities of the bodies after 1 s?
(d) What are the velocities of the bodies after body *A* has moved 1 m?

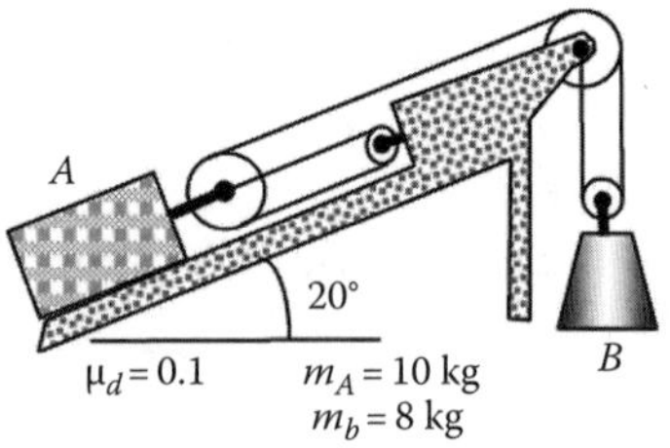

FIGURE P3.31

3.32 The two blocks shown in Figure P3.32 are initially at rest. The dynamic Coulomb-friction coefficient is $\mu_d = 0.2$. Neglect the pulley inertias.

Tasks:
(a) Select coordinates, draw the free-body diagrams, state the kinematic-constraint equation (s), and derive the equation of motion.
(b) How fast are the blocks going after 1 s?
(c) How fast are the blocks going after block *B* has moved 1 ft?

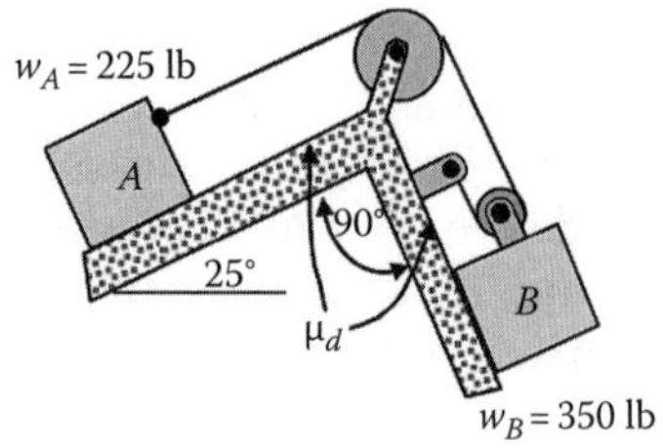

FIGURE P3.32

3.33 The system illustrated in Figure P3.33 is released from rest. The dynamic coefficient of friction is $\mu_d = 0.35$ on all surfaces. Neglect the pulley inertias.

Tasks:
(a) Select coordinates, draw the free-body diagrams, state the kinematic-constraint equation (s), and derive the equation of motion.
(b) Determine the velocities of the bodies after block *C* has moved down 8 in.
(c) Determine the velocities of the bodies after 3 s.

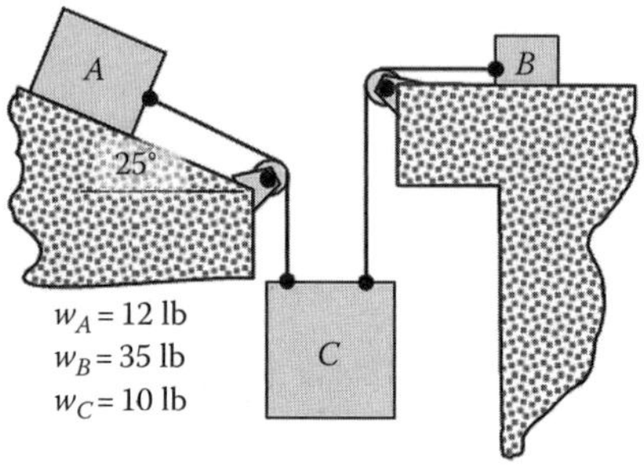

FIGURE P3.33

3.34 Two bodies *A* and *B* shown in Figure P3.34 have masses of 45 and 30 kg, respectively. The cord connecting the bodies is inextensible. The system is released from rest. The dynamic coefficient of friction is $\mu_d = 0.3$. Neglect pulley inertias.

Tasks:
(a) Select coordinates, draw the free-body diagrams, state the kinematic-constraint equation (s), and derive the equation of motion.
(b) Determine the velocities of both bodies after body *B* has moved 1 m.
(c) Determine the velocities of both bodies after 1 s.

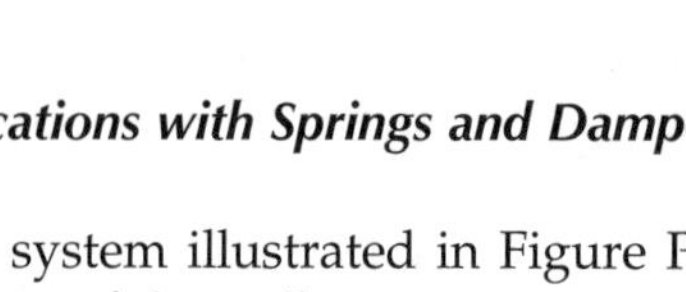

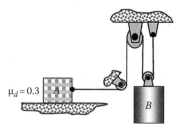

FIGURE P3.34

Pulley Applications with Springs and Dampers

3.35 For the system illustrated in Figure P3.35, neglect the inertia of the pulleys.

Tasks:
(a) Select coordinates, draw the free-body diagrams, state the kinematic-constraint equation(s), and derive the equation of motion in terms of the motion of mass *B*.

(b) Remove the damper and use conservation of energy to derive the equation of motion.
(c) From the results of part (a), determine the undamped natural frequency and the damping factor for the system.

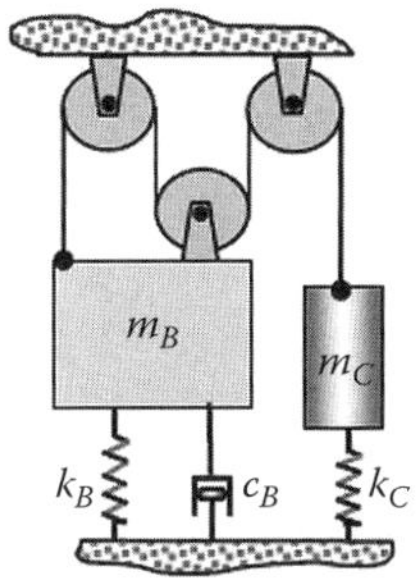

FIGURE P3.35

3.36 For the system illustrated in Figure P3.36, $m_B = 2m_A$. The system is released from rest with the spring undeflected. Neglect the pulley inertia.

Tasks:
(a) Select coordinates, draw the free-body diagrams, state the kinematic-constraint equation(s), and derive the equation of motion.
(b) Determine the equilibrium condition and the equation of motion for motion relative to the equilibrium position.
(c) Apply conservation of energy and obtain the equation of motion.
(d) Determine the natural frequency.

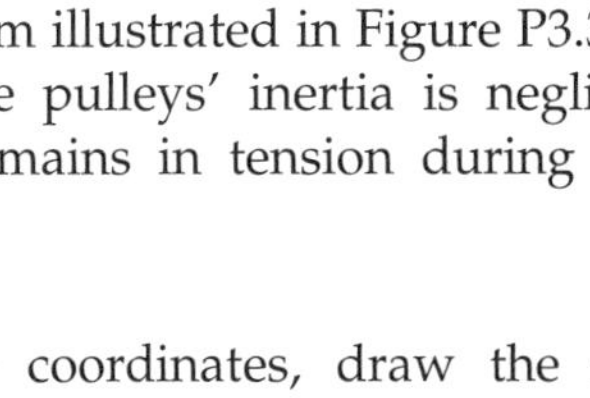

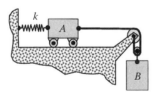

FIGURE P3.36

3.37 The system illustrated in Figure P3.37 is in equilibrium. The pulleys' inertia is negligible, and the spring remains in tension during motion of the mass.

Tasks:
(a) Select coordinates, draw the free-body diagrams, state the kinematic-constraint equation (s), and derive the equation of motion.
(b) Derive the equation of motion using conservation of energy.
Determine the natural frequency?

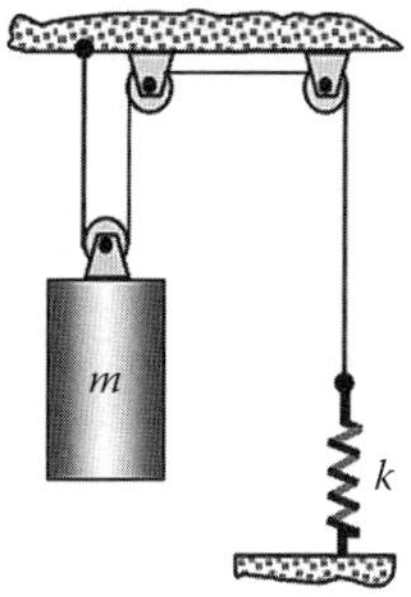

FIGURE P3.37

3.38 The system shown in Figure P3.38 consists of two masses connected by a single inextensible cord. Mass 1 on the left is connected to ground by a linear damper with damping coefficient c. Another cord extends from the ground over a pulley at the bottom of mass m_2 and back down to ground through a linear spring with stiffness coefficient k. The system is in equilibrium in the position shown, and the spring at A remains in tension for motion of the two masses. The mass and inertia of the pulleys are negligible.

Tasks:
(a) Select coordinates, draw the free-body diagrams, state the kinematic-constraint equation(s), and derive the equation of motion in terms of mass m_1's position.
(b) Remove the damper and derive the equation of motion using conservation of energy.

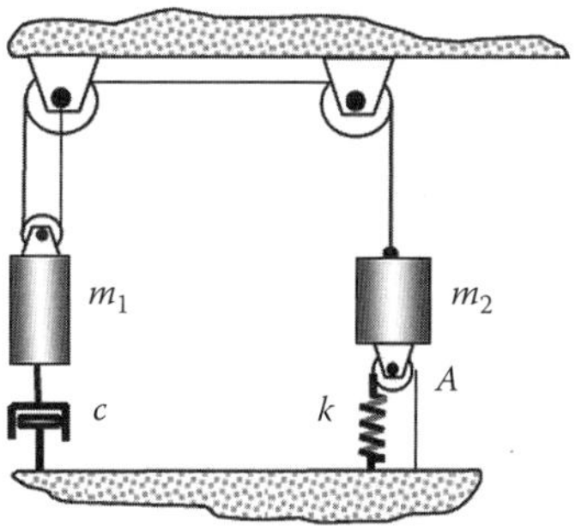

FIGURE P3.38

3.39 The system shown in Figure P3.39 consists of two masses connected by a single inextensible cord. Mass m_1 is connected to ground by a linear spring with stiffness coefficient k. Mass m_2 is connected to ground by a damper with linear damping coefficient c. The spring is undeflected in the position shown. The mass and inertia of the pulleys are negligible.

Tasks:

(a) Select coordinates, draw the free-body diagrams, state the kinematic-constraint equation (s), and derive the equation of motion in terms of mass m_1's displacement.

(b) Determine the equilibrium position and the equation of motion about the equilibrium position.

(c) Remove the damper and derive the equation of motion using conservation of energy.

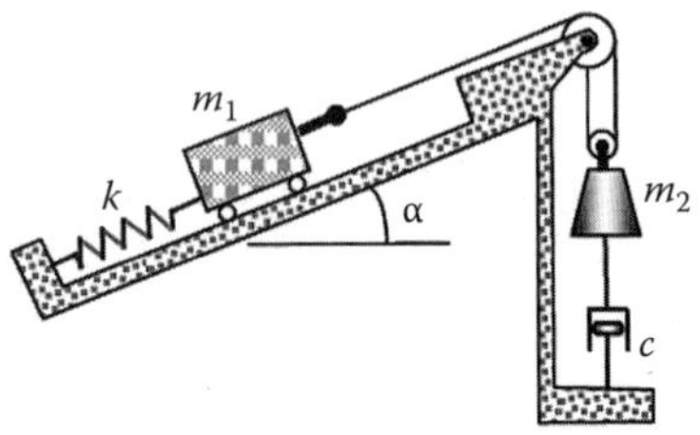

FIGURE P3.39

3.40 The system illustrated in Figure P3.40 is in equilibrium with negligible friction between mass m_1 and the contact surface. The mass and inertia of the pulleys are negligible.

Tasks:

(a) Select coordinates, draw the free-body diagrams, state the kinematic-constraint equation (s), and derive the equation of motion.

(b) Eliminate the damping and derive the equation of motion using conservation of energy.

(c) What are the undamped and damped natural frequencies and the damping factor?

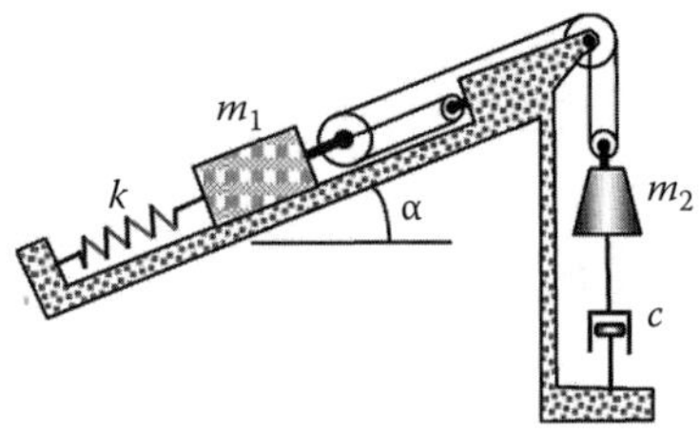

FIGURE P3.40

3.41 The system shown in Figure P3.41 is in equilibrium and consists of two masses connected by a single inextensible cord. Mass 1 is connected to ground by a linear damper with damping coefficient c. Another cord runs from the ceiling through a pulley at the top of mass 2 and back to the ceiling through a linear spring with stiffness coefficient k. The spring remains in tension during motion of the two masses. The mass and inertia of the pulleys are negligible.

Tasks:

(a) Select coordinates, draw the free-body diagrams, state the kinematic-constraint equation (s), and derive the equation of motion in terms of mass m_1.

(b) What are the undamped and damped natural frequencies the damping factor?

(c) Remove the damper and derive the equation of motion using conservation of energy.

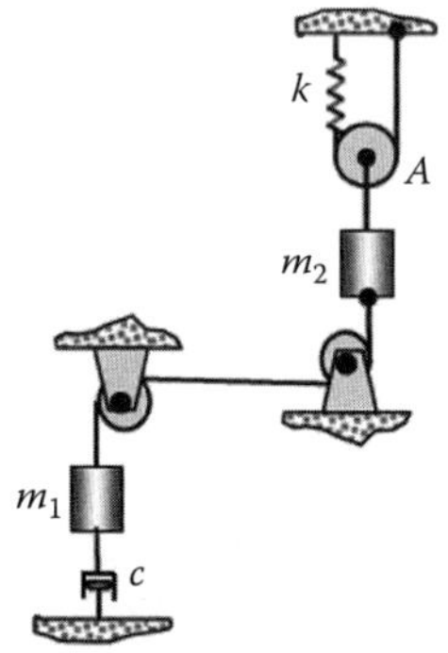

FIGURE P3.41

MDOF Equations of Motion

3.42 Derive the equations of motion for the system illustrated in Figure P3.42 and put your equations in standard matrix format.

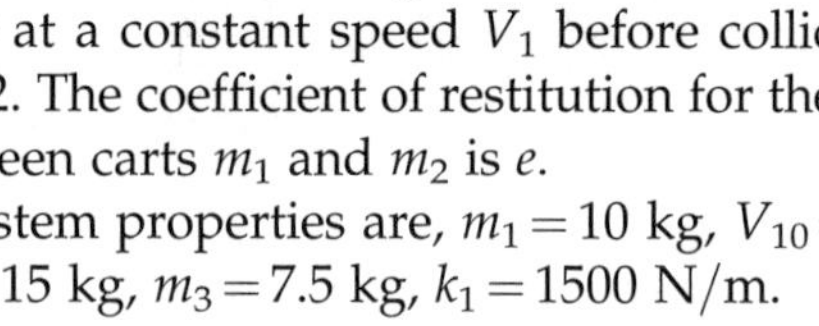

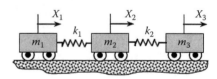

FIGURE P3.42

3.43 Cart 1 of mass m_1 in Figure P3.43 is traveling to the right at a constant speed V_1 before colliding with cart 2. The coefficient of restitution for the collision between carts m_1 and m_2 is e.

System properties are, $m_1 = 10$ kg, $V_{10} = 25$ m/s, $m_2 = 15$ kg, $m_3 = 7.5$ kg, $k_1 = 1500$ N/m.

Tasks:

(a) For $e = 1$, determine the velocity of mass m_2 immediately following the collision and solve for the response of the carts after the collision.

(b) For $e = 0.65$, determine the velocity of mass m_2 immediately following the collision and solve for the response of the carts after the collision.

(c) For $e = 0$, determine the velocity of mass m_2 immediately following the collision and solve for the response of the carts after the collision. Note: Recalculate the eigen properties for this case.

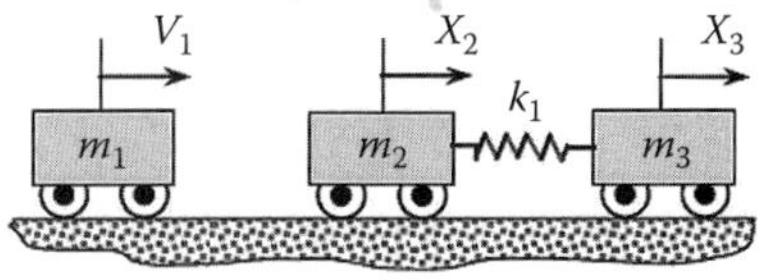

FIGURE P3.43

3.44 The unattached end of spring k_5 in Figure P3.44 is oscillating with motion defined by $X_5(t) = D\cos(\omega t)$, and the force $f(t) = f_O \sin(2\omega t)$ acts on mass m_2.

Tasks:

(a) Select coordinates and draw them on the figure.

(b) State the nature of the assumed motion and draw corresponding free-body diagrams.

(c) Derive the equations of motion and put them in standard matrix format.

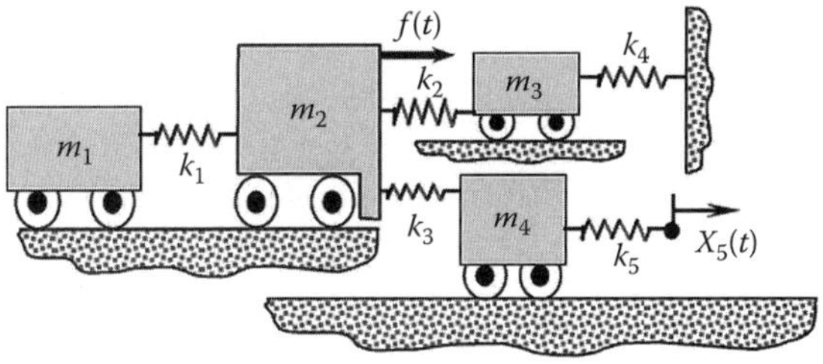

FIGURE P3.44

3.45 For the system illustrated in Figure P3.45, the left-hand wall is oscillating with motion defined by $X_0 = A\cos(\omega t)$.

Tasks:

(a) Select coordinates and draw them on the figure.

(b) State the nature of the assumed motion and draw corresponding free-body diagrams.

(c) Derive the equations of motion and put them in standard matrix format.

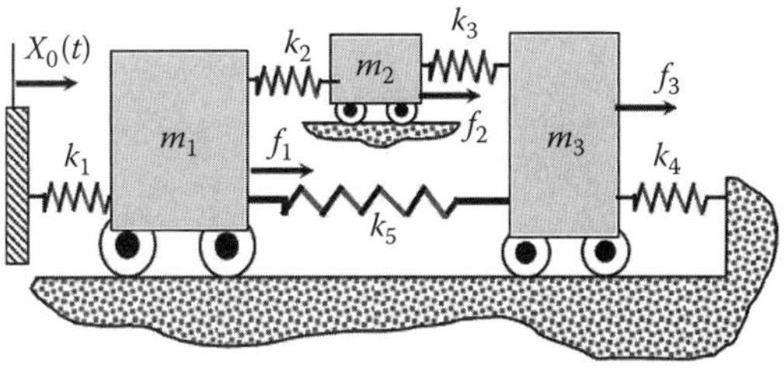

FIGURE P3.45

3.46 The system illustrated in Figure P3.46 is free to translate.

Tasks:

(a) Select coordinates and draw them on the figure.

(b) State the nature of the assumed motion and draw corresponding free-body diagrams.

(c) Derive the equations of motion and put them in standard matrix format.

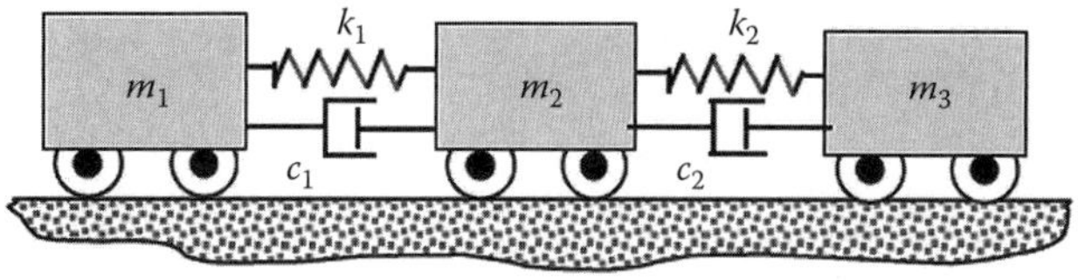

FIGURE P3.46

3.47 Note that the left-hand wall of negligible mass in Figure P3.47 is oscillating with motion defined by $X_0 = A\cos(\omega t)$.

Tasks:

(a) Select coordinates and draw them on the figure.

(b) State the nature of the assumed motion, and draw corresponding free-body diagrams.

(c) Derive the equations of motion and put them in standard matrix format.

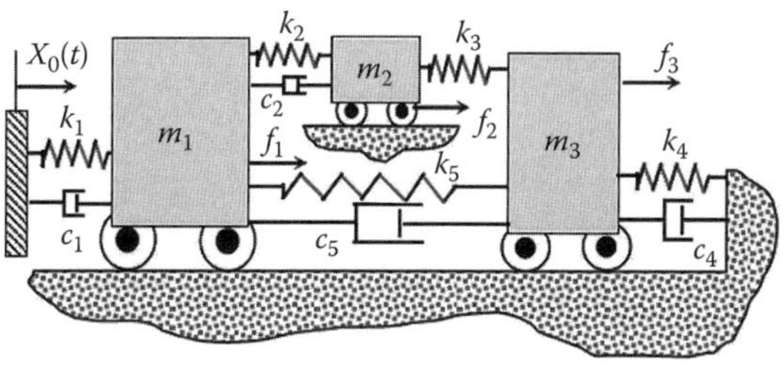

FIGURE P3.47

3.48 The unattached end of spring k_5 in Figure P3.48 is oscillating with motion defined by $X_5(t) = D\cos(\omega t)$, and force $f(t) = f_O \sin(2\omega t)$ acts on mass m_2.

Tasks:

(a) Select coordinates and draw them on the figure.

(b) State the nature of the assumed motion, and draw corresponding free-body diagrams.

(c) Derive the equations of motion and put them in standard matrix format.

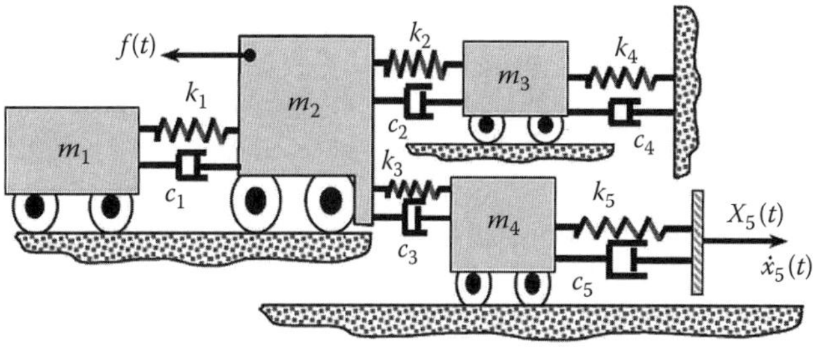

FIGURE P3.48

MDOF Response Applications

3.49 A vibrating system is defined by the equation,

$$\begin{bmatrix} m_1 & 0 \\ 0 & m_2 \end{bmatrix} \begin{Bmatrix} \ddot{X}_1 \\ \ddot{X}_2 \end{Bmatrix} + \begin{bmatrix} k_1+k_2 & -k_2 \\ -k_2 & k_2+k_3 \end{bmatrix} \begin{Bmatrix} X_1 \\ X_2 \end{Bmatrix} = \{0\}$$

with $m_1 = 1.25$ kg, $m_2 = 2.5$ kg, $k_1 = 5$ kN/m, $k_2 = 10$ kN/m, and $k_3 = 5$ kN/m. The initial conditions are $X_{10} = 0.05$ m, $X_{20} = 0.075$ m, $\dot{X}_{10} = \dot{X}_{20} = 0$.

Tasks:

(a) Calculate the eigenvalues and natural frequencies.

(b) Calculate the eigenvectors. Draw a plot of your eigenvectors showing the relative motion of the masses.

(c) Normalize the eigenvectors such that $[A^*]^T[M][A^*] = [I]$ and confirm that your normalized matrix of eigenvectors satisfies $[A^*]^T[K][A^*] = [\Lambda]$ where $[\Lambda]$ is a diagonal matrix of eigenvalues.

(d) Obtain the uncoupled modal equations of motion and the transformation from modal to physical coordinates.

(e) Obtain the modal-coordinate time response solutions.

(f) Obtain the physical-coordinate time response solutions.

3.50 The double-pendulum system is defined by the equation,

$$\begin{bmatrix} (m_1+m_2)l_1^2 & m_2 l_2 l_1 \\ m_2 l_2 l_1 & m_2 l_2^2 \end{bmatrix} \begin{Bmatrix} \ddot{\theta}_1 \\ \ddot{\theta}_2 \end{Bmatrix} + \begin{bmatrix} (w_1+w_2)l_1 & 0 \\ 0 & w_2 l_2 \end{bmatrix} \begin{Bmatrix} \theta_1 \\ \theta_2 \end{Bmatrix} = \{0\}$$

with $m_1 = 0.5$ kg, $m_2 = 0.25$ kg, $l_1 = 100$ cm, $l_2 = 50$ cm, $\theta_{20} = 5°, \dot{\theta}_{10} = \theta_{10} = \dot{\theta}_{20} = 0$.

Tasks:

(a) Calculate the eigenvalues and natural frequencies.

(b) Calculate the eigenvectors and draw them showing the relative motion of the masses.

(c) Normalize the eigenvectors such that $[A^*]^T[M][A^*] = [I]$ and confirm that your normalized matrix of eigenvectors satisfies $[A^*]^T[K][A^*] = [\Lambda]$, where $[\Lambda]$ is diagonal matrix of eigenvalues.

(d) Obtain the uncoupled modal equations of motion and the transformation from modal to physical coordinates.

(e) Obtain and the modal-coordinate time response.

(f) Obtain the physical-coordinate time response for the system.

3.51 A vibrating system is defined by the equation,

$$\begin{bmatrix} I_o + m_2R^2 & m_2R \\ m_2R & m_1+m_2 \end{bmatrix} \begin{Bmatrix} \ddot{\theta} \\ \ddot{X}_1 \end{Bmatrix} + \begin{bmatrix} k_2R^2 & k_2R \\ k_2R & k_1+k_2 \end{bmatrix} \begin{Bmatrix} \theta \\ X_1 \end{Bmatrix} = 0,$$

with $R = 0.25$ m, $m_2 = 1.5$ kg, $m_1 = 3$ kg, $k_1 = 5000$ N/m, $k_2 = 4750$ N/m and $I_o = (1/2)m_2R^2$.

Tasks:

(a) Calculate the eigenvalues and natural frequencies.

(b) Calculate the eigenvectors and draw them showing the relative motion of the coordinates.

(c) Normalize the eigenvectors such that $[A^*]^T[M][A^*] = [I]$ and confirm that your normalized matrix of eigenvectors satisfies $[A^*]^T[K][A^*] = [\Lambda]$, where $[\Lambda]$ is diagonal matrix with of eigenvalues.

3.52 The equation of motion for a system is

$$\begin{bmatrix} m_1 & 0 & 0 \\ 0 & m_2 & 0 \\ 0 & 0 & m_3 \end{bmatrix} \begin{Bmatrix} \ddot{X}_1 \\ \ddot{X}_2 \\ \ddot{X}_3 \end{Bmatrix} + \begin{bmatrix} k_1 & -k_1 & 0 \\ -k_1 & k_1+k_2 & -k_2 \\ 0 & -k_2 & k_2 \end{bmatrix} \times \begin{Bmatrix} X_1 \\ X_2 \\ X_3 \end{Bmatrix} = \begin{Bmatrix} 0 \\ f_0 \sin \omega t \\ 0 \end{Bmatrix}.$$

The system properties are $m_1 = 2$ kg, $m_2 = 3$ kg, $m_3 = 2$ kg, $k_1 = 100$ N/m, $k_2 = 150$ N/m, $f_0 = 20$ N, and $\omega = 15$ Hz, and the initial conditions are $X_{10} = 0.05\text{ m}, X_{20} = 0.075\text{ m}, X_{30} = 0.0\text{ m}\ \dot{X}_{10} = \dot{X}_{20} = \dot{X}_{30} = 0$.

Tasks:

(a) Determine the eigenvalues and normalized eigenvectors.

(b) Obtain the uncoupled modal equations of motion and the transformation from modal to physical coordinates.

(c) Obtain the modal-coordinate time response.

(d) Obtain the physical-coordinate time response.

(e) Obtain the steady-state modal and physical response.

3.53 The matrix equation of motion for a two-mass system is

$$\begin{bmatrix} m_1 & 0 \\ 0 & m_2 \end{bmatrix} \begin{Bmatrix} \ddot{x}_1 \\ \ddot{x}_2 \end{Bmatrix} + \begin{bmatrix} k_1+k_2 & -k_2 \\ -k_2 & k_2+k_3 \end{bmatrix} \begin{Bmatrix} x_1 \\ x_2 \end{Bmatrix} = \begin{Bmatrix} 2000t^2 \\ 0 \end{Bmatrix},$$

where $m_1 = 2$ kg, $m_2 = 4$ kg, $k_1 = 0$, $k_2 = k_3 = 2000$ N/m. The eigenvalues for the system are found to be $\omega^2_{n1} = 292.9$ (rad/s)^2 and $\omega^2_{n2} = 1701$ (rad/s)^2. The normalized matrix of eigenvectors is

$$[A^*] = \begin{bmatrix} 0.5 & 0.5 \\ 0.3536 & -0.3536 \end{bmatrix}.$$

The system starts from rest with $X_1(0) = X_2(0) = 0$. The first and second modes have 5% of critical damping.

Tasks:

(a) State the modal-coordinate differential equations and their time solutions.

(b) State the time solution to the physical coordinates.

3.54 For the model

$$\begin{bmatrix} m_1 & 0 \\ 0 & m_2 \end{bmatrix} \begin{Bmatrix} \ddot{x}_1 \\ \ddot{x}_2 \end{Bmatrix} + \begin{bmatrix} k_1+k_2 & -k_2 \\ -k_2 & k_2+k_3 \end{bmatrix} \begin{Bmatrix} x_1 \\ x_2 \end{Bmatrix} = \begin{Bmatrix} f_1(t) \\ f_2(t) \end{Bmatrix}$$

use the following data to determine the free motion of the two bodies:

$$m_1 = 0.5\text{ kg},\quad m_2 = 1\text{ kg},\quad k_1 = k_2 = 1\text{ N/m},$$
$$k_3 = 0;\quad x_2(0) = 1\text{ cm},$$
$$x_1(0) = \dot{x}_1(0) = \dot{x}_2(0) = 0,\quad f_1(t) = f_2(t) = 0.$$

3.55 For the model stated in Problem 3.54 use the following data to determine the free motion of the two bodies:

$$m_1 = 0.5\text{ kg},\ m_2 = 1\text{ kg},\ k_1 = 0,\ k_2 = 1\text{N/m},\ k_3 = 0;$$
$$x_1(0) = x_2(0)\dot{x}_1(0) = 0;\ \dot{x}_2(0) = 1\text{ cm/s},\ f_1(t) = f_2(t) = 0.$$

3.56 Develop the transient solution for the model stated in Problem 3.54 using the data from Problem 3.54 with the following initial conditions: $x_1(0) = x_2(0) = \dot{x}_1(0) = \dot{x}_2(0) = 0$, and the constant applied force vector $f_1(t) = 1$ N; $f_2(t) = 0$.

3.57 Develop the transient solution for the model stated in Problem P3.54 along with the system parameters defined in Problem 3.54 and the initial conditions, $x_1(0) = x_2(0) = \dot{x}_1(0) = \dot{x}_2(0) = 0$, the constant applied force vector, $f_1(t) = 1$ N; $f_2(t) = 0$, and the following two sets of modal damping factors, (a) $\zeta_1 = \zeta_2 = 0$; (b) $\zeta_1 = 0.0$, $\zeta_2 = 0.1$. Plot the solutions for both the modal and physical coordinates versus time so that 4 cycles of oscillations occur. Discuss your results.

3.58 Develop the transient solution for the model stated in Problem 3.54 with the initial conditions, $x_1(0) = x_2(0) = \dot{x}_1(0) = \dot{x}_2(0) = 0$, the constant applied force vector, $f_1(t) = 1$ N; $f_2(t) = 0$, the following modal damping factors, $\zeta_1 = \zeta_2 = 0.15$, and the parameters defined in Problem 3.54. Plot the solutions for both the modal and physical coordinates versus time so that 20 cycles of oscillations occur. Discuss your results.

3.59 A vibrating system is defined by the following matrix differential equation:

$$\begin{bmatrix} m_1 & 0 \\ 0 & m_2 \end{bmatrix} \begin{Bmatrix} \ddot{X}_1 \\ \ddot{X}_2 \end{Bmatrix} + \begin{bmatrix} k_1+k_2 & -k_2 \\ -k_2 & k_2 \end{bmatrix} \begin{Bmatrix} X_1 \\ X_2 \end{Bmatrix} = \begin{Bmatrix} f_1(t) \\ f_2(t) \end{Bmatrix},$$

with $m_1 = 2$ kg, $m_2 = 2$ kg, $k_1 = k_2 = 2$ N/m, $f_1(t) = 3\cos(2t)$, $f_2(t) = 5\cos(2t)$. The initial conditions are $X_{10} = 0.05$ m, $X_{20} = 0.075$ m, $\dot{X}_{10} = \dot{X}_{20} = 0$.

Tasks:

(a) Calculate the eigenvalues and natural frequencies.

(b) Calculate the eigenvectors. Draw a plot of your eigenvector showing the relative motion of the masses.

(c) Normalize the eigenvectors such that $[A^*]^T[M][A^*] = [I]$ and confirm that your normalized matrix of eigenvectors satisfies $[A^*]^T[K][A^*] = [\Lambda]$, where $[\Lambda]$ is a diagonal matrix with the eigenvalues completing the diagonal.

(d) Obtain the uncoupled equations of motion.

(e) Obtain the uncoupled modal time response for the system.

(f) Obtain the physical time response for the system.

(g) Obtain the steady-state response for the system.

3.60 The system shown in Figure P3.60 has the following properties: $w_1 = 10$ lb, $w_2 = 4$ lb, $k_1 = 25$ lb/in., $k_2 = 15$ lb/in., $k_3 = 30$ lb/in., $k_4 = 35$ lb/in., $f_0 = 10$ lb, and $\omega = 30$ Hz. The initial conditions for the problem are: $X_{10} = 1$ in., $X_{20} = 0.5$ in., $\dot{X}_{10} = \dot{X}_{20} = 0$.

Tasks:

(a) Select coordinates, define the nature of the assumed motion, and obtain the differential equation of motions and present them in matrix form.
(b) Determine the eigenvalues and natural frequency of the system.
(c) Obtain the eigenvectors for the system.
(d) Obtain the modal differential equation of motion.
(e) Obtain the time response for the physical coordinates.
(f) Obtain the steady-state response for the system.

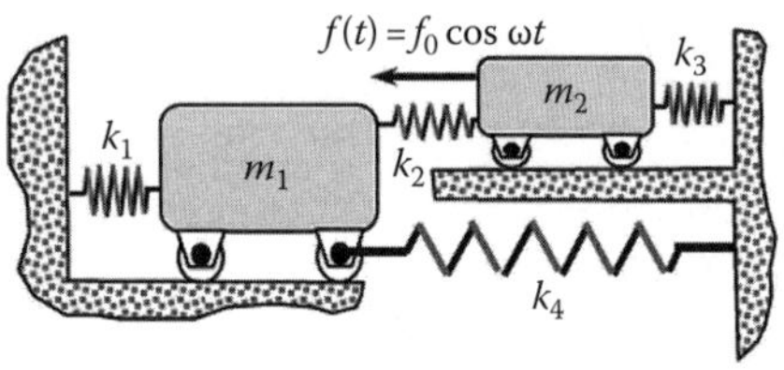

FIGURE P3.60

3.61 The system shown in Figure P3.61 has the following properties: $m_1 = 100$ kg, $m_2 = 25$ kg, $k_1 = 750$ N/m, $k_2 = 150$ N/m, $k_3 = 150$ N/m, $k_4 = 250$ N/m, $k_5 = 250$ N/m, $X_0(t) = X_0 \sin \omega t$, $X_0 = 10$ cm, and $\omega = 60$ Hz. The initial conditions for the problem are: $X_{10} = 10$ cm, $X_{20} = 0$, $\dot{X}_{10} = \dot{X}_{20} = 0$.

Tasks:

(a) Select coordinates, define the nature of the assumed motion, and obtain the differential equation of motions and present them in matrix form.
(b) Determine the eigenvalues and natural frequencies of the system.
(c) Obtain the eigenvectors for the system.
(d) Obtain the modal differential equation of motion.
(e) Obtain the response of the system in physical coordinates.
(f) Obtain the steady-state response of the system.

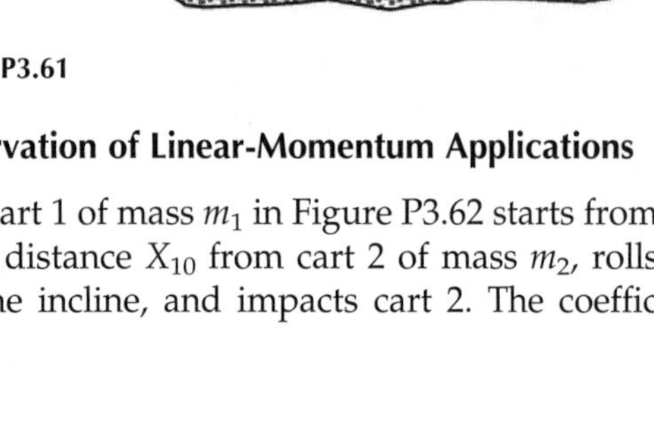

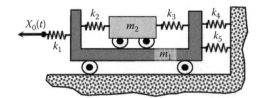

FIGURE P3.61

Conservation of Linear-Momentum Applications

3.62 Cart 1 of mass m_1 in Figure P3.62 starts from rest at a distance X_{10} from cart 2 of mass m_2, rolls down the incline, and impacts cart 2. The coefficient of restitution between the carts is e. Mass m_2 is in equilibrium at the time of impact. The properties of the system are $m_1 = 1500$ kg, $X_{10} = 10$ m, $m_2 = 5000$ kg, $k = 10^6$ N/m, $c = 1500$ N s/m, $e = 0.35$, and $\alpha = 20°$.

Tasks:

(a) Determine the velocity of the carts immediately following the collision.
(b) Obtain the equation of motion for the carts following the initial collision and prior to any subsequent collisions. What is the undamped natural frequency, the damped natural frequency, and the damping factor for cart 2?
(c) Determine the maximum gap, Δ, between the carts after the impact occurs.

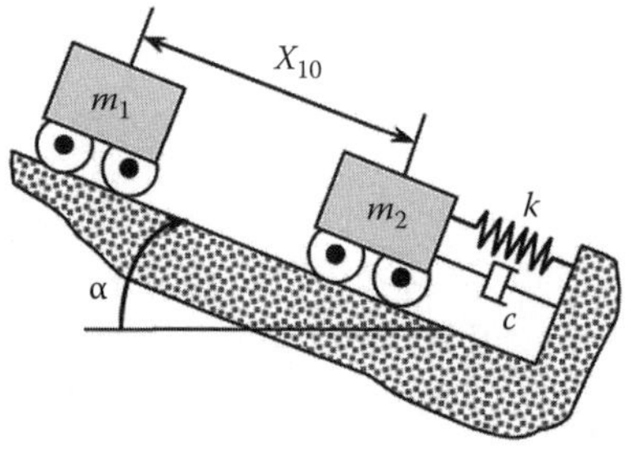

FIGURE P3.62

3.63 In Figure P3.63, the bullet of weight w_b is traveling at the constant speed v_0 when it impacts and is embedded within the cart of weight w_c. The mass of the bumper that impacts the wall is negligible.

Tasks:

(a) Determine the velocity of the cart immediately after impact.
(b) Derive the EOM for the block following the collision, after the bumper makes contact with the wall, and before it loses contact with the wall.
(c) Neglect damping and use the following system properties, $w_c = 5$ lb, $v_0 = 2820$ ft/s, $w_b = 0.229$oz, to determine the spring constant k such that the cart comes to rest in a distance of 1 in.
(d) Determine the required damping constant c so that the stopping time from part (c) is reduced by 15%.

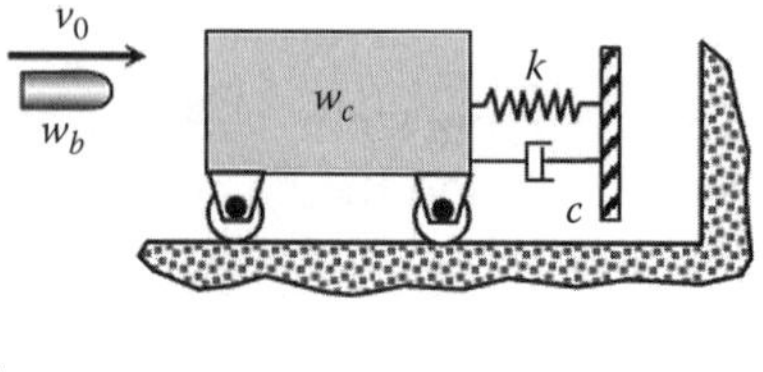

FIGURE P3.63

Conservation of Moment-of-Momentum Applications

3.64 Illustrated in Figure P3.64 is a hollow half sphere of radius R with gravity acting vertically downward. A small particle of mass m slides with negligible friction inside the half sphere. At an initial release angle of θ_0, the particle has an initial moment of momentum of $H_0 = mr_0^2\dot{\phi}_0$.

Tasks:

(a) Use the coordinates r, ϕ, and Z, draw a free-body diagram, and state the equations of motion.

(b) Using geometry, and taking advantage of conservation of moment of momentum about a vertical axis through the center of the half cylinder, derive an equation of motion in the variable θ (only).

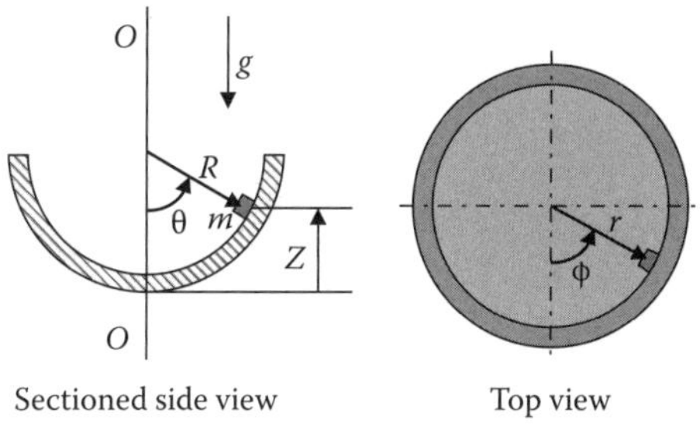

FIGURE P3.64

3.65 Figure P3.65 illustrates a particle of mass m that is sliding without friction on the inner surface of an inverted cone. The cone has the half angle α. The position of the particle is defined by S and denotes the position along the cone's surface from the apex. θ defines the counter clockwise rotation about the cone's axis, and the height above the cone's apex is Z. Assume that at an initial position S_0 the particle has an initial moment of momentum of $H_0 = mr_0^2\dot{\theta}_0$.

Tasks:

(a) Use the coordinates illustrated, draw a free-body diagram, and state the equations of motion.

(b) Using geometry, and taking advantage of conservation of moment of momentum about a vertical axis through the cone's axis, derive an equation of motion in the variable S (only).

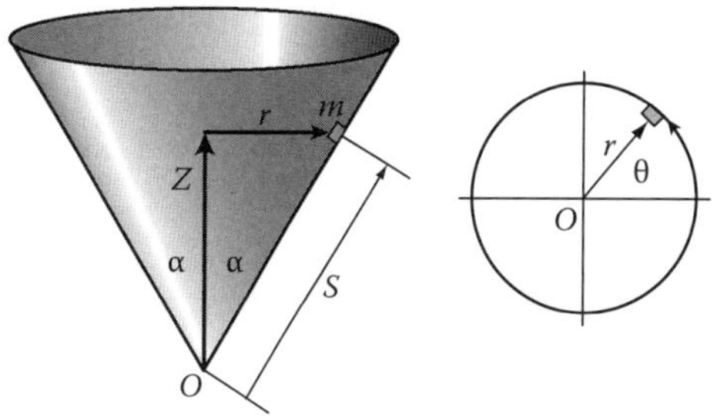

FIGURE P3.65

3.66 Figure P3.66 illustrates a particle of mass m connected to another particle of mass M by an inextensible cord of length l. Particle m is sliding on the outside of an inverted cone with cone angle α. The cord passes through a hole at the cone's vertex, and particle M moves vertically on a line through the cone's vertex. Assume that at an initial position S_0 the particle has an initial moment of momentum of $H_0 = mr_0^2\dot{\phi}_0$.

Tasks:

(a) Draw a free-body diagram for m, and state the equations of motion for the coordinates r, ϕ, and Z.

(b) Draw a free-body diagram for M and use it to derive M's EOM and then eliminate the cord tension.

(c) Using geometry, and taking advantage of conservation of moment of momentum about a vertical axis through the center of the half cylinder, derive an equation of motion in the variable S (only).

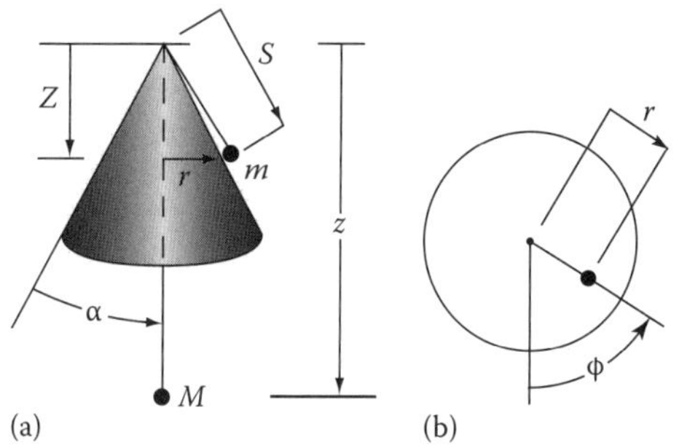

FIGURE P3.66

4

Planar Kinematics of Rigid Bodies

4.1 Introduction

The derivation of general kinematic equations for bodies in plane motion is the central objective of this chapter. *Planar* motion of a body means that all of a body's motion is confined to a plane, e.g., the plane of this page. Figure 4.1 illustrates a rigid body that is translating and rotating in the plane of the figure. The x, y coordinate system is fixed to the rigid body. The origin of the x, y system is point o, which is located in the X, Y system by the vector $\boldsymbol{R}_o = \boldsymbol{I}R_{oX} + \boldsymbol{J}R_{oY}$. The angle θ defines the orientation of both the body and the x, y coordinate system with respect to the X, Y system. Planar kinematics of a rigid body involves keeping track of the position of a specified point on the body (e.g., using R_{oX} and R_{oY} of Figure 4.1 to locate point o) and the body's orientation (θ angle of rotation of Figure 4.1) with respect to a fixed axis. Position, velocity, and acceleration vectors for points were the subjects covered in Chapter 2 on particle kinematics.

Formulating planar kinematics problems for rigid bodies requires that we introduce and define angular velocity and angular acceleration vectors for the rigid body. Figure 4.1b shows the rigid body at a later time $t + \Delta t$, during which the orientation angle has changed to $\theta + \Delta\theta$. The angular rotation rate of the body is $\dot{\theta} = \lim \Delta t \Rightarrow 0 | (\Delta\theta/\Delta t) = d\theta/dt$. Using the right-hand rule for an advancing screw, the angular velocity *vector* of the body is $\boldsymbol{\omega} = \boldsymbol{k}\dot{\theta}$. Here, $\boldsymbol{k}$ is the unit vector which is normal to the plane of the figure and pointing outward along the z-axis of Figure 4.2. Similarly, the angular acceleration *vector* for the body is $\dot{\boldsymbol{\omega}} = \boldsymbol{k}\ddot{\theta}$. Figure 4.2 shows the rigid body and the angular velocity vector in the X, Y, Z system.

The great majority of mechanisms in mechanical engineering are planar; hence, planar kinematics is both important and useful. However, many common problems of interest involve nonplanar motion; e.g., an airplane that is simultaneously pitching up and yawing to the left cannot be described by planar kinematics. The motion of a gyroscope provides another example of a rigid-body motion that cannot at all be described with planar kinematics. Three-dimensional vector kinematics is covered in more advanced dynamics texts.

No new or additional governing equations are introduced in this chapter; the kinematics equations developed in Chapter 2 are adequate to define the planar kinematics of rigid bodies. The problem objectives in this chapter normally involve the derivation of *general* governing kinematics equations for systems of rigid bodies, in particular, planar linkages. Most of the problems require the development of general governing equations versus requests for answers at a particular position. In the author's experience, students who learn to develop general solutions rarely have problems with specific questions, but the converse is not true. Also, in the design of mechanisms, one generally seeks information that can only be provided by a general solution format, namely, *peak* accelerations or velocities. The method of "instantaneous" centers is not covered in this chapter. In the author's experience, students consistently have trouble understanding why this method works for velocity but not for acceleration.

As noted in the preface, one persistent irritation with dynamics concerns the possibility of working (correctly) the same problem in many different ways. Most planar kinematics problems involving two or more bodies can be solved readily using Cartesian or polar coordinates of Chapter 2. They can also be solved with less effort by starting from geometric equations and then differentiating once with respect to time to obtain velocity equations and then differentiating again to obtain acceleration equations. Example problems provided in this chapter will demonstrate a range of alternative solution possibilities.

4.2 Rotation about a Fixed Axis

Before jumping into planar kinematics problems for rigid bodies, let us suppose we "warm up" by looking at the simple situation illustrated in Figure 4.3, namely, a wheel of radius r that is rotating about a fixed axis. The wheel's orientation with respect to the X-, Y-axes is defined by θ; its angular velocity and angular acceleration are defined by $\boldsymbol{\omega} = \boldsymbol{k}d\theta/dt = \boldsymbol{k}\dot{\theta}$, $\dot{\boldsymbol{\omega}} = \boldsymbol{k}d^2\theta/dt^2 = \boldsymbol{k}\ddot{\theta}$. Using

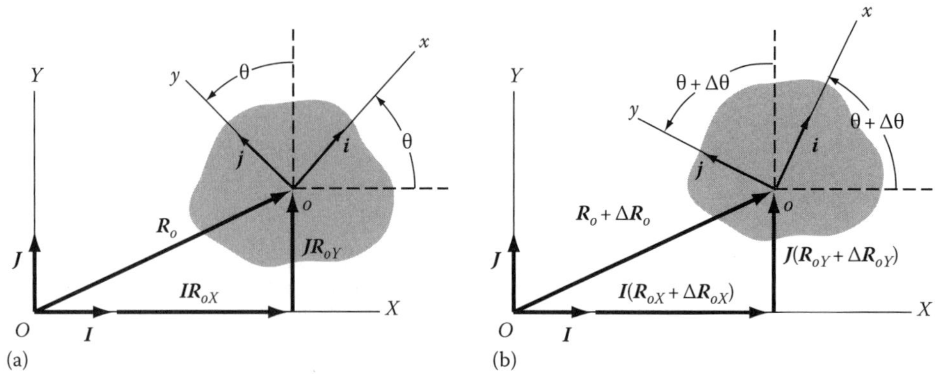

FIGURE 4.1
Planar motion of a rigid body moving in the plane of the page. (a) The body at time t and orientation θ. (b) The body at a slightly later time $t + \Delta t$ with a new position $R_o + \Delta R_o$ and new orientation $\theta + \Delta\theta$.

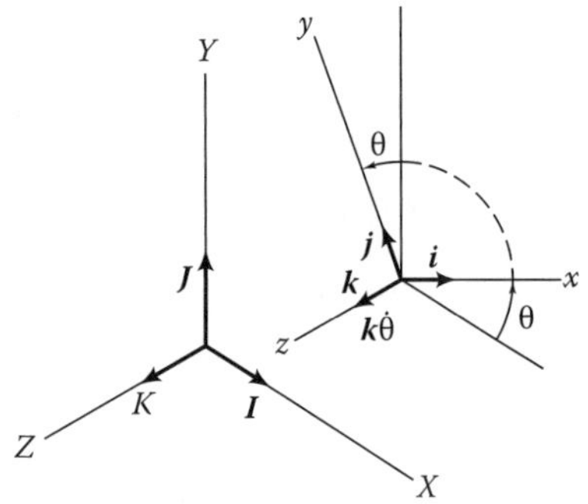

FIGURE 4.2
Rigid body moving in the X, Y plane with the angular velocity vector $\boldsymbol{\omega} = \boldsymbol{k}\dot{\theta}$ aligned with the z- and Z-axes.

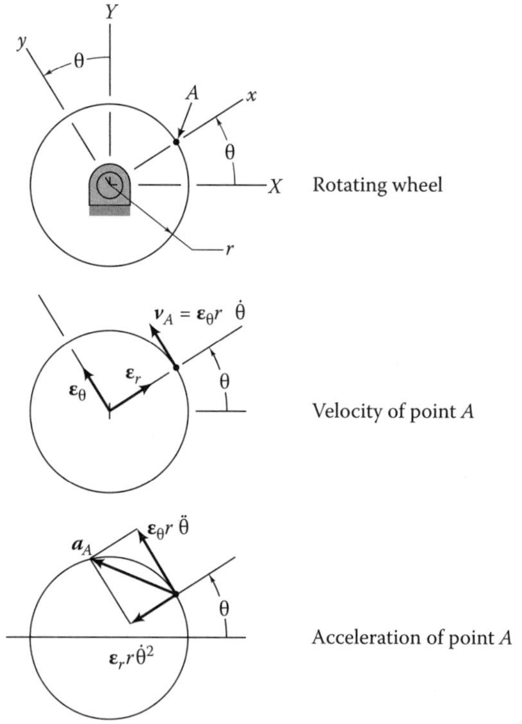

FIGURE 4.3
Simple fixed-axis rotation.

our knowledge from Chapter 2 (Equation 2.27) on polar-coordinate kinematics, the velocity of point A on the wheel is $v_A = \boldsymbol{\varepsilon}_\theta r\dot{\theta}$, as illustrated in Figure 4.2b. The acceleration of point A with respect to the X, Y coordinate system is (from Equation 2.30) $a_A = \boldsymbol{\varepsilon}_\theta\ddot{\theta} - \boldsymbol{\varepsilon}_r r\dot{\theta}^2$. The terms $r\ddot{\theta}$ and $-r\dot{\theta}^2$ are the circumferential and centripetal components of acceleration, respectively.

Now, think about applying these polar-coordinate kinematic equations to a simple rigid-body rotation problem. Figure 4.4 illustrates a stationary exercise bicycle. The radius of the drive sprocket is r_1, the radius of the driven sprocket is r_2, and the fly wheel has radius r_3. Assuming that the angular velocity of the drive sprocket is ω_1 rad/s, what is the angular velocity of the driven sprocket ω_2 (and the fly wheel)? Given that the chain connecting the driven and drive wheels is inextensible (cannot stretch), the circumferential velocity at the outside radii is the same for both wheels; i.e.,

$$v_\theta = r_1\omega_1 = r_2\omega_2 \Rightarrow \omega_2 = r_1\omega_1/r_2.$$

A comparable analysis answers the related question: If the drive sprocket has the angular acceleration $\dot{\omega}_1$, *what is* $\dot{\omega}_2$, *the angular acceleration of sprocket* 2? The acceleration along the path of the chain is constant; hence,

$$a_\theta = r_1\dot{\omega}_1 = r_2\dot{\omega}_2 \Rightarrow \dot{\omega}_2 = r_1\dot{\omega}_1/r_2$$

Note that the radial accelerations at the outer circumferences of the sprockets are not equal, and are defined by

$$\begin{aligned} a_{1r} &= -r_1\omega_1^2 \\ a_{2r} &= -r_2\omega_2^2 = -r_1^2\omega_1^2/r_2 \\ a_{3r} &= -r_3\omega_2^2 = -r_3\omega_1^2(r_1/r_2)^2. \end{aligned}$$

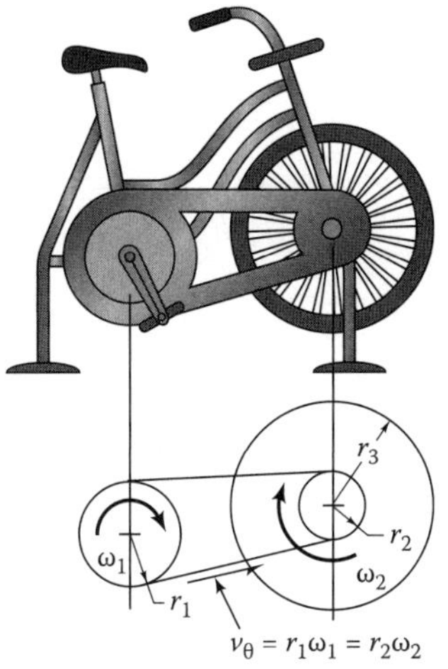

FIGURE 4.4
Sprocket arrangement for an exercise bike.

4.3 Velocity and Acceleration Relationships for Two Points in a Rigid Body

We derived general kinematic equations using two-coordinate systems in Section 2.8. Figure 4.5 illustrates the kinematic variables used for this development. Recall that the x, y coordinate system is moving with respect to the X, Y system, and its orientation with respect to the X, Y system is defined by the angle θ. The origin of the x, y system is located in the X, Y system by the vector $\boldsymbol{R}_o$. The point P is located in the X, Y system by the vector $\boldsymbol{r}$ and is located in the x, y system by the vector $\boldsymbol{\rho}$. From Equations 2.62 and 2.65, the vector relationships for the velocity and acceleration of point P, relative to the X, Y system are, respectively,

$$\dot{\boldsymbol{r}} = \dot{\boldsymbol{R}}_o + \hat{\dot{\boldsymbol{\rho}}} + \boldsymbol{\omega} \times \boldsymbol{\rho}, \tag{4.1a}$$

$$\ddot{\boldsymbol{r}} = \ddot{\boldsymbol{R}}_o + \hat{\ddot{\boldsymbol{\rho}}} + 2\boldsymbol{\omega} \times \hat{\dot{\boldsymbol{\rho}}} + \dot{\boldsymbol{\omega}} \times \boldsymbol{\rho} + \boldsymbol{\omega} \times (\boldsymbol{\omega} \times \boldsymbol{\rho}). \tag{4.1b}$$

In these equations, please recall the derivative notation and operations:

$$\dot{\boldsymbol{B}} = \left.\frac{d\boldsymbol{B}}{dt}\right|_{X,Y} = \boldsymbol{I}\dot{B}_X + \boldsymbol{J}\dot{B}_Y, \tag{4.2a}$$

$$\hat{\dot{\boldsymbol{B}}} = \left.\frac{d\boldsymbol{B}}{dt}\right|_{x,y} = \boldsymbol{i}\dot{B}_x + \boldsymbol{j}\dot{B}_y. \tag{4.2b}$$

In words, $\dot{B}$ is the derivative of $\boldsymbol{B}$ with respect to the X, Y coordinate system and is obtained by writing $\boldsymbol{B} = \boldsymbol{I}B_X + \boldsymbol{J}B_Y$, and then differentiating with respect to time, holding the unit vectors, $\boldsymbol{I}$ and $\boldsymbol{J}$ constant. Similarly, $\hat{\dot{\boldsymbol{B}}}$ is the derivative of $\boldsymbol{B}$ with respect to the x, y coordinate system. It is obtained by writing $\boldsymbol{B} = \boldsymbol{i}B_x + \boldsymbol{j}B_y$ and then differentiating with respect to time, holding the unit vectors $\boldsymbol{i}$ and $\boldsymbol{j}$ constant. Further, in this chapter, we will have occasion to use Equations 4.1 as they stand; however, we will more regularly need a simplified version to relate the velocities and accelerations vectors *for two points that are fixed in a rigid body*.

Figure 4.6 provides an alternative and simplified version of Figure 4.5, with the x, y coordinate system now fixed in a rigid body. The vectors $\boldsymbol{R}_A$, $\boldsymbol{R}_B$ have replaced $\boldsymbol{R}_o$ and $\boldsymbol{r}$. Because points A and B and the x, y coordinate system are fixed in the rigid body, the position vector $\boldsymbol{\rho} = \boldsymbol{R}_{AB}$ is also fixed in the x, y coordinate system; hence,

$$\hat{\dot{\boldsymbol{\rho}}} = \left.\frac{d\boldsymbol{\rho}}{dt}\right|_{x,y} = 0; \quad \hat{\ddot{\boldsymbol{\rho}}} = \left.\frac{d^2\boldsymbol{\rho}}{dt^2}\right|_{x,y} = 0.$$

In words, to an observer fixed to the rigid body, the vector $\boldsymbol{\rho} = \boldsymbol{r}_{AB}$ would be constant.

As applied to Figure 4.6, Equations 4.1 become

$$\dot{\boldsymbol{R}}_B = \dot{\boldsymbol{R}}_A + \boldsymbol{\omega} \times \boldsymbol{r}_{AB}$$
$$\ddot{\boldsymbol{R}}_B = \ddot{\boldsymbol{R}}_A + \dot{\boldsymbol{\omega}} \times \boldsymbol{r}_{AB} + \boldsymbol{\omega} \times (\boldsymbol{\omega} \times \boldsymbol{r}_{AB}),$$

where $\boldsymbol{\rho}$ has been replaced by $\boldsymbol{r}_{AB}$. Note that $\boldsymbol{r}_{AB}$ goes from point A to point B. Denoting $\boldsymbol{v}_A$, $\boldsymbol{v}_B$ and $\boldsymbol{a}_A$, $\boldsymbol{a}_B$ as

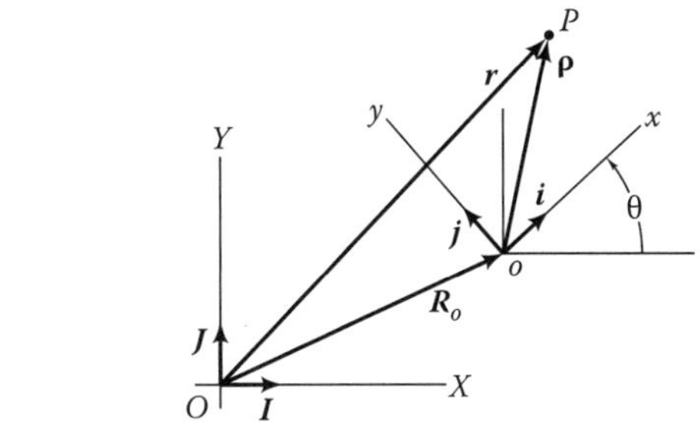

FIGURE 4.5
Two-coordinate arrangement for general planar kinematics.

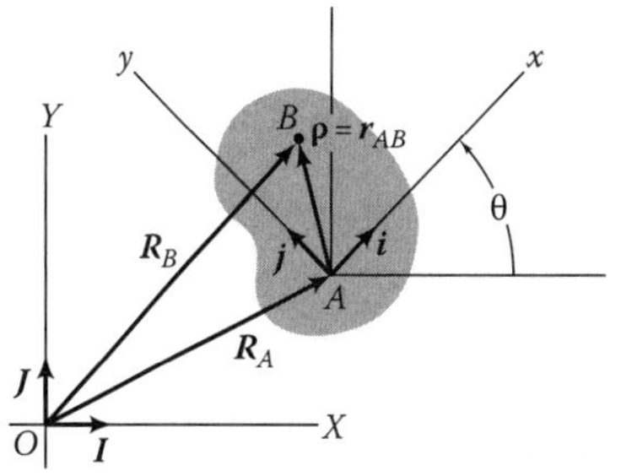

FIGURE 4.6
Two-coordinate arrangement with the x, y system fixed to the rigid body.

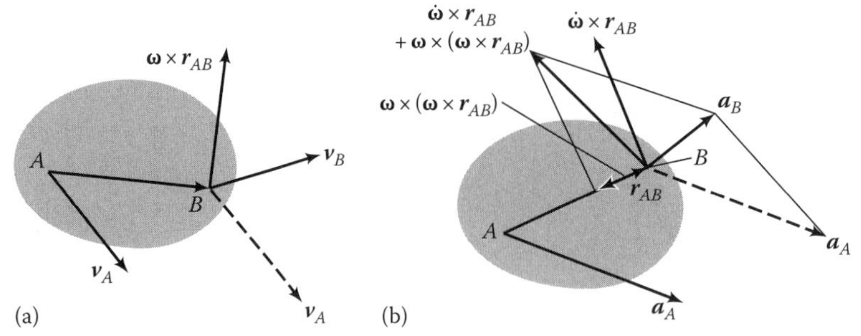

FIGURE 4.7
(a) Velocity and (b) acceleration relations for two points in a rigid body.

the velocities and accelerations, respectively, of points A and B with respect to the X, Y coordinate system, these equations can be written less formally as

$$\begin{aligned} v_B &= v_A + \boldsymbol{\omega} \times \boldsymbol{r}_{AB} \\ \boldsymbol{a}_B &= \boldsymbol{a}_A + \dot{\boldsymbol{\omega}} \times \boldsymbol{r}_{AB} + \boldsymbol{\omega} \times (\boldsymbol{\omega} \times \boldsymbol{r}_{AB}) \end{aligned} \tag{4.3}$$

Figure 4.7 provides a graphical presentation of Equation 4.3. In Figure 4.7a, v_A is shown at the left as the velocity of point A. The dashed representation for v_A on the right of the figure is added (vectorially) to $\boldsymbol{\omega} \times \boldsymbol{r}_{AB}$ to obtain v_B. Figure 4.7b shows $\boldsymbol{a}_A$, the acceleration of point A on the left. The centripetal acceleration term $\boldsymbol{\omega} \times (\boldsymbol{\omega} \times \boldsymbol{r}_{AB})$ points backward from point B to A. The circumferential acceleration term $\dot{\boldsymbol{\omega}} \times \boldsymbol{r}_{AB}$ is perpendicular to $\boldsymbol{r}_{AB}$. The vector $[\boldsymbol{\omega} \times (\boldsymbol{\omega} \times \boldsymbol{r}_{AB}) + \dot{\boldsymbol{\omega}} \times \boldsymbol{r}_{AB}]$ is shown as an intermediate vector sum, which is then added to $\boldsymbol{a}_A$ (dashed-line vector) to obtain $\boldsymbol{a}_B$

As a rule, students do not find the cross products of Equation 4.3 to be physically motivating and require practice in working problems to develop a sense of their direction and meaning. Your grasp of these terms may be helped by recalling from Chapter 2 that there is a direct correlation between the polar-coordinate definitions of velocity and acceleration and the cross product terms. Assuming that $\boldsymbol{\varepsilon}_r$ is directed along $\boldsymbol{r}_{AB}$ from A to B, the parallels between the cross products and polar terms are

$$\begin{aligned} &\boldsymbol{\omega} \times \boldsymbol{r}_{AB} \Rightarrow \boldsymbol{\varepsilon}_\theta r\dot{\theta}, \\ &\dot{\boldsymbol{\omega}} \times \boldsymbol{r}_{AB} \Rightarrow \boldsymbol{\varepsilon}_\theta r\ddot{\theta}, \\ &\boldsymbol{\omega} \times (\boldsymbol{\omega} \times \boldsymbol{r}_{AB}) \Rightarrow -\boldsymbol{\varepsilon}_r r\dot{\theta}^2. \end{aligned}$$

In words, the term $\boldsymbol{\omega} \times \boldsymbol{r}_{AB}$ is a direct parallel to the circumferential-velocity component $\boldsymbol{\varepsilon}_\theta r\dot{\theta}$, $\dot{\boldsymbol{\omega}} \times \boldsymbol{r}_{AB}$ is the circumferential acceleration component $\boldsymbol{\varepsilon}_\theta r\dot{\theta}$, and $\boldsymbol{\omega} \times (\boldsymbol{\omega} \times \boldsymbol{r}_{AB})$ is the centripetal acceleration component $-\boldsymbol{\varepsilon}_r r\dot{\theta}^2$.

Example Problem 4.1

For the instant illustrated in Figure XP4.1, point A on the bar has the following velocity and acceleration vectors with respect to the X, Y coordinate system:

$$\boldsymbol{v}_A = 5\boldsymbol{I} \text{ mm/s}, \quad \boldsymbol{a}_A = 0.5\boldsymbol{I} \text{ mm/s}^2 \tag{i}$$

At the same time, the bar has the following angular velocity and angular acceleration vectors:

$$\boldsymbol{\omega} = 0.1\boldsymbol{K} \text{ rad/s}, \quad \dot{\boldsymbol{\omega}} = 0.02\boldsymbol{K} \text{ rad/s}^2 \tag{ii}$$

Tasks:

a. Determine the velocity and acceleration vectors for points C and B.
b. Draw the velocity and acceleration vectors for points A, B, and C.

Solution

Applying the first of Equation 4.3 gives

$$\begin{aligned} \boldsymbol{v}_B &= \boldsymbol{v}_A + \boldsymbol{\omega} \times \boldsymbol{r}_{AB} = 5\boldsymbol{I} + 0.1\boldsymbol{K} \times 75\boldsymbol{I} = 5\boldsymbol{I} + 7.5\boldsymbol{J} \text{ mm/s} \\ \boldsymbol{v}_C &= \boldsymbol{v}_A + \boldsymbol{\omega} \times \boldsymbol{r}_{AC} = 5\boldsymbol{I} + 0.1\boldsymbol{K} \times (75\boldsymbol{I} + 100\boldsymbol{J}) \\ &= 5\boldsymbol{I} + 7.5\boldsymbol{J} - 10\boldsymbol{J} = -5\boldsymbol{I} + 7.5\boldsymbol{J} \text{ mm/s}. \end{aligned}$$

The acceleration vector of points B is obtained from the second of Equation 4.3 as

$$\begin{aligned} \boldsymbol{a}_B &= \boldsymbol{a}_A + \dot{\boldsymbol{\omega}} \times \boldsymbol{r}_{AB} + \boldsymbol{\omega} \times (\boldsymbol{\omega} \times \boldsymbol{r}_{AB}) \\ &= 0.5\boldsymbol{I} + 0.02\boldsymbol{K} \times 75.\boldsymbol{I} + 0.1\boldsymbol{K} \times [0.1\boldsymbol{K} \times 75.\boldsymbol{I}] \\ &= 0.5\boldsymbol{I} + 1.5\boldsymbol{J} - 0.75\boldsymbol{I} = -0.25\boldsymbol{I} + 1.5\boldsymbol{J} \text{ mm/s}^2. \end{aligned}$$

Similarly, for point C,

$$\begin{aligned} \boldsymbol{a}_C &= \boldsymbol{a}_A + \dot{\boldsymbol{\omega}} \times \boldsymbol{r}_{AC} + \boldsymbol{\omega} \times (\boldsymbol{\omega} \times \boldsymbol{r}_{AC}) \\ &= 0.5\boldsymbol{I} + 0.02\boldsymbol{K} \times (75\boldsymbol{I} + 100\boldsymbol{J}) + 0.1\boldsymbol{K} \\ &\quad \times [0.1\boldsymbol{K} \times (75\boldsymbol{I} + 100\boldsymbol{J})] \\ &= 0.5\boldsymbol{I} + (1.5\boldsymbol{J} - 2.\boldsymbol{I}) - (0.75\boldsymbol{I} + 1\boldsymbol{J}) = -2.25\boldsymbol{I} + 0.5\boldsymbol{J} \text{ mm/s}^2. \end{aligned}$$

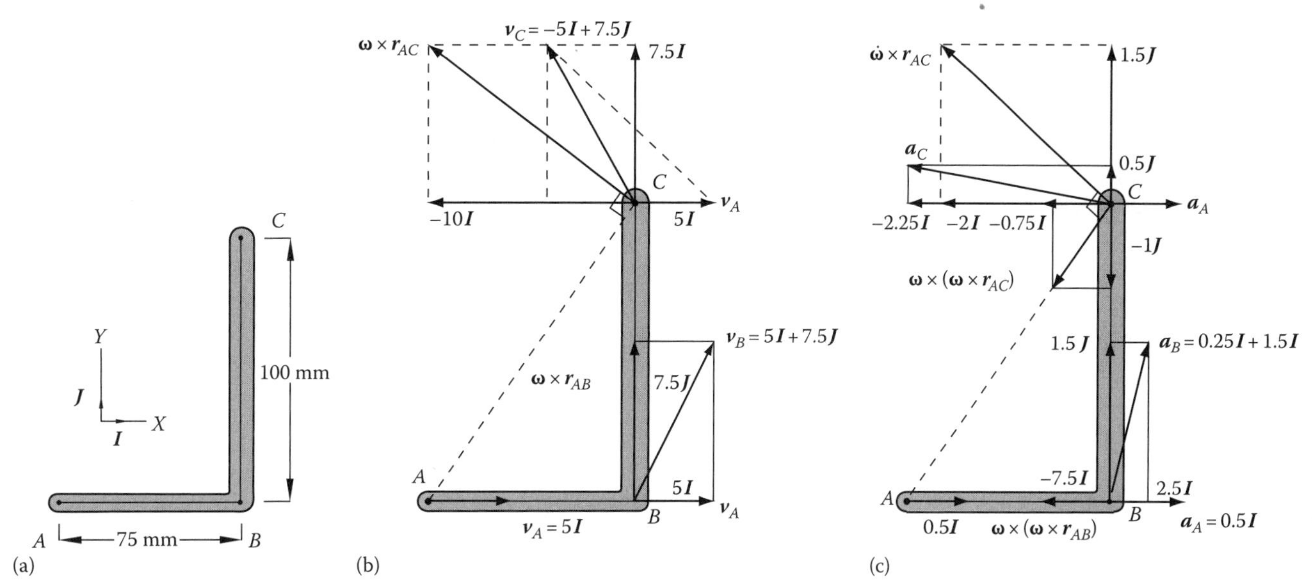

FIGURE XP4.1
(a) Bar with points *A*, *B*, *C*. (b) Velocity (mm/s). (c) Acceleration (mm/s^2).

This is a quite straightforward application of the velocity and acceleration equations for two points in a rigid body. You only need to be proficient in vector algebra to apply the equations.

Figure XP4.1b provides the velocity diagram, illustrating the addition to $\boldsymbol{v}_A$ of the cross-product terms for positions *B* and *C*. Figure XP4.1c provides the acceleration diagram showing the addition of $\boldsymbol{a}_A$, $\dot{\boldsymbol{\omega}} \times \boldsymbol{r}_{AB}$, and $\boldsymbol{\omega} \times (\boldsymbol{\omega} \times \boldsymbol{r}_{AB})$ to obtain $\boldsymbol{a}_B$, plus a similar result for $\boldsymbol{a}_C$. Drawing the velocity and acceleration diagrams illustrates the geometrical construction of the vectors while confirming the correctness of the vector algebra.

Example Problem 4.2

Figure XP4.2 illustrates a load that is being lifted by a pulley arrangement using inextensible cables. Point *A*, the end of the left cable has a velocity of 0.6 m/s and an acceleration of 0.13 m/s^2. Point *B*, the end of the right cable has a velocity of 1.2 m/s and an acceleration of −0.13 m/s^2. The central pulley has a radius of 0.2 m. Point *o* is at the center of the pulley, and point *P* is at the top of the pulley.

Tasks: Determine the velocity and acceleration of points *o* and *P*.

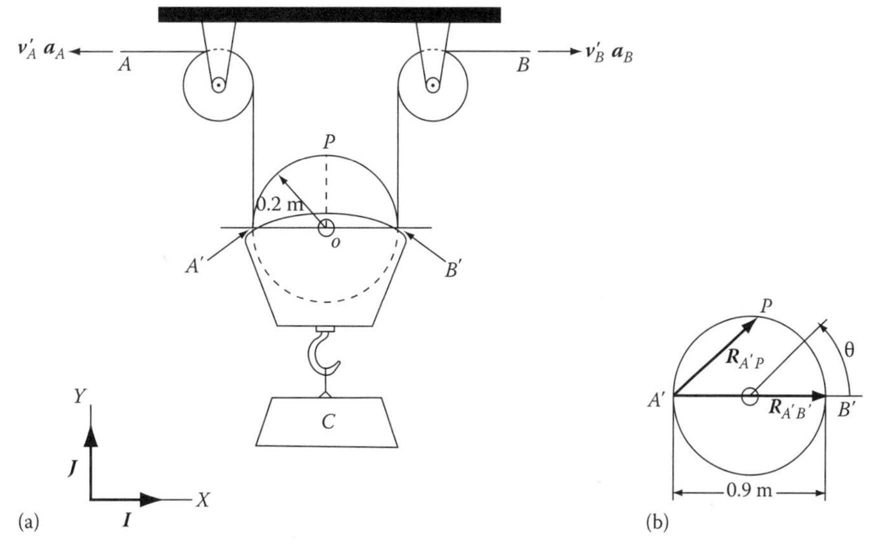

FIGURE XP4.2
(a) Pulley example and (b) position vectors.

Solution

We are going to use Equations 4.3, the velocity and acceleration vectors for two points on a rigid body to work through this example. The rigid body is the central pulley illustrated in Figure XP4.2b. Since the cable is inextensible, points A' and B' on the central pulley have the same velocities and *vertical* acceleration magnitudes as A and B, respectively. Specifically,

$$\begin{aligned} \mathbf{v}_{A'} &= 0.6\mathbf{J}\ \text{m/s}, \quad \mathbf{v}_{B'} = 1.2\mathbf{J}\ \text{m/s} \\ \mathbf{a}_{A'Y} &= 0.13\mathbf{J}\ \text{m/s}^2, \quad \mathbf{a}_{B'Y} = -0.13\mathbf{J}\ \text{m/s}^2. \end{aligned} \tag{i}$$

Velocity results: We start this development knowing $\mathbf{v}_{A'}$, $\mathbf{v}_{B'}$ and needing $\mathbf{v}_o$, $\mathbf{v}_P$. We could use the velocity relationship to find these unknowns providing that we knew $\boldsymbol{\omega} = \mathbf{K}\omega = \mathbf{K}\dot{\theta}$. We can determine $\boldsymbol{\omega}$ by applying $\mathbf{v}'_B = \mathbf{v}'_A + \boldsymbol{\omega} \times \mathbf{r}_{A'B'}$ for points A' and B' on the pulley. Substituting for $\mathbf{v}_{A'}$, $\mathbf{v}_{B'}$ from Equation (i) and $\mathbf{r}_{A'B'} = 0.4\mathbf{I}$ m gives

$$\begin{aligned} 1.2\mathbf{J}\ \text{m/s} &= 0.6\mathbf{J}\ \text{m/s} + \mathbf{K}\dot{\theta}\ \text{rad/s} \times 0.4\mathbf{I}\ \text{m} \\ &\Rightarrow \dot{\theta} = \frac{0.6}{0.4}\ \text{rad/s} = 1.5\ \text{rad/s}. \end{aligned} \tag{ii}$$

Proceeding with this result for $\boldsymbol{\omega}$ gives

$$\begin{aligned} \mathbf{v}_o &= \mathbf{v}_{A'} + (\boldsymbol{\omega} \times \mathbf{r}_{A'o}) = 0.6\mathbf{J}\ \text{m/s} + (1.5\mathbf{K}\ \text{rad/s} \times 0.2\mathbf{I}\ \text{m}) \\ &= 0.9\mathbf{J}\ \text{m/s} \\ \mathbf{v}_P &= \mathbf{v}_{A'} + (\boldsymbol{\omega} \times \mathbf{r}_{A'P}) \\ &= 0.6\mathbf{J}\ \text{m/s} + [1.5\mathbf{K}\ \text{rad/s} \times (0.2\mathbf{I} + 0.2\mathbf{J})\ \text{m}] \\ &= (0.6\mathbf{J} + 0.3\mathbf{J} - 0.3\mathbf{I})\ \text{m/s} = (0.9\mathbf{J} - 0.3\mathbf{I})\ \text{m/s} \end{aligned}$$

As expected, $\mathbf{v}_o$'s velocity is vertically upward. Because of the pulley's rotation, point P has a velocity component in the $-X$ direction. Note that we could have just as easily used the equations, $\mathbf{v}_o = \mathbf{v}_{B'} + (\boldsymbol{\omega} \times \mathbf{r}_{B'o})$ and $\mathbf{v}_P = \mathbf{v}_{B'} + (\boldsymbol{\omega} \times \mathbf{r}_{B'P})$ to get these results since we know $\mathbf{v}_{B'}$.

Acceleration development. The acceleration of points $A', B', \mathbf{a}_{A'}, \mathbf{a}_{B'}$ have the same vertical magnitudes, respectively, as $\mathbf{a}_A$, $\mathbf{a}_B$; specifically, $a_{A'Y} = a_A = .13\ \text{m/s}^2$ and $a_{B'Y} = a_B = -0.13\ \text{m/s}^2$. The acceleration vectors $\mathbf{a}_{A'}, \mathbf{a}_{B'}$ have horizontal components due to the pulley's rotation; hence, $\mathbf{a}_{A'} = 0.13\mathbf{J} + \mathbf{a}_{A'X}\mathbf{J}\ \text{m/s}^2$, and $\mathbf{a}_{B'} = -0.13\mathbf{J} + \mathbf{a}_{B'X}\mathbf{J}\ \text{m/s}^2$. We can verify this statement starting with $\mathbf{a}_o = a_o\mathbf{J}$, the acceleration of point o, even though a_o is unknown. Applying the acceleration equation from Equation 4.3 gives

$$\begin{aligned} \mathbf{a}_{A'} &= \mathbf{a}_o + \dot{\boldsymbol{\omega}} \times \mathbf{r}_{oA'} + \boldsymbol{\omega} \times (\boldsymbol{\omega} \times \mathbf{r}_{oA'}) \\ &= \mathbf{J}a_o + (\mathbf{K}\ddot{\theta} \times -0.2\mathbf{I}) + [\mathbf{K}\dot{\theta} \times (\mathbf{K}\dot{\theta} \times -0.2\mathbf{I})] \\ &= \mathbf{J}a_o - \mathbf{J}0.2\ddot{\theta} + \mathbf{I}0.2\ddot{\theta} = \mathbf{J}(a_o - 0.2\ddot{\theta}) + \mathbf{I}0.2\dot{\theta}^2\ \text{m/s}^2 \\ \mathbf{a}_{B'} &= \mathbf{a}_o + \dot{\boldsymbol{\omega}} \times \mathbf{r}_{oB'} + \boldsymbol{\omega} \times (\boldsymbol{\omega} \times \mathbf{r}_{oB'}) \\ &= \mathbf{J}a_o + (\mathbf{K}\ddot{\theta} \times 0.2\mathbf{I}) + [\mathbf{K}\dot{\theta} \times (\mathbf{K}\dot{\theta} \times 0.2\mathbf{I})] \\ &= \mathbf{J}a_o + \mathbf{J}0.2\ddot{\theta} + \mathbf{I}0.2\dot{\theta}^2 = \mathbf{J}(\mathbf{a}_o + 0.2\ddot{\theta}) - \mathbf{I}0.2\dot{\theta}^2\ \text{m/s}^2. \end{aligned} \tag{iii}$$

Note the horizontal components of $\mathbf{a}_{A'}$, $\mathbf{a}_{B'}$ arising from the centripetal acceleration term $r\dot{\theta}^2$ induced by the pulley rotation. With that result, we can proceed to solve for $\mathbf{a}_o$, $\mathbf{a}_P$.

The first step is solving for $\dot{\boldsymbol{\omega}}$. Applying $\mathbf{a}_{B'} = \mathbf{a}_{A'} + \dot{\boldsymbol{\omega}} \times \mathbf{r}_{A'B'} + \boldsymbol{\omega} \times (\boldsymbol{\omega} \times \mathbf{r}_{A'B'})$ gives

$$\begin{aligned} (-0.2\dot{\theta}^2\mathbf{I} - 0.13\mathbf{J}) &= (0.2\dot{\theta}^2\mathbf{I} + 0.13\mathbf{J}) + (\mathbf{K}\ddot{\theta} \times 0.4\mathbf{I}) \\ &\quad + [\mathbf{K}\dot{\theta} \times (\mathbf{K}\dot{\theta} \times 0.4\mathbf{I})] \\ &= 0.2\dot{\theta}^2\mathbf{I} + 0.13\mathbf{J} + 4\ddot{\theta}\mathbf{J} + 0.4\dot{\theta}^2\mathbf{I}\ \text{m/s}^2 \end{aligned}$$

Taking the $\mathbf{I}$ and $\mathbf{J}$ components separately gives

$$\mathbf{I}: -0.2\dot{\theta}^2 = 0.2\dot{\theta}^2 - 0.4\dot{\theta}^2,$$

$$\mathbf{J}: -0.13 = 0.13 + 0.4\ddot{\theta} \Rightarrow \ddot{\theta} = \frac{-0.26}{0.4}\ \text{rad/s}^2 = -0.65\ \text{rad/s}^2.$$

The X component result gives nothing; the Y component allows us to solve for $\ddot{\theta}$. At this point, we are in a position to directly solve for $\mathbf{a}_o$, $\mathbf{a}_P$ as

$$\begin{aligned} \mathbf{a}_o &= \mathbf{a}_{A'} + \dot{\boldsymbol{\omega}} \times \mathbf{r}_{A'o} + \boldsymbol{\omega} \times (\boldsymbol{\omega} \times \mathbf{r}_{A'o}) \\ &= (0.2\dot{\theta}^2\mathbf{I} + 0.13\mathbf{J}) + (\mathbf{K}\ddot{\theta} \times 0.2\mathbf{I}) + [\mathbf{K}\dot{\theta} \times (\mathbf{K}\dot{\theta} \times 0.2\mathbf{I})] \\ &= 0.2\dot{\theta}^2\mathbf{I} + 0.13\mathbf{J} + 0.2\ddot{\theta}\mathbf{J} - 0.2\dot{\theta}^2\mathbf{I} = (0.13 + 0.2 \times -0.65)\mathbf{J} \\ &= -1.17\mathbf{J}\ \text{m/s}^2 \\ \mathbf{a}_P &= \mathbf{a}_{A'} + \dot{\boldsymbol{\omega}} \times \mathbf{r}_{A'P} + \boldsymbol{\omega} \times (\boldsymbol{\omega} \times \mathbf{r}_{A'P}) \\ &= (0.2\dot{\theta}^2\mathbf{I} + 0.13\mathbf{J}) + \mathbf{K}\ddot{\theta} \times (0.2\mathbf{I} + 0.2\mathbf{J}) \\ &\quad + \mathbf{K}\dot{\theta} \times [\mathbf{K}\dot{\theta} \times (0.2\mathbf{I} + 0.2\mathbf{J})] \\ &= 0.2\dot{\theta}^2\mathbf{I} + 0.13\mathbf{J} + 0.2\ddot{\theta}(\mathbf{J} - \mathbf{I}) - 0.2\dot{\theta}^2(\mathbf{I} + \mathbf{J}) \\ &= -0.2\ddot{\theta}\mathbf{I} + (0.13 + 0.2\ddot{\theta} - 0.2\dot{\theta}^2)\mathbf{J} \\ &= (-0.2 \times -0.65)\mathbf{I} + [0.13 + (0.2 \times -0.65) - 0.2 \times 1.5^2]\mathbf{J} \\ &= -1.3\mathbf{I} - 1.5\mathbf{J}\ \text{m/s}^2. \end{aligned}$$

As expected, point o's acceleration is vertical. Also, P's vertical acceleration is entirely due to the $r\dot{\theta}^2$ term. This last step could have proceeded equally well from B' instead of A', since we also know $\mathbf{a}_{B'}$.

In this type of problem, the following sequential steps are used:

1. Starting with known velocities at two points, in this case A' and B', calculate $\boldsymbol{\omega}$. Then using a known velocity and $\boldsymbol{\omega}$ calculate any additional required velocities.
2. Starting with known accelerations at two points, calculate $\dot{\boldsymbol{\omega}}$. Then using one of the known accelerations plus $\boldsymbol{\omega}$ and $\dot{\boldsymbol{\omega}}$, calculate other required accelerations.

We detoured from that procedure in this solution when we developed Equation (iii) to emphasize the centripetal acceleration term arising from $r\dot{\theta}^2$ that are required to correctly define $\mathbf{a}_{A'}$, $\mathbf{a}_{B'}$.

4.4 Rolling without Slipping

4.4.1 A Wheel on a Plane

One of the basic problems of planar kinematics involves the regular observation experience of a wheel rolling on a surface. Figure 4.8 illustrates a geared wheel that is rolling in geared horizontal guide ways. This type of geared wheel is used on "cog" railways to climb slopes that are too steep for conventional locomotives. Clearly, the gear cannot slip with respect to the guides, emphasizing the kinematic requirement of rolling without slipping.

Figure 4.9 illustrates a wheel rolling on a horizontal surface. The wheel could be slipping or not. Think about the tires on your car. In normal steady-speed driving, they do not slip perceptibly; however, slipping can occur during rapid acceleration or during a rapid deceleration caused by braking. Slipping is more likely to occur when driving on wet or icy highways, surfaces for which the Coulomb coefficient of friction has been reduced. The question of whether or not a tire slips concerns forces and kinetics and will be covered in the next chapter. In this kinematics section, we simply "decree" that slipping will not occur (i.e., adequate friction or constraint forces are available to prevent slipping) and will investigate the kinematic relationships for velocity and acceleration that result.

The derivation and understanding of velocity and acceleration relationships for a wheel that is rolling without slipping is the fundamental objective of this subsection, and we will be using Figure 4.10 for this purpose. A rigid wheel is rolling on a rigid horizontal plane representing the ground. The wheel's center (point o) is located in the X, Y system by the vector $IX_o + Jr$. The wheel's orientation with respect to the X-, Y-axes is defined by the angle θ.

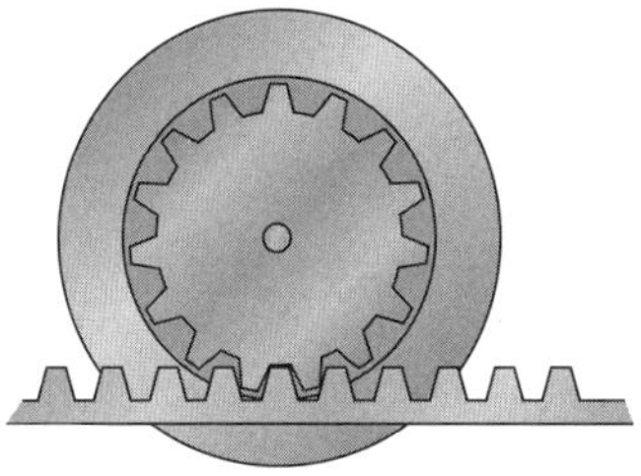

FIGURE 4.8
Gear rolling in geared horizontal guides.

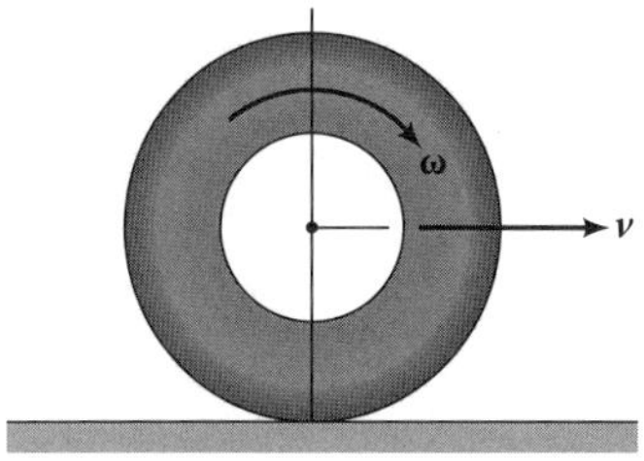

FIGURE 4.9
Wheel rolling on a horizontal surface.

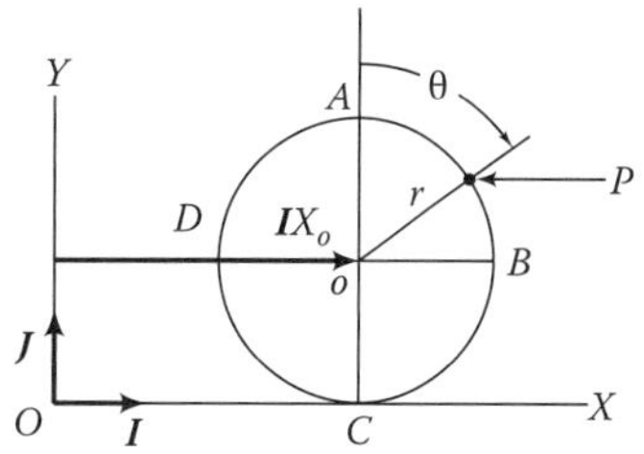

FIGURE 4.10
Wheel rolling without slipping on a horizontal surface.

4.4.1.1 Geometric Development

The wheel in Figure 4.10 advances to the right as θ increases. The first question is: Without slipping, how is the rotation angle θ related to the displacement of the wheel center X_o? Note first that the wheel's contact point between the wheel, denoted as C, advances to the right precisely the same distance as point o. If the wheel starts with $\theta_0 = X_0 = 0$, and rolls forward through one rotation *without slipping*, both the new contact point and the now displaced point o will have moved to the right a distance equal to the circumference of the wheel; i.e., $X_o = 2\pi r$. It may help you to think of the wheel as a paint roller and imagine the length of the paint strip that would be laid out on the plane during one rotation. Comparable to the result for a full rotation, *without slipping* the geometric constraint relating X_o and θ is

$$X_o = r\theta \tag{4.4}$$

Differentiating this result with respect to time gives the corresponding velocity and acceleration relationships:

$$\dot{X}_o = r\dot{\theta}, \quad \ddot{X}_o = r\ddot{\theta}. \tag{4.5}$$

These are the desired fundamental kinematic constraint equations for a wheel that is rolling without slipping.

The relationship between X_o and θ provided by Equation 4.4 means that a wheel which is rolling without slipping has one degree of freedom. Hence, either θ can be replaced by X_o or vice versa in the equations of motion. A slipping wheel will require both X_o and θ, and corresponding *independent* equations of motion for each coordinate.

We want to define the trajectory of the point P on the wheel located by θ. The coordinates of the displacement vector locating P are defined by

$$X = X_o + r\sin\theta = r\theta + r\sin\theta, \quad Y = r\cos\theta + r \tag{4.6}$$

Note that the $X_o = r\theta$ result of Equation 4.4 is used to eliminate X_o. The trajectory followed by point P as the wheel rolls to the right is the cycloid traced out in Figure 4.11. The letters A through D on the wheel indicate locations occupied by point P on the cycloid and also identified in Figure 4.10. The cycloid was obtained for $r=1$ by varying θ through two cycles of rotation to generate the coordinates $X(\theta)$ and $Y(\theta)$ and then plotting Y versus X.

Referring to the locations A through D of Figures 4.10 and 4.11, point P starts at A, reaches B after the wheel has rotated $\pi/2$ radians, reaches the contact location C after π radians, reaches D after $3\pi/2$ radians, and then returns to A after a full rotation. Note that P approaches the contact location C moving vertically downward, and departs contact moving vertically upwards. From this figure, we would expect point P (on the wheel) to have zero velocity at the contact location C, but to have a substantial vertical acceleration due to the abrupt change in the sign of the vertical velocity.

We can verify these expectations by differentiating the components of the position vector locating P (defined by Equation 4.6) to obtain components of the velocity and acceleration vectors for point P with respect to the X, Y coordinate system. The velocity components are

$$\dot{X} = r\dot{\theta} + r\dot{\theta}\cos\theta, \quad \dot{Y} = -r\dot{\theta}\sin\theta \tag{4.7}$$

Note that $\dot{X}$ and $\dot{Y}$ are components of the velocity vector of a point P on the wheel, located by the angle θ. The corresponding acceleration components are obtained by differentiating Equation 4.7 to get

$$\ddot{X} = r\ddot{\theta} + r\ddot{\theta}\cos\theta - r\dot{\theta}^2\sin\theta, \quad \ddot{Y} = -r\ddot{\theta}\sin\theta - r\dot{\theta}^2\cos\theta \tag{4.8}$$

Equations 4.7 and 4.8 can be used to evaluate the instantaneous velocity and acceleration components of any point on the wheel by specifying an appropriate value for θ.

Values for $\theta = 0$, $\pi/2$, π, and $3\pi/2$ correspond, respectively, to locations A, B, C, and D in Figure 4.10. The velocity and acceleration component results for these four locations are given below:

Position A (top of the wheel; $\theta = 0$):

$$\begin{aligned} &\dot{X} = r\dot{\theta} + r\dot{\theta}(1) = 2r\dot{\theta}, \quad \dot{Y} = -r\dot{\theta}(0) = 0 \\ &\ddot{X} = r\ddot{\theta} + r\ddot{\theta}(1) - r\dot{\theta}^2(0) = 2r\ddot{\theta}, \\ &\ddot{Y} = -r\ddot{\theta}(0) - r\dot{\theta}^2(1) = -r\dot{\theta}^2. \end{aligned} \tag{4.9}$$

Position B (right-hand side of the wheel; $\theta = \pi/2$):

$$\begin{aligned} &\dot{X} = r\dot{\theta} + r\dot{\theta}(0) = r\dot{\theta}, \quad \dot{Y} = -r\dot{\theta}(1) = -r\dot{\theta} \\ &\ddot{X} = r\ddot{\theta} + r\ddot{\theta}(0) - r\dot{\theta}^2(1) = r\ddot{\theta} - r\dot{\theta}^2, \\ &\ddot{Y} = -r\ddot{\theta}(1) - r\dot{\theta}^2(0) = -r\ddot{\theta}. \end{aligned} \tag{4.10}$$

Position C (bottom of the wheel at the contact location; $\theta = \pi$):

$$\begin{aligned} &\dot{X} = r\dot{\theta} + r\dot{\theta}(-1) = 0, \quad \dot{Y} = -r\dot{\theta}(0) = 0 \\ &\ddot{X} = r\ddot{\theta} + r\ddot{\theta}(-1) - r\dot{\theta}^2(0) = 0, \\ &\ddot{Y} = -r\ddot{\theta}(0) - r\dot{\theta}^2(-1) = -r\dot{\theta}^2. \end{aligned} \tag{4.11}$$

Position D (left-hand side of the wheel; $\theta = 3\pi/2$):

$$\begin{aligned} &\dot{X} = r\dot{\theta} + r\dot{\theta}(0) = r\dot{\theta}, \quad \dot{Y} = -r\dot{\theta}(-1) = r\dot{\theta} \\ &\ddot{X} = r\ddot{\theta} + r\ddot{\theta}(0) - r\dot{\theta}^2(-1) = r\ddot{\theta} + r\dot{\theta}^2, \\ &\ddot{Y} = -r\ddot{\theta}(-1) - r\dot{\theta}^2(0) = r\ddot{\theta}. \end{aligned} \tag{4.12}$$

Figure 4.12 shows the velocity components at points A through D and o. The zero vertical velocity for a point on the wheel at location A fits the cycloid trajectory, since the point has reached a vertical maximum and is moving in the horizontal direction. Point B has a horizontal

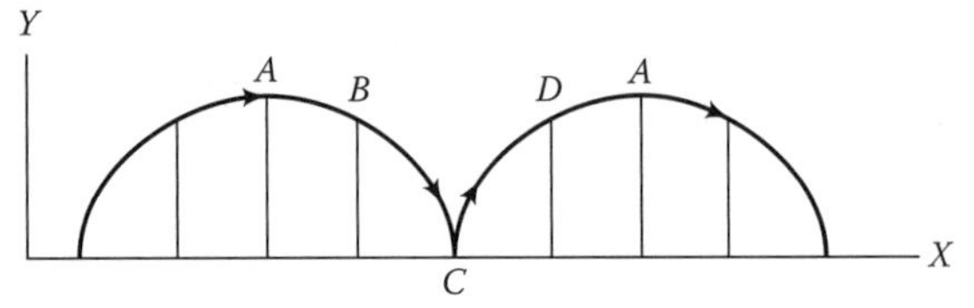

FIGURE 4.11
Cycloidal path traced out by a point on a wheel that is rolling without slipping.

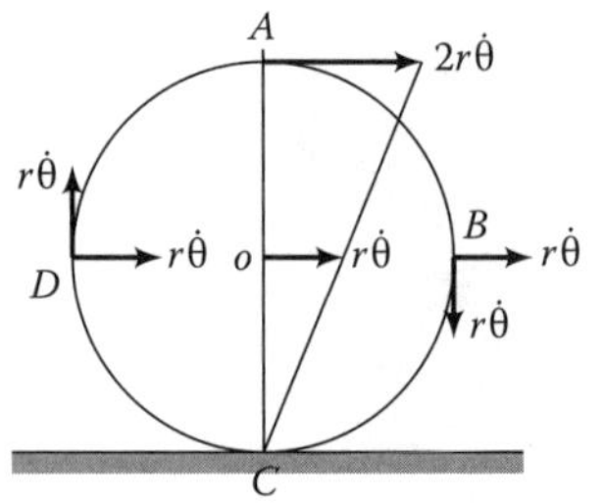

FIGURE 4.12
Velocity vectors for points on a wheel at locations A through D and o.

component of $r\dot{\theta}$ and a vertical velocity component of $-r\dot{\theta}$, which is again consistent with the cycloid trajectory, with movement to the right and downward. As noted previously, point C has zero velocity, and corresponds to the cusp of the cycloid. Point D has vertical and horizontal velocity components $r\dot{\theta}$, again consistent with the cycloid. Note in proceeding from the point of contact at C, vertically upward through o to A, that the velocity increases linearly with the vertical distance from C; zero at C, $r\dot{\theta}$ at o, and $2r\dot{\theta}$ at A.

Figure 4.13 illustrates the instantaneous acceleration components of points on the wheel at locations A through D. Note that each point has a "centripetal" acceleration component $r\dot{\theta}^2$ directed from the location at the wheel's periphery back toward the wheel center o. Precisely this acceleration component would be expected if the wheel were simply rotating about an axis through o at a constant rate $\dot{\theta}$. Absent the $r\dot{\theta}^2$ terms, there is a direct parallel between the velocity results of Figure 4.12 and the acceleration results of Figure 4.13. Note the analogous increase in the horizontal acceleration component as we move vertically upward from zero at location C, to $r\ddot{\theta}$ at o, and $2r\ddot{\theta}$ at A.

The acceleration of a point P at location C frequently causes confusion. Students regularly state that the acceleration of a point on the wheel at the contact location is zero. In comparison to the above results, this view is simply wrong. The logical fault in reaching this incorrect (zero-acceleration) conclusion is the idea that a zero-velocity vector necessarily implies a zero-acceleration vector. The point P in contact with the ground has zero velocity *at the instant of contact*. Constant-velocity components over a *finite* time period are required to give zero acceleration. *Note carefully that a point on the wheel at the contact location has a vertical acceleration of* $r\dot{\theta}^2$. This result was suggested by the cycloid trajectory, in that a point on the wheel approaches the contact location moving vertically downward and departs moving vertically upward. The abrupt change in the direction of the velocity vector is the source of this acceleration component.

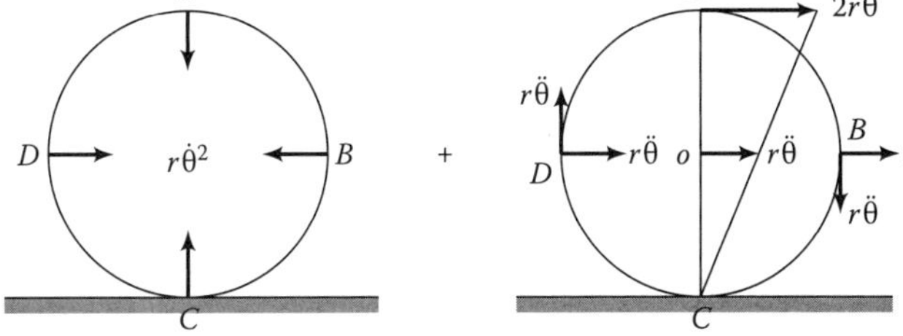

FIGURE 4.13
Acceleration vectors for points on a wheel for locations A through D and o.

The geometric approach used so far consists of the following sequential steps:

a. State (write out) the geometric X and Y component equations.
b. Differentiate the displacement component equations to obtain velocity component equations.
c. Differentiate the velocity component equations to obtain acceleration component equations.

This simple approach will prove to be very useful in formulating relationships for the linkage problems of Section 4.5 and will also be used heavily in Chapter 5 when developing kinematic relationships that are needed to complete dynamic models of planar linkages.

The geometric developments are intended to give you a physical and geometric understanding of the mechanics of rolling without slipping. Several of the rolling-without-slipping problems in this section are worked more easily using the vector equations of the preceding section. Hence, in the balance of this section, parallel developments will be demonstrated using the vector equations.

4.4.1.2 Vector Developments of Velocity Relationships

Returning to the wheel of Figure 4.10, suppose the wheel is rolling to the right, yielding the velocity vector $\boldsymbol{v}_o = \boldsymbol{I}\dot{X}_o$ for point o. Using the right-hand-rule convention for defining angular velocity vectors, the wheel's angular velocity vector is $\boldsymbol{\omega} = \boldsymbol{K}(-\dot{\theta}) = -\boldsymbol{K}\dot{\theta}$. (If the wheel were rolling to the left, $\dot{\theta}$ would be negative, and the angular velocity vector would be $\boldsymbol{\omega} = +\boldsymbol{K}|\dot{\theta}|$.)

C is a point on the wheel at the instantaneous contact location between the wheel and the ground, and it has a velocity of zero; i.e., $\boldsymbol{v}_C = \boldsymbol{I}0 + \boldsymbol{J}0$. This zero-contact-velocity specification can be used to develop a relationship between the coordinates $\dot{X}_o$ and $\dot{\theta}$. Applying the first of Equation 4.3 to points o and C gives

$$\boldsymbol{v}_C = \boldsymbol{v}_o + \boldsymbol{\omega} \times \boldsymbol{r}_{oC}.$$

Setting $\boldsymbol{v}_C$ to zero and substituting: $\boldsymbol{v}_o = \boldsymbol{I}\dot{X}_o$, $\boldsymbol{\omega} = -\boldsymbol{K}\dot{\theta}$, $\boldsymbol{r}_{oC} = -\boldsymbol{J}r$, gives

$$0 = \boldsymbol{I}\dot{X}_o + (-\boldsymbol{K}\dot{\theta} \times -\boldsymbol{J}r) \Rightarrow 0 = \boldsymbol{I}(\dot{X}_o - r\dot{\theta}).$$

Hence, the rolling-without-slipping kinematic result for velocity is (again)

$$\dot{X}_o = r\dot{\theta}. \quad (4.5a)$$

Returning to Figure 4.10, the velocity of point o can be written as $\boldsymbol{v}_o = \boldsymbol{I}\dot{X}_o = \boldsymbol{I}r\dot{\theta}$. We now want to find the velocity vectors for points A through D of Figure 4.10.

Starting with point A, and applying the velocity relationship from Equation 4.3 to points A and C on the wheel gives

$$v_A = v_C + \boldsymbol{\omega} \times \boldsymbol{r}_{CA}.$$

Substituting $v_C = 0, \boldsymbol{\omega} = -\boldsymbol{K}\dot{\theta}$, and $\boldsymbol{r}_{CA} = \boldsymbol{J}2r$ yields

$$v_A = 0 - \boldsymbol{K}\dot{\theta} \times \boldsymbol{J}2r = \boldsymbol{I}2r\dot{\theta}.$$

Looking back, this is precisely the velocity result obtained in Equation 4.9.

The velocity of a point on the wheel at location B can be found by applying Equation 4.3. Since we know the velocities at point o and C, two straightforward applications of Equation 4.3 are

$$v_B = v_o + \boldsymbol{\omega} \times \boldsymbol{r}_{oB}$$
$$v_B = v_C + \boldsymbol{\omega} \times \boldsymbol{r}_{CB}.$$

The first equation defines v_B by starting from a known velocity at point o; the second equation starts from a known velocity at C. The vectors v_o, v_C, and $\boldsymbol{\omega}$ have already been identified. The required new vectors are $\boldsymbol{r}_{oB} = \boldsymbol{I}r$ and $\boldsymbol{r}_{CB} = \boldsymbol{I}r + \boldsymbol{J}r$. Substitution into the above equations gives

$$v_B = \boldsymbol{I}r\dot{\theta} - \boldsymbol{K}\dot{\theta} \times \boldsymbol{I}r = \boldsymbol{I}r\dot{\theta} - \boldsymbol{J}r\dot{\theta}$$
$$v_B = 0 - \boldsymbol{K}\dot{\theta} \times (\boldsymbol{I}r + \boldsymbol{J}r) = \boldsymbol{I}r\dot{\theta} - \boldsymbol{J}r\dot{\theta}.$$

As expected, the same result is obtained with both equations, yielding the earlier geometric-based results of Equation 4.10. This answer is a vector-algebra confirmation of the velocity illustration in Figure 4.7.

The velocity of point D can be obtained by direct application of Equation 4.3, starting from a point on the wheel where the velocity is known. Given that the velocity vectors have been determined for points o, A, B, and C, any of the following vector equations can be used to determine v_D:

$$v_D = v_o + \boldsymbol{\omega} \times \boldsymbol{r}_{oD}$$
$$v_D = v_A + \boldsymbol{\omega} \times \boldsymbol{r}_{AD}$$
$$v_D = v_B + \boldsymbol{\omega} \times \boldsymbol{r}_{BD}$$
$$v_D = v_C + \boldsymbol{\omega} \times \boldsymbol{r}_{CD}$$

We know the velocity vectors for point o, A, B, and C, and can write expressions for the vectors $\boldsymbol{r}_{oD}$, $\boldsymbol{r}_{AD}$, $\boldsymbol{r}_{BD}$, and $\boldsymbol{r}_{CD}$. Applying (arbitrarily) the last equation with $\boldsymbol{r}_{CD} = -\boldsymbol{I}r + \boldsymbol{J}r$ gives

$$v_D = 0 - \boldsymbol{K}\dot{\theta} \times (-\boldsymbol{I}r + \boldsymbol{J}r) = \boldsymbol{I}r\dot{\theta} + \boldsymbol{J}r\dot{\theta}.$$

This outcome repeats the earlier, geometric-based results of Equation 4.12. To help develop confidence in the application of Equation 4.3 for velocity relationships, several homework problems involve precisely this type of exercise.

4.4.1.3 Vector Developments of Acceleration Results

In this section, we use the second of Equation 4.3, relating the acceleration vectors of two points on a rigid body, to derive the rolling-without-slipping acceleration result of Equation 4.5 ($\ddot{X}_o = r\ddot{\theta}$) and also to determine the acceleration vectors of points A through D of Figures 4.12 and 4.13. Starting with points o and C, we can apply the acceleration result of Equation 4.3 as

$$\boldsymbol{a}_C = \boldsymbol{a}_o + \dot{\boldsymbol{\omega}} \times \boldsymbol{r}_{oC} + \boldsymbol{\omega} \times (\boldsymbol{\omega} \times \boldsymbol{r}_{oC}).$$

Substituting $\boldsymbol{a}_o = \boldsymbol{I}\ddot{X}_o, \dot{\boldsymbol{\omega}} = -\boldsymbol{K}\ddot{\theta}$, and $\boldsymbol{r}_{oC} = -r\boldsymbol{J}$ gives

$$\boldsymbol{a}_C = (\boldsymbol{I}a_{CX} + \boldsymbol{J}a_{CY}) = \boldsymbol{I}\ddot{X}_o + (-\boldsymbol{K}\ddot{\theta} \times -\boldsymbol{J}r) - \boldsymbol{K}\dot{\theta} \times (\boldsymbol{K}\dot{\theta} \times -\boldsymbol{J}r),$$
$$\boldsymbol{I}a_{CX} + \boldsymbol{J}a_{CY} = \boldsymbol{I}(\ddot{X}_o - r\ddot{\theta}) + \boldsymbol{J}r\dot{\theta}^2.$$

Equating the $\boldsymbol{I}$ and $\boldsymbol{J}$ components gives

$$\begin{aligned} \boldsymbol{I}: a_{CX} &= (\ddot{X}_o - r\ddot{\theta}) = 0 \Rightarrow \ddot{X}_o = r\ddot{\theta} \\ \boldsymbol{J}: a_{CY} &= r\dot{\theta}^2. \end{aligned} \tag{4.13}$$

The acceleration component result $a_{CX} = 0$ follows from the no-slip condition at the contact point. The first of Equation 4.13 repeats the rolling-without-slipping acceleration condition of Equation 4.5. Together, Equation 4.13 repeat the geometric results for the acceleration of the point of contact provided by Equation. 4.11.

The accelerations of points A, B, and D can also be obtained from Equation 4.3, starting from any point on the wheel where the acceleration vector is known. Choosing point o (arbitrarily) gives

$$\boldsymbol{a}_A = \boldsymbol{a}_o + \dot{\boldsymbol{\omega}} \times \boldsymbol{r}_{oA} + \boldsymbol{\omega} \times (\boldsymbol{\omega} \times \boldsymbol{r}_{oA})$$
$$\boldsymbol{a}_B = \boldsymbol{a}_o + \dot{\boldsymbol{\omega}} \times \boldsymbol{r}_{oB} + \boldsymbol{\omega} \times (\boldsymbol{\omega} \times \boldsymbol{r}_{oB})$$
$$\boldsymbol{a}_D = \boldsymbol{a}_o + \dot{\boldsymbol{\omega}} \times \boldsymbol{r}_{oD} + \boldsymbol{\omega} \times (\boldsymbol{\omega} \times \boldsymbol{r}_{oD}).$$

Substituting for the variables on the right-hand side of these equations gives

$$\begin{aligned} \boldsymbol{a}_A &= \boldsymbol{I}r\ddot{\theta} + (-\boldsymbol{K}\ddot{\theta} \times \boldsymbol{J}r) - \boldsymbol{K}\dot{\theta} \times (-\boldsymbol{K}\dot{\theta} \times \boldsymbol{J}r) \\ &= \boldsymbol{I}2r\ddot{\theta} - \boldsymbol{J}r\dot{\theta}^2 \\ \boldsymbol{a}_B &= \boldsymbol{I}r\ddot{\theta} + (-\boldsymbol{K}\ddot{\theta} \times \boldsymbol{I}r) - \boldsymbol{K}\dot{\theta} \times (-\boldsymbol{K}\dot{\theta} \times \boldsymbol{I}r) \\ &= \boldsymbol{I}(r\ddot{\theta} - r\dot{\theta}^2) - \boldsymbol{J}r\ddot{\theta} \\ \boldsymbol{a}_D &= \boldsymbol{I}r\ddot{\theta} + (-\boldsymbol{K}\ddot{\theta} \times \boldsymbol{I}r) - \boldsymbol{K}\dot{\theta} \times (-\boldsymbol{K}\dot{\theta} \times -\boldsymbol{I}r) \\ &= \boldsymbol{I}(r\ddot{\theta} + r\dot{\theta}^2) + \boldsymbol{J}r\ddot{\theta} \end{aligned}$$

These are precisely the results obtained earlier in Equations 4.9 through 4.12 using geometry and differentiation.

Example Problem 4.3

As illustrated in Figure XP4.3, the lower wheel assembly is rolling without slipping on a plane that is inclined at 30° to the horizontal. It is connected to the top spool via an inextensible cable that is playing out cable. The center of the lower spool and its contact point are denoted, respectively by *o* and *C*. Point *A* denotes the top of the spool. At the instant of interest, the acceleration of the center of the lower spool is $\boldsymbol{a}_o = 6.44\boldsymbol{J}$ ft/s^2 and its velocity is $\boldsymbol{v}_o = 2.6\boldsymbol{J}$ ft/s.

Tasks: For the instant considered, determine the velocity and acceleration of points *C* and *A*. Determine the top spool's angular velocity and acceleration.

Solution

Starting with the desired velocity requirements, rolling without slipping requires zero velocity for the contact point *C*; i.e., $\boldsymbol{v}_C = 0$. We will use the $\boldsymbol{I}, \boldsymbol{J}$ coordinate system illustrated to define the velocity and acceleration vectors. The solution for $\boldsymbol{v}_A$ can be obtained using the velocity relationship for two points in a rigid body $\boldsymbol{v}_A = \boldsymbol{v}_P + \boldsymbol{\omega}_1 \times \boldsymbol{r}_{PA}$ provided that we know the velocity of point *P* and $\boldsymbol{\omega}_1 = \boldsymbol{K}\dot{\theta}$, the wheel assembly's angular velocity. Note that *o* is the only point on the spool that is moving in a straight line; hence, $\boldsymbol{v}_o = \boldsymbol{J}\dot{X}_o$. The velocity rolling-without-slipping condition of Equation 4.5 provides

$$\dot{X}_o = 2.6\,\text{ft/s} = r\dot{\theta} = \left(\frac{1.33}{2}\right)\text{ft} \times \dot{\theta}\ \text{rad/s}$$
$$\Rightarrow \dot{\theta} = 3.91\ \text{rad/s} \Rightarrow \boldsymbol{\omega}_1 = 3.91\boldsymbol{K}\ \text{rad/s}.$$

With $\boldsymbol{\omega}_1$ defined, starting from either *o* or *C*, $\boldsymbol{v}_A$ is

$$\boldsymbol{v}_A = \boldsymbol{v}_C + \boldsymbol{\omega}_1 \times r_{CA} = 0 + 3.91\boldsymbol{K}\,\text{rad/s} \times (0.665 + 2)\boldsymbol{I}\,\text{ft} = 10.42\boldsymbol{J}\ \text{ft/s}$$
$$\boldsymbol{v}_A = \boldsymbol{v}_o + \boldsymbol{\omega}_1 \times \boldsymbol{r}_{oA} = 2.6\boldsymbol{J}\,\text{ft/s} + 3.91\boldsymbol{K}\,\text{rad/s} \times 2\boldsymbol{I}\,\text{ft} = 10.42\boldsymbol{J}\ \text{ft/s}.$$

The angular velocity of the upper disk is obtained by noting that the velocity of the lower spool at the cable contact point is $\boldsymbol{v}_A = 10.42$ ft/s. Since the cable is inextensible, the cable contact point on the upper spool has the same velocity, and

$$v_B = v_A = 10.42\,\text{ft/s} = \omega_2 \times 1\,\text{ft} \Rightarrow \omega_2 = 10.42\ \text{rad/s},$$
$$\text{and}\quad \boldsymbol{\omega}_2 = 10.42\boldsymbol{K}\ \text{rad/s}^2.$$

Proceeding to the acceleration developments, we showed earlier in Equation 4.11 that $\boldsymbol{a}_C = \boldsymbol{I}r\dot{\boldsymbol{\theta}}^2 = 0.665\,\text{ft} \times (3.91\ \text{rad/s})^2 = 10.16\boldsymbol{I}\,\text{ft/s}^2$. Specifically, as illustrated in Figure 4.13, the wheel's contact acceleration is perpendicular to the plane and arises due to centripetal acceleration. To find $\boldsymbol{a}_A$ we use $\boldsymbol{a}_A = \boldsymbol{a}_P + \dot{\boldsymbol{\omega}}_1 \times \boldsymbol{r}_{PA} + \boldsymbol{\omega}_1 \times (\boldsymbol{\omega}_1 \times \boldsymbol{r}_{PA})$, starting from the acceleration of a known point *P*. The first step in using this equation is finding $\dot{\boldsymbol{\omega}}_1$. Again, point *o* is moving down the plane in a straight line; hence, $\boldsymbol{a}_o = \boldsymbol{J}\ddot{X}_o$. The rolling-without-slipping conditions of Equation 4.5 relates *o*'s acceleration to the spool's angular acceleration as

$$\ddot{X}_o = 6.44\ \text{ft/s}^2 = r\ddot{\theta} \Rightarrow \ddot{\theta} = 6.44\ (\text{ft/s}^2)/0.665\ \text{ft}$$
$$= 9.68\ \text{rad/s}^2 \Rightarrow \dot{\boldsymbol{\omega}}_1 = 9.68\boldsymbol{K}\ \text{rad/s}^2.$$

Starting at *o*, $\boldsymbol{a}_A$ is

$$\begin{aligned}\boldsymbol{a}_A &= \boldsymbol{a}_o + \dot{\boldsymbol{\omega}}_1 \times \boldsymbol{r}_{oA} + \boldsymbol{\omega}_1 \times (\boldsymbol{\omega}_1 \times \boldsymbol{r}_{oA})\\ &= 6.44\ \frac{\text{ft}}{\text{s}^2}\boldsymbol{J} + \left(9.68\ \frac{\text{rad}}{\text{s}^2}\boldsymbol{K} \times 2\ \text{ft}\boldsymbol{I}\right)\\ &\quad + \left[3.91\ \frac{\text{rad}}{\text{s}}\boldsymbol{K} \times \left(3.91\frac{\text{rad}}{\text{s}}\boldsymbol{K} \times 2\ \text{ft}\boldsymbol{I}\right)\right]\\ &= (6.44 + 19.36)\boldsymbol{J} - 30.6\boldsymbol{I} = -30.6\boldsymbol{I} + 25.8\boldsymbol{J}\ \text{ft/s}^2.\end{aligned}$$

Starting at *C*, $\boldsymbol{a}_A$ is

$$\begin{aligned}\boldsymbol{a}_A &= \boldsymbol{a}_C + \dot{\boldsymbol{\omega}}_1 \times \boldsymbol{r}_{CA} + \boldsymbol{\omega}_1 \times (\boldsymbol{\omega}_1 \times \boldsymbol{r}_{CA})\\ &= 10.16\boldsymbol{I}\ \frac{\text{ft}}{\text{s}^2} + \left(9.68\ \frac{\text{rad}}{\text{s}^2}\boldsymbol{K} \times 2.665\ \text{ft}\boldsymbol{I}\right)\\ &\quad + \left[3.91\ \frac{\text{rad}}{\text{s}}\boldsymbol{K} \times \left(3.91\ \frac{\text{rad}}{\text{s}}\boldsymbol{K} \times 2.665\ \text{ft}\boldsymbol{I}\right)\right]\\ &= 25.8\boldsymbol{J} + (10.61 - 40.74)\boldsymbol{I} = -30.6\boldsymbol{I} + 25.8\boldsymbol{J}\ \text{ft/s}^2\end{aligned}$$

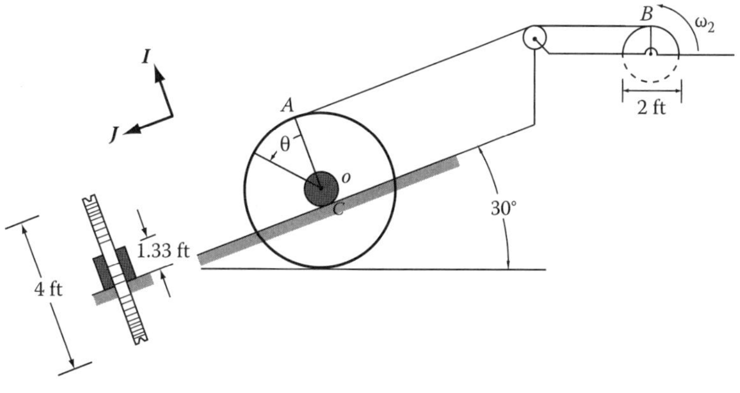

FIGURE XP4.3
Rolling wheel assembly.

Solving for $\dot{\boldsymbol{\omega}}_2$, the angular acceleration of the upper spool follows the same logic that was used in finding $\boldsymbol{\omega}_2$. Specifically, the acceleration of the wheel assembly at A is $\mathbf{a}_A = 25.8\mathbf{J} - 30.6\mathbf{I}$ ft/s^2. The acceleration of spool 2 at its contact point B is $\mathbf{a}_B = 1\dot{\omega}_2\mathbf{J} - 1\omega_2^2\mathbf{I}$ ft/s^2. Because the cable is inextensible, the $\mathbf{J}$ components of these accelerations (along the cable) must be equal. Hence, $\dot{\omega}_2 = (25.8\text{ ft/s}^2)/1\text{ ft} = 25.8\text{ rad/s}^2$, and $\dot{\boldsymbol{\omega}}_2 = 25.8\mathbf{K}$ rad/s^2.

Example Problem 4.4

Figure XP4.4 illustrates a cylinder of radius $r_2 = 1$ m that is rolling without slipping on a horizontal surface. At the instant of interest, the surface is moving to the left with $\mathbf{v}_s = -\mathbf{I}$ 1.25 m/s and has an acceleration to the right of $\mathbf{a}_s = \mathbf{I}$ 1.0 m/s^2. An inextensible cord is wrapped around an inner cylinder of radius $r_1 = 0.5$ m and anchored to a wall at D.

Tasks: Determine the angular velocity and acceleration of the cylinder and the velocity and acceleration of points *A*, *o*, *B*, and *C*.

Solution

We will start with an analysis of velocity to determine $\dot{\theta}$. Point C, the contact point between the outer surface of the cylinder and the moving surface, has the velocity $\mathbf{v}_c = \mathbf{I}v_s$. Point A, the contact point of the inner cylinder with the cord, has zero velocity; $\mathbf{v}_A = 0$. The cylinder can be visualized as rolling without slipping (to the right) on the cord line *A–D*. We could use the vector equation $\mathbf{v}_I = \mathbf{v}_J + \boldsymbol{\omega} \times r_{JI}$ to state multiple correct equations between the velocities of points *A*, *B*, *o*, and *C*, and most of these equations would not be helpful in determining $\dot{\theta}$. The points to use are A and C because we know their velocities, and we do not know the velocities of the remaining points. Since points C and A are on the cylinder (rigid body), we can apply Equation 4.3 as

$$\mathbf{v}_C = \mathbf{v}_A + \boldsymbol{\omega} \times \mathbf{r}_{AC} \quad \text{or}$$

$$-\mathbf{I}\,(1.25\text{ m/s}) = 0 + \mathbf{K}\dot{\theta}\text{ (rad/s)} \times -0.5\mathbf{J} = \mathbf{I}0.5\dot{\theta}\text{ (m/s)}$$

$$\therefore\ \dot{\theta} = -1.25/5 = -2.5\text{ rad/s}, \quad \boldsymbol{\omega} = -\mathbf{K}2.5\text{ rad/s} \qquad \text{(i)}$$

Hence, the cylinder is rotating in a clockwise direction.

We can determine the velocity of point o at the center of the cylinder using either $\mathbf{v}_o = \mathbf{v}_C + \boldsymbol{\omega} \times \mathbf{r}_{Co}$ or $\mathbf{v}_o = \mathbf{v}_A + \boldsymbol{\omega} \times \mathbf{r}_{Ao}$, because we know $\mathbf{v}_A = 0$ and $\mathbf{v}_C = -\mathbf{I}$ 1.25 m/s. Proceeding from point A,

$$\begin{aligned}\mathbf{v}_o &= \mathbf{v}_A + \boldsymbol{\omega} \times \mathbf{r}_{Ao} = 0 - \mathbf{K}2.5\text{ (rad/s)} \times \mathbf{J}0.5\text{ (m)}\\ &= \mathbf{I}1.25\text{ m/s}.\end{aligned}$$

Hence, point o at the center of the cylinder moves horizontally to the right. Similarly,

$$\begin{aligned}\mathbf{v}_B &= \mathbf{v}_A + \boldsymbol{\omega} \times \mathbf{r}_{AB} = 0 - \mathbf{K}2.5\text{ (rad/s)} \times \mathbf{J}1.5\text{ (m)}\\ &= \mathbf{I}3.75\text{ m/s}.\end{aligned}$$

We have now completed the velocity analysis determining $\dot{\theta}, \boldsymbol{\omega}, \mathbf{v}_o, \mathbf{v}_B, \mathbf{v}_A$.

Moving to acceleration analysis, we first need to find $\ddot{\theta}$. Point A has a zero horizontal acceleration component; i.e., $\mathbf{a}_A = \mathbf{I}0 + \mathbf{J}a_{AY}$ m/s^2. From the nonslipping condition, point C has the horizontal acceleration component, $a_{CX} = 1.0$ m/s^2; hence, $\mathbf{a}_c = \mathbf{I}1 + \mathbf{J}a_{CY}$ m/s^2.

Applying the second of Equation 4.3 to points C and A yields

$$\begin{aligned}\mathbf{a}_C &= \mathbf{a}_A + \dot{\boldsymbol{\omega}} \times \mathbf{r}_{AC} + \boldsymbol{\omega} \times (\boldsymbol{\omega} \times \mathbf{r}_{AC}), \quad \text{or}\\ \mathbf{I}1\text{ m/s}^2 + \mathbf{J}a_{CY} &= \mathbf{J}a_{AY} + [\mathbf{K}\ddot{\theta}\text{ (rad/s}^2) \times -0.5\text{ m}\mathbf{J}] + \mathbf{K}\dot{\theta}\text{ (rad/s)}\\ &\quad \times [\mathbf{K}\dot{\theta}\text{ (rad/s)} \times -0.5\text{ m}\mathbf{J}]\\ &= [\mathbf{I}0.5\ddot{\theta} + \mathbf{J}(a_{AY} + 0.5\dot{\theta}^2)]\text{ (m/s}^2).\end{aligned}$$

Taking the $\mathbf{I}$ and $\mathbf{J}$ components separately gives

$$\begin{aligned}\mathbf{I}: 1\text{ ms}^2 &= 0.5\ddot{\theta}\text{ m/s}^2 \Rightarrow \ddot{\theta} = 2\text{ rad/s}^2 \Rightarrow \dot{\boldsymbol{\omega}} = 2\mathbf{K}\text{ rad/s}^2\\ \mathbf{J}: a_{CY}\text{ (m/s}^2) &= [a_{AY} + 0.5\dot{\theta}^2]\text{ m/s}^2\\ &= [a_{AY} + 0.5 \times (-2.5)^2]\text{ m/s}^2\\ &= (a_{AY} + 3.125)\text{ m/s}^2\end{aligned} \qquad \text{(ii)}$$

The $\mathbf{I}$ component result is immediately useful, determining $\ddot{\theta}$. The $\mathbf{J}$ component result is not helpful, since it only provides a relationship between the two unknowns a_{AY} and a_{CY}.

We still need to calculate these components. Point o has no vertical motion, i.e., $\mathbf{a}_o = \mathbf{I}a_{oX}$ Hence, we can proceed as follows:

$$\begin{aligned}\mathbf{a}_C &= \mathbf{a}_o + \dot{\boldsymbol{\omega}} \times \mathbf{r}_{oC} + \boldsymbol{\omega} \times (\boldsymbol{\omega} \times \mathbf{r}_{oC}), \quad \text{or}\\ \mathbf{I}1\text{ m/s}^2 + \mathbf{J}a_{CY} &= \mathbf{I}a_{oX} + [\mathbf{K}\ddot{\theta}\text{ (rad/s}^2) \times -1\text{ m}\mathbf{J}]\\ &\quad + \{\mathbf{K}\dot{\theta}\text{ (rad/s)} \times [\mathbf{K}\dot{\theta}\text{ (rad/s)} \times -1\text{ m}\mathbf{J}]\\ &= \mathbf{I}(a_{oX} + \ddot{\theta}) + \mathbf{J}\dot{\theta}^2\text{ (m/s}^2).\end{aligned}$$

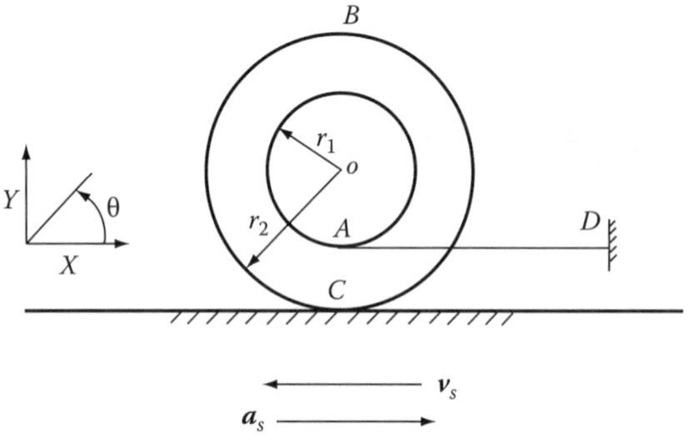

FIGURE XP4.4
Cylinder rolling without slipping on a horizontal plane (moving surface) while being restrained by a cord from *D*.

The $\boldsymbol{I}$ and $\boldsymbol{J}$ components give

$$\begin{aligned} \boldsymbol{I}&: 1 = (a_{oX} + \ddot{\theta}) \Rightarrow a_{oX} = (1 - 2) = -1 \text{ m/s}^2 \\ \boldsymbol{J}&: a_{CY} = \dot{\theta}^2 = (2.5)^2 \text{ m/s}^2 = 6.25 \text{ m/s}^2. \end{aligned} \quad \text{(iii)}$$

We substituted $\ddot{\theta} = 2$ rad/s^2 and $\dot{\theta} = -2.5$ rad/s into the $\boldsymbol{I}$ and $\boldsymbol{J}$ component equations, respectively. You can verify that we could also use $\boldsymbol{a}_A = \boldsymbol{a}_o + \dot{\boldsymbol{\omega}} \times \boldsymbol{r}_{oA} + \boldsymbol{\omega} \times (\boldsymbol{\omega} \times \boldsymbol{r}_{oA})$ successfully to determine a_{AY}.

From Equation (ii), $a_{CY} = a_{AY} + 3.125$ (m/s^2); hence using the second of Equation (iii), $a_{AY} = 6.25 - 3.125 = 3.125$ (m/s^2). Note that $a_{AY} = r_{oA}\dot{\theta}^2 = 0.5 \times (2.5)^2 = 3.125$ m/s^2. In this example, point A corresponds to the "contact point C" of Figure 4.13. It is in contact with cord line AD, and its only acceleration is vertical due to the centripetal acceleration term.

We now have

$$\begin{aligned} \boldsymbol{a}_A &= \boldsymbol{J}3.125 \text{ (m/s}^2) \\ \boldsymbol{a}_o &= -\boldsymbol{I}1 \text{ (m/s}^2) \\ \boldsymbol{a}_C &= -\boldsymbol{I}1 + \boldsymbol{J}6.25 \text{ (m/s}^2) \end{aligned}$$

The acceleration of point B can be obtained starting from points A, o, or C, since we have the acceleration of all these points. Starting from o,

$$\begin{aligned} \boldsymbol{a}_B &= \boldsymbol{a}_o + \dot{\boldsymbol{\omega}} \times \boldsymbol{r}_{oB} + \boldsymbol{\omega} \times (\boldsymbol{\omega} \times \boldsymbol{r}_{oB}), \quad \text{or} \\ &= -\boldsymbol{I}1 \text{ (m/s}^2) + [\boldsymbol{K}2 \text{ (rad/s}^2) \times \boldsymbol{J}1 \text{ m}] \\ &\quad - \boldsymbol{K}2.5 \text{ (rad/s)} \times [-\boldsymbol{K}2.5 \text{ (rad/s)} \times \boldsymbol{J}1 \text{ m}] \\ &= \boldsymbol{I}(-1 - 2) - \boldsymbol{J}6.25 \text{ m/s}^2 = -\boldsymbol{I}3 - \boldsymbol{J}6.25 \text{ m/s}^2) \end{aligned}$$

In reviewing this example, note that the key to the solution is (i) first find $\dot{\theta}$, and (ii) then find $\ddot{\theta}$. We used a velocity relation between A and C to find $\dot{\theta}$ and an acceleration relation between the same two points to find $\ddot{\theta}$. These points work because *we know the X components* of $\boldsymbol{v}_C$, $\boldsymbol{a}_C$, $\boldsymbol{v}_A$, $\boldsymbol{a}_A$. We can write valid equations relating the velocity and acceleration for any two of the points; however, only the combination of A and C will produce directly useful results in calculating $\dot{\theta}$ and $\ddot{\theta}$.

Also, note that to determine the Y components of $\boldsymbol{a}_A$, $\boldsymbol{a}_B$, $\boldsymbol{a}_C$, we needed to use an acceleration relationship involving $\boldsymbol{a}_O$, the acceleration of center of the cylinder. Valid acceleration relationships can certainly be stated between the points A, B, and C. However, the results are not helpful; e.g., we obtained $a_{CY} = a_{AY} + 3.125$ m/s^2 in Equation (ii), which involves two unknowns a_{CY} and a_{AY}. Point o "works" because we know that its vertical acceleration is zero.

4.4.2 A Wheel Rolling inside or on a Cylindrical Surface

4.4.2.1 Wheel Rolling inside a Cylindrical Surface

Figure 4.14b illustrates a wheel of radius r rolling without slipping on the inside of a cylindrical surface with radius R. O denotes the center of the cylindrical surface; o denotes the center of the wheel. The angle θ defines the rotation angle of the line O–o; ϕ defines the rotation of the wheel with respect to ground. The point of contact of the wheel with the larger radius is C. We want to use the nonslip condition at C to develop a relationship between the angular velocities $\dot{\theta}$ and $\dot{\phi}$.

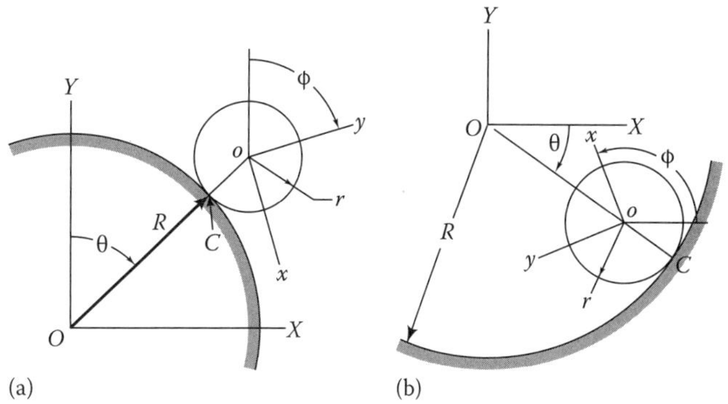

FIGURE 4.14
(a) Wheel rolling on a cylindrical surface. (b) Wheel rolling inside a cylindrical surface.

We can use the $\boldsymbol{\varepsilon}_r$, $\boldsymbol{\varepsilon}_\theta$ unit vectors to define the velocity of point o with respect to ground as

$$v_o = \boldsymbol{\varepsilon}_\theta (R - r)\dot{\theta} \quad (4.14)$$

Since o and C are both on the rigid rotating wheel, we can use the first of Equation 4.3 as

$$\begin{aligned} v_C = 0 = v_o + \boldsymbol{\omega} \times \boldsymbol{r}_{oC} &= (R - r)\dot{\theta}\boldsymbol{\varepsilon}_\theta + \boldsymbol{k}\dot{\phi} \times r\boldsymbol{\varepsilon}_r \\ &= [(R - r)\dot{\theta} - r\dot{\phi}]\boldsymbol{\varepsilon}_\theta, \end{aligned}$$

where $\boldsymbol{\omega} = \boldsymbol{k}\dot{\phi}$ is the angular velocity of the wheel with respect to ground. The nonslip condition at C gives

$$v_C = 0 \Rightarrow (R - r)\dot{\theta} = r\dot{\phi}, \quad (4.15a)$$

and differentiation gives $(R - r)\ddot{\theta} = r\ddot{\phi}$. Equation 4.15a provides the desired relation between $\dot{\theta}$ and $\dot{\phi}$.

This result can also be obtained by using Equation 4.3 to obtain the alternative result for v_o,

$$v_o = v_C + \boldsymbol{\omega} \times \boldsymbol{r}_{Ao} = 0 + \boldsymbol{k}\dot{\phi} \times (-r\boldsymbol{\varepsilon}_r) = r\dot{\phi}\boldsymbol{\varepsilon}_\theta.$$

In this case, we are starting at C to find v_o. Equating this result to Equation 4.14 again gives $(R - r)\dot{\theta} = r\dot{\phi}$. Recalling that $v_C = v_o + \boldsymbol{\omega} \times \boldsymbol{r}_{oC}$ applies for two points fixed in a rigid body, one could question the application of this equation for the contact point C, arguing that C is the contact point *on the cylindrical surface* versus a point *on the body*. In this application of Equation 4.3, we are considering a point on the wheel that is temporarily (instantaneously) at C.

4.4.2.2 *Wheel Rolling on the outside of a Cylindrical Surface*

Figure 4.14a illustrates a wheel of radius r rolling without slipping on the outside of a cylindrical surface of radius R. The angles θ and ϕ define the rotation angles of line O–o and the wheel with respect to ground. Following the same procedure, the velocity at the center of the wheel is

$$v_o = \boldsymbol{\varepsilon}_\theta (R+r)\dot{\theta}.$$

Since point o and the wheel contact point at C are both on the rigid wheel, we can again use the first of Equation 4.3 to obtain

$$\begin{aligned} v_C = 0 = v_o + \boldsymbol{\omega} \times \boldsymbol{r}_{oC} &= (R+r)\dot{\theta}\boldsymbol{\varepsilon}_\theta + (-\boldsymbol{k}\dot{\phi}) \times (-r\boldsymbol{\varepsilon}_r) \\ &= [(R+r)\dot{\theta} - r\dot{\phi}]\boldsymbol{\varepsilon}_\theta. \end{aligned}$$

Hence,

$$(R+r)\dot{\theta} = r\dot{\phi}, \tag{4.15b}$$

and $(R+r)\ddot{\theta} = r\ddot{\phi}$.

By denoting the radius of the cylinder-center motion as $\bar{R} = R - r$ in Equation 4.15a and $\bar{R} = R + r$ in Equation 4.15b, the results $\bar{R}\dot{\theta} = r\dot{\phi}$ and $\bar{R}\ddot{\theta} = r\ddot{\phi}$ apply for both cases. Kinematic relationships for rolling without slipping between cylindrical surfaces are not needed as frequently as rolling on a plane surface but arise regularly in dealing with geared systems.

4.5 Planar Mechanisms

4.5.1 Introduction

The wheel that rolled without slipping in the preceding section provides an appropriate "thought model" for the present topic of planar mechanisms. The wheel example had the following two characteristics that carry apply here:

a. The wheel had two coordinates X_o and θ. However, the rolling-without-slipping requirement yielded the linear relationship $X_o = r\theta$; hence, the wheel only had one degree of freedom.
b. We were able to obtain the desired velocity and acceleration results by either a geometric or traditional vector approach.

Similar outcomes will be observed for the mechanisms in this section, and solutions will be developed initially using both the geometric and vector approaches.

4.5.2 A Slider-Crank Mechanism

4.5.2.1 *Geometric Approach*

Planar mechanisms are connected linkages, designed to achieve specified motion of a point or link within the mechanism. A planar mechanism will normally involve a single input coordinate, with the possibility of multiple output coordinates. Figure 4.15 illustrates a "slider-crank" mechanism. The illustration could represent either (i) the crankshaft, connecting rod, and piston of a one-cylinder, internal-combustion engine with power developed by combustion and removed through the crankshaft or (ii) a section through the cylinder of a positive-displacement pump or compressor with power supplied from the crankshaft resulting in flow and elevated pressure. In either case, the figure shows three variables (θ, ϕ, and X_P). A mental inspection of this figure should convince you that only one degree of freedom is involved, since specifying any one of the coordinates will yield the remaining two variables. Given this observation, we would expect two relationships between the coordinates such that, having specified one variable, we could calculate the remaining two. In fact, the following relationships may be obtained by inspection:

$$X_P = l_1 \cos\theta + l_2 \cos\Phi, \quad l_1 \sin\theta = l_2 \sin\Phi \tag{4.16}$$

The first equation is obtained from the geometric relationships for the X (horizontal) components of the linkages; the second equation follows from the Y (vertical) components.

Looking at Figure 4.15, you could think of a civil engineer or surveyor doing a transit survey around a triangular piece of property bounded by sides AB, BC, and CA. If the surveyor follows the path, A to B to C to A, adding and subtracting vertical and horizontal

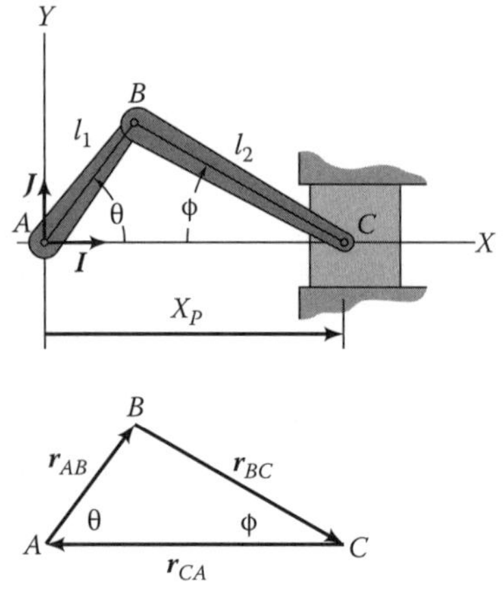

FIGURE 4.15
Slider-crank mechanism.

distances, the net change in position should be zero. Equation 4.16 reflect the scalar requirements that the change in position be zero. The first of Equation 4.16 states that the horizontal change in position is zero; the second equation states that the vertical change is zero.

The vector diagram in Figure 4.15 shows the position vectors $\boldsymbol{r}_{AB}$, $\boldsymbol{r}_{BC}$, and $\boldsymbol{r}_{CA}$. For these vectors, the transit idea requires

$$\boldsymbol{r}_{AB} + \boldsymbol{r}_{BC} + \boldsymbol{r}_{CA} = 0 \tag{4.17}$$

Substituting,

$$\begin{aligned} \boldsymbol{r}_{AB} &= \boldsymbol{I} l_1 \cos\theta + \boldsymbol{J} l_1 \sin\theta \\ \boldsymbol{r}_{BC} &= \boldsymbol{I} l_2 \cos\phi - \boldsymbol{J} l_2 \sin\phi \\ \boldsymbol{r}_{CA} &= -\boldsymbol{I} X_P, \end{aligned}$$

into Equation 4.17 gives

$$\begin{aligned} \boldsymbol{I} &: l_1 \cos\theta + l_2 \cos\phi - X_P = 0 \\ \boldsymbol{J} &: l_1 \sin\theta - l_2 \sin\phi = 0. \end{aligned}$$

Obviously, the formal vector approach of Equation 4.17 yields Equation 4.16. However, most students quickly develop a talent for "seeing" the scalar kinematic constraint equations by inspection, without recourse to the formal vector development. The "closed-loop" vector formalism remains a comfortable and certain approach when starting out.

Let us now review the mechanism of Figure 4.15, considering the statement that a planar mechanism has one input coordinate and possibly several output coordinates. The "customary" input coordinate for this mechanism is θ with ϕ and X_P being output coordinates. Figure 4.16 represents a mechanism for which X_P is the input coordinate delivered by a hydraulic cylinder, and θ and ϕ are the output coordinates. A real mechanism for which ϕ is the input coordinate, and θ and X_P are the output coordinates is unlikely.

Retaining θ as the input coordinate, we could consider the following engineering-analysis task: For a given constant rotation rate $\dot{\theta} = \omega$, find the velocity and acceleration of the piston for one cycle of θ. Results from this task could be required to determine the peak load on the pin used to hold the piston. Thinking about this task, you might look back at Equation 4.16 and think, "Well, if I differentiate Equation 4.16 with respect to time, I could get two equations relating $\dot{X}_P$, ϕ, and $\dot{\theta}$." The resulting scalar equations from this differentiation would be

$$\begin{aligned} \dot{X}_P + l_2 \sin\phi\dot{\phi} &= -l_1 \sin\theta\dot{\theta} = -l_1\,\omega \sin\theta \\ l_2 \cos\phi\dot{\phi} &= l_1 \cos\theta\dot{\theta} = l_1\,\omega\cos\theta. \end{aligned} \tag{4.18a}$$

In matrix format, these equations are

$$\begin{bmatrix} 1 & \sin\phi \\ 0 & \cos\phi \end{bmatrix} \begin{Bmatrix} \dot{X}_P \\ l_2\dot{\phi} \end{Bmatrix} = l_1\omega \begin{Bmatrix} -\sin\theta \\ \cos\theta \end{Bmatrix}. \tag{4.18b}$$

The equations have been rearranged with the unknowns ($\dot{X}_P$, and $\dot{\phi}$) on the left and the known rotation rate $\dot{\theta} = \omega$ on the right. Note that ϕ, a parameter in these equations, is obtained by solving the last of Equation 4.16 as $\phi = \sin^{-1}(l_1 \sin\theta / l_2)$. Both ϕ and X_P can be determined from Equation 4.16 for any specified value of θ.

Differentiating Equation 4.18a with respect to time gives

$$\begin{aligned} \ddot{X}_P + l_2 \sin\phi\ddot{\phi} &= -l_1 \sin\theta\ddot{\theta} - l_1 \cos\theta\dot{\theta}^2 - l_2\cos\phi\dot{\phi}^2 \\ &= -l_1\,\omega^2 \cos\theta - l_2 \cos\phi\dot{\phi}^2 \\ l_2 \cos\phi\ddot{\phi} &= l_1 \cos\theta\ddot{\theta} - l_1 \sin\theta\dot{\theta}^2 + l_2 \sin\phi\dot{\phi}^2 \\ &= -l_1\,\omega^2 \sin\theta + l_2 \sin\phi\dot{\phi}^2. \end{aligned} \tag{4.19a}$$

The facts that $\ddot{\theta} = 0$, and $\dot{\theta} = \omega =$ constant were used to simplify these equations, and their matrix statement is

$$\begin{bmatrix} 1 & \sin\phi \\ 0 & \cos\phi \end{bmatrix} \begin{Bmatrix} \ddot{X}_P \\ l_2\ddot{\phi} \end{Bmatrix} = -l_1\omega^2 \begin{Bmatrix} \cos\theta \\ \sin\theta \end{Bmatrix} + l_2\dot{\phi}^2 \begin{Bmatrix} -\cos\phi \\ \sin\phi \end{Bmatrix}. \tag{4.19b}$$

The unknowns are $\ddot{X}_P$ and $\ddot{\phi}$. The right-hand side terms are known. Equation 4.16 defines ϕ; Equation 4.18 defines both $\dot{\phi}$ (that is needed in Equation 4.19b) and $\dot{X}_P$ (that is not).

Your engineering task could be accomplished by the following sequential steps:

1. Vary θ over the range of $[0, 2\pi]$, yielding discrete values θ_i
2. For each θ_i value, solve Equation 4.16 to determine corresponding values for ϕ_i and X_{Pi}, via

$$\begin{aligned} \phi_i &= \sin^{-1}\left(\frac{l_1}{l_2}\sin\theta_i\right) \\ X_{P_i} &= l_1\cos\theta_i + l_2\cos\phi_i \end{aligned} \tag{i}$$

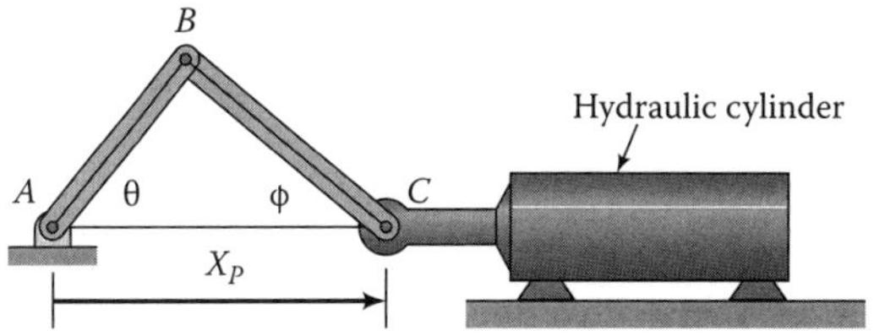

FIGURE 4.16
Slider-crank mechanism with displacement input from a hydraulic cylinder.

3. Enter Equation 4.18a with known values for θ_i and ϕ_i to determine $\dot{\phi}_i$, and $\dot{X}_{Pi}$ via

$$\begin{aligned} \dot{\phi}_i &= \frac{l_1}{l_2}\omega\cos\theta_i \\ \dot{X}_{Pi} &= -l_2\dot{\phi}_i\sin\phi_i - l_1\omega\sin\theta_i \end{aligned} \quad \text{(ii)}$$

4. Enter Equation 4.19a with known values for θ_i, ϕ_i, and $\dot{\phi}_i$, to determine $\ddot{\phi}_i$ and $\ddot{X}_{Pi}$ via

$$\begin{aligned} \ddot{\phi}_i &= \left(-l_1\omega^2\sin\theta_i + l_2\dot{\phi}_i^2\sin\dot{\phi}_i\right)/(l_2\cos\phi_i) \\ \ddot{X}_{Pi} &= -l_2\ddot{\phi}_i\sin\phi - l_1\omega^2\cos\theta_i - l_2\dot{\phi}_i^2\cos\phi_i \end{aligned} \quad \text{(iii)}$$

The solution to these equations can be readily developed with standard spread-sheet software, and Figure 4.17 presents a solution for $\ddot{X}_P$ for one θ cycle with $l_1 = 250$ mm, $l_2 = 300$ mm, $\omega = 14.6$ rad/s.

Figure 4.17 shows peak accelerations of approximately -10 g's for $\theta = \phi = 0$, corresponding to a fully extended position. The peak positive acceleration of approximately 8 g's arises for $\theta = \pi$, $\phi = 0$ with the piston in its left-most position.

Returning to the mechanism of Figure 4.16, let us consider reformulating the problem with $X_P(t)$ given as the input, with (θ and ϕ), ($\dot{\theta}$ and $\dot{\phi}$), and ($\ddot{\theta}$ and $\ddot{\phi}$) as the desired output coordinates. This second problem can be worked, using exactly the same equations but with an exchange of knowns and unknowns. From Equation 4.16 the desired equations for the coordinates are

$$l_1\cos\theta + l_2\cos\phi = X_P, \quad l_1\sin\theta - l_2\sin\phi| = 0 \quad (4.20a)$$

Now, the input coordinate is X_P, and the coordinates to be determined are θ and ϕ. The equations are nonlinear and harder to solve for θ and ϕ than Equation 4.16 was for X_P and ϕ. However, they can be easily solved (numerically) for θ and ϕ. (You could also start with θ as the independent variable, solve for X_P, ϕ, and then produce a table of X_{Pi}, ϕ_i, θ_i.) From Equation 4.20a, the velocity relationships are

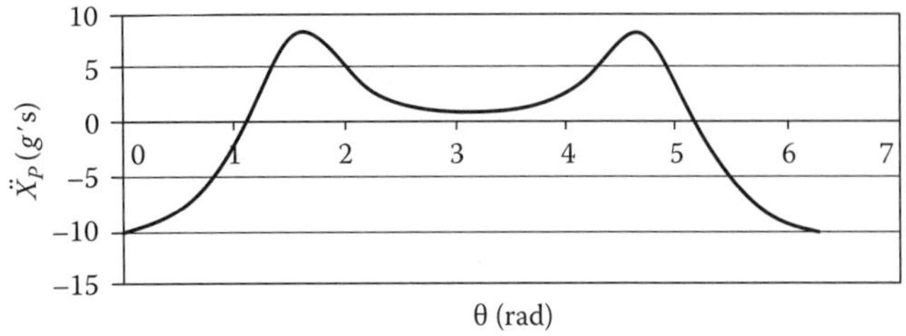

FIGURE 4.17
$\ddot{X}_P(g\text{'s}) = \ddot{X}/9.81\ \text{m/s}^2$ versus θ.

$$l_2\sin\phi\dot{\phi} + l_1\sin\theta\dot{\theta} = -\dot{X}_P, \quad l_2\cos\phi\dot{\phi} - l_1\cos\theta\dot{\theta} = 0. \quad (4.20b)$$

Finally, differentiating Equation 4.20b produces the required acceleration component equations:

$$\begin{aligned} l_2\sin\phi\ddot{\phi} + l_1\sin\theta\ddot{\theta} &= -\ddot{X}_P - l_1\cos\theta\dot{\theta}^2 - l_2\cos\phi\dot{\phi}^2 \\ l_2\cos\phi\ddot{\phi} - l_1\cos\theta\ddot{\theta} &= -l_1\sin\theta\dot{\theta}^2 + l_2\sin\phi\dot{\phi}^2. \end{aligned} \quad (4.20c)$$

The problem solution is obtained for specified values of $X_P(t)$, $\dot{X}_P(t)$, $\ddot{X}_P(t)$ by proceeding *sequentially* through Equations 4.20a, 4.20b, and 4.20c. Note that Equation 4.20a defining θ and ϕ are nonlinear, while Equation 4.20b are linear in the unknowns $\dot{\phi}$ and $\dot{\theta}$, and Equation 4.20c are linear in the unknowns $\ddot{\phi}$ and $\ddot{\theta}$.

The essential first step in developing kinematic equations for planar mechanisms via geometric relationships is drawing a picture of the mechanism. A general orientation is required, yielding equations that can be subsequently differentiated. Looking back at Figures 4.15 and 4.16, note that a general position is shown for the coordinates. Also, note in reviewing the mechanisms of these figures, that two ostensibly different mechanisms are modeled by the same scalar equations for coordinates, and velocity and acceleration terms. From a geometric viewpoint, there are simply not that many *fundamentally different* models.

As noted previously, there are generally several different and correct ways to work kinematics problems. To the author, the geometric approach outlined above has always appeared to be the simplest and most straightforward procedure, and most students share this view. However, "easy" like beauty is frequently in the eye of the beholder, and some students and engineers prefer a vector approach using the kinematic results of Chapter 2 and Section 4.3. To demonstrate this approach, we will now work through the slider-crank problem using the velocity and acceleration results of Section 4.3.

4.5.2.2 Vector Approach for Velocity and Acceleration Results

Figure 4.18 provides an illustration of the slider-crank mechanism in a disassembled condition. Applying the velocity result of Equation 4.3 separately to links 1 and 2, gives

$$\boldsymbol{v}_B = \boldsymbol{v}_A + \boldsymbol{\omega}_1 \times \boldsymbol{r}_{AB}, \quad \boldsymbol{v}_C = \boldsymbol{v}_B + \boldsymbol{\omega}_2 \times \boldsymbol{r}_{BC}.$$

Equation 4.3 applies for two points in a rigid body. The first of these equations is for points A and B in rigid

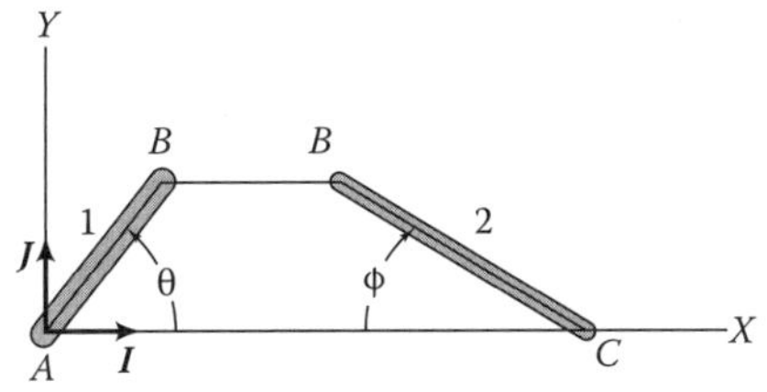

FIGURE 4.18
Disassembled view of the slider-crank mechanism for vector analysis.

body 1. The second equation is for points C and B in rigid body 2. We are going to get two scalar governing equations from these vector equations by equating the two answers that these equations provide for $\boldsymbol{v}_B$, obtaining

$$\boldsymbol{v}_B = \boldsymbol{v}_A + \boldsymbol{\omega}_1 \times \boldsymbol{r}_{AB} = \boldsymbol{v}_C - \boldsymbol{\omega}_2 \times \boldsymbol{r}_{BC}.$$

To complete the velocity relationships, we need to "plug in" components for the vectors in this equation. Since point A is fixed in the X, Y system, observe that $\boldsymbol{v}_A = 0$. Similarly, given that point C can only move horizontally, $\boldsymbol{v}_C = \boldsymbol{I}\dot{X}_P$. The vector $\boldsymbol{\omega}_1$ is the angular velocity of link 1 with respect to the X, Y system. Using the right-hand rule, this definition implies $\boldsymbol{\omega}_1 = \boldsymbol{K}\dot{\theta}$. Similarly, $\boldsymbol{\omega}_2$ is the angular velocity of body 2 with respect to the X, Y system, and the right-hand rule definition gives $\boldsymbol{\omega}_2 = -\boldsymbol{K}\dot{\phi}$. The position vector $\boldsymbol{r}_{AB}$ goes from point A to point B and is defined by $\boldsymbol{r}_{AB} = l_1(\boldsymbol{I}\cos\theta + \boldsymbol{J}\sin\theta)$. Similarly, in link 2, $\boldsymbol{r}_{BC}$ extends from point B to point C and is defined by $\boldsymbol{r}_{BC} = l_2(\boldsymbol{I}\cos\phi - \boldsymbol{J}\sin\phi)$. Substituting these results gives

$$\begin{aligned} 0 &+ \boldsymbol{K}\dot{\theta} \times l_1(\boldsymbol{I}\cos\theta + \boldsymbol{J}\sin\theta) \\ &= \boldsymbol{I}\dot{X}_p - (-\boldsymbol{K}\dot{\phi}) \times l_2(\boldsymbol{I}\cos\phi - \boldsymbol{J}\sin\phi). \end{aligned}$$

Carrying out the cross products and gathering terms gives

$$\begin{aligned} \boldsymbol{I} &: -l_1\,\dot{\theta}\sin\theta = \dot{X}_P + l_2\sin\phi\dot{\phi} \\ \boldsymbol{J} &: l_1\,\dot{\theta}\cos\theta = l_2\,\dot{\phi}\cos\phi. \end{aligned}$$

These equations coincide with our earlier velocity results of Equation 4.18a.

Returning to Figure 4.18, we can use the acceleration form of Equation 4.3 to write

$$\begin{aligned} \boldsymbol{a}_B &= \boldsymbol{a}_A + \dot{\boldsymbol{\omega}}_1 \times \boldsymbol{r}_{AB} + \boldsymbol{\omega}_1 \times (\boldsymbol{\omega}_1 \times \boldsymbol{r}_{AB}) \\ \boldsymbol{a}_C &= \boldsymbol{a}_B + \dot{\boldsymbol{\omega}}_2 \times \boldsymbol{r}_{BC} + \boldsymbol{\omega}_2 \times (\boldsymbol{\omega}_2 \times \boldsymbol{r}_{BC}). \end{aligned}$$

The first equation applies to points A and B of link 1 and relates the acceleration of these points with respect to the X, Y system. The second equation relates the acceleration of points B and C of link 2 with respect to the X, Y system. We can equate the separate definitions provided by these equations for $\boldsymbol{a}_B$ to obtain

$$\begin{aligned} \boldsymbol{a}_B &= \boldsymbol{a}_A + \dot{\boldsymbol{\omega}}_1 \times \boldsymbol{r}_{AB} + \boldsymbol{\omega}_1 \times (\boldsymbol{\omega}_1 \times \boldsymbol{r}_{AB}) \\ &= \boldsymbol{a}_C - \dot{\boldsymbol{\omega}}_2 \times \boldsymbol{r}_{BC} - \boldsymbol{\omega}_2 \times (\boldsymbol{\omega}_2 \times \boldsymbol{r}_{BC}). \end{aligned}$$

Since point A is fixed, $\boldsymbol{a}_A = 0$. Also, since point C is constrained to move in the horizontal plane, $\boldsymbol{a}_C = \boldsymbol{I}\ddot{X}_P$. The remaining undefined variables are $\dot{\boldsymbol{\omega}}_1 = \boldsymbol{K}\ddot{\theta}$, the angular acceleration of link 1 with respect to the X, Y system, and $\dot{\boldsymbol{\omega}}_2 = \boldsymbol{K}\ddot{\phi}$, the angular acceleration of link 2 with respect to the X, Y system. The right-hand rule for defining angular velocities and angular accelerations has been used in these definitions. Substituting the remaining variables gives

$$\begin{aligned} 0 &+ \boldsymbol{K}\ddot{\theta} \times l_1(\boldsymbol{I}\cos\theta + \boldsymbol{J}\sin\theta) + \boldsymbol{K}\dot{\theta} \times [\boldsymbol{K}\dot{\theta} \times l_1(\boldsymbol{I}\cos\theta + \boldsymbol{J}\sin\theta)] \\ &= \boldsymbol{I}\ddot{X}_P - (-\boldsymbol{K}\ddot{\phi}) \times l_2(\boldsymbol{I}\cos\phi - \boldsymbol{J}\sin\phi) \\ &\quad - (-\boldsymbol{K}\dot{\phi}) \times [-\boldsymbol{K}\dot{\phi} \times l_2(\boldsymbol{I}\cos\phi - \boldsymbol{J}\sin\phi)]. \end{aligned}$$

Completing the cross products and algebra gives the following component equations:

$$\begin{aligned} \boldsymbol{I} &: -l_1\,\ddot{\theta}\sin\theta - l_1\,\dot{\theta}^2\cos\theta = \ddot{X}_P + l_2\,\ddot{\phi}\sin\phi + l_2\dot{\phi}^2\cos\phi \\ \boldsymbol{J} &: l_1\,\ddot{\theta}\cos\theta - l_1\,\dot{\theta}^2\sin\theta = l_2\,\ddot{\phi}\cos\phi - l_2\,\dot{\phi}^2\sin\phi, \end{aligned}$$

which replicate the earlier results of Equation 4.19a.

The judgment as to whether the geometric or vector approach is better for deriving velocity and acceleration equations is a matter of personal preference. Properly applied, either approach gives the correct answer. However, irrespective of the approach used to develop the velocity and acceleration equations, the geometric Equation 4.16 must be derived to work out a general solution. With the geometric approach, they start the process. With the vector approach, they tend to be an afterthought.

4.5.3 A Four-Bar-Linkage Example

Figure 4.19a illustrates a four-bar-linkage mechanism, involving three coordinates α, β, and γ. The stationary wall connecting points A and D is the nonvisible "fourth" bar in this four-bar linkage. As with the earlier slider-crank mechanism, this mechanism only has one degree of freedom. The mechanism could reasonably be driven from A, making α the input coordinate, and β and γ the output coordinates. Alternatively, it could be driven from D making γ the input coordinate and α and β the output coordinates.

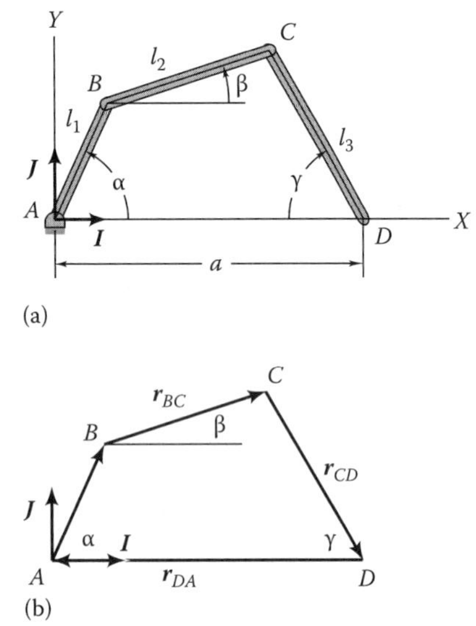

FIGURE 4.19
(a) Four-bar linkage. (b) Vector diagram for linkage.

Consider the following engineering-analysis task: For a constant rotation rate, $\dot{\alpha} = \omega$ determine the angular velocities $\dot{\beta}$, $\dot{\gamma}$ and angular accelerations $\ddot{\beta}$, $\ddot{\gamma}$ for one rotation of α. We are going to work through this task with geometric and vector equations, starting with the geometric relations.

4.5.3.1 Geometric Approach

Inspecting Figure 4.19a yields the following geometric relationships:

$$\begin{aligned} X&: l_1 \cos\alpha + l_2 \cos\beta + l_3 \cos\gamma = a \\ Y&: l_1 \sin\alpha + l_2 \sin\beta - l_3 \sin\gamma = 0. \end{aligned} \tag{4.21}$$

These results are obtained by starting at A and proceeding to D, separately adding the X and Y components. Figure 4.19b shows a closed-loop vector representation that can be formally used to obtain Equation 4.21. The results from Figure 4.19b can be stated, $\boldsymbol{r}_{AB} + \boldsymbol{r}_{BC} + \boldsymbol{r}_{CD} + \boldsymbol{r}_{DA} = 0$. Substituting

$$\boldsymbol{r}_{AB} = l_1(I \cos\alpha + J \sin\alpha), \quad \boldsymbol{r}_{BC} = l_2(I \cos\beta + J \sin\beta)$$
$$\boldsymbol{r}_{CD} = l_3(\boldsymbol{I} \cos\gamma - \boldsymbol{J} \sin\gamma), \quad \boldsymbol{r}_{DA} = -\boldsymbol{I}a$$

into this equation gives the same result as Equation 4.21.

Equation 4.21 can be restated to make the specified nature of the input coordinate α and output coordinates (β, γ) clearer as

$$\begin{aligned} l_2 \cos\beta + l_3 \cos\gamma &= a - l_1 \cos\alpha \\ l_2 \sin\beta - l_3 \sin\gamma &= -l_1 \sin\alpha. \end{aligned} \tag{4.22}$$

These nonlinear equations can be solved analytically. The solution is instructive and will be discussed further below. Differentiating Equation 4.22 gives the velocity relationships

$$\begin{aligned} -l_2 \sin\beta\dot{\beta} - l_3 \sin\gamma\dot{\gamma} &= l_1 \sin\alpha\dot{\alpha} = l_1\,\omega \sin\alpha \\ l_2 \cos\beta\dot{\beta} - l_3 \cos\gamma\dot{\gamma} &= -l_1 \cos\alpha\dot{\alpha} = -l_1\,\omega \cos\alpha. \end{aligned} \tag{4.23}$$

In matrix format, these equations become

$$\begin{bmatrix} \sin\beta & \sin\gamma \\ -\cos\beta & \cos\gamma \end{bmatrix} \begin{Bmatrix} l_2\dot{\beta} \\ l_3\dot{\gamma} \end{Bmatrix} = l_1\omega \begin{Bmatrix} -\sin\alpha \\ \cos\alpha \end{Bmatrix}.$$

They are linear in $\dot{\beta}$ and $\dot{\gamma}$ and can be solved for any specified value of α and then β, γ from Equation 4.22. Using Cramer's rule (Appendix A), their solution is

$$\begin{aligned} l_2\dot{\beta} &= \frac{l_1\omega}{\cos\gamma\sin\beta + \sin\gamma\cos\beta} \begin{vmatrix} -\sin\alpha & \sin\gamma \\ \cos\alpha & \cos\gamma \end{vmatrix} \\ &= \frac{-l_1\omega\sin(\alpha+\gamma)}{\sin(\beta+\gamma)} \\ l_3\dot{\gamma} &= \frac{l_1\omega}{\cos\gamma\sin\beta + \sin\gamma\cos\beta} \begin{vmatrix} \sin\beta & -\sin\alpha \\ -\cos\beta & \cos\alpha \end{vmatrix} \\ &= \frac{l_1\omega\sin(\beta-\alpha)}{\sin(\beta+\gamma)}, \end{aligned} \tag{4.24}$$

where the determinant of the coefficient matrix is $D = \sin(\beta+\gamma)$. Note that the solution is undefined for $\sin(\beta+\gamma) = 0 \Rightarrow \beta+\gamma = 0, \pi$.

Differentiating Equation 4.23 gives the following acceleration equations:

$$\begin{aligned} -l_2 \sin\beta\ddot{\beta} - l_3 \sin\gamma\ddot{\gamma} &= l_1 \sin\alpha\ddot{\alpha} + l_1 \cos\alpha\dot{\alpha}^2 \\ &\quad + l_2 \cos\beta\dot{\beta}^2 + l_3 \cos\gamma\dot{\gamma}^2 \\ l_2 \cos\beta\ddot{\beta} - l_3 \cos\gamma\ddot{\gamma} &= -l_1 \cos\alpha\ddot{\alpha} + l_1 \sin\alpha\dot{\alpha}^2 \\ &\quad + l_2 \sin\beta\dot{\beta}^2 - l_3 \sin\gamma\dot{\gamma}^2. \end{aligned}$$

Setting $\ddot{\alpha} = 0$, and $\dot{\alpha} = \omega$ reduces them to

$$\begin{aligned} -l_2 \sin\beta\ddot{\beta} - l_3 \sin\gamma\ddot{\gamma} &= l_1 \cos\alpha\omega^2 + l_2 \cos\beta\dot{\beta}^2 + l_3 \cos\gamma\dot{\gamma}^2 \\ l_2 \cos\beta\ddot{\beta} - l_3 \cos\gamma\ddot{\gamma} &= -l_1 \sin\alpha\omega^2 + l_2 \sin\beta\dot{\beta}^2 - l_3 \sin\gamma\dot{\gamma}^2, \end{aligned} \tag{4.25}$$

or, in matrix format,

$$\begin{bmatrix} -\sin\beta & -\sin\gamma \\ \cos\beta & -\cos\gamma \end{bmatrix} \begin{Bmatrix} l_2\ddot{\beta} \\ l_3\ddot{\gamma} \end{Bmatrix}$$
$$= \begin{Bmatrix} l_1\omega^2\cos\alpha + l_2\dot{\beta}^2\cos\beta + l_3\dot{\gamma}^2\cos\gamma \\ -l_1\omega^2\sin\alpha + l_2\dot{\beta}^2\sin\beta - l_3\dot{\gamma}^2\sin\gamma \end{Bmatrix} = \begin{Bmatrix} g_1 \\ g_2 \end{Bmatrix}$$

Using Cramer's rule, the solution is

$$l_2\ddot{\beta} = \frac{-g_1 \cos\gamma + g_2 \sin\gamma}{\sin(\beta+\gamma)}, \quad l_3\ddot{\gamma} = \frac{-g_1 \cos\beta - g_2 \sin\beta}{\sin(\beta+\gamma)}. \tag{4.26}$$

Note that the determinant for the coefficient matrix is (again) $\sin(\beta+\gamma)$; hence, $(\ddot{\beta}, \ddot{\gamma})$ are undefined for $\sin(\beta+\gamma) = 0$ as with $(\dot{\beta}, \dot{\gamma})$ in Equation 4.24.

We have now obtained the last required set of kinematic equations, and the engineering task outlined above would be accomplished by executing the following sequential steps:

1. Vary α over the range to be specified, yielding discrete values α_i.
2. For each α_i value, solve Equation 4.22 to determine corresponding values for β_i and γ_i. As discussed below, these equations can be solved analytically.
3. Enter Equation 4.24 with known values for α_i β_i and $\dot{\gamma}_i$ to determine $\dot{\beta}_i$, and $\dot{\gamma}_i$.
4. Enter Equation 4.26 with known values for α_i, β_i, γ_i, $\dot{\beta}_i$, and $\dot{\gamma}_i$ to determine $\ddot{\beta}_i$ and $\ddot{\gamma}_i$. Again, these linear equations can be solved readily for the unknown angular accelerations.

As noted above, Equation 4.22 can be solved analytically, and the solution is worth reviewing for several reasons. First, it shows that two different solutions are generally possible. Specifically, for a given input parameter set, l_1, l_2, l_3, a and a specified rotation angle α, two solutions for β and γ may be possible. Moving to solve for β and γ, we restate Equation 4.22 as

$$l_2 \cos\beta = (a - l_1 \cos\alpha) - l_3 \cos\gamma = h_1 - l_3 \cos\gamma$$
$$l_2 \sin\beta = -l_1 \sin\alpha + l_3 \sin\gamma = h_2 + l_3 \sin\gamma.$$

Note that $h_1 = a - l_1 \cos\alpha$ and $h_2 = -l_1 \sin\alpha$ are defined in terms of α and are known quantities.

Squaring both of these equations and adding them together gives

$$\begin{aligned} l_2^2(\cos^2\beta + \sin^2\beta) &= h_1^2 - 2h_1 l_3 \cos\gamma + l_3^2 \cos^2\gamma \\ &\quad + h_2^2 + 2h_2 l_3 \sin\gamma + l_3^2 \sin^2\gamma \\ \therefore\ l_2^2 &= h_1^2 + h_2^2 + l_3^2 - 2h_1 l_3 \cos\gamma + 2h_2 l_3 \sin\gamma. \end{aligned}$$

Rearranging gives

$$\begin{aligned} 2h_1 l_3 \cos\gamma &= h_1^2 + h_2^2 + l_3^2 - l_2^2 + 2h_2 l_3 \sin\gamma \\ &= 2d + 2h_2 l_3 \sin\gamma \\ \therefore\ h_1^2 l_3^2 \cos^2\gamma &= h_1^2 l_3^2 (1 - \sin^2\gamma) \\ &= d^2 + 2d h_2 l_3 \sin\gamma + h_2^2 l_3^2 \sin^2\gamma, \end{aligned}$$

where $2d = h_1^2 + h_2^2 + l_3^2 - l_2^2$. Restating this result gives the following quadratic equation in $\sin\gamma$

$$\sin^2\gamma + \frac{2dh_2}{l_3(h_1^2 + h_2^2)} \sin\gamma + \frac{(d^2 - h_1^2 l_3^2)}{l_3^2(h_1^2 + h_2^2)} = 0.$$

The equation $\sin^2\gamma + B\sin\gamma + C = 0$ has the two roots:

$$\begin{aligned} \sin\gamma &= -\frac{B}{2} + \frac{1}{2}\sqrt{B^2 - 4C}, \\ \sin\gamma &= -\frac{B}{2} - \frac{1}{2}\sqrt{B^2 - 4C}. \end{aligned} \tag{4.27}$$

Depending on values for B and C, this equation can have one real root, two real roots, or two complex roots. Two real roots implies two distinct solutions, and this possibility is illustrated by Figure 4.20, where the same α value gives an orientation that differs from Figure 4.19a.

As we will confirm below, the one real-root solution corresponding to $B^2 = 4C$ defines an extreme "locked" position for the mechanism, as illustrated in Figure 4.21. Note that this position corresponds to $\beta_1 = -\gamma_1$ netting $\sin(\beta_1 + \gamma_1) = 0$, which also caused the angular velocities and angular accelerations to be undefined in Equations 4.24 and 4.26, respectively.

We can solve for the limiting α value in Figure 4.21 by substituting $\beta_1 = -\gamma_1$ into Equation 4.22 to get

$$l_2 \cos\gamma_1 + l_3 \cos\gamma_1 = (l_2 + l_3)\cos\gamma_1 = a - l_1 \cos\alpha_1$$
$$-l_2 \sin\gamma_1 - l_3 \sin\gamma_1 = -(l_2 - l_3)\sin\gamma_1 = -l_1 \sin\alpha_1.$$

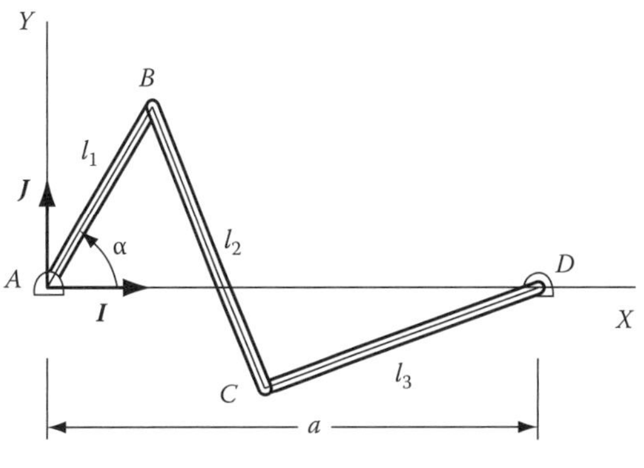

FIGURE 4.20
Alternate configuration for the linkage of Figure 4.19a.

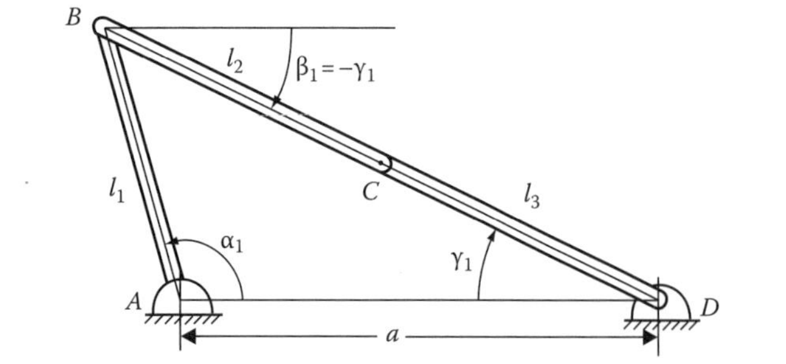

FIGURE 4.21
Locked position for the linkage of Figure 4.19a with $l_2 = l_3 = 1.45\ l_1$, $a = 2.80\ l_1$.

Squaring both equations and adding them together gives

$$a^2 - 2al_1 \cos\alpha_1 + l_1^2 = (l_2 + l_3)^2 \Rightarrow \cos\alpha_1 = \frac{a^2 + l_1^2 - (l_2 + l_3)2}{2al_1}. \qquad (4.28)$$

If the parameters l_1, l_2, l_3, a are such that a solution exists for α_1, then a locked position can occur.

For Figure 4.21, the limiting value for α_1, corresponds to

$$\alpha_1 = \pi \Rightarrow (a + l_1)^2 = (l_1 + l_3)^2 \Rightarrow a + l_1 = l_2 + l_3.$$

For $l_2 + l_3 > a + l_1$ there is no limiting rotation angle α_1 and the left-hand link can rotate freely through 360°.

As noted above, limiting conditions can also be obtained from Equation 4.27. Substituting for B and C into $B^2 - 4C = 0$ gives $d^2 = l_3^2(h_1^2 + h_2^2)$. Substituting for d, h_1, h_2 (plus some algebra) into this equation gives the two limiting conditions:

$$\begin{aligned} a_2 - 2al_1 \cos\alpha_1 + l_1^2 &= (l_2 + l_3)^2 \\ a_2 - 2al_1 \cos\alpha_1 + l_1^2 &= (l_2 - l_3)^2. \end{aligned} \qquad (4.29)$$

The first result coincides with Equation 4.28. The second corresponds to a situation where the linkage is driven from the right link with γ as the input and will not be discussed further. Assuming that the parameters l_1, l_2, l_3, a produce a limiting value α_1 from Equation 4.28, then a larger value for α will yield $(B^2 - 4C) < 0$, producing two complex roots for $\sin\gamma$ in Equation 4.27 that represent nonphysical solutions.

Figure 4.22 illustrates a solution for β, γ, $\dot{\beta}$, $\dot{\gamma}$ with $l_1 = 0.35$ m, $l_2 = .816$ m, $l_3 = 1$ m, $a = 0.6$ m, and $\omega = 3$ rad/s for α_i. over $[0, 2\pi]$. The solution illustrated corresponds to the first solution (positive square root) in Equation 4.27.

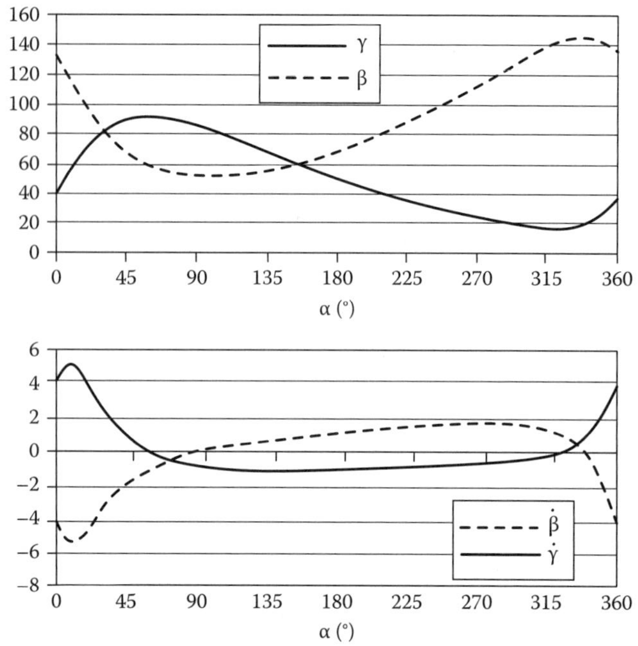

FIGURE 4.22
Numerical solution for (β, γ), $(\dot{\beta}, \dot{\gamma})$ versus α for $l_1 = 0.35$ m, $l_2 = .816$ m, $l_3 = 1$ m, $a = 0.6$ m and $\omega = 3$ rad/s.

Caution is advisable in using Equation 4.27 to solve for γ_i. to make sure that the solution is in the proper quadrant. Also, in solving for β_i with Equation 4.22, use $\beta_i = \tan^{-1}(\sin\beta_i / \cos\beta_i)$ using the known value for α_i and the calculated value for γ_i.

4.5.3.2 Vector Approach for Velocity and Acceleration Relationships

For comparison to the geometric approach, we will now develop the velocity and acceleration relationships for this example using the vector Equation 4.3. Figure 4.23 provides a disassembled view of the three rigid bodies that make up the mechanism. First, looking to develop relationships for $\dot{\alpha}$, $\dot{\beta}$, and $\dot{\gamma}$, we need a vector velocity equation for each link. Starting on the left with link 1, and looking from point A to B gives

$$\boldsymbol{v}_B = \boldsymbol{v}_A + \boldsymbol{\omega}_1 \times \boldsymbol{r}_{AB}.$$

Next, for link 3, looking from point D back to point C we can write

$$\boldsymbol{v}_C = \boldsymbol{v}_D + \boldsymbol{\omega}_3 \times \boldsymbol{r}_{DC}.$$

Finally, for link 2, looking from point B to point C gives

$$\boldsymbol{v}_C = \boldsymbol{v}_B + \boldsymbol{\omega}_2 \times \boldsymbol{r}_{BC}.$$

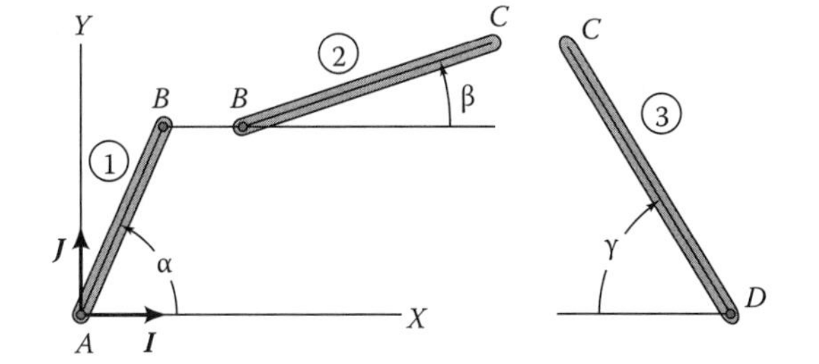

FIGURE 4.23
Disassembled view of the four-bar linkage of Figure 4.19 for vector analysis.

Substituting into this last equation for v_B and v_C, and observing that $v_A = v_D = 0$ gives the desired vector result

$$\boldsymbol{\omega}_3 \times \boldsymbol{r}_{DC} = \boldsymbol{\omega}_1 \times \boldsymbol{r}_{AB} + \boldsymbol{\omega}_2 \times \boldsymbol{r}_{BC}.$$

Equation 4.21 define the position vectors of this equation. Using the right-hand rule, the angular velocity vectors are defined as $\boldsymbol{\omega}_1 = \boldsymbol{K}\dot{\alpha}$, $\boldsymbol{\omega}_2 = \boldsymbol{K}\dot{\beta}$, and $\boldsymbol{\omega}_2 = -\boldsymbol{K}\dot{\gamma}$. Substituting these results give

$$-\boldsymbol{K}\dot{\gamma} \times l_3(-\boldsymbol{I}\cos\gamma + \boldsymbol{J}\sin\gamma) = \boldsymbol{K}\dot{\alpha} \times l_1(\boldsymbol{I}\cos\alpha + \boldsymbol{J}\sin\alpha) + \boldsymbol{K}\dot{\beta} \times l_2(\boldsymbol{I}\cos\beta + \boldsymbol{J}\sin\beta).$$

Carrying out the cross products gives the following component equations:

$$\boldsymbol{I}: l_3\dot{\gamma}\sin\gamma = -l_1\dot{\alpha}\sin\alpha - l_2\dot{\beta}\sin\beta$$
$$\boldsymbol{J}: l_3\dot{\gamma}\cos = l_1\dot{\alpha}\cos\alpha + l_2\cos\dot{\beta}\cos\beta.$$

You can confirm that these vector-based equations coincide with the geometric relationships of Equation 4.22.

The acceleration relationships are obtained via the same logic from the vector equations

$$\boldsymbol{a}_B = \boldsymbol{a}_A + \dot{\boldsymbol{\omega}}_1 \times \boldsymbol{r}_{AB} + \boldsymbol{\omega}_1 \times (\boldsymbol{\omega}_1 \times \boldsymbol{r}_{AB})$$
$$\boldsymbol{a}_C = \boldsymbol{a}_D + \dot{\boldsymbol{\omega}}_3 \times \boldsymbol{r}_{DC} + \boldsymbol{\omega}_3 \times (\boldsymbol{\omega}_3 \times \boldsymbol{r}_{DC})$$
$$\boldsymbol{a}_C = \boldsymbol{a}_B + \dot{\boldsymbol{\omega}}_2 \times \boldsymbol{r}_{BC} + \boldsymbol{\omega}_2 \times (\boldsymbol{\omega}_2 \times \boldsymbol{r}_{BC}).$$

Substituting from the first and second equations for $\boldsymbol{a}_B$ and $\boldsymbol{a}_C$ into the last equation gives

$$\boldsymbol{a}_D + \dot{\boldsymbol{\omega}}_3 \times \boldsymbol{r}_{DC} + \boldsymbol{\omega}_3 \times (\boldsymbol{\omega}_3 \times \boldsymbol{r}_{BC})$$
$$= \boldsymbol{a}_A + \dot{\boldsymbol{\omega}}_1 \times \boldsymbol{r}_{AB} + \boldsymbol{\omega}_1 \times (\boldsymbol{\omega}_1 \times \boldsymbol{r}_{AB})$$
$$+ \dot{\boldsymbol{\omega}}_2 \times \boldsymbol{r}_{BC} + \boldsymbol{\omega}_2 \times (\boldsymbol{\omega}_2 \times \boldsymbol{r}_{AC}).$$

Noting that $\boldsymbol{a}_D$ and $\boldsymbol{a}_A$ are zero, and substituting $\dot{\boldsymbol{\omega}}_1 = \boldsymbol{K}\ddot{\alpha}$, $\dot{\boldsymbol{\omega}}_2 = \boldsymbol{K}\ddot{\beta}$, and $\dot{\boldsymbol{\omega}}_3 = \boldsymbol{K}\ddot{\gamma}$ gives

$$-\boldsymbol{K}\ddot{\gamma} \times l_3(-\boldsymbol{I}\cos\gamma + \boldsymbol{J}\sin\gamma)$$
$$- \boldsymbol{K}\dot{\gamma} \times [-\boldsymbol{K}\dot{\gamma} \times l_3(-\boldsymbol{I}\cos\gamma + \boldsymbol{J}\sin\gamma)]$$
$$= \boldsymbol{K}\ddot{\alpha} \times l_1(\boldsymbol{I}\cos\alpha + \boldsymbol{J}\sin\alpha)$$
$$+ \boldsymbol{K}\dot{\alpha} \times [\boldsymbol{K}\dot{\alpha} \times l_1(\boldsymbol{I}\cos\alpha + \boldsymbol{J}\sin\alpha)]$$
$$+ \boldsymbol{K}\ddot{\beta} \times l_2(\boldsymbol{I}\cos\beta + \boldsymbol{J}\sin\beta)$$
$$+ \boldsymbol{K}\dot{\beta} \times [\boldsymbol{K}\dot{\beta} \times l_2(\boldsymbol{I}\cos\beta + \boldsymbol{J}\sin\beta)].$$

Carrying out the cross products and gathering terms gives the component-equation results

$$\boldsymbol{I}: l_3\ddot{\gamma}\sin\gamma + l_3\dot{\gamma}^2\cos\gamma$$
$$= -l_1\ddot{\alpha}\sin\alpha - l_1\dot{\alpha}^2\cos\beta - l_2\ddot{\beta}\sin\beta - l_2\dot{\beta}^2\cos\beta$$
$$\boldsymbol{J}: l_3\ddot{\gamma}\cos\gamma - l_3\dot{\gamma}^2\sin\gamma$$
$$= l_1\ddot{\alpha}\cos\alpha - l_1\dot{\alpha}^2\sin\alpha + l_2\ddot{\beta}\cos\beta - l_2\dot{\beta}^2\sin\beta.$$

Comparison to Equation 4.25 confirms that the same results have been obtained for the acceleration component equations with vectors as were obtained earlier from the geometric equations.

In comparing the geometric and vector approaches, you have a choice for deriving kinematic equations for velocity and acceleration components; specifically, you can use either the geometric component equations and differentiate, or you can apply the vector algebraic equations. However, if general governing equations are required, the geometric relationships of Equation 4.22 must be developed.

The four-bar mechanism in Figure 4.19 was analyzed alternately with geometry and vectors. The discussion of the two alternatives obscures the most important step in kinematics analysis, namely, drawing the mechanism in a general position and selecting coordinates to define its orientation. Example Problem 4.5 is included to emphasize this important step.

Example Problem 4.5

Figure XP4.5a illustrates an oil pumping rig that is typically used for shallow oil wells. An electric motor drives the rotating arm *OA* at a constant, clockwise angular velocity $\omega = 20$ rpm. A cable attaches the pumping rod at *D* to the end of the rocking arm *BE*. Rotation of the driving link produces a vertical oscillation that drives a positive-displacement pump at the bottom of the well.

Tasks:

a. Draw the rig in a general position and select coordinates to define the bars' general position. State the kinematic constraint equations defining the angular positions of bars *AB* and *BCE* in terms of bar *OA*'s angular position.

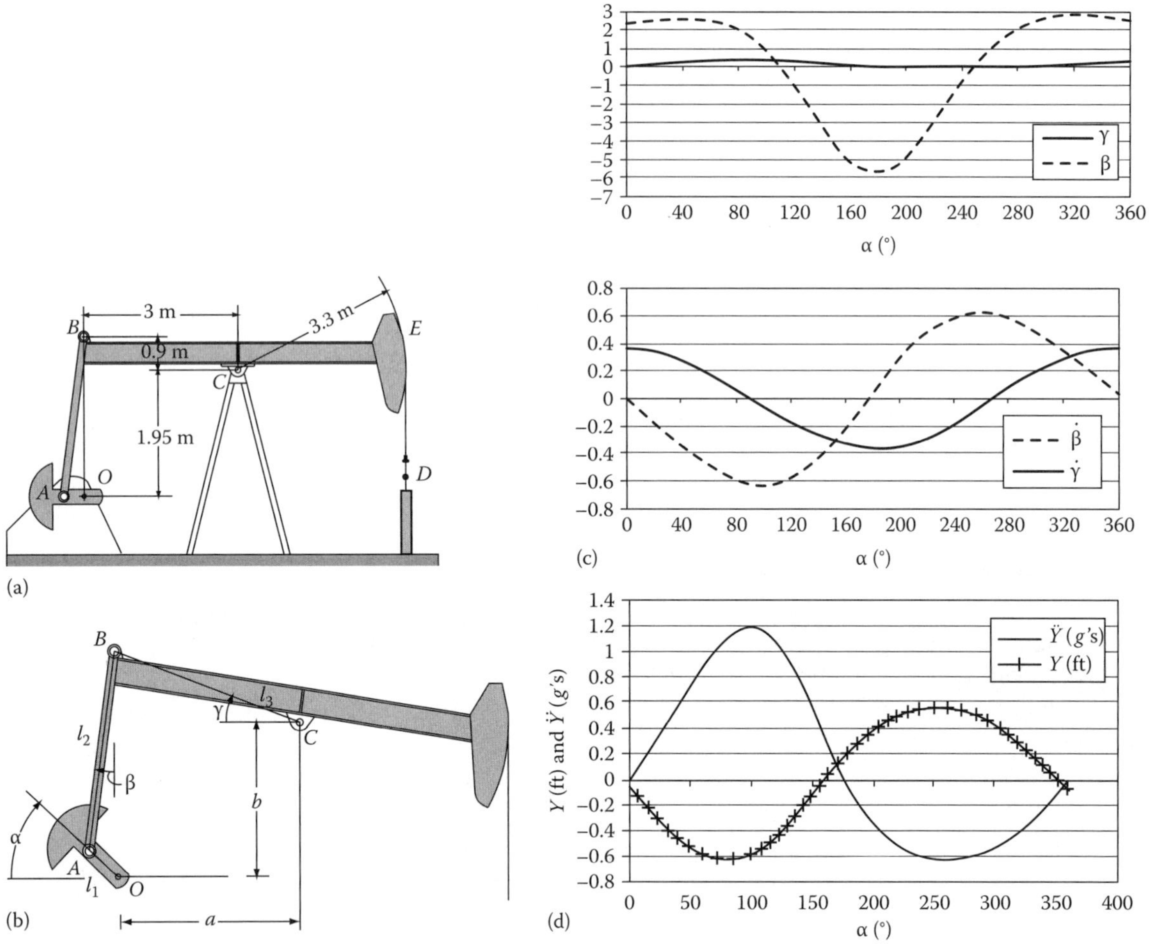

FIGURE XP4.5
(a) Oil well pumping rig. (Adapted from Meriam, J. and Kraige, L., *Engineering Mechanics, Dynamics*, vol. 2, 3rd edn., Wiley, New York, 1992.) (b) Pumping rig in a general position with coordinates. (c) γ, β, and $\dot{\gamma}$, $\dot{\beta}$ versus α. (d) Vertical acceleration and change in position of the pumping rod.

b. Outline a solution procedure to determine the orientations of bars *AB* and *BCE* in terms of bar *OA*'s angular position.
c. Derive general expressions for the angular velocities of bars *AB* and *BCE* in terms of bar *OA*'s angular position and angular velocity. Solve for the unknown angular velocities.
d. Derive general expressions for the angular accelerations of bars *AB* and *BCE* in terms of bar *OA*'s angular position, velocity, and acceleration. Solve for the unknown angular accelerations.
e. Derive general expressions for the change in vertical position and vertical acceleration of point *D* as a function of bar *OA*'s angular position.

Solution

The sketch of Figure XP4.5b starts Task a and will be used to state the governing relationships between the bars' rotation angles. This figure shows the angles α, β, γ defining the angular positions of lines *OA*, *AB*, and *BC*, respectively. The angle α is the (known) input variable, while β and γ are the (unknown) output variables. The length l_3 extends from *B* to *C*. Stating the components of the bars in the *X*- and *Y*-directions gives

$$\begin{aligned} &-l_1\cos\alpha + l_2\sin\beta + l_3\cos\gamma = a \\ &\quad\Rightarrow l_2\sin\beta + l_3\cos\gamma = a + l_1\cos a = h_1(\alpha) \\ &l_1\sin\alpha + l_2\cos\beta - l_3\sin\gamma - = b \\ &\quad\Rightarrow l_2\cos\beta - l_3\sin\gamma = b - l_1\sin\alpha = h_2(\alpha) \end{aligned} \qquad \text{(i)}$$

and concludes Task a. As a first step in solving for (β, γ), we state the equations as

$$l_2\sin\beta = h_1 - l_3\cos\gamma, \quad l_2\cos\beta = h_2 + l_3\sin\gamma$$

Squaring these equations and adding them together gives

$$\begin{aligned} l_2^2(\sin^2\beta + \cos^2\beta) = l_2^2 &= h_1^2 - 2h_1l_3\cos\gamma + l_3^2\cos^2\gamma \\ &\quad + h_2^2 + 2h_2l_3\sin\gamma + l_3^2\sin^2\gamma \\ \therefore\ 2h_1l_3\cos\gamma &= h_1^2 + h_2^2 + l_3^2 - l_2^2 + 2h_2l_3\sin\gamma \\ &= 2d + 2h_2l_3\sin\gamma. \end{aligned}$$

where $2d = h_1^2 + h_2^2 + l_3^2 - l_2^2$. Substituting $\cos\gamma = \sqrt{1-\sin^2\gamma}$ leads to

$$h_1^2 l_3^2(1-\sin^2\gamma) = d^2 + 2dh_2 l_3 \sin\gamma + h_2^2 l_3^2 \sin^2\gamma$$

$$\therefore \sin^2\gamma + \frac{2dh_2}{l_3(h_1^2+h_2^2)}\sin\gamma + \frac{d^2 - h_1^2 l_3^2}{l_3^2(h_1^2+h_2^2)} = 0. \quad \text{(ii)}$$

For a specified value of α, solving this quadratic equation gives $\sin\gamma \Rightarrow \gamma = \sin^{-1}\gamma$, and back substitution into Equation (i) nets β. These steps concludes Task b, and Figure XP4.5c illustrates the results for the lengths of Figure XP4.5a.

Proceeding to Task c, we can differentiate Equation (i) with respect to time to obtain

$$\begin{aligned} l_2\cos\beta\dot\beta - l_3\sin\gamma\dot\gamma &= -l_1\sin\alpha\dot\alpha = -l_1\omega\sin\alpha \\ -l_2\sin\beta\dot\beta - l_3\cos\gamma\dot\gamma &= -l_1\cos\alpha\dot\alpha = -l_1\omega\cos\alpha. \end{aligned} \quad \text{(iii)}$$

In matrix format, these equations become

$$\begin{bmatrix} \cos\beta & -\sin\gamma \\ \sin\beta & \cos\gamma \end{bmatrix} \begin{Bmatrix} l_2\dot\beta \\ l_3\dot\gamma \end{Bmatrix} = l_1\omega \begin{Bmatrix} -\sin\alpha \\ \cos\alpha \end{Bmatrix}.$$

Using Cramer's rule (Appendix A), their solution can be stated:

$$\begin{aligned} l_2\dot\beta &= \frac{l_1\omega}{\cos\gamma\sin\beta + \sin\gamma\cos\beta}\begin{vmatrix} -\sin\alpha & -\sin\gamma \\ \cos\alpha & \cos\gamma \end{vmatrix} \\ &= \frac{l_1\omega\sin(\gamma-\alpha)}{\cos(\beta-\gamma)} \\ l_3\dot\gamma &= \frac{l_1\omega}{\cos\gamma\sin\beta + \sin\gamma\cos\beta}\begin{vmatrix} \cos\beta & -\sin\alpha \\ \sin\beta & \cos\alpha \end{vmatrix} \\ &= \frac{l_1\omega\cos(\alpha-\beta)}{\sin(\beta-\gamma)}, \end{aligned} \quad \text{(iv)}$$

concluding Task c. Figure XP 4.5c illustrates $\dot\gamma$, $\dot\beta$ versus α

Moving to Task d, we can differentiate Equation (iii) with respect to time to obtain:

$$\begin{aligned} l_2\cos\beta\ddot\beta - l_3\sin\gamma\ddot\gamma &= -l_1\omega^2\cos\alpha + l_2\sin\beta\dot\beta^2 \\ &\quad + l_3\cos\gamma\dot\gamma^2 = g_1 \\ -l_2\sin\beta\ddot\beta - l_3\cos\gamma\ddot\gamma &= l_1\omega_2\sin\alpha + l_2\cos\beta\dot\beta^2 \\ &\quad - l_3\sin\gamma\dot\gamma^2 = -g_2. \end{aligned}$$

In matrix format, these equations become

$$\begin{bmatrix} \cos\beta & -\sin\gamma \\ \sin\beta & \cos\gamma \end{bmatrix} \begin{Bmatrix} l_2\ddot\beta \\ l_2\ddot\gamma \end{Bmatrix} = \begin{Bmatrix} g_1 \\ g_2 \end{Bmatrix}.$$

The solution can be stated

$$\begin{aligned} l_2\ddot\beta &= \frac{1}{\cos\gamma\sin\beta + \sin\gamma\cos\beta}\begin{vmatrix} g_1 & -\sin\gamma \\ g_2 & \cos\gamma \end{vmatrix} \\ &= \frac{g_1\cos\gamma + g_2\sin\gamma}{\cos(\beta-\gamma)} \\ l_3\ddot\gamma &= \frac{1}{\cos\gamma\sin\beta + \sin\gamma\cos\beta}\begin{vmatrix} \cos\beta & g_1 \\ \sin\beta & g_2 \end{vmatrix} \\ &= \frac{g_2\cos\beta - g_1\sin\beta}{\cos(\beta-\gamma)}, \end{aligned}$$

concluding Task d.

In regard to Task e, as long as the circular arc at the end of the rocker arm is long enough, the tangent point of the cable with the circular-faced end of the rocker arm will be at a horizontal line running through *C*. Hence, the change in the horizontal position of point *D* is the amount of cable rolled off the arc due to a change in the rocker arm γ, i.e., $\delta\gamma = -3.3\delta\gamma$. Similarly, the vertical acceleration of the sucker rod at *D* is the circumferential acceleration of a point on the arc i.e., $a_\theta = r\ddot\theta \Rightarrow \ddot Y = 3.3\ddot\gamma$. Figure XP4.5d illustrates $\delta\gamma$, $\ddot Y$ as a function of alpha. The distance traveled by the pump rod in one cycle is $0.633-(-0.633) = 1.27$ m. The peak positive acceleration is $1.17g$, and the minimum is $-0.63g$. Note that $|-0.63g| < lg$, indicating from a rigid-body viewpoint that the cable will remain in tension during its downward motion.

4.5.4 Another Slider-Crank Mechanism

4.5.4.1 Geometric Approach

Figure 4.24 illustrates an alternative slider-crank mechanism. The crank of link *AB* is driven by varying θ, causing point *B* to follow a circular path around *A*. As *B* follows this circular path, rod *BD* is pulled back and forth through the pivoted bushing at *C*. The distance from point *C* to *B* is *S*. The input coordinate is θ; the output coordinates are *S* and ϕ.

The following engineering-analysis task applies for this mechanism: For $\dot\theta = \omega =$ constant, determine ϕ and *S* and their first and second derivatives for one cycle of θ.

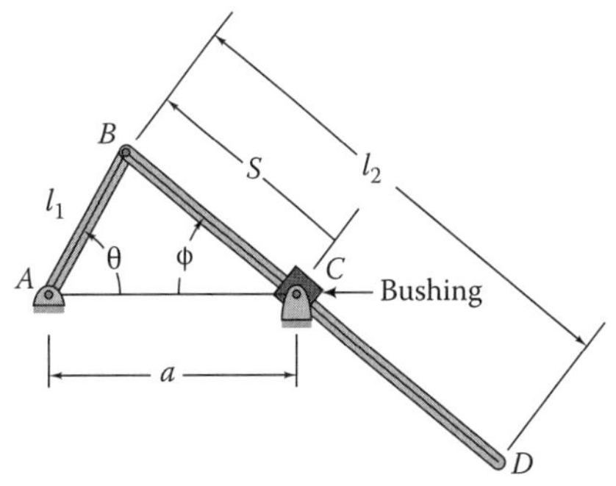

FIGURE 4.24
Alternative slider-crank mechanism.

An inspection of Figure 4.24 yields the following coordinate results:

$$X: l_1 \cos\theta + S\cos\phi = a$$
$$Y: l_1 \sin\theta = S\sin\phi.$$

Reordering these equations to

$$\begin{aligned} S\cos\phi &= a - l_1\cos\theta \\ S\sin\phi &= l_1\sin\theta, \end{aligned} \tag{4.30}$$

emphasizes that θ is the input coordinate, with ϕ and S the output coordinates. These equations are nonlinear but can be readily solved for ϕ and S in terms of θ, via

$$\begin{aligned} S^2(\cos^2\phi + \sin^2\phi) &= (a - l_1\cos\theta)^2 + (l_1\sin\theta)^2 \\ &\Rightarrow S^2 = a^2 - 2al_1\cos\theta + l_1^2, \end{aligned} \tag{4.31}$$

and

$$\tan\phi = \frac{\sin\phi}{\cos\phi} = \frac{l_1\sin\theta}{a - l_1\cos\theta}. \tag{4.32}$$

The velocity relations are obtained by differentiating Equation 4.30:

$$\begin{aligned} \dot{S}\cos\phi - S\sin\phi\dot{\phi} &= l_1\sin\theta\dot{\theta} = l_1\omega\sin\theta \\ \dot{S}\sin\phi + S\cos\phi\dot{\phi} &= l_1\cos\theta\dot{\theta} = l_1\omega\cos\theta. \end{aligned} \tag{4.33}$$

The matrix statement of these equations is

$$\begin{bmatrix} \cos\phi & -\sin\phi \\ \sin\phi & \cos\phi \end{bmatrix} \begin{Bmatrix} \dot{S} \\ S\dot{\phi} \end{Bmatrix} = l_1\omega \begin{Bmatrix} \sin\theta \\ \cos\theta \end{Bmatrix}.$$

This coefficient matrix is orthogonal, i.e., $[A]^{-1} = [A]^T$. Hence, the unknowns can be determined by premultiplying the equation by $[A]^T$. They can also be solved using Cramer's rule to obtain

$$\dot{S} = l_1\omega\sin(\theta + \phi), \quad S\dot{\phi} = l_1\omega\cos(\theta + \phi). \tag{4.34}$$

The acceleration relationships are obtained by differentiating Equation 4.34 with respect to time, getting

$$\begin{aligned} \ddot{S} &= l_1\omega\cos(\theta + \phi)(\dot{\theta} + \dot{\phi}) \\ &= l_1\omega\cos(\theta + \phi)(\omega + \dot{\phi}) \\ S\ddot{\phi} &= -\dot{S}\dot{\phi} - l_1\omega\sin(\theta + \phi)(\dot{\theta} + \dot{\phi}) \\ &= -\dot{S}\dot{\phi} - l_1\omega\sin(\theta + \phi)(\omega + \dot{\phi}). \end{aligned} \tag{4.35}$$

These are the last required set of kinematic equations, and the engineering-analysis task outlined above would be accomplished by executing the following sequential steps:

1. Vary θ over the range $[0, 2\pi]$, yielding discrete values θ_i.
2. For each θ_i value, evaluate Equations 4.31 and 4.32 to determine corresponding values for ϕ_i and S_i as

$$\begin{aligned} S_i &= \left(a^2 - 2al_1\cos\theta_i + l_1^2\right)^{1/2}, \\ \phi_i &= \tan^{-1}\left[\frac{l_1\sin\theta_i}{a - l_1\cos\theta_i}\right]. \end{aligned} \tag{i}$$

3. Enter Equation 4.34 with known values for θ_i, ϕ_i, and S_i to compute $\dot{\phi}_i$ and $\dot{S}_i$ from

$$\begin{aligned} \dot{S}_i &= l_i\omega\sin(\theta_i + \phi_i), \\ \dot{\phi}_i &= [l_1\omega\cos(\theta_i + \phi_i)]/S_i. \end{aligned} \tag{ii}$$

4. Enter Equation 4.35 with known values for θ_i, ϕ_i, $\dot{\theta}_i$, $\dot{\phi}_i$, and $\dot{S}_i$, to determine $\ddot{\phi}_i$ and $\ddot{S}_i$

$$\begin{aligned} \ddot{S}_i &= l_1\omega(\omega + \dot{\phi}_i)\cos(\theta_i + \phi_i) \\ \ddot{\phi}_i &= [-\dot{S}_i\dot{\phi}_i - l_1\omega(\omega + \dot{\phi}_i)\sin(\theta_i + \phi_i)]/S_i. \end{aligned} \tag{iii}$$

These equations can be readily evaluated using a spreadsheet analysis. Figure 4.25 illustrates solutions for $\ddot{S}$, $\ddot{\phi}$ with $l_1, = 250$ mm, $a = 600$ mm, $\omega = 1$ Hz $= 6.28$ rad/s.

For the mechanism of Figure 4.24, θ is the independent (input) coordinate with ϕ and S as dependent (output) coordinates. Figure 4.26 illustrates a geometrically

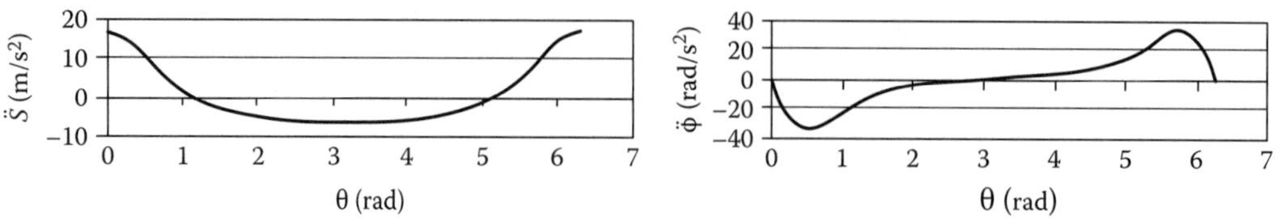

FIGURE 4.25
$\ddot{S}$ and $\ddot{\phi}$ versus θ for $l_1, = 250$ mm, $a = 600$ mm, $\omega = 1$ Hz $= 6.28$ rad/s.

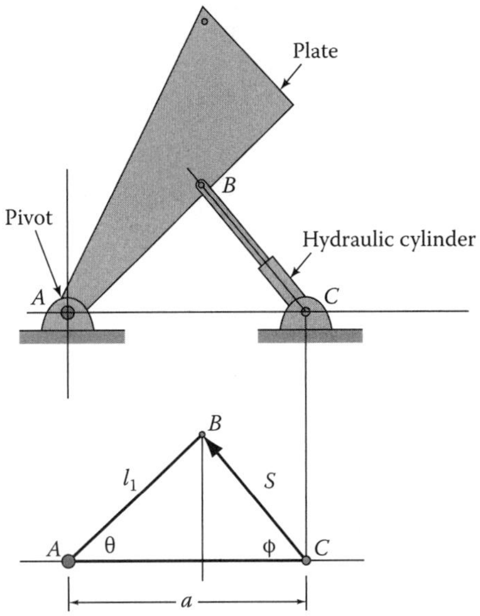

FIGURE 4.26
Alternative version of the mechanism in Figure 4.24 with S as the input.

similar mechanism for which a hydraulic cylinder is used to control $S(t)$, making S the input coordinate, and θ and ϕ the dependent output coordinates. Equation 4.22 continue to be valid; however, they would now be used to solve for θ_i and ϕ_i for a specified value of S_i. Equation 4.33 can be reordered as

$$S\dot{\phi}\sin\phi + l_1\dot{\theta}\sin\theta = \dot{S}\cos\phi$$
$$S\dot{\phi}\cos\phi - l_1\dot{\theta}\cos\theta = -\dot{S}\sin\phi,$$

to define $\dot{\phi}$ and $\dot{\theta}$. Similarly, rearranging Equation 4.35 provides the following defining equations for $\ddot{\phi}$ and $\ddot{\theta}$:

$$-S\sin\phi\ddot{\phi} - l_1\sin\theta\ddot{\theta} = -\ddot{S}\cos\phi + S\cos\phi\dot{\phi}^2 + 2\dot{S}\dot{\phi}\sin\phi + l_1\cos\theta\dot{\theta}^2$$
$$S\cos\phi\ddot{\phi} - l_1\cos\theta\ddot{\theta} = -\ddot{S}\sin\phi + S\sin\phi\dot{\phi}^2 - 2\dot{S}\dot{\phi}\cos\phi - l_1\sin\theta\dot{\theta}^2.$$

The basic geometry of Figures 4.24 and 4.26 tends to show up regularly in planar mechanisms.

Returning to Figure 4.24, the new aspect of the current mechanism, versus our earlier examples, is that the length of side BC is variable. This characteristic is of minor concern in writing the scalar geometric equations, but prevents us from using Equation 4.3 in a vector development to relate the velocities and accelerations of points B and C. Equation 4.3 apply for two points in a rigid body, and points B and C clearly do not meet this requirement.

4.5.4.2 Vector, Two-Coordinate-System Approach for Velocity and Acceleration Relationships

The two-coordinate, velocity, and acceleration relationships provided by Equation 4.1 are the basis for the second approach for analyzing the mechanism of Figure 4.24. You may wish to review Sections 2.8 and 4.3 before proceeding further. As illustrated in Figure 4.27, the two-coordinate-system model of Figure 4.5 has been "mapped" into the mechanism of Figure 4.27 by attaching the X, Y coordinate system to ground, and aligning the second "moving" x, y system with rod BD. The origin of the x, y system is at the pivot-point C, and the vector $\boldsymbol{\rho} = jS$ extends from C to point B.

The basic velocity and acceleration equations for the two coordinate systems are

$$\begin{aligned} \dot{r} &= \dot{R}_o + \hat{\dot{\boldsymbol{\rho}}} + \boldsymbol{\omega}\times\boldsymbol{\rho} \\ \ddot{r} &= \ddot{R}_o + \hat{\ddot{\boldsymbol{\rho}}} + 2\boldsymbol{\omega}\times\hat{\dot{\boldsymbol{\rho}}} + \dot{\boldsymbol{\omega}}\times\boldsymbol{\rho} + \boldsymbol{\omega}\times(\boldsymbol{\omega}\times\boldsymbol{\rho}) \end{aligned} \quad (4.1)$$

We want to analyze the linkage shown in Figure 4.27 by defining the individual entries in Equation 4.1 in terms of the figure's variables. Starting with easy terms, the vector $\boldsymbol{R}_o$ locates the origin of the x, y system (point C) in the X, Y system. Since C is fixed in the X, Y system, we can immediately state $\dot{\boldsymbol{R}}_o = \ddot{\boldsymbol{R}}_o = 0$. The vector $\boldsymbol{\omega}$ is defined as the angular velocity of the system relative to the X, Y system. From Figure 4.27, using the right-hand-screw convention,

$$\boldsymbol{\omega} = -\boldsymbol{K}\dot{\phi} = -\boldsymbol{k}\dot{\phi}.$$

Given that $\dot{\boldsymbol{\omega}} = \left.\dfrac{d\boldsymbol{\omega}}{dt}\right|_{X,Y}$, we obtain by direct differentiation

$$\dot{\boldsymbol{\omega}} = -\boldsymbol{K}\ddot{\phi} = -\boldsymbol{k}\ddot{\phi}.$$

From Figure 4.27,

$$\boldsymbol{\rho} = jS.$$

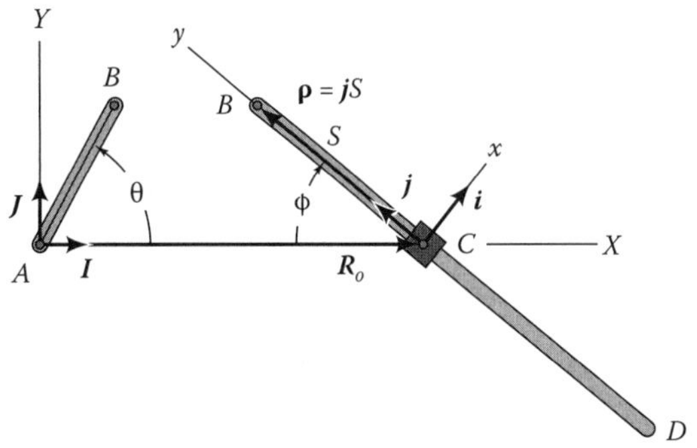

FIGURE 4.27
Two-coordinate arrangement for rod BD of the slider-crank mechanism in Figure 4.24.

Differentiating this vector holding $\boldsymbol{j}$ constant gives

$$\hat{\dot{\boldsymbol{\rho}}} = \boldsymbol{j}\dot{S}.$$

Differentiating again gives

$$\hat{\ddot{\boldsymbol{\rho}}} = \boldsymbol{j}\ddot{S}.$$

Substituting these results into the definitions provided by Equation 4.1 gives

$$\begin{aligned}\dot{\boldsymbol{r}}_B &= \boldsymbol{v}_B = 0 + \boldsymbol{j}\dot{S} - \boldsymbol{k}\dot{\phi} \times \boldsymbol{j}S \\ \ddot{\boldsymbol{r}}_B &= \boldsymbol{a}_B = 0 + \boldsymbol{j}\ddot{S} + 2(-\boldsymbol{k}\dot{\phi}) \times \boldsymbol{j}\dot{S} - \boldsymbol{k}\ddot{\phi} \\ &\quad \times \boldsymbol{j}S - \boldsymbol{k}\dot{\phi} \times (-\boldsymbol{k}\dot{\phi} \times \boldsymbol{j}S)\end{aligned} \qquad (4.30')$$

Carrying through the cross products and completing the algebra nets

$$\begin{aligned}\boldsymbol{v}_B &= \boldsymbol{i}S\dot{\phi} + \boldsymbol{j}\dot{S} \\ \boldsymbol{a}_B &= \boldsymbol{i}(S\ddot{\phi} + 2\dot{S}\dot{\phi}) + \boldsymbol{j}(\ddot{S} - S\dot{\phi}^2)\end{aligned} \qquad (4.31')$$

Returning to Figure 4.27, we can apply Equation 4.3 to state the velocities and accelerations of points A and B as

$$\begin{aligned}\boldsymbol{v}_B &= \boldsymbol{v}_A + \boldsymbol{\omega}_1 \times \boldsymbol{r}_{AB} \\ \boldsymbol{a}_B &= \boldsymbol{a}_A + \dot{\boldsymbol{\omega}}_1 \times \boldsymbol{r}_{AB} + \boldsymbol{\omega}_1 \times (\boldsymbol{\omega}_1 \times \boldsymbol{r}_{AB}).\end{aligned}$$

We can use Equation 4.3 for points A and B of link 1 because they are *two points fixed in a rigid body*. We were forced to use the general expression, Equation 4.1 to relate the velocities of points C and D because *they are not fixed in a rigid body*. Since A is fixed, $\boldsymbol{v}_A = \boldsymbol{a}_A = 0$. From the right-hand-rule convention, $\boldsymbol{\omega}_1 = \boldsymbol{K}\dot{\theta}$, $\dot{\boldsymbol{\omega}}_1 = \boldsymbol{K}\ddot{\theta}$. From Figure 4.27, $\boldsymbol{r}_{AB} = l_1(\boldsymbol{I}\cos\theta + \boldsymbol{J}\sin\theta)$. Substituting these results, we obtain

$$\begin{aligned}\boldsymbol{v}_B &= 0 + \boldsymbol{K}\dot{\theta} \times l_1(\boldsymbol{I}\cos\theta + \boldsymbol{J}\sin\theta) \\ \boldsymbol{a}_B &= 0 + \boldsymbol{K}\ddot{\theta}_1 \times l_1(\boldsymbol{I}\cos\theta + \boldsymbol{J}\sin\theta) + \boldsymbol{K}\dot{\theta} \\ &\quad \times [\boldsymbol{K}\dot{\theta} \times l_1(\boldsymbol{I}\cos\theta + \boldsymbol{J}\sin\theta)].\end{aligned}$$

Carrying out the cross products and algebra gives

$$\begin{aligned}\boldsymbol{v}_B &= l_1\dot{\theta}(-\boldsymbol{I}\sin\theta + \boldsymbol{J}\cos\theta) \\ \boldsymbol{a}_B &= l_1\ddot{\theta}(-\boldsymbol{I}\sin\theta + \boldsymbol{J}\cos\theta) - l_1\dot{\theta}^2(\boldsymbol{I}\cos\theta + \boldsymbol{J}\sin\theta)\end{aligned} \qquad (4.32')$$

Equations 4.31 and 4.32 provide alternate statements for $\boldsymbol{v}_B$ and $\boldsymbol{a}_B$, and our effort might reasonably be concluded by equating these results. However, the results in

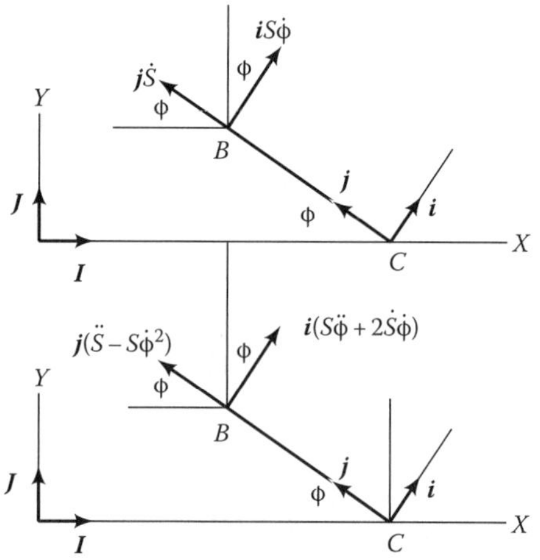

FIGURE 4.28
Velocity and acceleration definitions for the velocity of point B from a two-coordinate development.

Equation 4.32 are given in terms of $\boldsymbol{I}$ and $\boldsymbol{J}$ unit vectors, versus $\boldsymbol{i}$ and $\boldsymbol{j}$ for Equation 4.31. In other words, the vectors of Equation 4.31 are defined in terms of their components in the x, y coordinate system, while the components of Equation 4.32 are stated in the X, Y system.

We need the components of the vectors in the same system, and Figure 4.28 can be used to transform the x, y components of $\boldsymbol{v}_B$ and $\boldsymbol{a}_B$ in Equation 4.31 to components in the X, Y system. From Figure 4.28, we obtain

$$\begin{aligned}\boldsymbol{v}_B &= S\dot{\phi}(\boldsymbol{I}\sin\dot{\phi} + \boldsymbol{J}\cos\phi) + \dot{S}(-\boldsymbol{I}\cos\phi + \boldsymbol{J}\sin\phi) \\ \boldsymbol{a}_B &= (S\ddot{\phi} - S\dot{\phi}^2)(\boldsymbol{I}\sin\phi + \boldsymbol{J}\cos\phi) \\ &\quad + (\ddot{S} + 2\dot{S}\dot{\phi})(-\boldsymbol{I}\cos\phi + \boldsymbol{J}\sin\phi)\end{aligned} \qquad (4.33')$$

Equating this definition for $\boldsymbol{v}_B$ with the result from the first of Equation 4.32 gives

$$\begin{aligned}\boldsymbol{I}&: -l_1\dot{\theta}\sin\theta = -\dot{S}\cos\phi + S\dot{\phi}\sin\phi \\ \boldsymbol{J}&: l_1\dot{\theta}\cos\theta = \dot{S}\sin\phi + S\dot{\phi}\cos\phi,\end{aligned}$$

which repeats the earlier results from geometric approaches. Equating the definition for $\boldsymbol{a}_B$. provided by the second of Equation 4.33 with the second of Equation 4.32 gives

$$\begin{aligned}\boldsymbol{I}&: -l_1\ddot{\theta}\sin\theta - l_1\dot{\theta}^2\cos\theta \\ &= -(\ddot{S} - S\dot{\phi}^2)\cos\phi + (S\ddot{\phi} + 2\dot{S}\dot{\phi})\sin\phi \\ \boldsymbol{J}&: l_1\ddot{\theta}\cos\theta - l_1\dot{\theta}^2\sin\theta \\ &= (\ddot{S} - S\dot{\phi}^2)\sin\phi + (S\ddot{\phi} + 2\dot{S}\dot{\phi})\cos\phi,\end{aligned}$$

which repeats our earlier results.

As you can see, we have worked the same problem with two different approaches. You will have to reach your own conclusion concerning the question: What is the best way to work the problem? Given that geometric relationships are required in terms of the coordinates for a general solution, the geometric approach is the quickest (if you can differentiate). Also, the geometric approach eliminates the task of coordinate transformations that arise using the two-coordinate-system formulations. While disadvantaged in planar kinematics, the two-coordinate approach works better in 3D kinematics than the geometric approach.

4.5.4.3 *Solution for the Velocity and Acceleration of Point D*

Suppose that we are asked to determine general expressions for the velocity and acceleration for point D at the end of rod BD in Figure 4.24. From the developments above, we would start this task having already obtained a solution for ϕ, $\dot{\phi}$, and $\ddot{\phi}$ as a function of θ. We could follow the geometric approach and write out the X and Y coordinates for point D in the X, Y coordinate system, and differentiate once with respect to time to get the velocity components, and then differentiate again to get the acceleration components. However, a simpler approach (given that we know ϕ, $\dot{\phi}$, and $\ddot{\phi}$) is the direct vector formulation. Applying Equation 4.3 to points A and B gives

$$\boldsymbol{v}_B = \boldsymbol{v}_A + \boldsymbol{\omega}_1 \times \boldsymbol{r}_{AB}$$
$$\boldsymbol{a}_B = \boldsymbol{a}_A + \dot{\boldsymbol{\omega}}_1 \times \boldsymbol{r}_{AB} + \boldsymbol{\omega}_1 \times (\boldsymbol{\omega}_1 \times \boldsymbol{r}_{AB}).$$

We have already worked through these equations, obtaining solutions for $\boldsymbol{v}_B$ and $\boldsymbol{a}_B$ in Equation 4.30. We can also apply Equation 4.3 to points B and D, since they are points on a rigid body (unlike point C), obtaining

$$\boldsymbol{v}_D = \boldsymbol{v}_B + \boldsymbol{\omega}_2 \times \boldsymbol{r}_{BD}$$
$$\boldsymbol{a}_D = \boldsymbol{a}_B + \dot{\boldsymbol{\omega}}_2 \times \boldsymbol{r}_{BD} + \boldsymbol{\omega}_2 \times (\boldsymbol{\omega}_2 \times \boldsymbol{r}_{BD}).$$

Substituting from Equation 4.30 for $\boldsymbol{v}_B$ and $\boldsymbol{a}_B$ plus $\boldsymbol{\omega}_2 = -\boldsymbol{K}\dot{\phi}$, $\dot{\boldsymbol{\omega}}_2 = -\boldsymbol{K}\ddot{\phi}$, and $\boldsymbol{r}_{BD} = l_2(\boldsymbol{I}\cos\phi - \boldsymbol{J}\sin\phi)$ gives

$$\boldsymbol{v}_D = l_1\dot{\theta}(-\boldsymbol{I}\sin\theta + \boldsymbol{J}\cos\theta) - \boldsymbol{K}\dot{\phi} \times l_2(\boldsymbol{I}\cos\phi - \boldsymbol{J}\sin\phi)$$
$$\boldsymbol{a}_D = l_1\ddot{\theta}(-\boldsymbol{I}\sin\theta + \boldsymbol{J}\cos\theta) - l_1\dot{\theta}^2(\boldsymbol{I}\cos\theta + \boldsymbol{J}\sin\theta)$$
$$- \boldsymbol{K}\ddot{\phi} \times l_2(\boldsymbol{I}\cos\phi - \boldsymbol{J}\sin\phi)$$
$$- \boldsymbol{K}\dot{\phi} \times [-\boldsymbol{K}\dot{\phi} \times l_2(\boldsymbol{I}\cos\phi - \boldsymbol{J}\sin\phi)].$$

Carrying out the cross products and gathering terms yields

$$\boldsymbol{v}_D = -\boldsymbol{I}(l_1\dot{\theta}\sin\theta + l_2\dot{\phi}\sin\phi) + \boldsymbol{J}(l_1\dot{\theta}\cos\theta - l_2\dot{\phi}\cos\phi)$$
$$\boldsymbol{a}_D = \boldsymbol{I}(-l_1\ddot{\theta}\sin\theta - l_1\dot{\theta}^2\cos\theta - l_2\ddot{\phi}\sin\phi - l_2\dot{\phi}^2\cos\phi)$$
$$+ \boldsymbol{J}(l_1\ddot{\theta}\cos\theta - l_1\dot{\theta}^2\sin\theta - l_2\ddot{\phi}\cos\phi + l_2\dot{\phi}^2\sin\phi).$$

These are general equations for $\boldsymbol{v}_D$ and $\boldsymbol{a}_D$. Substituting $\dot{\theta} = \omega$ and $\ddot{\theta} = 0$ completes the present effort, with ϕ, $\dot{\phi}$, and $\ddot{\phi}$ defined by Equations 4.32, 4.34, and 4.35, respectively.

Example Problem 4.6

Figure XP4.6a provides a top view of a power-gate actuator. An electric motor drives a lead screw mounted in the arm connecting points C and D. Lengthening arm CD closes the gate; shortening it opens the gate. During closing action, arm CD extends from a length of 3.3 to 4.3 ft in about 17 s to proceed from a fully open to fully closed positions. The gate reaches its steady extension rate quickly at the outset and decelerates rapidly when the gate nears the closed position.

Tasks:

a. Draw the gate actuator in a general position and derive the governing equations that define the orientations of bars CD and BC as a function of the length of arm CD.
b. Assume that arm CD extends at a constant rate (gate closing) and determine a relationship for the angular velocities of arms CD and BE.
c. Continuing to assume that bar CD extends at a constant rate, determine a relationship for the angular accelerations of arms CD and BE.

Solution

Figure XP4.6b illustrates a general position for the gate and satisfies Task a. For Task b, Figure XP4.6b provides the following horizontal and vertical component relations:

$$\text{horizontal}: S\sin\theta = 0.416 + 3.8886\sin\phi;$$
$$3.886 = \sqrt{3.875^2 + 0.3^2} \qquad \text{(i)}$$
$$\text{vertical}: S\cos\theta = 3.886\cos\phi - 0.546$$

For this example, S is the input and θ, ϕ are the unknown output variables. The easy way to get the desired results is to first solve these equations for θ_i, S_i with specified ϕ_i, and then build a table for S_i, θ_i, ϕ_i.

Differentiating Equation (i) with respect to time gives

$$\dot{S}\sin\theta + S\cos\theta\dot{\theta} = 3.886\cos\phi\dot{\phi},$$
$$\dot{S}\cos\theta - S\sin\theta\dot{\theta} = -3.886\sin\phi\dot{\phi} \qquad \text{(ii)}$$

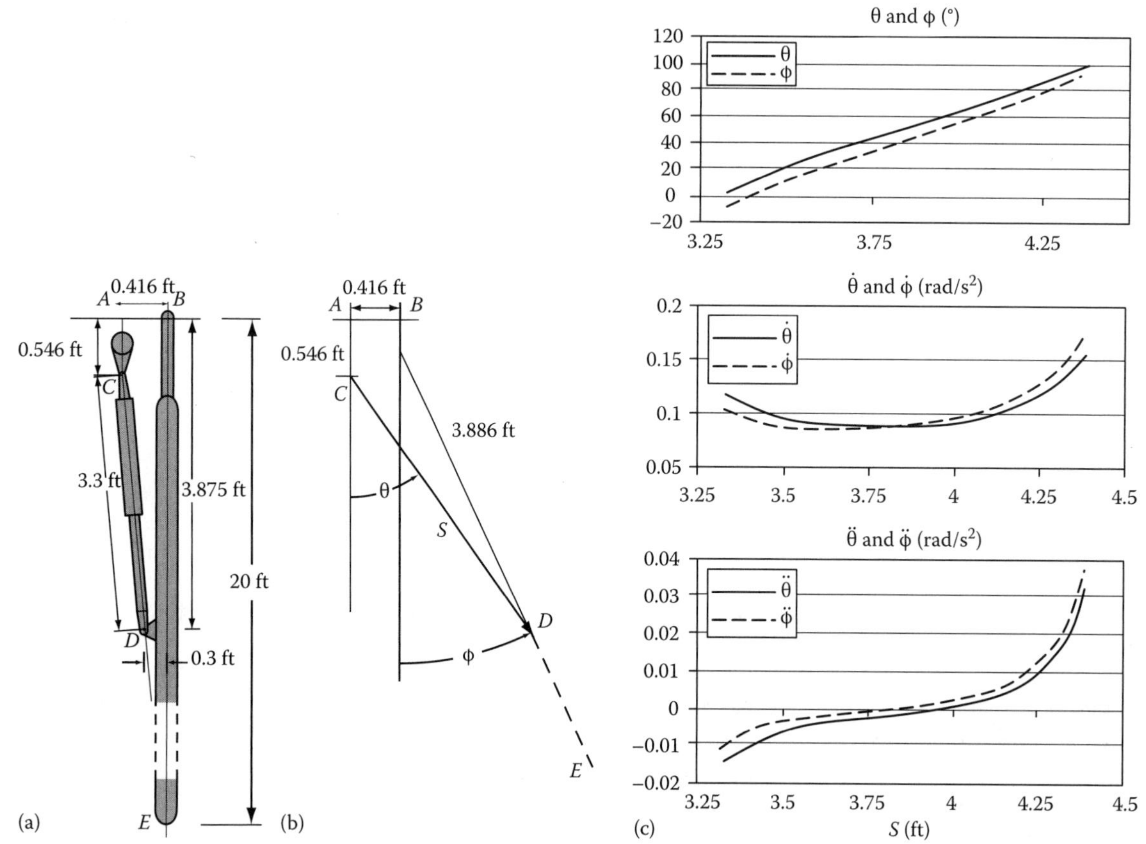

FIGURE XP4.6
(a) Top view of gate in the fully opened position. (b) General position with coordinates. (c) Angular position, velocity, and acceleration versus S.

Rearranging Equations (ii) and putting them in matrix format gives

$$\begin{bmatrix} 3.886\cos\phi & -S\cos\theta \\ -3.886\sin\phi & S\sin\theta \end{bmatrix} \begin{Bmatrix} \dot{\phi} \\ \dot{\theta} \end{Bmatrix} = \dot{S}\begin{Bmatrix} \sin\theta \\ \cos\theta \end{Bmatrix}. \qquad \text{(iii)}$$

Differentiating Equation (ii) gives

$$\begin{aligned} &\ddot{S}\sin\theta + 2\dot{S}\dot{\theta}\cos\theta + S\cos\theta\ddot{\theta} - S\sin\theta\dot{\theta}^2 \\ &\quad = 3.886\cos\phi\ddot{\phi} - 3.886\sin\phi\dot{\phi}^2 \\ &\ddot{S}\cos\theta - 2\dot{S}\dot{\theta}\sin\theta - S\sin\theta\ddot{\theta} - S\cos\theta\dot{\theta}^2 \\ &\quad = -3.886\sin\phi\ddot{\phi} - 3.886\cos\phi\dot{\phi}^2 \end{aligned}$$

Rearranging and placing the unknown acceleration terms on the left-hand side gives

$$\begin{aligned} 3.886\cos\phi\ddot{\phi} - S\cos\theta\ddot{\theta} &= \ddot{S}\sin\theta + 2\dot{S}\dot{\theta}\cos\theta - S\sin\theta\dot{\theta}^2 \\ &\quad + 3.886\sin\phi\dot{\phi}^2 = g_1 \\ -3.886\cos\phi\ddot{\phi} + S\cos\theta\ddot{\theta} &= \ddot{S}\cos\theta - 2\dot{S}\dot{\theta}\sin\theta - S\cos\theta\dot{\theta}^2 \\ &\quad + 3.886\cos\phi\dot{\phi}^2 = g_2 \end{aligned} \qquad \text{(iv)}$$

In matrix format, Equation (iv) becomes

$$\begin{bmatrix} 3.886\cos\phi & -S\cos\theta \\ -3.886\sin\phi & S\sin\theta \end{bmatrix} \begin{Bmatrix} \ddot{\phi} \\ \ddot{\theta} \end{Bmatrix} = \begin{Bmatrix} g_1 \\ g_2 \end{Bmatrix} \qquad \text{(v)}$$

Figure XP4.6c illustrates the solution for θ, ϕ, $\dot{\theta}$, $\dot{\phi}$, and $\ddot{\theta}$, $\ddot{\phi}$ versus S. Note the rapid closure rate associated with $\dot{\phi}$ as the gate approaches a closed position.

4.5.4.4 Closing Comments

The mechanism of Figure 4.24 was introduced for reasons of instruction versus wide application. However, as you will see in the exercise problems, many practical mechanisms exist which are geometrically similar to this mechanism, involving a triangular-shaped loop with one side of variable length. The following lessons should be learned from working through solutions for this problem:

a. The relationships between coordinates for a planar mechanism must be developed from geometry and trigonometry.
b. The relationships between velocities and accelerations can be developed by geometric or

two-coordinate-system vector approaches. Personal judgments determine which of these approaches is the easiest or best.

c. The lesson to be learned from the last step in determining v_D and a_D of Figure 4.37 is that the "best" approach for determining *relationships* between coordinates may not be the best approach for determining the velocity and acceleration *of a particular point*.

4.6 Summary and Discussion

No new equations or relationships were introduced in this section; the kinematics results of Chapter 2 were adequate for planar rigid body kinematics. The techniques for applying these results to rigid-body kinematics can be challenging. The results must be mastered for competency in kinematics analysis and will be essential in performing kinetics analysis in Chapter 5.

The rolling-without-slipping circumstances of Section 4.4 are of fundamental and continuing value in planar kinematics and also in Chapter 5 on planar kinetics. The rolling-without-slipping example provides the first instance of a rigid-body, kinematic constraint equation, echoing the earlier particle kinematic constraint results of Section 3.3. Wheels of vehicles ranging from bicycles to railroad cars nominally roll without slipping. Geared systems frequently involve rolling of a circular gear within or on a gear of a larger diameter. Mastery of this material and an understanding of the phenomenon are essential for an engineer doing dynamic analysis.

Section 4.5 presents representative planar mechanisms that arise in engineering practice. All of these mechanisms involve multiple rigid bodies and multiple coordinates but have only one degree of freedom. Hence, all of the examples involve kinematic constraint equations between the coordinates. Mechanisms have one input variable and one or more output variables. Depending on the requirements, the same basic geometric mechanism can see a reversal of the role of input and output variables. The slider-crank mechanism of Figures 4.15 and 4.16 illustrate this potential reversal. Developing general relationships between the coordinates, their velocities (time derivative of the coordinates), and accelerations (second time derivative of coordinates) is the central task of planar kinematics. The coordinate relations are normally nonlinear equations, while the velocity and acceleration relations are linear. Kinematic analysis involves the sequential solution for (i) the output coordinates, (ii) their derivatives with respect to time, and (iii) their second time derivatives.

As noted in the introduction of this chapter, there are many correct ways to perform planar kinematics analysis of rigid bodies. The rolling-without-slipping material of Section 4.4.1, illustrated this point with separate analyses alternatives via either (i) geometry followed by differentiation or (ii) the two-coordinate relationships of Section 2.9. Alternative analyses are presented for the planar mechanisms using (i) geometry followed by differentiation and (iii) the two-coordinate relationships of Section 2.9. Solutions involving two-coordinate-system relationships normally need coordinate transformations to get all the vector-component results (Section 2.4) stated in the same coordinate system. The "best" approach to use depends on personal experience and comfort levels. However, general kinematic equations always require geometry to develop relationships between coordinates, and your development of these equations should begin with the drawing of the mechanism in a general position.

3D rigid-body kinematics is beyond the scope of this chapter. The same vector equations apply, but 3D motion requires at least three ordered rotations to define the orientation of a rigid body versus the single rotation required for planar motion. This step is a "mind expanding" experience for most students.

Problems

Relative Motion Problems

4.1 In Figure P4.1, points A and B on the planar body shown have the following properties:

$$\boldsymbol{v}_A = 46.64 \text{ m/s } \angle 68.31^\circ; \quad \boldsymbol{a}_A = 8909.5 \text{ m/s}^2 \angle 118^\circ$$

$$\boldsymbol{v}_B = 30 \text{ m/s } \angle 120^\circ; \quad \boldsymbol{a}_B = 200 \text{ m/s}^2 \angle 330^\circ$$

The points on the body have the following coordinates: $X_A = (15, -5)$ cm, $X_B = (8, 8)$ cm, $X_P = (-5, 8)$ cm.

Tasks: State all results in vector form.

(a) Solve for the angular velocity of the body.
(b) Solve the velocity of point P.
(c) Solve for the angular acceleration of the body.
(d) Solve for the acceleration of point P.

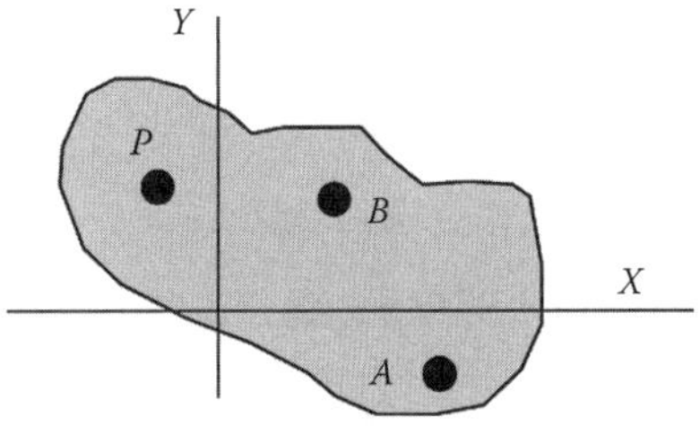

FIGURE P4.1

4.2 The planar body shown in Figure P4.1 has the angular velocity $\boldsymbol{\omega} = \mathbf{K}$ (600 rpm) and angular acceleration $\boldsymbol{\alpha} = \boldsymbol{K}(-3\ \text{rad/s}^2)$. Point A has a velocity of 10 ft/s acting at ∠180° from the horizontal and an acceleration of 2 ft/s^2 acting at ∠50°. At the instant shown, the points on the body have the coordinates: $X_A = (15, -5)$ in., $X_B = (8, 8)$ in., $X_P = (-5, 8)$ in.

Tasks:

(a) Determine point P's velocity and draw a figure showing the components.

(b) Determine point P's acceleration and draw a figure showing the components.

Rolling without Slipping

4.3 The disk's center O has the velocity and acceleration shown in Figure P4.3. The disk rolls without slipping on the horizontal surface.

Task: Determine the velocity of A and the acceleration of B for the instant presented.

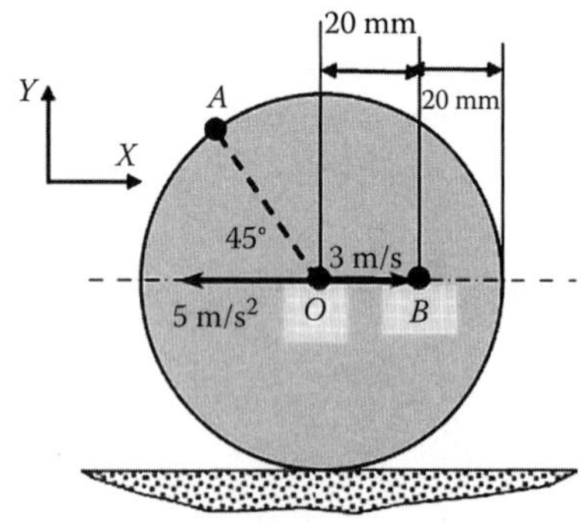

FIGURE P4.3

4.4 The wheel assembly illustrated in Figure P4.4 consists of an outer disk of radius R that is rigidly attached to the shaft of radius r. It is rolling without slipping, and at the instant shown, has velocity and acceleration of $\boldsymbol{v}_o$ and $\boldsymbol{a}_o$, respectively, to the right.

Tasks:

(a) Determine the angular velocity and acceleration of the wheel.

(b) Determine the velocity and acceleration of point E at the instant shown.

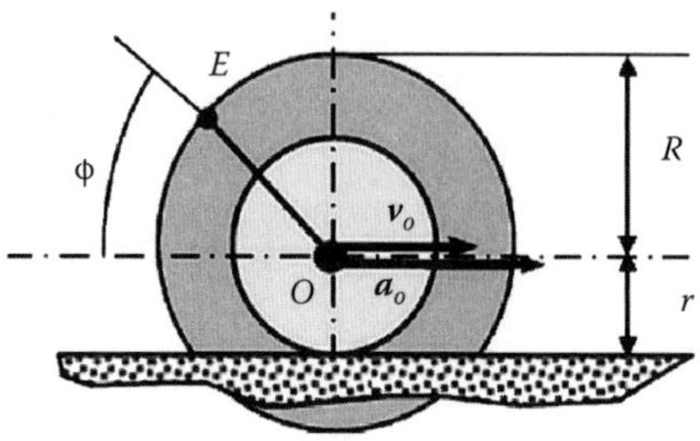

FIGURE P4.4

4.5 The cylindrical assembly shown in Figure P4.5 is rolling without slipping down the inclined plane. The acceleration and velocity of the assembly center at O are 2.45 m/s^2 and 0.5 m/s and in the $+X$ direction.

Task: Determine the velocity and acceleration of points A, B, and D with respect to ground.

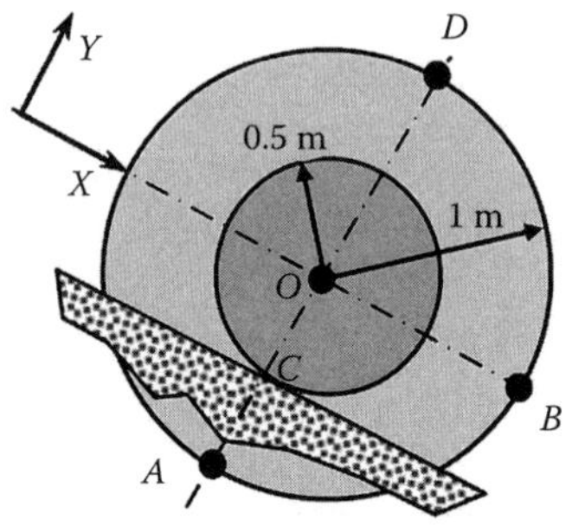

FIGURE P4.5

4.6 The cylinder shown in Figure P4.5 is rolling without slipping along the inclined plane. At the instant of interest, the wheel has a clockwise angular velocity of 3.13 rad/s and a counterclockwise angular acceleration of 9.81 rad/s^2.

Task: Determine the velocity and acceleration of points O and A with respect to ground.

4.7 The spool shown in Figure P4.7 unravels from the cord and its center moves vertically downward. At the instant shown, the spool has a clockwise angular velocity of 3.25 rad/s and an angular acceleration of 4.25 rad/s^2.

Task: Find the velocity and acceleration of points G and B with respect to ground.

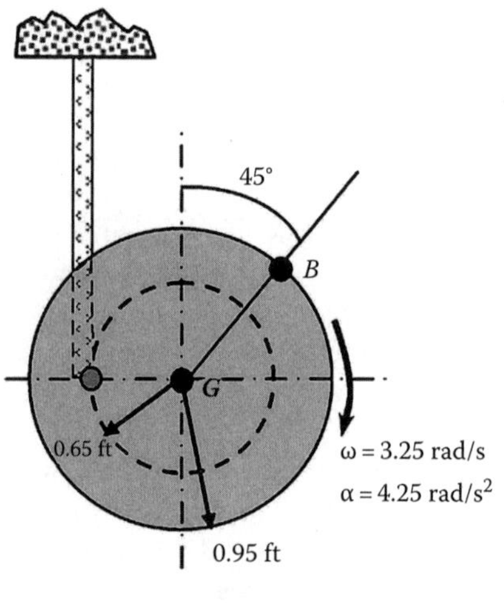

FIGURE P4.7

4.8 The spool shown in Figure P4.8 ($R=3$ ft, $r=1.5$ ft) rolls without slipping on the horizontal plane. The cord is being pulled from the spool at the speed $v_A=15$ ft/s, and is accelerating at a rate of 0.5 ft/s^2. At the instant of interest, point B on the wheel is located 45° below the horizontal.

Tasks:

(a) Determine the disk's angular velocity.
(b) Determine the disk's angular acceleration.
(c) Determine the velocity and acceleration of the disk's center with respect to ground.
(d) Determine point B's velocity and acceleration components with respect to ground.

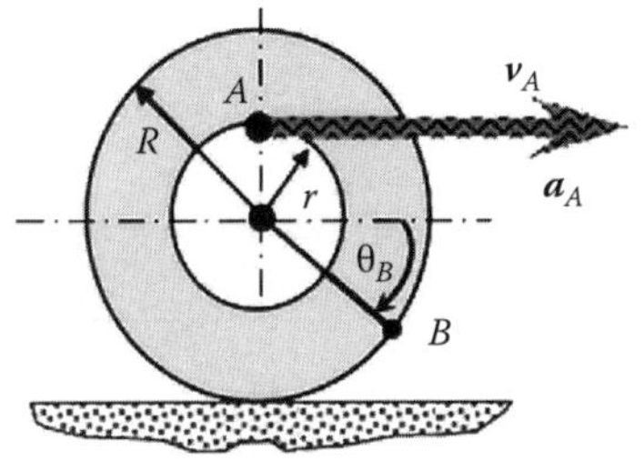

FIGURE P4.8

4.9 As shown in Figure P4.9, the center G of the flat plate moves to the left at a constant velocity v_G. The plate is on top of the two cylinders, and rolling without slipping exists at all contact points. Let $r=4$ in., $R=10$ in.

Tasks:

(a) Determine the angular velocities of the two cylinders in terms of v_G.
(b) Determine the velocity of point E on the small cylinder if the plate's velocity is 2 ft/s.
(c) Determine the resulting angular acceleration of the large cylinder if the plate is brought to rest in 5 s by applying a constant deceleration from $v_G=2$ ft/s.

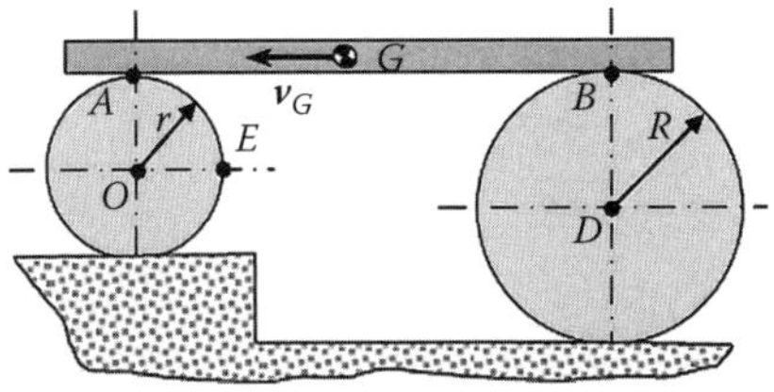

FIGURE P4.9

4.10 As shown in the Figure P4.10, the cable feed from pulley D has a constant velocity v, and surface A of the lower support plate has velocity v_A and acceleration a_A. Rolling without slipping exists at the contact point between the drum and the surface.

Tasks: (Solutions need to be stated in terms of the known variables shown in the figure.)

(a) Obtain the expression for the drum's angular velocity.
(b) Determine the velocity of point O.
(c) Obtain the velocity for point B on the drum that is located an angle ϕ below the horizontal.
(d) Draw the velocity vectors (components and resultants) for points B, C, and E on a separate figure for the drum.
(e) Knowing that the velocity at E is constant, obtain the expression for the angular acceleration of the drum. *Hint*: Do not assume that a_O is zero.

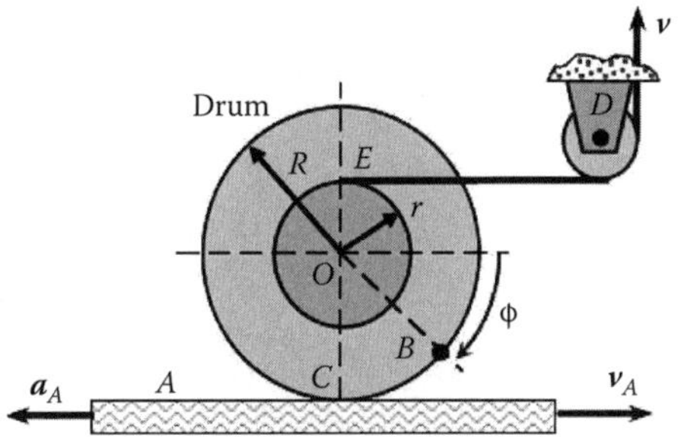

FIGURE P4.10

4.11 As shown in Figure P4.11, the sliding rails A and B engage the rims of the double wheel without slipping.

Tasks:

(a) For the indicated velocities of A and B, determine the angular velocity ω of the wheel and point P's velocity.
(b) For the indicated velocities and accelerations of A and B, find point P's acceleration.

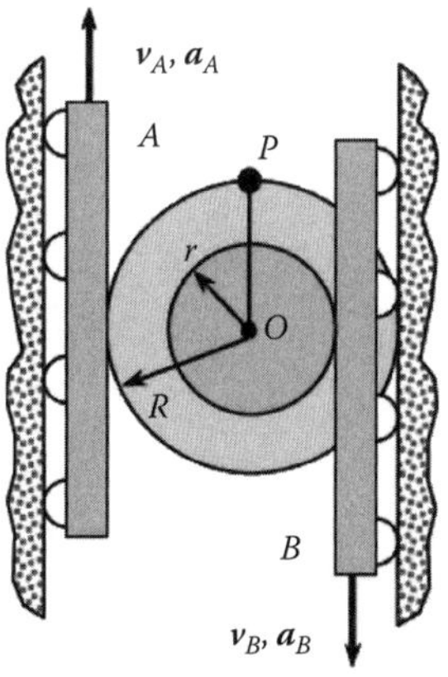

FIGURE P4.11

4.12 At the instant pictured in Figure P4.12, the load L has an upward acceleration of 5 m/s^2, and point A of the hoisting cable has an upward velocity of

2.1 m/s (not necessarily constant). Point O is the center of the pulley shaft mounted on the hook housing.

Tasks:

(a) Determine the angular velocity and angular acceleration of the pulley.

(b) Determine the acceleration of point B on the pulley.

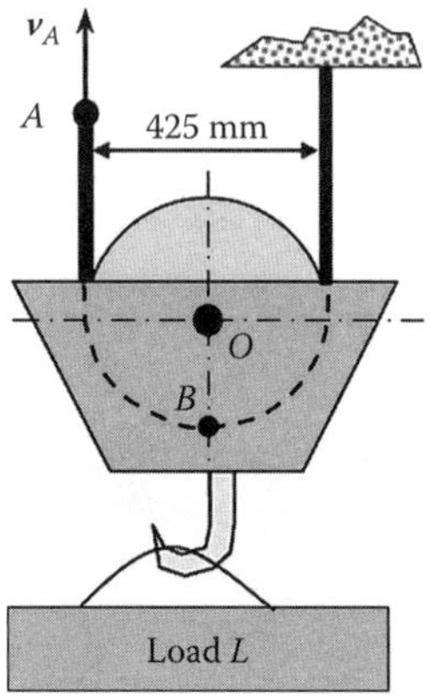

FIGURE P4.12

4.13 The circular disk rolls without slipping between two moving plates A and B as shown in Figure P4.13. Plate B has a velocity of 2 m/s to the left and an acceleration of 8 m/s^2 to the right. Point O, the center of the disk, has a velocity of 1 m/s to the left and an acceleration of 4 m/s^2 to the right. The radius of the disk is 0.5 m.

Tasks:

(a) Determine the angular velocity and angular acceleration of the disk.

(b) Determine the velocity and acceleration of plate A.

(c) Determine the velocity and acceleration of point D.

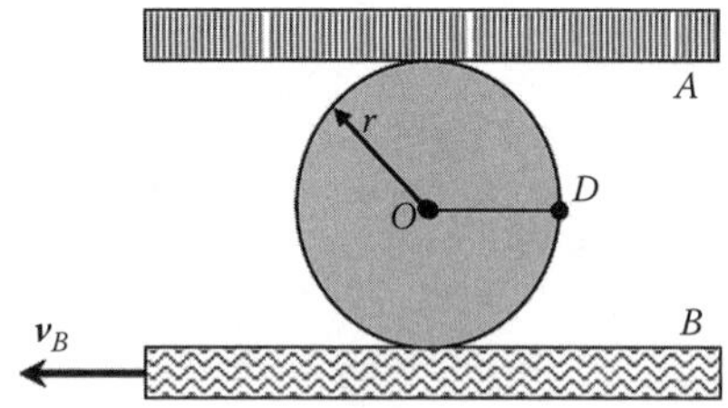

FIGURE P4.13

4.14 The load L shown in Figure P4.14 is lowered by the two pulleys. The pulley at O is composed of two additional pulleys that rotate as a single unit. For the instant presented, drum A has a counterclockwise angular velocity of 6 rad/s that is decreasing by 4.5 rad/s^2. Simultaneously, drum B has a clockwise angular velocity of 5 rad/s that is increasing by 1.75 rad/s^2.

Task: Calculate the acceleration of points C and D and the load L.

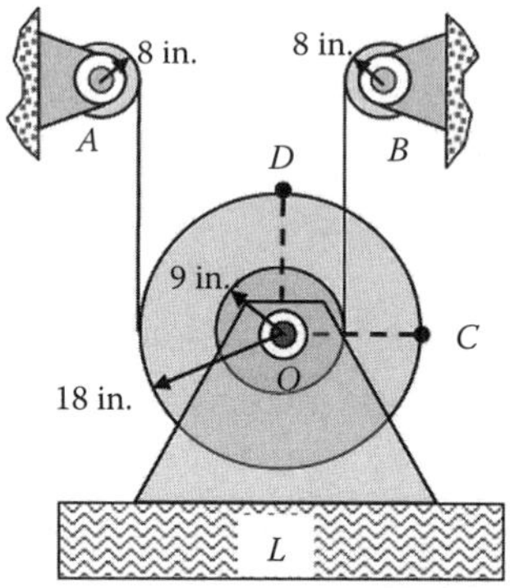

FIGURE P4.14

4.15 As shown in the Figure P4.15, the cable feed from pulley D has a velocity v, and surface A is fixed. Rolling without slipping exists at the contact point between the drum and the surface.

Tasks: (Solutions need to be stated in terms of the known variables shown in the figure)

(a) Using a vector relationship, obtain the expression for the angular velocity of the drum.

(b) Obtain the velocity for point B on the drum located at angle ϕ below the horizontal.

(c) Draw the velocity vectors (components and resultants) for points B and E on a separate figure of the drum.

(d) Using vector relationships, obtain the expression for the angular acceleration of the drum if point O has the acceleration shown.

(e) Draw a figure showing the acceleration components of point E.

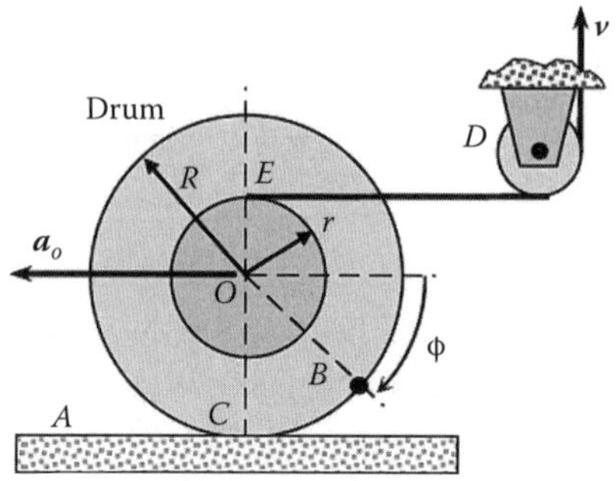

FIGURE P4.15

4.16 The railroad utility vehicle starts from rest and, under a constant acceleration, reaches a cruising speed of 35 mph in 120 ft. The trucks tires are used to propel it while the 8 in. diameter guide wheels keep the truck on the tracks (Figure P4.16).

Tasks:

(a) Determine the equation that relates the angular motion, i.e., displacement, velocity, and acceleration, of the guide wheel to the linear motion of the truck.

(b) Calculate the number of revolutions and angular speed of the guide wheel when the truck reaches its cruising speed.

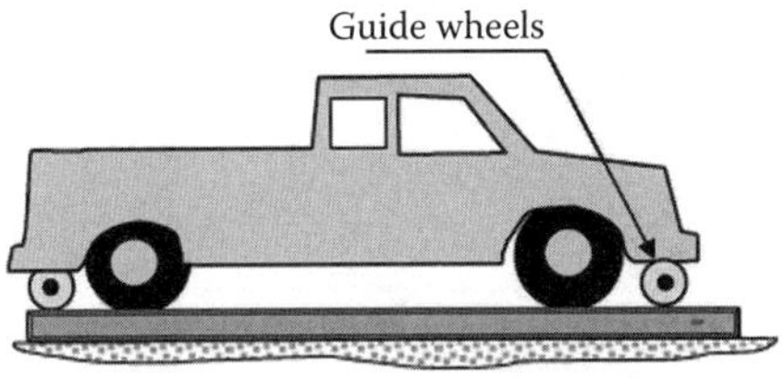

FIGURE P4.16

4.17 At the instant shown, the cable feed in Figure P4.17 has a constant velocity v, and point B on the drum is at an angle ϕ below the horizontal. The drive roller of the conveyor system has a constant angular velocity ω_1, and the pulleys on each end have the same diameter, $2b$. Rolling without slipping exists at the contact point C.

Tasks:

(a) Find and state the drum's angular velocity in terms of the angular speed of the conveyer pulley and v.

(b) Determine the velocity of point O.

(c) Obtain the velocity for points A and B in terms of the conveyor pulley angular velocity and draw a kinematic diagram showing their components.

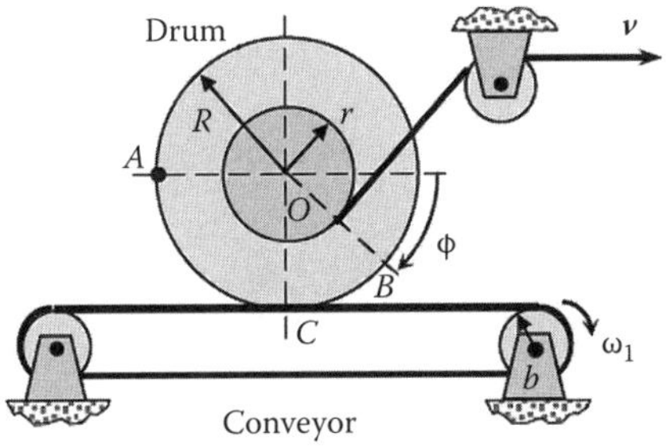

FIGURE P4.17

4.18 The electric motor in Figure P4.18 operates at 1800 rpm. The gear attached to the motor has a diameter d of 8 in.

Tasks:

(a) Determine the angular velocity of the fan ω_S in terms of the motor speed ω_m and the two gear diameters, d and D.

(b) Determine the required diameter, D, of the mating gear so that point A has a tip speed of 102 ft/s. The blade length L is 24 in.

(c) The motor starts at rest, is accelerated at a constant rate, and reaches its operating speed in 15 s. Determine the acceleration components of point A on the fan blade at $t = 10$ s.

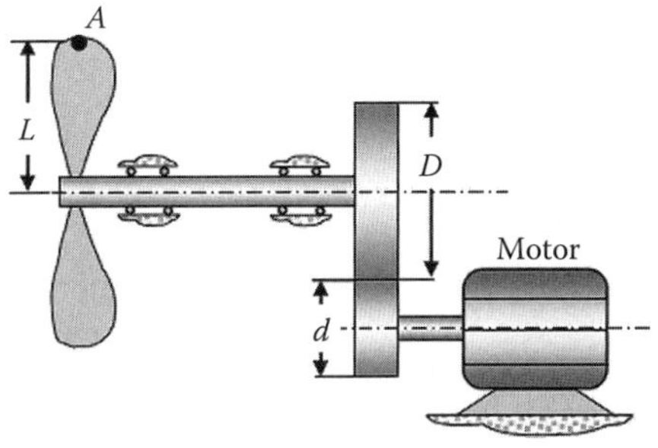

FIGURE P4.18

4.19 A belt-and-pulley system is shown in Figure P4.19. Pulley A operates at speed 3600 rpm and has a diameter of 10 in. Pulley B has a diameter of 6 in. Assume that the belt is inelastic and that no slip occurs.

Tasks:

(a) Determine the angular velocity of pulley B.

(b) For $r_C = 2.75$ in., determine the velocity of point C on pulley B.

(c) If pulley A has an angular motion defined by $\theta_A(t) = 2t^3$ rad, determine the acceleration of point C on pulley B at $t = 0.75$ s.

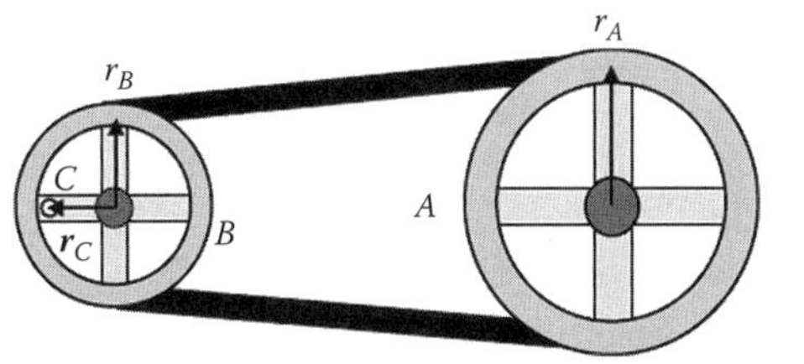

FIGURE P4.19

4.20 The drum shown in Figure P4.20 has a constant angular speed of ω_O. The inner ring is fixed. The cylinder rolls without slipping on both the inner ring and the drum at all points of contact.

Tasks:

(a) Determine the angular velocity of the cylinder about its center A.

(b) Determine the angular velocity of the cylinder's center about point O.
(c) Determine the velocity and acceleration of point A.

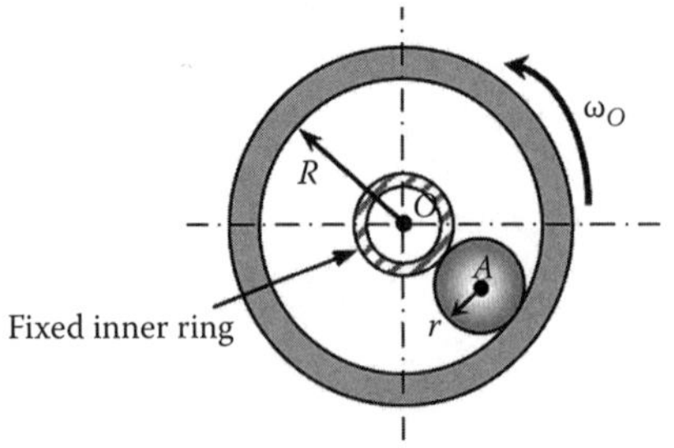

FIGURE P4.20

Planar Mechanisms

4.21 Link OA in Figure P4.21 is driven at the constant counterclockwise angular speed of 3600 rpm that causes the block at B to translate along the X-axis.

Tasks: For, $l_1 = 1.5$ in., $l_2 = 6$ in., and $\theta = 30°$, do the following:
(a) Select coordinates and obtain the kinematic constraint expressions.
(b) Determine the velocity of point B.
(c) Determine the acceleration of point B.
(d) Determine the velocity and acceleration components of point G, the midpoint of link AB.

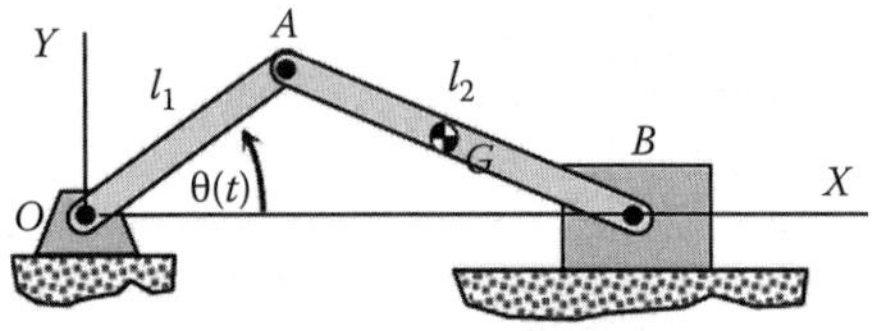

FIGURE P4.21

4.22 The hydraulic cylinder at the left of Figure P4.22 controls the position of point A on link AC as denoted by $X(t)$. Link AC is connected to link BO by a hinge connection at B. Link BO is connected to ground by a hinge at O.

Tasks:
(a) Select coordinates for motion of links AC and OB and state the general geometric equations defining the orientation of links AC and BO in terms of the position of A.
(b) State general equations defining the angular velocities of links AC and BO in terms of the position and velocity of A. Identify your unknowns and present a matrix format for their solution.
(c) State the general equations defining the angular accelerations of links AC and BO in terms of the position, velocity, and acceleration of A. Identify your unknowns and present the matrix format for their solution.
(d) Provide general expressions for the velocity and acceleration of point C in terms of $X(t)$ and θ and their derivatives.

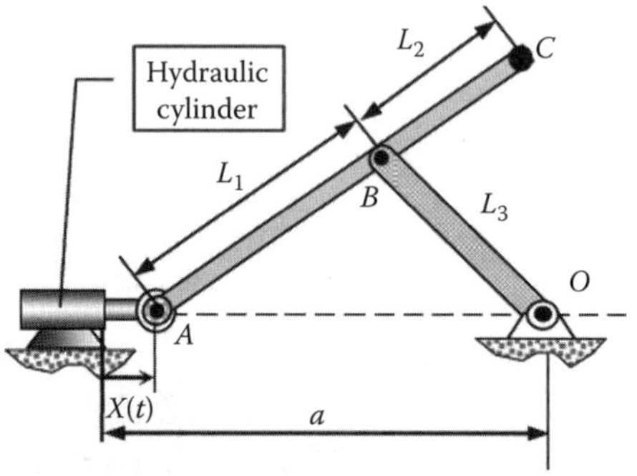

FIGURE P4.22

4.23 For the mechanism illustrated in Figure P4.23, link AB is driven counterclockwise, and its position is defined by the general function $\theta(t)$. Bar BC slides through the collar at D. The collar at D is free to pivot about D.

Tasks:
(a) Redraw the figure in a general configuration, select coordinates for the unknowns, and obtain the kinematic constraint equations.
(b) Determine general expressions for the angular velocity and acceleration of bar BC. Present the equations for solutions of your unknowns in matrix format.
(c) Provide a general expression for the velocity and acceleration of point C in terms of θ, and your defined variables and their derivatives.
(d) Determine the resulting velocity and acceleration of point C given the following data: $r = 20$ cm, $L = 1.5$ m, $a = 1.2$ m, at $\theta = 130°$, $\dot{\theta} = 180$ rpm, $\ddot{\theta} = 0.25$ rad/s^2.

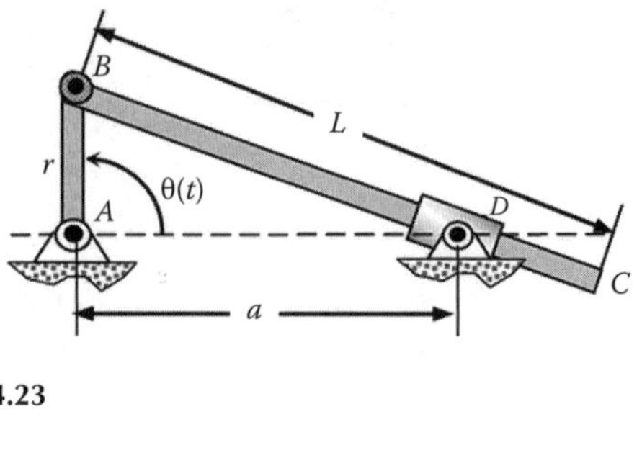

FIGURE P4.23

4.24 As shown in Figure P4.24, the angular position of link *OA* is a known function of time and rotates in the clockwise direction (measured from dotted vertical reference line). Link *OA* is connected to link *CB* through the sliding collar *A*, and *CB* is connected to the ground by a hinge at *C*.

Tasks:

(a) Redraw the mechanism in a general configuration, select coordinates, and obtain the kinematic constraint expressions. Also, obtain the general equations defining the orientation of link *CB* and the distance from *C* to *A*.

(b) State the general equation defining the angular velocity of link *CB* and put them into matrix format for solution.

(c) State the general equation defining the angular acceleration of link *CB* and put them into matrix format for solution.

(d) State general equations defining the velocity and acceleration of point *B*.

(e) Determine the resulting velocity and acceleration of point *B* given the following: $L_1 = 0.5$ ft, $L_2 = 1.8$ ft, $a = 0.65$ ft, at $\theta = 65°$, $\dot{\theta} = 120$ rpm, $\ddot{\theta} = 0.25$ rad/s^2.

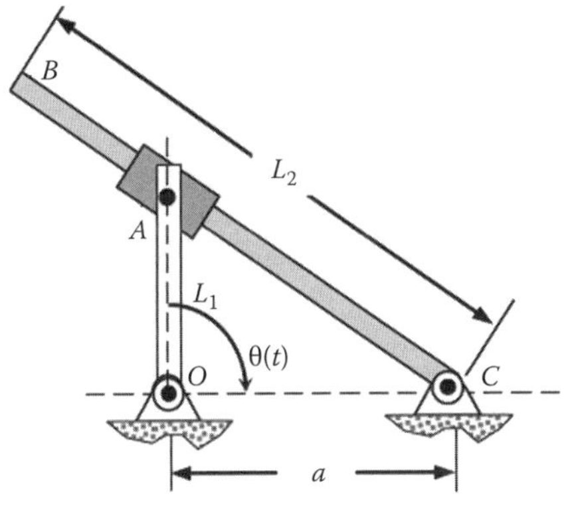

FIGURE P4.24

4.25 As shown in Figure P4.25, link *AO*'s rotation is controlled by the piston rod of hydraulic cylinder *BC* that is elongating at the constant rate $\dot{s} = v_C$ for an interval of motion.

Tasks:

(a) Select coordinates and obtain the kinematic constraint equations for the assembly.

(b) Determine angular position and angular velocity of link *AO* in terms of the input motion of the hydraulic cylinder.

(c) Determine the velocity and acceleration of point *A* in terms of *S* and its derivatives.

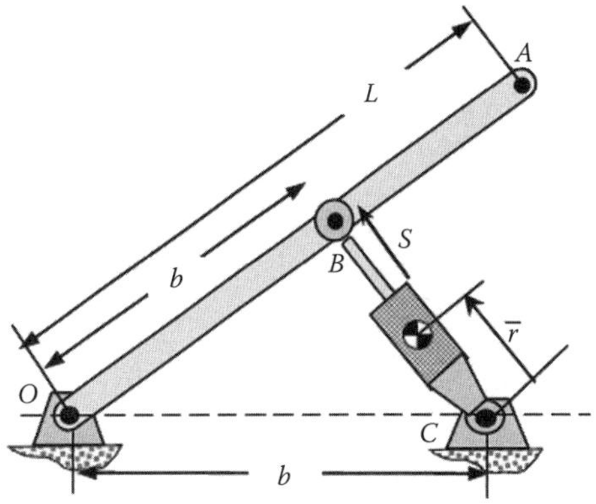

FIGURE P4.25

4.26 Figure P4.26 illustrates two links. For the time interval of interest, link *DC* has a constant counterclockwise angular velocity of $\dot{\theta} = \omega_O$. The collar at *C* is free to slide along link *AB* and can pivot freely about point *C*.

Tasks:

(a) Select coordinates and obtain the kinematic constraint equations.

(b) Derive general expressions for the angular velocity of link *AB* and the displacement between pins *A* and *C*. Present the equations for solution of your unknowns in matrix format.

(c) Derive general expressions for the angular acceleration of link *AB* and for the displacement between pins *A* and *C*. Present the equations for solution of your unknowns in matrix format.

(d) Obtain expressions for the velocity and acceleration vectors for point *B* in terms of variables obtained in (a).

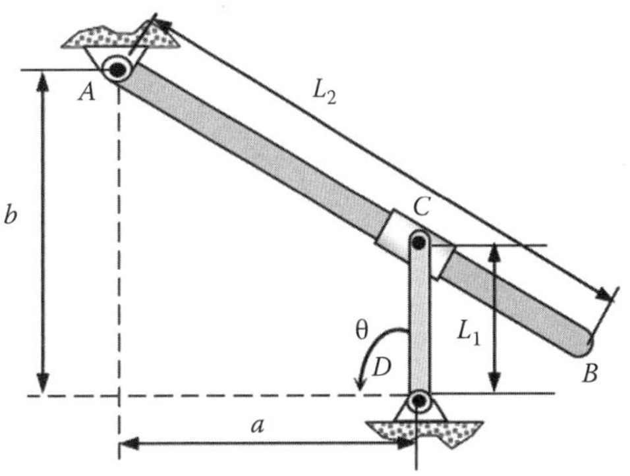

FIGURE P4.26

4.27 For the system illustrated in Figure P4.27, θ is given as a function of time. Link *AB* is free to slide through the collar at *C*. The collar at *C* is free to pivot about the point at *C*.

Tasks:

(a) Select coordinates, and obtain the kinematic constraint expressions.

(b) Derive the expressions for the angular velocity of link *AB* and the distance between points *A* and *C* as a function of θ and $\dot{\theta}$.

(c) Derive the expressions for the angular acceleration of link *AB* and the second time derivative of displacement between points *A* and *C* as a function of θ, $\dot{\theta}$, and $\ddot{\theta}$.

(d) Determine point *B*'s velocity and acceleration.

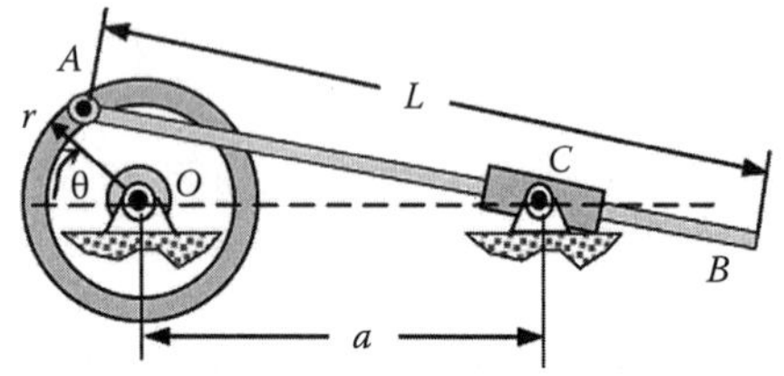

FIGURE P4.27

4.28 The scotch yoke shown in Figure P4.28 is driven by $X = X_0 \sin(\omega t)$. (At the instant illustrated, $X = 0$.) Pin A is rigidly attached to link *OB*.

Tasks:

(a) Redraw the mechanism in a general orientation, select coordinates, and derive the kinematic constraint equations.

(b) Derive expressions for the angular velocity and acceleration of link *OB*.

(c) Given that $\omega = 30$ rad/s and $X_0 = 5$ in., determine the velocity and acceleration of point *B* at $t = 0.5$ s. Let $OA = 0.75$ ft, $L = 1$ ft, $d = 0.25$ ft.

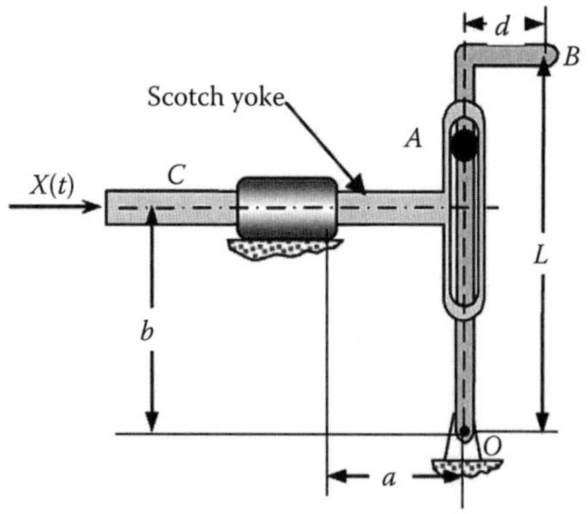

FIGURE P4.28

4.29 A four-bar linkage is illustrated in Figure P4.29. For the interval of interest, bar *OA* is rotating with a constant angular velocity of ω rad/s, counterclockwise.

Tasks:

(a) Redraw the mechanism in a general orientation, select coordinates, and obtain the kinematic constraint expressions.

(b) Derive general expressions for the angular velocities of links *AB* and *BC*.

(c) Derive general expressions for the angular accelerations of links *AB* and *BC*.

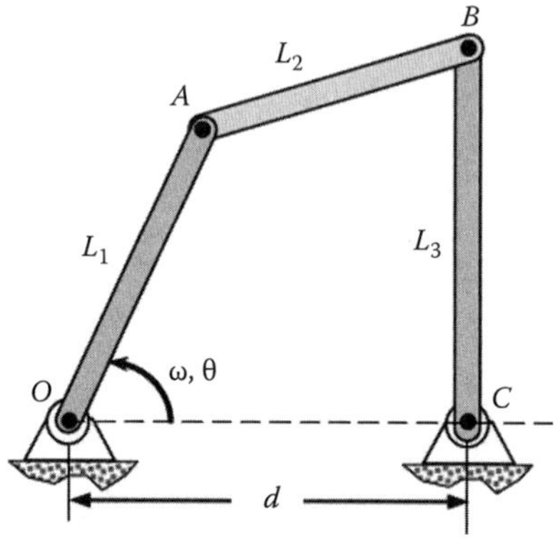

FIGURE P4.29

4.30 The four-bar mechanism shown in Figure P4.30 is driven from the right-hand side such that link *ED* has a constant counterclockwise angular velocity of $\dot{\theta} = \omega_0$ rad/s.

Tasks:

(a) Redraw the mechanism in a general orientation, select coordinates, and state the general equations for the orientation of the system.

(b) State the general equations for the angular velocities of bars *DE*, *DB*, and *BA*. Put your equations into matrix format for solutions of the angular velocities of links *BA* and *BD* as unknowns.

(c) State the general equations for the angular accelerations of bars *DE*, *DB*, and *BA*. Put your equations into a matrix format for solutions of the angular accelerations of links *DA* and *DB* as unknowns.

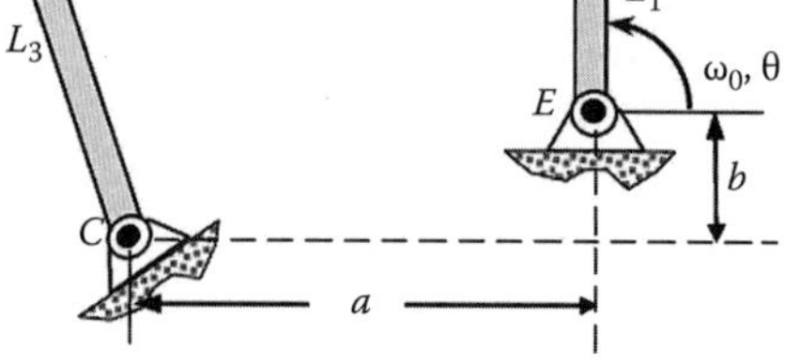

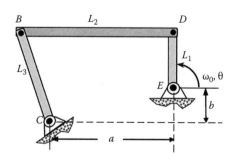

FIGURE P4.30

4.31 Illustrated in Figure P4.31 is a disk driven with a constant counterclockwise rotation rate of $\dot{\theta} = \omega_0$ rad/s.

Tasks:

(a) Redraw the mechanism in a general orientation, select coordinates, and obtain the general equations for the orientation of the assembly.

(b) Derive general expressions in matrix format for solution of the unknown angular velocities of links *AB* and *BC*.

(c) Derive general expressions in matrix format for solution of the unknown angular acceleration of links *AB* and *BC*.

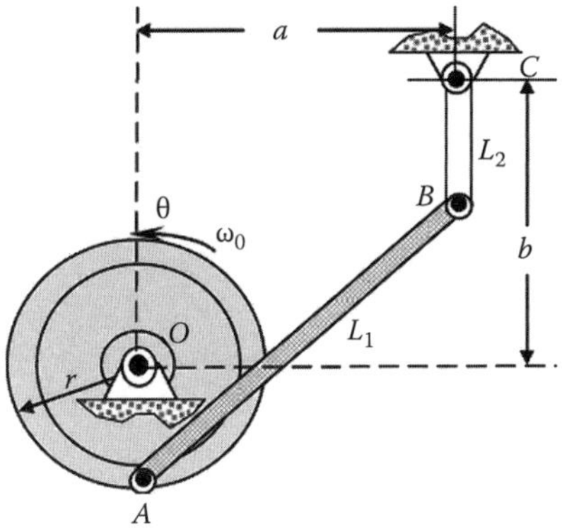

FIGURE P4.31

4.32 For the system illustrated in Figure P4.32, θ, is given as a general function of time.

Tasks:

(a) Select coordinates and obtain the kinematic constraint equations.

(b) Obtain the angular velocity of the plate *ABD* and the link BO_2 as a function of $\dot{\theta}$ and θ.

(c) Obtain angular acceleration of the plate *ABD* and the link BO_2 as a function of $\theta, \dot{\theta}$, and $\ddot{\theta}$.

(d) Determine the velocity and acceleration vectors for point *D* in terms of the variables obtained in tasks (a), (b), and (c).

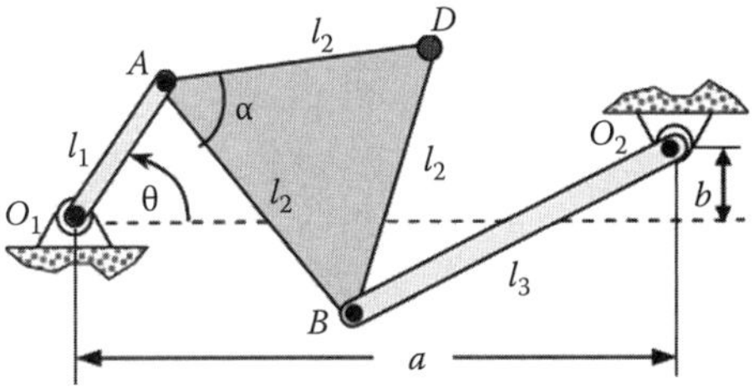

FIGURE P4.32

4.33 The system illustrated in Figure P4.33 consists of two links driven at *B* by a hydraulic cylinder. For the motion of interest, the hydraulic cylinder is causing point *B* to move straight up (vertically) at a constant velocity $\dot{S} = v_0$.

Tasks:

(a) Redraw the mechanism in a general orientation, select coordinates, and state general equations for the system.

(b) State general equations for the angular velocities of bars *AC* and *AD*. Put your equations into matrix format for solution of unknown angular velocities of links *AC* and *AD*.

(c) State the general equations for the angular accelerations of bars *AC* and *AD*. Put your equations into matrix format for solution of the unknown angular accelerations of links *AC* and *AD*.

(d) Find the general expressions for the velocity and acceleration of point *D*. Express your answers in terms of ***I*** and ***J*** components.

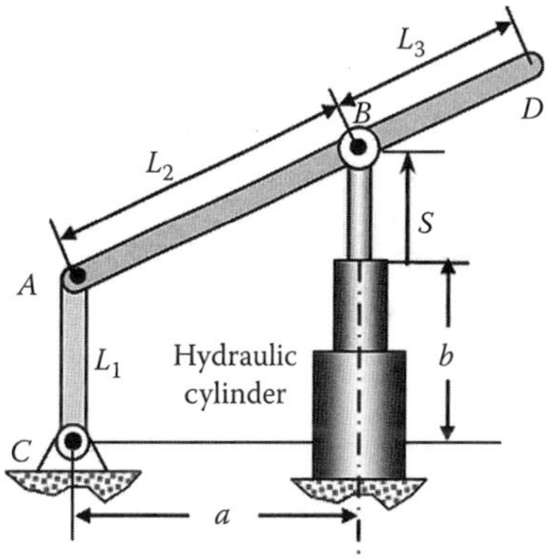

FIGURE P4.33

4.34 For the problem illustrated in Figure P4.34, θ_1 and θ_2 are known functions of time. Points *A* and *B* are joined by a telescoping cylinder.

Tasks:

(a) Redraw the mechanism in a general orientation, select coordinates, and obtain equations to solve for the orientation of the system.

(b) Derive general velocity expressions for the orientation and change in length of *AB* as a function of θ_1, θ_2, $\dot{\theta}_1$, and $\dot{\theta}_2$. Solve for each unknown explicitly.

(c) Derive general acceleration expressions for the orientation and second derivative of length *AB* as a function of θ_1, θ_2, $\dot{\theta}_1$, $\dot{\theta}_2$, $\ddot{\theta}_1$, and $\ddot{\theta}_2$. Solve for each unknown explicitly.

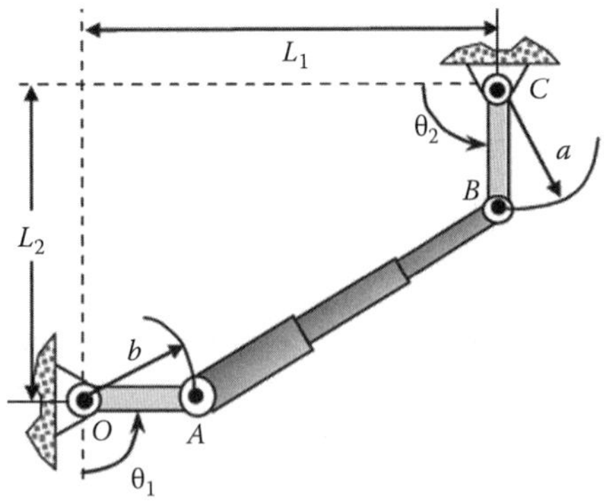

FIGURE P4.34

4.35 For the system illustrated in Figure P4.35, θ is given as a function of time.

Tasks:

(a) Select coordinates and obtain the kinematic constraint equations.
(b) Determine the angular velocity of link *BC* and the velocity of point *C*.
(c) Determine the angular acceleration of link *BC* and the acceleration of point *C*.

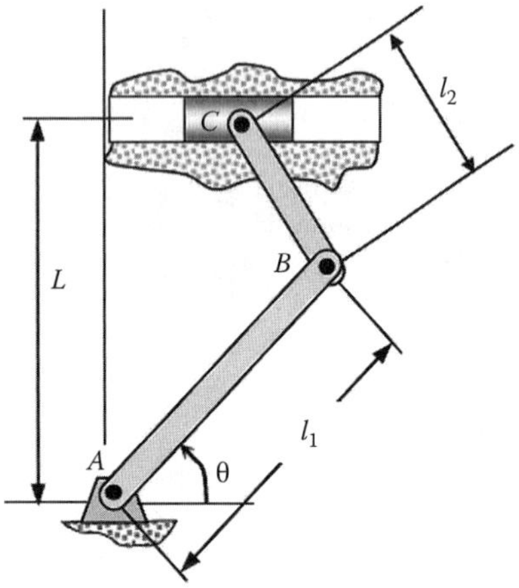

FIGURE P4.35

4.36 Link *AB* in Figure P4.36 is rotating with constant, counterclockwise angular velocity, $\dot{\theta} = \omega_0$. *L* is the length of link *CD*, and *R* is the length of link *AB*.

Tasks:

(a) Select coordinates and obtain the kinematic constraint expressions.
(b) Determine the angular velocity of link *CD*.
(c) Determine the angular acceleration of link *CD*.
(d) Determine the velocity and acceleration of point *D*.

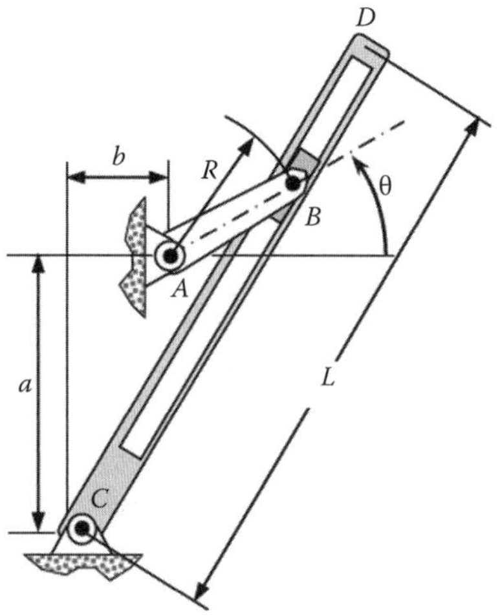

FIGURE P4.36

4.37 For the mechanism illustrated in Figure P4.37, the angle θ is a general function of time, θ(*t*) and acts in the clockwise direction.

Tasks:

(a) Redraw the mechanism in a general orientation, select coordinates, and obtain the kinematic constraint expressions.
(b) Determine general expressions for the angular velocity and angular accelerations of bar *BC* and for the time derivative of the distance between points *B* and *C*. State the equations for solution of your unknowns in matrix format.
(c) Provide a general expression for the velocity of point *C*.

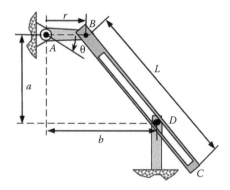

FIGURE P4.37

4.38 As illustrated in Figure P4.38, the disk at *O* is driven at a constant angular speed ω_O in the clockwise direction. Rolling without slipping exists between disks *O* and *A*. The pin at *B* is rigidly attached to the disk at *A*.

Tasks:

(a) Redraw the mechanism in a general orientation, select coordinates, and derive the kinematic constraints for the mechanism.

(b) Derive expressions for the velocity and acceleration of the scotch yoke through the bushing at C in terms of ω_O.

Disk at O has an angular velocity of 1800 rpm. Determine, with respect to ground, the velocity and acceleration of the scotch yoke moving through the collar at C at $t=1.35$ s. Let $AB=8$ in., $R=10$ in., $r=4$ in., $a=15$ in., $b=12$ in.

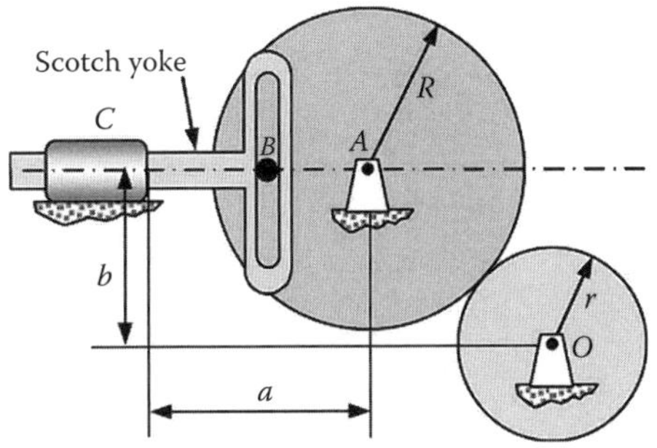

FIGURE P4.38

4.39 For the system illustrated in Figure P4.39, link AC rotates in the counterclockwise direction at the angular speed $\dot{\theta} = \omega_0 =$ constant.

Tasks:

(a) Redraw the mechanism in a general orientation, select coordinates, and obtain the kinematic constraint expressions.

(b) Obtain general relationships for the angular velocity and angular acceleration of bar OB and present the solutions in matrix format.

(c) Provide general expressions for the velocity and acceleration of point B (provide answer in vector format).

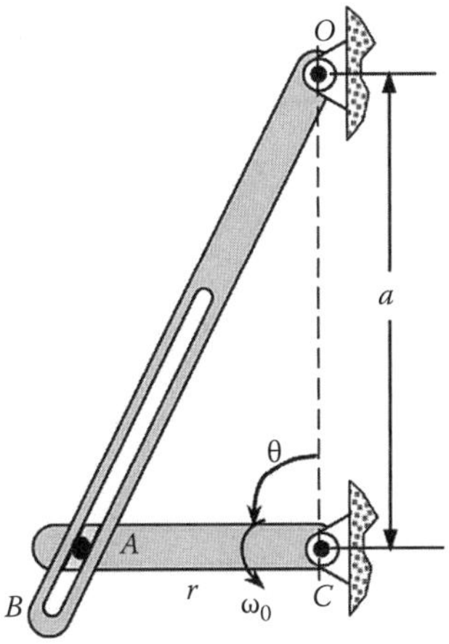

FIGURE P4.39

4.40 The mechanism shown in Figure P4.40 causes unit B to move intermittently as disk A rotates at a constant angular speed ω_0. Pin C is fixed to disk A and is located a distance r from the pivot O of disk A.

Tasks:

(a) Redraw the mechanism in a general orientation, select coordinates, and obtain the kinematic constraint expressions.

(b) Determine relations that describe the velocity and acceleration of pin C relative to O_2 in the corresponding slot for the given angular velocity of disk A.

(c) Obtain an expression for the angular velocity of disk B during engagement.

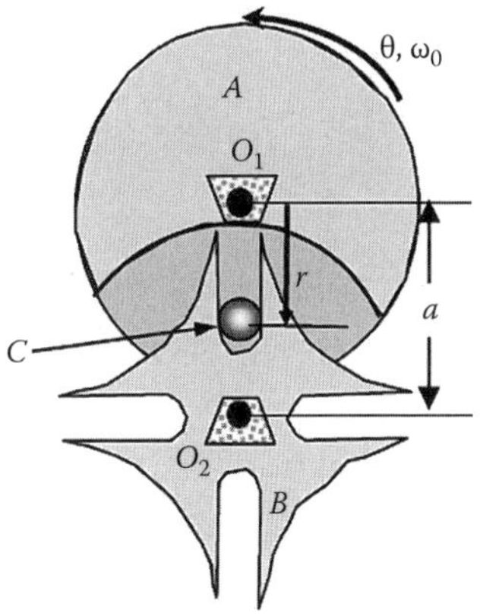

FIGURE P4.40

4.41 As illustrated in Figure P4.41, the disk's axis of rotation at A is offset from the disk's center of geometry O by a distance e. The uniform bar BC pivots about point C and remains in contact with the disk. The disk rotates at a constant angular speed ω_O.

Tasks:

(a) Redraw the mechanism in a general orientation, select coordinates, and obtain the kinematic constraint expressions.

(b) Determine the angular velocity of the bar BC.

(c) Determine the angular acceleration of the bar BC.

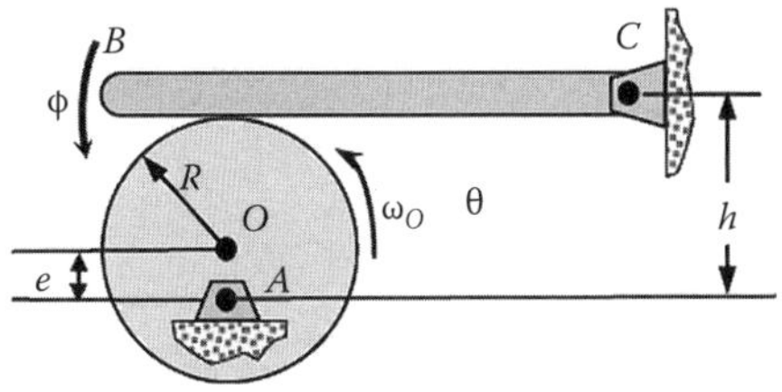

FIGURE P4.41

4.42 The mechanism shown in Figure P4.42 is driven by link OA that rotates at a constant angular speed ω_0 in the clockwise direction.

Tasks:

(a) Select coordinates and obtain the kinematic constraint expressions.

(b) Determine expressions that describe the angular velocities of links BC, and DC and state in matrix format.

(c) Determine expressions that describe the angular acceleration of links BC, and DC and state in matrix format.

(d) Determine expressions for the velocity and acceleration of point D and point A. Express the equations for solutions of your unknowns in matrix form.

Hint: Use two sets of kinematic constraints.

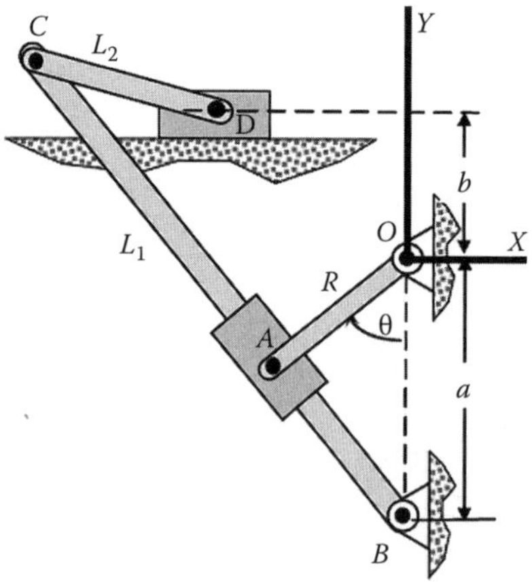

FIGURE P4.42

5

Planar Kinetics of Rigid Bodies

5.1 Introduction

This chapter revisits all of the principles and objectives covered in Chapter 3 for particles, and now applies them to planar motion of rigid bodies. The objectives include deriving equations of motion using Newtonian and energy approaches, plus developing the principles of conservation of momentum for planar motion of a rigid body. You have presumably just completed Chapter 4 on *planar* kinematics of rigid bodies and understand the idea of planar motion for a rigid body.

One of the attractions of dynamics is that governing equations proceed from a limited number of principles and equations. You enter this chapter knowing (or at least having been exposed to) Newton's second law of motion for a particle, $\sum \boldsymbol{f} = m\ddot{\boldsymbol{r}}$, where $\ddot{\boldsymbol{r}}$ is the particle's acceleration with respect to an inertial coordinate system. The particle kinematics presented in Chapter 2 provides the tools for determining the required $\ddot{\boldsymbol{r}}$ in Newton's second law. You will learn in this chapter how to proceed from Newton's equation *for a single particle* to derive governing equations of motion *for a rigid body*, containing an infinite number of particles. Similarly, you worked problems in Chapter 3 involving the kinetic energy for a particle, $T = mv^2/2$, where v is the velocity magnitude of the particle with respect to an inertial coordinate system. In this chapter, we will use the particle definition $T = mv^2/2$ to develop the defining equation for the kinetic energy of a rigid body.

As with Chapter 3, most of the problems in this chapter involve development of models including deriving governing differential equations of motion and kinematic constraint equations. The derivations will proceed from (i) Newtonian methods using free-body diagrams and (ii) work–energy principles. The kinematic constraint equations of Chapter 4 will frequently be required to develop a complete set of governing equations. For most examples, both approaches are employed showing their advantages and disadvantages.

The next section introduces the mass moment of inertia for a rigid body and the parallel-axis formula for moments of inertia. You should have previously been introduced to the mass moment of inertia and the parallel-axis formula in statics and calculus; hence, the introduction is brief.

Nomenclature

(Mainly these terms are introduced in this chapter. Check Chapter 3 Nomenclature for additional terms.)

$\boldsymbol{b}_{og}$	See Figure 5.3. Position vector from point o in a body to point g, the mass center	[L]
I_o	Moment of inertia of a body about an axis through o	[ML^2]
I_g	Moment of inertia of a body about an axis through the body's mass center	[ML^2]
E	Young's modulus of elasticity	[F/L^2]
G	Shear modulus of elasticity; see Equation 5.41	[F/L^2]
$\boldsymbol{H}_o$	Rigid-body moment of momentum; see Equation 5.282	[ML^2/T]
k_g	Radius of gyration. If the moment of inertia about an axis through point A through the body is I_A, and the body's mass is m, $I_A = mk_g^2$	[L]
k_θ	Torsional spring constant, see Equation 5.41	[F-L]
1DOF	One degree of freedom	
2DOF	Two degree of freedom	
MDOF	Multi-degree of freedom	
EOM	Equation of motion	

5.2 Inertia Properties and the Parallel-Axis Formula

5.2.1 Centroids and Moments of Inertia

Inertia properties are not exciting but they are essential for the development of governing equations for rigid bodies and are a good way to start thinking about the difference between a particle and a rigid body. Figure 5.1 illustrates a rigid body with an imbedded x, y, z

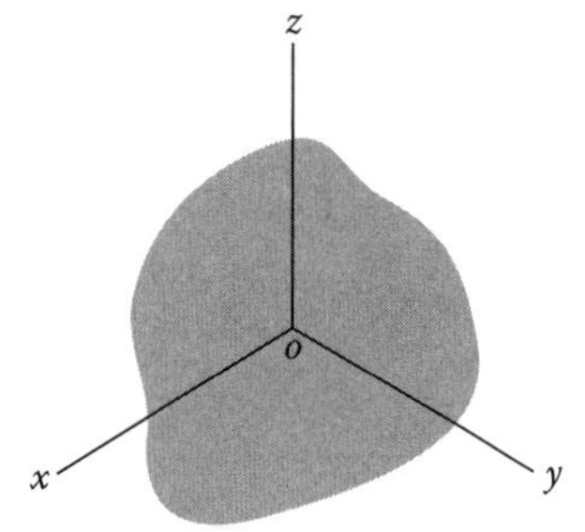

FIGURE 5.1
Rigid body with an imbedded x, y, z coordinate system.

coordinate system. The origin of the x, y, z system is denoted as o. Although the topic of this chapter is *planar* kinetics of rigid body, the body in Figure 5.1 has a general form, as opposed to a "planar-looking" body such as a thin book or a sheet of plywood. A differential element within the body is a cube with dimensions dx, dy, dz, and the mass of a cube located at the point x, y, z is $\gamma\, dx\, dy\, dz$ where $\gamma(x, y, z)$ is the body's density at this point. The mass of the body is defined by

$$m = \int_V \gamma\, dx\, dy\, dz \tag{5.1}$$

Using $\boldsymbol{\rho} = \boldsymbol{i}x + \boldsymbol{i}y + \boldsymbol{k}z$ as the position vector for a point in the rigid body, the body's mass center is located in the x, y, z system by the vector $\boldsymbol{b}_{og}$, defined by

$$\boldsymbol{mb}_{og} = \boldsymbol{m}(\boldsymbol{i}\boldsymbol{b}_{ogx} + \boldsymbol{j}\boldsymbol{b}_{ogy} + \boldsymbol{k}\boldsymbol{b}_{ogz}) = \int_V \boldsymbol{\rho}\ \boldsymbol{\gamma}\, \boldsymbol{dx}\, \boldsymbol{dy}\, \boldsymbol{dz} = \int_m \boldsymbol{\rho}\ \boldsymbol{dm}. \tag{5.2}$$

The mass moment of inertia about a z-axis through o is defined by

$$I_{zzo} = \int_V (x^2 + y^2)\gamma\, dx\, dy\, dz = \int_m (x^2 + y^2) dm. \tag{5.3}$$

Figure 5.2 illustrates a triangular plate of constant unit thickness and a constant mass *per unit area* in the x, y plane of $\bar{\gamma}$. Applying Equation 5.1 to this example gives

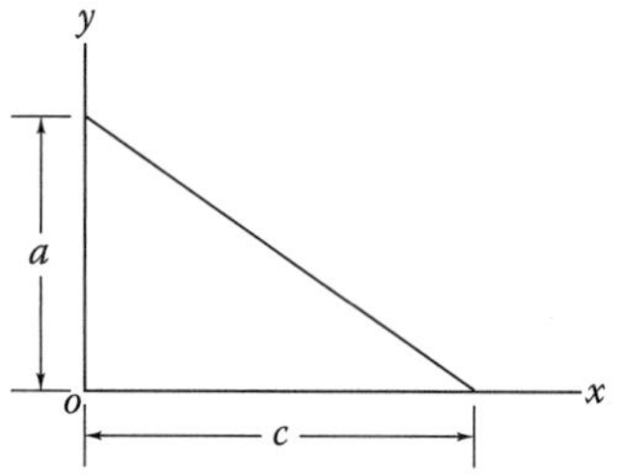

FIGURE 5.2
Triangular mass of unit depth and uniform mass per unit area $\bar{\gamma}$.

$$m = \bar{\gamma}\int_0^c y\, dx = \bar{\gamma}\int_0^c \left(a - \frac{a}{x}x\right) dx = \bar{\gamma}\Big|_0^c \left(ax - \frac{ax^2}{2c}\right) = \frac{ac\bar{\gamma}}{2}.$$

This result is straightforward and scarcely surprising, the area of the plate being $ac/2$ times $\bar{\gamma}$, the density per unit area. We will now apply Equation 5.2 to find the mass center location for the plate as

$$m\boldsymbol{b}_{og} = m(\boldsymbol{i}b_{ogx} + \boldsymbol{j}b_{ogy}) = \int_A \boldsymbol{\rho}\bar{\gamma}\, dx\, dy = \int_A (\boldsymbol{i}x + \boldsymbol{j}y)\bar{\gamma}\, dx\, dy.$$

Hence,

$$mb_{ogx} = \bar{\gamma}\int_A x\, dx\, dy, \quad mb_{ogy} = \bar{\gamma}\int_A y\, dx\, dy.$$

These area integrals are stated more efficiently as

$$\begin{aligned} mb_{ogx} &= \bar{\gamma}\int_0^c xy\ dx = \bar{\gamma}\int_0^c x\left(a - \frac{ax}{c}\right)dx \\ &= \bar{\gamma}\Big|_0^c\left(\frac{ax^2}{2} - \frac{ax^3}{3c}\right) = \bar{\gamma}\frac{ac^2}{6} = m\frac{c}{3} \\ mb_{ogy} &= \bar{\gamma}\int_0^c yx\ dy = \bar{\gamma}\int_0^c y\left(c - \frac{cy}{a}\right)dy \\ &= \bar{\gamma}\Big|_0^a\left(\frac{cy^2}{2} - \frac{cy^3}{3a}\right) = \bar{\gamma}\frac{ca^2}{6} = m\frac{a}{3}. \end{aligned} \tag{5.4}$$

Hence, the mass center is located in the x, y system by the vector $\boldsymbol{b}_{og} = \boldsymbol{i}c/3 + \boldsymbol{j}a/3$, which should strike a familiar chord in your memory.

Proceeding from Equation 5.3 for the moment-of-inertia definition, we obtain

$$\begin{aligned} I_{zzo} &= \int_A \bar{\gamma}(x^2 + y^2)dx\ dy \\ &= \bar{\gamma}\int_0^a\left[\int_0^{c(1-\frac{y}{a})} (x^2 + y^2)dx\right]dy = \frac{m(c^2 + a^2)}{6}. \end{aligned} \tag{5.5}$$

A considerable amount of work is hidden in getting across the last equality sign.

For fixed rotation about an axis, the "radius of gyration" is sometimes used to define a moment of inertia, and is defined by $I_o = mk_g^2$. The radius of gyration is the radius at which all of the mass could be concentrated to obtain the same moment of inertia. From Equation 5.5, the radius of gyration for the present example about the z-axis is

$$k_g = \frac{(c^2 + a^2)^{1/2}}{\sqrt{6}}.$$

The radius of gyration definition is used more often for composite bodies.

This book concerns dynamics as opposed to calculus and does not include exercises for calculating inertia properties. Appendix C provides a summary of inertia properties for many common rigid bodies. The problems included in this chapter will regularly include circular disks, uniform rods, and rectangular plates. If you work through many of the exercise problems, you should soon commit the inertia properties of these bodies to memory.

A particle has all of its mass concentrated at a point and has negligible dimensions of length, breadth, depth, etc. Rigid bodies have finite dimensions, yielding properties such as area, volume, and moment of inertia. Observe that continuing to reduce the dimensions of the triangular plate in Figure 5.2 will cause the moment of inertia defined by Equation 5.5 to rapidly approach zero, which is consistent with a particle.

5.2.2 The Parallel-Axis Formula

Our chore in this section is to determine a relationship between the moments of inertia of a rigid body as defined in two separate coordinate systems whose axes are parallel. Figure 5.3a illustrates two coordinate systems whose origins and orientations are fixed with respect to a rigid body. The x-, y-, z-axes are parallel, respectively, to the $\bar{x}$-, $\bar{y}$-, $\bar{z}$-axes. The mass center of the body is located at the origin of the $\bar{x}$, $\bar{y}$, $\bar{z}$ coordinate system and is located in the x, y, z coordinate system by the vector $\boldsymbol{b}_{og} = \boldsymbol{i}b_{ogx} + \boldsymbol{j}b_{ogy} + \boldsymbol{k}b_{ogz}$. The question of interest is: Suppose that we know the moment of inertia about the $\bar{z}$-axis, what is it about the z-axis? You could obviously reverse the question and start, knowing the moment of inertia about the z-axis, and want to determine it for the $\bar{z}$-axis.

Figure 5.3b provides an end view along the z-axis. A point that is located in the $\bar{x}$, $\bar{y}$, $\bar{z}$ coordinate system by the vector $\bar{\boldsymbol{\rho}} = \bar{\boldsymbol{i}}\bar{x} + \bar{\boldsymbol{j}}\bar{y} + \bar{\boldsymbol{k}}\bar{z}$ is located in the x, y, z system by $\boldsymbol{\rho} = \bar{\boldsymbol{\rho}} + \boldsymbol{b}_{og}$, or $\boldsymbol{\rho} = \boldsymbol{i}x + \boldsymbol{j}y + \boldsymbol{k}z = \boldsymbol{b}_{og} + \bar{\boldsymbol{\rho}} = \boldsymbol{i}(b_{ogx} + \bar{x}) + \boldsymbol{j}(b_{ogy} + \bar{y}) + \boldsymbol{k}z$; hence,

$$x = b_{ogx} + \bar{x}, \quad y = b_{ogy} + \bar{y}.$$

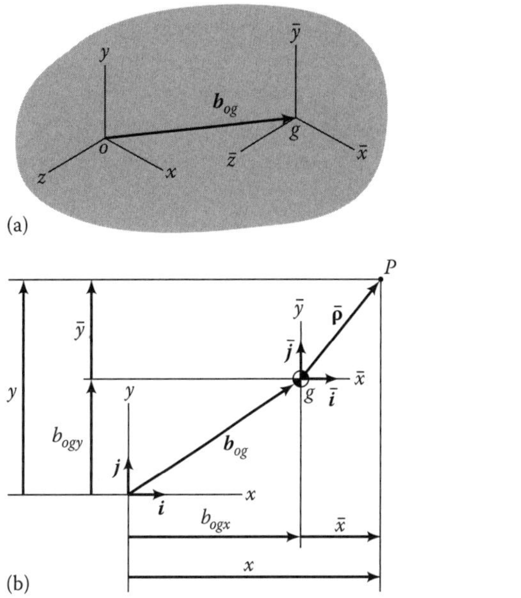

FIGURE 5.3
(a) Two parallel coordinate systems fixed in a rigid body. (b) End view looking in along the z-axis.

The moment of inertia about a z-axis through o is defined to be

$$I_{zzo} = \int_m (x^2 + y^2)\, dm.$$

Substituting for x and y gives

$$\begin{aligned} I_{zzo} &= \int_m \left(b_{ogx}^2 + 2b_{ogx}\bar{x} + \bar{x}^2 + b_{ogy}^2 + 2b_{ogy}\bar{y} + \bar{y}^2\right) dm \\ &= \int_m (\bar{x}^2 + \bar{y}^2) dm + \left(b_{ogx}^2 + b_{ogy}^2\right) m \\ &\quad + 2b_{ogx}\int_m \bar{x}\, dm + 2b_{ogy}\int_m \bar{y}\, dm \\ &= I_{\bar{z}\bar{z}} + m(b_{ogx}^2 + b_{ogy}^2) + 2b_{ogx}\int_m \bar{x}\, dm + 2b_{ogy}\int_m \bar{y}\, dm. \end{aligned} \quad (5.6)$$

Because the mass center is at the origin of the $\bar{x}$, $\bar{y}$, $\bar{z}$ coordinate system, the last two integrals in Equation 5.6 are zero, and we obtain

$$I_{zz} = I_{\bar{z}\bar{z}} + m(b_{ogx}^2 + b_{ogy}^2) = I_g + m(b_{ogx}^2 + b_{ogy}^2). \quad (5.7)$$

Note that this expression is only valid when the mass center of the body is at the origin of the $\bar{x}$, $\bar{y}$, $\bar{z}$ *coordinate system*. We use Equation 5.7 to relate the moments of inertia for various "z-axes" that are perpendicular to the x, y plane. To simplify the notation, we will simply use subscripts on the moment of inertia to identify the point

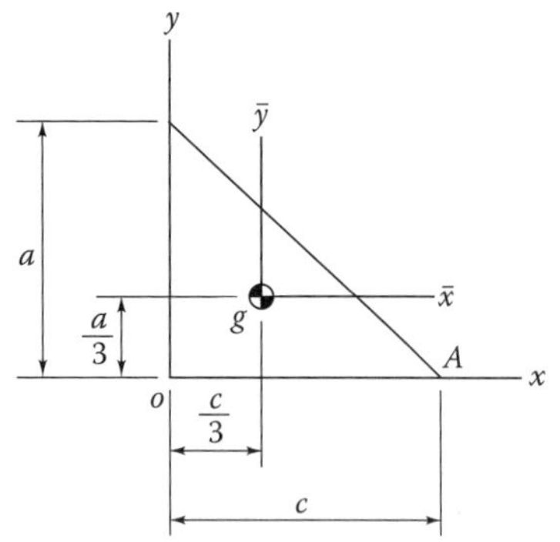

FIGURE 5.4
Two coordinate systems in the triangular plate of Figure 5.2.

through which an axis passes; for example, in Equation 5.7, the subscript g denotes the mass center.

Figure 5.4 provides an illustration of the plate in Figure 5.2, with the origin of the $\bar{x}$, $\bar{y}$, $\bar{z}$ system at the plate's mass center. We will use this figure to demonstrate the parallel-axis formula. We previously determined $I_{zz}=I_o$, the z moment of inertia for an axis through the origin of the x, y, z system. (The result is given in Equation 5.5.) Suppose that we now want the moment of inertia about a z-axis (perpendicular to the plate) through point A at the right-hand corner. To meet this objective using Equation 5.7, we need to go from o, the origin of the x, y, z system, to g at the mass center and then from g to A. The vector from point o to g is $\boldsymbol{b}_{og}=\boldsymbol{i}c/3+\boldsymbol{j}a/3$; hence, applying Equation 5.7 to find I_g, the moment of inertia about an axis through the mass center, gives

$$I_g = I_o - m|\boldsymbol{b}_{og}|^2 = \frac{m}{6}(c^2+a^2) - \frac{m}{9}(a^2+c^2) = \frac{m}{18}(c^2+a^2).$$

The vector from point g to A is $\boldsymbol{b}_{gA}=\boldsymbol{i}2c/3-\boldsymbol{j}a/3$. Hence, applying Equation 5.7 gives

$$\begin{aligned} I_A = I_g + m|\boldsymbol{b}_{gA}|^2 &= \frac{m}{18}(c^2+a^2) + m\left(\frac{4c^2}{9}+\frac{a^2}{9}\right) \\ &= \frac{m}{18}(9c^2+3a^2) \end{aligned}$$

You should confirm that trying to use Equation 5.7 to go *directly* from the moment of inertia at point o to find I_A gives a wrong answer.

Example Problem 5.1

Figure XP5.1a illustrates a welded assembly consisting of a uniform bar with dimensions $l_1=25$ cm, $d_1=2.5$ cm

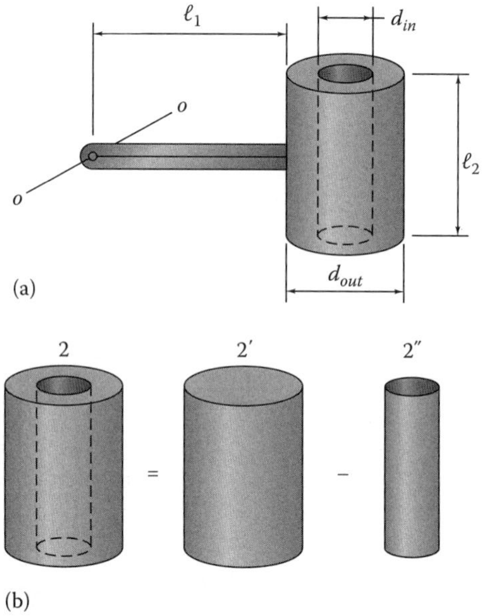

FIGURE XP5.1
(a) Assembly rotating about axis o–o. (b) Modeling approach for a hollow cylinder.

attached to a hollow cylinder with length $l_2=20$ cm and inner and outer diameters $d_{out}=150$ mm; $d_{in}=75$ mm, respectively. The assembly is made from steel with density $\gamma=7750$ kg/m^3. The assembly rotates about the o–o axis, and the following engineering-analysis tasks apply:

a. Determine the moment of inertia of the assembly about the o–o axis.
b. Determine the assembly's radius of gyration for rotation about the o–o axis.
c. Determine the assembly's mass center location.

Solution

We will break the assembly into two pieces and analyze the bar and hollow cylinder separately. From Appendix C, the moment of inertia for a slender bar about a transverse axis at its end is $I_{end}=ml^2/3$, and the moment of inertia for a transverse axis through the mass center is $I_g=ml^2/12$. These results are related to each other via the parallel-axis formula as

$$I_{end} = I_g + m\left(\frac{1}{2}\right)^2 = \frac{ml^2}{12} + \frac{ml^2}{4} = \frac{ml^2}{3}. \tag{i}$$

You should work at committing the bar's inertia-property definition to memory.

The bar's mass is

$$\begin{aligned} m_1 &= \left(\frac{\pi d_1^2}{4}\right) l_1 \gamma = \frac{3.14159 \times 0.025^2\,(\mathrm{m}^2)}{4} 0.25\,(\mathrm{m}) \times 7750\left(\frac{\mathrm{kg}}{\mathrm{m}^3}\right) \\ &= 0.951\ \mathrm{kg}. \end{aligned} \tag{ii}$$

Hence, from Equation (i),

$$I_{1o} = \frac{m_1 l_1^2}{3} = \frac{0.951\ (\mathrm{kg}) \times 0.25^2\ (\mathrm{m}^2)}{3} = 0.0198\ \mathrm{kg\ m}^2 \quad \text{(iii)}$$

Note that all dimensions are converted to meters in accordance with the kg/m^3 dimensions of density.

Figure XP5.1b shows that the hollow cylinder can be "constructed" by subtracting a solid cylinder (denoted 2″) with the inner diameter d_{in} from a solid cylinder (denoted 2′) with the outer radius d_{out}. Starting with the inner cylinder

$$\begin{aligned} m_{2''} &= \left(\frac{\pi d_{in}^2}{4}\right) \times l_2 \times \gamma \\ &= \left(\frac{3.14159 \times 0.075^2\ (\mathrm{m}^2)}{4}\right) \times 0.2\ (\mathrm{m}) \times 7750 \left(\frac{\mathrm{kg}}{\mathrm{m}^3}\right) \\ &= 6.847\ \mathrm{kg}. \end{aligned} \quad \text{(iv)}$$

From Appendix C, the moment of inertia for a transverse axis through the cylinder's mass center is

$$\begin{aligned} I_{2''g} &= m_{2''} \times \left(\frac{r_{in}^2}{4} + \frac{l_2^2}{12}\right) \\ &= 6.847\ (\mathrm{kg}) \times \left[\frac{0.0375^2\ (\mathrm{m})^2}{4} + \frac{0.2^2\ (\mathrm{m}^2)}{12}\right] \\ &= 0.0252\ \mathrm{kg\ m}^2. \end{aligned} \quad \text{(v)}$$

For the solid outer cylinder,

$$\begin{aligned} m_{2'} &= \left(\frac{\pi d_{out}^2}{4}\right) \times l_2 \times \gamma \\ &= \left[\frac{3.14159 \times 0.15^2\ (\mathrm{m}^2)}{4}\right] \times 0.2\ (\mathrm{m}) \times 7750 \left(\frac{\mathrm{kg}}{\mathrm{m}^3}\right) \\ &= 28.12\ \mathrm{kg}, \end{aligned} \quad \text{(vi)}$$

and

$$\begin{aligned} I_{2'g} &= m_{2'} \times \left(\frac{r_{out}^2}{4} + \frac{l_2^2}{12}\right) \\ &= 28.12\ (\mathrm{kg}) \times \left[\frac{0.075^2\ (\mathrm{m}^2)}{4} + \frac{0.2^2\ (\mathrm{m}^2)}{12}\right] \\ &= 0.1332\ \mathrm{kg\ m}^2. \end{aligned} \quad \text{(vii)}$$

From Equations (iv) through (vii),

$$\begin{aligned} m_2 &= m_2' - m_2'' = 28.12 - 6.847 = 21.27\ \mathrm{kg} \\ I_2 &= I_2' - I_2'' = 0.1332 - 0.0252 = 0.1080\ \mathrm{kg\ m}^2. \end{aligned}$$

These values conclude the individual results for the hollow cylinder.

Combining the results for the bar and the hollow cylinder via the parallel-axis formula, the assembly moment of inertia is

$$\begin{aligned} I_0 &= I_{10} + I_2 + m_2 \times b_{2g}^2 \\ &= (0.0198 + 0.1080)\ (\mathrm{kg\ m}^2) + 21.27\ (\mathrm{kg}) \times 0.325^2\ (\mathrm{m}^2) \\ &= 2.374\ \mathrm{kg\ m}^2, \end{aligned}$$

and concludes Task a. The radius of gyration k_g is obtained from $I_o = mk_g^2$ where I_o is the assembly moment of inertia; hence,

$$\begin{aligned} 2.347\ (\mathrm{kg\ m}^2) &= (0.951 + 21.27)\ (\mathrm{kg}) \times k_g^2\ (\mathrm{m}^2) \\ &= 22.22 k_g^2\ (\mathrm{kg\ m}^2) \Rightarrow kg = 0.325\ \mathrm{m}, \end{aligned} \quad \text{(viii)}$$

and concludes Task b.

The assembly mass location is found from

$$\begin{aligned} md &= (m_1 + m_2)d = 22.22d \\ &= (m_1 d_1 + m_2 d_2) = 0.951\ (\mathrm{kg}) \times \frac{0.25\ (\mathrm{m})}{2} \\ &\quad + 21.27\ (\mathrm{kg}) \times 0.325\ (\mathrm{m}) = 7.150\ (\mathrm{kg\ m}); \end{aligned}$$

hence,

$$d = 0.322\ \mathrm{m}.$$

Note that the mass center location defined by d is not related to the radius of gyration k_g.

5.3 Governing Force and Moment Equations for a Rigid Body

As noted in Section 5.1, we start this chapter with the knowledge that Newton's second law of motion $\sum f = m\ddot{r}$ can be used to define the governing differential equations of motion for a *particle*. In this section, we will derive corresponding governing equations of motion for planar motion of a *rigid body*. The derivation will proceed from $\sum f = m\ddot{r}$ for a particle, plus the particle kinematics results of Chapter 4, to derive governing force and moment differential equations of motion for planar motion of a rigid body.

5.3.1 Force Equation

Figure 5.5 illustrates a rigid body acted on by several external forces. To derive the force differential equations of motion for the body, we start with the required

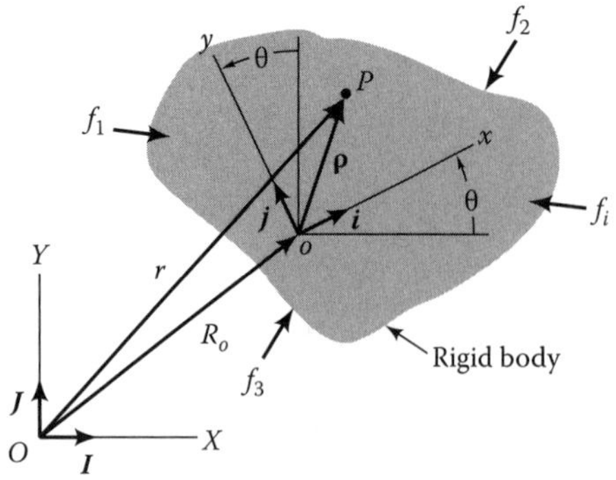

FIGURE 5.5
Rigid body acted on by external forces. The x, y, z coordinate system is fixed in the rigid body; the X, Y, Z system is an inertial coordinate system.

kinematics. Figure 5.5 illustrates an inertial X, Y, Z coordinate system and an x, y, z coordinate system that is fixed in the rigid body. The origin of the x, y, z system is denoted as o. The x, y and X, Y planes lie in the plane of Figure 5.5, and all motion occurs in this common plane. The Z- and z-axes are parallel and directed out of the plane of the figure. The angle θ defines the orientation of the rigid body (and the x, y, z coordinate system) with respect to the X, Y, Z system. Consistent with the positive right-hand screw rule, $\boldsymbol{\omega} = \boldsymbol{k}\dot{\theta}$ is the angular velocity of the rigid body (and the x, y, z coordinate system) with respect to X, Y, Z coordinate system.

The vector $\boldsymbol{R}_o = \boldsymbol{I}R_{oX} + \boldsymbol{J}R_{oY}$ locates the origin of the x, y, z system in the X, Y, Z system. A point P in the rigid body is located in the X, Y, Z system by the vector $\boldsymbol{r} = \boldsymbol{I}r_X + \boldsymbol{J}r_Y + \boldsymbol{K}r_Z$. The same point is located in the x, y, z system by the vector $\boldsymbol{\rho} = \boldsymbol{i}x + \boldsymbol{j}y + \boldsymbol{k}z$. Hence, $\boldsymbol{r}$ can be stated as

$$r = \boldsymbol{R}_o + \boldsymbol{\rho}$$

Suppose that we are interested in a particle at point P with dimensions dx, dy, dz and mass $dm = \gamma\, dx\, dy\, dz$ where γ is the mass density of the rigid body. Applying Newton's second law to the particle gives

$$\boldsymbol{f}_P = dm\,\ddot{\boldsymbol{r}} = dm \left.\frac{d^2\boldsymbol{r}}{dt^2}\right|_{X,Y,Z}, \tag{5.8}$$

where
- $\boldsymbol{f}_P$ is the *resultant* force
- $\ddot{\boldsymbol{r}}$ is the acceleration of the particle with respect to the inertial X, Y, Z system

Most of the forces contributing to the *resultant* force $\boldsymbol{f}_P$ arise from internal reaction forces acting between the particle at point P and its infinite number of neighboring points. To obtain the governing force differential equation of motion for the rigid body, we propose to integrate Equation 5.8 over the full mass of the body. Integration of the left-hand side of the equation over the body will cause all of the *internal* forces to cancel, yielding the resultant of the *external* forces acting on the rigid body. The integral expression of Equation 5.8 is

$$\sum \boldsymbol{f}_i = \int_V \ddot{\boldsymbol{r}}\gamma\, dx\, dy\, dz = \int_m \ddot{\boldsymbol{r}}\, dm, \tag{5.9}$$

where $\sum \boldsymbol{f}_i$ is the summation of *external* forces acting on the body illustrated in Figure 5.5. In reviewing Equation 5.9, you might ask, "If all of the motion is supposed to occur in the common x, y and X, Y planes, why does the integral in Equation 5.9 include the z coordinate?" The answer to this question is: While the body's motion is confined to the common plane, the physical mass of the body will normally extend out of the plane.

Our next step in obtaining a useful force/acceleration equation involves a restatement of $\ddot{\boldsymbol{r}}$. Point P (the particle location) and point o (the origin of the x, y, z system) are both fixed in the rigid body; hence, we can use the last of Equations 4.3 from Section 4.3 relating the accelerations of two points in a rigid body as

$$\boldsymbol{a}_P = \boldsymbol{a}_o + \dot{\boldsymbol{\omega}} \times \boldsymbol{r}_{oP} + \boldsymbol{\omega} \times (\boldsymbol{\omega} \times \boldsymbol{r}_{oP}).$$

Since $\boldsymbol{r}$ and $\boldsymbol{R}_o$ locate points P and o, respectively, in the X, Y, Z system, and ρ is the vector from point o to P, the appropriate statement of this equation (in terms of our present variables) is

$$\ddot{\boldsymbol{r}} = \ddot{\boldsymbol{R}}_o + \dot{\boldsymbol{\omega}} \times \boldsymbol{\rho} + \boldsymbol{\omega} \times (\boldsymbol{\omega} \times \boldsymbol{\rho}). \tag{5.10}$$

This is the desired relationship to be substituted into the integral of Equation 5.9.

Recall that $dm = \gamma\, dx\, dy\, dz$; hence, the integration of Equation 5.9 extends over the volume of the rigid body. However, in Equation 5.10, $\ddot{\boldsymbol{R}}_o$, $\dot{\boldsymbol{\omega}}$, and $\boldsymbol{\omega}$ are constant with respect to the x, y, z integration variables and can be brought outside the integral sign. Hence, the result of substituting Equation 5.10 into Equation 5.9 can be stated

$$\sum \boldsymbol{f}_i = m\ddot{\boldsymbol{R}}_o + \dot{\boldsymbol{\omega}} \times \int_m \boldsymbol{\rho}\, dm + \boldsymbol{\omega} \times \left(\boldsymbol{\omega} \times \int_m \boldsymbol{\rho}\, dm\right). \tag{5.11}$$

The vector $\boldsymbol{b}_{og}$ locates the center of mass of the rigid body in the x, y, z system and is defined by

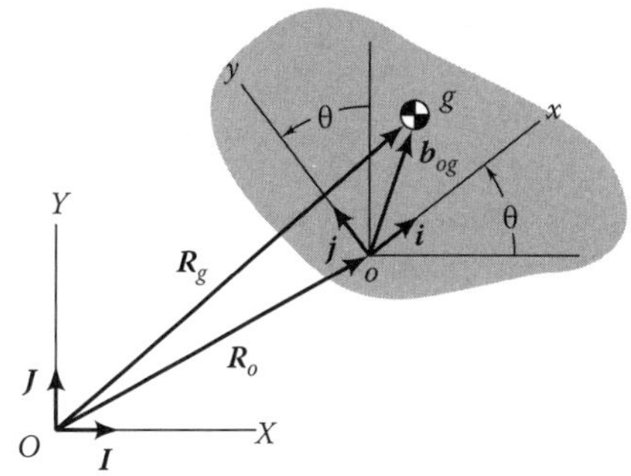

FIGURE 5.6
A rigid body with a mass center located in the body-fixed x, y, z system by the vector $\boldsymbol{b}_{og}$ and located in the X, Y, Z system by $\boldsymbol{R}_g$.

$$m\boldsymbol{b}_{og} = \int_m \boldsymbol{\rho}\, dm = m(\boldsymbol{i}b_{ogx} + \boldsymbol{j}b_{ogy} + \boldsymbol{k}b_{ogz}). \tag{5.12}$$

Substituting from Equation 5.12 into Equation 5.11 gives

$$\sum \boldsymbol{f}_i = m\left[\ddot{\boldsymbol{R}}_o + \dot{\boldsymbol{\omega}} \times \boldsymbol{b}_{og} + \boldsymbol{\omega} \times (\boldsymbol{\omega} \times \boldsymbol{b}_{og})\right]. \tag{5.13}$$

As we will now demonstrate, this result can be stated in a more physically meaningful form. In Figure 5.6, point g denotes the mass center of the rigid body and is located by vectors $\boldsymbol{b}_{og}$ and $\boldsymbol{R}_g$ in the x, y, z and X, Y, Z systems, respectively. As noted previously, point o denotes the origin of the x, y, z system. Hence, Equation 4.3 relating the acceleration of two points in a rigid body applies to points o and g, and can be stated

$$\boldsymbol{a}_g = \boldsymbol{a}_o + \dot{\boldsymbol{\omega}} \times \boldsymbol{r}_{og} + \boldsymbol{\omega} \times (\boldsymbol{\omega} \times \boldsymbol{r}_{og})$$

Introducing the notation of the present development gives

$$\ddot{\boldsymbol{R}}_g = \ddot{\boldsymbol{R}}_o + \dot{\boldsymbol{\omega}} \times \boldsymbol{b}_{og} + \boldsymbol{\omega} \times (\boldsymbol{\omega} \times \boldsymbol{b}_{og}).$$

Hence, Equation 5.13 can be written (finally) as

$$\sum \boldsymbol{f}_i = m\ddot{\boldsymbol{R}}_g. \tag{5.14}$$

In words, Equation 5.14 states that a rigid body can be treated like a particle, in that the summation of external forces acting on the rigid body equals the mass of the body times the acceleration of its mass center with respect to an inertial coordinate system. We have struggled through a considerable amount of math to arrive at this simple but quite significant result.

Equation 5.14 is a vector equation. Writing the components in terms of the X, Y system gives the two scalar equations:

$$\sum f_{iX} = m\ddot{R}_{gX}, \quad \sum f_{iY} = m\ddot{R}_{gY}. \tag{5.15}$$

As with particle kinematics, Equation 5.14 could also be stated in terms of polar or path-coordinate components.

5.3.2 Moment Equation

Figure 5.7 illustrates the rigid body of Figure 5.5 now acted on by several external forces $\boldsymbol{f}_i$ and moments $\boldsymbol{M}_i$. Vector $\boldsymbol{a}_i$ locates the point of action of force $\boldsymbol{f}_i$ in the body-fixed x, y, z coordinate system. As you may recall from statics, the moments acting on the rigid body are "free" vectors, meaning that the resultant moment on the rigid body is independent of their points of application. We now want to derive a moment equation for the rigid body that will be comparable to the force equation provided by Equation 5.14. Returning to Equation 5.8, suppose that we take the moment of both sides of Equation 5.8 about o the origin of the x, y, z coordinate system. In Figure 5.5, the position vector $\boldsymbol{\rho}$ extends from o to a particle at point P. For moments about o, $\boldsymbol{\rho}$ is the moment arm, and the equation is

$$\boldsymbol{\rho} \times \boldsymbol{f}_P = \boldsymbol{\rho} \times dm\,\ddot{\boldsymbol{r}}. \tag{5.16}$$

As with the force equation, the resultant force $\boldsymbol{f}_P$ acting on the differential mass dm is composed mainly of internal forces acting between dm and its infinite number of neighboring points. However, when we integrate over the mass of the body, the moments due to internal forces cancel leaving only the resultant *external* moment due to the external forces and moments. Hence, integrating Equation 5.16 over the mass of the rigid body yields

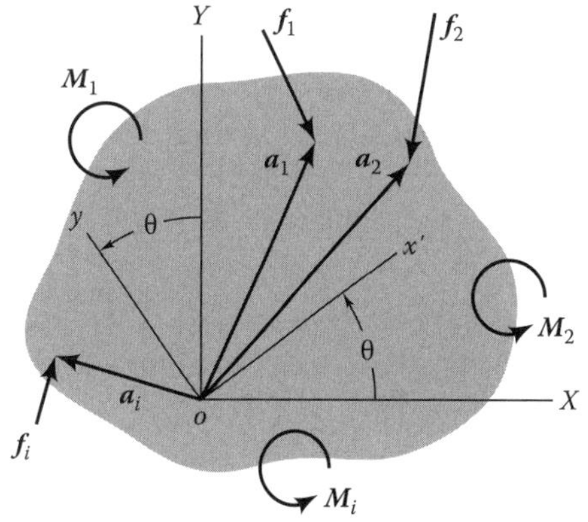

FIGURE 5.7
A rigid body acted on by external forces $\boldsymbol{f}_i$ and external moments $\boldsymbol{M}_i$.

$$\sum (\boldsymbol{a}_i \times \boldsymbol{f}_i) + \sum \boldsymbol{M}_i = \boldsymbol{M}_o = \int_V (\boldsymbol{\rho} \times \ddot{\boldsymbol{r}})\gamma\, dx\, dy\, dz = \int_m (\boldsymbol{\rho} \times \ddot{\boldsymbol{r}})\, dm. \quad (5.17)$$

The vector $\boldsymbol{M}_o$ on the left is the *resultant* external moment acting on the rigid body about o, the origin of the x, y, z coordinate system. Turning our attention to the integral in Equation 5.17, we can use Equation 5.10 to obtain

$$\boldsymbol{\rho} \times \ddot{\boldsymbol{r}} = (\boldsymbol{\rho} \times \ddot{\boldsymbol{R}}_o) + \boldsymbol{\rho} \times (\dot{\boldsymbol{\omega}} \times \boldsymbol{\rho}) + \boldsymbol{\rho}[\boldsymbol{\omega} \times (\boldsymbol{\omega} \times \boldsymbol{\rho})].$$

The vector identity,

$$\boldsymbol{A} \times [\boldsymbol{B} \times (\boldsymbol{B} \times \boldsymbol{A})] = \boldsymbol{B} \times [\boldsymbol{A} \times (\boldsymbol{B} \times \boldsymbol{A})],$$

allows us to rewrite the last term of this equation to obtain

$$\boldsymbol{\rho} \times \ddot{\boldsymbol{r}} = (\boldsymbol{\rho} \times \ddot{\boldsymbol{R}}_o) + \boldsymbol{\rho} \times (\dot{\boldsymbol{\omega}} \times \boldsymbol{\rho}) + \boldsymbol{\omega} \times [\boldsymbol{\rho} \times (\boldsymbol{\omega} \times \boldsymbol{\rho})]. \quad (5.18)$$

We are now ready to substitute this result back into Equation 5.17. Recall in Equation 5.18 that $\rho = \boldsymbol{i}x + \boldsymbol{j}y + \boldsymbol{k}z$ is a function of the variables of integration and that $\ddot{\boldsymbol{R}}_o$, $\dot{\boldsymbol{\omega}}$, and $\boldsymbol{\omega}$ are not; hence substitution from Equation 5.18 into Equation 5.17 gives

$$\boldsymbol{M}_o = m(\boldsymbol{b}_{og} \times \ddot{\boldsymbol{R}}_o) + \int \boldsymbol{\rho} \times (\dot{\boldsymbol{\omega}} \times \boldsymbol{\rho})dm + \boldsymbol{\omega} \times \int [\boldsymbol{\rho} \times (\boldsymbol{\omega} \times \boldsymbol{\rho})]dm. \quad (5.19)$$

The first term in this equation uses the completed integral definition of Equation 5.12 for $\boldsymbol{b}_{og}$. Note that using the vector identity to obtain Equation 5.18 allowed us to move $\boldsymbol{\omega}$ outside the last integral in Equation 5.19.

We need a single scalar equation defining moments about the z-axis from the vector Equation 5.19. We can start this process by stating the first term in Equation 5.19 as

$$\boldsymbol{b}_{og} \times m\ddot{\boldsymbol{R}}_o = m\begin{vmatrix} \boldsymbol{i} & \boldsymbol{j} & \boldsymbol{k} \\ b_{ogx} & b_{ogy} & b_{ogz} \\ \ddot{R}_{ox} & \ddot{R}_{oy} & 0 \end{vmatrix} = \begin{matrix} -\boldsymbol{i}mb_{ogz}\ddot{R}_{oy} \\ +\boldsymbol{j}mb_{ogz}\ddot{R}_{ox} \\ +\boldsymbol{k}m(b_{ogx}\ddot{R}_{oy} - b_{ogy}\ddot{R}_{ox}) \end{matrix} \quad (5.20)$$

In carrying out the cross product, note that $\ddot{R}_o$ is stated in terms of its components in the x, y, z coordinate system, versus the customary X, Y, Z system.

To proceed further, we need to evaluate the integrals in Equation 5.19. Defining the vectors in Equation 5.19 in terms of their components gives

$$\boldsymbol{\rho} = \boldsymbol{i}x + \boldsymbol{j}y + \boldsymbol{k}z, \quad \boldsymbol{\omega} = \boldsymbol{k}\dot{\theta}, \quad \dot{\boldsymbol{\omega}} = \boldsymbol{k}\ddot{\theta}.$$

Hence,

$$\boldsymbol{\omega} \times \boldsymbol{\rho} = \boldsymbol{k}\dot{\theta} \times (\boldsymbol{i}x + \boldsymbol{j}y + \boldsymbol{k}z) = \dot{\theta}(\boldsymbol{j}x - \boldsymbol{i}y)$$
$$\dot{\boldsymbol{\omega}} \times \boldsymbol{\rho} = \ddot{\theta}(\boldsymbol{j}x - \boldsymbol{i}y),$$

and

$$\boldsymbol{\rho} \times (\boldsymbol{\omega} \times \boldsymbol{\rho}) = \begin{vmatrix} \boldsymbol{i} & \boldsymbol{j} & \boldsymbol{k} \\ x & y & z \\ -y\dot{\theta} & x\dot{\theta} & 0 \end{vmatrix} = \begin{matrix} -\boldsymbol{i}\dot{\theta}xz \\ -\boldsymbol{j}\dot{\theta}yz \\ +\boldsymbol{k}\dot{\theta}(x^2 + y^2) \end{matrix} \quad (5.21)$$

Similarly,

$$\boldsymbol{\rho} \times (\dot{\boldsymbol{\omega}} \times \boldsymbol{\rho}) = -\boldsymbol{i}\ddot{\theta}xz - \boldsymbol{j}\ddot{\theta}yz + \boldsymbol{k}\ddot{\theta}(x^2 + y^2) \quad (5.22)$$

Substituting from Equations 5.20 through 5.22 into Equation 5.19 gives the z-component equation

$$\boldsymbol{k}M_{oz} = \boldsymbol{k}m(\boldsymbol{b}_{og} \times \ddot{\boldsymbol{R}}_o)_z + \boldsymbol{k}\ddot{\theta}\int_m (x^2 + y^2)dm + \boldsymbol{k}\dot{\theta} \times \boldsymbol{k}\dot{\theta}\int_m (x^2 + y^2)dm. \quad (5.23)$$

The last term is zero because $\boldsymbol{k} \times \boldsymbol{k} = 0$. This outcome is the result of our restrictions to planar motion. In *general*, 3D rigid-body kinetics, the cross product is not zero, leading to interesting "gyroscopic" phenomena.

The integral in Equation 5.23 is the I_{zz} moment of inertia for an axis through o; hence, Equation 5.23 can be stated (finally) as

$$\sum M_{oz} = I_o\ddot{\theta} + m(\boldsymbol{b}_{og} \times \ddot{\boldsymbol{R}}_o)_z. \quad (5.24)$$

This is the governing moment differential equation of motion for θ. In using Equation 5.24, recall that M_{oz} is the z component of the resultant moment acting on the rigid body about o, I_o is the moment of inertia for the z-axis through point o (the origin of the x, y, z coordinate system), $\boldsymbol{b}_{og}$ goes from point o to g and locates the mass center in the x, y system, and $\ddot{\boldsymbol{R}}_o$ is the acceleration of point o with respect to the X, Y system. Equation 5.24 and the two components of

Equation 5.14 complete the governing equations for planar motion of a rigid body.

5.3.2.1 Reduced Forms for the Moment Equation

We can and will use the moment Equation 5.24 "as is" in solving many problems; however, two simplified forms of the equation will also prove to be very useful. First, consider the consequence of locating the rigid body's mass center at o, the origin of the x, y, z coordinate system. With this position, the vector $\boldsymbol{b}_{og}$ is zero, and Equation 5.24 reduces to

$$M_{gz} = I_g \ddot{\theta} \quad (5.25)$$

This equation is *only* correct for moments taken about the mass center of the rigid body.

The second useful form for Equation 5.24 arises when point o is fixed in the (inertial) X, Y, Z coordinate system yielding $\ddot{\boldsymbol{R}}_o = 0$. This outcome reduces Equation 5.24 to

$$M_{oz} = I_o \ddot{\theta} \quad (5.26)$$

This reduced equation holds when point o is fixed in inertial space, that is, when point o's acceleration is zero. The next section concerns problems involving fixed-axis rotation for which Equation 5.26 specifically applies.

5.4 Kinetic Energy for Planar Motion of a Rigid Body

In Section 3.6.1, the following general equation was developed

$$\text{Work}_{n.c.} = (T_2 + V_2) - (T_1 + V_1) \quad (5.27)$$

for a particle. Given that a rigid body is an infinite collection of particles, this equation also applies to motion of a rigid body. However, before applying Equation 5.27 to planar-dynamics problems, we need to develop an expression for the kinetic energy of a rigid body in planar motion. Developing this result is our present objective.

In Section 5.3, we developed force and moment equations for planar motion for a rigid body, proceeding from $\sum f = m\ddot{r}$ for a particle. Our present objective is the development of an expression for the kinetic energy of a rigid body, starting from $T = mv^2/2$ for a particle. Figure 5.8 shows a rigid body with an imbedded x, y, z coordinate system and an inertial X, Y coordinate system. The origin of the x, y, z system is located in the X, Y plane by $\boldsymbol{R}_o$. The body's mass center is located in the x, y, z system by the position vector $\boldsymbol{b}_{og}$ defined earlier as

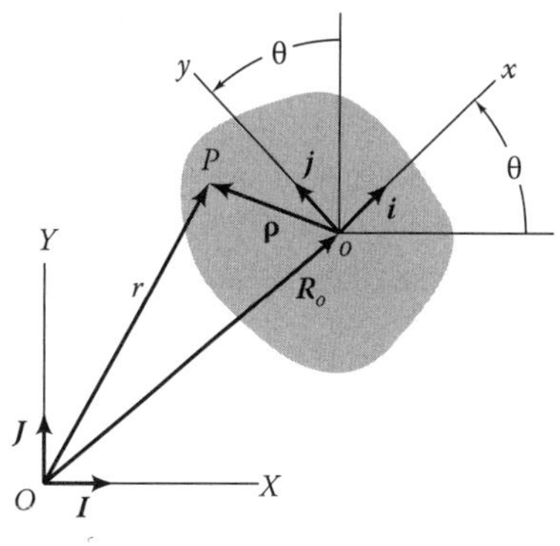

FIGURE 5.8
Rigid body with an imbedded x, y coordinate system. The origin of the x, y system, point o, is located in the inertial X, Y system by $\boldsymbol{R}_o$.

$$m\boldsymbol{b}_{og} = m(\boldsymbol{i}b_{ogx} + \boldsymbol{j}b_{ogy} + \boldsymbol{k}b_{ogz}) = \int_V \boldsymbol{\rho}\gamma\, dx\, dy\, dz \int_m \boldsymbol{\rho}\, dm, \quad (5.2)$$

where $\gamma(x, y, z)$ is the mass density of the body at point P. A point P in the body is located in the X, Y plane by the position vector $\boldsymbol{r}$ and in the x, y system by $\boldsymbol{\rho} = \boldsymbol{i}x + \boldsymbol{j}y + \boldsymbol{k}z$.

The kinetic energy of the mass is

$$T = \frac{1}{2}\int_m v^2\, dm = \frac{1}{2}\int_m (\dot{\boldsymbol{r}}\cdot\dot{\boldsymbol{r}})\, dm, \quad (5.28)$$

where $\dot{\boldsymbol{r}}$ is the velocity of a particle of mass dm at point P with respect to the X, Y coordinate system. Since points o and P are both fixed in the rigid body, we can use the first of Equation 4.3 to state $\boldsymbol{v}_P = \boldsymbol{v}_o + \boldsymbol{\omega} \times \boldsymbol{r}_{oP}$, or in terms of our current variables, $\dot{\boldsymbol{r}} = \dot{\boldsymbol{R}}_o + \boldsymbol{\omega} \times \boldsymbol{\rho}$, and

$$\dot{\boldsymbol{r}}\cdot\dot{\boldsymbol{r}} = |\dot{\boldsymbol{R}}_o|^2 + 2\dot{\boldsymbol{R}}_o\cdot\boldsymbol{\omega} \times \boldsymbol{\rho} + |\boldsymbol{\omega} \times \boldsymbol{\rho}|^2. \quad (5.29)$$

We are shortly going to substitute from Equation 5.29 into the integral of Equation 5.28; however, recall that the integration in Equation 5.28 is carried out for the x, y, z coordinates. $\dot{\boldsymbol{R}}_o$, the velocity of point o, is not a function of the variables of integration, and can come outside the integral. Continuing,

$$\boldsymbol{\omega} \times \boldsymbol{\rho} = \boldsymbol{k}\dot{\theta} \times (\boldsymbol{i}x + \boldsymbol{j}y + \boldsymbol{k}z) = \dot{\theta}(\boldsymbol{j}x - \boldsymbol{j}y), \quad (5.30)$$

and

$$|\boldsymbol{\omega} \times \boldsymbol{\rho}|^2 = (\boldsymbol{\omega} \times \boldsymbol{\rho})\cdot(\boldsymbol{\omega} \times \boldsymbol{\rho}) = \dot{\theta}^2(x^2 + y^2)$$
$$\int (\boldsymbol{\omega} \times \boldsymbol{\rho})dm = \boldsymbol{\omega} \times \int \boldsymbol{\rho}\, dm = \boldsymbol{\omega} \times m\boldsymbol{b}_{og} \tag{5.31}$$

Substituting into Equation 5.28 gives

$$T = \frac{m|\dot{\boldsymbol{R}}_o|^2}{2} + \dot{\boldsymbol{R}}_o\cdot(\boldsymbol{\omega} \times m\boldsymbol{b}_{og}) + \frac{I_o\dot{\theta}^2}{2}, \tag{5.32}$$

where,

$$I_o = I_{zzo} = \int (x^2 + y^2)\, dm,$$

is the moment of inertia about a z-axis through point o, the origin of the x, y, z system. This general equation has a limited range of useful applications. A much more useful definition for the kinetic energy is obtained if the origin of the x, y, z system coincides with the body's mass center, yielding $\boldsymbol{b}_{og} = 0$ and reducing Equation 5.32 to

$$T = \frac{m|\dot{\boldsymbol{R}}_g|^2}{2} + \frac{I_g\dot{\theta}^2}{2}. \tag{5.33}$$

This equation states that the kinetic energy of a rigid body is the sum of the following terms:

a. The translational energy of the body assuming that all of its mass is concentrated at the mass center.
b. The rotational kinetic energy for rotation about the mass center.

Equation 5.33, rather than Equation 5.32, is almost always the preferred definition for the kinetic energy of plane motion for a rigid body.

Think about driving a nail through point o in Figure 5.63, so that the only possible motion is pure rotation about o. For this circumstance, $\dot{\boldsymbol{R}}_o = 0$ in Equation 5.32, and the following simplified definition applies

$$T = \frac{I_o\dot{\theta}^2}{2}. \tag{5.34}$$

Equation 5.34 defines the kinetic energy of the body for pure rotation about an axis through a point o fixed in space. This equation is the *energy* analog of the *moment* Equation 5.26 for rotation about a fixed axis and will be useful in working the examples that follow.

5.5 Fixed-Axis-Rotation Applications of the Force, Moment, and Energy Equations

In this section, we derive equations of motion for bodies involving rotation about a fixed axis. The differential equations of motion will frequently resemble the equations of motion that were presented in Chapter 3 for motion of a particle in a straight line.

5.5.1 Rotor in Frictionless Bearings: Moment Equation

We begin with the rotor illustrated in Figure 5.9. A circular disk of radius r and mass m is supported by a rigid shaft, which is in turn supported by two frictionless bearings. The mass center of the disk coincides with the axis of the bearings. The clockwise moment $M(t)$ is applied to the shaft. The appropriate engineering task is to *derive the differential equation of motion for the rotor*.

To obtain the governing equation of motion we can either state $M_{oz} = I_o\ddot{\theta}$ for an axis through the center of the shaft or $M_{oz} = I_o\ddot{\theta}$ since the mass center lies on an axis through the shaft.

For the present example, $M_{oz} = M(t)$ is the only moment acting on the system, and we only need to substitute $I_o = mr^2/2$ from Appendix C into $M_{oz} = I_o\ddot{\theta}$ to obtain the EOM

$$\frac{mr^2}{2}\ddot{\theta} = M(t). \tag{5.35}$$

We assume that the radius and mass of the shaft are small in comparison to the radius and mass of the disk. If the mass and radius of the shaft were provided, we could easily calculate its moment of inertia about the axis of rotation and add it to the disk's moment of inertia. The moment $M(t)$ is positive because it is acting in the $+\theta$ direction. This is basically the same second-order differential equation obtained for a particle of

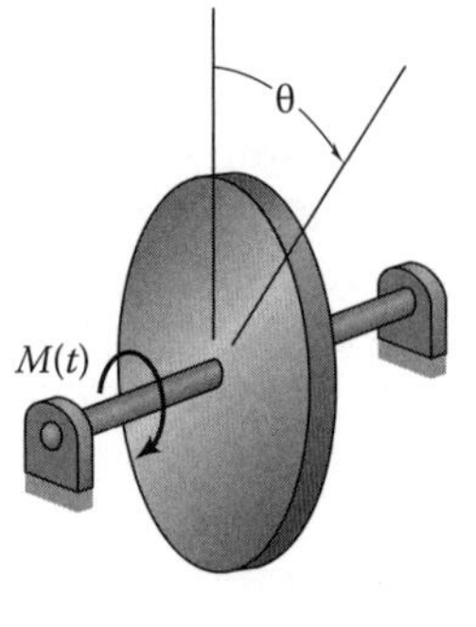

FIGURE 5.9
A disk mounted on a massless shaft, supported by two frictionless bearings and acted on by the applied torque $M(t)$.

mass m acted on by the force $f(t)$, namely, $m\ddot{x} = f(t)$, where x locates the particle in an inertial coordinate system.

5.5.2 Rotor in Frictionless Bearings: Energy Equation

We just derived the equation of motion from the moment equation and will eventually perform the same task here, using the work–energy equation. However, to ''warm up,'' we will first consider the following engineering task. *Assume that the rotor has an initial angular velocity of* $\dot{\theta}(0) = \omega_0$ *and is acted on by a constant drag moment* $\overline{M}$*; how many revolutions will it take to come to rest?*

There is no change to the potential energy, and the final kinetic energy is zero; hence, applying $T_o = I_o\dot{\theta}^2/2$ gives

$$\text{Work}_{n.c.} = \Delta(T + V) \Rightarrow \text{Work}_{n.c.} = (0 + 0) - \left(\frac{I_o\omega_o^2}{2} + 0\right). \qquad (5.36)$$

To complete the task, we need to calculate the work done by the resistance torque. We know that the differential work due to an applied force f acting through the differential distance ds is $dWk = f \cdot ds$. We can proceed from this result to replace a moment M by a force acting at a fixed radius $\bar{r}$, such that $M = f\bar{r}$. When the moment M rotates through the differential angle $d\theta$, the force will act through the arc distance $ds = \bar{r}\,d\theta$, and the differential work will be

$$dWk = f\,ds = f\,\bar{r}\,d\theta = M\,d\theta. \qquad (5.37)$$

This differential work result is important and will be used throughout the balance of the book.

Using Equation 5.37, Equation 5.36 becomes

$$\int_o^{\Delta\theta} -\overline{M}\,d\theta = -\overline{M}\,\Delta\theta = -\left(\frac{I_o\omega_o^2}{2}\right)$$

$$\Rightarrow \Delta\text{revolutions} = \frac{\Delta\theta}{2\pi} = \frac{I_o\omega_o^2}{4\pi\overline{M}}, \qquad (5.38)$$

and completes this engineering task. The work integral is negative because it decreases the energy of the system. This solution is a simple, advantageous, and direct application of the work–energy equation.

Using the work–energy equation can be easier than the moment equation, because a free-body diagram is not required. Further, using the moment equation to complete the same task would require either (i) integrating $(mr^2/2)\ddot{\theta} = M(t) = -\overline{M}$ once with respect to time to find t_f the time at which the rotor would reach zero velocity and then integrating again to find $\theta(t_f)$ or (ii) using the energy-integral substitution $\ddot{\theta} = d(\dot{\theta}^2/2)/d\theta$ and integrating with respect to θ to get Equation 5.38. The work–energy equation starts with this expression.

Consider the following additional task. *Assume that the rotor is acted on by the positive (in the direction of* $+\theta$*) applied moment* $M(t)$*, and derive the equation of motion.* For this task, the work–energy equation gives

$$\int_o^{\theta} M(t)\,dx = \frac{I_{oz}\dot{\theta}^2}{2} - T_0 = \frac{I_{oz}\dot{\theta}^2}{2} - \frac{I_o\omega_o^2}{2}, \qquad (5.39)$$

where T_0 is the initial kinetic energy. Equation 5.37 was used as $d\text{Work} = M\,d\theta$ to develop the work integral on the left-hand side of this equation. In Equation 5.38, $M(t) = \overline{M}$ is constant, and we could complete the integral. In the present case, $M(t)$ is a general function of time (not θ), and the integral cannot be completed. However, we can still differentiate Equation 5.39 with respect to θ to obtain the EOM

$$I_{oz}\frac{d}{d\theta}\left(\frac{\dot{\theta}^2}{2}\right) = I_{oz}\ddot{\theta} = M(t),$$

which completes the task.

5.5.3 Rotor in Bearings with Viscous Drag: Moment Equation

Assuming that the rotor of Figure 5.9 is supported in journal bearings and that the flow is laminar, the drag torque at each bearing is proportional to the rotor's angular velocity $\dot{\theta}$. Specifically, $M_d = -C_d\dot{\theta}$. Figure 5.10 provides the free-body diagram showing the two drag moments. The moment equation $\sum M_o = I_o\ddot{\theta}$ gives

$$\frac{mr^2}{2}\ddot{\theta} = \sum M_z = M(t) - 2C_d\dot{\theta}. \qquad (5.40)$$

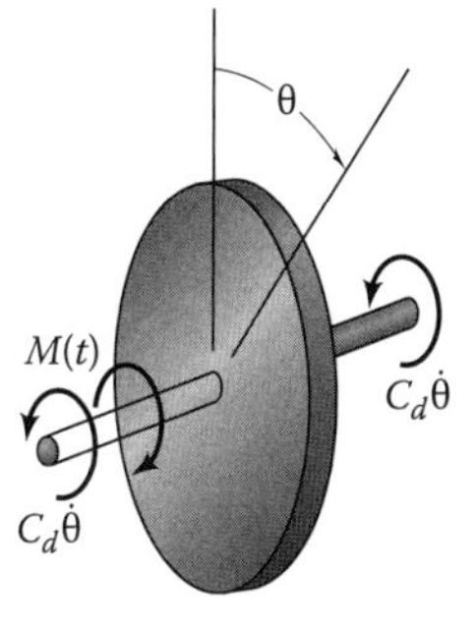

FIGURE 5.10
Free-body diagram for the rotor of Figure 5.9 with a drag torque $C_d\dot{\theta}$ acting at each bearing.

The drag moment terms have negative signs because they are acting in $-\theta$ direction in opposition to the $+\dot{\theta}$ rotation. The EOM,

$$\left(\frac{mr^2}{2}\right)\ddot{\theta} + 2C_d\dot{\theta} = M(t),$$

has the same form as a particle of mass m acted on by the force $f(t)$ plus a linear dashpot with a damping coefficient c; namely, $m\ddot{x} + c\dot{x} = f(t)$.

5.5.4 Rotor in Bearings with Viscous Drag: Energy Equation

Applying, $\text{Work}_{n.c.} = (T + V) - (T_0 + V_0)$ to derive the EOM gives

$$\int_o^\theta \left[M(t) - 2C_d\dot{\theta}\right]dx = \frac{I_{oz}\dot{\theta}^2}{2} - T_o.$$

Differentiation with respect to θ gives

$$I_{oz}\frac{d}{d\theta}\left(\frac{\dot{\theta}^2}{2}\right) I_{oz}\ddot{\theta} = M(t) - 2C_d\dot{\theta} \Rightarrow I_{oz}\ddot{\theta} + 2C_d\dot{\theta} = M(t),$$

coinciding with Equation 5.40. Using the work–energy equation has no particular advantage in developing these equations of motion, since the free-body diagram of Figure 5.10 is needed to state the nonconservative moment acting on the rotor.

5.5.5 Torsional Vibration Example: Moment Equation

Figure 5.11 illustrates a 1DOF vibration problem involving a disk of mass m and radius R, supported by a slender circular rod of radius r and length l. The top end of the rod is fixed to a rigid horizontal surface, and the bottom end is fixed at the disk's mass center, perpendicular to the plane of the disk. The applied moment $M(t)$ acts on the lower disk in the $+\theta$ direction. Consider twisting the disk about the vertical axis through the rod and visualize the resistance moment that would be developed by the rod. Twisting the rod about its axis through an angle θ will create a reaction moment, related to θ by

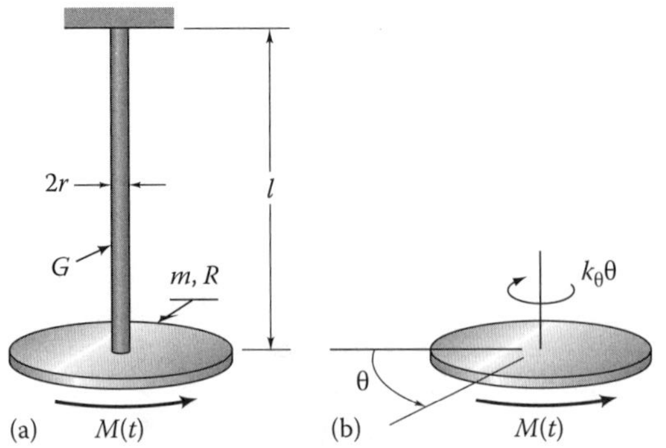

FIGURE 5.11
(a) Circular disk of mass m and radius R, supported by a slender rod of length l, radius r, and shear modulus G. (b) Free-body diagram for $\theta > 0$.

$$M_\theta = -k_\theta\theta = -\frac{GJ}{l}\theta = -\frac{G}{l}\frac{\pi r^4}{2}\theta. \tag{5.41}$$

In this equation, we have used a result from strength of materials for the torsional properties of a circular rod; specifically, $k_\theta = GJ/l$, where G is the shear modulus of the rod, and $J = \pi r^4/2$ is the rod's area polar moment of inertia. Recall that the SI units for G is N/m^2; hence, k_θ has the units: N-m/rad, that is, moment per unit torsional rotation of the rod.

Returning to Figure 5.11, applying $M_{oz} = I_o\ddot{\theta}$ yields

$$\frac{mR^2}{2}\ddot{\theta} = \sum M_z = M(t) + M_\theta = M(t) - \frac{G}{l}\frac{\pi r^4}{2}\theta,$$

The signs of the moments on the right-hand side of this equation are positive or negative, depending on whether they are, respectively, in the $+\theta$ direction or $-\theta$ direction. The distributed inertia of the rod was neglected in stating this moment equation. Since $I_o = mR^2/2$, the EOM is

$$\frac{mR^2}{2}\ddot{\theta} + \frac{\pi G r^4}{2l}\theta = M(t). \tag{5.42}$$

This result is analogous to the EOM for a particle of mass m, acted on by an external force $f(t)$ and supported by a linear spring with stiffness coefficient k; namely, $m\ddot{x} + kx = f(t)$. Equation 5.42 can be rewritten as

$$\ddot{\theta} + \omega_n^2\theta = \frac{2M(t)}{mR^2},$$

where the undamped natural frequency is $\omega_n = \sqrt{\pi G r^4/lmR^2}$.

Figure 5.12 illustrates the disk-rod assembly of Figure 5.11 with the disk now immersed in a viscous fluid. Rotation of the disk at a positive rotational velocity $\dot{\theta}$ within the fluid causes the drag moment, $-C_d\dot{\theta}$, on the disk. The negative sign for the drag term applies because it acts in the $-\theta$ direction. The complete moment equation is

$$\frac{mR^2}{2}\ddot{\theta} = \sum M_z = M(t) - \frac{G}{l}\frac{\pi r^4}{2}\theta - C_d\dot{\theta},$$

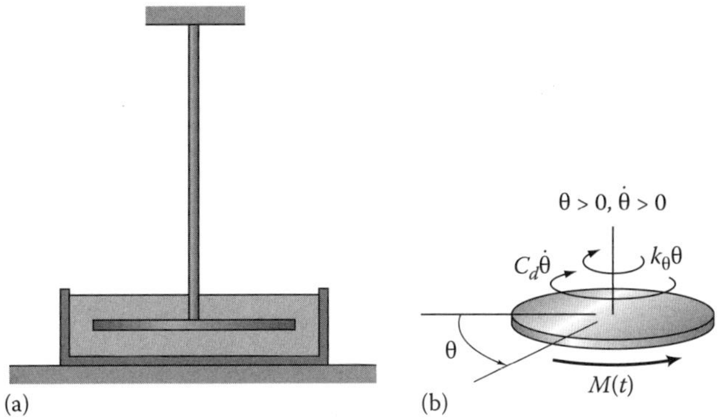

FIGURE 5.12
(a) Torsion example of Figure 5.11 now immersed in a viscous fluid. (b) Free-body diagram for $\theta > 0$, $\dot{\theta} > 0$.

with the governing EOM

$$\frac{mR^2}{2}\ddot{\theta} + C_d\dot{\theta} + \frac{\pi G r^4}{2l}\theta = M(t). \tag{5.43}$$

Equation 5.43 can be restated as

$$\ddot{\theta} + 2\zeta\omega_n\dot{\theta} + \omega_n^2\theta = \frac{2M(t)}{mR^2},$$

where
$\omega_n^2 = (\pi G r^2)/(mlR^2)$
ζ is the damping factor, defined by $2\zeta\omega_n = 2C_d/mR^2$

Equation 5.43 has the same form as a particle of mass m supported by a parallel arrangement of a spring with stiffness coefficient k and a linear damper with damping coefficient c; namely, $m\ddot{x} + c\dot{x} + kx = f(t)$. You might wish to compare these results to Equations 3.21 and 3.22.

The models developed from Figures 5.11 and 5.12 show the same damped and undamped vibration possibilities for rotational motion of a disk as we reviewed earlier for linear motion of a particle. The same possibilities exist to define damped and undamped natural frequencies, damping factors, etc.

5.5.6 Torsional Vibration Example: Energy Equation

For the model of Figure 5.11, applying $\text{Work}_{n.c.} = (T + V) - (T_1 + V_1)$ to derive the EOM gives

$$\int_0^\theta M(t)\,dx = \left(\frac{I_o\dot{\theta}^2}{2} + V\right) - (T_0 + V_0), \tag{5.44}$$

where $d\text{Work} = M\,d\theta$, was used to define the differential work integrand. The potential energy of the system is stored in the shaft due to the torsional rotation θ. Recall that the reaction moment is defined from

$$M_\theta = -k_\theta\theta = -\frac{GJ}{l}\theta = -\frac{G}{l}\frac{\pi r^4}{2}\theta,$$

where
G is the shear modulus of the rod
$J = \pi r^4/2$ is the rod's cross-sectional area polar moment of inertia

The requirement that a potential force (or moment) be derivable as the negative gradient of a potential function gives

$$M_\theta = -k_\theta\theta = -\frac{dV_\theta}{d\theta} \Rightarrow V_\theta = k_\theta\left(\frac{\theta^2}{2}\right). \tag{5.45}$$

This potential energy stored by twisting a circular shaft has an obvious similarity to the potential energy stored by stretching or compressing a conventional spring. Substituting for V_θ into Equation 5.44 gives

$$\int_0^\theta M(t)\,d\theta = \left(\frac{I_o\dot{\theta}^2}{2} + \frac{k_\theta\theta^2}{2}\right) - (T_0 + V_0).$$

Differentiating with respect to θ gives

$$I_o\frac{d}{d\theta}\left(\frac{\dot{\theta}^2}{2}\right) + k_\theta\theta = M(t) \Rightarrow \frac{mR^2}{2}\ddot{\theta} + \frac{\pi G r^4}{2l}\theta = M(t),$$

which coincides with Equation 5.42.

Adding the drag torque arising from the viscous fluid in Figure 5.12 gives

$$\int_0^\theta \left[M(t) - C_D\dot{\theta}\right]d\theta = \left(\frac{I_o\dot{\theta}^2}{2} + \frac{k_\theta\theta^2}{2}\right) - (T_0 + V_0),$$

and differentiation with respect to θ gives

$$I_o\frac{d}{d\theta}\left(\frac{\dot{\theta}^2}{2}\right) + k_\theta\theta = M(t) - C_d\dot{\theta}$$

$$\Rightarrow \frac{mR^2}{2}\ddot{\theta} + C_d\dot{\theta} + \frac{\pi G r^4}{2l}\theta = M(t),$$

which coincides with Equation 5.43. In deriving equations of motion, the work–energy equation is most advantageous when energy is conserved. For nonconservative examples, a free-body diagram is needed to define the nonconservative moment acting on the body.

5.5.7 Pulley/Weight Example: Free-Body Approach

Figure 5.13a illustrates two weights connected together by an inextensible cord that is looped over a pulley. The pulley has mass M and the radius of gyration k_{Og} netting $I_o = Mk_{Og}^2$. Weight 2 weighs more than weight 1 and is being supported prior to release. The pulley is assumed to rotate about an axis through O without friction. After w_2 is released, we assume that the cord does not slip on the pulley. Our immediate tasks are selecting coordinates, drawing a free-body diagram for the weights and pulley following release, and using them to derive a governing EOM.

Figure 5.13 illustrates the coordinates y_1, y_2 that have been selected to locate w_1, w_2, respectively. The pulley's rotation with respect to ground is θ. Tensions in the left- and right-hand side cords are T_{c1}, T_{c2}. Stating the moment equation for the pulley about an axis through O gives

$$\sum M_O = I_O\ddot{\theta} = T_{c2}r - T_{c1}r. \tag{5.46}$$

Applying $\sum f = m\ddot{r}$ separately to the two bodies gives

$$\begin{aligned} &\text{Body 1:} \quad \sum f_{y1} = w_1 - T_{c1} = m_1\ddot{y}_1 \\ &\text{Body 2:} \quad \sum f_{y2} = w_2 - T_{c2} = m_2\ddot{y}_2 \end{aligned} \tag{5.47}$$

We now have three equations and five unknowns, $\ddot{\theta}, T_{c1}, T_{c2}, \ddot{y}_1, \ddot{y}_2$. We clearly need kinematic constraints. Returning to Figure 5.13b, w_2's downward acceleration equals the circumferential acceleration of a point on the pulley at its contact with the cord. A similar outcome holds for w_1; hence,

$$\ddot{y}_2 = r\ddot{\theta}, \quad \ddot{y}_1 = -r\ddot{\theta} \tag{5.48}$$

Substituting these results into Equation 5.47 gives

$$\begin{aligned} T_{c1} &= w_1 - m_1\ddot{y}_1 = w_1 - m_1(-r\ddot{\theta}) = w_1 + m_1r\ddot{\theta} \\ T_{c2} &= w_2 - m_2\ddot{y}_2 = w_2 - m_2(r\ddot{\theta}) = w_2 - m_2r\ddot{\theta} \end{aligned} \tag{5.49}$$

Substituting for T_{c1}, T_{c2} into Equation 5.46 gives

$$\begin{aligned} &I_O\ddot{\theta} = T_{c2}r - T_{c1}r = r(w_2 - m_2r\ddot{\theta}) - r(w_1 - m_1r\ddot{\theta}) \\ &\therefore \ (I_O + m_1r^2 + m_2r^2)\ddot{\theta} = I_{eff}\ddot{\theta} = rw_2 - rw_1. \end{aligned} \tag{5.50}$$

Note that I_{eff}, the effective inertia, is the same result that we would have obtained if the particles m_1, m_2 had been "stuck" to the rims of the pulley. Also, note that w_2's downward acceleration is

$$\ddot{y}_2 = r\ddot{\theta} = \frac{r^2(w_2 - w_1)}{(I_O + m_1r^2 + m_2r^2)} = \frac{w_2 - w_1}{((I_O/r^2) + m_1 + m_2)} \tag{5.51}$$

A problem like this could have been assigned in Chapter 3 for particle dynamics with the assumption that the pulley's inertia was negligible. *With negligible inertia for the pulley*, $I_O \cong 0$, a moment equation about O gives

$$\sum M_O = I_O\ddot{\theta} = 0 = T_{c2}r - T_{c1}r \Rightarrow T_{c1} = T_{c2} = T_c.$$

Recall that the cord tensions on opposite sides of pulleys in Chapter 3 were assumed to be equal. Obviously, when inertia is introduced, this outcome stops being correct. Equations 5.47 are still valid, but they now give us two equations in the unknowns T_c, $\ddot{y}_1$, $\ddot{y}_2$. The kinematic constraint Equations 5.49 are still correct, but θ and $\ddot{\theta}$ are normally not present in the "inertialess," particle-dynamics version of this problem. Instead, the inextensible nature of the cord is used to give

$$y_1 + y_2 + c = \text{cord length} \Rightarrow \ddot{y}_1 = \ddot{y}_2.$$

Substituting $T_{c1} = T_{c2} = T_c$ and $\ddot{y}_1 = -\ddot{y}_2$ gives

$$\begin{aligned} T_c &= w_1 - m_1\ddot{y}_1 = w_1 - m_1(-\ddot{y}_2) = w_2 - m_2\ddot{y}_2 \\ &\Rightarrow \ddot{y}_2(m_1 + m_2) = w_2 - w_1 \\ \therefore \ \ddot{y}_2 &= \frac{w_2 - w_1}{(m_1 + m_2)} \end{aligned}$$

Comparing this result to Equation 5.51 shows the impact of I_O in reducing $\ddot{y}_2$. This example is provided

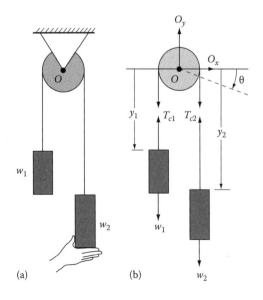

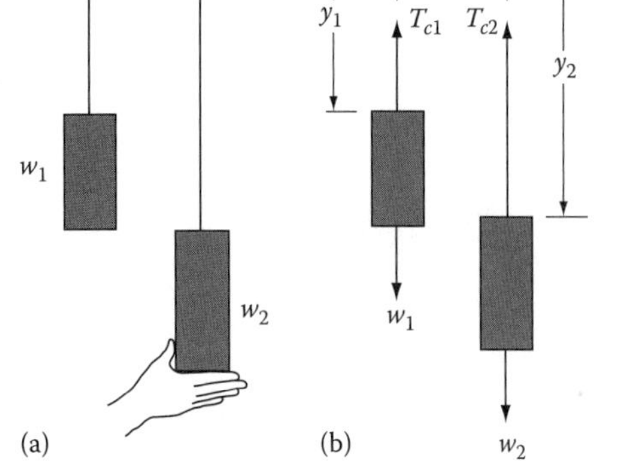

FIGURE 5.13
Weight–pulley example. (a) Prior to release. (b) Coordinates and free-body diagram.

to clarify your understanding concerning the differences between particle and rigid body dynamics.

5.5.8 Pulley/Weight Example: Energy Approach

Energy is conserved for this example, so we can use $T+V=T_0+V_0$ to obtain the EOM. The kinetic energy is

$$T = I_O\frac{\dot{\theta}^2}{2} + m_1\frac{\dot{y}_1^2}{2} + m_2\frac{\dot{y}_2^2}{2}$$

Assuming that the datum for gravity potential energy is on a horizontal plane passing through the center of the pulley gives $V_g = -w_1y_1 - w_2y_2$. We want to use θ as the coordinate. Rotating the disk through θ moves the two weights as follows:

$$y_1 = y_{10} - r\theta \Rightarrow \dot{y}_1 = -r\dot{\theta}, \quad y_2 = y_{20} + r\theta \Rightarrow \dot{y}_2 = r\dot{\theta}$$

Substitution gives

$$T = I_O\frac{\dot{\theta}^2}{2} + m_1\frac{(-r\dot{\theta})^2}{2} + m\frac{(r\dot{\theta})^2}{2} = (I_O + m_1r^2 + m_2r^2)\frac{\dot{\theta}^2}{2}$$
$$V = V_g = -w_1(y_{10} - r\theta) - w_2(y_{20} + r\theta).$$

Hence, $T+V=T_0+V_0$ gives

$$(I_O + m_1r^2 + m_2r^2)\frac{\dot{\theta}^2}{2} - w_1(y_{10} - r\theta) - w_2(y_{20} + r\theta)$$
$$= T_0 + V_0$$

Differentiating with respect to θ gives

$$(I_O + m_1r^2 + m_2r^2)\frac{d}{d\theta}\left(\frac{\dot{\theta}^2}{2}\right) + w_1r - w_2r = 0$$
$$\Rightarrow (I_O + m_1r^2 + m_2r^2)\ddot{\theta} = w_2r - w_1r$$

This results coincides with Equation 5.50. Applying conservation of energy to develop the equations of motion avoids the free-body diagram; hence, there is no reason to note that $T_{cl} \neq T_{c2}$.

5.5.9 An Example Involving a Disk and a Particle: Newtonian Approach

Figure 5.14a provides a mild kinematic complication in applying the moment Equation 5.26, illustrating a circular disk of mass M and radius r, supported in frictionless bearings. The disk rotates about its mass center. A light, inextensible cord is wrapped around the radius of the disk and is connected to a particle of mass m. The engineering tasks are as follows: *Select coordinates, draw a free-body diagram, and use it to derive the system EOM.*

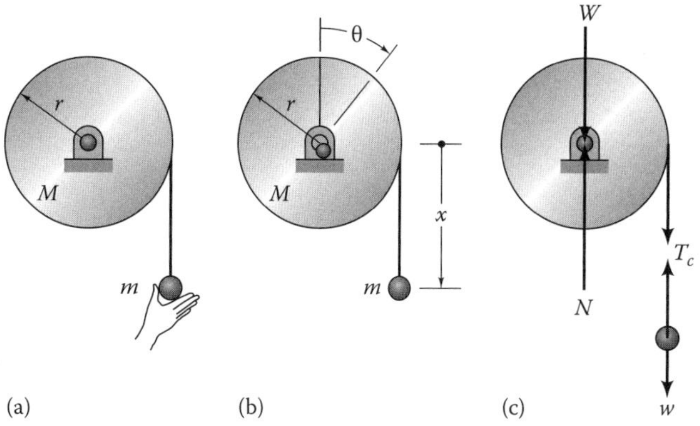

FIGURE 5.14
Disk of mass M and radius r connected to a particle of mass m by a light and inextensible cord. (a) Prior to release. (b) Coordinates. (c) Free-body diagram.

As shown in Figure 5.14b, the rotation of the disk is defined by the angle θ. The change in position of the mass m is defined by x. First, note that, although there are two bodies, there is only one degree of freedom. Rotating the disk through θ radians will reel $r\theta$ radians of cord off the disk and will lower the mass m a distance $r\theta$; hence, δx (the change in position for m) and θ and their derivatives are related by

$$\delta x = r\theta, \quad \delta\dot{x} = \dot{x} = r\dot{\theta}, \quad \delta\ddot{x} = \ddot{x} = r\ddot{\theta}. \tag{5.52}$$

The second equation is obtained by differentiating the first equation and can be interpreted physically as the requirement that $\dot{x}$, the vertical downward velocity of the mass m, must equal $v_\theta = r\dot{\theta}$, the circumferential velocity of the disk's rim at the point where the cord is leaving the disk. The acceleration result in Equation 5.52 is obtained and interpreted similarly; namely, $\ddot{x}$, the downward acceleration of the mass m, equals $a_\theta = r\ddot{\theta}$, the circumferential acceleration of the disk at the point where the cord is leaving the disk.

Figure 5.14c provides the free-body diagram for the two bodies. The disk's EOM is obtained by writing a moment equation about its axis of rotation. The EOM for mass m follows from $\sum f = m\ddot{r}$. The governing equations are

$$\frac{Mr^2}{2}\ddot{\theta} = \sum M_{oz} = T_cr, \quad m\ddot{x} = \sum f = w - T_c, \tag{5.53}$$

where T_c is the tension in the cord. (The mass of the cord has been neglected in stating these equations.) In the first of Equation 5.53, the moment term T_cr is positive because it acts in the $+\theta$ direction. In the force equation, the sign of w is positive because it acts in the $+x$-*direction*; T_c has a negative sign because it is directed in the $-x$ direction.

Equations 5.53 provides two equations for the three unknowns: $\ddot{x}$, $\ddot{\theta}$, and T_c. Eliminating T_c gives

$$\frac{Mr^2}{2}\ddot{\theta} + rm\ddot{x} = wr. \tag{5.54}$$

Substituting from the last of Equation 5.52 for $\ddot{x} = r\ddot{\theta}$ gives the final EOM

$$\left(\frac{M}{2} + m\right) r^2\ddot{\theta} = I_{eff}\ddot{\theta} = wr.$$

The contributions of M and m to the effective moment of inertia I_{eff} are both positive. A negative contribution would signal an error. Nature does not have negative moments of inertia.

When faced with this problem and employing $M_{oz} = I_o\ddot{\theta}$, students are occasionally tempted to simply apply wr, the moment due to the weight w, directly to the inertia. This viewpoint is wrong in that it neglects the motion of mass m, which is also being accelerated by the weight w. Note that $rm\ddot{x}$ in Equation 5.54 would be incorrectly omitted due to this erroneous approach.

5.5.10 An Example Involving a Disk and a Particle: Work–Energy Approach

In this example, energy is conserved. Defining the gravity potential-energy datum as a horizontal plane through the pivot point of the bearing gives

$$T + V = T_0 + V_o \Rightarrow \frac{I_o\dot{\theta}^2}{2} + \frac{m\dot{x}^2}{2} - mg(\bar{x} + \delta x) = T_0 + V_0.$$

We need $I_o = Mr^2/2$ from Appendix C, and the kinematics of Equation 5.52, $\delta x = r\theta$, $\delta\dot{x} = r\dot{\theta}$, to obtain

$$\frac{\dot{\theta}^2 r^2}{2}\left(\frac{M}{2} + m\right) - mgr\theta - mg\bar{x} = T_0 + V_0.$$

Differentiating with respect to θ gives

$$\left(\frac{M}{2} + m\right) r^2\ddot{\theta} = wr,$$

which coincides with our earlier Newtonian-based developments. This example is an ideal application for the energy approach in developing the EOM, since energy is conserved. We still need the kinematics result of Equation 5.52 but do not require a free-body diagram.

5.5.11 Two Driven Pulleys Connected by a Belt

Figure 5.15a illustrates two pulleys that are connected by a light and inextensible belt. The pulley at the left has mass m_1, radius of gyration k_{g1} about the pulley's axis of rotation $(I_1 = m_1k_{g1}^2)$, and is acted on by the counterclockwise moment M_o. The pulley at the right has mass m_2 and a radius of gyration k_{2g} about its axis of rotation. The belt runs in grooves in pulleys 1 and 2 with radii, respectively of r_1 and r_2. The tasks for this example are as follows: *Select coordinates, draw a free-body diagram for the system, and use it to derive the EOM in terms of* θ *and its derivatives*.

As shown in Figure 5.15b, we choose angles of rotation for pulleys 1 and 2, respectively, as θ and ϕ. The free-body diagram for the two pulleys is shown in Figure 5.15c. Applying the fixed-axis moment equation $M_{oz} = I_0\ddot{\theta}$ separately to the two bodies gives

$$\begin{aligned} I_1\ddot{\theta} &= \sum M_{1oz} = M_o(t) + r_1(T_{c2} - T_{c1}) \\ I_2\ddot{\phi} &= \sum M_{2oz} = r_2(T_{c1} - T_{c2}), \end{aligned} \tag{5.55}$$

where T_{c1} and T_{c2} are the tension components in the upper and lower segments of the belt. In the first of Equation 5.55, the term $M_o(t)$ has a positive sign because it is acting in the $+\theta$ direction; $r_1(T_{c2} - T_{c1})$ has a positive sign because it acts in the $+\theta$ direction. Similarly, $r_2(T_{c1} - T_{c2})$ has a positive sign in the second of Equation 5.55 because it is acting in the $+\phi$ direction.

Returning to Equation 5.55, we can eliminate the tension terms in the two equations, obtaining

$$I_1\ddot{\theta} = M_o - \frac{r_1}{r_2}I_2\ddot{\phi} \Rightarrow I_1\ddot{\theta} + \frac{r_1}{r_2}I_2\ddot{\phi} = M_o(t). \tag{5.56}$$

We now have one equation for the two unknowns $\ddot{\theta}$ and $\ddot{\phi}$ and need an additional kinematic equation. Given that

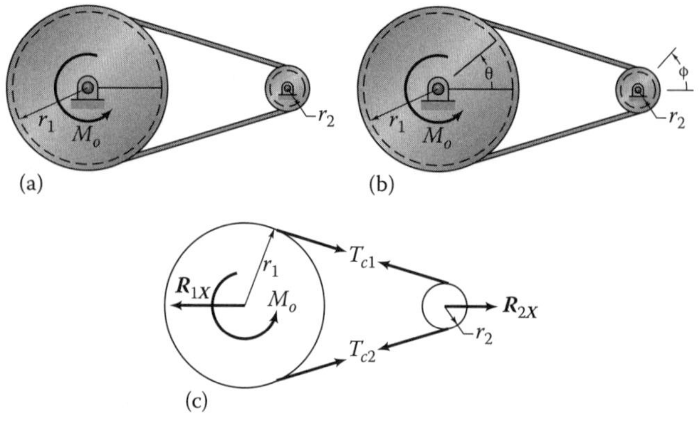

FIGURE 5.15
(a) Two pulleys connected by an inextensible belt. (b) Coordinates. (c) Free-body diagram.

the belt connecting the pulleys is inextensible (cannot stretch) the velocity of the belt v leaving both pulleys must be equal; hence,

$$v = r_1\dot{\theta} = r_2\dot{\phi} \Rightarrow r_1\ddot{\theta} = r_2\ddot{\phi}. \quad (5.57)$$

Substituting this result back into Equation 5.56 gives the desired final result

$$\left[I_1 + \left(\frac{r_1}{r_2}\right)^2 I_2\right]\ddot{\theta} = I_{eff}\ddot{\theta} = M_o(t), \quad (5.58)$$

where $I_1 = m_1k_{g1}^2$, $I_2 = m_2k_{g2}^2$. Coupling the two pulleys' motion by the belt acts to increase the effective inertia I_{eff} in resisting the applied moment. Note that I_1 and $(r_1/r_2)^2I_2$ are both positive in contributing to I_{eff}. A negative sign in either term would be a certain indication that an error had been made. As with mass, negative moments of inertia simply do not occur in nature. This example is relatively straightforward and illustrates a common situation where the equations of motion are not adequate (by themselves) to define the problem. We could have anticipated from the outset that an additional kinematic relationship would be required between variables because we had two coordinates (leading to two equations of motion) but only one degree of freedom.

5.5.12 Two Driven Pulleys Connected by a Belt: Work–Energy Approach

Energy is not conserved, but there is no energy dissipation for this example. Hence, work–energy provides a simple approach for deriving the differential equations of motion. There is no change in the potential energy of this system; hence, $\text{Work}_{n.c.} = \Delta(T+V)$ yields

$$\text{Work}_{n.c.} = T - T_0 = \left(\frac{I_1\dot{\theta}^2}{2} + \frac{I_1\dot{\phi}^2}{2}\right) - 0.$$

From Equation 5.57, $\dot{\phi} = \dot{\theta}(r_1/r_2)$, and

$$\text{Work}_{n.c.} = \frac{\dot{\theta}^2}{2}\left[I_1 + \left(\frac{r_1}{r_2}\right)^2 I_2\right] = \frac{I_{eff}\dot{\theta}^2}{2}.$$

Using $d\text{Work} = M\,d\theta$, to define the work integral on the left-hand side gives

$$\int_0^{\Delta\theta} M_o(t)\,d\theta = \frac{I_{eff}\dot{\theta}^2}{2}.$$

Differentiating with respect to θ gives

$$I_{eff}\ddot{\theta} = M_o,$$

which coincides with Equation 5.58.

5.6 Compound-Pendulum Applications

5.6.1 The Simple Compound Pendulum: EOM, Linearization, Stability

Figure 5.16 a illustrates a compound pendulum consisting of a uniform bar of mass m and length l supported by a frictionless pivot at o. The term "compound" is used to distinguish the present rigid-body pendulum from the "simple" pendulum of Section 3.4.2.3, which consisted of a particle at the end of a massless string. The angle θ defines the orientation of the pendulum with respect to the vertical. In the preceding examples of this section, axes through the mass center g and the fixed axis of rotation o coincided. As we shall see, for the compound pendulum, the EOM is obtained more quickly by stating the moment equation about o than about g.

The engineering tasks are as follows: (i) Select a coordinate, draw a free-body diagram, and use it to derive the equation of motion, and (ii) For small motion about equilibrium, determine the undamped natural frequency. Figure 5.16b illustrates the coordinate choice θ, and Figure 5.16c provides the free-body diagram with the pendulum acted on by its weight w and the reaction force components at o. By taking moments about o, the moment due to the reaction component o_θ is eliminated, yielding

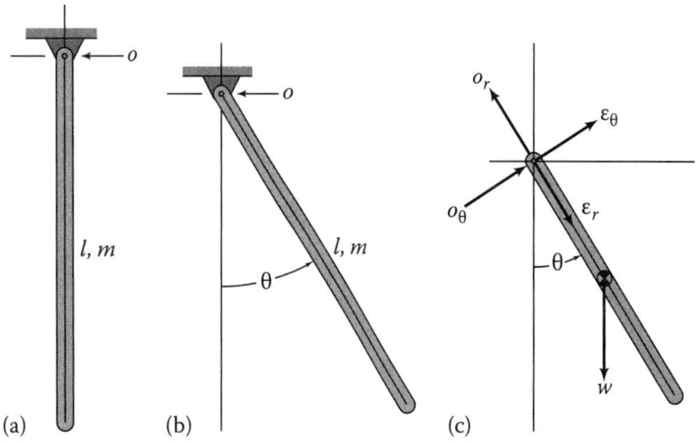

FIGURE 5.16
Compound pendulum. (a) At rest in equilibrium. (b) General position with coordinate θ. (c) Free-body diagram.

$$I_o\ddot{\theta} = \sum M_o = -w\frac{1}{2}\sin\theta, \quad I_o = \frac{ml^2}{3} \tag{5.59}$$

The minus sign on the right-hand side applies because the moment is acting in the $-\theta$ direction; hence, the governing EOM is

$$\frac{ml^2}{3}\ddot{\theta} + \frac{mgl}{2}\sin\theta = 0 \Rightarrow \ddot{\theta} + \frac{3g}{2l}\sin\theta = 0. \tag{5.60}$$

In Section 3.4.2.3, while analyzing the simple pendulum, we used a Taylor series expansion to justify the linearizing approximation, $\sin\theta \cong \theta$. You may want to go back and review this development. The approximation is valid for small θ ($\theta \leq$ about $15 = .262$ rad) and reduces Equation 5.60 to

$$\ddot{\theta} + \frac{3g}{2l}\theta = 0 \Rightarrow \ddot{\theta} + \omega_n^2\theta = 0. \tag{5.61}$$

This differential equation is the *rotation* analog of the 1DOF *displacement*, vibration model $\ddot{Y} + \omega_n^2 Y = 0$, and the compound pendulum's natural frequency is

$$\omega_n = \sqrt{\frac{3g}{2l}}.$$

Consider the following additional engineering task in connection with the model of Figure 5.16b and c: *Assuming that the pendulum is released from rest at* $\theta = \pi/2$ rad $= 90°$, *define the reaction-force components* o_θ, o_r *as a function of* θ (*only*). The free-body diagram of Figure 5.16c can be used to find the reaction forces. Using polar coordinates, and writing $\sum f = m\ddot{\boldsymbol{R}}_g$ for the bar's mass center gives

$$\begin{aligned}\sum f_r &= -o_r + w\cos\theta = ma_r = m(\ddot{r} - r\dot{\theta}^2) = -m\frac{l}{2}\dot{\theta}^2\\ \sum f_\theta &= o_\theta - w\sin\theta = ma_\theta = m(r\ddot{\theta} + 2\dot{r}\dot{\theta}) = m\frac{l}{2}\ddot{\theta}.\end{aligned} \tag{5.62}$$

In the acceleration terms, $\dot{r} = \ddot{r} = 0$ because $r = (l/2)$ is a constant. Equations 5.62 define the reaction-force components o_r, o_θ, but not as a function of θ alone. To complete the definitions, we will need to use the nonlinear differential equation of motion, Equation 5.60, defining $\ddot{\theta}$ as a function of θ. Recall that the linearized model of Equation 5.61 is only valid for small motion; hence, the nonlinear model of Equation 5.60 is required. Direct substitution for $\ddot{\theta}$ from Equation 5.60 into the second of Equations 5.62 defines o_θ as

$$o_\theta = w\sin\theta + \frac{ml}{2}\ddot{\theta} = w\sin\theta - \frac{ml}{2}\left(\frac{3g}{2l}\right)\sin\theta = \frac{w}{4}\sin\theta. \tag{5.63}$$

Finding a comparable relationship for o_r is more complicated, because the first of Equations 5.62 involves $\dot{\theta}^2$. We will see below that $\dot{\theta}^2(\theta)$ can be readily obtained from conservation of energy; however, we will first integrate the differential equation of motion via the energy-integral substitution to obtain $\dot{\theta}^2(\theta)$, proceeding from

$$\ddot{\theta} = \frac{d}{d\theta}\left(\frac{\dot{\theta}^2}{2}\right) = -\frac{3g}{2l}\sin\theta.$$

Multiplying by $d\theta$ nets an exact differential on both sides of the equation. Integration gives

$$\frac{\dot{\theta}^2}{2} - \frac{\dot{\theta}_o^2}{2} = \int_{\frac{\pi}{2}}^{\theta} -\frac{3g}{2l}\sin u\, du \Rightarrow \frac{\dot{\theta}^2}{2} = \frac{3g}{2l}\Big|_{\frac{\pi}{2}}^{\theta}\cos x = \frac{3g}{2l}\cos\theta. \tag{5.64}$$

Substituting this result into the first of Equation 5.62 gives

$$o_r = w\cos\theta + ml\frac{\dot{\theta}^2}{2} = w\cos\theta + ml\left(\frac{3g}{2l}\right)\cos\theta = \frac{5w}{2}\cos\theta. \tag{5.65}$$

This result shows that the dynamic reaction force will be 2.5 times greater than the static weight w when the rod reached its lowest position ($\theta = 0$).

Suppose that we had chosen to take moments about g, the mass center of the rod in Figure 5.16c, obtaining

$$\sum M_g = I_g\ddot{\theta} = -o_\theta\frac{l}{2}.$$

The moment in this equation is negative because it is acting in the $-\theta$ direction. Substituting for o_θ from Equation 5.62 gives

$$-\frac{l}{2}\left(m\frac{l}{2}\ddot{\theta} + w\sin\theta\right) = \frac{ml^2}{12}\ddot{\theta},$$

where $I_g = ml^2/12$ from Appendix C. Rearranging the equation gives

$$-\frac{wl}{2}\sin\theta = \left(\frac{ml^2}{12} + \frac{ml^2}{4}\right)\ddot{\theta} = \frac{ml^2}{3}\ddot{\theta} = I_o\ddot{\theta}. \tag{5.66}$$

As expected, this EOM coincides with Equation 5.60. However, writing the moment equation about g involves more work. Note in the intermediate step of Equation 5.66 that we are accomplishing the parallel-axis formula in moving from $I_g = ml^2/12$ to $I_o = ml^2/3$, via $I_o = I_g + m(b_{ogx}^2 + b_{ogy}^2) = I_g + m(l/2)^2$, where $\mathbf{b}_{go}$ is the vector from the mass center to the pivot point. Also note that both $ml^2/12$ and $ml^2/4$ are positive in developing I_o.

Taking a work–energy approach to derive the EOM starts by defining the datum for potential energy as lying on a horizontal plane through the pendulum's pivot point. Starting with the bar in a horizontal position corresponding to $\theta = (\pi/2)$ and using conservation of energy gives

$$T + V = T_0 + V_0 \Rightarrow \frac{ml^2}{3}\frac{\dot{\theta}^2}{2} - w\frac{l}{2}\cos\theta = 0 + 0 \quad (5.67)$$

Differentiating with respect to θ gives

$$\frac{ml^2}{3}\frac{d}{d\theta}\left(\frac{\dot{\theta}^2}{2}\right) + w\frac{l}{2}\sin\theta = 0 \Rightarrow \ddot{\theta} + \frac{3g}{2l}\sin\theta = 0,$$

coinciding with Equation 5.60. Again, starting with conservation of energy, the EOM is obtained without recourse to a free-body diagram. Earlier, we needed $\dot{\theta}^2(\theta)$ to obtain $o_\theta(\theta)$. Equation 5.67 can be used directly to obtain $\dot{\theta}^2(\theta) = (3g)/(2l) \times \cos\theta$, repeating the results of Equation 5.64 without the effort of integrating the EOM.

We will now consider the pendulum's *static stability*. Equilibrium is defined from Equation 5.59 by $\sum M_o = 0 \Rightarrow \ddot{\theta} = 0$. This outcome is obtained at $\theta = 0$ and $\theta = \pi$. In the above development, we derived Equation 5.61 to define the motion of the pendulum for small motion about the equilibrium position $\theta = 0$. For the initial conditions, $\dot{\theta}(0) = \dot{\theta}_0$; $\theta(0) = \theta_0$, Equation 5.61's solution is

$$\theta(t) = \theta_0 \cos\omega_n t + \frac{\dot{\theta}_0}{\omega_n}\sin\omega_n t,$$

consisting of a *stable* oscillation at the natural frequency. Hence, $\theta = 0$ is said to be a *stable equilibrium point* for the body. Remember that Equation 5.61 is an approximate model for small motion about $\theta = 0$; hence, this solution is only valid if $\theta(t)$ remains less than approximately 15°.

Now consider the equilibrium position represented by the position $\theta = \pi$. We just worked through the solution for $\dot{\theta}^2$ as a function of θ in Equation 5.64 when the body is released from rest at $\theta = \pi/2$ and found that the pendulum accelerated from this initial position with $\dot{\theta}^2$ increasing steadily with increasing θ. We would expect a similar outcome if the pendulum is released from rest in the vertical position defined by $\theta = \pi$. With these thoughts in mind, one could reasonably ask, "How can you call $\theta = \pi$ an equilibrium position?" To answer this question, consider the EOM

$$\ddot{\theta} = -\frac{3g}{2l}\sin\theta. \quad (5.60)$$

We showed that $\theta = 0$ is a stable equilibrium point. The question here is: For small motion about the equilibrium position $\theta = \pi$, is the pendulum's motion stable? The keywords in this question is *small*. Mathematically, we can examine small motion of Equation 5.60 by expanding $\sin\theta$ in a Taylor's series about π, obtaining

$$\begin{aligned}\sin\theta &= \sin(\pi + \delta\theta) = \sin\pi\cos\delta\theta + \cos\pi\sin\delta\theta \\ &= -\sin\delta\theta = -\delta\theta + \frac{(\delta\theta)^3}{6} - \frac{(\delta\theta)^5}{120} + \cdots. \end{aligned} \quad (5.68)$$

This result is very similar to the expansion for $\sin\theta$ about zero in Equation 3.106. Retaining only the linear term in Equation 5.68, noting that $\delta\ddot{\theta} = \ddot{\theta}$, and substituting back into Equation 5.60 gives

$$\delta\ddot{\theta} - \frac{3g}{2l}\delta\theta = 0. \quad (5.69)$$

Observe the negative sign in the coefficient for $\delta\theta$. If this were a harmonic oscillator consisting of a spring supporting a mass, a comparable negative sign would imply a negative stiffness, yielding a differential equation of the form $m\ddot{x} - kx = 0$. Substituting the assumed solution $\delta\theta = Ae^{st}$ into Equation 5.69 gives

$$(s^2 - \omega_n^2)Ae^{st} = 0 \Rightarrow s = \pm\omega_n, \quad \omega_n^2 = \frac{3g}{2l}$$

Hence the solution to Equation 5.69 is

$$\delta\theta(t) = A_1 e^{\omega_n t} - A_2 e^{-\omega_n t}$$

The first term grows exponentially with time. Hence, any small disturbance of the pendulum from the equilibrium position $\theta = \pi$ will grow exponentially with time, and $\theta = \pi$ is an *unstable equilibrium point*. In deriving equations of motion, remember that a "negative stiffness" coefficient implies *static* instability. A negative stiffness coefficient is to be expected for the present *unstable* inverted pendulum. It is not an expected outcome for stable mechanical systems, and you should check the sign of this coefficient to insure that it is consistent with your expectation of the physics of the situation.

We have analyzed the equilibrium of the pendulum from a differential-equation viewpoint. The static equilibrium issue can also be approached from an energy viewpoint. Starting with the potential energy function,

$$V = -w\frac{1}{2}\cos\theta. \tag{5.70}$$

The two equilibrium points arise when V has a maximum and minimum with respect to θ; that is,

$$\frac{dV}{d\theta} = \frac{wl}{2}\sin\theta = 0 \Rightarrow \theta = 0,\ \pi.$$

Stable equilibrium occurs when the potential energy is a minimum, that is, when $\theta = 0$. The pendulum's potential energy is a maximum at $\theta = \pi$ predicting (by default) that the pendulum is statically unstable at this position.

The results on small motion about an equilibrium position for the present problem will be of continuing use and value throughout this book. We had previously used the simple pendulum example in Section 3.4 to examine small motion about a stable equilibrium position, leading to oscillatory motion at the pendulum's natural frequency. The present compound pendulum introduces the notion of an unstable equilibrium position, for which the pendulum is *statically* unstable.

Using the moment equation to derive the EOM for the compound pendulum provides several lessons. First, recall that the EOM was determined more simply by stating the moment equation about o, the axis of rotation, than g, the mass center, since the moment equation about o sufficed (by itself). Stating the moment equation about g pulled the reaction-force component o_θ into the development and required the additional use of $\sum f = m\ddot{R}_g$ for the mass center. This is a generally applicable lesson; namely, with fixed-axis rotation, using the moment equation for the axis of rotation is always simpler.

For the present example, a requirement for the calculation of reaction forces necessitates stating $\sum f = m\ddot{R}_g$ for the mass center. Look back at this step in the present example and note that the polar-coordinate form was selected instead of Cartesian coordinates. This should seem like the obvious choice to you given the geometry of the problem. The analysis can be executed using Cartesian coordinates, but the solution is more tedious.

Example Problem 5.2

Figure XP5.2 illustrates a plate of mass m, length $2a$, and width a. It is supported by a frictionless pivot at o and by a ledge at B. The engineering tasks associated with this problem follow. Assuming that the support at B is suddenly removed, carry out the following steps:

a. Select a coordinate, draw the body in a general position, and draw a free-body diagram.
b. Use the free-body diagram to derive the EOM.
c. Use work–energy to derive the EOM.
d. Develop relationships that define the components of the reaction force as a function of the rotation angle only.
e. Derive the governing differential equation of motion for small motion about the plate's equilibrium position. Determine the natural frequency of the plate for small motion about this position.

Figure XP5.2b illustrates the plate in a general position with two coordinates: θ, Θ. The angle θ defines the plate's orientation with respect to the horizontal, while Θ defines the rotation from the horizontal of a line from o through g, the plate's mass center. The two angles are

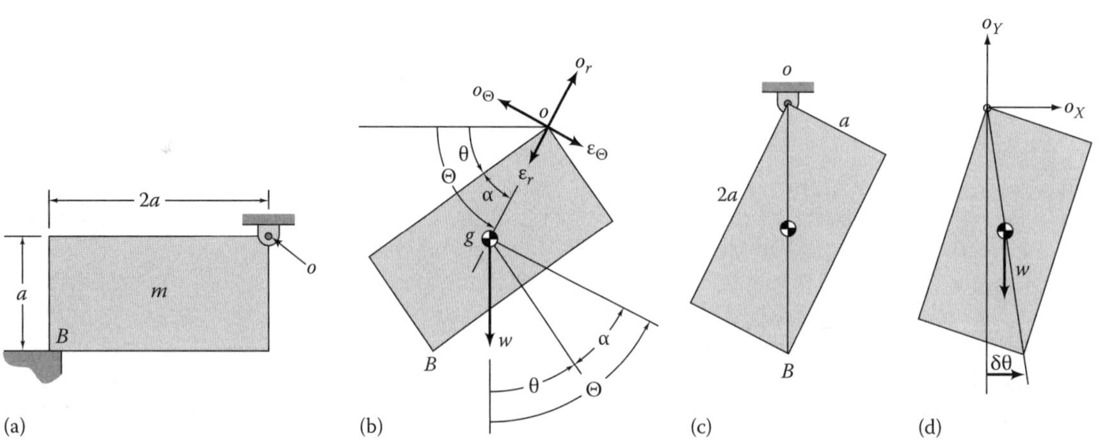

FIGURE XP5.2
(a) Rectangular plate supported at o by a frictionless pivot and at B by a ledge. (b) Free-body diagram after the support at B has been removed. (c) Plate in its stable equilibrium position. (d) Free-body diagram for motion about equilibrium.

related by $\Theta = (\theta + \alpha)$. The angle α lies between the top surface of the plate and a line running from o through g, and is defined by

$$\alpha = \tan^{-1}[(a/2)/a] = \tan^{-1}\left(\frac{1}{2}\right) = 26.57^\circ$$

Note that $\dot{\theta} = \dot{\Theta}$.

Now look at the forces acting on the plate in the free-body diagram. The weight acts straight down through g and can be added to the free-body diagram without much thought. However, there are choices involved in drawing the components of the reaction force at o. For example, the components could be drawn in the horizontal and vertical directions, or drawn parallel to the edges of the rotated plate. The component-definition choice shown in Figure 5.23b was made looking forward to Task b, which will require a statement of $\sum \boldsymbol{f} = m\ddot{\boldsymbol{R}}_g$ for the plate's masscenter. The geometry of this problem argues for the polar-coordinate version of $\sum \boldsymbol{f} = m\ddot{\boldsymbol{R}}_g$, and the choice of Figure XP5.2b has o_r and o_Θ aligned, respectively, with the $\boldsymbol{\varepsilon}_r$, $\boldsymbol{\varepsilon}_\Theta$ unit vectors. Frequently, success and comparative ease in working dynamics problems hinges on the decisions that are made in selecting kinematics and in drawing the free-body diagram. Good decisions at this stage lead to a comparatively easy and direct solution. If you are having trouble working a problem, go back and review these choices, looking for a simpler way to state the problem.

Moving *to* Task b, the immediate choice in working this problem is: Should the moment equation be stated about point g (the mass center) or o (the fixed axis of rotation)? Stating the moment about g has the undesirable outcome of bringing the (unknown) reaction component o_Θ into the moment equation. Stating the moment about o avoids this complication, yielding

$$I_o\ddot{\theta} = \sum M_{ox} = \frac{wa\sqrt{5}}{2}\cos(\theta + \alpha). \quad \text{(i)}$$

The distance from o to g is $a\sqrt{5}/2$, and w develops the external moment acting through the moment arm $a\sqrt{5}/2\cos(\theta + \alpha)$. The moment is positive because it is acting in the $+\theta$ direction.

To proceed with Equation (i), we require I_o the moment of inertia for an axis through o. However, Appendix C only provides I_g. Applying the parallel-axis formula of Equation 5.7 gives

$$I_o = I_g + m(b_{ogx}^2 + b_{ogy}^2) = \frac{m}{12}(a^2 + 4a^2) + 5m\frac{a^2}{4} = \frac{5ma^2}{3}. \quad \text{(ii)}$$

Substituting this result into Equation (i) gives

$$\ddot{\theta} = \frac{3\sqrt{5}}{10}\frac{g}{a}\cos(\theta + \alpha), \quad \text{(iii)}$$

and we have completed Task b.

Moving to Task c, obtaining the equation of motion from conservation of energy, we will choose a horizontal plane running through the pivot point o as the datum for gravity potential energy; hence,

$$T + V = T_0 + V_0 \Rightarrow I_o\frac{\dot{\theta}^2}{2} - w\frac{a\sqrt{5}}{2}\sin(\theta + \alpha) = 0 - w\frac{a}{2} \quad \text{(iv)}$$

The potential energy function is negative because it is below the datum. Substituting for I_o into Equation (iv) and differentiating with respect to θ gives

$$\frac{5ma^2}{3}\frac{d}{d\theta}\left(\frac{\dot{\theta}^2}{2}\right) - w\frac{a\sqrt{5}}{2}\cos(\theta + \alpha) = 0$$

$$\Rightarrow \frac{5ma^2}{3}\ddot{\theta} - w\frac{a\sqrt{5}}{2}\cos(\theta + \alpha) = 0 \quad \text{(v)}$$

$$\therefore\ \ddot{\theta} = \frac{3\sqrt{5}}{10}\frac{g}{a}\cos(\theta + \alpha),$$

which coincides with Equation (iii), again without drawing a free-body diagram.

For Task d, getting governing equations for the reaction components, we will need to state $\sum \boldsymbol{f} = m\ddot{\boldsymbol{R}}_g$ for the plate's mass center. The nature of the current problem would argue strongly for a polar-coordinate version of these vector equations. Recall that $\Theta = (\theta + \alpha)$ implies $\dot{\Theta} = \dot{\theta}$, and $\ddot{\Theta} = \ddot{\theta}$, since α is a constant. Hence, from the free-body diagram of Figure XP5.2b, the polar statement of $\sum \boldsymbol{f} = m\ddot{\boldsymbol{R}}_g$ for the mass center gives

$$\begin{aligned}\sum f_r &= w\sin(\theta + \alpha) - o_r \\ &= ma_{gr} = m(\ddot{r}_g - r_g\dot{\theta}^2) = -ma\frac{\sqrt{5}}{2}\dot{\theta}^2 \\ \sum f_\Theta &= w\cos(\theta + \alpha) - o_\Theta \\ &= ma_{g\Theta} = m(r_g\ddot{\theta} - 2\dot{r}_g\dot{\theta}) = ma\frac{\sqrt{5}}{2}\ddot{\theta}.\end{aligned} \quad \text{(vi)}$$

In these equations $\dot{r}_g = \ddot{r}_g = 0$, because $r_g = a\sqrt{5}/2$ is constant.

The solution for o_Θ as a function of θ (alone) is obtained directly by substituting for $\ddot{\theta}$ from Equation (iii) into the second of Equation (vi), obtaining

$$\begin{aligned}o_\Theta &= w\cos(\theta + \alpha) - ma\frac{\sqrt{5}}{2}\cdot\frac{3\sqrt{5}}{10}\frac{g}{a}\cos(\theta + \alpha) \\ &= \frac{w}{4}\cos(\theta + \alpha).\end{aligned}$$

This result states that o_Θ starts at $(w/4)\cos\alpha$ for $\theta = 0$, $(\Theta = \alpha)$ and is zero when g is directly beneath the pivot point o at $\Theta = (\theta + \alpha) = \pi/2$.

We could use the energy-integral substitution $\ddot{\theta} = d(\dot{\theta}^2/2)/d\theta$ to integrate Equation (iii), but we will take the quicker route of using the energy-equation Equation (iv) to get

$$\frac{5ma^2}{3}\frac{\dot{\theta}^2}{2}=\frac{wa\sqrt{5}}{2}\sin(\theta+\alpha)-w\frac{a}{2}\Rightarrow\frac{\dot{\theta}^2}{2}$$
$$=\frac{3g}{2a\sqrt{5}}\sin(\theta+\alpha)-\frac{3g}{10a}.$$

Substituting this result back into the first of Equation (vi) gives

$$o_r=w\sin(\theta+\alpha)+m\frac{a\sqrt{5}}{2}\times\left[\frac{3g}{a\sqrt{5}}\sin(\theta+\alpha)-\frac{3g}{5a}\right]$$
$$=\frac{5w}{2}\sin(\theta+\alpha)-\frac{3w}{2}\sqrt{\frac{a}{5}},$$

and Task d *is* now completed.

The equilibrium condition for the pendulum is obtained by setting the right-hand side of Equation (iii) equal to zero, obtaining

$$\ddot{\theta}=\frac{3\sqrt{5}}{10}\frac{g}{a}\cos(\theta+\alpha)=0\Rightarrow\cos(\alpha+\bar{\theta})=0$$
$$\Rightarrow\bar{\theta}+\alpha=\frac{\pi}{2},\ \frac{3\pi}{2}.$$

$\bar{\theta}+\alpha=\pi/2$ defines the lower stable position with the mass center directly under the pivot point. The first step toward developing the EOM for small motion about the equilibrium condition is taken by substituting $\bar{\theta}+\alpha=\pi/2+\delta\theta$ into Equation (iii), obtaining

$$\delta\ddot{\theta}=\frac{3\sqrt{5}}{10}\frac{g}{a}\cos\left(\frac{\pi}{2}+\delta\theta\right),$$

and

$$\cos\left(\frac{\pi}{2}+\delta\theta\right)=\cos\left(\frac{\pi}{2}\right)\cos\delta\theta-\sin\left(\frac{\pi}{2}\right)\sin\delta\theta$$
$$=-\sin\delta\theta\cong-\delta\theta.$$

Hence, for small motion about the bottom equilibrium position,

$$\delta\ddot{\theta}+\frac{3\sqrt{5}}{10}\frac{g}{a}\delta\theta=0, \qquad \text{(vii)}$$

and the natural frequency is

$$\omega_n^2=\frac{3\sqrt{5}}{10}\frac{g}{a}\Rightarrow\omega_n=.819\sqrt{\frac{g}{a}}.$$

Task d is now completed.

We have proceeded from the original nonlinear equation of motion to get Equation (viii). Figure XP5.2c shows the plate initially hanging in its equilibrium position. Figure XP5.2d provides a free-body diagram for the plate rotated through $\delta\theta$ away from the equilibrium position. Stating the moment equation about o for Figure XP5.2d gives

$$I_o\delta\ddot{\theta}=\sum M_{oz}=-\frac{wa\sqrt{5}}{2}\sin\delta\theta\Rightarrow\delta\ddot{\theta}+\frac{3\sqrt{5}}{10}\frac{g}{a}\delta\theta=0.$$

This result coincides with Equation (viii).

The swinging-plate example introduces the following new features: (1) The parallel-axis theorem is required to obtain I_o in Equation 5.65. (2) More importantly, the plate is not in a general position in Figure XP5.2a, and no coordinate has been selected. Problems related to this subject will require that you select a coordinate and draw the body in a general position before proceeding to the free-body diagram, which must include the correct arrangement of external, body, and reaction forces. Note in writing the EOM that the positive-sign choice for the direction of rotation applies to the angular velocity, the angular acceleration, and the moments.

5.6.2 The Compound Pendulum with Damping

You may have had occasion to play with or observe a compound pendulum. The pendulum in a grandfather clock provides a common demonstration of this device. A spring or set of counterweights provides the energy to drive the clock. The fact that a "wound" clock eventually runs down is indicative of energy dissipation due to various forms of damping within the clock.

In our prior discussion, we assumed that the pivot joint was frictionless. Now, suppose that the pivot joint is a sleeve filled with a lubricant, constituting a fluid-film bearing. Figure 5.17 illustrates this situation. The relative rotation between the pendulum arm and the fixed pivot pin creates a fluid wedge that supports the pendulum. A resistance moment proportional to the pendulum rotation rate is also developed, defined by

$$M_\theta=-C_d\dot{\theta}.$$

The negative sign in this equation states that the resistance (drag) moment is directed opposite to

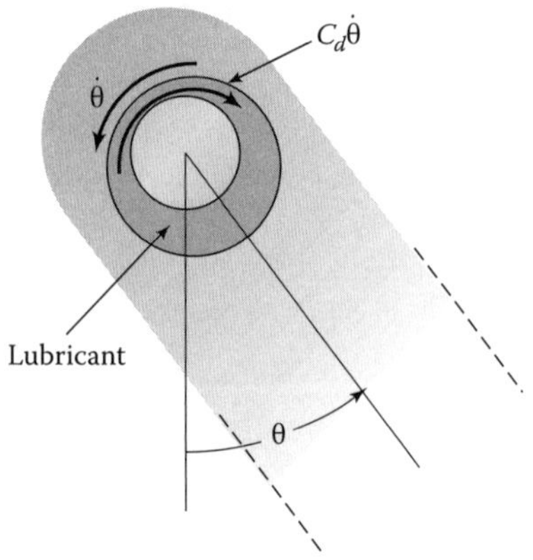

FIGURE 5.17
Pivot sleeve in a compound pendulum containing a viscous lubricant that develops a resistance torque due to rotation of the pendulum.

the rotation-velocity direction defined by $\dot{\theta}$. Adding this resistance moment to the moments on the right of Equation 5.33 gives

$$\frac{ml^2}{3}\ddot{\theta} = -w\frac{1}{2}\sin\theta - C_d\dot{\theta}, \qquad (5.71)$$

or

$$\ddot{\theta} + \frac{3C_d}{ml^2}\dot{\theta} + \frac{3g}{2l}\sin\theta = 0.$$

This equation is nonlinear because of the sin θ term. Retaining only the linear term in the Taylor series expansion for sin θ (given in Equation 3.106) gives the linear differential equation

$$\ddot{\theta} + 2\zeta\omega_n\dot{\theta} + \omega_n^2\theta = 0, \qquad (5.72)$$

where, the damping factor and natural frequency are

$$\omega_n^2 = \frac{3g}{2l}, \quad 2\zeta\omega_n = \frac{3C_d}{ml^2}$$

From Appendix B, in terms of the arbitrary initial conditions, $\dot{\theta}(0) = \dot{\theta}_0$; $\theta(0) = \theta_0$, Equation 5.72's solution is

$$\theta(t) = e^{-\zeta\omega_n t}\left[\theta_0\cos\omega_d t + \frac{(\dot{\theta}_0 + \zeta\omega_n\theta_0)}{\omega_d}\sin\omega_d t\right],$$

where the damped natural frequency $\omega_d = \omega_n\sqrt{1-\zeta^2}$. As expected, this solution consists of an exponentially decaying harmonic motion at the damped natural frequency ω_d.

We can now consider deriving the EOM from $\text{Work}_{n.c.} = (T + V) - (T_0 + V_0)$. Making the pivot point the datum for gravity potential energy gives

$$\left(I_o\frac{\dot{\theta}^2}{2} - w\frac{l}{2}\cos\theta\right) - \left(0 - w\frac{l}{2}\right) = \text{Work}_{\text{nonconservative}}$$

We found in Equation 5.37 that the differential work done by a moment acting through the differential angle $d\text{Work} = M\ d\theta$; hence, the work done by the moment $M = -C_d\dot{\theta}$ produces

$$\left(I_o\frac{\dot{\theta}^2}{2} - w\frac{l}{2}\cos\theta\right) - \left(0 - w\frac{l}{2}\right) = -\int_0^\theta C_d\dot{\theta}\,du \qquad (5.73)$$

Differentiating with respect to θ gives

$$\frac{ml^2}{3}\frac{d}{d\theta}\left(\frac{\dot{\theta}^2}{2}\right) + w\frac{l}{2}\sin\theta = -C_d\dot{\theta}$$

$$\Rightarrow \frac{ml^2}{3}\ddot{\theta} + C_d\dot{\theta} + w\frac{l}{2}\sin\theta = 0.$$

This result coincides with Equation 5.71. Note that we need the free-body diagram of Figure 5.17 to define the drag moment and state the work integral in Equation 5.73. Hence, in this case, there is no particular advantage in using the energy equation to derive the equation of motion.

5.6.3 Compound Pendulum/Spring and Damper Connections: Linearization and Equilibrium

Various vibration problems arise for a compound pendulum with additional spring connections from the pendulum to ground. Typically, vibration amplitudes are small for these problems, and linearization is pursued to obtain governing equations that apply for small motion about an equilibrium position. We considered linearization of the pendulum equation earlier in this section. Linearization of spring and damper forces for small motion is the first subject of this section.

5.6.3.1 Compound Pendulum with a Spring Attachment to Ground: Moment Equation

Figure 5.18a illustrates a compound pendulum with a spring connecting its lowest point to ground. The spring has length $l/3$ and is undeflected at $\theta = 0$. The following tasks apply:

a. Select coordinates, draw a free-body diagram, and derive the equation of motion.
b. For small motion about the equilibrium position, develop the linearized equation of motion.

Figure 5.18b provides the free-body diagram illustrating the stretched spring. The deflected spring length is defined from

$$\begin{aligned} l_s^2 &= \left(\frac{l}{3} + l\sin\theta\right)^2 + [l(1-\cos\theta)]^2 \\ &= l^2\left(\frac{1}{9} + \frac{2\sin\theta}{3} + \sin^2\theta + 1 - 2\cos\theta + \cos^2\theta\right) \\ &= \frac{l^2}{9}(19 + 6\sin\theta - 18\cos\theta). \end{aligned} \qquad (5.74)$$

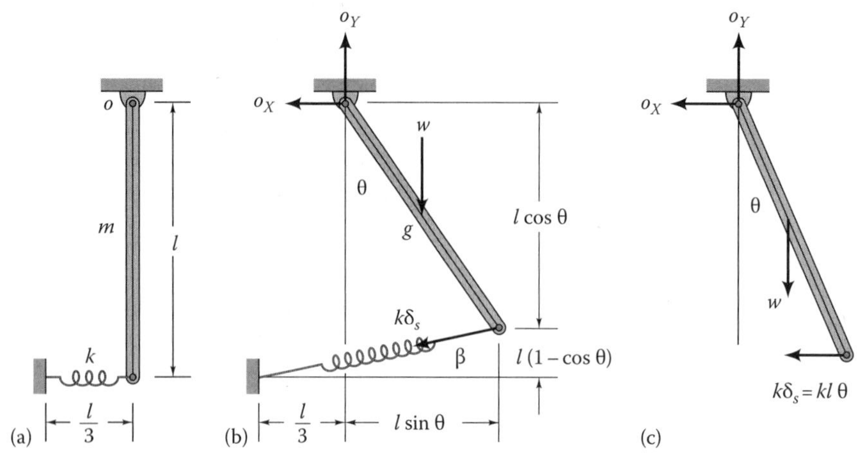

FIGURE 5.18
Compound pendulum with spring attachment to ground. (a) At rest in equilibrium. (b) General position. (c) Small-angle free-body diagram.

The spring force is

$$f_s = k\delta_s = k\left(l_s - \frac{l}{3}\right),$$

and it acts at the angle β from the horizontal defined by

$$\sin\beta = \frac{l(1-\cos\theta)}{l_s}, \quad \cos\beta = \frac{((1/3)+l\sin\theta)}{l_s}. \tag{5.75}$$

The pendulum equation of motion is obtained by a moment equation about the pivot point, yielding

$$\begin{aligned}\sum M_o = I_o\ddot{\theta} &= -w\frac{1}{2}\sin\theta - k\delta_s\cos\beta \times l\cos\theta \\ &\quad - k\delta_s\sin\beta \times l\sin\theta \\ &= -w\frac{l}{2}\sin\theta - k\delta_s l\cos(\theta-\beta) \end{aligned}\tag{5.76}$$

Substituting for $\delta_s = l_s - l/3$, $\cos\beta$, $\sin\beta$ (plus a considerable amount of algebra) yields

$$\begin{aligned}&\frac{ml^2}{3}\ddot{\theta} + \frac{wl}{2}\sin\theta + \frac{kl^2}{3}(\cos\theta + 3\sin\theta) \\ &\times\left[1 - \frac{1}{(19+6\sin\theta-18\cos\theta)^{1/2}}\right] = 0.\end{aligned}\tag{5.77}$$

This type of nonlinearity is referred to as "geometric." The spring is linear, but the finite θ rotation causes a nonlinear reaction force.

Our interest concerns the linearized equation of motion and particularly the linearized spring force. For small θ, expanding $\delta_s = l_s - l/3$ with l_s defined by Equation 5.74 in a Taylor's series expansion gives $\delta_s \cong l\theta$. Also, for small θ, a Taylor series expansion gives $\beta \cong \sin\beta \cong -3\theta^2/2 \cong 0$. Hence, for small θ, the spring force acts perpendicular to the pendulum axis, $\cos(\theta-\beta) \cong \cos\theta \cong 1$ and the moment Equation 5.76 reduces to

$$\frac{ml^2}{3}\ddot{\theta} = -w\frac{l}{2}\theta - k(l\theta)\times l \Rightarrow \frac{ml^2}{3}\ddot{\theta} + \left(\frac{wl}{2} + kl^2\right)\theta = 0. \tag{5.78}$$

For small θ, the spring deflection is $\delta_s = l\theta$, the spring force $f_s \cong -k\delta_s = -kl\theta$ acts perpendicular to the pendulum, and the moment of the spring force about o is $kl^2\theta$. Figure 5.18c provides the small-angle free-body diagram for the pendulum–spring system. Note that the spring force is independent of its initial spring length. From Equation 5.78, the natural frequency is

$$\omega_n = \sqrt{\frac{3g}{2l} + \frac{3k}{m}},$$

showing (as expected) an increase in the pendulum natural frequency due to the spring's stiffness.

5.6.3.2 Compound Pendulum with a Spring Attachment to Ground: Energy Approach

We will now use work–energy approaches to complete the following tasks:

a. *Assuming that the pendulum is released from rest at $\theta = \pi/2$, what is its angular velocity when it reaches the lowest position?*
b. *Use the energy equation to derive the equation of motion.*

Only conservative forces are present; hence, energy is conserved, and $T + V = T_0 + V_0$ applies. A horizontal plane through the pivot point serves as a datum for gravity potential energy. The clear difficulty with this example is determining the deflected length of the spring l_s as a function of θ to plug into the potential-energy function. Fortunately, the work was completed in Equation 5.74, and we can state

$$l_s = \frac{l}{3}(19 + 6\sin\theta - 18\cos\theta)^{1/2}. \tag{5.79}$$

The potential energy stored in the spring is $V_s = k\delta_s^2/2$ where the spring's deflection is $\delta_s = l_s - l/3$. At $\theta = \pi/2$ the spring length is $l_s(\theta = \pi/2) = 5l/3$. Substituting into $T + V = T_0 + V_0$ gives

$$\frac{ml^2}{3}\frac{\dot{\theta}^2}{2} + \frac{k}{2}\left[\frac{l}{3}(19 + 6\sin\theta - 18\cos\theta)^{1/2} - \frac{l}{3}\right]^2 - \frac{wl}{2}\cos\theta = \frac{k}{2}\left(\frac{5l}{3} - \frac{l}{3}\right)^2. \tag{5.80}$$

This equation applies for any value of θ. Evaluating Equation 5.80 at $\theta = 0$, concludes Task a as

$$\dot{\theta}(\theta = 0) = \left(\frac{3g}{l} + \frac{16k}{3m}\right)^{1/2}.$$

Task b is accomplished by differentiating Equation 5.80 with respect to θ, obtaining

$$\frac{ml^2}{3}\ddot{\theta} + \frac{wl}{2}\sin\theta + \frac{kl^2}{3}(\cos\theta + 3\sin\theta) \times \left[1 - \frac{1}{(19 + 6\sin\theta - 18\cos\theta)^{1/2}}\right] = 0, \tag{5.81}$$

where $\ddot{\theta} = d(\dot{\theta}^2/2)d\theta$. This unpleasant nonlinear differential equation would need to be integrated numerically from initial conditions to obtain time solutions $\dot{\theta}(t)$, $\theta(t)$.

Equation 5.81 coincides with Equation 5.77 obtained earlier from a moment equation about an axis through the pivot point. Developing Equation 5.77 from the moment equation required the following steps:

a. A free-body diagram was developed, including the spring force to obtain a moment equation about the pivot point o.
b. The deflected-spring length l_s was calculated leading to the spring force $f_s = k(l_s - l/3)$.
c. Components of f_s were needed in the horizontal and vertical directions to be used in defining the spring-force moment about the pendulum support point o.

For this example, the energy equation provides a much quicker path to the equation of motion. Also, satisfying Task a (starting with Equation 5.77) would require integrating it using the energy-integral substitution $\ddot{\theta} = d(\dot{\theta}^2/2)/d\theta$, versus simply starting with $T + V = T_0 + V_0$. With either approach, the main struggle involves getting Equation 5.74 that defines $l_s(\theta)$.

For the small displacement of the spring, starting with $T + V = T_0 V_0$, the EOM is

$$\frac{ml^2}{3}\frac{\dot{\theta}^2}{2} + \frac{k}{2}(l\theta^2) - \frac{wl}{2}\cos\theta = \frac{k}{2}\left(\frac{5l}{3} - \frac{l}{3}\right)^2. \tag{5.82}$$

Differentiating Equation 5.82 with respect to θ gives

$$\frac{ml^2}{3}\ddot{\theta} + kl^2\theta + \frac{wl}{2}\sin\theta = 0 \Rightarrow \frac{ml^2}{3}\ddot{\theta} + \left(kl^2 + \frac{wl}{2}\right)\theta \cong 0. \tag{5.83}$$

The last step in Equation 5.83 follows from $\sin\theta \cong \theta$ for small θ, and coincides with Equation 5.78. Note particularly that we did not use the "customary" $\cos\theta \cong 1$ small-θ assumption in stating $V_g = -wl\cos\theta/2$ for Equation 5.82. Had we used this approximation, when we differentiated with respect to 2, the w contribution would have vanished in Equation 5.83. The appropriate small-angle approximation would have been $\cos\theta \cong 1 - \theta^2/2$; however, it is simpler to impose the small-angle approximation at the end as we did in Equation 5.83.

5.6.3.3 Compound Pendulum with a Damper Attachment to Ground: Moment Equation

The issue of forces from dampers due to small motion about an equilibrium position also arises. Figure 5.19a shows a compound pendulum with an attached linear damper; Figure 5.19b provides the corresponding general position (large θ) free-body diagram. The damper reaction force, $f_d = -c\dot{\delta}_s$, acts at the angle β from the horizontal. From Equation 5.74,

$$\dot{\delta}_s = \dot{l}_s = l(\cos\theta/3 + \sin\theta)\dot{\theta} \Big/ \sqrt{\frac{19}{9} + \frac{2\sin\theta}{3} - 2\cos\theta}$$

We could plow through the same nonlinear development for this system as we did for the spring. However, for small θ, $\beta \cong 0$, and the damping force acts perpendicular to the pendulum axis and reduces to $f_d = -cl\dot{\theta}$, where $v_\theta = l\dot{\theta}$ is the pendulum's circumferential velocity at the attachment point. Figure 5.19c provides a "small θ" free-body diagram, yielding

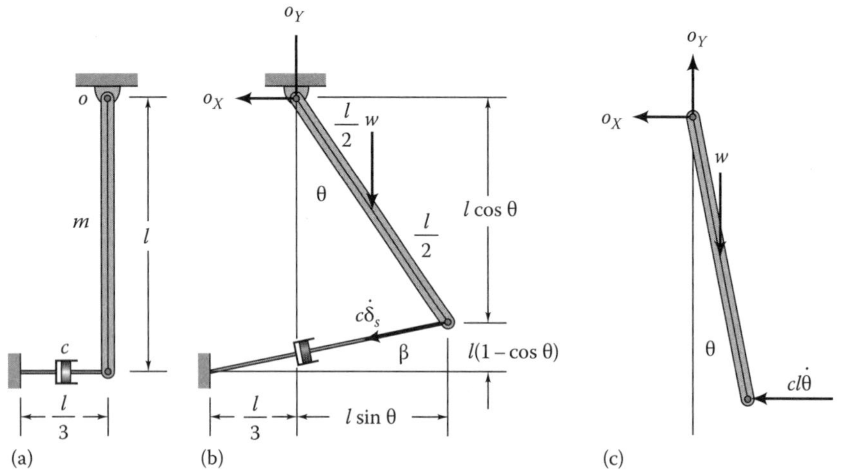

FIGURE 5.19
Compound pendulum. (a) At rest in equilibrium. (b) General-position free-body diagram. (c) Small-rotation free-body diagram.

$$\sum M_o = I_o\ddot{\theta} = -w\frac{l}{2}\theta - l \times cl\dot{\theta} \Rightarrow \frac{ml^2}{3}\ddot{\theta} + cl^2\dot{\theta} + \frac{wl}{2}\theta = 0,$$

with $I_o = ml^2/3$ defined in Appendix C. The natural frequency and damping factor are

$$\omega_n = \sqrt{\frac{3g}{2l}}, \quad 2\zeta\omega_n = cl^2 \times \frac{3}{ml^2} = \frac{3c}{m}.$$

As with the spring, for small θ, the damping force $f_d = -cl\dot{\theta}$ is independent of the initial damper length. Several small-motion pendulum/spring/damper examples are considered in the balance of this section.

Example Problem 5.3

Figure XP5.3a shows a pendulum with mass $m = 0.5$ kg and length $l = 1$ m supported by a pivot point located $l/3$ from the pendulum's end. Two linear springs with stiffness coefficient $k = 15$ N/m are attached to the pendulum a distance $l/3$ down from the pivot point, and a linear damper with damping coefficient $c = 0.5$ N s/m is attached to the pendulum's end. The springs are preloaded with the same tensile force when the pendulum is vertical. The following tasks apply:

a. Select coordinates, draw a free-body diagram, and derive the EOM.
b. Determine the natural frequency and damping factor.
c. Use work–energy to derive the EOM.

A "small θ" free-body diagram is given in Figure XP5.3b. Note that the spring forces shown do not involve the preloads. Only the change in the spring force due to θ impacts the EOM. Taking moments about the pivot point gives

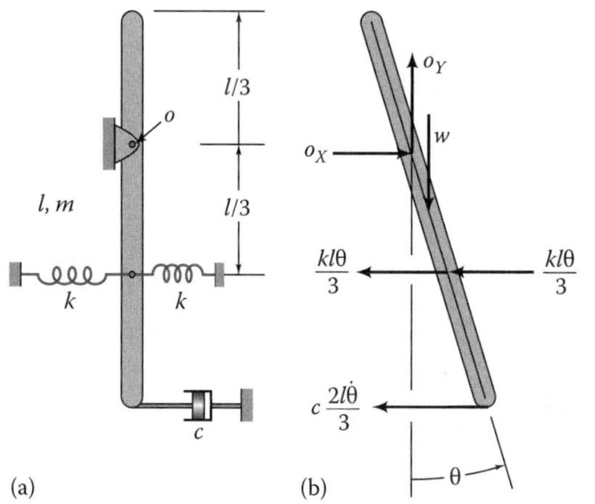

FIGURE XP5.3
(a) Pendulum attached to ground by two linear springs and a viscous damper. (b) Coordinate and free-body diagram.

$$\begin{aligned} I_o\ddot{\theta} &= \sum M_o = -w\frac{l}{6}\sin\theta - 2 \times k\frac{l\theta}{3} \times \frac{l}{3} - c\frac{2l\dot{\theta}}{3} \times \frac{2l}{3} \\ \therefore \quad & \frac{ml^2}{9}\ddot{\theta} + \frac{4cl^2}{9}\dot{\theta} + \frac{wl}{6}\sin\theta + \frac{2kl^2}{9}\theta = 0, \end{aligned} \quad \text{(i)}$$

where from Appendix C and the parallel-axis formula, $I_o = \frac{ml^2}{12} + m\left(\frac{l}{6}\right)^2 = \frac{ml^2}{9}$. For sin θ ≅ θ, the linearized EOM is

$$\ddot{\theta} + \frac{4c}{m}\dot{\theta} + \left(\frac{3g}{2l} + \frac{2k}{m}\right)\theta = 0.$$

The natural frequency and damping factor are

$$\omega_n = \left(\frac{3g}{2l} + \frac{2k}{m}\right)^{1/2} = \left(\frac{3 \times 9.81}{2 \times 1} + \frac{30}{0.5}\right)^{1/2}$$
$$= 8.64\ \frac{\text{rad}}{\text{s}} \Rightarrow f_n = 1.38\ \text{Hz}$$
$$2\zeta\omega_n = \frac{4c}{m} \Rightarrow \zeta = \frac{4c}{2\omega_n m} = \frac{4 \times 5}{2 \times 8.64 \times 0.5} = 0.231,$$

concluding Task b.

Applying work$_{n.c.} = \Delta(T + V)$ gives

$$\int_o^\theta -c\frac{2l\dot{\theta}}{3} \times \frac{2l}{3}\,dx = \left[I_o\frac{\dot{\theta}^2}{2} + 2\frac{k}{2}\left(\frac{l\theta}{3}\right)^2 - w\frac{l}{6}\cos\theta\right] - \left[0 - w\frac{l}{6}\right],$$

where the term on the left accounts the work done by the moment of the damping force $-c(2l\dot{\theta})/3$ about the pivot point o. The terms on the right account for the kinetic energy of the bar, the potential energy stored in the springs, and the gravity potential energy of the bar for a datum plane through o. Differentiating with respect to θ gives

$$-c\left(\frac{2l}{3}\right)^2\dot{\theta} = I_o\frac{d}{d\theta}\left(\frac{\dot{\theta}^2}{2}\right) + \frac{2kl^2}{9}\theta + w\frac{l}{6}\sin\theta$$
$$\Rightarrow \frac{ml^2}{9}\ddot{\theta} + \frac{4cl^2}{9}\dot{\theta} + \frac{2kl^2}{9}\theta + \frac{wl}{6}\sin\theta = 0,$$

which coincides with Equation (i) and concludes Task c.

The examples of this section are provided to help you (i) develop an understanding for the "geometrical" nonlinearities that can arise for finite rotation of a pendulum, (ii) better understand the mechanics of linearization about an equilibrium position in terms of force definition and direction, and finally (iii) conclude that, for motion about an equilibrium position, the equation of motion is independent of the static preload in the springs.

5.6.3.4 Bars Supported by Springs: Preload and Equilibrium

The amount of preload developed by a spring and its influence on the equation of motion frequently causes confusion in developing equations of motion for a pendulum with spring attachments. Figure 5.20a illustrates a bar of mass m and length l in equilibrium with linear springs having stiffness coefficients k_1, k_2 counteracting the weight w. The springs act at a distance $2l/3$ from the pivot support point and have been preloaded (stretched or compressed) to maintain the bar in its equilibrium position. The engineering-analysis tasks for this example are as follows: *Draw a free-body diagram, derive the EOM, and determine the natural frequency.*

Figure 5.20b provides an equilibrium free-body diagram for the bar at the rotation angle $\bar{\theta}$. The bar's weight w is supported by the reaction $\bar{O}_Y$ at the pivot support point, by a compression force $k_1\delta_1$ in the lower spring, and a tension force $k_2\delta_2$ in the upper spring. The moment equilibrium-condition requirement is found by taking moments about the pivot point, obtaining

$$\sum M_o = 0 = w\frac{l}{2}\sin\bar{\theta} - k_1\delta_1\frac{2l}{3} - k_2\delta_2\frac{2l}{3}. \tag{5.84}$$

Figure 5.20c provides a free-body diagram for a general displaced position defined by the rotation angle $\bar{\theta} + \delta\theta$. For small $\delta\theta$, the spring-support point moves the

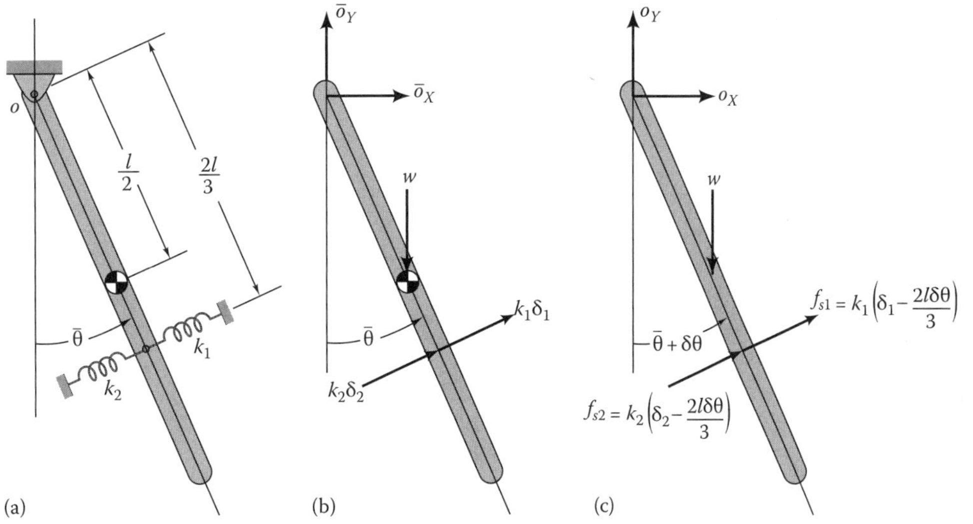

FIGURE 5.20
Uniform bar. (a) In equilibrium at angle $\bar{\theta}$. (b) Equilibrium free-body diagram. (c) Displaced-position free-body diagram.

perpendicular distance $\delta_s = (2l/3)\delta\theta$. Hence, the stretch of the upper spring decreases from δ_1 to $\delta_1 - (2l/3)\delta\theta$, and the compression of the lower spring decreases from δ_1 to $\delta_2 - (2l/3)\delta\theta$. The spring reaction forces are

$$f_{s1} = k_1\left(\delta_1 - \frac{2l}{3}\delta\theta\right), \quad f_{s2} = k_2\left(\delta_2 - \frac{2l}{3}\delta\theta\right).$$

Taking moments about o gives

$$\begin{aligned}\sum M_o = I_o\delta\ddot{\theta} &= -w\frac{l}{2}\sin(\bar{\theta}+\delta\theta) + \frac{2l}{3}k_1\left(\delta_1 - \frac{2l}{3}\theta\right) \\ &\quad + \frac{2l}{3}k_2\left(\delta_2 - \frac{2l}{3}\theta\right) \\ &\cong -w\frac{l}{2}(\sin\bar{\theta} + \cos\bar{\theta}\delta\theta) + \frac{2l}{3}k_1\delta_1 \\ &\quad + \frac{2l}{3}k_2\delta_2 - \left(\frac{2l}{3}\right)^2(k_1+k_2)\delta\theta \\ &= \left(-w\frac{1}{2}\sin\bar{\theta} + \frac{2l}{3}k_1\delta_1 + \frac{2l}{3}k_2\delta_2\right) \\ &\quad - w\frac{l}{2}\cos\bar{\theta}\delta\theta - \left(\frac{2l}{3}\right)^2(k_1+k_2)\delta\theta,\end{aligned}$$

after dropping second-order terms in $\delta\theta$. Rearranging provides the EOM,

$$\begin{aligned}&\frac{ml^2}{3}\delta\ddot{\theta} + \left[\left(\frac{2l}{3}\right)^2(k_1+k_2) + w\frac{l}{2}\cos\bar{\theta}\right]\delta\theta = 0 \\ &= -w\frac{1}{2}\sin\bar{\theta} + \frac{2l}{3}k_1\delta_1 + \frac{2l}{3}k_2\delta_2. \end{aligned} \tag{5.85}$$

The right-hand side of Equation 5.85 is zero from the equilibrium result of Equation 5.84. If the bar is in equilibrium in a vertical position ($\bar{\theta} = 0$, $\cos\bar{\theta} = 1$), the weight contribution to the EOM reverts to the compound pendulum results of Equation 3.60. For a horizontal equilibrium position, ($\bar{\theta} = \pi/2$, $\cos\bar{\theta} = 0$), and the weight term is eliminated. The natural frequency is

$$\begin{aligned}\omega_n^2 &= \frac{3}{ml^2}\left[\left(\frac{2l}{3}\right)^2(k_1+k_2) + w\frac{l}{2}\cos\bar{\theta}\right] \\ \Rightarrow \omega_n &= \sqrt{\frac{4(k_1+k_2)}{3m} + \frac{3g}{2l}\cos\bar{\theta}}.\end{aligned} \tag{5.86}$$

Some students find the equilibrium results puzzling, thinking, ''Wait a minute, suppose the bar is in equilibrium by some other combination of spring forces?'' To pursue this question, consider the alternative equilibrium situation provided by Figure 5.21a. Now the lower spring is also assumed to be in tension with a

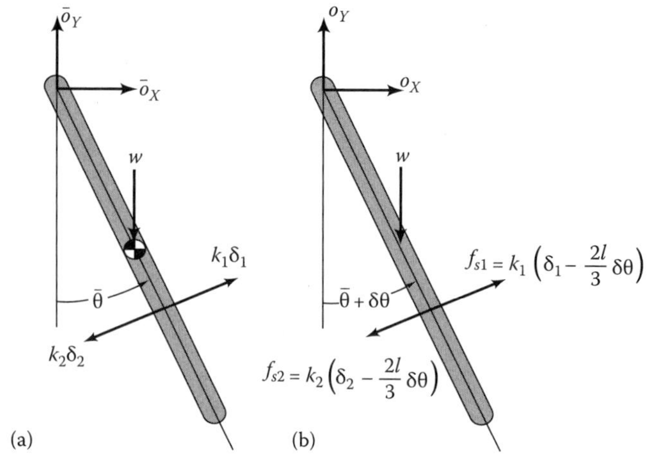

FIGURE 5.21
Uniform bar. (a) Alternative static equilibrium free-body diagram. (b) Displaced position free-body diagram.

static stretch δ_2, developing the tension force $k_2\delta_2$ at equilibrium. Taking moments about o gives the static equilibrium requirement

$$\sum M_o = 0 = -w\frac{l}{2}\sin\bar{\theta} + k_1\delta_1\frac{2l}{3} - k_2\delta_2\frac{2l}{3}. \tag{5.87}$$

The $\delta\theta$ rotation increases the stretch in the lower spring from δ_2 to $\delta_2 + (2l/3)\delta\theta$, decreases the stretch in the upper spring from δ_1 to $\delta_1 - (2l/3)\delta\theta$, and the reaction forces are

$$f_{s1} = k_1\left(\delta_1 - \frac{2l}{3}\delta\theta\right), \quad f_{s2} = k_2\left(\delta_2 + \frac{2l}{3}\delta\theta\right).$$

From Figure 5.21b,

$$\begin{aligned}\sum M_o = I_o\delta\ddot{\theta} &= -w\frac{l}{2}\sin(\bar{\theta}+\delta\theta) + \frac{2l}{3}k_1\left(\delta_1 - \frac{2l}{3}\delta\theta\right) \\ &\quad - \frac{2l}{3}k_2\left(\delta_2 + \frac{2l}{3}\delta\theta\right) \\ &\cong -w\frac{l}{2}(\sin\bar{\theta} + \cos\bar{\theta}\delta\theta) + \frac{2l}{3}k_1\delta_1 - \frac{2l}{3}k_2\delta_2 \\ &\quad - \left(\frac{2l}{3}\right)^2(k_1+k_2)\delta\theta \\ &= \left(-w\frac{l}{2}\sin\bar{\theta} + \frac{2l}{3}k_1\delta_1 - \frac{2l}{3}k_2\delta_2\right) \\ &\quad + w\frac{l}{2}\cos\bar{\theta} - \left(\frac{2l}{3}\right)^2(k_1+k_2)\delta\theta,\end{aligned}$$

and the EOM is (again)

$$\frac{ml^2}{3}\delta\ddot{\theta}+\left[\left(\frac{2l}{3}\right)^2(k_1+k_2)+w\frac{l}{2}\cos\bar{\theta}\right]\delta\theta=0$$
$$=-w\frac{1}{2}\sin\bar{\theta}+\frac{2l}{3}k_1\delta_1-\frac{2l}{3}k_2\delta_2. \quad (5.88)$$

The right-hand side is zero from the equilibrium requirement of Equation 5.87, and Equation 5.88 repeats the EOM of Equation 5.85.

The lesson from this second development is: For small motion about equilibrium, the same EOM is obtained *irrespective* of the initial equilibrium forces in the (linear) springs. The spring-force contributions to the differential equation arise from the *change* in the equilibrium forces due to a *change in position*. This is the same basic outcome that we obtained for a mass m supported by linear springs in Figure 3.6. The change in equilibrium angle $\bar{\theta}$ changes w's contribution to the EOM, because $-w(l/2)\sin\theta$, the moment due to w, is a *nonlinear* function of θ.

Figure 5.22 illustrates a situation where the equilibrium angle does impact the linear support spring's contribution to the EOM. As shown in Figure 5.22a, the bar is in equilibrium, being supported by the pivot at o and horizontal and vertical springs at its lower end. Figure 5.22b illustrates the equilibrium free-body diagram with both springs in tension. Taking moments about o gives the equilibrium requirement

$$\sum M_o=0=-w\frac{l}{2}\sin\bar{\theta}+k_x\delta_x l\cos\bar{\theta}+k_y\delta_y l\sin\bar{\theta}. \quad (5.89)$$

Figure 5.22c illustrates the bar rotated through the angle $\bar{\theta}+\delta\theta$. The $\delta\theta$ rotation displaces the bar's end the small distance $l\delta\theta$. As illustrated in Figure 5.22d, this displacement produces horizontal and vertical displacements, respectively, of $l\delta\theta\cos\theta$ and $l\delta\theta\sin\bar{\theta}$. With these displacements, the vertical and horizontal spring deflections are, respectively, $\delta_x-l\delta\theta\cos\bar{\theta}$ and $\delta_y-l\delta\theta\sin\bar{\theta}$. Figure 5.22c provides the displaced-position free-body diagram for reaction forces conforming to these displacements.

Taking moments about o gives

$$\sum M_o=I_o\delta\ddot{\theta}=-w\frac{1}{2}\sin(\bar{\theta}+\delta\theta)+k_x(\delta_x-l\delta\theta\cos\bar{\theta})\times l\cos(\bar{\theta}+\delta\theta)+k_y(\delta_y-l\delta\theta\sin\bar{\theta})\times l\sin(\bar{\theta}+\delta\theta)$$

For small δ_x, δ_y, $\delta\theta$, expanding this equation produces

$$\begin{aligned}I_o\delta\ddot{\theta}&=-w\frac{1}{2}(\sin\bar{\theta}\cos\delta\theta+\cos\bar{\theta}\sin\delta\theta)\\&\quad+k_xl(\delta_x-l\delta\theta\cos\bar{\theta})(\cos\bar{\theta}\cos\delta\theta-\sin\bar{\theta}\sin\delta\theta)\\&\quad+k_yl(\delta_y-l\delta\theta\sin\bar{\theta})(\sin\bar{\theta}\cos\delta\theta+\cos\bar{\theta}\sin\delta\theta)\\&\cong\left(-w\frac{l}{2}\sin\bar{\theta}+k_x\delta_xl\cos\bar{\theta}+k_y\delta_yl\sin\bar{\theta}\right)\\&\quad-w\frac{l}{2}\cos\bar{\theta}-l(k_x\cos^2\bar{\theta}-k_y\sin^2\bar{\theta})l\delta\theta.\end{aligned}$$

Hence, the EOM is

$$I_o\delta\ddot{\theta}+\left[w\frac{l}{2}\cos\bar{\theta}+l^2(k_x\cos^2\bar{\theta}+k_y\sin\bar{\theta}^2)\right]\delta\theta=0$$
$$=-w\frac{l}{2}\sin\bar{\theta}+k_x\delta_xl\cos\bar{\theta}+k_y\delta_yl\sin\bar{\theta} \quad (5.90)$$

The right-hand side is zero because of Equation 5.89's equilibrium requirement. Even though the springs are linear, their contribution to the effective stiffness $(k_x\cos^2\bar{\theta}+k_y\sin\bar{\theta}^2)$ is a nonlinear function of the

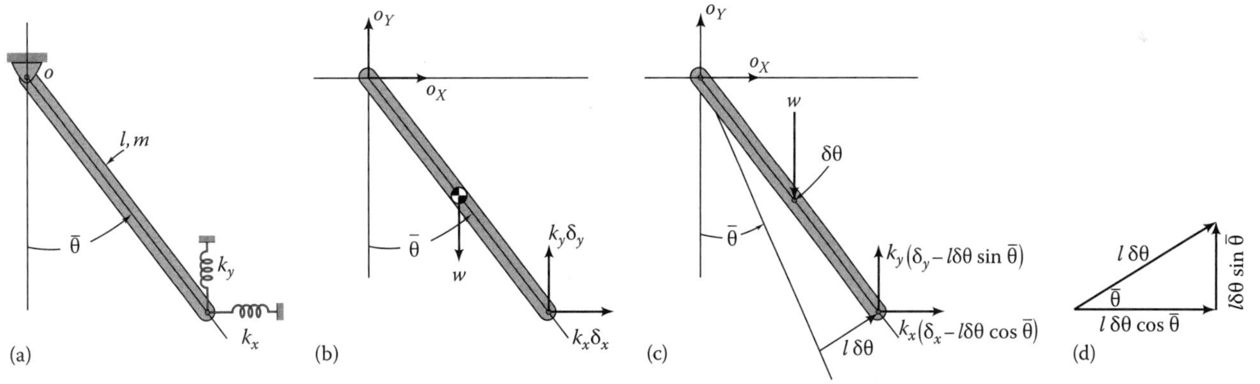

FIGURE 5.22
Bar supported by a pivot point and horizontal and vertical springs. (a) Equilibrium position and condition. (b) Equilibrium free-body diagram. (c) Displaced free-body diagram. (d) Components of the bar's end displacement.

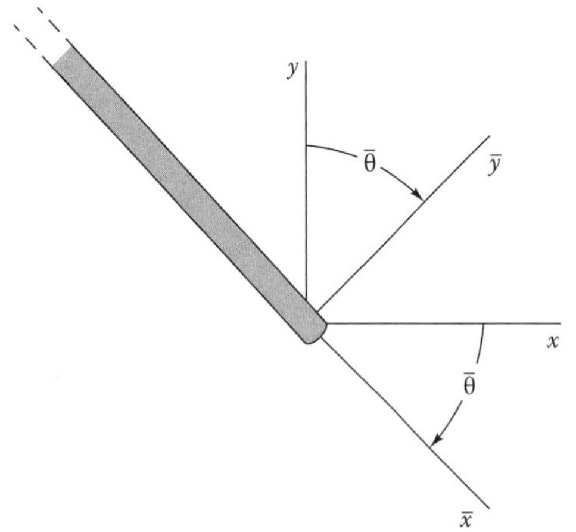

FIGURE 5.23
x, y and $\bar{x}$, $\bar{y}$ coordinate systems.

equilibrium angle $\bar{\theta}$. This is another example of a geometric nonlinearity.

To get an alternative viewpoint of this outcome, consider the two coordinate systems in Figure 5.23. The x, y system corresponds to the coordinate system of Figure 5.22. For the $\bar{x}$, $\bar{y}$ system, $\bar{x}$ is along the axis of the bar; $\bar{y}$ is perpendicular to the bar's axis and is pointed in the direction of the $l\delta\theta$ displacement in Figure 5.22c and d. The stiffness matrix for the springs in the x, y system defines the reaction force components as

$$\begin{Bmatrix} f_{sx} \\ f_{sy} \end{Bmatrix} = -\begin{bmatrix} k_x & 0 \\ 0 & k_y \end{bmatrix}\begin{Bmatrix} x \\ y \end{Bmatrix} \quad \text{or} \quad (f_s)_i = -[k_i](r_i). \tag{5.91}$$

We would like to obtain the stiffness matrix for the $\bar{x}$, $\bar{y}$ system. Figure 5.23 defines the following coordinate transformation [see Equation 2.15]:

$$\begin{Bmatrix} B_x \\ B_y \end{Bmatrix} = \begin{bmatrix} \cos\bar{\theta} & \sin\bar{\theta} \\ -\sin\bar{\theta} & \cos\bar{\theta} \end{bmatrix}\begin{Bmatrix} B_{\bar{x}} \\ B_{\bar{y}} \end{Bmatrix} \quad \text{or} \quad (B)_i = [A](B)_{\bar{i}}. \tag{5.92}$$

Using Equation 5.92 to define $\boldsymbol{f}_s$ and $\boldsymbol{r}$ in terms of their $\bar{x}$, $\bar{y}$ components gives $[A](f_s)_{\bar{i}} = -[k_i][A](r)_{\bar{i}} \Rightarrow (f_g)_{\bar{i}} = -[A]^T[k_i][A](r)_{\bar{i}}$, and the stiffness matrix for the $\bar{x}$, $\bar{y}$ coordinate system is

$$[k_{\bar{i}}] = [A]^T[k_i][A] = \begin{bmatrix} k_x\cos^2\bar{\theta} + k_y\sin^2\bar{\theta} & (k_x - k_y)\sin\bar{\theta}\cos\bar{\theta} \\ (k_x - k_y)\sin\bar{\theta}\cos\bar{\theta} & k_x\sin^2\bar{\theta} + k_y\cos^2\bar{\theta} \end{bmatrix}$$

Using this stiffness matrix, the reaction force due to the displacements, $\delta\bar{x} = 0$, $\delta\bar{y} = l\,\delta\theta$ is

$$\begin{Bmatrix} f_{s\bar{x}} \\ f_{s\bar{y}} \end{Bmatrix} = -\begin{bmatrix} k_x\cos^2\bar{\theta} + k_y\sin^2\bar{\theta} & (k_x - k_y)\sin\bar{\theta}\cos\bar{\theta} \\ (k_x - k_y)\sin\bar{\theta}\cos\bar{\theta} & k_x\sin^2\bar{\theta} + k_y\cos^2\bar{\theta} \end{bmatrix}\begin{Bmatrix} 0 \\ l\delta\theta \end{Bmatrix} = -\begin{Bmatrix} l\delta\theta\left[(k_x - k_y)\sin\bar{\theta}\cos\bar{\theta}\right] \\ l\delta\theta(k_x\cos^2\bar{\theta} + k_y\sin^2\bar{\theta}) \end{Bmatrix}$$

The $f_{s\bar{x}}$ component, $-l\delta\theta[(k_x - k_y)\sin\bar{\theta}\cos\bar{\theta}]$, is along the rod's axis and does not enter the moment equation. The $f_{s\bar{y}}$ component, $-l\delta\theta(k_x\cos^2\theta + k_y\sin^2\bar{\theta})$ provides the spring's stiffness contributions in Equation 5.90.

5.6.3.5 *Closing Comments and (Free) Advice*

The examples of this section are generally representative of 1DOF, fixed-axis problems. In applying the moment equation $M_{oz} = I_o\ddot{\theta}$ to derive the EOM, selecting coordinates and drawing a correct free-body diagram is the most important step. You should always take advantage of a fixed-rotation axis in stating a moment equation. If the mass center location g does not coincide with the fixed axis of rotation location o, stating the moment equation about o is always preferable. Problems that require the calculation of reactions will always require the statement of $\sum \boldsymbol{f} = m\ddot{\boldsymbol{R}}_g$ for the mass center of the body. Choose your coordinate system carefully in stating this equation. Most fixed-axis rotation problems proceed more quickly if you use polar coordinates.

The moment equation will always yield a second-order differential equation in the rotation angle. Some of the equations developed here were linear; some were nonlinear. The compound-pendulum and swinging-plate problems were illustrative of governing nonlinear differential equations of motion that could be profitably linearized about an equilibrium position. Linearization of nonlinear differential equations about an equilibrium position is a continuing important step in engineering analysis, and we will have repeated demonstrations of this procedure throughout this book. We showed in the compound-pendulum problem that the pendulum had two equilibrium positions, one stable and one unstable. We could have shown a similar outcome for the swinging plate problem. Almost all linear vibration problems involve an implicit assumption of small motion about a stable equilibrium position to justify their linear state. Dynamic systems that are statically unstable are generally quite undesirable and are less common.

5.6.4 Prescribed Acceleration of a Compound Pendulum's Pivot Support Point

Moment equations for the fixed-axis rotation problems of the preceding section were taken about *a fixed* pivot point, employing the moment equation

$$M_{oz} = I_o\ddot{\theta}, \tag{5.26}$$

where o identifies the axis of rotation. The problems involved here concern situations where the pivot point is accelerating, and the general moment equation

$$M_{oz} = I_o\ddot{\theta} + m(\boldsymbol{b}_{og} \times \ddot{\boldsymbol{R}}_o)_z, \qquad (5.24)$$

is required. In applying Equation 5.24, recall the following points:

a. Moments are being taken about the body-fixed axis o, and I_o is the moment of inertia through axis o.
b. The vector $\boldsymbol{b}_{og}$ goes from a z-axis through o to a z-axis through the mass center at g.
c. The positive rotation and moment sense in Equation 5.24 correspond to a counterclockwise rotation for θ.

The last term in the moment equation is positive because the positive right-hand-rule convention for the cross product in this term coincides with the $+\theta$ sense. For a rigid body with a positive clockwise rotation angle, this last term requires a negative sign.

Figure 5.24a illustrates an accelerating pickup truck with an unsecured tail gate. The angle θ defines the tail gate's rotation angle. Given that the pickup has a constant acceleration of $g/3$, neglecting friction at the pivot, and assuming that the tailgate can be modeled as a uniform plate of mass m, carry out the following tasks:

a. Select coordinates.
b. Derive the governing equation of motion.
c. Assuming that the tailgate starts from rest at $\theta = 0$, what will $\dot{\theta}$ be at $\theta = \pi/2$?
d. Determine the reactions at pivot point o as a function of θ (only).

Figure 5.24b provides the free-body diagram for the tail gate and

$$\boldsymbol{b}_{og} = \frac{l}{2}(-\boldsymbol{I}\sin\theta + \boldsymbol{J}\cos\theta)$$
$$\ddot{\boldsymbol{R}}_o = \boldsymbol{I}\ddot{X} = \boldsymbol{I}\frac{g}{3}$$
$$\boldsymbol{b}_{og} \times \ddot{\boldsymbol{R}}_o = -\boldsymbol{K}\frac{gl}{6}\cos\theta.$$

Hence, Equation 5.24 gives

$$\sum M_o = w\frac{l}{2}\sin\theta = \frac{ml^2}{3}\ddot{\theta} - m\frac{gl}{6}\cos\theta \Rightarrow \frac{ml^2}{3}\ddot{\theta}$$
$$= \frac{wl}{2}\sin\theta + \frac{wl}{6}\cos\theta. \qquad (5.93)$$

We have now completed Task b. As expected, the pickup's acceleration is causing the tail gate to swing open more rapidly than it would under the influence of gravity alone. We can use the energy-integral substitution to integrate this nonlinear EOM as

$$\ddot{\theta} = \frac{d}{d\theta}\left(\frac{\dot{\theta}^2}{2}\right) = \frac{3g}{2l}\sin\theta + \frac{g}{2l}\cos\theta. \qquad (5.94)$$

Multiplying by $d\theta$ reduces both sides of this equation to exact differentials. Integrating from the initial condition $\dot{\theta}(\theta = 0) = 0$ gives

$$\frac{\dot{\theta}^2}{2} = \frac{g}{l}\Big|_0^{\theta}\left(-\frac{3}{2}\cos u + \frac{1}{2}\sin u\right) = \frac{g}{l}\left[\frac{1}{2}\sin\theta + \frac{3}{2}(1-\cos\theta)\right]. \qquad (5.95)$$

Hence, at $\theta = \pi/2$, $\dot{\theta}(\pi/2) = 2\sqrt{g/l}$, and we have completed Task c.

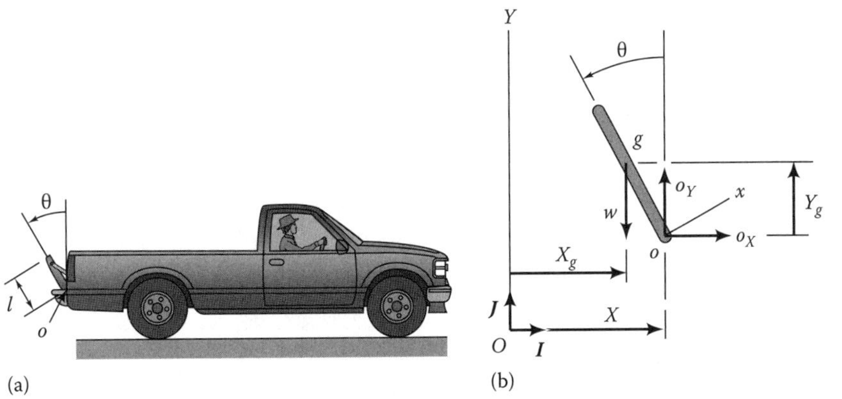

FIGURE 5.24
(a) An accelerating pickup truck with a loose tail gate. (b) Free-body diagram for the tail gate.

Moving to Task d, stating $\sum f = m\ddot{\mathbf{R}}_g$ for the mass center gives

$$\begin{aligned}\sum f_X &= o_X = m\ddot{X}_g \\ \sum f_Y &= o_Y - w = m\ddot{Y}_g\end{aligned} \tag{5.96}$$

We need to determine $\ddot{X}_g$, $\ddot{Y}_g$. From Figure 5.24b,

$$X_g = X - \frac{l}{2}\sin\theta; \quad Y_g = \frac{l}{2}\cos\theta.$$

Differentiating twice with respect to time gives

$$\begin{aligned}\ddot{X}_g &= \ddot{X} - \frac{l}{2}\cos\theta\ddot{\theta} + \frac{l}{2}\sin\theta\dot{\theta}^2 = \frac{g}{3} - \frac{l}{2}\cos\theta\ddot{\theta} + \frac{l}{2}\sin\theta\dot{\theta}^2 \\ \ddot{Y}_g &= -\frac{l}{2}\sin\theta\ddot{\theta} - \frac{l}{2}\cos\theta\dot{\theta}^2.\end{aligned}$$

where $\ddot{X}$ has been replaced with $g/3$, the pickup truck's acceleration. Substituting into Equations 5.96 gives

$$\begin{aligned}o_X &= m\left(\frac{g}{3} - \frac{l}{2}\cos\theta\ddot{\theta} + \frac{l}{2}\sin\theta\dot{\theta}^2\right) \\ o_Y - w &= -m\left(\frac{l}{2}\sin\theta\ddot{\theta} + \frac{l}{2}\cos\theta\dot{\theta}^2\right),\end{aligned} \tag{5.97}$$

Substituting from Equations 5.94 and 5.95 for $\ddot{\theta}$ and $\dot{\theta}^2$, respectively, (and some algebra) gives

$$\begin{aligned}o_X &= w\left(\frac{11}{24} + \frac{3}{2}\sin\theta - \frac{9}{8}\sin 2\theta - \frac{3}{8}\cos 2\theta\right) \\ o_Y &= w\left(\frac{11}{8} - \frac{3}{2}\cos\theta + \frac{9}{8}\cos 2\theta - \frac{3}{8}\sin 2\theta\right),\end{aligned}$$

and completes Task d.

The decision to use the general moment Equation 5.24 and sum moments about the pivot point o instead of the mass center g saves a great deal of effort in arriving at the differential equation of motion. To confirm this statement, consider the following moment equation about g

$$\sum M_g = o_X\frac{l}{2}\cos\theta + o_y\frac{l}{2}\sin\theta = I_g\ddot{\theta}$$

Substituting from Equation 5.65 for o_X, o_Y gives

$$\begin{aligned}I_g\ddot{\theta} &= \frac{l}{2}\cos\theta m\left(\frac{g}{3} - \frac{l}{2}\cos\theta\ddot{\theta} + \frac{l}{2}\sin\theta\dot{\theta}^2\right) \\ &\quad + \frac{l}{2}\sin\theta\left[w - m\left(\frac{l}{2}\sin\theta\ddot{\theta} + \frac{l}{2}\cos\theta\dot{\theta}^2\right)\right].\end{aligned}$$

Gathering like terms gives

$$\begin{aligned}&\left[I_g + \frac{ml^2}{4}(\sin^2\theta + \cos^2\theta)\right]\ddot{\theta} \\ &= \frac{wl}{6}\cos\theta + \frac{wl}{2}\sin\theta + \frac{ml}{4}\dot{\theta}^2(\sin\theta\cos\theta - \sin\theta\cos\theta).\end{aligned}$$

Simplifying these equations gives Equation 5.93, the original differential equation of motion. This development is "messy," but actually understates the comparative development complexity in starting with the moment equation about g, since the original solution did not require the reaction components o_X, o_Y.

The lesson from this short section is as follows: In problems where a pivot support point has a prescribed acceleration, stating the moment equation (correctly) about the pivot point will lead to the governing equation of motion much more quickly and easily than taking moments about the mass center.

Note that energy is not conserved with base acceleration. Students sometimes incorrectly try to derive the equation of motion using conservation of energy. An external source of energy is needed to accelerate the support point.

5.7 General Applications of Force, Moment, and Energy Equations for Planar Motion of a Rigid Body

This section's contents largely involve deriving an EOM for a body or bodies with one degree of freedom. We will consider and compare the Newtonian approach following from free-body diagrams and work–energy approaches. Formulating governing equations of motion for most of the systems will require an understanding of the kinematics covered in Chapter 4.

5.7.1 Rolling-without-Slipping Examples: Newtonian and Energy Approaches

Section 4.4 dealt (at length) with the kinematics involved in a wheel that rolls without slipping, and this section will consider the derivation of equations of motion for several examples that involve rolling without slipping.

5.7.1.1 A Cylinder Rolling Down an Inclined Plane: Free-Body Diagram Approach

Figure 5.25a illustrates a uniform disk (or cylinder) of mass m and radius r, rolling down a plane that is inclined from the horizontal by the angle α. The static

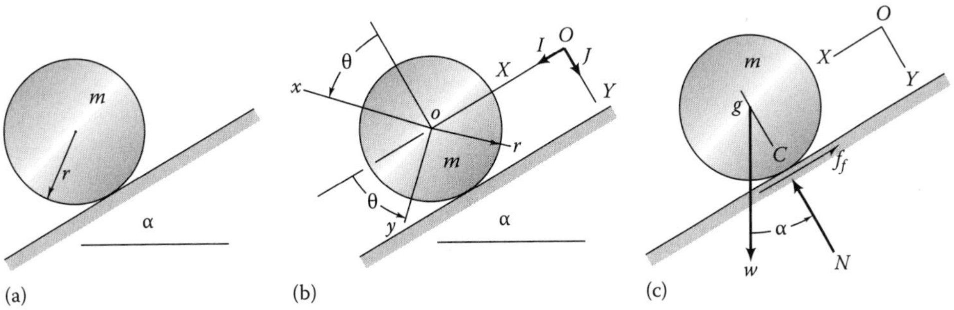

FIGURE 5.25
(a) Uniform disk of mass m and radius r rolling (without slipping) down an inclined plane. (b) Coordinate choices. (c) Free-body diagram.

Coulomb coefficient of friction for the plane is μ_s. We face the following tasks:

a. Select coordinates and draw the body in a general position.
b. Draw a free-body diagram for the cylinder in a general position and use it to derive the governing equation of motion.

Our choice of kinematics is illustrated in Figure 5.25b. The disk rotation angle θ is drawn counterclockwise from an axis that is perpendicular to the plane. The x, y coordinate system is fixed to the disk, and its origin o coincides with the disk's mass center g. The origin of the inertial X, Y system is at O. The center of the disk is located a distance X down the plane from point O. The Y-axis is perpendicular to and directed into the plane. The rolling-without-slipping kinematic result provided by Equations 4.4 and 4.5 gives

$$X = r\theta, \quad \dot{X} = r\dot{\theta}, \quad \ddot{X} = r\ddot{\theta}. \tag{5.98}$$

In the free-body diagram of Figure 5.25c, w acts vertically downward, the reaction force N acts perpendicular to the surface of the plane, and the friction force f_f acts parallel to the plane to oppose slipping and induce rotation of the disk. In looking at the free-body diagram, you might think, "Why didn't they put in $\mu_s N$ for the friction force?" The answer is that the friction force is "just" large enough to prevent slipping. It would only reach $\mu_s N$ in the limit when slipping was about to start. Think about starting the plate in a horizontal position and then progressively increasing α. At some limiting value of α, the disk would stop rolling and start slipping. Similarly, if we keep α constant, and progressively reduce μ_s, slipping will begin at some lower limit for friction. Without friction, the disk will slide down the plane, not roll.

Applying $\sum \boldsymbol{f} = m\ddot{\boldsymbol{R}}_g$ for the disk's mass center in terms of the X, Y coordinate system gives

$$\sum f_X = w\sin\alpha - f_f = m\ddot{X}, \; \sum f_Y = w\cos\alpha - N = m\ddot{Y} = 0. \tag{5.99}$$

Components of forces acting on the body are positive or negative, depending on whether they are in the $\pm X$ direction or $\pm Y$ direction, respectively. The acceleration component $\ddot{Y}$ is zero in the second of these equations, because the mass center's distance from the plane is constant; hence, $N = w\cos\alpha$

Applying $\sum M_g = I_g\ddot{\theta}$ gives

$$\sum M_g = f_f r = I_g\ddot{\theta} = \frac{mr^2}{2}\ddot{\theta}, \quad I_g = \frac{mr^2}{2} \tag{5.100}$$

Note that a moment acting in the counterclockwise direction about the mass center of the disk is positive because it is acting in the $+\theta$ direction (not because it would be in the counterclockwise direction per se, or because of the right-hand rule). Also, note that w and N do not appear in Equation 5.100, because they act through the mass center.

Equation 5.100 and the first of Equation 5.99 constitute two equations in the three unknowns: $\ddot{\theta}$, $\ddot{X}$, and f_f. The rolling-without-slipping acceleration results from Equation 5.98 provides the third equation. Solving for f_f from Equation 5.100, and substituting into the first of Equation 5.99 gives

$$m\ddot{X} = w\sin\alpha - m\frac{r}{2}\ddot{\theta}.$$

Now, substituting $\ddot{\theta} = \ddot{X}/r$ gives

$$m\ddot{X} + m\frac{r}{2}\cdot\frac{\ddot{X}}{r} = \frac{3m}{2}\ddot{X} = w\sin\alpha, \tag{5.101}$$

which completes the differential equation with X as the dependent variable.

We could have taken a different tack, using θ as the dependent variable, by solving the first of Equation 5.98 for f_f and then substituting into Equation 5.99 to obtain

$$I_g\ddot{\theta} = r(w\sin\alpha - m\ddot{X}),$$

Substituting $\ddot{X} = r\ddot{\theta}$ gives

$$(I_g + mr^2)\ddot{\theta} = \frac{3mr^2}{2}\ddot{\theta} = wr\sin\alpha. \quad (5.102)$$

Equations 5.101 and 5.102 are basically the same equation.

Having derived the EOM, let us go back and look at the results. If the friction force is zero, then Equations 5.99 and 5.100 predict that (i) the wheel will not rotate; that is, $\ddot{\theta} = 0$, and (ii) the wheel will skid down the plane with the constant acceleration $\ddot{X} = g\sin\alpha$. However, when the wheel rolls without slipping, the acceleration is reduced to $\ddot{X} = (2/3)g\sin\alpha$. Without friction, all of the work done by gravity acts to increases the translational energy of the wheel. With friction, part of the work also acts to increase the rotational energy of the wheel. Without slipping, the friction force does no work, and energy is conserved.

Returning to the issue of slipping, we can use the first of Equation 5.98 and Equation 5.101 to solve for the friction force as

$$f_f = w\sin\alpha - m\ddot{X} = w\sin\alpha - m\left(\frac{2w\sin\alpha}{3m}\right) = \frac{w}{3}\sin\alpha. \quad (5.103)$$

As noted earlier, the normal reaction force is defined from the second of Equation 5.99 as $N = w\cos\alpha$; hence, the "required" μ_s to prevent slipping is

$$\mu_s\ (\text{required}) = \frac{f_f}{N} = \frac{w\sin\alpha}{3} \times \frac{1}{w\cos\alpha} = \frac{\tan\alpha}{3}. \quad (5.104)$$

If this calculated value is less than or equal to μ_s, the wheel will roll without slipping. If it greater than μ_s, the wheel will slip, and $f_f = \mu_d N$. When the wheel rolls without slipping, the kinematic constraint Equation 5.100 holds, and there is only one degree of freedom. However, when the wheel slips, there are two degrees of freedom, and, from Equations 5.99 and 5.100, the equations of motion are

$$w\sin\alpha - \mu_d w\cos\alpha = m\ddot{X}, \quad \frac{mr^2}{2}\ddot{\theta} = r\mu_d w\cos\alpha. \quad (5.105)$$

These equations are obtained by substituting $f_f = \mu_d N = \mu_d w \cos\alpha$ into the governing equations.

The following *general* moment equation was developed in Section 5.3:

$$M_{oz} = I_o\ddot{\theta} + m(\boldsymbol{b}_{og} \times \ddot{\boldsymbol{R}}_o)_z, \quad (5.24)$$

where moments are taken about a point o in the body, the vector $\boldsymbol{b}_{og}$ goes from point o to the mass center location at g, and $\ddot{\boldsymbol{R}}_o$ is the acceleration of point o. This equation has the two reduced forms provided by $M_g = I_g\ddot{\theta}$ (moments taken about g) and $M_o = I_o\ddot{\theta}$ (moments taken about a fixed axis of rotation). Looking back at Figure 5.25c, one might think (correctly) that taking moment about C, the point of contact between the wheel and the plane would be advantageous, since both f_f and N act through this point and develop a zero moment. Stating Equation 5.24 for moments about point C gives

$$M_{Cz} = w\sin\boldsymbol{\alpha} \times r = I_C\ddot{\theta} + m(\boldsymbol{b}_{Cg} \times \ddot{\boldsymbol{R}}_C)_z,$$

where the weight of the disk provides the only external moment about C. From Appendix C, $I_g = mr^2/2$ and the parallel-axis formula gives $I_C = I_g + mr^2 = 3mr^2/2$. In terms of the X, Y system, the vector $\boldsymbol{b}_{Cg}$ proceeds from C to g and is defined by $\boldsymbol{b}_{Cg} = -\boldsymbol{J}r$. The acceleration of the cylinder at point C is $\ddot{\boldsymbol{R}}_C = -\boldsymbol{J}r\dot{\theta}^2$, directed from C back toward the disk's center. (You can look back at Equation 4.11 or Equation 4.13 to refresh your memory concerning this result.) Hence, the vectors $\boldsymbol{b}_{Cg}$ and $\ddot{\boldsymbol{R}}_C$ are parallel, $\boldsymbol{b}_{Cg} \times \ddot{\boldsymbol{R}}_C = 0$ and the EOM is

$$\frac{3mr^2}{2}\ddot{\theta} = wr\sin\theta,$$

which coincides with our earlier result of Equation 5.102.

For this example, using Equation 5.24 and calculating moments about the contact point is quicker than using a moment equation about the mass center (plus the first of Equations 5.99) to develop the EOM. Beginners regularly misapply Equation 5.24 by failing to include the last term. Generally, $\sum M_{oz} \neq I_o\ddot{\theta}$ unless the point o coincides with either the mass center g or a fixed axis of rotation. The present example provides an opportunity to make this mistake (omit the last term) and still get the correct answer, since it turns out to be zero anyway. Most dynamics problems are not so forgiving.

5.7.1.2 A Cylinder Rolling Down an Inclined Plane: Work–Energy Approach

Absent slipping, energy is conserved. We can use a horizontal plane through O, the origin of the X, Y system for the gravity potential-energy datum and apply $(T + V) = (T_0 + V_0)$ as

$$\frac{m\dot{X}^2}{2} + \frac{I_g\dot{\theta}^2}{2} - wX\sin\alpha = \frac{m\dot{X}_0^2}{2} + \frac{I_g\dot{\theta}_0^2}{2} + 0.$$

The potential energy term on the left has a negative sign because the mass center is *below* the datum. Substituting the rolling-without-slipping kinematic conditions $X = r\theta$, $\dot{X} = r\dot{\theta}$ yields

$$\frac{(I_g + mr^2)\dot{\theta}^2}{2} - wr\theta \sin\alpha = \frac{m\dot{X}_0^2}{2} + \frac{I_g\dot{\theta}_0^2}{2}. \qquad (5.106)$$

Substituting $I_g = mr^2/2$ and differentiating with respect to θ and gives the EOM,

$$\frac{3mr^2}{2}\frac{d}{d\theta}\left(\frac{\dot{\theta}^2}{2}\right) = \frac{3mr^2}{2}\ddot{\theta} = wr\sin\alpha.$$

This result coincides with the earlier (Newtonian-based) result of Equation 5.102 with considerably less effort.

Another straightforward question is, *If the cylinder is released from rest at* $X = \theta = 0$, *what will its angular velocity be when* $\theta = \bar{\theta}$? Plugging into Equation 5.106 gives the answer

$$\frac{3mr^2}{4}\dot{\theta}^2(\theta = \bar{\theta}) - wr\bar{\theta}\sin\alpha = 0 \Rightarrow \dot{\theta}(\theta = \bar{\theta}) = \sqrt{\frac{4g\sin\alpha\bar{\theta}}{3r}}.$$

Obtaining these results from the energy equation is quicker than the earlier Newtonian developments. We could go on to solve for the normal and friction reaction forces, but we would need (i) the free-body diagram of Figure 5.25c and (ii) the force equations provided by Equations 5.99. These additional requirements would reduce the energy approach's advantages.

5.7.1.3 An Imbalanced Cylinder Rolling Down an Inclined Plane: Newtonian Approach

Figure 5.26a illustrates a complication to the earlier rolling disk of Figure 5.25a; namely, the mass center of the disk is now displaced a distance e from the disk's geometric center. The disk's radius of gyration about the cylinder's geometric center at point o is k_{og}; hence, $I_o = mk_{og}^2$. The engineering tasks are as follows:

a. *Select coordinates, and draw the body in a general position.*
b. *Draw a free-body diagram for the cylinder in a general position and use it to derive the governing EOM.*

Figure 5.26b illustrates the choice of coordinates. Point o is located in the inertial X, Y coordinate system by IX. The rotation of the disk relative to an axis perpendicular to the inclined plane is defined by the clockwise rotation angle β. The rolling-without-slipping condition gives

$$X = r\beta, \quad \dot{X} = r\dot{\beta}, \quad \ddot{X} = r\ddot{\beta}. \qquad (5.107)$$

The disk's mass center is located in the X, Y system by

$$X_g = X + e\sin\beta = r\beta + e\sin\beta, \quad Y_g = e\cos\beta. \quad (5.108)$$

Hence, without slipping, the disk has two variables, X and β, but only one degree of freedom.

Applying $\sum f = m\ddot{R}_g$ for the disk's mass center gives

$$\begin{aligned}\sum f_X &= w\sin\alpha - f_f = m\ddot{X}_g\\ \sum f_Y &= N - w\cos\alpha = m\ddot{Y}_g.\end{aligned} \qquad (5.109)$$

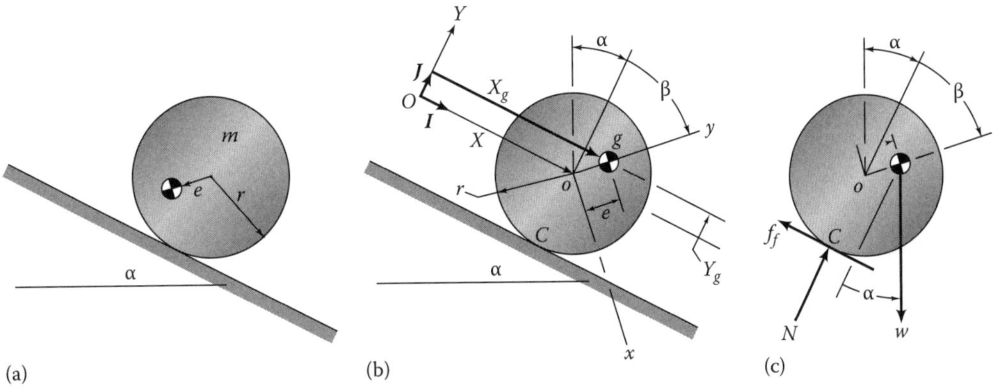

FIGURE 5.26
(a) Disk with its mass center displaced a distance e from its geometrical center, rolling down an inclined plane. (b) Coordinate choices. (c) Free-body diagram.

In reviewing these equations, note that $\ddot{Y}_g$ is not zero, and that $\ddot{X}_g$, $\ddot{Y}_g$ remain to be defined in terms of β and its derivatives.

A person studying the free-body diagram of Figure 5.26c could reasonably choose to take moments about g (the mass center), or C (the contact point between the disk and the plane); we examine these two options before leaving this example. The moment equation about g is

$$\sum M_g = Ne\sin\beta + f_f(r + e\cos\beta) = I_g\ddot{\beta}. \tag{5.110}$$

Note that positive moments are in the $+\beta$, clockwise direction. Also, from the parallel-axis formula, $I_g = I_o - me^2 = m(k_g^2 - e^2)$.

Equations 5.109 and 5.110 provide three equations in the five unknowns: $\ddot{\beta}$, $\ddot{X}_g$, $\ddot{Y}_g$, N, f_f, and completing the governing equations requires that we define $\ddot{X}_g$, $\ddot{Y}_g$ in terms of β and its derivatives. Differentiating Equations 5.108 once with respect to time gives

$$\dot{X}_g = r\dot{\beta} + e\cos\beta\dot{\beta}, \quad \dot{Y}_g = -e\sin\beta\dot{\beta}. \tag{5.111}$$

Differentiating a second time gives

$$\begin{aligned} \ddot{X}_g &= r\ddot{\beta} + e\cos\beta\ddot{\beta} - e\sin\beta\dot{\beta}^2, \\ \ddot{Y}_g &= -e\sin\beta\ddot{\beta} - e\cos\beta\dot{\beta}^2, \end{aligned} \tag{5.112}$$

which provides our final two equations. The governing equation of motion is obtained from the following steps:

a. Substitute for $\ddot{X}_g$, $\ddot{Y}_g$ from Equation 5.112 into Equations 5.109, obtaining

$$\begin{aligned} f_f &= w\sin\alpha - m(r\ddot{\beta} + e\cos\beta\ddot{\beta} - e\sin\beta\dot{\beta}^2) \\ N &= w\cos\alpha - m(e\sin\beta\ddot{\beta} + e\cos\beta\dot{\beta}^2). \end{aligned} \tag{5.113}$$

b. Substitute for N and f_f into Equation 5.110, obtaining

$$\begin{aligned} I_g\ddot{\beta} = &\left[w\cos\alpha - m\left(e\sin\beta\ddot{\beta} + e\cos\beta\dot{\beta}^2\right)\right]e\sin\beta \\ &+ \left[w\sin\alpha - m\left(r\ddot{\beta} + e\cos\beta\ddot{\beta} - e\sin\beta\dot{\beta}^2\right)\right] \\ &\times (r + e\cos\beta). \end{aligned}$$

After a fair amount of algebra, β's governing EOM is

$$\begin{aligned} &m(k_{og}^2 + r^2 + 2re\cos\beta)\ddot{\beta} - mer\sin\beta\dot{\beta}^2 \\ &\quad = w[r\sin\alpha + e\sin(\beta + \alpha)]. \end{aligned} \tag{5.114}$$

This equation reduces to Equation 5.102 if $e = 0$, and $I_o = mr^2/2$.

Looking at Figure 5.26c, taking moments about C—the contact point between the wheel and the plane—offers some clear advantages, since the friction force f_f and the normal force N have zero moments about C. The moment equation is

$$\begin{aligned} M_{Cz} &= w\cos\alpha e\sin\beta + w\sin\alpha(e\cos\beta + r) \\ &= wr\sin\alpha + we\sin(\alpha + \beta) = I_C\ddot{\beta} - m(\boldsymbol{b}_{Cg} \times \ddot{\boldsymbol{R}}_C)_z. \end{aligned} \tag{5.115}$$

The minus sign for $m(b_{Cg} \times \ddot{R}_C)_z$ in this equation denotes the rotation for β is in the clockwise direction, not the counterclockwise direction used in defining the moment equation (Equation 5.24). We need the parallel-axis formula to determine I_C in Equation 5.115, the body's moment of inertia about point C. The following steps accomplish this goal:

$$\begin{aligned} I_g &= I_o - m|\boldsymbol{b}_{og}|^2 = I_o - me^2 = m\left(k_{og}^2 - e^2\right), \\ I_C &= I_g + m|\boldsymbol{b}_{gC}|^2 = I_g + m\left[(r + e\cos\beta)^2\right. \\ &\quad \left. + (e\sin\beta)^2\right] = m\left(k_{og}^2 + r^2 + 2re\cos\beta\right). \end{aligned} \tag{5.116}$$

The first step goes from an axis through o to an axis through g; the second step goes from g to C. Remember that the parallel-axis formula applies for two axes, one of which must be an axis through g. We could not use the formula (correctly) to go directly from an axis through o to an axis through C.

Returning to Equation 5.115, the vector $\boldsymbol{b}_{Cg}$ goes from C to g and is

$$\boldsymbol{b}_{Cg} = \boldsymbol{I}e\sin\beta + \boldsymbol{J}(r + e\cos\beta), \tag{5.117}$$

and the acceleration of point C is $\ddot{\boldsymbol{R}}_C = \boldsymbol{J}r\dot{\beta}^2$. Hence,

$$\boldsymbol{b}_{Cg} \times \ddot{\boldsymbol{R}}_C = \boldsymbol{K}er\sin\beta\dot{\beta}^2. \tag{5.118}$$

Substituting Equation 5.118 into Equation 5.115 gives

$$\begin{aligned} &m\left(k_{og}^2 + r^2 + 2re\cos\beta\right)\ddot{\beta} - mer\sin\beta\dot{\beta}^2 \\ &\quad = wr\sin\alpha + we\sin(\beta + \alpha), \end{aligned} \tag{5.119}$$

which coincides with Equation 5.114. Taking moments about C minimizes the algebra involved in arriving at the governing equation, but requires the parallel-axis formula.

5.7.1.4 An Imbalanced Cylinder Rolling Down an Inclined Plane: Work–Energy Approach

Without slipping, $T+V=T_0+V_0$ applies and can be used to state

$$\frac{I_g}{2}\dot{\beta}^2+\frac{m}{2}\left(\dot{X}_g^2+\dot{Y}_g^2\right)-wX\sin\alpha+we\cos(\alpha+\beta)=(T_0+V_0), \quad (5.120)$$

where $I_g=I_o-me^2$. From Figure 5.26b, this example has the three coordinates X_g, Y_g, β. To eliminate unwanted coordinates, we need the rolling-without-slipping condition $X=r\beta$, plus the kinematic conditions,

$$X_g = X + e\sin\beta = r\beta + e\sin\beta \Rightarrow \dot{X}_g = r\dot{\beta} + e\cos\beta\dot{\beta}$$
$$Y_g = e\cos\beta \Rightarrow \dot{Y}_g = -e\sin\beta\dot{\beta},$$

from Equations 5.108 and 5.111. Substituting into Equation 5.120 gives

$$m\left(k_{og}^2+r^2+2er\cos\beta\right)\frac{\dot{\beta}^2}{2}-w[r\beta\sin\alpha-e\cos(\alpha+\beta)] = (T_0+V_0). \quad (5.121)$$

Differentiating with respect to β gives

$$m\left(k_{og}^2+r^2+2er\cos\beta\right)\ddot{\beta}-mer\sin\beta\dot{\beta}^2 = w[r\sin\alpha+e\sin(\alpha+\beta)]. \quad (5.122)$$

This result coincides with our earlier Newtonian developments in Equation 5.114 and is obtained with far less effort.

5.7.1.5 A Half Cylinder Rolling on a Horizontal Plane: Newtonian Approach

Figure 5.27a illustrates a half cylinder held vertically upright on its left edge. The body is released from rest in this position and rolls without slipping, to the right. The following tasks apply:

a. *Select coordinates and draw the body in a general position.*
b. *Draw a free-body diagram for the general position and use it to derive the EOM.*
c. *Determine the static Coulomb friction coefficient* μ_s *required to prevent slipping at the initial time that the body is released.*
d. *Find the natural frequency for small motion about the equilibrium position.*

Figure 5.27b illustrates the body in a general position. The coordinate X locates the center of the half disk o with respect to ground, while β defines a positive clockwise rotation. The point of contact between the cylinder and the plane is C. The rolling-without-slipping requirement gives $X=r\beta$. The mass center is located in an inertial X, Y coordinate system by X_g, Y_g that are defined by

$$X_g = X - e\sin\beta = r\beta - e\sin\beta, \quad Y_g = r - e\cos\beta \quad (5.123)$$

Figure 5.27c provides a free-body diagram, with w acting vertically down, and the normal reaction force N acting vertically up. The friction force f_f acts to oppose slipping. Applying $\sum f = m\ddot{R}_g$ provides

$$\sum f_{Xg}=f_f=m\ddot{X}_g, \quad \sum f_{Yg}=N-w=m\ddot{Y}_g. \quad (5.124)$$

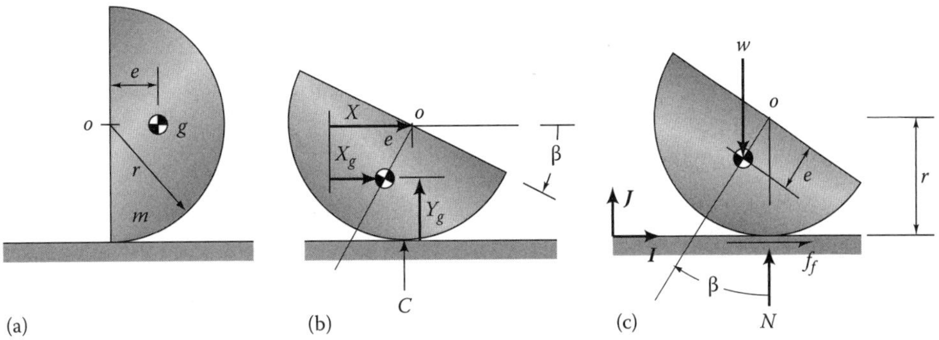

FIGURE 5.27
Half cylinder shown: (a) vertically upright and at rest on one edge, (b) in a general position with coordinates X, β, (c) free-body diagram.

The positive and negative signs for the force terms in these equations follow from $\pm X_g$ and $\pm Y_g$. Taking moments about g gives

$$\sum M_g = -Ne\sin\beta - f_f(r - e\cos\beta) = I_g\ddot{\beta}, \quad (5.125)$$

From Appendix C, the mass center of the body g is located at the distance $e = 4r/3\pi$ from point o, and $I_o = mr^2/2$. The parallel-axis formula gives

$$I_g = I_o - m|\boldsymbol{b}_{go}|^2 = \frac{mr^2}{2} - m\left(\frac{4r}{3\pi}\right)^2. \quad (5.126)$$

Equations 5.124 and 5.125 provide three equations for the five unknowns $\ddot{X}_g$, $\ddot{Y}_g$, $\ddot{\beta}$, f_f, N. To complete the development, we need to define $\ddot{X}_g$, $\ddot{Y}_g$ in terms of X, β and their derivatives. Differentiating, Equation 5.123 with respect to time gives

$$\dot{X}_g = r\dot{\beta} - e\cos\beta\dot{\beta}, \quad \dot{Y}_g = e\sin\beta\dot{\beta}. \quad (5.127)$$

Differentiating again with respect to time,

$$\ddot{X}_g = r\ddot{\beta} - e\cos\beta\ddot{\beta} + e\sin\beta\dot{\beta}^2, \quad \ddot{Y}_g = e\sin\beta\ddot{\beta} + e\cos\beta\dot{\beta}^2. \quad (5.128)$$

Equations 5.124, 5.125, and 5.128 provide five equations in the five unknowns $\ddot{X}_g$, $\ddot{Y}_g$, $\ddot{\beta}$, f_f, N. Proceeding to eliminate variables, we can substitute for $\ddot{X}_g$, $\ddot{Y}_g$ into Equation 5.124 for f_f and N to obtain

$$\begin{aligned} f_f &= m\ddot{X}_g = m(r\ddot{\beta} - e\cos\beta\ddot{\beta} + e\sin\beta\dot{\beta}^2) \\ N &= w + m\ddot{Y}_g = w + m(e\sin\beta\ddot{\beta} + e\cos\beta\dot{\beta}^2). \end{aligned} \quad (5.129)$$

Substituting these results into Equation 5.125 gives

$$\begin{aligned} &-\left[w + m\left(e\sin\beta\ddot{\beta} + e\cos\beta\dot{\beta}^2\right)\right]e\sin\beta \\ &- m\left(r\ddot{\beta} - e\cos\beta\ddot{\beta} + e\sin\beta\dot{\beta}^2\right)(r - e\cos\beta) = I_g\ddot{\beta}. \end{aligned}$$

Rewriting the above equation,

$$\begin{aligned} &-we\sin\beta + \ddot{\beta}(-e^2\sin^2\beta - r^2 + 2re\cos\beta - e^2\cos\beta) \\ &+\dot{\beta}^2 m\left[(-e^2\cos\beta\sin\beta - e\sin\beta)(r - e\cos\beta)\right] = I_g\ddot{\beta}. \end{aligned}$$

Simplifying and continuing to gather terms gives

$$\left[I_g + m(r^2 + e^2 - 2er\cos\beta)\right]\ddot{\beta} + mer\sin\beta\dot{\beta}^2 = -we\sin\beta. \quad (5.130)$$

After substituting $I_o = I_g + me^2 = mr^2/2$ and $e = 4r/(3\pi)$ we obtain the final EOM

$$mr^2\left(\frac{3}{2} - \frac{8\cos\beta}{3\pi}\right)\ddot{\beta} + \frac{4mr^2}{3\pi}\sin\beta\dot{\beta}^2 = -\frac{4wr}{3\pi}\sin\beta. \quad (5.131)$$

This equation completes Task b.

In considering the free-body diagram of Figure 5.27c, C is an attractive point for which moments could be stated, since both f_f and N have zero moments about this point. Stating moments about C yields, from Equation 5.24,

$$\sum M_{Cz} = -we\sin\beta = I_C\ddot{\beta} - m(\boldsymbol{b}_{Cg} \times \ddot{\boldsymbol{R}}_C)_z, \quad (5.132)$$

where

I_C is the body's moment of inertia about C
$\boldsymbol{b}_{Cg}$ is the position vector proceeding from C to g
$\ddot{\boldsymbol{R}}_C$ is the acceleration of C

The minus sign on the right arises because β is a clockwise rotation in contrast to the clockwise development used in getting Equation 5.24. The elements of Equation 5.132 are

$$\begin{aligned} I_C &= I_g + m|\boldsymbol{b}_{gC}|^2 = (I_o - me^2) + m(r^2 - 2er\cos\beta + e^2) \\ &= \frac{3mr^2}{2} - 2mer\cos\beta \\ \boldsymbol{b}_{Cg} &= -\boldsymbol{I}e\sin\beta + \boldsymbol{J}(r - e\cos\beta), \\ \ddot{\boldsymbol{R}}_C &= \boldsymbol{J}r\dot{\beta}^2 \Rightarrow \boldsymbol{b}_{Cg} \times \ddot{\boldsymbol{R}}_C = -\boldsymbol{K}e\sin\beta r\dot{\beta}^2 \end{aligned}$$

You may need to look back at Equation 4.11 or 4.13 regarding the result for $\ddot{\boldsymbol{R}}_C$. Substituting these results into Equation 5.132 gives

$$\left(\frac{3mr^2}{2} - 2mer\cos\beta\right)\ddot{\beta} + mer\sin\beta\dot{\beta}^2 = -we\sin\beta,$$

which coincides with Equation 5.131. Taking moments about the contact point C clearly requires less algebra than taking moments about g.

To attack Task c, we need to solve for N and f_f from Equation 5.129 at $t = 0$, $\beta = -\pi/2$. At the initial time and position, $\dot{\beta}(t = 0) = 0$. From Equation 5.131,

$$\begin{aligned} &mr^2\left(\frac{3}{2} - \frac{8 \times 0}{3\pi}\right)\ddot{\beta}(0) + \frac{4mr^2}{3\pi}(-1)\dot{\beta}(0)^2 = -\frac{4wr}{3\pi}(-1) \\ &\therefore\ \ddot{\beta}(0) = \frac{4wr}{3\pi} \times \frac{2}{3mr^2} = \frac{8g}{9\pi r}. \end{aligned}$$

Substituting into Equations 5.129 gives

$$f_f(0) = mr\ddot{\beta}(0) = mr\frac{8g}{9\pi r} = 0.283w$$
$$N(0) = w + me\sin\left(-\frac{\pi}{2}\right)\ddot{\beta}(0) = w - me\frac{8g}{9\pi r} = 0.880w.$$

Hence, the static coefficient of friction required to *initially* prevent slipping is

$$\mu_s(t=0)_{\text{required}} = \frac{f_f(0)}{N(0)} = 0.322.$$

Given that the Coulomb friction coefficient for a smooth steel surface rubbing against a smooth steel surface is on the order of 0.6, this is a distinctly realizable outcome.

Proceeding to Task d, Equation 5.131 is the EOM and will be used to develop the linearized equation for small motion about an equilibrium position. Observe that $\beta = 0$ defines an equilibrium position for Equation 5.131, since it yields $\dot{\beta} = \ddot{\beta} = 0$ as a solution. Small motion about this equilibrium position would be produced by perturbing the body (slightly) away from this position. Assuming small motion such that $\beta = \delta\beta$, $\sin\beta \cong \beta$, $\cos\beta \cong 1$ and dropping second-order terms in $\delta\beta$ gives

$$mr^2\left(\frac{3}{2} - \frac{8}{3\pi}\right)\delta\ddot{\beta} + \frac{4wr}{3\pi}\delta\beta = 0. \tag{5.133}$$

Hence, the undamped natural frequency is defined by

$$\omega_n^2 = \frac{4g}{3\pi} \times \frac{6\pi}{r(9\pi - 16)} = \frac{8g}{r(9\pi - 16)} = 0.652\frac{g}{r} \Rightarrow \omega_n = 0.807\sqrt{\frac{g}{r}}.$$

Sometimes, students "balk" at dropping the $\dot{\beta}^2$ term in moving from Equations 5.131 through 5.133. The usual question is: O.K., I can see how you linearize $\sin\beta$ and $\cos\beta$, but how do you know that the $\dot{\beta}^2$ terms isn't large? This question can be addressed by working back from the solution to Equation 5.133. If the body is released from rest at $\delta\beta(0) = B$, the solution can be stated as

$$\delta\beta = B\cos\omega_n t \Rightarrow \delta\dot{\beta} = -\omega_n B\sin\omega_n t \Rightarrow \delta\ddot{\beta} = -B\omega_n^2\cos\omega_n t$$

Recall that B is a "small" parameter on the order of 15° (0.26 rad). Substituting this solution into Equation 5.131, plus $\sin\beta \cong \delta\beta = B\cos\omega_n t$ and $\cos\beta \cong 1$ gives

$$-mr^2\left(\frac{9\pi - 16}{6\pi}\right)\omega_n^2 B\cos\omega_n t + \frac{4mr^2}{3\pi}(-B\cos\omega_n t)\left(\omega_n^2 B^2\sin^2\omega_n t\right) = -\frac{4wr}{3\pi}B\cos\omega_n t.$$

This equation shows that the $\dot{\beta}^2$ term is small (order of B^2) compared to the $\ddot{\beta}$ and w terms (order of B) in the original nonlinear differential equation. The $\sin\beta$ term is also on the order of B; hence, $\sin\beta\dot{\beta}^2$ is on the order of B^3. Dividing through by the inertia term on the left gives

$$-\omega_n^2 B\cos\omega_n t - \frac{8\omega_n^2 B^3}{(9\pi - 16)}\sin^2\omega_n t\cos\omega_n t = -\frac{8g}{r(9\pi - 16)}\cos\omega_n t = -\omega_n^2 B\cos\omega_n t.$$

Note that ω_n^2 appears in each term (as it should from dimensional considerations). This form of the equation shows, even more clearly, the "smallness" of the $\delta\dot{\beta}^2$ term in comparison to the $\delta\ddot{\beta}$ and $\delta\beta$ terms. It also shows that the relative "smallness" of the dynamic terms $\dot{\beta}$ and $\ddot{\beta}$ *does not depend on the magnitude of* ω_n.

5.7.1.6 A Half Cylinder Rotating on a Horizontal Plane: Energy Approach

Our tasks in this development are as follows:

a. Use work–energy approaches to derive the EOM.
b. Assuming that the half-disk is released from rest in the position shown in Figure 5.27a and rolls to the right without slipping, what is its maximum velocity?

For rolling without slipping, $T + V = T_0 + V_0$ holds and nets

$$\frac{I_g}{2}\dot{\beta}^2 + \frac{m}{2}\left(\dot{X}_g^2 + \dot{Y}_g^2\right) + w(r - e\cos\beta) = (T_0 + V_0), \tag{5.134}$$

where the gravity potential-energy datum is the horizontal plane. To get a differential equation in β, we will need the velocity relations,

$$\dot{X}_g = r\dot{\beta} - e\cos\beta\dot{\beta}, \quad \dot{Y}_g = e\sin\beta\dot{\beta}. \tag{5.104′}$$

We also need $I_g = I_o - me^2$ from the parallel-axis equation and $I_o = mr^2/2$, $e = 4r/3\pi$, from Appendix C. Substituting these results into Equation 5.134 gives

$$(I_o + mr^2 - 2mer\cos\beta)\frac{\dot{\beta}^2}{2} + w(r - e\cos\beta) = (T_0 + V_0), \quad \text{or}$$
$$mr^2\left(\frac{3}{2} - \frac{8}{3\pi}\cos\beta\right)\frac{\dot{\beta}^2}{2} + wr\left(1 - \frac{4}{3\pi}\cos\beta\right) = (T_0 + V_0) \tag{5.135}$$

Differentiating with respect to β gives

$$mr^2\left(\frac{3}{2}-\frac{8\cos\beta}{3\pi}\right)\ddot{\beta}+\frac{4mr^2}{3\pi}\sin\beta\dot{\beta}^2+\frac{4wr}{3\pi}\sin\beta=0, \tag{5.136}$$

where $\ddot{\beta}=d(\dot{\beta}^2/2)/d\beta$. This result coincides with Equation 5.131, obtained via the Newtonian approach, but it is developed with a fraction of the effort.

Equation 5.136 is highly nonlinear and cannot be solved analytically for a time solution of the form $\beta(t)$, $\dot{\beta}(t)$. Equation 5.135 from the energy equation provides a solution of the form $\dot{\beta}(\beta)$. For the position shown in Figure 5.27a, substituting $\dot{\beta}(0)=0$, $\beta(0)=-\pi/2$ into Equation 5.135 gives

$$mr^2\left(\frac{3}{2}-\frac{8}{3\pi}\cos\beta\right)\frac{\dot{\beta}^2}{2}+wr\left(1-\frac{4}{3\pi}\cos\beta\right)=0+wr$$

$$\therefore\ \dot{\beta}^2=\frac{8g}{3\pi}\cos\beta/r\left(\frac{3}{2}-\frac{8}{3\pi}\cos\beta\right)=\frac{16g}{r}\left(\frac{\cos\beta}{9\pi-16\cos\beta}\right)$$

Evaluating this equation at $\beta=0$ gives the maximum angular velocity result,

$$\dot{\beta}|_{\beta=0}=4\sqrt{\frac{g}{r(9\pi-16)}}.$$

The absence of energy dissipation yields a solution for which the half cylinder will continue to roll back and forth (indefinitely) from edge to edge.

5.7.1.7 A Cylinder, Restrained by a Spring and Rolling on a Plane: Newtonian Approach

Figure 5.28a illustrates a solid cylinder of mass m and radius r that is connected to a wall by a linear spring with spring coefficient k. The cylinder rolls without slipping. The spring is initially undeflected when $x=0$. The following engineering-analysis tasks apply:

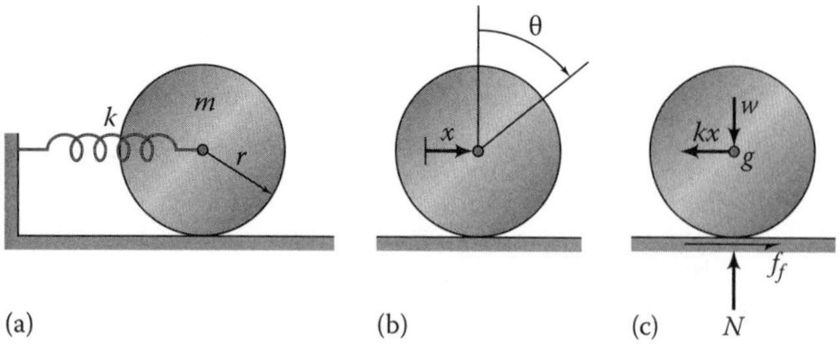

FIGURE 5.28
(a) Spring-restrained cylinder. (b) Kinematic variables. (c) Free-body diagram.

a. Select coordinates and draw the body in a general position.
b. Draw a free-body diagram for the cylinder in a general position and use it to derive the EOM.
c. Determine the natural frequency for small-amplitude vibrations.

Figure 5.28b shows the kinematic variables with x defining the change in position of the cylinder's mass center and θ defining the cylinder's rotation. The rolling-without-slipping kinematic conditions,

$$x=r\theta\Rightarrow\ddot{x}=r\ddot{\theta}, \tag{5.137}$$

apply.

Figure 5.28c provides the free-body diagram for the body in a general position. Applying $\sum f=m\ddot{R}_g$ for the cylinder's mass center nets

$$\sum f_X=f_f-kx=m\ddot{x}. \tag{5.138}$$

Stating the moment equation about the cylinder's mass center gives

$$\sum M_g=-f_f r=I_g\ddot{\theta}. \tag{5.139}$$

We now have three equations in the three unknowns $\ddot{x}$, $\ddot{\theta}$, f_f. Substituting for f_f from Equation 5.138 into Equation 5.139 gives

$$-r(kx+m\ddot{x})=I_g\ddot{\theta}=\frac{mr^2}{2}\ddot{\theta},$$

where, $I_g=mr^2/2$. We can use Equations 5.137 to eliminate x and $\ddot{x}$, obtaining

$$-r(kr\theta+mr\ddot{\theta})=\frac{mr^2}{2}\ddot{\theta}\Rightarrow\frac{3mr^2}{2}\ddot{\theta}+kr^2\theta=0 \tag{5.140}$$

This result concludes Task b, and from Equation 5.140, the natural frequency is

$$\omega_n^2=kr^2\Big/\frac{3mr^2}{2}=\frac{2k}{3m}\Rightarrow\omega_n=\sqrt{\frac{2k}{3m}}.$$

This result concludes Task c. The cylinder inertia has caused a substantial reduction in the natural frequency as compared to a simple spring-mass system that would yield $\omega_n=\sqrt{k/m}$.

For the past several examples, we have found it advantageous (quicker) to take moments about the contact point between the cylinder and ground, since the friction force f_f and the normal force N both have zero moments about this point. These circumstances also hold for the present example. Try applying the general moment equation for the contact point for Figure 5.28c.

5.7.1.8 A Cylinder, Restrained by a Spring and Rolling on a Plane: Energy Approach

For rolling without slipping, $T+V=T_0+V_0$ applies and can be used to state

$$\frac{m\dot{x}^2}{2}+\frac{I_g\dot{\theta}^2}{2}+\frac{kx^2}{2}=\text{constant.}$$

Substituting the rolling-without-slipping kinematic conditions, $x=r\theta$, $\dot{x}=r\dot{\theta}$ gives

$$m\frac{(r\dot{\theta})^2}{2}+\frac{mr^2}{2}\frac{\dot{\theta}^2}{2}+\frac{k}{2}(r\theta)^2=\text{constant.}$$

Differentiating with respect to θ gives

$$\frac{3mr^2}{2}\ddot{\theta}+kr^2\theta=0,$$

where $\ddot{\theta}=d(\dot{\theta}^2/2)/d\theta$. This result coincides with Equation 5.140. Again, an energy approach is clearly quicker than the Newtonian approach in deriving the equation of motion.

5.7.1.9 A Cylinder Rolling inside a Cylindrical Surface

Figure 5.29a illustrates a cylinder of mass m and radius r at rest at the bottom of a cylindrical surface with radius R. O denotes the origin of the stationary X, Y coordinate system. The following tasks apply:

a. Select coordinates and draw the cylinder in a general position.
b. Draw a free-body diagram for the cylinder in a general position and use it to derive the equation of motion.
c. For small motion about the bottom equilibrium position, determine the natural frequency.
d. Use work–energy principles to derive the equation of motion.
e. Use the coordinates of Figure 5.27b, assume that the cylinder is released from rest at $\theta=\pi/2$, and define the normal reaction force as a function of θ (only).

As illustrated in Figure 5.29b, an x, y coordinate system is fixed to the cylinder, and its origin o coincides with the cylinder's mass center g. The angle θ defines the counterclockwise rotation of the line O–g, while ϕ defines the cylinder's clockwise rotation with respect to ground. The contact point between the two cylinders is C. The rolling-without-slipping kinematic constraint relating θ and ϕ was developed in Figure 4.14b as

$$v_C=0\Rightarrow(R-r)\dot{\theta}=r\dot{\phi}, \qquad (4.15a)$$

Figure 5.29c provides the free-body diagram. In applying $\sum f=m\ddot{R}_g$, we will use the polar-coordinate unit vectors. Starting with the $\boldsymbol{\varepsilon}_\theta$ direction,

$$\sum f_\theta=f_f-w\sin\theta=ma_{g\theta}=m(R-r)\ddot{\theta}. \qquad (5.141)$$

Stating the moment equation about the mass center g gives

$$\sum M_g=I_g\ddot{\phi}=-rf_f. \qquad (5.142)$$

The moment due to the friction force is negative because it is acting in the $-\phi$ direction. We now have two equations in the three unknowns $\ddot{\theta}$, $\ddot{\phi}$, f_f. From Equation 4.15a, the required kinematic constraint equation is

$$(R-r)\dot{\theta}=r\dot{\phi}\Rightarrow(R-r)\ddot{\theta}=r\ddot{\phi}. \qquad (5.143)$$

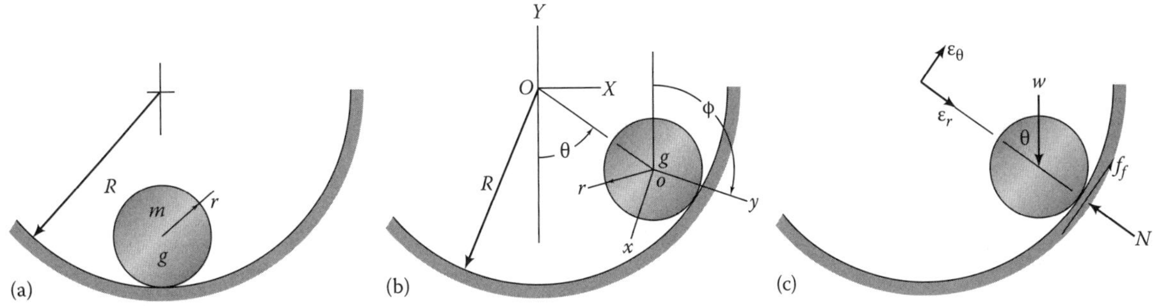

FIGURE 5.29
Cylinder rolling inside a cylindrical surface. (a) Cylinder resting at the bottom of the cylinder. (b) Kinematic variable. (c) Free-body diagram.

Substituting for f_f from Equation 5.141 into Equation 5.142 gives

$$-r\left[w\sin\theta + m(R-r)\ddot{\theta}\right] = I_g\ddot{\phi}.$$

Now substituting for $\ddot{\phi}$ from Equation 5.143 gives

$$\begin{aligned} -r\left[w\sin\theta + m(R-r)\ddot{\theta}\right] &= \frac{mr^2}{2}\frac{(R-r)}{r}\ddot{\theta} \\ \Rightarrow \frac{3m}{2}(R-r)\ddot{\theta} + w\sin\theta &= 0, \end{aligned} \quad (5.144)$$

where, from Appendix C, $I_g = mr^2/2$. Equation 5.144 is the EOM requested in Task b. For small θ, $\sin\theta \cong \theta$ and the linearized EOM is

$$\ddot{\theta} + \frac{2g}{3(R-r)}\theta = 0.$$

The natural frequency is $\omega_n = \sqrt{2g/3(R-r)}$, which concludes Task c.

Without slipping, energy is conserved and $T + V = T_0 + V_0$ can be used to obtain the EOM. We take a horizontal plane through O as the datum reference for gravity conservation of energy to obtain

$$V_g = -w(R-r)\cos\theta.$$

The kinetic energy is

$$T = \frac{m}{2}|\dot{\mathbf{R}}_g|^2 + \frac{I_g}{2}\dot{\phi}^2 = \frac{m}{2}\left[(R-r)\dot{\theta}\right]^2 + \frac{I_g}{2}\dot{\phi}^2$$

where the velocity of the cylinder's mass center is $\dot{R}_g = (R-r)\dot{\theta}$. Substituting the kinematic constraint $(R-r)\dot{\theta} = r\dot{\phi}$ from Equation 5.143 gives

$$\begin{aligned} &\frac{m}{2}\left[(R-r)\dot{\theta}\right]^2 + m\frac{r^2}{2}\cdot\frac{(R-r)^2}{r^2}\frac{\dot{\theta}^2}{2} - w(R-r)\cos\theta = 0 \\ &\Rightarrow \frac{3m}{2}(R-r)\frac{\dot{\theta}^2}{2} - w\cos\theta = 0 \end{aligned} \quad (5.145)$$

Differentiating with respect to θ yields

$$\frac{3m}{2}(R-r)\ddot{\theta} + w\sin\theta = 0,$$

where $\ddot{\theta} = d(\dot{\theta}^2/2)/d\theta$. This result coincides with Equation 5.144 and was obtained with substantially less work.

Looking at Figure 5.27c, the normal reaction force N is obtained from the r component of $\sum f = m\ddot{\mathbf{R}}_g$ as

$$\sum f_r = w\cos\theta - N = ma_{gr} = -m(R-r)\dot{\theta}^2. \quad (5.146)$$

The requested solution for N as a function of θ (only) will require that we find $\dot{\theta}^2(\theta)$. We could proceed from Equation 5.144 and use the energy-integral substitution $\ddot{\theta} = d(\dot{\theta}^2/2)/d\theta$ to get

$$\frac{d}{d\theta}\left(\frac{\dot{\theta}^2}{2}\right) = -\frac{2gr}{3(R-r)}\sin\theta.$$

Integrating this equation, using the initial condition, $\dot{\theta}(\theta = \pi/2) = 0$ will get $\dot{\theta}^2(\theta)$, but it is easier to start from the $T + V = T_0 + V_0$ result provided in Equation 5.145 as

$$\frac{3m}{2}(R-r)^2\frac{\dot{\theta}^2}{2} - wr\cos\theta = 0 \Rightarrow \dot{\theta}^2 = \frac{4gr\cos\theta}{3(R-r)^2}. \quad (5.147)$$

Substituting for $\dot{\theta}^2$ into Equation 5.146 defines N as

$$N = w\cos\theta + m(R-r)\times\frac{4gr\cos\theta}{3(R-r)^2} = w\cos\theta\left[1 + \frac{4r}{3(R-r)}\right] \quad (5.148)$$

and meets the requirements of Task c. The peak reaction force would occur for the bottom position at $\theta = 0 \Rightarrow \cos\theta = 1$.

Note from Equation 5.148 that $N = 0$ at $\theta = \pi/2$; hence, from our initial conditions, the wheel will slip initially until N becomes large enough for the Coulomb friction force $f_f = \mu_d N$ to prevent slipping. Until f_f is large enough to prevent slipping, the model provided by Equations 5.144 and the solutions provided by Equations 5.147 and 5.148 are invalid. With slipping, the appropriate model as provided from Equations 5.141 and 5.142 with $f_f = \mu_d N$ is

$$\begin{aligned} \mu_d N - w\sin\theta &= m(R-r)\ddot{\theta} \\ -r\mu_d N &= I_g\ddot{\phi} \\ N &= w\cos\theta + m(R-r)\dot{\theta}^2. \end{aligned}$$

Eliminating N gives the two coupled nonlinear equations of motion:

$$\begin{aligned} m(R-r)\ddot{\theta} &= \mu_d\left[w\cos\theta + m(R-r)\dot{\theta}^2\right] - w\sin\theta \\ I_g\ddot{\phi} &= -r\mu_d\left[w\cos\theta + m(R-r)\dot{\theta}^2\right]. \end{aligned} \quad (5.149)$$

These equations apply during slipping, provided that the direction of the friction force f_f in Figure 5.29 does not change. (The friction force would have a different sign if the cylinder were released from rest at $\theta = -\pi/2$.)

With slipping, the cylinder has two degrees of freedom, θ and ϕ. Note from Equations 5.149 that initially at $\theta = \pi/2$ and $\dot{\theta}(0) = 0$, $\ddot{\theta} < 0$ and $\ddot{\phi} = 0$. As the cylinder moves down the surface, $\dot{\theta}$ and $\dot{\phi}$ increase in magnitude but remain negative. The friction force f_f acts to slow down the magnitude increase in $\dot{\theta}$ and accelerate the magnitude increase in $\dot{\phi}$. When the kinematic condition $(R - r)\dot{\theta} = r\dot{\phi}$ is met, slipping stops, Equations 5.149 become invalid, and Equation 5.144 applies.

5.7.1.10 Pulley-Assembly Example: Newtonian Approach

Figure 5.30a illustrates a pulley assembly consisting of a pulley (disk) of mass m_1 and an attached mass of mass m_2. As illustrated, the assembly is in equilibrium. It is supported by an inextensible cord in series with a linear spring with stiffness coefficient k. On the right, the cord's end is rigidly attached to a horizontal surface. On the left, the cord is attached to the spring which is attached to the same surface. The pulley has mass m_l and $I_g = m_1 r^2/2$. The cord does not slip on the pulley. Think about pulling the assembly down and releasing it. You would expect the pulley assembly to oscillate directly up and down. The tasks associated with this system are as follows:

a. Select coordinates and draw the body in a general position.
b. Draw a free-body diagram for the body in a general position and use it to derive the EOM.
c. Determine the system's natural frequency.

As illustrated in Figure 5.30b, y defines the change in position of the pulley, and θ defines the pulley's rotation angle. You could reasonably look at this system and wonder why it is included in a subsection concerning "rolling without slipping." However, the statement that the "cord does not slip on the pulley" introduces the rolling-without-slipping condition. The pulley can be visualized as rolling without slipping on the vertical surface defined by the right-hand-side cord line. The δy and θ coordinates are related via

$$\delta y = r\theta \Rightarrow \delta\dot{y} = r\dot{\theta} \Rightarrow \delta\ddot{y} = r\ddot{\theta}. \tag{5.150}$$

Figure 5.30c provides the appropriate free-body diagrams with the pulley and the lower assembly separated. The reaction force N acts between the two masses at the pivot connection point. The pulley's mass center and the lower assembly have the same vertical acceleration $\delta\ddot{y}$; hence, their equations of motion are

$$\sum f_y = N + w_2 - (\overline{T}_{c1} + \delta T_{c1}) - (\overline{T}_{c2} + \delta T_{c2}) = m_1 \delta\ddot{y}$$
$$\sum f_y = w_1 - N = m_2 \delta\ddot{y}.$$

We can eliminate N by adding these last equations to obtain

$$\begin{aligned} w_1 + w_2 - (\overline{T}_{c1} - \overline{T}_{c2}) - (\delta T_{c1} - \delta T_{c2}) \\ = -(\delta T_{c1} - \delta T_{c2}) = (m_1 + m_2)\ddot{y}. \end{aligned} \tag{5.151}$$

This equation is simplified by the equilibrium requirement for the pulley and body $\sum f_y = 0 = w_1 + w_2 - (\overline{T}_{c1} + \overline{T}_{c2})$. Taking moments about the pulley's mass center gives

$$\begin{aligned} \sum M_g &= (\overline{T}_{c2} + \delta T_{c2})r - (\overline{T}_{c1} + \delta T_{c1})r \\ &= (\delta T_{c2} - \delta T_{c1})r = I_g \ddot{\theta} = \frac{m_1 r^2}{2}\ddot{\theta} \end{aligned} \tag{5.152}$$

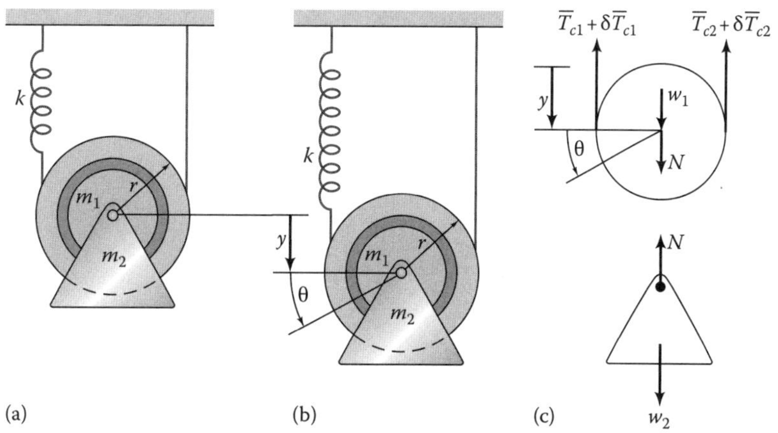

FIGURE 5.30
(a) Pulley assembly in equilibrium. (b) General displaced position defined by (y, θ). (c) Free-body diagram.

At equilibrium, the moment equation gives $\overline{T}_{c1}=\overline{T}_{c2}$, but with motion, $\delta T_{c1} \neq \delta T_{c2}$. They cannot be equal and still cause the pulley's angular acceleration.

Equations 5.150 through 5.152 provide three equations for the four unknowns: $\ddot{\theta}$, δT_{c1}, δT_{c2}, $\delta\ddot{y}$. Obviously, we need another kinematic result. Following the developments of Section 3.3.1, pulling the pulley down a distance δy will pull the cord end attached to the spring down a distance $2\delta y$. Hence, the cord tension δT_{c1} is defined by $\delta T_{c1}=k(2\delta y)=2k\delta y$. Substituting this result and eliminating δy and $\delta\ddot{y}$ gives

$$r\delta T_{c2} - r[2k(r\theta)] = \frac{m_1 r^2}{2}\ddot{\theta}$$
$$- 2k(r\theta) - \delta T_{c2} = (m_1 + m_2)r\ddot{\theta}.$$

Multiplying the second of these equations by r and adding the resulting equation to the first eliminates δT_{c2} and gives

$$r(-2kr\theta) - 2kr^2\theta = (m_1 + m_2)r^2\ddot{\theta} + \frac{m_1 r^2}{2}\ddot{\theta}.$$

Gathering terms and simplifying gives

$$\left(\frac{3m_1}{2} + m_2\right)r^2\ddot{\theta} + 4kr^2\theta = 0. \tag{5.153}$$

The natural frequency is

$$\omega_n = \sqrt{\frac{8k}{(3m_1 + 2m_2)}}\ \mathrm{s}^{-1}.$$

This example is "tricky" in that the rolling-without-slipping constraint and the second pulley constraint to define the spring deflection, $\delta_s=2\delta y$, require some thought.

5.7.1.11 Pulley-Assembly Example: Energy Approach

Starting from $T+V=T_0+V_0$ gives

$$I_g\frac{\dot{\theta}^2}{2} + (m_1 + m_2)\frac{\dot{y}^2}{2} + k\frac{\delta_s^2}{2} = T_0 + V_0 = 0,$$

where δ_s is the spring deflection. The potential energy function does not include w_1 and w_2 because the assembly is initially in equilibrium, and the spring is linear. Substituting: (i) $I_g=m_1r^2/2$, (ii) the rolling-without-slipping condition $\delta\dot{y}=r\dot{\theta}$, and (iii) the pulley condition $\delta_s=2\delta y=2r\theta$ gives

$$\frac{m_1 r^2}{2} \times \frac{\dot{\theta}^2}{2} + (m_1 + m_2)\frac{(r\dot{\theta})^2}{2} + \frac{k}{2}(2r\theta)^2 = 0$$
$$\Rightarrow \left(\frac{3m_1}{2} + m_2\right)r^2\frac{\dot{\theta}^2}{2} + 4kr^2\frac{\theta^2}{2} = 0.$$

Differentiating with respect to θ gives

$$\left(\frac{3m_1}{2} + m_2\right)r^2\ddot{\theta} + 4kr^2\theta = 0,$$

where $\ddot{\theta}=d(\dot{\theta}/2)/d\theta$. This result coincides with Equation 5.153 and was (again) obtained with substantially less effort than required for the Newtonian approach.

5.7.1.12 Closing Comments

As noted at the outset of this section, all the examples considered require a rolling-without-slipping kinematic condition for solution. Looking back through these examples, note that the Newtonian solution developments tend to progress through the following steps:

a. Select coordinates to define the position and orientation of the body, noting at the outset that the body has two coordinates, which will reduce to one if rolling without slipping applies.
b. Draw a sketch of the body in a general position displaying your choice of coordinates.
c. Draw a free-body diagram of the body in a general position.
d. Apply $\sum \boldsymbol{f} = \ddot{\boldsymbol{R}}_g$ to the mass center of the rigid body obtaining two scalar equations.
e. Either apply $\sum M_g = I_g\ddot{\theta}$ for the mass center or select another point on the body for which moments are to be taken and then apply $\sum M_{oz} = I_o\ddot{\theta} + m(\boldsymbol{b}_{og} \times \ddot{\boldsymbol{R}}_o)_z$. In this equation, be sure that positive moments agree with the positive direction selected for the rotation coordinate. If positive moments are in the counterclockwise direction, it applies directly; however, if your rotation coordinate has a positive clockwise sense, it needs to be modified as

$$\sum M_{oz} = I_o\ddot{\beta} - m(\boldsymbol{b}_{og} \times \ddot{\boldsymbol{R}}_o)_z.$$

f. Step *e* completes the equations of motion, and there will invariably be more unknowns than equations when this step is completed. Additional kinematics equations must be developed to complete the model, including (at a minimum) the rolling-without-slipping condition.

g. Using the kinematic relationships, eliminate enough unknowns to obtain a single, second-order differential equation in one of your coordinates.

With rolling without slipping, energy is conserved in all of the examples considered in this subsection. Hence, most of the equations of motion are readily (and more easily) obtained from work–energy principles. The following procedure was used in obtaining the equations of motion from energy principles:

a. Select coordinates to define the position and orientation of the body, noting at the outset that the body has two coordinates, which will reduce to one if rolling without slipping applies.
b. Draw a sketch of the body in a general position displaying your choice of coordinates.
c. Write out the kinematic constraint equations that apply including the rolling-without-slipping equation.
d. "Map" the variables for the specific example into $T+V=T_0+V_0$, using the kinematic constraint equations to get an equation in one coordinate and its time derivative.
e. Differentiate the energy equation from step d with respect to the single coordinate, using the energy-integral substitution $d(\dot{x}^2/2)/dx=\ddot{x}$ to obtain the EOM.

Recall that we do not always know whether slipping occurs or not. For rolling without slipping on a plane, the displacement and rotation coordinates are related through the kinematic condition, and there is only one degree of freedom and one governing EOM. If slipping occurs, the displacement and rotation coordinates are independent, the problem has two degrees of freedom and two differential equations of motion. The (slipping) friction force is defined by $f_f=\mu_d N$. These comments also apply to the example involving a cylinder rolling without slipping on a cylindrical surface.

For the examples of this section, proceeding from an energy equation provides the easiest and quickest path to derive the EOM, mainly because one can avoid drawing a free-body diagram. That happy circumstance is diminished considerably when energy dissipation is present. The next section provides some contrary results regarding the relative ease in deriving equations of motion from Newtonian and energy approaches.

5.7.2 One Degree of Freedom, Planar-Motion Applications, Newtonian and Energy Approaches

The rolling-without-slipping examples of the preceding section have two coordinates but only one degree of freedom. The rigid bodies of this section's examples have two coordinates corresponding to translation and rotation, but, because of a kinematic constraint, have only a rotation-angle degree of freedom. We will use both Newtonian and energy approaches to derive the equation of motion.

5.7.2.1 A Uniform Bar, Acted on by an External Force, Moving in Slots, and Constrained by Springs

Figure 5.31a illustrates a uniform bar of length l and mass m. The vertical force $f=Jf(t)$ acts at the bar's center. Light rollers at the bar's ends move with negligible friction in slots. The bar's ends are connected to ground by linear springs that are undeflected in the position shown. The springs at ends A and B have stiffness coefficients k_1 and k_2, respectively. The relevant engineering task are as follows:

a. Draw the bar in a general position and select coordinates to define its position and orientation.
b. Draw a free-body diagram for the bar's general position and use it to derive the equation of motion.

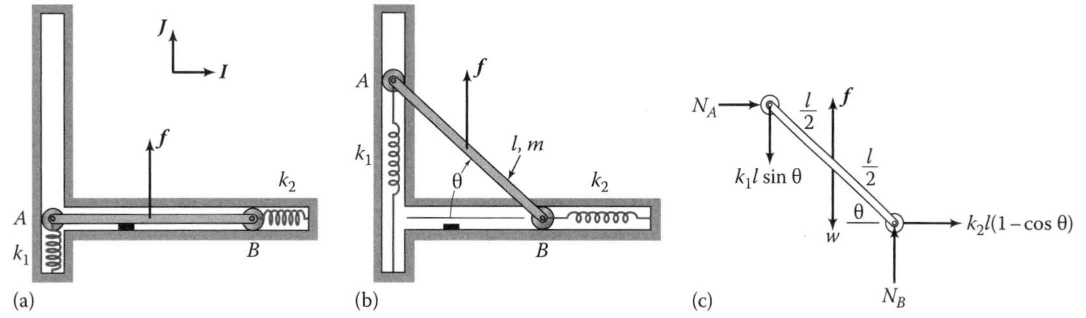

FIGURE 5.31
(a) Starting position for a bar acted on by a vertical force $f=Jf$. (b) Bar in a general position with coordinates. (c) Free-body diagram.

c. Use work–energy principles to derive the equation of motion.
d. Assuming that the bar starts at rest, and $f(t)=\bar{f}$, determine the bar's angular velocity when it reaches a vertical orientation. What is the minimum constant force that will cause the bar to reach this position?

Figure 5.31b illustrates the bar in a general position with θ serving as the coordinate. The bar's mass center is located by X_g, Y_g that are defined in terms of θ by

$$X_g = \frac{l}{2}\cos\theta, \quad Y_g = \frac{l}{2}\sin\theta. \tag{5.154}$$

Figure 5.31c provides the free-body diagram. Given the reaction N_A at A and N_B at B, the moment equation is most readily stated about the mass center, obtaining

$$\begin{aligned} I_g\ddot{\theta} = \sum M_g &= N_A\frac{l}{2}\sin\theta - k_1 l\sin\theta\cdot\frac{l}{2}\cos\theta \\ &- N_B\frac{l}{2}\cos\theta - k_2 l(1-\cos\theta)\cdot\frac{l}{2}\sin\theta. \end{aligned} \tag{5.155}$$

Stating $\sum \boldsymbol{f} = m\ddot{\boldsymbol{R}}_g$ for the mass center gives

$$\begin{aligned} \sum f_X &= N_A + k_2 l(1-\cos\theta) = m\ddot{X}_g \\ \sum f_Y &= f(t) + N_B - w - k_1 l\sin\theta = m\ddot{Y}_g. \end{aligned} \tag{5.156}$$

We now have three equations in the five unknowns N_A, N_B, $\ddot{\theta}$, $\ddot{X}_g$, $\ddot{Y}_g$.

The additional kinematic equations are obtained from the geometric relations. Differentiating Equations 5.154 once with respect to time gives

$$\dot{X}_g = -\frac{l}{2}\sin\theta\dot{\theta}, \quad \dot{Y}_g = \frac{l}{2}\cos\theta\dot{\theta}. \tag{5.157}$$

A second differentiation gives

$$\ddot{X}_g = -\frac{l}{2}(\sin\theta\ddot{\theta} + \cos\theta\dot{\theta}^2), \quad \ddot{Y}_g = \frac{l}{2}(\cos\theta\ddot{\theta} - \sin\theta\dot{\theta}^2). \tag{5.158}$$

Substitution for $\ddot{X}_g$, $\ddot{Y}_g$ into Equations 5.156 defines N_A, N_B as

$$\begin{aligned} N_A &= -k_2 l(1-\cos\theta) - \frac{ml}{2}(\sin\theta\ddot{\theta} + \cos\theta\dot{\theta}^2) \\ N_B &= -f(t) + w + k_1 l\sin\theta + \frac{ml}{2}(\cos\theta\ddot{\theta} - \sin\theta\dot{\theta}^2). \end{aligned}$$

Substitution for N_A, N_B into the moment Equation 5.155 gives

$$\begin{aligned} \frac{ml^2}{12}\ddot{\theta} &= \frac{l}{2}\sin\theta\left[-k_2 l(1-\cos\theta) - \frac{ml}{2}(\sin\theta\ddot{\theta} + \cos\theta\dot{\theta}^2)\right] \\ &- \frac{k_1 l^2}{2}\sin\theta\cos\theta \\ &- \frac{l}{2}\cos\theta\left[-f(t) + w + k_1 l\sin\theta + \frac{ml}{2}(\cos\theta\ddot{\theta} - \sin\theta\dot{\theta}^2)\right] \\ &- \frac{k_2 l^2}{2}(1-\cos\theta)\sin\theta, \end{aligned}$$

where $I_g = ml^2/12$. Gathering terms and simplifying gives the EOM

$$\begin{aligned} &\frac{ml^2}{3}\ddot{\theta} + k_2 l^2\sin\theta(1-\cos\theta) + k_1 l^2\sin\theta\cos\theta \\ &+ \frac{wl}{2}\cos\theta = \frac{fl}{2}\cos\theta. \end{aligned} \tag{5.159}$$

Developing Equation 5.159 was straightforward but algebraically messy. We started with the two components of $\sum \boldsymbol{f} = m\ddot{\boldsymbol{R}}_g$ and the moment equation $\sum M_g = I_g\ddot{\theta}$ to arrive at three equations in five unknowns. We then used kinematics to define $\ddot{X}_g$, $\ddot{Y}_g$. Back substitution leads to Equation 5.159 and concludes Task b.

We will now use energy principles to derive the equation of motion. Because of the external force, energy is not conserved, and the general work–energy equation, $\text{Work}_{n.c.} = (T_2 + V_2) - (T_1 + V_1)$, applies. We will start by calculating the nonconservative work term. For reasons that will become evident, we will use the force vector $\boldsymbol{f}_1 = \boldsymbol{J}f(t)$ illustrated in Figure 5.32a. It is located in the fixed X, Y system by $\boldsymbol{r}_1 = l/2(\boldsymbol{I}\cos\theta + \boldsymbol{J}\sin\theta)$. The differential work done by f_1 acting through a differential change in position is

$$\begin{aligned} d\text{Work}_{n.c} &= \boldsymbol{f}_1\cdot d\boldsymbol{r}_1 = \boldsymbol{J}f(t)\cdot\frac{l}{2}(-\boldsymbol{I}\sin\theta + \boldsymbol{J}\cos\theta)d\theta \\ &= \frac{f(t)l}{2}\cos\theta\, d\theta = Q_{\theta 1}\, d\theta, \end{aligned}$$

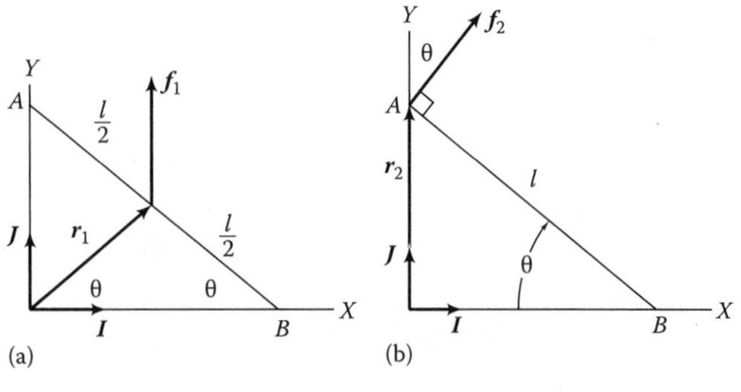

FIGURE 5.32
(a) External force $\boldsymbol{f}_1 = \boldsymbol{J}f_1$ located in the center of the bar by the vector $\boldsymbol{r}_1 = l/2(\boldsymbol{I}\cos\theta + \boldsymbol{J}\sin\theta)$. (b) External force $\boldsymbol{f}_2 = f_2(\boldsymbol{I}\cos\theta + \boldsymbol{J}\sin\theta)$ located at $\boldsymbol{r}_2 = \boldsymbol{J}l\sin\theta$.

where $Q_{\theta 1}$ is *the generalized force* associated with θ and f_1. The kinetic energy is

$$T = \frac{m}{2}\left(\dot{X}_g^2 + \dot{Y}_g^2\right) + \frac{I_g}{2}\dot{\theta}^2. \tag{5.160}$$

Substituting from Equation 5.157 for $\dot{X}_g$, $\dot{Y}_g$ gives

$$\begin{aligned} T &= \frac{m}{2}\cdot\frac{l^2}{4}(\sin^2\theta + \cos^2\theta)\dot{\theta}^2 + \frac{I_g}{2}\dot{\theta}^2 \\ &= \left(\frac{ml^2}{4} + \frac{ml^2}{12}\right)\frac{\dot{\theta}^2}{2} = \frac{ml^2}{3}\frac{\dot{\theta}^2}{2}. \end{aligned} \tag{5.161}$$

Establishing the gravity potential-energy datum in the center of the lower guide slot yields $V_g = wl/2\sin\theta$. The spring potential energy function is

$$V_s = \frac{k_1}{2}\delta_1^2 + \frac{k_2}{2}\delta_2^2 = \frac{k_1}{2}(l\sin\theta)^2 + \frac{k_2}{2}(l - l\cos\theta)^2.$$

Plugging all of these results into the work–energy equation gives

$$\begin{aligned} \int_0^\theta \frac{f(t)l}{2}\cos x\,dx = &\left[\frac{ml^2}{6}\dot{\theta}^2 + \frac{wl}{2}\sin\theta + \frac{k_1l^2}{2}\sin^2\theta\right. \\ &\left. + \frac{k_2l^2}{2}(1-\cos\theta)^2\right], \end{aligned} \tag{5.162}$$

where $T_0 = V_0 = 0$. Differentiating with respect to θ gives

$$\begin{aligned} &\frac{ml^2}{3}\ddot{\theta} + \frac{wl}{2}\cos\theta + k_1l^2\sin\theta\cos\theta \\ &+ k_2l^2(1-\cos\theta)\sin\theta = \frac{f(t)l}{2}\cos\theta = Q_{\theta 1}, \end{aligned} \tag{5.163}$$

where $\ddot{\theta} = d(\dot{\theta}^2/2)/d\theta$. Equation 5.163 coincides with Equation 5.159 and concludes Task c. As with the earlier rolling-without-slipping examples, getting the equation of motion is quicker using an energy approach than the Newtonian approach. We will show shortly that the potential advantage for this example is even greater than this initial development suggests.

Moving to Task d, for $f(t) = \bar{f}$, when the bar reaches a vertical position at $\theta = \pi/2$, its angular velocity is obtained from Equation 5.162 as

$$\begin{aligned} \frac{\bar{f}l}{2}\int_0^{\frac{\pi}{2}}\cos x\,dx &= \frac{\bar{f}l}{2}\sin x\Big|_0^{\frac{\pi}{2}} = \frac{\bar{f}l}{2} \\ &= \left[\frac{ml^2}{6}\dot{\theta}^2 + \frac{wl}{2}\sin\theta + \frac{k_1l^2}{2}\sin^2\theta\right. \\ &\quad\left. + \frac{k_2l^2}{2}(1-\cos\theta)^2\right]\Bigg|_{\theta=\frac{\pi}{2}} \\ &= \left[\frac{ml^2}{6}\dot{\theta}\left(\theta = \frac{\pi}{2}\right)^2 + \frac{wl}{2} + \frac{k_1l^2}{2} + \frac{k_2l^2}{2}\right]. \end{aligned}$$

Hence,

$$\dot{\theta}\left(\theta = \frac{\pi}{2}\right) = \left\{\frac{6}{ml^2}\left[\frac{\bar{f}l}{2} - \left(\frac{wl}{2} + \frac{k_1l^2}{2} + \frac{k_2l^2}{2}\right)\right]\right\}^{\frac{1}{2}}.$$

From this equation, the minimum required constant force $\bar{f}$ for the body to just reach the vertical is obtained by setting $\dot{\theta}(\theta = \pi/2) = 0$, producing $f = w + k_1l + k_2l$.

Note the generalized force term $Q_{\theta 1}$ on the right-hand side of Equation 5.163. Changing the external-force definition modifies only this term, leaving the left-hand side of the equation unchanged. Consider the force definition of Figure 5.32b. The external force $\boldsymbol{f}_2 = f_2(\boldsymbol{I}\sin\theta + \boldsymbol{J}\cos\theta)$ is located by the vector $\boldsymbol{r}_2 = \boldsymbol{J}l\sin\theta$. Hence, the differential work is

$$\begin{aligned} d\text{Work} &= \boldsymbol{f}_2\cdot d\boldsymbol{r}_2 = f_2(\boldsymbol{I}\sin\theta + \boldsymbol{J}\cos\theta)\cdot\boldsymbol{J}l\cos\theta\,d\theta \\ &= f_2l\cos^2\theta\,d\theta = Q_{\theta 2}d\theta. \end{aligned}$$

Substituting $Q_{\theta 2}$ for $Q_{\theta 1}$ in Equation 5.163 provides the correct equation for the loading situation of Figure 5.32b. Performing the same task with Newtonian mechanics would be more difficult, requiring changes in the force and moment equations and a rework of the algebra to arrive at the final equation. Ease in changing the equation of motion due to a change in the external-force definition is a distinct advantage for the work–energy formulation.

5.7.2.2 Adding Viscous Damping to the Slots Supporting the Uniform Bar

Figure 5.33a illustrates a variation of this example with lubricated guides (of negligible mass) now replacing the end rollers. This arrangement makes viscous damping in the slots more believable. The tasks associated with Figure 5.33a are as follows:

a. Assume that viscous damping is present in the slots and is quantified by the damping coefficient c, draw a free-body diagram, and use it to derive the EOM.
b. Use work–energy principles to derive the EOM.

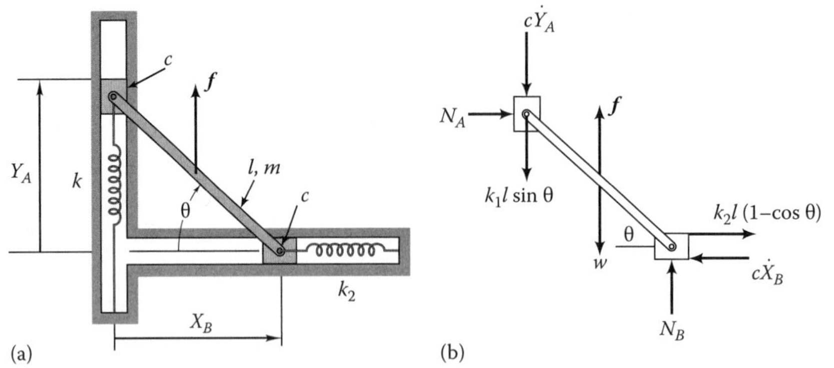

FIGURE 5.33
(a) The uniform bar of Figure 5.31 with lubricated guides now replacing the end rollers. (b) Free-body diagram for viscous damping.

We are going to continue using the kinematic variables introduced in Figure 5.31b.

The free-body diagram of Figure 5.33b accounts for viscous damping within the slots. Note that the viscous damping forces at A and B are defined by $f_{dA} = -cv_A = -cJ\dot{Y}_A$ and $f_{dB} = -cv_B = -Ic\dot{X}_B$, respectively. These equations state, for example, that the viscous damping force at A is proportional to the magnitude of $v_A = J\dot{Y}_a$ and acts in a direction opposite to v_A. From Figure 5.33a, points A and B are located via

$$Y_A = l\sin\theta, \quad X_B = l\cos\theta.$$

Differentiating with respect to time gives

$$\dot{Y}_A = l\cos\theta\dot{\theta}, \quad \dot{X}_B = -l\sin\theta\dot{\theta} \tag{5.164}$$

Stating the moment equation about the mass center of the bar in Figure 5.33b gives

$$\begin{aligned} I_g\ddot{\theta} = \sum M_g &= N_A\frac{l}{2}\sin\theta - k_1 l\sin\theta\cdot\frac{l}{2}\cos\theta \\ &\quad - N_B\frac{l}{2}\cos\theta - k_2 l(1-\cos\theta)\cdot\frac{l}{2}\sin\theta \\ &\quad - c\dot{Y}_A\frac{l}{2}\cos\theta - c\dot{X}_B\frac{l}{2}\sin\theta. \end{aligned}$$

Applying $\sum f = m\ddot{R}_g$ to the free-body diagram of Figure 5.33b provides

$$\begin{aligned} \sum f_X &= N_A - k_2 l(1-\cos\theta) - c\dot{X}_B = m\ddot{X}_g \\ \sum f_Y &= f + N_B - w - k_1 l\sin\theta - c\dot{Y}_A = m\ddot{Y}_g. \end{aligned}$$

The acceleration components $\ddot{X}_g, \ddot{Y}_g$ continue to be defined by Equations 5.158. Taking moments about g gives

$$\begin{aligned} I_g\ddot{\theta} = \sum M_g &= N_A\frac{l}{2}\sin\theta - k_1 l\sin\theta\cdot\frac{l}{2}\cos\theta \\ &\quad - N_B\frac{l}{2}\cos\theta - k_2 l(1-\cos\theta)\cdot\frac{l}{2}\sin\theta \\ &\quad - c\dot{X}_A\frac{l}{2}\cos\theta - c\dot{Y}_b\frac{l}{2}\sin\theta. \end{aligned}$$

Substituting from Equation 5.164 for $\dot{Y}_A$, $\dot{X}_B$ into the moment and force equations gives

$$\begin{aligned} &I_g\ddot{\theta} + \frac{cl^2}{2}(\cos^2\theta + \sin^2\theta)\dot{\theta} = \frac{N_A l}{2}\sin\theta - \frac{k_1 l^2}{2}\sin\theta\cos\theta \\ &\quad - \frac{N_B l}{2}\cos\theta - \frac{k_2 l^2}{2}(1-\cos\theta)\sin\theta, \\ &N_A = cl\sin\theta\dot{\theta} - k_2 l(1-\cos\theta) - \frac{ml}{2}(\sin\theta\ddot{\theta} + \cos\theta\dot{\theta}^2), \\ &N_B = -f + w + k_1 l\sin\theta + cl\cos\theta\dot{\theta} + \frac{ml}{2}(\cos\theta\ddot{\theta} - \sin\theta\dot{\theta}^2). \end{aligned} \tag{5.165}$$

Substitution for N_A, N_B into the moment equation and simplifying terms gives the final EOM

$$\begin{aligned} &\frac{ml^2}{3}\ddot{\theta} + cl^2\dot{\theta} + k_2 l^2\sin\theta(1-\cos\theta) \\ &\quad + k_1 l^2\sin\theta\cos\theta + \frac{wl}{2}\cos\theta = \frac{fl}{2}\cos\theta, \end{aligned} \tag{5.166}$$

and concludes Task a. The damping term on the left is the only addition to the prior moment Equation 5.159.

Using the work–energy equation to derive the equation of motion requires that we account for the nonconservative forces acting on the bar. The three nonconservative forces and their vector locations in the X, Y system are

i. $Jf(t), \boldsymbol{r_1} = l/2(\boldsymbol{I}\cos\theta + \boldsymbol{J}\sin\theta) \Rightarrow \boldsymbol{dr_1} = l/2(-\boldsymbol{I}\sin\theta + \boldsymbol{J}\cos\theta)d\theta$,
ii. $-\boldsymbol{J}c\dot{Y}_A$, $\boldsymbol{r_2} = \boldsymbol{J}Y_A = \boldsymbol{J}l\sin\theta \Rightarrow \boldsymbol{dr_2} = \boldsymbol{J}l\cos\theta\, d\theta$,
iii. $-\boldsymbol{I}c\dot{X}_B$, $\boldsymbol{r_3} = \boldsymbol{I}X_B = \boldsymbol{I}l\cos\theta \Rightarrow \boldsymbol{dr_3} = -\boldsymbol{I}l\sin\theta\, d\theta$.

The velocity components for points A and B are

$$\dot{Y}_A = l\cos\theta\dot{\theta}, \quad \dot{X}_B = -l\sin\theta\dot{\theta}. \tag{5.164}$$

Hence, the nonconservative differential work due to the three forces is

$$\begin{aligned}
d\text{Work}_{n.c.} &= \boldsymbol{J}f(t)\cdot\boldsymbol{dr_1} - \boldsymbol{J}c\dot{Y}_A\cdot\boldsymbol{dr_2} - \boldsymbol{I}c\dot{X}_B\cdot\boldsymbol{dr_3} \\
&= \boldsymbol{J}f(t)\cdot\frac{l}{2}(-\boldsymbol{I}\sin\theta + \boldsymbol{J}\cos\theta)d\theta - \boldsymbol{J}c(l\cos\theta\dot{\theta}) \\
&\quad \bullet \boldsymbol{J}l\cos\theta\, d\theta - \boldsymbol{I}c(-l\sin\theta\dot{\theta})\cdot\boldsymbol{I}(-l\sin\theta)d\theta \\
&= \left[\frac{f(t)l}{2}\cos\theta - cl^2\cos^2\theta\dot{\theta} - cl^2\sin^2\theta\dot{\theta}\right]d\theta \\
&= \left[\frac{f(t)l}{2}\cos\theta - cl^2\dot{\theta}\right]d\theta = Q_{\theta 3}d\theta.
\end{aligned}$$

Substituting $Q_{\theta 3}$ into Equation 5.163 gives

$$\begin{aligned}
&\frac{7ml^2}{12}\ddot{\theta} + \frac{wl}{2}\cos\theta + k_1l_1^2\sin\theta\cos\theta + k_2l_2^2(1-\cos\theta)\sin\theta \\
&\quad = \frac{f(t)l}{2}\cos\theta - cl^2\dot{\theta},
\end{aligned}$$

which coincides with Equation 5.166.

The energy approach continues to be quicker than the Newtonian but it required the free-body diagram of Figure 5.33b to define the damping forces at A and B. The energy approach's advantage becomes even clearer if we choose to account for the mass of the sliders, moving them from negligible to m_s. From Equations 5.160 and 5.161, the kinetic energy function for the bar and the two slider masses is now

$$\begin{aligned}
T &= \frac{m}{2}\left(\dot{X}_g^2 + \dot{Y}_g^2\right) + \frac{I_g}{2}\dot{\theta}^2 + m_s\frac{\dot{Y}_A^2}{2} + m_s\frac{\dot{X}_B^2}{2} \\
&= \frac{ml^2}{3}\frac{\dot{\theta}^2}{2} + \frac{m_s}{2}l^2(\sin^2\theta + \cos^2\theta)\dot{\theta}^2 \\
&= \left(\frac{m}{3} + \frac{m_s}{2}\right)l^2\frac{\dot{\theta}^2}{2}.
\end{aligned}$$

The potential energy equation has not changed, so Equation 5.162 becomes

$$\begin{aligned}
\int_0^{\theta}\frac{f(t)l}{2}\cos x\,dx &= \left[\left(\frac{m}{3} + \frac{m_s}{2}\right)l^2\frac{\dot{\theta}^2}{2} + \frac{wl}{2}\sin\theta\right. \\
&\quad \left. + \frac{k_1l^2}{2}\sin^2\theta + \frac{k_2l^2}{2}(1-\cos\theta)^2\right],
\end{aligned} \tag{5.165'}$$

and differentiation gives

$$\begin{aligned}
&\left(\frac{m}{3} + \frac{m_s}{2}\right)\ddot{\theta} + \frac{wl}{2}\cos\theta + k_1l^2\sin\theta\cos\theta \\
&\quad + k_2l^2(1-\cos\theta)\sin\theta = \frac{fl}{2}\cos\theta = Q_{\theta 1}.
\end{aligned} \tag{5.166'}$$

The Newtonian approach would require new free-body diagrams for the sliders and the bar, including the reaction forces between the bar and the sliders. New $\sum f = m\ddot{\boldsymbol{R}}_g$ equations would be needed for the slider masses to eliminate the reaction forces on the road to developing the equation of motion.

5.7.2.3 A Bar Leaning and Sliding on a Smooth Floor and against a Smooth Vertical Wall

Figure 5.34a illustrates a uniform bar of mass m and length l that is initially leaning against a wall. It begins sliding against a smooth floor and down a smooth vertical wall. The following tasks apply:

a. Draw the bar in a general position and select coordinates to define its position and orientation.
b. Neglect damping, draw a free-body diagram for the bar in a general position, and use it to derive the EOM.
c. Assuming that the bar starts from rest as shown in Figure 5.34a, for what position will it lose contact with the wall? What model applies after the bar leaves the wall?
d. Use energy principles to derive the bar's EOM while it is in contact with the wall.

As illustrated in Figure 5.34b, θ defines the bar's orientation with respect to the wall, and its mass center is located by

$$X_g = \frac{l}{2}\sin\theta, \quad Y_g = \frac{l}{2}\cos\theta \tag{5.167}$$

Figure 5.34c provides the free-body diagram showing the weight w and the normal reaction forces N_A and N_B acting at the wall and floor, respectively. Proceeding with Task b, and applying $\sum f = m\ddot{\boldsymbol{R}}_g$ for the mass center gives

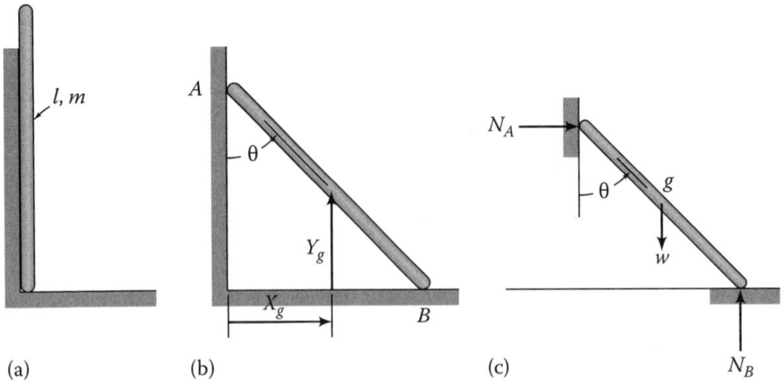

FIGURE 5.34
(a) Uniform bar of mass m and length l initially leaning against a vertical wall. (b) Coordinates X_g, Y_g, θ to locate the bar. (c) Free-body diagram.

$$\sum f_X = N_A = m\ddot{X}_g, \quad \sum f_Y = N_B - w = m\ddot{Y}_g. \quad (5.168)$$

Taking moments about the mass center gives

$$\sum M_g = N_B \frac{l}{2} \sin\theta - N_A \frac{l}{2} \cos\theta = I_g \ddot{\theta} = \frac{ml^2}{12}\ddot{\theta}. \quad (5.169)$$

The positive sense for moments is counterclockwise in accordance with the $+\theta$ rotation direction. We have three equations for the five unknowns N_A, N_B, $\ddot{\theta}$, $\ddot{X}_g$, $\ddot{Y}_g$ signaling the need for kinematics equations. Differentiating Equations 5.167 once with respect to time gives

$$\dot{X}_g = \frac{l}{2}\cos\theta\dot{\theta}, \quad \dot{Y}_g = -\frac{l}{2}\sin\theta\dot{\theta} \quad (5.170)$$

Differentiating again gives

$$\ddot{X}_g = \frac{l}{2}(\cos\theta\ddot{\theta} - \sin\theta\dot{\theta}^2), \quad \ddot{Y}_g = -\frac{l}{2}(\sin\theta\ddot{\theta} + \cos\theta\dot{\theta}^2)$$

Substituting for $\ddot{X}_g$, $\ddot{Y}_g$ into Equation 5.168 defines N_A and N_B as

$$N_A = \frac{ml}{2}(\cos\theta\ddot{\theta} - \sin\theta\dot{\theta}^2), \quad N_B = w - \frac{ml}{2}(\sin\theta\ddot{\theta} + \cos\theta\dot{\theta}^2) \quad (5.171)$$

Substituting N_A and N_B into the moment Equation 5.169 gives

$$\frac{ml^2}{12}\ddot{\theta} = \frac{wl}{2}\sin\theta - \frac{ml^2}{4}\ddot{\theta}(\cos^2\theta + \sin^2\theta) + \frac{ml^2}{4}\dot{\theta}^2(\cos\theta\sin\theta - \cos\theta\sin\theta),$$

or more simply

$$\left(\frac{ml^2}{12} + \frac{ml^2}{4}\right)\ddot{\theta} = \frac{ml^2}{3}\ddot{\theta} = \frac{wl}{2}\sin\theta \Rightarrow \ddot{\theta} = \frac{3g}{2l}\sin\theta. \quad (5.172)$$

This equation concludes Task b. Equation 5.172 is nonlinear and would customarily be solved by numerical integration from specified initial conditions to obtain time solutions, $\theta(t)$, $\dot{\theta}(t)$.

The bar loses contact with the wall when N_A becomes zero. Looking at the first of Equation 5.171, N_A is defined in terms of θ, $\dot{\theta}$, $\ddot{\theta}$. To conclude Task c, we need a definition for N_A in terms of θ alone. The $\ddot{\theta}$ term is easily eliminated via Equation 5.172. As will see below, the easiest way to get $\dot{\theta}^2(\theta)$ is to start from $T + V = T_0 + V_0$. However, we can also get it by integrating Equation 5.172, starting with the energy-integral substitution $\ddot{\theta} = d(\dot{\theta}^2/2)/d\theta$ to obtain

$$\frac{d}{d\theta}\left(\frac{\dot{\theta}^2}{2}\right) = \frac{3g}{2l}\sin\theta \Rightarrow \frac{\dot{\theta}^2}{2} - 0 = \int_0^\theta -\frac{3g}{2l}\cos dx$$
$$\Rightarrow \dot{\theta}^2 = \frac{3g}{l}(1 - \cos\theta),$$

where $\dot{\theta}(\theta = 0) = 0$. Substituting for $\dot{\theta}^2$ and $\ddot{\theta}$ into the first of Equation 5.171 for N_A gives

$$\begin{aligned} N_A &= \frac{ml}{2}(\cos\theta\ddot{\theta} - \sin\theta\ddot{\theta}^2) \\ &= \frac{ml}{2}\left\{\cos\theta\left(\frac{3g}{2l}\sin\theta\right) - \sin\theta\left[\frac{3g}{l}(1 - \cos\theta\right]\right\} \\ &= 3w\sin\theta\left(\frac{3\cos\theta}{2} - 1\right) \end{aligned}$$

For $N_A = 0$, this equation has the following two solutions: (i) $\sin\bar{\theta} = 0 \Rightarrow \bar{\theta} = 0$ that occurs at the initial position and (ii) $\cos\bar{\theta} = 2/3 \Rightarrow \bar{\theta} = 48.2°$ that is the desired solution. For values of θ on the interval (0, 48.2°), N_A is positive. For greater values of θ, N_A is negative, which is not possible, since the wall will not support a tensile reaction force.

When the bar loses contact with the wall, it has two degrees of freedom, for example, X_g and θ, and the appropriate force and moment equations are

$$\Sigma f_X = 0 = m\ddot{X}_g$$

$$\Sigma f_Y = N_B - w = m\ddot{Y}_g = -\frac{ml}{2}(\sin\theta\ddot{\theta} + \cos\theta\dot{\theta}^2)$$

$$\Sigma M_g = N_B\frac{l}{2}\sin\theta = I_g\ddot{\theta} = \frac{ml^2}{12}\ddot{\theta}.$$

The kinematic equation $Y_g = (l/2)\cos\theta$ still applies and is used in the second of these equations to define $\ddot{Y}_g$. Eliminating N_B and simplifying gives the final equations of motion

$$\ddot{X}_g = 0$$

$$\frac{ml^2}{4}\left(\frac{5}{6} - \frac{\cos 2\theta}{2}\right)\ddot{\theta} + \frac{ml^2}{8}\sin 2\theta\dot{\theta}^2 = \frac{wl}{2}\sin\theta. \quad (5.173)$$

The first of these equations is linear and easily solved. The second is nonlinear and must be integrated numerically to obtain time solutions $\theta(t)$, $\dot{\theta}(t)$. For Equations 5.173, taking the initial time $t = 0$ as the time that the bar leaves the wall, the required initial conditions $\dot{X}_g(0)$, $X_g(0)$, $\dot{\theta}(0)$, $\theta(0)$ are obtained from the "final" conditions of the solution for the bar's motion when it leaves the wall, that is,

$$\theta(0) = \bar{\theta} = 48.2°,$$

$$\dot{\theta}(0) = \dot{\theta}(\bar{\theta}) = \sqrt{\frac{3g}{l}(1 - \cos\bar{\theta})} = \sqrt{\frac{3g}{l}\left(1 - \frac{2}{3}\right)} = \sqrt{\frac{g}{l}}$$

$$X_g(0) = X_g(\bar{\theta}) = \frac{l}{2}\sin\bar{\theta} = \frac{l}{2} \times \frac{\sqrt{5}}{3} = \frac{\sqrt{5}l}{6}$$

$$\dot{X}_g(0) = \dot{X}_g(\bar{\theta}) = \frac{l}{2}\cos\bar{\theta} \times \dot{\theta}(\bar{\theta}) = \frac{l}{2} \times \frac{2}{3} \times \sqrt{\frac{g}{l}} = \frac{\sqrt{gl}}{3},$$

where from Equation 5.170, $\dot{X}_g = l/2\cos\theta\dot{\theta}$.

We will now proceed to Task d: Use energy principles to derive the equation of motion while the bar is in contact with the wall. For a smooth wall and floor, $T + V = T_0 + V_0$ applies. Using the floor as the gravity potential-energy datum yields

$$T + V = T_0 + V_0 \Rightarrow \frac{m}{2}\left(\dot{X}_g^2 + \dot{Y}_g^2\right) + \frac{I_g\dot{\theta}^2}{2} + \frac{wl\cos\theta}{2} = 0 + \frac{wl}{2}. \quad (5.174)$$

From Equation 5.171, $\dot{X}_g = (l/2)\cos\theta\dot{\theta}$, $\dot{Y}_g = -(l/2)\sin\theta\dot{\theta}$. Substituting these results into Equation 5.174, plus $I_g = ml^2/12$ gives

$$\frac{m}{2} \times \frac{l^2}{4}(\sin^2\theta + \cos^2\theta)\dot{\theta}^2 + \frac{ml^2}{12}\frac{\dot{\theta}^2}{2} = \frac{ml^2}{3}\frac{\dot{\theta}^2}{2}$$

$$= \frac{wl}{2}(1 - \cos\theta) \Rightarrow \dot{\theta}^2 = \frac{3g}{l}(1 - \cos\theta) \quad (5.175)$$

Differentiating with respect to θ gives

$$\ddot{\theta} = \frac{3g}{2l}\sin\theta,$$

which coincides with Equation 5.172. Again, proceeding from $T + V = T_0 + V_0$ produces the EOM more quickly than a free-body, Newtonian approach.

5.7.2.4 A Bar Leaning and Sliding on a Floor and against a Vertical Wall with Coulomb Friction

Figure 5.35 provides the free-body diagram for the bar in Figure 5.34 if the same dynamic Coulomb-friction coefficient μ_d applies to both the wall and the floor. In Figure 5.35, the directions for the Coulomb friction force components $\mu_d N_A$, $\mu_d N_B$ only apply for $\dot{\theta} > 0$; however, for the present example, this condition is always satisfied. The following engineering tasks apply:

a. Use the free-body diagram of Figure 5.35 to derive the equation of motion.
b. Use work–energy principles to derive the equation of motion.

The following force and moment equations can be obtained from the free-body diagram:

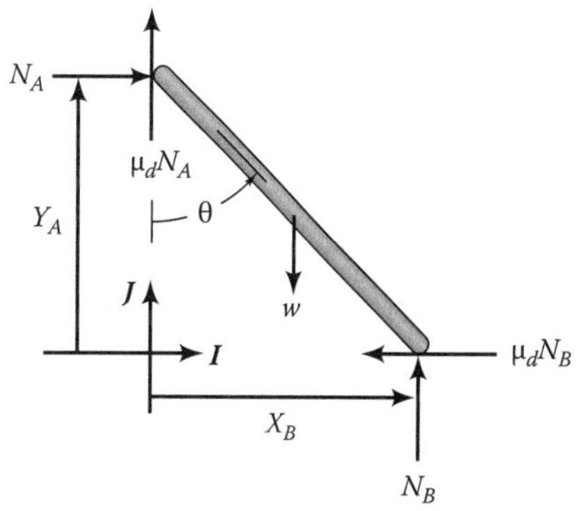

FIGURE 5.35
Free-body diagram for the bar of Figure 5.34 with Coulomb friction at the sliding surfaces.

$$\sum f_X = N_A - \mu_d N_B = m\ddot{X}_g = \frac{ml}{2}(\cos\theta\ddot{\theta} - \sin\theta\dot{\theta}^2)$$
$$\sum f_Y = N_B + \mu_d N_A - w = m\ddot{Y}_g = -\frac{ml}{2}(\sin\theta\ddot{\theta} + \cos\theta\dot{\theta}^2)$$
$$\sum M_g = N_B\frac{l}{2}\sin\theta - N_A\frac{l}{2}\cos\theta - \mu_d N_B\frac{l}{2}\cos\theta - \mu_d N_A\frac{l}{2}\sin\theta = I_g\ddot{\theta} = \frac{ml^2}{12}\ddot{\theta}. \quad (5.176)$$

Using Cramer's rule to solve the first two equation for N_A and N_B gives

$$N_A = \frac{\mu_d w}{(1+\mu_d^2)} + \frac{ml}{2(1+\mu_d^2)} \times \left[\ddot{\theta}(\cos\theta - \mu_d\sin\theta) - \dot{\theta}^2(\sin\theta + \mu_d\cos\theta)\right]$$
$$N_B = \frac{w}{(1+\mu_d^2)} - \frac{ml}{2(1+\mu_d^2)} \times \left[\ddot{\theta}(\sin\theta + \mu_d\cos\theta) + \dot{\theta}^2(\cos\theta - \mu_d\sin\theta)\right] \quad (5.177)$$

Substituting these results into the moment equation (plus lots of algebra) gives the EOM

$$\left[\frac{ml^2}{3} - \frac{m\mu_d^2 l^2}{2(1+\mu_d^2)}\right]\ddot{\theta} - \frac{\mu_d ml^2}{2(1+\mu_d^2)}\dot{\theta}^2 = \frac{wl}{2}\sin\theta - \frac{\mu_d wl}{(1+\mu_d^2)}(\cos\theta + \mu_d\sin\theta), \quad (5.178)$$

and concludes Task a. Note that Equation 5.178 reduces to Equation 5.172, the original equation of motion, if $\mu_d = 0$.

This equation is nonlinear and would have to be solved numerically to obtain time solutions $\theta(t)$, $\dot{\theta}(t)$. It can be reduced to

$$\left[\frac{ml^2}{3} - \frac{m\mu_d^2 l^2}{2(1+\mu_d^2)}\right]\frac{d}{d\theta}\left(\frac{\dot{\theta}^2}{2}\right) - \frac{\mu_d ml^2}{(1+\mu_d^2)}\left(\frac{\dot{\theta}^2}{2}\right) = \frac{wl}{2}\sin\theta - \frac{\mu_d wl}{(1+\mu_d^2)}(\cos\theta + \mu_d\sin\theta),$$

where $\ddot{\theta} = d(\dot{\theta}^2/2)/d\theta$. Following Appendix B, this constant-coefficient, first-order, linear differential equation can be solved speedily to produce a $\dot{\theta}^2(\theta)$ solution; however, using the solution to find the $\theta = \bar{\theta}$ value when the bar leaves the wall would be tedious.

To derive the EOM using work–energy principles, we will need the free-body diagram of Figure 5.35 to define the nonconservative forces acting on the bar. The friction force vectors at the ends of the bars and the position vectors that locate them in the X, Y coordinate system are ($f_1 = J\mu_d N_a$, $r_1 = Jl\cos\theta$), ($f_2 = -I\mu_d N_B$, $r_2 = Il\sin\theta$). The nonconservative differential work is

$$d\text{Work}_{n.c.} = \boldsymbol{f_1}\cdot\boldsymbol{dr_1} + \boldsymbol{f_2}\cdot\boldsymbol{dr_2} = \boldsymbol{J}\mu_d N_A\cdot -\boldsymbol{J}l\sin\theta\, d\theta + (-\boldsymbol{I}\mu_d N_B)\cdot\boldsymbol{I}l\cos\theta\, d\theta = -\mu_d l(N_A\sin\theta + N_B\cos\theta)d\theta. \quad (5.179)$$

N_A, N_B are defined in Equation 5.177. Substituting into the differential work expression of Equation 5.179 (plus algebra) gives

$$d\text{Work}_{n.c.} = \left[-\frac{\mu_d wl}{(1+\mu_d^2)}(\cos\theta + \mu_d\sin\theta) + \frac{\mu_d ml^2\ddot{\theta}}{2(1+\mu_d^2)} + \frac{\mu_d ml^2\dot{\theta}^2}{2(1+\mu_d^2)}\right]d\theta = Q_\theta\, d\theta.$$

Substituting this result into the work–energy equation produces

$$\int_0^\theta Q_\theta\, dx = (T+V) - (T_0+V_0) = \frac{ml^2}{6}\dot{\theta}^2 + \frac{wl}{2}\cos\theta - \frac{wl}{2}.$$

Differentiating with respect to θ nets

$$\frac{ml^2}{3}\ddot{\theta} - \frac{wl}{2}\sin\theta = Q_\theta = -\frac{\mu_d wl}{(1+\mu_d^2)}(\cos\theta + \mu_d\sin\theta) + \frac{\mu_d^2 ml^2\ddot{\theta}}{2(1+\mu_d^2)} + \frac{\mu_d ml^2\dot{\theta}^2}{2(1+\mu_d^2)},$$

which coincides with Equation 5.178 when rearranged as

$$\left[\frac{ml^2}{3} - \frac{m\mu_d^2 l^2}{2(1+\mu_d^2)}\right]\ddot{\theta} - \frac{\mu_d ml^2\dot{\theta}^2}{2(1+\mu_d^2)} = \frac{wl}{2}\sin\theta - \frac{\mu_d wl}{(1+\mu_d^2)}(\cos\theta + \mu_d\sin\theta).$$

In this example, Coulomb friction markedly reduces the advantages of using the work–energy equation to derive equations of motion. We had to use a free-body diagram and apply $\sum \boldsymbol{f} = m\ddot{\boldsymbol{R}}_g$ to define the nonconservative forces.

5.7.2.5 Summary and Discussion

The examples of this section provide the following lessons for 1DOF systems:

a. Absent nonconservative forces, deriving the equations of motion is quicker via the work–energy equation (followed by differentiation), than a direct application of the force and moment equations. This speed advantage follows mainly because (i) no free-body diagram is required, (ii) the kinematics is simpler only requiring the development of velocity components, versus acceleration components when applying $\sum f = m\ddot{\boldsymbol{R}}_g$, and (iii) the algebra is generally much simpler.
b. Nonconservative forces can be readily accounted for in the work–energy approach; however, $\text{Work}_{n.c.} = \Delta(T+V)$ must be employed versus the simpler $T+V=T_0+V_0$. Also, a free-body diagram is needed to define the nonconservative forces. For the Coulomb friction example above, the acceleration components of the mass center were also required. For the examples of this subsection, the algebra remained generally simpler for the work–energy approach than the Newtonian approach.
c. Applying the work–energy equation for examples that include nonconservative forces requires a continuing mastery of Newtonian dynamics. Free-body diagrams continue to be required plus, in some cases, acceleration components.

5.7.3 Multi-Body, Single-Coordinate Applications of the Work–Energy Equation

In contrast to the preceding material of this section, all of the examples in this subsection involve two or more rigid bodies; however, they only have one degree of freedom and only require one coordinate. The presence of more than one body resembles the 1DOF mechanisms that we will consider in Section 5.7.5; however, those examples will involve multiple coordinates that are related via nonlinear kinematic constraint equations, and they will entail more than one coordinate. The position and orientation of the present examples can be defined by a single coordinate.

5.7.3.1 Two Bars with an Applied Force and a Connecting Spring

Figure 5.36a illustrates two bars connected to each other by a frictionless pivot at A. Each bar has mass m and length l. The lower bar pivots without friction about point O. Point B at the top end of the upper bar moves in a vertical slot without friction. A linear spring with spring coefficient k connects points O and B. The spring is undeflected at $\beta = \beta_0$. The vertical external force $\boldsymbol{f} = \boldsymbol{J}f$ acts at point B. The following tasks apply:

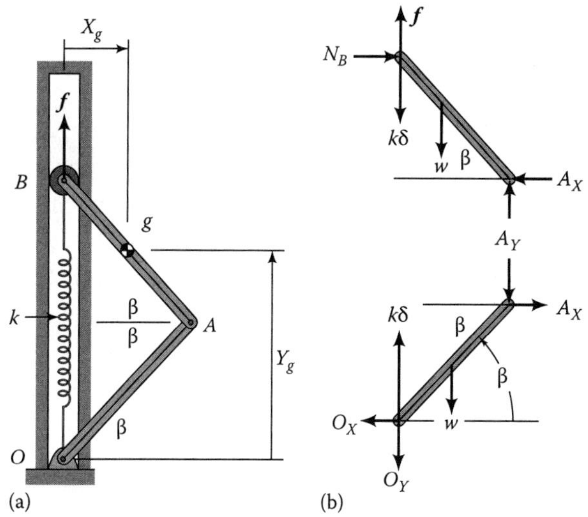

FIGURE 5.36
Two bars, each with mass m and length l connected by a spring. (a) Coordinates. (b) Free-body diagram.

a. Develop a general work–energy equation.
b. If the bodies start from rest at $\beta = \beta_0$ and $f(t) = \bar{f} =$ constant, determine the angular velocity of the bodies when $\beta = \pi/2$.
c. Use the work–energy equation to derive the EOM.

Because of the external force, energy is not conserved, and the general work–energy equation $\text{Work}_{n.c.} = \Delta(T+V)$ applies. The kinetic energy of the two bars is

$$T = \frac{I_O\dot{\beta}^2}{2} + \left[\frac{I_g\dot{\beta}^2}{2} + \frac{m}{2}\left(\dot{X}_g^2 + \dot{Y}_g^2\right)\right]. \tag{5.180}$$

The first term accounts for the kinetic energy of the lower bar that is rotating about an axis through O. The second term accounts for the upper bar that is both translating and rotating. The velocity components of the upper bar's mass center are

$$\begin{aligned} X_g &= \frac{l}{2}\cos\beta \Rightarrow \dot{X}_g = -\frac{l}{2}\sin\beta\dot{\beta}, \\ Y_g &= \frac{3l}{2}\sin\beta \Rightarrow \dot{Y}_g = \frac{3l}{2}\cos\beta\dot{\beta}. \end{aligned} \tag{5.181}$$

Substituting these results (plus $I_O = ml^2/3$, $I_g = ml^2/12$) into Equation 5.180 gives

$$T = \frac{ml^2}{6}\dot{\beta}^2 + \frac{ml^2}{24}\dot{\beta}^2 + \frac{m}{2}\left[\frac{l^2}{4}\sin^2\beta + \frac{9l^2}{4}\cos^2\beta\right]\dot{\beta}^2$$
$$= ml^2\left(\frac{1}{6}+\frac{1}{24}+\frac{1}{8}\right)\dot{\beta}^2 + ml^2\cos^2\beta\dot{\beta}^2$$
$$= \frac{ml^2}{3}\dot{\beta}^2 + ml^2\cos^2\beta\dot{\beta}^2 = ml^2\dot{\beta}^2\left(\frac{1}{3}+\cos\beta^2\right)$$

For the gravity potential-energy datum through O, the mechanical potential energy is

$$V = \frac{wl}{2}\sin\beta + \frac{3wl}{2}\sin\beta + \frac{k\delta^2}{2}$$
$$= 2wl\sin\beta + \frac{k}{2}(2l\sin\beta - 2l\sin\beta_0)^2.$$

The differential work due to the external force is

$$d\text{Work}_{n.c.} = \boldsymbol{J}f(t)\boldsymbol{\cdot}d(\boldsymbol{J}2l\sin\beta) = 2f(t)l\cos\beta\, d\beta = Q_\beta d\beta.$$

Substituting into the work–energy equation gives

$$\int_{\beta_0}^{\beta} 2f(t)l\cos x\,dx = ml^2\dot{\beta}^2\left(\frac{1}{3}+\cos^2\beta\right)$$
$$+\left[2wl\sin\beta + \frac{k}{2}(2l\sin\beta - 2l\sin\beta_0)^2\right] - (T_0 + V_0) \tag{5.182}$$

and concludes Task a.

Task b is accomplished by substituting $f(t)=\bar{f}$ and $\beta=\pi/2$ into Equation 5.182 to obtain

$$\int_{\beta_0}^{\pi/2} 2\bar{f}l\cos\beta\,d\beta = 2\bar{f}l(1-\sin\beta) = \frac{ml^2}{3}\dot{\beta}^2(\beta=\pi/2)$$
$$+2wl(1-\sin\beta_0) + 2kl^2(1-\sin\beta_0)^2,$$

and the requested answer is

$$\dot{\beta}\left(\beta=\frac{\pi}{2}\right) = \left\{\frac{6}{ml}\left[\bar{f}(1-\sin\beta_0) - w(1-\sin\beta_0)\right.\right.$$
$$\left.\left. -kl(1-\sin\beta_0)^2\right]\right\}^{1/2}.$$

The EOM is obtained by differentiating Equation 5.182 with respect to β, obtaining

$$2ml^2\ddot{\beta}\left(\frac{1}{3}+\cos^2\beta\right) - ml^2\dot{\beta}^2\sin 2\beta + 2wl\cos\beta$$
$$+4kl^2(\sin\beta - \sin\beta_0)\cos\beta = Q_\beta = 2f(t)l\cos\beta$$

This result concludes Task c.

This example is ideally suited to the work–energy approach. Working the same problem with Newtonian methods would follow the procedures of Section 5.7.2. The free-body diagram of Figure 5.36b would be required for the separate bodies, including the reaction forces at A, B, and O. The four equations of motion would consist of a moment equation about O for the lower body, a moment equation about the mass center for the upper body, and the X and Y components of $\sum \boldsymbol{f} = m\ddot{\boldsymbol{R}}_g$ for the upper bar. Two additional kinematic equations would be obtained by differentiating $\dot{X}_g$, $\dot{Y}_g$ in Equation 5.181 to obtain $\ddot{X}_g$, $\ddot{Y}_g$. After this step, there are four equations in the four unknowns $\ddot{\theta}$, N_A, A_x, A_Y. The development of the equation of motion is completed by algebra to eliminate N_A, A_x, A_Y. For this example, Newtonian approaches require a major struggle to duplicate the easily obtained energy approach result.

Consider the modified version of this example provided by Figure 5.37. The upper bar now has length $2l$ and mass $2m$. The system now has the two coordinates β and ϕ that are related by the kinematic condition

$$2l\sin\phi = l\cos\beta \Rightarrow \sin\phi = \frac{\cos\beta}{2},$$
$$\cos\phi = \frac{\sqrt{4-\cos^2\beta}}{2}. \tag{5.183}$$

Differentiating the first equation with respect to time gives

$$2l\cos\phi\dot{\phi} = -l\sin\beta\dot{\beta} \tag{5.184}$$

The position and velocity components of the upper bar's mass center are now defined by

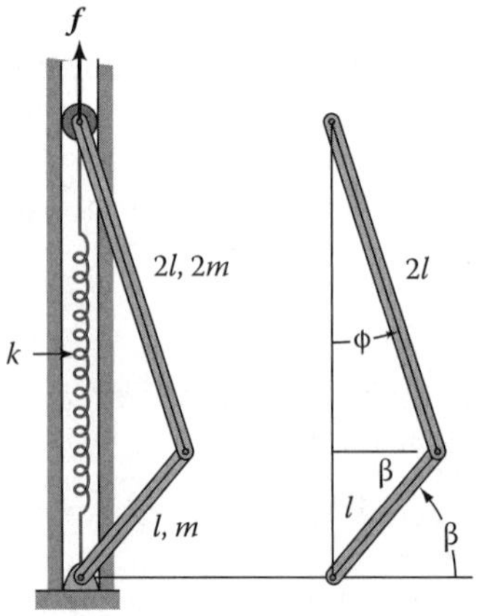

FIGURE 5.37
Modification to Figure 5.36 with the upper bar now having mass $2m$ and length $2l$.

$$X_g = \frac{l}{2}\sin\beta = l\sin\phi \Rightarrow \dot{X}_g = \frac{l}{2}\cos\beta\dot{\beta} = l\cos\phi\dot{\phi}$$
$$Y_g = l\sin\beta + l\cos\phi \Rightarrow \dot{Y}_g = l\cos\beta\dot{\beta} - l\sin\phi\dot{\phi}, \tag{5.185}$$

and the kinetic energy expression is

$$T = \frac{I_O\dot{\beta}^2}{2} + \left[\frac{I_g\dot{\phi}^2}{2} + \frac{m}{2}\left(\dot{X}_g^2 + \dot{Y}_g^2\right)\right].$$

Substituting for I_O, I_g and the components $\dot{X}_g$, $\dot{Y}_g$ gives

$$T = \frac{ml^2}{6}\dot{\beta}^2 + \frac{ml^2}{3}\dot{\phi}^2 + ml^2(\dot{\phi}^2 + \cos^2\beta\dot{\beta}^2 - 2\cos\beta\sin\phi\dot{\beta}\dot{\phi}).$$

The potential-energy function becomes

$$V = \left(\frac{wl}{2}\sin\beta + 2wl\cos\phi\right) + \frac{k}{2}[(l\sin\beta + 2l\cos\phi) - (l\sin\beta_0 + 2l\cos\phi_0)]^2. \tag{5.186}$$

The applied force and its position vector are $\boldsymbol{f} = \boldsymbol{J}f(t)$, $\boldsymbol{r} = \boldsymbol{J}(l\sin\beta + 2l\cos\phi)$; hence, the differential nonconservative work is

$$d\text{Work}_{n.c.} = \boldsymbol{f}\cdot d\boldsymbol{r} = f(l\cos\beta\, d\beta - 2l\sin\phi\, d\phi).$$

Substituting these results into the work–energy equation gives

$$\int_{\beta_0}^{\beta} f(t)l\cos x\,dx - \int_{\phi_0}^{\phi} 2f(t)l\sin x\,dx = \frac{ml^2\dot{\beta}^2}{6} + \frac{ml^2\dot{\phi}^2}{3} + ml^2(\dot{\phi}^2 + \cos^2\beta\dot{\beta}^2 - 2\cos\beta\sin\phi\dot{\beta}\dot{\phi}) + \frac{k}{2}[(l\sin\beta + 2l\cos\phi) - (l\sin\beta_0 + 2l\cos\phi_0)]^2 + \left(\frac{wl}{2}\sin\beta + wl\cos\phi\right) - \left(\frac{wl}{2}\sin\beta_0 + wl\cos\phi_0\right) \tag{5.187}$$

This is an awkward but correct work–energy equation.

Getting the EOM is tedious. To get a differential equation for β, we need to substitute from Equations 5.184 and 5.185 into Equation 5.177 to eliminate ϕ and $\dot{\phi}$. In the second integral, we also need to take the differential of $2l\sin\phi = l\cos\beta$ to obtain

$$2l\cos\phi\,d\phi = -l\sin\beta\,d\beta \Rightarrow d\phi = -\frac{\sin\beta\,d\beta}{2\cos\phi} = -\frac{\sin\beta\,d\beta}{\sqrt{4-\cos^2\beta}}$$

Making these substitutions into Equation 5.187 gives

$$ml^2\dot{\beta}^2\left[\frac{1}{6} + \cos^2\beta + \frac{\sin\beta(4\sin\beta + 3\cos^2\beta)}{3(4-\cos^2\beta)}\right] + \frac{k}{2}\left[\left(l\sin\beta + l\sqrt{4-\cos^2\beta}\right) - (l\sin\beta_0 + 2l\cos\phi_0)\right]^2 + \left(\frac{wl}{2}\sin\beta + wl\frac{\sqrt{4-\cos^2\beta}}{2}\right) - \left(\frac{wl}{2}\sin\beta_0 + wl\cos\phi_0\right)$$
$$= \int_{\beta_0}^{\beta} fl\left[\cos x + \frac{\sin x\cos x}{4-\cos^2 x}\right]d\beta = \int_{\beta_0}^{\beta} Q_\beta\,d\beta$$

Differentiating this unpleasant-looking equation with respect to β will give the governing differential equation. This development shows that the equation of motion *can* be developed after the problem has been changed to include a second coordinate. However, developing the equation of motion from the work–energy equation is no longer the speedy, easy, and graceful exercise that it was with one coordinate. The clear advantage of the work–energy approach in the one-coordinate case is largely eliminated. The present two-coordinate, one-degree-of-freedom example is a good application for Lagrange's equations using a Lagrange multiplier to account for the kinematic constraint. It can also be attacked using the procedures of Section 5.7.5 related to mechanisms.

5.7.3.2 Hinged Bar/Plate Example

Figure 5.38a presents another two-body, single-coordinate example. The bar on the left has mass m and length b and rotates freely about a pivot at O. It is hinged at B to a square plate with sides of length b and mass M. A small roller at A supports the plate on a horizontal plane. The horizontal force $\boldsymbol{f} = \boldsymbol{I}f(t)$ acts at corner D. A linear spring with spring-coefficient k connects points O and A and is undeflected at $\theta = \theta_0$. Point C is at the midpoint of side AD. The following tasks apply:

a. Develop a general work–energy equation.
b. If the bodies starts from rest at $\theta = 0$, and $f(t) = \bar{f}$, determine $\dot{\theta}$ when $\theta = \pi/2$.
c. Develop the EOM.

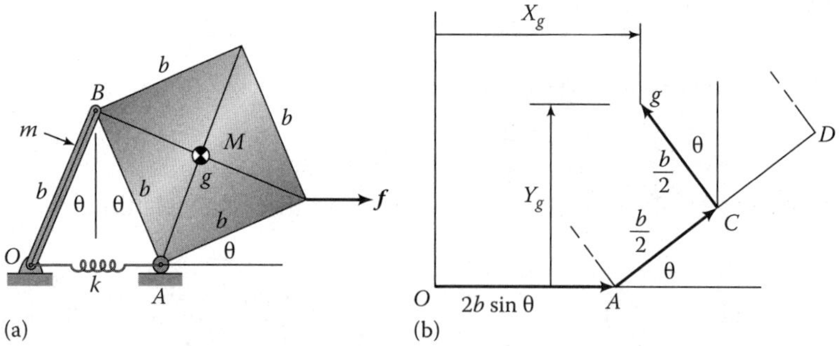

FIGURE 5.38
(a) Hinged bar and plate combination. (b) Geometry locating the mass center.

Because of the external force, energy is not conserved, and $\text{Work}_{n.c.} = \Delta(T+V)$ applies. The kinetic energy of the two bodies is

$$T = \frac{I_O\dot{\theta}^2}{2} + \left[\frac{I_g\dot{\theta}^2}{2} + \frac{M}{2}\left(\dot{X}_g^2 + \dot{Y}_g^2\right)\right], \tag{5.188}$$

where the first term accounts for the kinetic energy of the bar, and the second term takes care of the plate. Figure 5.38b illustrates the geometry required to locate the plate's mass center. In this figure, C is located at the midpoint of side AD. Observe from Figure 5.38b that

$$X_g = 2b\sin\theta + \frac{b}{2}\cos\theta - \frac{b}{2}\sin\theta \Rightarrow \dot{X}_g = \frac{b\dot{\theta}}{2}(3\cos\theta - \sin\theta)$$
$$Y_g = \frac{b}{2}\sin\theta + \frac{b}{2}\cos\theta \Rightarrow \dot{Y}_g = \frac{b\dot{\theta}}{2}(\cos\theta - \sin\theta). \tag{5.189}$$

These component results for X_g and Y_g are obtained by proceeding vectorially along the path: $O \Rightarrow A \Rightarrow C \Rightarrow g$.

Substituting $I_O = mb^2/3$ and $I_g = Mb^2/6$ from Appendix C, plus the kinematics results of Equation 5.189, into Equation 5.188 (plus considerable amounts of algebra), gives

$$T = b^2\dot{\theta}^2\left[\frac{m}{6} + M\left(\frac{5}{6} + \cos 2\theta - \sin 2\theta\right)\right]$$

Using a plane through the pivot point O as the datum for the gravity potential-energy function gives the potential-energy function

$$V = \frac{wb}{2}\cos\theta + \frac{Wb}{2}(\sin\theta + \cos\theta) + \frac{k}{2}(2b\sin\theta - 2b\sin\theta_0)^2.$$

The external force $f = \boldsymbol{I}f(t)$ is located in the X, Y system by $\boldsymbol{r} = [\boldsymbol{I}(2b\sin\theta + b\cos\theta) + \boldsymbol{J}b\sin\theta]$. The differential nonconservative work due to this force is

$$d\text{Work}_{n.c.} = \boldsymbol{f}\cdot d\boldsymbol{r} = \boldsymbol{I}f(t)\cdot[\boldsymbol{I}b(2\cos\theta - \sin\theta) - \boldsymbol{J}b\sin\theta]d\theta$$
$$= bf(t)(2\cos\theta - \sin\theta)d\theta = Q_\theta\, d\theta.$$

Substituting these results into $\text{Work}_{n.c.} = \Delta(T+V)$ gives

$$b\int_{\theta_0}^{\theta} f(t)(2\cos x - \sin x)dx$$
$$= b^2\dot{\theta}^2\left[\frac{m}{6} + M\left(\frac{19}{12} + \cos 2\theta - \sin 2\theta\right)\right] + \frac{wb}{2}\cos\theta$$
$$+ \frac{Wb}{2}(\sin\theta + \cos\theta) + 2kb^2(\sin\theta - \sin\theta_0)^2 - (T_0 + V_0), \tag{5.190}$$

and concludes Task a.

Moving to Task b, the integral on the left can be completed for $f(t) = \bar{f}$. Integrating over the interval $(\theta = \theta_0 = 0,\ \theta = \pi/2)$ and setting $T_0 = 0$, $V_0 = (w+W)b/2 + k(2b\sin\theta_0)^2/2$ yields from Equation 5.190

$$2b\bar{f} = \dot{\theta}_f^2\frac{b^2}{6}\left(m + \frac{7M}{2}\right) - 2kb^2(1 - \sin\theta_0)^2$$
$$- \frac{wb}{2} + 2kb^2\sin^2\bar{\theta}.$$

Solving for the angular velocity gives

$$\dot{\theta}_f = \dot{\theta}\left(\theta = \frac{\pi}{2}\right)$$
$$= \left\{\left[\frac{12\bar{f}}{b} - 12k(1 - 2\sin\bar{\theta}) + \frac{3w}{b}\right]\Big/\left(m + \frac{7M}{2}\right)\right\}^{1/2},$$

and concludes Task c.

Differentiating Equation 5.190 with respect to θ gives the EOM

$$\ddot{\theta}\left[\frac{m}{3}+2M\left(\frac{5}{6}+\cos 2\theta-\sin 2\theta\right)\right]-2\dot{\theta}^2M(\sin 2\theta-\cos 2\theta)$$
$$-\frac{w}{2b}\sin\theta+\frac{W}{2b}\cos\theta+4k(\sin\theta-\sin\bar{\theta})\cos\theta$$
$$=\frac{f(t)}{b}(2\cos\theta-\sin\theta)$$

where $\ddot{\theta}=d(\dot{\theta}^2/2)/d\theta$.

Again, this example is ideally suited for developing the EOM via the work–energy equation. It provides a much more direct development than a Newtonian approach. However, a small modification to the geometry to require an additional coordinate (e.g., changing the square plate dimension from b to $3b/2$) would seriously degrade the energy-equation's advantage as a basis for developing the EOM.

5.7.3.3 A Parallel, Double-Bar Arrangement for Retracting a Cylinder

Figure 5.39a illustrates a bar assembly for retracting a cylinder. The four bars in the assembly have mass m and length l. The cross bar at the top also has mass m. The cylinder has radius r and mass M. A time-varying force with magnitude $f(t)$ is acting through a cable that extends over a pulley and is attached at the center of the assembly's connecting bar. The cylinder rolls without slipping on a horizontal plane, and its rotation angle is β. Figure 5.39b provides a side view of the assembly, showing the left-hand bars connected to the wall at point O, at the distance r above the horizontal plane. The bars of the folding assembly are connected to each other via a frictionless sleeve at A. The right-hand bars are connected to the cylinder at B. The pulley through which the force acts is pivoted at the distance l above O. Including the mass of the transverse connecting bar, this system involves six rigid bodies (five bars and the cylinder) but has only one degree of freedom θ. Definitions of the position and orientation of the bars by θ are similar to the prior examples of this section. You could expect to define the cylinder's rotation angle β and position in terms of θ via the rolling-without-slipping kinematic condition. The fact that only one coordinate is required to define the positions and orientations of the rigid bodies of this system makes a development based on the work–energy equation particularly attractive.

The following tasks apply:

a. Develop a general work–energy equation.
b. If the bodies starts from rest at $\theta=0$, and $f(t)=\bar{f}=$ constant, determine $\dot{\theta}_f=\dot{\theta}\ (\theta=\pi/2)$.
c. Use $\text{Work}_{n.c.}=\Delta(T+V)$ to develop the EOM.

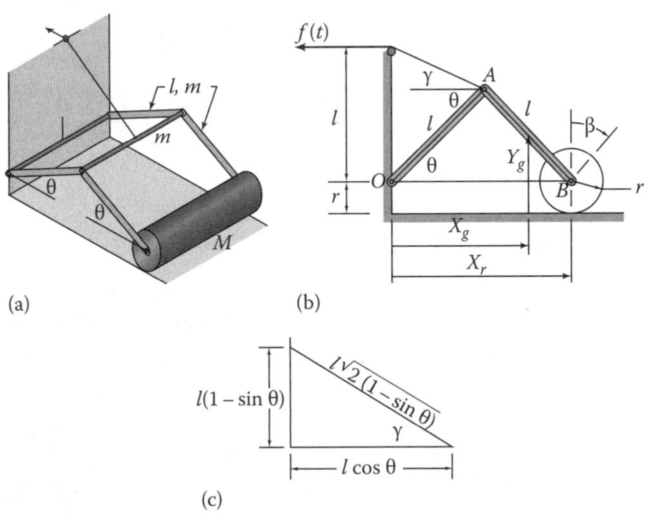

FIGURE 5.39
(a) Parallel, double-bar arrangement for retracting a cylinder. (b) Side view. (c) Load-angle geometry.

Because of the external force $f(t)$, energy is not conserved, and we will need the general work–energy equation $\text{Work}_{n.c.}=\Delta(T+V)$. Looking back at Figure 5.39a, you would expect to define the kinetic energy of the left-hand bars by $T=I_O\dot{\theta}^2/2$ because they are rotating about a fixed axis. The right-hand bars and the cylinder are translating and rotating and will require the general kinetic energy equation $T=I_g\dot{\theta}^2/2+m\left(\dot{X}_g^2+\dot{Y}_g^2\right)/2$. The top transverse bar's velocity is $v_{\dot{\theta}}=l\dot{\theta}$. From the geometry of Figure 5.39b, the kinetic energy is

$$T=2\times\frac{I_O\dot{\theta}^2}{2}+2\times\left[\frac{I_g\dot{\theta}^2}{2}+\frac{m}{2}\left(\dot{X}_g^2+\dot{Y}_g^2\right)\right]$$
$$+\frac{m}{2}(l\dot{\theta})^2+\left(\frac{I_B\dot{\beta}^2}{2}+\frac{M\dot{X}_g^2}{2}\right). \qquad (5.191)$$

The first, second, third, and fourth terms define the kinetic energy of the left-hand bars, the right-hand bars, the transverse bar, and the cylinder, respectively. From Figure 5.39b,

$$X_g=\frac{3l}{2}\cos\theta\Rightarrow\dot{X}_g=-\frac{3l}{2}\sin\theta\dot{\theta}$$
$$Y_g=\frac{l}{2}\sin\theta\Rightarrow\dot{Y}_g=\frac{l}{2}\cos\theta\dot{\theta} \qquad (5.192)$$
$$X_r=2l\cos\theta\Rightarrow\dot{X}_r=-2l\sin\theta\dot{\theta}$$

The rolling-without-slipping condition is

$$r\dot{\beta} = \dot{X}_r = -2l\sin\theta\dot{\theta} \Rightarrow \dot{\beta} = -\frac{2l}{r}\sin\theta\dot{\theta}. \quad (5.193)$$

Substituting from Equations 5.192 and 5.193 plus $I_O = ml^2/3$, $I_g = ml^2/12$, and $I_B = Mr^2/2$ into Equation 5.191 gives

$$\begin{aligned} T &= \dot{\theta}^2\left[\frac{ml^2}{3} + \frac{ml^2}{12} + ml^2(1 + 2\sin^2\theta)\right. \\ &\quad \left. + \frac{ml^2}{2} + \frac{Mr^2}{4}\cdot\frac{4l^2\sin^2\theta}{r^2} + \frac{M}{2}\cdot 4l^2\sin^2\theta\right] \\ &= \dot{\theta}^2 l^2\left[\frac{23m}{12} + (2m + 3M)\sin^2\theta\right]. \end{aligned}$$

Defining the datum for gravity potential energy by a horizontal plane through O and B yields

$$V = 4 \times w\frac{l}{2}\sin\theta + wl\sin\theta = 3wl\sin\theta.$$

Working out the differential work function due to the force $f(t)$ acting at A is a little complicated. The force acts at the angle γ from the horizontal. Figure 5.39c shows the right triangle defining γ and provides

$$\sin\gamma = \frac{(1 - \sin\theta)}{\sqrt{2(1 - \sin\theta)}}, \quad \cos\gamma = \frac{\cos\theta}{\sqrt{2(1 - \sin\theta)}}.$$

Hence, the force vector is $\boldsymbol{f} = f(t)(-\boldsymbol{I}\cos\gamma + \boldsymbol{J}\sin\gamma)$. The force acts at A in position $\boldsymbol{r} = l(\boldsymbol{I}\cos\theta + \boldsymbol{J}\sin\theta)$, and the differential nonconservative work is

$$\begin{aligned} d\text{Work}_{n.c.} &= \boldsymbol{f}\cdot d\boldsymbol{r} = f(t)(-\boldsymbol{I}\cos\gamma + \boldsymbol{J}\sin\gamma) \\ &\quad \cdot l(-\boldsymbol{I}\sin\theta + \boldsymbol{J}\cos\theta)\, d\theta \\ &= \frac{f(t)l[\sin\theta\cos\theta + \cos\theta(1 - \sin\theta)]}{\sqrt{2(1 - \sin\theta)}}\, d\theta \\ &= \frac{f(t)l\cos\theta\, d\theta}{\sqrt{2(1 - \sin\theta)}} = Q_\theta\, d\theta. \end{aligned}$$

Substituting for T, V, and $d\text{Work}_{n.c.}$ into the work–energy equation gives

$$\int_{\theta_0}^{\theta} \frac{f(t)l\cos x\, dx}{\sqrt{2(1 - \sin x)}} = \dot{\theta}^2 l^2\left[\frac{23m}{12} + (2m + 3M)\sin^2\theta\right] + 3wl\sin\theta, \quad (5.194)$$

and concludes Task a.

Task b is initiated by setting $f(t) = \bar{f}$ and completing the integral in Equation 5.194 for θ over the interval $[0, \pi/2]$. In fact, the work done by a constant force $\bar{f}$ for this change in θ can be calculated by inspection. The constant force acts through the distance $l\sqrt{2}$ as the assembly folds up developing the nonconservative work $\text{Work}_{n.c.} = \bar{f}l\sqrt{2}$. Substituting this result and $T_0 = V_0 = 0$ into Equation 5.194 gives

$$\dot{\theta}_f = \dot{\theta}\left(\theta = \frac{\pi}{2}\right) = \left[\left(\bar{f}\sqrt{2} - 3w\right)/l\left(\frac{47m}{12} + 3M\right)\right]^{1/2}$$

and concludes Task b.

Differentiating Equation 5.194 with respect to θ gives

$$\begin{aligned} &\ddot{\theta}l^2\left[\frac{23m}{6} + 2(2m + 3M)\sin^2\theta\right] \\ &\quad + 2\dot{\theta}^2 l^2(2m + 3M)\sin\theta\cos\theta \\ &\quad + 3wl\cos\theta = Q_\theta = \frac{f(t)l\cos\theta}{\sqrt{2(1 - \sin\theta)}}, \end{aligned}$$

where $\ddot{\theta} = d(\dot{\theta}^2/2)/d\theta$, and concludes Task c.

5.7.3.4 *Closing Comments*

Keep in mind that the examples of this subsection are not representative of dynamics problems in general. Specifically, multi-body, single-coordinate examples are not that common in dynamics. Multi-body problems normally involve more than one coordinate, and the work–energy equation can only be used directly to derive an equation of motion for one coordinate.

The example of Figure 5.37 shows quickly that two coordinates (with one degree of freedom) seriously complicates the development of an equation of motion using the work–energy equation. This type of problem is much more easily attacked with Lagrange multipliers and Lagrange equations of motion. The techniques for developing and using generalized forces carries over directly to Lagrange's equations for multi-degree-of-freedom examples.

5.7.4 Examples Having More Than One Degree of Freedom

The examples of Sections 5.5 and 5.6 (rotation about a fixed axis) had one coordinate and one degree of freedom, and the examples of the immediately preceding section also had one degree of freedom. The rolling-without-slipping examples of Section 5.7.1 had two coordinates but (because of the rolling-without-slipping kinematic condition) had only one degree of freedom. In this section, we return to the topic of Section 3.5; namely, systems with more than one degree of freedom. We start with torsional vibration problems that are defined by comparatively simple linear equations of motion and then proceed through several examples with nonlinear governing equations of motion that can be linearized to obtain linear vibration equations. Most of the examples to be considered have two degrees of freedom; one torsional vibration example has three. Obviously, planar mechanical systems can have many degrees of freedom; however, the procedures for attacking systems with two degrees of freedom are generally transferrable to systems with multiple degrees of freedom.

5.7.4.1 Torsional Vibration Examples

We worked through a 1DOF, torsional vibration example in Section 5.5.5, starting with the model of Figure 5.11. Figure 5.40a illustrates a 2DOF extension to this example. The upper disk has mass m_1, radius R_1, and is connected to "ground" by a circular shaft of radius r_1 length l_1, and shear modulus G_1. The lower disk has mass m_2, radius R_2, and is connected to disk 1 by a circular shaft of radius r_2, length l_2, and shear modulus G_2. As shown in Figure 5.40a, the rotation angles θ_1 and θ_2 define the two disks' orientations. The shafts have zero elastic deflections and moments when these angles are zero. Twisting the upper disk through θ_1 will develop the reaction moment,

$$\overline{M}_1 = -k_{\theta 1}\theta_1 = -\frac{G_1 J_1}{l_1}\theta_1 = -\frac{G_1}{l_1}\frac{\pi r_1^4}{2}\theta_1, \tag{5.195}$$

acting on top of the upper disk. The negative sign in this equation implies that the moment is acting in a direction opposite to a $+\theta_1$ rotation. The reaction moment acting on the bottom of disk 1 and the top of disk 2 is proportional to the difference between θ_1 and θ_2. Assuming

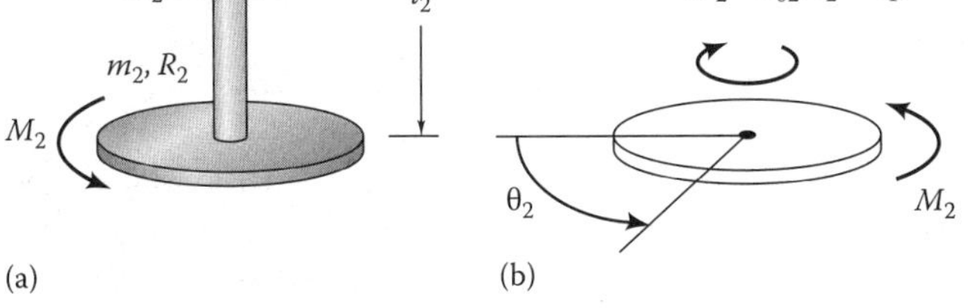

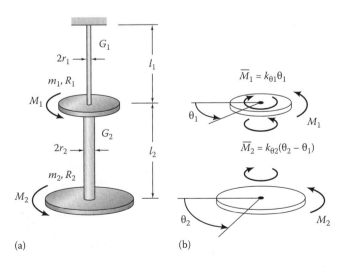

FIGURE 5.40
(a) Two-disk, torsional vibration example. (b) Coordinates and free-body diagram for $\theta_2 > \theta_1$.

$\theta_2 > \theta_1$, the reaction moment acting *on disk* 1 from the lower shaft is

$$\overline{M}_2 = k_{\theta 2}(\theta_2 - \theta_1) = \frac{G_2 J_2}{l_2}(\theta_2 - \theta_1) = \frac{G_2}{l_2}\cdot\frac{\pi r_2^4}{2}(\theta_2 - \theta_1). \tag{5.196}$$

The positive sign for the moment implies that it is acting in the $+\theta_1$ direction, that is, acting to rotate disk 1 in a $+\theta_1$ direction. The negative of this moment acts on the top of disk 2. As shown in Figure 5.40, the applied moments $M_1(t)$ and $M_2(t)$ are acting, respectively, on disks 1 and 2. Individually summing moments about the axis of symmetry for the two bodies yields

$$\begin{aligned} I_{g1}\ddot{\theta}_1 &= \frac{m_1 R_1^2}{2}\ddot{\theta}_1 = M_1(t) - \overline{M}_1 + \overline{M}_2 \\ &= M_1(t) - k_{\theta 1}\theta_1 + k_{\theta 2}(\theta_2 - \theta_1) \\ I_{g2}\ddot{\theta}_2 &= \frac{m_2 R_2^2}{2}\ddot{\theta}_2 = M_2(t) - \overline{M}_2 = M_2(t) - k_{\theta 2}(\theta_2 - \theta_1). \end{aligned}$$

The matrix statement of these equations is

$$\begin{bmatrix} I_{g1} & 0 \\ 0 & I_{g2} \end{bmatrix}\begin{Bmatrix} \ddot{\theta}_1 \\ \ddot{\theta}_2 \end{Bmatrix} + \begin{bmatrix} (k_{\theta 1} + k_{\theta 2}) & -k_{\theta 2} \\ -k_{\theta 2} & k_{\theta 2} \end{bmatrix}\begin{Bmatrix} \theta_1 \\ \theta_2 \end{Bmatrix} = \begin{Bmatrix} M_1(t) \\ M_2(t) \end{Bmatrix}. \tag{5.197}$$

Note that the inertia matrix is diagonal, and the stiffness matrix is symmetric. A stiffness matrix that is not symmetric, or cannot be made symmetric by multiplying one of its rows by a constant, indicates a potentially unstable system. The present system is neutrally stable; that is, the inertia and torsional stiffness elements can store energy but they cannot dissipate energy or introduce energy into the system. For this type of system, stiffness matrices should always be symmetric. To be blunt, for a neutrally stable system, if your development yields a nonsymmetric stiffness matrix, you have made a mistake in developing the equations of motion.

Equations 5.197 have precisely the same form that we obtained for the two-mass system of Section 3.5. The solution of natural frequencies and mode shapes for two-degree-of freedom vibration problems is covered in detail in Section 3.5.2, and you may wish to review this material before proceeding. Calculating the undamped natural frequencies and mode shapes starts by substituting the assumed solution, $(\theta_1, \theta_2)^T = (a_1, a_2)^T \cos\omega_n t$ into the homogeneous version of Equation 5.197 to obtain

$$\begin{bmatrix} [-\omega_n^2 I_{g1} + (k_{\theta 1} + k_{\theta 2})] & -k_{\theta 2} \\ -k_{\theta 2} & (-\omega_n^2 I_{g2} + k_{\theta 2}) \end{bmatrix}\begin{Bmatrix} a_1 \\ a_2 \end{Bmatrix}\cos\omega t = 0. \tag{5.198}$$

For a nontrivial solution ($a_1, a_2 \neq 0$) the determinant of the coefficient matrix in Equation 5.198 must be zero, yielding the characteristic equation

$$\omega_n^4 I_{g1} I_{g2} - \omega_n^2 \left[(k_{\theta 1} + k_{\theta 2}) I_{g2} + k_{\theta 2} I_{g1} \right] + k_{\theta 1} k_{\theta 2} = 0. \tag{5.199}$$

A numerical example is defined by

$$\begin{aligned} G &= 8.27 \times 10^{10} \text{ Pa}, \quad l_1 = l_2 = .3 \text{ m}, \\ r_1 &= r_2 = .0119 \text{ m} \Rightarrow k_{\theta 1} = k_{\theta 2} = 6184 \text{ N m/rad} \\ m_1 &= 5 \text{ kg}, \quad R_1 = .15 \text{ m} \Rightarrow I_{g1} = .05625 \text{ kg}^2 \text{ m}^2 \\ m_2 &= 10 \text{ kg}, \quad R_2 = .15 \text{ m} \Rightarrow I_{g2} = .1125 \text{ kg}^2 \text{ m}^2 \end{aligned} \tag{5.200}$$

Plugging these numbers into Equation 5.199 gives

$$\omega_n^4 - 2.748 \times 10^5 \, \omega_n^2 + 6.043 \times 10^9 = 0 \ (\text{s}^{-8}),$$

with the eigenvalues $\omega_{nl}^2 = 2.410 \times 10^4 \text{ s}^{-2}$; $\omega_{n2}^2 = 2.507 \times 10^5 \text{ s}^{-2}$. The natural frequencies are $\omega_{n1} = 155.$ rad/s and $\omega_{n2} = 501.$ rad/s.

Many torsional examples arise in turbomachinery, that is, turbines, compressors, pumps, etc. A turbomachine rotor is approximately modeled by many "lumped" disks that are connected by massless shafts. We can get an idea of this type of model from Figure 5.41 that shows a two-disk rotor model in a horizontal position supported by frictionless bearings. In contrast to the model of Figure 5.40, note that there is no connection from disk 1 to "ground." Hence, the EOM for the present problem are obtained from Equation 5.197 by setting $k_{\theta 1}$ equal to zero to obtain

$$\begin{bmatrix} I_{g1} & 0 \\ 0 & I_{g2} \end{bmatrix} \begin{Bmatrix} \ddot{\theta}_1 \\ \ddot{\theta}_2 \end{Bmatrix} + \begin{bmatrix} k_{\theta 2} & -k_{\theta 2} \\ -k_{\theta 2} & k_{\theta 2} \end{bmatrix} \begin{Bmatrix} \theta_1 \\ \theta_2 \end{Bmatrix} = \begin{Bmatrix} M_1(t) \\ M_2(t) \end{Bmatrix}. \tag{5.201}$$

Substituting the assumed solution, $(\theta_1, \theta_2)^T = (a_1, a_2)^T \cos \omega_n t$ into the homogeneous version of Equation 5.201 nets

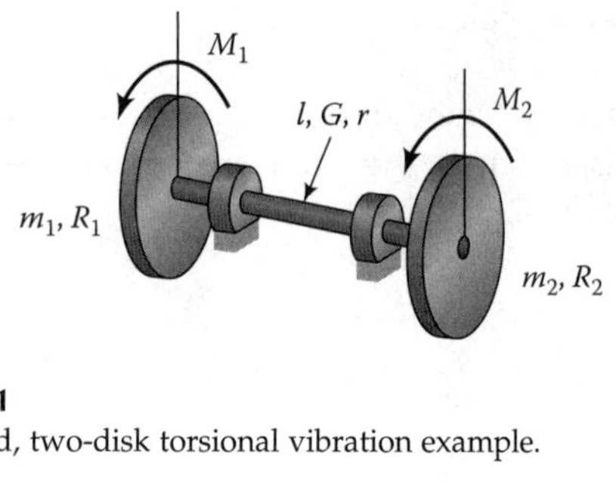

FIGURE 5.41
Unrestrained, two-disk torsional vibration example.

$$\begin{bmatrix} (-\omega_n^2 I_{g1} + k_{\theta 2}) & -k_{\theta 2} \\ -k_{\theta 2} & (-\omega_n^2 I_{g2} + k_{\theta 2}) \end{bmatrix} \begin{Bmatrix} a_1 \\ a_2 \end{Bmatrix} = 0. \tag{5.202}$$

For $k_{\theta 1} = 0$, Equation 5.199 reduces to

$$\omega_n^4 I_{g1} I_{g2} - \omega_n^2 k_{\theta 2} (I_{g1} + I_{g2}) = 0 \Rightarrow \omega_{nl}^2 = 0; \ \omega_{n2}^2 = \frac{k_{\theta 2}(I_{g1} + I_{g2})}{I_{g1} I_{g2}}.$$

For the numbers in Equation 5.200, the eigenvalues are $\omega_{n1}^2 = 0$, $\omega_{n2}^2 = 1.649 \times 10^5 \text{ s}^{-2}$, and the natural frequencies are $\omega_{n1} = 0$, $\omega_{n2} = 406$ rad/s. The $\omega_{nl}^2 = 0$ result differs from prior results in this book. The physical interpretation of this outcome is assisted by reviewing the associated matrix of eigenvectors and the modal differential equations. Substituting $\omega_{n1}^2 = 0$ into Equation 5.202, gives

$$\begin{bmatrix} 6184 & -6184 \\ -6184 & 6184 \end{bmatrix} \begin{Bmatrix} a_{11} \\ a_{21} \end{Bmatrix} = 0.$$

The determinant of the coefficient matrix is clearly zero, and the first eigenvector can be defined from either scalar equation by setting $a_{11} = 1$ to obtain $a_{21} = 1$. The first eigenvector is

$$\begin{Bmatrix} a_{11} \\ a_{21} \end{Bmatrix} = \begin{Bmatrix} 1 \\ 1 \end{Bmatrix}.$$

The second mode is obtained by substituting the second eigenvalue into Equation 5.202—plus substituting for I_{g1}, I_{g2} and $k_{\theta 2}$ from Equation 5.200—to obtain

$$\begin{bmatrix} -3092 & -6184 \\ -6184 & -12368 \end{bmatrix} \begin{Bmatrix} a_{12} \\ a_{22} \end{Bmatrix} = 0.$$

Again, the determinant of the coefficient matrix is zero, and the second eigenvector is

$$\begin{Bmatrix} a_{12} \\ a_{22} \end{Bmatrix} = \begin{Bmatrix} 1 \\ -0.5 \end{Bmatrix}.$$

The matrix of eigenvectors is

$$[A] = \begin{bmatrix} 1 & 1 \\ 1 & -0.5 \end{bmatrix}.$$

The first step in obtaining the modal differential equations is taken by introducing the modal coordinates, via the coordinate transformation, $(\theta_i) = [A](q_i)$,

$$\begin{Bmatrix} \theta_1 \\ \theta_2 \end{Bmatrix} = \begin{bmatrix} A_{11} & A_{12} \\ A_{12} & A_{22} \end{bmatrix} \begin{Bmatrix} q_1 \\ q_2 \end{Bmatrix} = \begin{bmatrix} 1 & 1 \\ 1 & -0.5 \end{bmatrix} \begin{Bmatrix} q_1 \\ q_2 \end{Bmatrix}.$$

Substituting $(\theta_i) = [A](q_i)$ into Equation 5.201 and then premultiplying by the transpose matrix $[A]^T$ gives the uncoupled modal differential equations:

$$(I_{g1} + I_{g2})\ddot{q}_1 = M_1(t) + M_2(t)$$
$$(I_{g1} + 0.25I_{g2})\ddot{q}_1 + 10822q_2 = M_1(t) - 0.5M_2(t).$$

Observe that the first modal coordinate q_1 (with the zero eigenvalue and zero natural frequency) defines rigid-body rotation of the rotor with zero relative rotation between θ_1 and θ_2. The second modal coordinate q_2 defines relative motion with the two disks moving in opposite directions. Motion of the rotor consists of a rigid-body rotation, defined by $q_1(t)$, plus relative rotations between the disks defined by $q_2(t)$.

So far, the torsion examples present an exact parallel to the two-degree-of-freedom, displacement vibration examples of Section 3.5.2. The displacement, two-mass vibration model of Equation 3.124 would have precisely the same type of combined rigid-body/relative motion as the example of Equation 5.201, if the displacement spring attachments to ground are eliminated ($k_1 = k_3 = 0$). However, as demonstrated in Figure 5.42, torsional vibration systems present one complication that has no parallel in displacement-vibration models.

Figure 5.42a shows two rotors, each with two disks connected by a shaft. The rotations of disks 1 and 2 on rotor 1 are defined by the counterclockwise rotations θ_1, θ_2. The rotations of disks 2′ and 3 of rotor 2 are defined by the clockwise rotations θ'_2, θ_3. The radii of disks 2 and 2′ are, respectively, R_2, R'_2. The two disks are "geared" together with no relative slipping between their edges; hence, θ_1 and θ_2 are related by the kinematic constraint equations:

$$R_2\theta_2 = R'_2\theta'_2 \Rightarrow R_2\dot{\theta}_2 = R'_2\dot{\theta}'_2 \Rightarrow R_2\ddot{\theta}_2 = R'_2\ddot{\theta}'_2. \quad (5.203)$$

As illustrated in Figure 5.42b, the second of these equations requires that the individual disk circumferential velocity components be equal at their point of contact.

Figure 5.42c illustrates the reaction force $\bar{f}$ acting between disks 2 and 2′. Stating the equations of motion for the disks of the two rotors gives

Rotor 1 (+ counterclockwise rotations and moments)

$$\begin{aligned} I_{g1}\ddot{\theta}_1 + k_{\theta 1}(\theta_1 - \theta_2) &= 0 \\ I_{g2}\ddot{\theta}_2 + k_{\theta 1}(\theta_2 - \theta_1) &= -\bar{f}R_2 \end{aligned} \quad (5.204)$$

Rotor 2 (+ clockwise rotations and moments)

$$\begin{aligned} I'_{g2}\ddot{\theta}'_2 + k_{\theta 2}(\theta'_2 - \theta_3) &= \bar{f}R'_2 \\ I_{g3}\ddot{\theta}_3 + k_{\theta 2}(\theta_3 - \theta'_2) &= 0. \end{aligned} \quad (5.205)$$

Equations 5.203 through 5.205 comprise five equations in the five unknowns $(\ddot{\theta}_1, \ddot{\theta}_2, \ddot{\theta}'_2, \ddot{\theta}_3, \bar{f})$. Equating $\bar{f}$ in the second of Equations 5.204 and the first of Equations 5.205 eliminates this variable. Also, Equation 5.203 can be used to eliminate θ'_2, $\ddot{\theta}'_2$ yielding the following three coupled equations:

$$I_{g1}\ddot{\theta}_1 + k_{\theta 1}(\theta_1 - \theta_2) = 0$$
$$\left[I_{g2} + (R_2/R'_2)^2 I'_{2g}\right]\ddot{\theta}_2 - k_{\theta 1}\theta_1 + \theta_2\left[k_{\theta 1} + k_{\theta 2}(R_2/R'_2)^2\right] - k_{\theta 2}(R_2/R'_2)\theta_3 = 0$$
$$I_{g3}\ddot{\theta}_3 + k_{\theta 2}\left[\theta_3 - (R_2/R'_2)\theta_2\right] = 0.$$

These equations define a 3DOF model with the matrix statement

$$\begin{bmatrix} I_{g1} & 0 & 0 \\ 0 & I_{g2} + I'_{g2}(R_2/R'_2)^2 & 0 \\ 0 & 0 & I_{g3} \end{bmatrix} \begin{Bmatrix} \ddot{\theta}_1 \\ \ddot{\theta}_2 \\ \ddot{\theta}_3 \end{Bmatrix} + \begin{bmatrix} k_{\theta 1} & -k_{\theta 1} & 0 \\ -k_{\theta 1} & \left[k_{\theta 1} + k_{\theta 2}(R_2/R'_2)^2\right] & -k_{\theta 2}(R_2/R'_2) \\ 0 & -k_{\theta 2}(R_2/R'_2) & k_{\theta 2} \end{bmatrix} \begin{Bmatrix} \theta_1 \\ \theta_2 \\ \theta_3 \end{Bmatrix} = 0. \quad (5.206)$$

Aside from having an additional degree of freedom, the motion predicted by this equation is quite similar to the 2DOF vibration problem of Equation 5.201. Similar (R_2/R'_2) factors were developed in the equations of

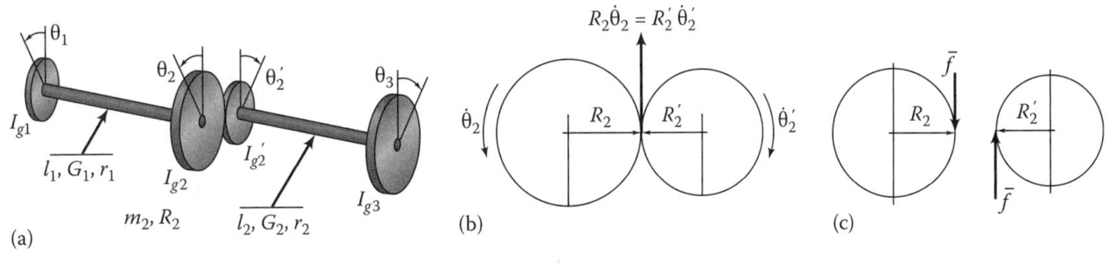

FIGURE 5.42
(a) Two, two-disk rotors with coupled motion. (b) Velocity kinematic constraint between disks 2 and 2′. (c) Reaction-force components between disks 2 and 2′.

motion for the pulley connected by a belt in Figure 5.15. Equations 5.206 yield a cubic frequency equation in the eigenvalue ω_n^2. Because neither of the component rotors is connected to ground, one of the eigenvalues is zero, and the remaining two roots (and their associated eigenvectors) can be determined analytically.

We have not discussed damping for torsional vibration problems. To be direct, there is a great deal of uncertainty involved in the appropriate torsional damping levels provided by turbomachinery components. Most torsional vibration problems involve very low levels of damping, and steady operation with a torsional excitation on or near a torsional natural frequency can result in a failed shaft. A diagonal damping matrix would normally be appropriate for torsional vibration problems, because the damping is concentrated at a disk or support bearing. Hence the damping is a direct function of the local angular velocity versus the relative angular velocity between adjacent disks.

5.7.4.2 Beams as Springs: Bending Vibration Examples

Beams are regularly used as elements in machines and structures. In turbomachinery (pumps, compressors, turbines, etc.), a machine's flexible rotor can be modeled as a collection of rigid bodies (short solid cylinders or disks) connected by massless beams. Figure 5.43 illustrates a rotor model consisting of two rigid disks attached to each other by flexible but massless beams. The springs connecting the model to ground represent the bearings that support the rotor in its housing. This type of model is used to calculate a rotor's multiple natural frequencies, and a considerable effort is expended in designing the machine to avoid steady running speeds that are near one of its natural frequencies. The coincidence of a rotor's running speed ω with a lateral natural frequency ω_{ni} of the rotor/bearing system is called a critical speed.

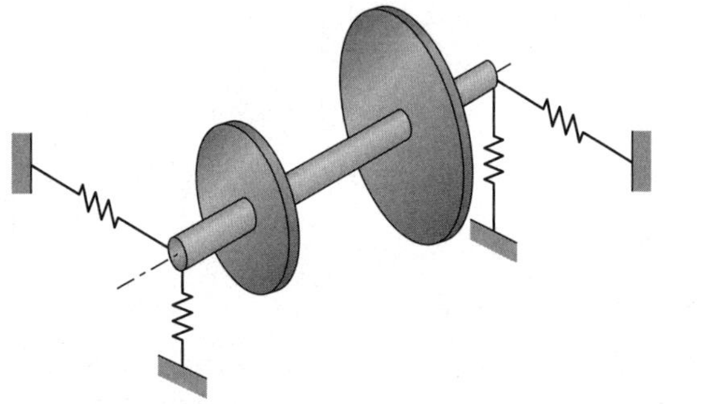

FIGURE 5.43
Lumped-parameter rotor model including two disks.

Beams are also regularly encountered in civil-engineering structures such as bridges, steel-framed buildings, etc. A model to predict the response of a bridge to a moving vehicle or a building's motion due to earthquake excitation would normally involve multiple beam elements. Offshore drilling platforms include beam construction, and these structures must be analyzed and designed to withstand heavy dynamic loads due to wave action.

We will start with the simple cantilever beam of Figure 5.44a that is supporting a disk at its end via an inextensible cable. The beam is a slender "Euler beam." Leonhard Euler (1707–1803) was an incredibly innovative and productive mathematician, who developed the differential equation that defines the deflected shape of a slender beam. As shown, the disk is in equilibrium. We want the EOM resulting from pulling the disk down from its equilibrium position and releasing it. Figure 5.44b provides the free-body diagram for the coordinate δ. The weight is not shown in the free-body diagram because the displacement is away from the equilibrium position. Applying $\sum f = m\ddot{\boldsymbol{R}}_g$, the EOM is

$$m\ddot{\delta} = -k_p\delta \tag{5.207}$$

We now need to solve for k_p. As you may recall from your mechanics of materials course, the displacement at the end of a cantilever beam due to a load P is $\delta = Pl^3/3EI_a$, where l is the beam length, and EI_a is the "section modulus." Further, E is the modulus of elasticity and I_a is the beam's cross-sectional area moment of inertia about its neutral axis. (For a rectangular beam cross section with base b and height h, $I_a = bh^3/12$.) $l^3/3EI_a$ is the flexibility coefficient for the end of the

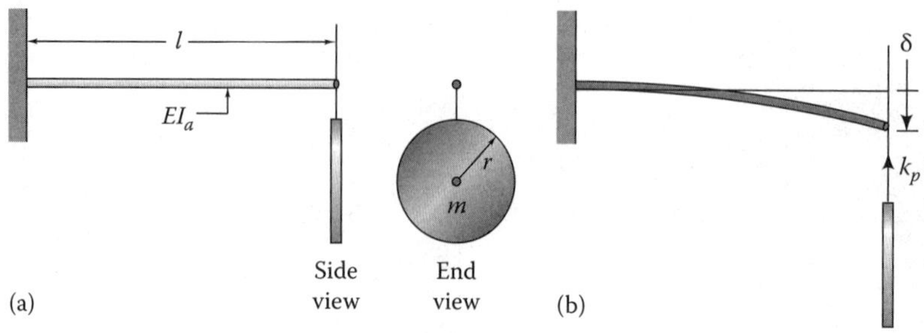

FIGURE 5.44
(a) Cantilever beam with length l and section modulus EI_a supporting a disk of mass m (in equilibrium). (b) Free-body diagram.

beam, defined as the displacement of the beam's end due to a unit load. The stiffness coefficient is obtained by inverting the flexibility coefficient equation to obtain $P = \delta 3EI/l^3$. Hence, the lateral stiffness at the end of a moment-free (pinned) cantilever beam is

$$k_p = \frac{3EI_a}{l^3}, \qquad (5.208)$$

and the EOM is

$$m\ddot{\delta} + \frac{3EI}{l^3}\delta = 0 \qquad (5.209)$$

Note in deriving this equation that we are implicitly assuming small displacements such that the cord remains in tension during motion.

Suppose that the beam has length $l = 750$ mm, a circular cross section with diameter $d = 25$ mm. It is made from steel with modulus of elasticity $E = 2.1 \times 10^{11}$ Pa (N/m^2). The disk has radius $r = 250$ mm, thickness $b = 25$ mm and is also made from steel (density $= \gamma = 7830$ kg/m^3). We would like to calculate the natural frequency. As a first step, the bending section modulus and mass are

$$EI_a = E \times \frac{\pi r^4}{4} = 2.1 \times 10^{11}\,(\text{N/m}^2) \times \frac{3.1416(.0125)^4}{4}\,(\text{m}^4)$$
$$= 4030.\ \text{N m}^2$$
$$m = \pi r^2 b\gamma = 3.1416 \times .25^2\ (\text{m}^2) \times .025\ (\text{m}) \times 7830\ (\text{kg/m}^3) = 38.4\ \text{kg}.$$

Hence, the natural frequency is

$$\omega_n = \left(\frac{3EI_a}{ml^3}\right)^{1/2} = \left[\frac{3 \times 4030.\ (\text{N m}^2)}{38.4\ (\text{kg}) \times .75^3\ (\text{m}^3)}\right]^{1/2}$$
$$= 27.3\ \text{rad/s} = 27.3\ \text{s}^{-1}. \qquad (5.210)$$

This simple example is a good refresher concerning a beam as a spring for which the beam has zero moment about its end, irrespective of its displacement. However, deflecting or rotating the end of a beam normally produces both a reaction force and a reaction moment. Determining the stiffness matrix for a cantilever beam incorporating both reaction forces and moments is our current objective.

Figure 5.45a illustrates a slender Euler beam with length l and section modulus EI_a. The beam is cantilevered at its left end (zero slope and rotation) and is supporting a rigidly attached circular disk (radius r, mass m, and transverse mass moment of inertia $I_g = mr^2/4$) at its right end. The displacement and rotation of the disk are defined, respectively, by R_X and β_Y. Figure 5.45b provides the free-body diagram for the disk including the applied force f_X (that can include the weight) and moment M_Y and the reaction force and moment pair, $\bar{f}_X$, $\overline{M}_Y$. Applying $\sum \mathbf{f} = m\ddot{\mathbf{R}}_g$, and $\sum M_{gY} = I_g\ddot{\beta}_Y$, the disk's equations of motion are

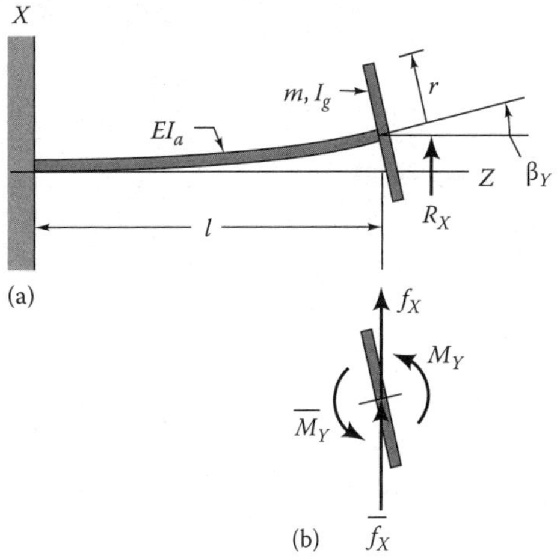

FIGURE 5.45
(a) Cantilevered beam supporting a circular disk at its right end. (b) Disk free-body diagram.

$$m\ddot{R}_X = f_X + \bar{f}_X, \quad I_g\ddot{\beta}_Y = M_Y + \overline{M}_Y. \qquad (5.211)$$

The reaction force and moments are positive in these equations.

Defining the reaction force $\bar{f}_X$ and moment $\overline{M}_Y$ in terms of the displacement and rotation coordinates R_X, β_Y is the principal difficulty in completing Equations 5.211. Figure 5.46a illustrates the beam with a concentrated load f applied at its end, yielding (from strength of materials) the displacement and rotation,

$$R_{X1} = \frac{fl^3}{3EI_a}, \quad \beta_{Y1} = \frac{fl^2}{2EI_a}. \qquad (5.212a)$$

Similarly, Figure 5.46b illustrates the result of applying the moment M to the beam's end, yielding

$$R_{X2} = \frac{Ml^2}{2EI_a}, \quad \beta_{Y2} = \frac{Ml}{EI_a}. \qquad (5.212b)$$

Combining these results as $R_X = R_{X1} + R_{X2}$ and $\beta_Y = \beta_{Y1} + \beta_{Y2}$ gives

$$\begin{Bmatrix} R_X \\ \beta_Y \end{Bmatrix} = \begin{bmatrix} l^3/3EI_a & l^2/2EI_a \\ l^2/2EI_a & l/EI_a \end{bmatrix} \begin{Bmatrix} f \\ M \end{Bmatrix}. \qquad (5.213)$$

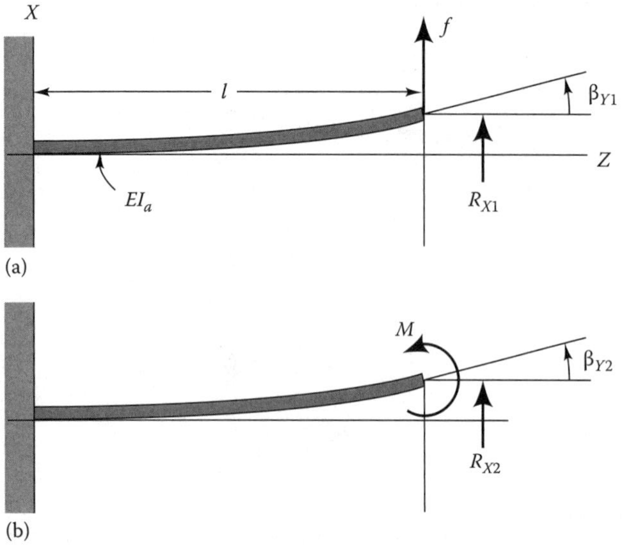

FIGURE 5.46
(a) Cantilevered beam with an applied end force. (b) Applied end moment.

The coefficient matrix is a "flexibility" matrix $[F]$. An F_{ij} flexibility-matrix entry is the displacement (or rotation) at point i due to a unit load (or moment) at point j. Multiplying through by $[F]^{-1}$ gives

$$\begin{Bmatrix} f \\ M \end{Bmatrix} = \begin{bmatrix} 12EI_a/l^3 & -6EI_a/l^2 \\ -6EI_a/l^2 & 4EI_a/l \end{bmatrix} \begin{Bmatrix} R_X \\ \beta_Y \end{Bmatrix}.$$

This coefficient matrix is the stiffness matrix $[k]=[F]^{-1}$. This equation is a little difficult to interpret in terms of *applied* loads (f_X, M_Y) and (R_X, β_Y), since we are (properly) accustomed to thinking of the (f_X, M_Y) as the input variables and (R_X, β_Y) as the output variables. However, it is very useful in defining the reaction forces and moments $(\bar{f}_X, \bar{M}_Y)$ as output variables due to displacements and rotations. With this viewpoint, the reaction force and moment of Figure 5.45b for the X, Z plane are

$$\begin{Bmatrix} \bar{f}_X \\ \bar{M}_Y \end{Bmatrix} = -[k_{XZ}] = -\begin{bmatrix} 12EI_a/l^3 & -6EI_a/l^2 \\ -6EI_a/l^2 & 4EI_a/l \end{bmatrix} \begin{Bmatrix} R_X \\ \beta_Y \end{Bmatrix}. \quad (5.214)$$

Substituting this result into Equations 5.211 gives the matrix equation of motion

$$\begin{bmatrix} m & 0 \\ 0 & I_g \end{bmatrix} \begin{Bmatrix} \ddot{R}_X \\ \ddot{\beta}_Y \end{Bmatrix} + \begin{bmatrix} 12EI_a/l^3 & -6EI_a/l^2 \\ -6EI_a/l^2 & 4EI_a/l \end{bmatrix} \begin{Bmatrix} R_X \\ \beta_Y \end{Bmatrix} = \begin{Bmatrix} f_X \\ M_Y \end{Bmatrix}. \quad (5.215)$$

An entry k_{ij} for the stiffness matrix is the negative reaction force (or moment) at point i due to a displacement (or rotation) at station j, *with all other displacements and rotations equal to zero*. As noted previously, the stiffness matrix for a neutrally stable structure is always symmetric. The following example problem illustrates a solution for the natural frequencies and mode shapes for the model defined by Equation 5.215.

Example Problem 5.4

The cantilevered beam of Figure 5.45 has length $l=750$ mm, a circular cross section with diameter $d=25$ mm. It is made from steel with modulus of elasticity $E=2.1\times10^{11}$ Pa (N/m^2). The disk has radius $r=250$ mm, thickness $b=25$ mm and is also made from steel (density $=\gamma=$ 7830 kg/m^3). The following tasks apply:

a. Determine the inertia and stiffness matrices and state the matrix equation of motion.
b. Determine the eigenvalues, natural frequencies, and eigenvectors.

Solution

As a first step, the bending section modulus is

$$EI_a = E\times\frac{\pi r^4}{4} = 2.1\times10^{11}\ \text{N/m}^2 \times \frac{3.1416(0.0125)^4}{4}\ \text{m}^4$$
$$= 4030\ \text{N m}^2.$$

Continuing, the stiffness coefficients are

$$k_{11} = \frac{12EI_a}{l^3} = 12\times4030\ \text{N m}^2\times\frac{1}{0.75^3\ \text{m}^3} = 1.15\times10^5\ \text{N/m}$$
$$k_{12}=k_{21}=\frac{6EI_a}{l^2} = 6\times4030\ \text{N m}^2\times\frac{1}{0.75^2\ \text{m}^2} = 4.30\times10^4\ \text{N}.$$
$$k_{22} = \frac{4EI_a}{l} = 4\times4030\ \text{N m}^2\times\frac{1}{0.75\ \text{m}} = 2.15\times10^4\ \text{N m}$$

The inertia-matrix entries are

$$m=\pi r^2 b\gamma = 3.1416\times0.25^2\,(\text{m}^2)\times0.025\,(\text{m})\times7830\,(\text{kg/m}^3)$$
$$=38.4\ \text{kg}$$
$$I_g = \frac{mr^2}{4} = \frac{38.4\,(\text{kg})\times0.25^2\,(\text{m}^2)}{4} = 0.60\ \text{kg m}^2.$$

Substituting these results into Equation 5.215 defines the model as

$$\begin{bmatrix} 38.4 & 0 \\ 0 & 0.60 \end{bmatrix} \begin{Bmatrix} \ddot{R}_X \\ \ddot{\beta}_Y \end{Bmatrix} + \begin{bmatrix} 1.15\times10^5 & -4.30\times10^4 \\ -4.30\times10^4 & 2.15\times10^4 \end{bmatrix}$$
$$\times \begin{Bmatrix} R_X \\ \beta_Y \end{Bmatrix} = \begin{Bmatrix} f_X \\ M_Y \end{Bmatrix}. \quad \text{(i)}$$

You may want to review Section 3.5.2 if you need a refresher on eigenanalysis. Substituting the assumed solution $(r_X, \beta_Y)^T = (a_1, a_2)^T \cos \omega t$ into the homogeneous version of Equation (i) nets

$$\begin{bmatrix} -38.4\omega^2 + 1.15 \times 10^5 & -4.30 \times 10^4 \\ -4.30 \times 10^4 & -0.60\omega^2 + 2.15 \times 10^4 \end{bmatrix} \times \begin{Bmatrix} a_1 \\ a_2 \end{Bmatrix} \cos \omega t = 0. \quad \text{(ii)}$$

For a nontrivial solution, the determinant of this coefficient matrix must be zero, leading to the characteristic equation

$$23.04\omega^4 - 8.946 \times 10^5\omega^2 + 6.235 \times 10^8 = 0.$$

The eigenvalues and natural frequencies defined by the roots of this equation are

$$\begin{aligned} \omega_{n1}^2 &= 709.9 \text{ s}^{-2} \Rightarrow \omega_{n1} = 26.6 \text{ s}^{-1} \\ \omega_{n2}^2 &= 3.812 \times 10^4 \text{ s}^{-2} \Rightarrow \omega_{n2} = 195.2 \text{ s}^{-1}. \end{aligned} \quad \text{(iii)}$$

Alternately substituting ω_{n1}^2 and ω_{n2}^2 into Equation (ii), the corresponding eigenvectors are

$$\begin{aligned} \begin{Bmatrix} a_1 \\ a_2 \end{Bmatrix}_1 &= \begin{Bmatrix} R_X \\ \beta_Y \end{Bmatrix}_1 = \begin{bmatrix} 1 \\ 2.04 \end{bmatrix}, \\ \begin{Bmatrix} a_1 \\ a_2 \end{Bmatrix}_2 &= \begin{Bmatrix} R_X \\ \beta_Y \end{Bmatrix}_2 = \begin{Bmatrix} 1 \\ -31.4 \end{Bmatrix}. \end{aligned} \quad \text{(iv)}$$

As illustrated in Figure XP5.4, the displacement and rotation are in phase for the first mode and out of phase for the second.

Figure 5.47a provides another variation of the cantilever-beam example, showing the disk constrained by rollers that prevent rotation. For $\beta_Y = \ddot{\beta}_Y = 0$, Equation 5.215 gives

$$m\ddot{R}_X + \frac{12EI_a}{l^3} R_X = f_X, \quad M_Y = -\frac{6EI_a}{l^2} R_X. \quad (5.216)$$

The first equation is the equation of motion for R_X. The second equation defines the reaction moment that the constraint rollers must provide to keep the disk from rotating. Figure 5.47b provides a separate free-body diagram for the reaction force and moment developed by displacing the end of the beam while holding the slope at

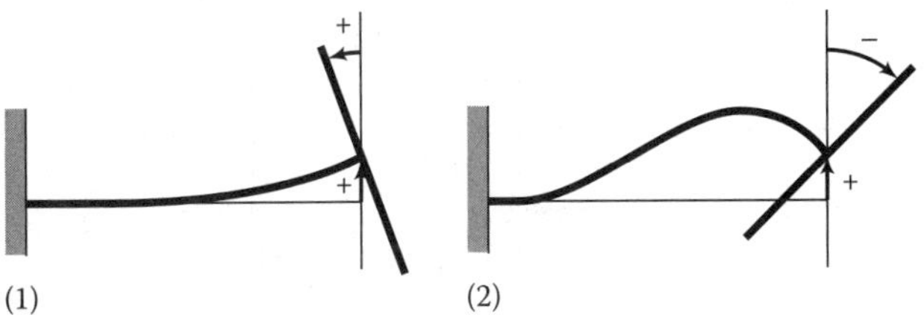

FIGURE XP5.4
Calculated mode shape of Equation (iv); not to scale.

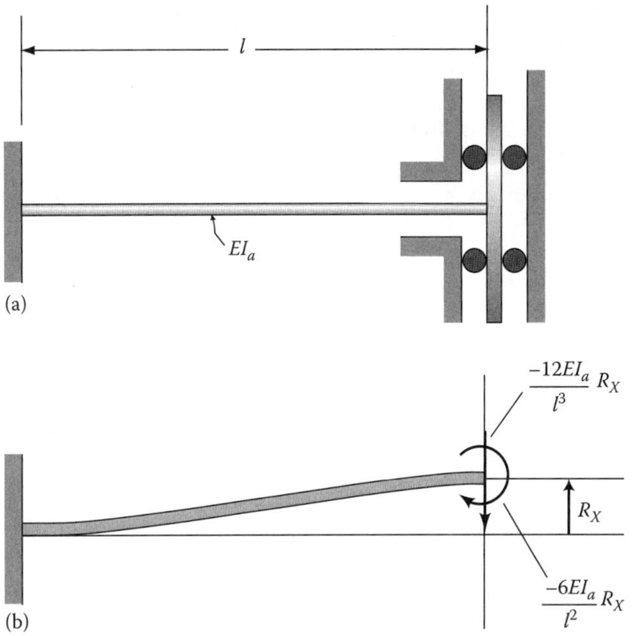

FIGURE 5.47
(a) Cantilever beam with the end disk allowed to move up and down but prevented from rotating. (b) Free-body diagram.

zero. The stiffness of the end of a zero-rotation cantilevered beam is

$$k_c = \frac{12EI_a}{l^3}, \quad (5.217)$$

and it is four times greater than the stiffness at the free end of a cantilevered beam as given in Equation 5.208. Using the data of Example Problem 5.4, the natural frequency is

$$\omega_n = \left(\frac{12EI_a}{ml^3}\right)^{1/2} = \left[\frac{12 \times 4030 \text{ (N/m}^2)}{38.4 \text{ (kg)} \times 0.75^3 \text{ (m}^3)}\right]^{1/2} = 54.6 \text{ s}^{-1}.$$

The moment restraint on the disk has doubled the natural frequency result for the beam of Figure 5.44 where the disk was hung from the end of the beam in a zero-end-moment condition. The message from this example problem is that boundary conditions make a big difference in the natural frequencies of rigid bodies supported by beams.

Example Problem 5.5

The framed structure has two square floors. The first floor has mass $m_1 = 2000$ kg and is supported to the foundation by *four* solid columns with square cross sections. These columns are cantilevered from the foundation and are welded to the bottom of the first floor. The second floor has mass $m_2 = 2000$ kg and is supported from the first floor by *four* solid columns with square cross sections. These columns are welded to the top of the first floor and are hinged to the second floor. The bottom and top columns have length $l_1 = 6$ m and $l_2 = 4$ m. The top and bottom beams cross-sectional dimensions are $b_1 = 100$ mm and $b_2 = 75$ mm. They are made from steel with a modulus of elasticity $E = 2.1 \times 10^{11}$ N/m^2. A model is required to account for motion of the foundation due to earthquake excitation defined by $x(t)$.

Tasks:

a. Select coordinates, draw a free-body diagram, derive the equations of motion.
b. State the equations of motion in matrix format and solve for the eigenvalues and eigenvectors. Draw the eigenvectors.

Solution

Figure XP5.5b illustrates the coordinates $x_1(t)$, $x_2(t)$ selected to locate the first and second floors with respect to ground. All beams connecting the foundation and the first floor are cantilevered at both ends, similar to the beam in Figure 5.47 with a stiffness $k_c = 12EI_a/l^3$. The free-body diagram of Figure XP5.5c was drawn assuming that the first floor has moved further than the ground ($x_1 > x$) and defines the reaction force

$$\bar{f}_{\text{firststory-ground}} = -4k_c(x_1 - x), \quad k_c = \frac{12EI_a}{l^3},$$

due to all four cantilevered beams acting at the bottom of floor 1.

Each beam connecting the floors has a cantilevered end attached to floor 1 and a pinned end attached to floor 2, similar to the pinned-end beam of Figure 5.44, with a stiffness coefficient $k_p = 3EI_a/l^3$. The free-body diagram in Figure XP5.5c was developed assuming that the second floor has moved further than the first floor ($x_2 > x_1$) and provides the following reaction force acting on the bottom of the second floor

$$\bar{f}_{\text{second-ground}} = -4k_p(x_2 - x_1), \quad k_p = \frac{3EI_a}{l^3}.$$

The negative of this force is acting at the top of floor 1. Summing forces for the two floors gives

$$\text{floor1}: m_1\ddot{x}_1 = \sum f_{x1} = -4k_c(x_1 - x) + 4k_p(x_2 - x_1)$$
$$\Rightarrow m_1\ddot{x}_1 + 4(k_c + k_p)x_1 - 4k_p x_2 = 4k_c x$$
$$\text{floor2}: m_2\ddot{x}_2 = \sum f_{x2} = -4k_p(x_2 - x_1) \Rightarrow m_2\ddot{x}_1 - 4_p k_p x_1$$
$$+ 4k_p x_2 = 0$$

Putting these equations in matrix form gives

$$\begin{bmatrix} m_1 & 0 \\ 0 & m_2 \end{bmatrix}\begin{Bmatrix} \ddot{x}_1 \\ \ddot{x}_2 \end{Bmatrix} + 4\begin{bmatrix} k_c + k_p & -k_p \\ -k_p & k_p \end{bmatrix}\begin{Bmatrix} x_1 \\ x_2 \end{Bmatrix} = \begin{Bmatrix} 4k_c x(t) \\ 0 \end{Bmatrix} \tag{i}$$

This outcome is similar to Equation 3.126 for two masses connected by springs.

Filling in the numbers gives

$$EI_{a1} = E \times \frac{b_1 h_1^3}{12} = 2.1 \times 10^{11}\,\frac{\text{N}}{\text{m}^2} \times \frac{0.1^4}{12}\,\text{m}^4$$
$$= 1.75 \times 10^6\ \text{N m}^2$$
$$EI_{a2} = E \times \frac{b_2 h_2^3}{12} = 2.1 \times 10^{11}\,\frac{\text{N}}{\text{m}^2} \times \frac{(0.075)^4}{12}\,\text{m}^4$$
$$= 5.54 \times 10^5\ \text{N m}^2.$$

Continuing, the stiffness coefficients are

$$k_c = \frac{12EI_{a1}}{l^3} = \frac{12(1.75 \times 10^6)\ \text{N m}^2}{(6m^3)} = 97200\ \frac{\text{N}}{\text{m}}$$
$$k_p = \frac{3EI_{a2}}{l^2} = \frac{3(5.54 \times 10^5\ \text{N m}^2)}{(4m)^3} = 26000\ \frac{\text{N}}{\text{m}}.$$

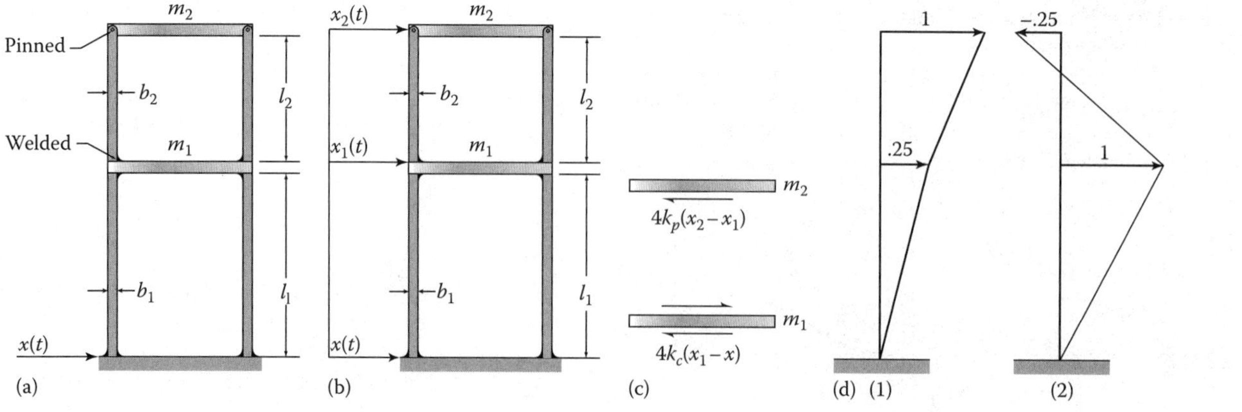

FIGURE XP5.5
(a) Front view of a two-story framed structure excited by base excitation. (b) Coordinates. (c) Free-body diagram for $x_1 > x > 0$, $x_2 > x_1$. (d) Eigenvectors.

Plugging these results into Equation (i) gives

$$\begin{bmatrix} 2000 & 0 \\ 0 & 2000 \end{bmatrix} \begin{Bmatrix} \ddot{x}_1 \\ \ddot{x}_2 \end{Bmatrix} + \begin{bmatrix} 493000 & -104000 \\ -104000 & 104000 \end{bmatrix} \times \begin{Bmatrix} x_1 \\ x_2 \end{Bmatrix} = \begin{Bmatrix} 398000x(t) \\ 0 \end{Bmatrix}. \qquad \text{(ii)}$$

Substituting the assumed solution $(x_1, x_2)^T = (a_1, a_2)^T \cos \omega t$ into the homogeneous version of this equation gives

$$\begin{bmatrix} [-2000\omega^2 + 493000] & -104000 \\ -104000 & [-2000\omega^2 + 104000] \end{bmatrix} \times \begin{Bmatrix} a_1 \\ a_2 \end{Bmatrix} \cos \omega t = \begin{Bmatrix} 0 \\ 0 \end{Bmatrix} \qquad \text{(iii)}$$

Since, $\cos \omega t \neq 0$, and a_1, and a_2 are also not zero, a nontrivial solution, for Equation (iii) requires that the determinant of the coefficient matrix equal zero, producing

$$4 \times 10^6 \omega^4 - 1.194 \times 10^9 \omega^2 + 4.0546 \times 10^{10} = 0$$
$$\Rightarrow \omega^4 - 298.5\omega^2 + 10110 = 0.$$

This characteristic equation defines the two eigenvalues and natural frequencies:

$$\omega_{n1}^2 = 38.95\ (\text{rad/s})^2 \Rightarrow \omega_{n1} = 6.24\ \text{rad/s},$$
$$\omega_{n2}^2 = 259.\ (\text{rad/s})^2 \Rightarrow \omega_{n2} = 16.1\ \text{rad/s}$$

Alternately substituting $\omega_{n1}^2 = 38.95\ (\text{rad/s})^2$ and $\omega_{n2}^2 = 259.5\ (\text{rad/s})^2$ into Equation (iii) gives the eigenvectors

$$(a)_1 = \begin{Bmatrix} 1 \\ 4.0 \end{Bmatrix}, \quad (a)_2 = \begin{Bmatrix} 1 \\ -0.25 \end{Bmatrix}$$

Figure XP5.5d illustrates these two eigenvectors, showing the relative motion of the two floors somewhat better than the two-mass eigenvectors of Figure 3.51.

Figure 5.48a illustrates a cantilever beam supporting a uniform cylinder of radius r_d and length l_d. Figure 5.48b provides the free-body diagram. The reaction force and moment from the cantilever act at the edge of the cylinder, a distance $l_d/2$ from the disk's mass center. The finite length of the cylinder is the main complication in this example versus the thin disk of Figure 5.45. The following tasks apply: *Select coordinates, draw a free-body diagram, and derive the equations of motion.*

For the thin disk of Figure 5.45, the reaction force $\bar{f}_X$ and moment $\overline{M}_Y$ act (approximately) at the disk mass center. For the present example, they act at the left-hand side of the cylinder, a distance $l_d/2$ from the mass center. Moreover, the displacement of the cylinder's mass center R_{Xg} differs from the displacement of the left-hand end of the beam R_X. For small β_Y, they are related via

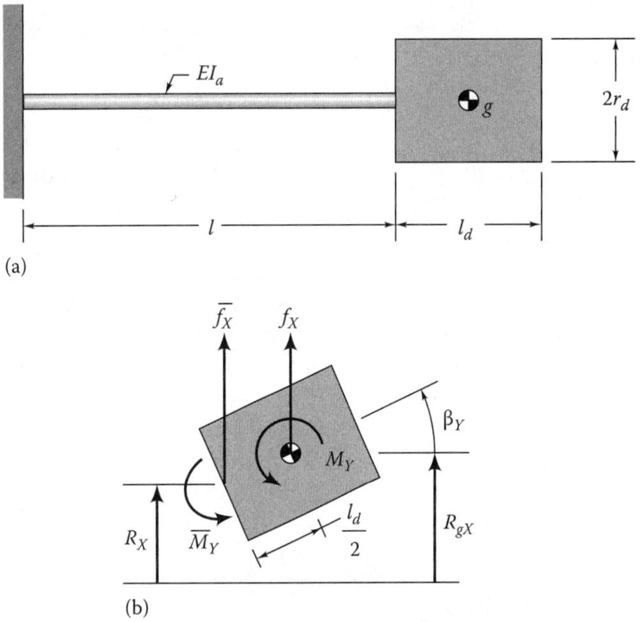

FIGURE 5.48
(a) Shaft supporting a cylinder of length l_d and radius r_d. (b) Free-body diagram.

$$R_{gX} = R_X + \beta_Y \times \frac{l_d}{2} \Rightarrow \ddot{R}_{gX} = \ddot{R}_X + \frac{l_d}{2}\ddot{\beta}_Y$$

where β_Y is the rotation angle of both the right-hand end of the beam and the cylinder. The reaction force $\bar{f}_X$ and moment $\overline{M}_Y$ continue to be defined by Equation 5.214. From the free-body diagram of Figure 5.48b, the equations of motion are

$$\sum f_X = f_X + \bar{f}_X = m\ddot{R}_{gX} = m\left(\ddot{R}_X + \frac{l_d}{2}\ddot{\beta}_Y\right)$$
$$\sum M_g = M_Y + \overline{M}_y - \bar{f}_X \frac{l_d}{2} = I_g \ddot{\beta}_Y. \qquad (5.218)$$

Substituting for $\bar{f}_X$ from the first of these equations into the second nets

$$I_g \ddot{\beta}_Y = M_Y + \overline{M}_y + \frac{l_d}{2}\left[f_X - m\left(\ddot{R}_X + \frac{l_d}{2}\ddot{\beta}_Y\right)\right]$$

$$\text{or} \quad \left[I_g + m\left(\frac{l_d}{2}\right)^2\right]\ddot{\beta}_Y + \frac{ml_d}{2}\ddot{R}_X = M_Y + \overline{M}_Y + \frac{l_d}{2} f_X$$

From the first of Equation 5.218 and this last equation, plus the definitions for the reaction force $\bar{f}_X$ and moment $\overline{M}_Y$ provided by Equation 5.214, the matrix equation of motion for the variables (R_X, β_Y) is

$$\begin{bmatrix} m & ml_d/2 \\ ml_d/2 & \left[I_g + m\left(\frac{l_d}{2}\right)^2\right] \end{bmatrix} \begin{Bmatrix} \ddot{R}_X \\ \ddot{\beta}_Y \end{Bmatrix} + \begin{bmatrix} 12EI_a/l^3 & -6EI_a/l^2 \\ -6EI_a/l^2 & 4EI_a/l \end{bmatrix} \times \begin{Bmatrix} R_X \\ \beta_Y \end{Bmatrix} = \begin{Bmatrix} f_X \\ M_Y + f_X \frac{l_d}{2} \end{Bmatrix}. \quad (5.219)$$

In comparison to Equation 5.215, the inertia matrix continues to be symmetric but is now coupled via the off-diagonal terms. The moment term on the right includes the moment of the external force $f_X l_d/2$ about the connection point of the beam to the disk.

Suppose that we wanted to use (R_{gX}, β_Y) as variables. Substituting $R_X = R_{Xg} - l_d\beta_y/2$, the reaction force would now be defined as

$$\begin{aligned} \bar{f}_X &= -k_{11}\left(R_{gX} - \frac{l_d}{2}\beta_Y\right) - k_{12}\beta_Y \\ &= -k_{11}R_{gX} - \left(k_{12} - \frac{l_d}{2}k_{11}\right)\beta_Y, \end{aligned}$$

with k_{11}, k_{12} defined by Equation 5.214. Substituting this result into the first of Equation 5.218, gives

$$m\ddot{R}_{gX} + k_{11}R_{gX} + \left(k_{12} - \frac{l_d}{2}k_{11}\right)\beta_Y = f_X,$$

and the moment equation becomes

$$\begin{aligned} I_g\ddot{\beta}_Y &= M_Y + \overline{M}_y - \bar{f}_X\frac{l_d}{2} \\ &= M_Y + \left[-k_{21}\left(R_{gX} - \frac{l_d}{2}\beta_Y\right) - k_{22}\beta_Y\right] \\ &\quad - \frac{l_d}{2}\left[-k_{11}R_{gX} - \left(k_{12} - \frac{l_d}{2}k_{11}\right)\beta_Y\right] \\ &= M_Y - \left(k_{21} - \frac{l_d}{2}\right)R_{gX} - \left[k_{22} - \frac{l_d}{2}k_{21} + \left(\frac{l_d}{2}\right)^2 k_{11}\right]\beta_Y \end{aligned}$$

with k_{21}, k_{22} defined by Equation 5.214. The matrix vibration equation is

$$\begin{bmatrix} m & 0 \\ 0 & I_g \end{bmatrix} \begin{Bmatrix} \ddot{R}_{gX} \\ \ddot{\beta}_Y \end{Bmatrix} + \begin{bmatrix} k_{11} & \left(k_{12} - \frac{l_d}{2}k_{11}\right) \\ \left(k_{21} - \frac{l_d}{2}k_{11}\right) & k_{22} - \frac{l_d}{2}k_{21} + \left(\frac{l_d}{2}\right)^2 k_{11} \end{bmatrix} \times \begin{Bmatrix} R_{gX} \\ \beta_Y \end{Bmatrix} = \begin{Bmatrix} f_X \\ M_Y \end{Bmatrix}. \quad (5.220)$$

With these variables, the inertia matrix is unchanged, but the stiffness matrix has been modified. The stiffness matrix of Equation 5.220 defines the reaction forces and moments acting at the cylinder mass center due to the mass center's displacement and rotation. The models provided by Equations 5.219 and 5.220 are equally valid; they simply use different coordinates to model the system.

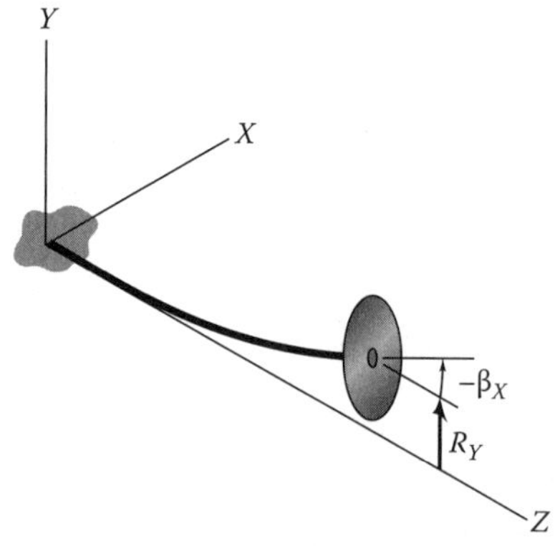

FIGURE 5.49
Shaft supporting a disk in the Y–Z plane.

We have worked in the X, Z plane in this development related to beams as stiffness elements. As illustrated in Figures 5.45 and 5.46, positive f_X force and M_Y moment components produce positive displacements R_X and rotations β_Y. As presented in Figure 5.49, shifting to the Y, Z plane produces different signs for (i) a rotation due to an applied force and (ii) the displacement due to an applied moment. The differences arise because of the right-hand rule for defining positive rotations about the X- and Y-axes. For Figure 5.49, the right-hand rule requires that a clockwise rotation about the X-axis be positive when looking out along X.

Figures 5.50a shows a free-body diagram with an applied force and moment pair (f_Y, M_X) and reaction force and moment pair $(\bar{f}_Y, \overline{M}_X)$ acting on the end disk of Figure 5.49. Applying $\sum \boldsymbol{f} = m\ddot{\boldsymbol{R}}_g$ and $\sum M_{gX} = I_g\ddot{\beta}_X$, the disk's equations of motion are

$$m\ddot{R}_Y = f_Y + \bar{f}_Y, \quad I_g\ddot{\beta}_X = M_X + \overline{M}_X. \quad (5.221)$$

As with the earlier development in the X–Z plane, defining the reaction force and moment pair $(\bar{f}_Y, \overline{M}_X)$ in terms of the displacement R_Y and rotation β_X is the main task in completing the equations of motion. Figures 5.50b and c illustrate the displacement and rotation due to applied forces and moments. Note that the positive applied force f produces a negative β_X rotation. Similarly, a positive moment M will develop a negative R_Y displacement. Hence Equations 5.212a and b become

$$R_{X1} = \frac{fl^3}{3EI_a}, \quad \beta_{Y1} = -\frac{fl^2}{2EI_a}, \quad R_{X2} = -\frac{Ml^2}{2EI_a}, \quad \beta_{Y2} = \frac{M}{EI_a}.$$

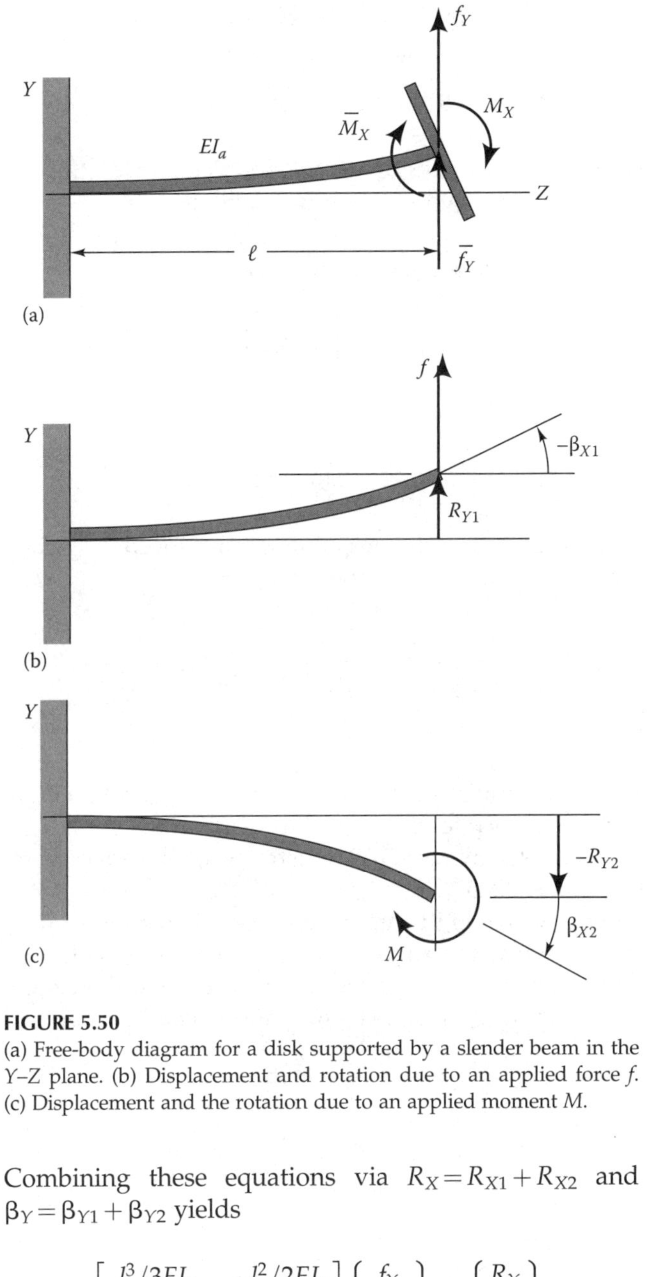

FIGURE 5.50
(a) Free-body diagram for a disk supported by a slender beam in the Y–Z plane. (b) Displacement and rotation due to an applied force f. (c) Displacement and the rotation due to an applied moment M.

Combining these equations via $R_X = R_{X1} + R_{X2}$ and $\beta_Y = \beta_{Y1} + \beta_{Y2}$ yields

$$\begin{bmatrix} l^3/3EI_a & -l^2/2EI_a \\ -l^2/2EI_a & l/EI_a \end{bmatrix} \begin{Bmatrix} f_Y \\ M_X \end{Bmatrix} = \begin{Bmatrix} R_Y \\ \beta_X \end{Bmatrix}.$$

This flexibility matrix has negative off-diagonal terms versus the positive terms of Equation 5.213. Inverting it yields the $[k_{YZ}]$ stiffness matrix, defining the reaction force and moment of Figure 5.50a as

$$\begin{Bmatrix} \bar{f}_Y \\ \bar{M}_X \end{Bmatrix} = -[k_{YZ}] \begin{Bmatrix} R_Y \\ \beta_X \end{Bmatrix} = -\begin{bmatrix} 12EI_a/l^3 & 6EI_a/l^2 \\ 6EI_a/l^2 & 4EI_a/l \end{bmatrix} \begin{Bmatrix} R_Y \\ \beta_X \end{Bmatrix}.$$

The off-diagonal terms are positive versus negative for the corresponding $[k_{XZ}]$ stiffness matrix of Equation 5.214. Substituting this result into Equations 5.221 gives the matrix equation of motion,

$$\begin{bmatrix} m & 0 \\ 0 & I_g \end{bmatrix} \begin{Bmatrix} \ddot{R}_Y \\ \ddot{\beta}_X \end{Bmatrix} + \begin{bmatrix} 12EI_a/l^3 & 6EI_a/l^2 \\ 6EI_a/l^2 & 4EI_a/l \end{bmatrix} \begin{Bmatrix} R_Y \\ \beta_X \end{Bmatrix} = \begin{Bmatrix} f_Y \\ M_X \end{Bmatrix}.$$

Except for the off-diagonal terms in the stiffness matrix, this result coincides with Equation 5.215 for the X–Z plane. You can repeat Example Problem 5.3 to show that the same *physical* results (eigenvalues and eigenvectors) are obtained with the present model and the model of Equation 5.215.

5.7.4.3 The Jeffcott/Laval Rotor Model

Figure 3.18 included a discussion of a 1DOF model for excitation due to a rotating imbalance. That simple model's principal virtue is the simplicity involved in the one degree of freedom. However, one degree of freedom limits the model's value in analyzing real rotating machines. Figure 5.51 illustrates a 2DOF model consisting of a disk supported by two identical flexible beams. The beam's end supports correspond to rigid bearings. The disk's geometric center is located by the vector $\boldsymbol{R} = \boldsymbol{I}R_X + \boldsymbol{J}R_Y$ in the inertial X, Y, Z system. The disk and the x, y, z coordinate system are rotating at the constant speed ω about the bearings' axis. The x, y, z coordinate system is fixed to the disk with z parallel with the bearings' axis. The disk's mass center is displaced from its geometric center by the imbalance vector $\boldsymbol{a} = \boldsymbol{i}a$. We want to predict the steady-state rotor response defined by the components of $\boldsymbol{R}$ due to the rotating imbalance vector $\boldsymbol{a}$.

The model of Figure 5.51 is referred to by most English-speaking engineers as the Jeffcott (1919) model developed by an H.H. Jeffcott. This model was developed earlier by a German named Föppl (1895). Föppl's model and analysis did not include damping; Jeffcott's did. Most engineers on the European continent refer to the Jeffcott–Föppl model as the "Laval model" after the Swedish engineer Carl Laval (1845–1913) one of the early inventors of steam turbines.

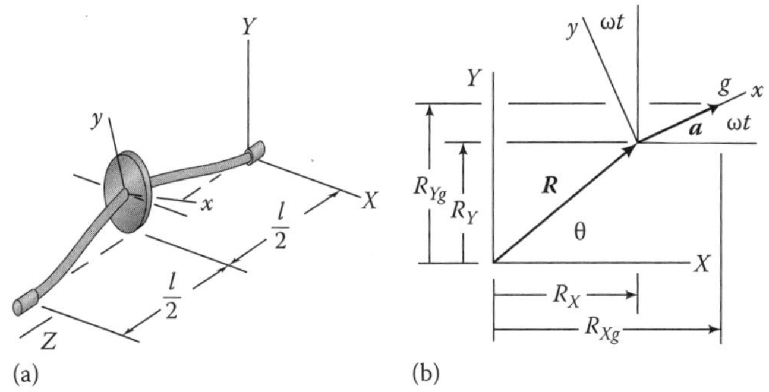

FIGURE 5.51
The Föppl–Jeffcott flexible-rotor model.

The disk's mass center is located in the inertial X, Y, Z system by

$$R_{Xg} = R_X + a\cos\omega t, \quad R_{Yg} = R_Y + a\sin\omega t.$$

Hence, the components of the mass center acceleration are

$$\ddot{R}_{Xg} = \ddot{R}_X - a\omega^2\cos\omega t, \quad \ddot{R}_{Yg} = \ddot{R}_Y - a\omega^2\sin\omega t.$$

Applying $\sum f = m\ddot{\boldsymbol{R}}_g$ for the disk produces

$$\begin{aligned} m(\ddot{R}_X - a\omega^2\cos\omega t) &= -kR_X \Rightarrow m\ddot{R}_X + kR_X = ma\omega^2\cos\omega t \\ m(\ddot{R}_Y - a\omega^2\sin\omega t) &= w - kR_Y \Rightarrow m\ddot{R}_Y + kR_Y \\ &= ma\omega^2\sin\omega t - w. \end{aligned}$$

The stiffness coefficient k can be developed from Equation 5.208. The mass imbalance ma produces a harmonic excitation at the rotation speed ω that is proportional to running-speed squared ω^2. With external viscous damping, the model becomes

$$\begin{aligned} m\ddot{R}_X + c\dot{R}_X + kR_X &= ma\omega^2\cos\omega t \\ m\ddot{R}_Y + c\dot{R}_Y + kR_Y &= ma\omega^2\sin\omega t - w. \end{aligned} \tag{5.222}$$

Neglecting w, the steady-state solutions for these equations can be developed directly; however, the task can be simplified via complex analysis. Multiplying the second of Equations 5.222 by $j = \sqrt{-1}$ and adding it to the first yields

$$\begin{aligned} m(\ddot{R}_X + j\ddot{R}_Y) + c(\dot{R}_X + j\dot{R}_Y) + k(R_X + jR_Y) \\ = ma\omega^2(\cos\omega t + j\sin\omega t), \quad \text{or} \\ \ddot{\boldsymbol{R}} + 2\zeta\omega_n\dot{\boldsymbol{R}} + \omega_n^2\boldsymbol{R} = a\omega^2 e^{j\omega t}, \end{aligned}$$

where $\boldsymbol{R} = R_X + jR_Y$, $\omega_n^2 = k/m$, and $2\zeta\omega_n = c/m$. Substituting the assumed solution $\boldsymbol{R} = \boldsymbol{A}e^{j\omega t}$ gives

$$\begin{aligned} \left[(\omega_n^2 - \omega^2) + j2\zeta\omega\omega_n\right]\boldsymbol{A} &= a\omega^2 \\ \Rightarrow \frac{\boldsymbol{A}}{a} = \frac{\omega^2}{\left[(\omega_n^2 - \omega^2) + j2\zeta\omega\omega_n\right]} &= |J(\omega)|e^{j\psi}, \end{aligned}$$

and the solution is

$$\boldsymbol{R} = a|J(\omega)|e^{j(\omega t + \psi)}.$$

In terms of the nondimensional rotating speed, $r = \omega/\omega_n$, the amplification factor and phase are

$$|J(r)| = \frac{r^2}{\left[(1 - r^2)^2 - 4\zeta^2 r^2\right]^{1/2}}, \quad \tan\psi = -\frac{2\zeta r}{(1 - r^2)}. \tag{5.223}$$

This amplification-factor result coincides with Equation 3.55 developed earlier for the 1DOF rotor model of Section 3.2.5. Figure 5.52 illustrates $|J(r)|$ and ψ from Equations 5.223.

Observe that high amplitudes arise when $r = \omega/\omega_n = 1$, that is, when the running speed ω coincides with ω_n. This circumstance is called a "critical speed." At high speeds ($r \gg 1$), note that $|J(r)|$ approaches 1 implying that $A \cong a$.

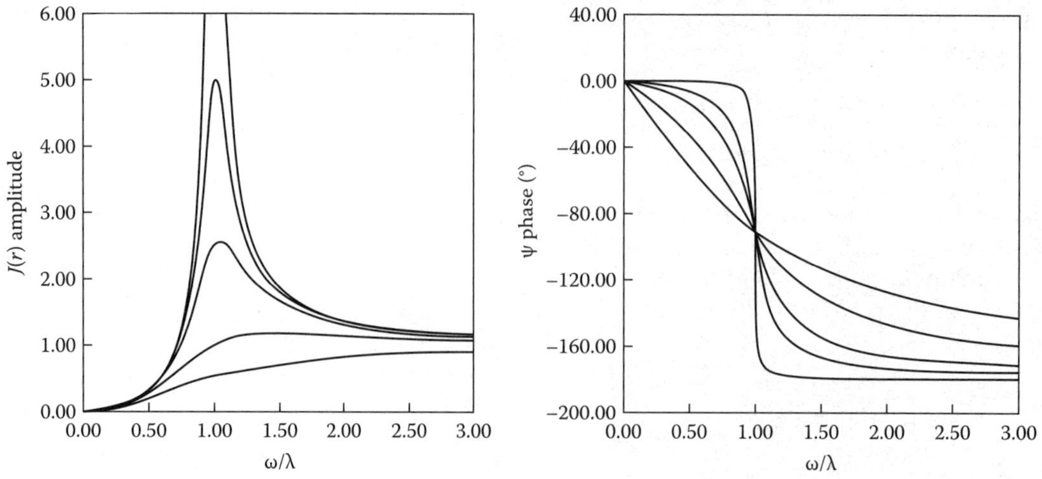

FIGURE 5.52
Amplitude and phase response for the Jeffcott–Föppl rotor model. (Redrawn from Childs, D.W., *Turbomachinery Rotordynamics*, Wiley, New York, 1993.)

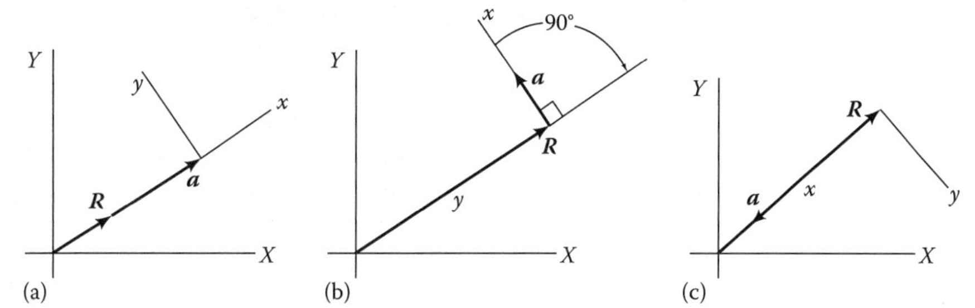

FIGURE 5.53
Phase relations for $\boldsymbol{a}$ and $\boldsymbol{R}$ at (a) $\omega \ll \omega_n$, (b) $\omega = \omega_n$, and (c) $\omega \gg \omega_n$.

As illustrated in Figure 5.53, Equation 5.223's phase definition has definite physical meaning. At low speeds ($r \ll 1$), $\psi = 0$, and the rotor response vector $\boldsymbol{A}$ is in phase with the rotating imbalance vector $\boldsymbol{a} = ae^{j\omega t}$. At the critical speed ($r = \omega/\omega_n = 1$), $\psi = -\pi/2 = -90°$, and $\boldsymbol{A}$ is 90° behind $\boldsymbol{a}$. At high speeds ($r \gg 1$), $\boldsymbol{A} \cong -\boldsymbol{a}$. At low speeds, the disk's mass center rotates about the bearings' axis. At high speeds, the disk mass center lies approximately on the bearings' axis, and the geometric center rotates about it.

5.7.4.4 A Translating Mass with an Attached Compound Pendulum

Figure 5.54a shows a cart-pendulum assembly in equilibrium. The tasks for this example are as follows:

a. Select coordinates and draw the two bodies in a general position.
b. Draw free-body diagrams and use them to derive the EOMs.
c. For $l = 1$ m, $m = 1$ kg, $M = 100$ kg, $k = 1470$ N/m, calculate the system's natural frequencies and eigenvectors.

Figure 5.54b illustrates a general position and the selected coordinates. Figure 5.54c and d provide free-body diagrams. The cart has finite dimensions and (consequently) has a moment of inertia. However, it does not rotate in the X, Y plane, and the governing equations of motion are

$$\begin{aligned}\sum f_X &= f(t) + o_X - kX = M\ddot{X} \\ \sum f_Y &= o_Y + w + R_A + R_B = 0.\end{aligned} \tag{5.224}$$

The resultant force in the vertical direction is zero because the cart does not leave the plane and moves only in the horizontal direction (zero vertical acceleration).

Looking at the pendulum's free-body diagram, we can state the moment equation about either the pivot point o or the mass center g. Stating the moment equation about g means that we could use the simpler moment equation $\sum M_g = I_g\ddot{\theta}$. However, this approach will draw the unknown and unwanted reaction components o_X, o_Y into the moment equation. For this application, the general moment equation, $M_{oz} = I_o\ddot{\theta} + m\ (\boldsymbol{b}_{og} \times \ddot{\boldsymbol{R}}_o)_z$

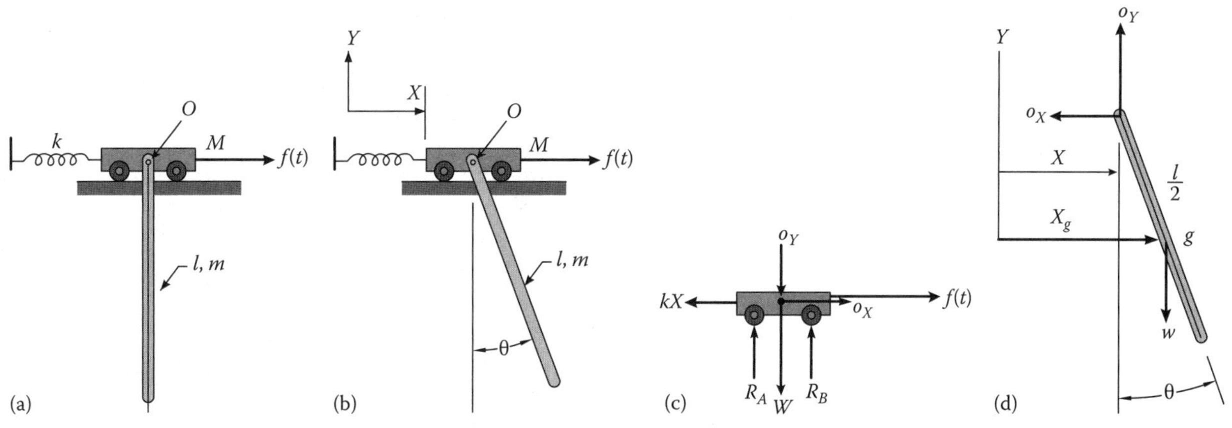

FIGURE 5.54
Translating cart of mass M supported by frictionless rollers and supporting a compound pendulum of length l and mass m. (a) Equilibrium position. (b) General position. (c) Cart free-body diagram. (d) Pendulum free-body diagram.

is quicker. An inspection of the pendulum in Figure 5.54c gives

$$\boldsymbol{b}_{og} = \frac{l}{2}(\boldsymbol{I}\sin\theta - \boldsymbol{J}\cos\theta),$$

$$\ddot{\boldsymbol{R}}_o = \boldsymbol{I}\ddot{X} \Rightarrow \boldsymbol{b}_{og} \times \ddot{\boldsymbol{R}}_o = \boldsymbol{K}\frac{l}{2}\cos\theta\ddot{X}.$$

Hence, stating the moment equation about o gives

$$\sum M_o = -w\frac{l}{2}\sin\theta = \frac{ml^2}{3}\ddot{\theta} + m\frac{l}{2}\cos\theta\ddot{X}. \quad (5.225)$$

We now have two equations (the first of Equations 5.224 and 5.225) for the three unknowns: $\ddot{X}$, $\ddot{\theta}$, o_X. The X component of the $\sum \boldsymbol{f} = m\ddot{\boldsymbol{R}}_g$ equation for the pendulum gives the last required EOM as

$$\sum f_X = -o_X = m\ddot{X}_g. \quad (5.226)$$

However, this equation introduces the new unknown $\ddot{X}_g$, which can be eliminated, starting from the geometric relationship

$$X_g = X + \frac{l}{2}\sin\theta.$$

Differentiating this equation twice with respect to time gives

$$\ddot{X}_g = \ddot{X} + \frac{l}{2}\cos\theta\ddot{\theta} - \frac{l}{2}\sin\theta\dot{\theta}^2.$$

Substituting for $\ddot{X}_g$ into Equation 5.226 gives

$$-o_X = m\left(\ddot{X} + \frac{l}{2}\cos\theta\ddot{\theta} - \frac{l}{2}\sin\theta\dot{\theta}^2\right).$$

Now, substituting this o_X definition into the first of Equation 5.224 gives

$$(m+M)\ddot{X} + kX + M\left(\frac{l}{2}\cos\theta\ddot{\theta} - \frac{l}{2}\sin\theta\dot{\theta}^2\right) = f(t). \quad (5.227)$$

Equations 5.225 and 5.227 comprise the two governing equations in $\ddot{X}$, $\ddot{\theta}$. Their matrix statement is

$$\begin{bmatrix} \frac{ml^2}{3} & \frac{ml}{2}\cos\theta \\ \frac{ml}{2}\cos\theta & (M+m) \end{bmatrix} \begin{Bmatrix} \ddot{\theta} \\ \ddot{X} \end{Bmatrix} = \begin{Bmatrix} -\frac{wl}{2}\sin\theta \\ f(t) - kX + \frac{ml^2}{2}\dot{\theta}^2\sin\theta \end{Bmatrix}. \quad (5.228)$$

We have now completed Task a. Given initial conditions, these nonlinear equations can be integrated with respect to time. Note that the inertia matrix on the left is symmetric. For a stable system, this matrix should be symmetric or capable of being made symmetric by appropriate multiplication or division of one of the governing equations of motion.

Assuming "small" motion for this system means that second-order terms in X and θ are dropped. Introducing the small-angle approximations $\sin\theta \cong \theta$; $\cos\theta \cong 1$, and dropping second order and higher terms in θ and $\dot{\theta}$ yields

$$\begin{bmatrix} \frac{ml^2}{3} & \frac{ml}{2} \\ \frac{ml}{2} & (M+m) \end{bmatrix} \begin{Bmatrix} \ddot{\theta} \\ \ddot{X} \end{Bmatrix} + \begin{bmatrix} \frac{mgl}{2} & 0 \\ 0 & k \end{bmatrix} \begin{Bmatrix} \theta \\ X \end{Bmatrix} = \begin{Bmatrix} 0 \\ f(t) \end{Bmatrix}. \quad (5.229)$$

which concludes Task b.

Calculating the undamped natural frequencies and mode shapes from the homogeneous form of Equation 5.229 starts with the assumed solution, $(\theta, X)^T = (a_1, a_2)^T \cos\omega_n t$. (The solution of natural frequencies and mode shapes for two-degree-of freedom vibration problems is covered in detail in Section 3.5, and you may wish to review this material before proceeding.) Substituting this assumed solution into the homogeneous version of Equation 5.229 gives

$$\begin{bmatrix} -\omega_n^2\frac{ml^2}{3} + \frac{mgl}{2} & -\omega_n^2\frac{ml}{2} \\ -\omega_n^2\frac{ml}{2} & [-\omega_n^2(M+m)+k] \end{bmatrix} \begin{Bmatrix} a_1 \\ a_2 \end{Bmatrix} \cos\omega_n t = 0. \quad (5.230)$$

For a nontrivial solution (i.e., $a_1, a_2 \neq 0$), the determinant of the coefficient matrix in Equation 5.230 must be zero, yielding the characteristic equation

$$\omega_n^4\left[\frac{(m+M)m}{3} - \frac{m^2}{4}\right]l^2 - \omega_n^2\left[\frac{(m+M)mg}{2} + \frac{kml}{3}\right]l + \frac{kmgl}{2} = 0.$$

Plugging the numbers provided above into this equation gives

$$\omega_n^4 - 29.78\omega_n^2 + 220.18 = 0,$$

with the eigenvalue solutions $\omega_{n1}^2 = 13.61\ \text{s}^{-2}$; $\omega_{n2}^2 = 16.18\ \text{s}^{-2}$. The associated eigenvectors are

$$\begin{Bmatrix} a_1 \\ a_2 \end{Bmatrix}_1 = \begin{Bmatrix} 1 \\ 18.45 \end{Bmatrix}, \quad \begin{Bmatrix} a_1 \\ a_2 \end{Bmatrix}_2 = \begin{Bmatrix} 1 \\ -16.6 \end{Bmatrix}.$$

These modes show the cart and pendulum moving in phase in the first mode and out of phase in the second. In both cases, the pendulum motion is much larger than the cart, because the cart's mass is a 100 times larger than the pendulum's.

The two natural frequencies are $\omega_{n1} = 3.69\ \text{s}^{-1}$; $\omega_{n2} = 4.02\ \text{s}^{-1}$. By comparison, the natural frequency for the cart (alone) is $\omega_{nc} = \sqrt{k/m} = 3.87\ \text{s}^{-1}$, and the natural frequency for the pendulum with the cart held motionless is $\omega_{np} = \sqrt{3g/2l} = 3.83\ \text{s}^{-1}$. Hence, the first natural frequency of the coupled system is lower than either of the uncoupled natural frequencies, and the second natural frequency is higher.

5.7.4.5 A Swinging Bar Supported at Its End by a Cord

Figure 5.55a shows a swinging bar *AB*, supported by a cord connecting end *A* to the support point *O* in equilibrium. Figure 5.55b illustrates a general position with the two coordinates ϕ and θ. The cord has length l_1; the bar has length l_2 and mass m_2. This system has two degrees of freedom. The analysis tasks are

a. Derive the governing EOM.
b. For small motion about the equilibrium position, $\theta = \phi = 0$, state the linearized equations of motion.

Figure 5.55c provides the free-body diagram for the bar, showing the tension force T_c in the cord and the weight of the bar acting vertically. We would expect three equations of motion for the bar; namely, two scalar equations arising from $\sum \boldsymbol{f} = m\ddot{\boldsymbol{R}}_g$ and one moment equation. The force equations are

$$\sum f_X = w_2 - T_c \cos\phi = m_2 \ddot{X}_g$$
$$\sum f_Y = -T_c \sin\phi = m_2 \ddot{Y}_g.$$

The acceleration components can be obtained by stating the components of $\boldsymbol{R}_g$ as

$$X_g = l_1 \cos\phi + \frac{l_2}{2}\cos\theta, \quad Y_g = l_1 \sin\phi + \frac{l_2}{2}\sin\theta.$$

Differentiating these equations twice with respect to time gives

$$\ddot{X}_g = -l_1 \cos\phi\dot{\phi}^2 - l_1 \sin\phi\ddot{\phi} - \frac{l_2}{2}\cos\theta\dot{\theta}^2 - \frac{l_2}{2}\sin\theta\ddot{\theta}$$
$$\ddot{Y}_g = -l_1 \sin\phi\dot{\phi}^2 - l_1 \cos\phi\ddot{\phi} - \frac{l_2}{2}\sin\theta\dot{\theta}^2 + \frac{l_2}{2}\cos\theta\ddot{\theta}. \tag{5.231}$$

Substituting these results into the component force equations gives

$$w_2 - T_c\cos\phi = m_2\left(-l_1 \cos\phi\dot{\phi}^2 - l_1 \sin\phi\ddot{\phi} - \frac{l_2}{2}\cos\theta\dot{\theta}^2 - \frac{l_2}{2}\sin\theta\ddot{\theta}\right)$$
$$-T_c\sin\phi = m_2\left(-l_1 \sin\phi\dot{\phi}^2 + l_1 \cos\phi\ddot{\phi} - \frac{l_2}{2}\sin\theta\dot{\theta}^2 + \frac{l_2}{2}\cos\theta\ddot{\theta}\right).$$

We can eliminate T_c from these equations by the following steps: (i) multiply the first by $\sin\phi$, (ii) multiply the second by $-\cos\phi$, and (iii) add the results to obtain

$$w_2 \sin\phi = m_2\left[-l_1\ddot{\phi} - \frac{l_2}{2}\cos(\theta - \phi)\ddot{\theta} + \frac{l_2}{2}\sin(\theta - \phi)\dot{\theta}^2\right]. \tag{5.232}$$

This is the first of our required equations.

One could reasonably state a moment equation for the bar about either *g*, the mass center, or the end *A*. Stating the moment about *A* has the advantage of eliminating

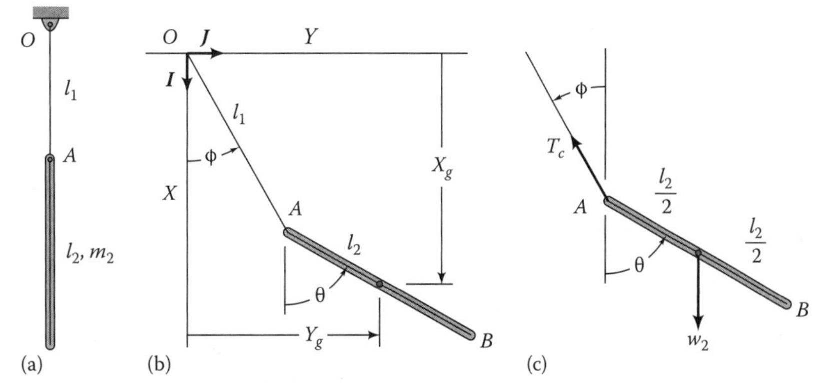

FIGURE 5.55
Swinging bar supported at its end by a cord: (a) Equilibrium position, (b) general position with coordinates, and (c) free-body diagram.

the reaction force T_c, and we will use the following version of the moment Equation 5.24

$$I_A\ddot{\theta} + m_2(\boldsymbol{b}_{Ag} \times \ddot{\boldsymbol{R}}_A)_z = M_{Az} = -\frac{w_2 l_2}{2}\sin\theta. \quad (5.233)$$

The moment due to w_2 is negative because it acts opposite to the $+\theta$ direction. The vector $\boldsymbol{b}_{Ag}$ goes from A to g and is defined by

$$\boldsymbol{b}_{Ag} = \frac{l_2}{2}(\boldsymbol{I}\cos\theta + \boldsymbol{J}\sin\theta) \quad (5.234)$$

To complete Equation 5.233, we need to define $\ddot{\boldsymbol{R}}_A$. Throughout this section, we have been accomplishing this requirement by stating the geometric equations and then differentiating twice. For the present example, we can use the polar-coordinate definition of Equation 2.30, as illustrated in Figure 5.56. The cord length is constant; hence, the radial acceleration component consists of the centrifugal acceleration term $-\boldsymbol{\varepsilon}_r l_1\dot{\phi}^2$. Similarly, the circumferential acceleration term reduces to $\boldsymbol{\varepsilon}_\phi l_2\ddot{\phi}$. Resolving these terms into their components along the X- and Y-axes gives

$$\ddot{\boldsymbol{R}}_A = l_1\left[-\boldsymbol{I}(\ddot{\phi}\sin\phi + \dot{\phi}^2\cos\phi) + \boldsymbol{J}(\ddot{\phi}\cos\phi - \dot{\phi}^2\sin\phi)\right] \quad (5.235)$$

The cross product of $\mathbf{b}_{Ag}$ from Equation 5.234 and $\ddot{\boldsymbol{R}}_A$ from Equation 5.235 gives

$$(\boldsymbol{b}_{Ag} \times \ddot{\boldsymbol{R}}_A)_z = \frac{l_1 l_2}{2}\left[\ddot{\phi}\cos(\theta-\phi) + \dot{\phi}^2\sin(\theta-\phi)\right]$$

Substituting this result into Equation 5.233 gives

$$\frac{m_2 l_2^2}{3}\ddot{\theta} + \frac{w_2 l_2}{2}\sin\theta + \frac{m_2 l_1 l_2}{2} \times \left[\ddot{\phi}\cos(\theta-\phi) + \dot{\phi}^2\sin(\theta-\phi)\right] = 0, \quad (5.236)$$

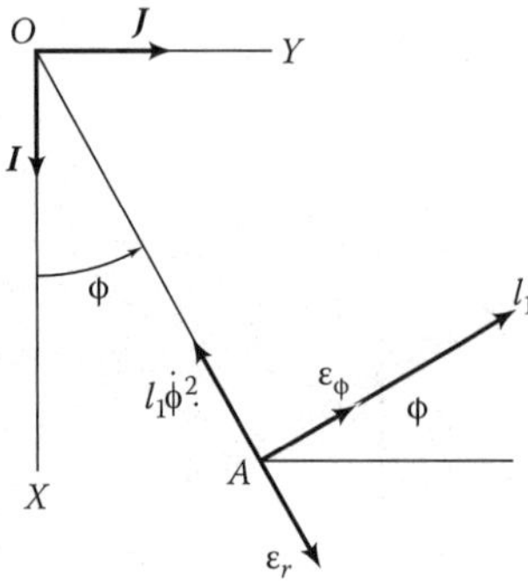

FIGURE 5.56
Polar kinematics for the cord to determine $\ddot{\boldsymbol{R}}_A$.

with $I_A = ml^2/3$. This is the desired moment equation for the bar and is the last equation needed to complete the model. Equations 5.232 and 5.236 are stated in the following matrix equation

$$\begin{bmatrix} m_2 l_1 & \frac{m_2 l_2}{2}\cos(\theta-\phi) \\ \frac{m_2 l_1 l_2}{2}\cos(\theta-\phi) & \frac{m_2 l_2^2}{3} \end{bmatrix}\begin{Bmatrix}\ddot{\phi}\\ \ddot{\theta}\end{Bmatrix} = \begin{Bmatrix} -w_2\sin\phi + m_2\frac{l_2}{2}\sin(\theta-\phi)\dot{\theta}^2 \\ -\frac{w_2 l_2}{2}\sin\theta - \frac{m_2 l_1 l_2}{2}\dot{\phi}^2\sin(\theta-\phi)\end{Bmatrix}. \quad (5.237)$$

This matrix equation completes Task a. The inertia-coupling matrix can be made symmetric by multiplying the first row by l_1. Eliminating second-order terms in θ and ϕ gives the linearized model

$$\begin{bmatrix} m_2 l_1^2 & \frac{m_2 l_2 l_1}{2} \\ \frac{m_2 l_2 l_1}{2} & \frac{m_2 l_2^2}{3}\end{bmatrix}\begin{Bmatrix}\delta\ddot{\phi}\\ \delta\ddot{\theta}\end{Bmatrix} + \begin{bmatrix} w_2 l_1 & 0 \\ 0 & \frac{w_2 l_2}{2}\end{bmatrix}\begin{Bmatrix}\delta\phi\\ \delta\theta\end{Bmatrix} = 0. \quad (5.238)$$

This inertia matrix has been made symmetric by multiplying the first row by l_1.

Matrix Equation 5.237 is nonlinear, and its solution would generally require numerical integration. The inertia coupling of the acceleration terms $\ddot{\phi}$, $\ddot{\theta}$ poses a new complication for numerical integration. We have encountered inertia coupling before via mass or inertia matrices; however, these matrices were constant, as is the inertia matrix of Equation 5.238. Stating Equation 5.238 as $[J](\ddot{x})_i + [K](x)_i = 0$, we could do a one-time premultiplication of the inertia-matrix inverse to obtain $(\ddot{x})_i = [J]^{-1}[K](x) = [B](x)_i$. Choosing the state variables as $\phi = x_1$, $\dot{\phi} = x_2$, $\theta = x_3$, $\dot{\theta} = x_4$, the state-variable model would then look like

$$\begin{aligned}\dot{x}_1 &= x_2\\ \dot{x}_2 &= B_{11}x_1 + B_{12}x_2\\ \dot{x}_3 &= x_4\\ \dot{x}_4 &= B_2 x_1 + B_{22}x_2.\end{aligned}$$

Numerical integration of these equations would involve repeated calculation of the right-hand functions in these equations.

Since the inertia matrix of Equation 5.237 is a function of ϕ and θ, it would vary continuously during motion, and a one-time inversion of the inertia matrix will not work. For Equation 5.237 with its 2×2 coefficient matrix, we could actually use Cramer's rule to solve for $\ddot{\phi}$, $\ddot{\theta}$ analytically and eliminate this complication; however, for typical problems with larger dimensions an analytical matrix inversion is not an option. Numerical integration requires repeated solution of the linear algebraic Equation 5.237 for $\ddot{\phi}$, $\ddot{\theta}$. Solution of the linear algebraic Equations 5.237 for $\ddot{\phi}$, $\ddot{\theta}$ produces numerical values denoted as $G_2(x_1, x_2)$ and $G_4(x_1, x_2)$. Using these variables, the state-variable form for Equation 5.237 would be

$$
\begin{aligned}
\dot{x}_1 &= x_2 = G_1(x_1, x_2)\\
\dot{x}_2 &= G_2(x_1, x_2)\\
\dot{x}_3 &= x_4 = G_3(x_1, x_2)\\
\dot{x}_4 &= G_4(x_1, x_2).
\end{aligned}
$$

Each integration step requires several evaluations of these functions. For example, using a fourth-order Runge–Kutta integrator would require four solutions of the linear equations and evaluations for $G_2(x_1, x_2)$ and $G_4(x_1, x_2)$ per integration step. $G_1(x_1, x_2)$ and $G_3(x_1, x_2)$ would also need to be calculated four times per integration step; however, their evaluation does not require an algebraic solution of Equation 5.237.

Figure 5.57 illustrates a numerical solution for $\phi(t)$ and $\theta(t)$ for the data: $l_1 = l_2 = 1$, $m = 5$ kg and initial conditions: $\phi(0) = 0.5233$ rad $= 30°$, $\theta(0) = 1.0467$ rad $= 60°$, $\dot{\theta}(0) = \dot{\phi}(0) = 0$. Even though the initial angles are not small ($\leq 15°$), the nonlinear response shows the contributions of the two modes of vibrations.

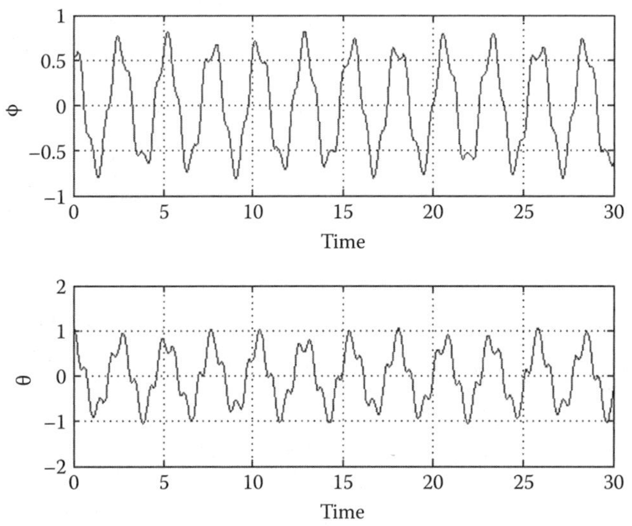

FIGURE 5.57
Numerical solution to Equation 5.237 for $\phi(0) = 0.5233$ rad $= 30°$, $\theta(0) = 1.0467$ rad $= 60°$, $\dot{\theta}(0) = \dot{\phi}(0) = 0$.

5.7.4.6 A Double Compound Pendulum

Figure 5.58a illustrates a double compound pendulum in an equilibrium position. Figure 5.58b shows a general position with coordinates. The upper pendulum *OA* is supported by a frictionless pivot at *O* and supports the second pendulum *AB* with a frictionless pivot joint at *A*. This system has two rigid bodies and two degrees of freedom. The angles θ and ϕ define the orientations of bars *OA* and *AB*, respectively, and are the two degrees of freedom. This system clearly resembles the immediately preceding "swinging bar" example, with bar *OA* of Figure 5.58a replacing cord *OA* of Figure 5.55a. The following tasks apply:

a. Derive governing differential equations of motion.
b. Obtain the linearized equations of motion that apply for small motion about the equilibrium position defined by $\phi = \theta = 0$.

Figure 5.58c provides the free-body diagram for the two bodies. Comparing the free-body diagrams for the swinging bar of Figure 5.55c with this figure shows that the reaction at *A* is defined differently. In Figure 5.55c, the reaction is directed along the axis of the cord, versus the general reaction components A_X, A_Y of Figure 5.58c.

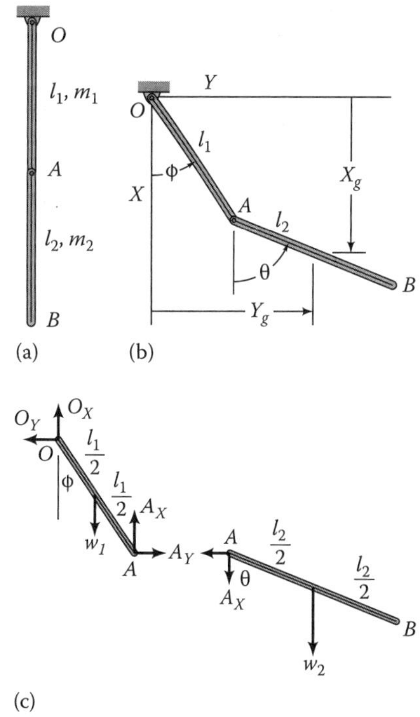

FIGURE 5.58
Double compound pendulum. (a) Equilibrium position. (b) General position with coordinates. (c) Free-body diagram.

Bar OA has only one degree of freedom; hence, it has only one EOM. The logical approach to get this equation is to state a moment equation about the pivot point O. This approach keeps the unwanted reactions O_X, O_Y out of the model equations. The moment equation is

$$\sum M_{Oz} = -w_1 \frac{l_1}{2} \sin\phi + A_Y l_1 \cos\phi + A_X l_1 \sin\phi = I_O \ddot{\phi}. \tag{5.239}$$

In this equation, moments are positive or negative, respectively, if they are in the $+\phi$ direction or $-\phi$ direction. We now have one equation of motion and three unknowns $\ddot{\phi}$, A_X, A_Y.

We can eliminate A_X, A_Y from Equation 5.239 by applying $\sum f = m\ddot{\boldsymbol{R}}_g$ for the mass center of bar AB, obtaining the two component equations:

$$\sum f_X = w_2 + A_X = m_2 \ddot{X}_g, \quad \sum f_Y = -A_Y = m_2 \ddot{Y}_g.$$

You should compare these equations to the force equations from the preceding example. The kinematics definitions provided $\ddot{X}_g$, $\ddot{Y}_g$ by Equations 5.231 continue to apply; hence, the force equations become

$$\begin{aligned} w_2 + A_X &= m_2\Big(-l_1 \cos\phi\dot{\phi}^2 - l_1 \sin\phi\ddot{\phi} \\ &\quad -\frac{l_2}{2}\cos\theta\dot{\theta}^2 - \frac{l_2}{2}\sin\theta\ddot{\theta}\Big) \\ -A_Y &= m_2\Big(-l_1 \sin\phi\dot{\phi}^2 + l_1 \cos\phi\ddot{\phi} \\ &\quad -\frac{l_2}{2}\sin\theta\dot{\theta}^2 + \frac{l_2}{2}\cos\theta\ddot{\theta}\Big) \end{aligned} \tag{5.240}$$

We now have four equations in the four unknowns A_X, A_Y, $\ddot{\theta}$, $\ddot{\phi}$. We can use Equations 5.240 to solve *for* A_X and A_Y and then eliminate these variables from Equation 5.239 by substitution to obtain

$$\begin{aligned} \frac{m_1 l_1^2}{3}\ddot{\phi} &= -\frac{w_1 l_1}{2}\sin\phi + l_1\cos\phi \\ &\quad \left[-m_2\left(l_1 \sin\phi\dot{\phi}^2 - l_1\cos\phi\ddot{\phi} + \frac{l_2}{2}\sin\theta\dot{\theta}^2 - \frac{l_2}{2}\cos\theta\ddot{\theta}\right)\right] \\ &\quad + l_1\sin\phi\Big[-w_2 + m_2\Big(l_1\cos\phi\phi^2 - l_1\sin\phi\ddot{\phi} \\ &\quad -\frac{l_2}{2}\cos\theta\dot{\theta}^2 - \frac{l_2}{2}\sin\theta\ddot{\theta}\Big)\Big], \end{aligned}$$

where $I_O = m_1 l_1^2/3$. Combining terms and simplifying gives

$$\begin{aligned} &l_1^2\left(\frac{m_1}{3} + m_2\right)\ddot{\phi} + \frac{m_2 l_1 l_2}{2}\cos(\theta-\phi)\ddot{\theta} \\ &\quad + \frac{m_2 l_1 l_2}{2}\sin(\theta-\phi)\dot{\theta}^2 + \left(\frac{w_1 l_1}{2} + w_2 l_2\right)\sin\phi = 0. \end{aligned} \tag{5.241}$$

Returning to the free-body diagram for bar AB we would expect to get three independent equations, a moment equation, and the two component equations of $\sum f = m\ddot{\boldsymbol{R}}_g$. As in the preceding swinging-bar example, we could reasonably state the moment equation about either the pivot location A or the mass center g. The moment equation about A does not contain the unknowns A_X, A_Y. The moment Equation 5.236 about A of the preceding example continues to apply, plus the kinematic results of Equations 5.231. Hence, the final governing moment equation in the two unknowns $\ddot{\theta}$, $\ddot{\phi}$ is provided by Equation 5.236 as

$$\begin{aligned} &\frac{m_2 l_2^2}{3}\ddot{\theta} + \frac{w_2 l_2}{2}\sin\theta + \frac{m_2 l_1 l_2}{2} \\ &\quad \times\left[\ddot{\phi}\cos(\theta-\phi) + \phi^2\sin(\theta-\phi)\right] = 0. \end{aligned} \tag{5.242}$$

Equations 5.241 and 5.242 can be combined into the matrix equation

$$\begin{aligned} &\begin{bmatrix} l_1^2\left(\frac{m_1}{3} + m_2\right) & \frac{m_2 l_1 l_2}{2}\cos(\theta-\phi) \\ \frac{m_2 l_2 l_1}{2}\cos(\theta-\phi) & \frac{m_2 l_2^2}{3} \end{bmatrix} \begin{Bmatrix} \ddot{\phi} \\ \ddot{\theta} \end{Bmatrix} \\ &\quad = -\begin{Bmatrix} \frac{m_2 l_1 l_2}{3}\sin(\theta+\phi)\dot{\theta}^2 - \left(\frac{w_1 l_1}{2} + w_2 l_2\right)\sin\phi \\ \frac{w_2 l_2}{2}\sin\theta + \frac{m_2 l_1 l_2}{2}\dot{\phi}^2\sin(\theta-\phi) \end{Bmatrix}. \end{aligned} \tag{5.243}$$

For small motion about the equilibrium position defined by $\theta = \phi = 0$, the EOMs are

$$\begin{aligned} &\begin{bmatrix} l_1^2\left(\frac{m_1}{3} + m_2\right) & \frac{m_2 l_2 l_1}{2} \\ \frac{m_2 l_2 l_1}{2} & \frac{m_2 l_2^2}{3} \end{bmatrix} \begin{Bmatrix} \delta\ddot{\phi} \\ \delta\ddot{\theta} \end{Bmatrix} + \begin{bmatrix} \left(\frac{w_1 l_1}{2} + w_2 l_2\right) & 0 \\ 0 & \frac{w_2 l_2}{2} \end{bmatrix} \\ &\quad \times \begin{Bmatrix} \delta\phi \\ \delta\theta \end{Bmatrix} = 0. \end{aligned} \tag{5.244}$$

This equation concludes Task b.

5.7.5 Planar Mechanisms

Section 4.5 dealt (at length) with the kinematics of planar mechanisms. The mechanisms included several connected links and bodies whose orientations and positions were defined by displacements and angles; however, each mechanism had only one degree of freedom. The relevant engineering-task objectives dealt with the derivation of kinematic equations relating the coordinates and their derivatives. We return to mechanisms in this section to derive a set of governing equations that define the mechanisms' equations of motion that will include force equations, moment equations, and kinematic constraint equations.

In the rolling-without-slipping examples of Section 5.7.1, we were able to combine equations of motion and kinematic equations to obtain a single equation of motion. Similarly, in Section 5.7.4, we could combine the force, moment, and kinematic equations to obtain two equations of motion, corresponding to two degrees of freedom. Models for planar mechanisms do not work that way. The "model" for a mechanism will consist of coupled *algebraic* and differential equations that define angular accelerations, displacement accelerations, and reaction-force components. The kinematic constraint equations for mechanisms are nonlinear and cannot be used (effectively) to reduce the model to a single second-order differential equation.

5.7.5.1 Slider-Crank Mechanism

Figure 5.59a illustrates a slider-crank mechanism, driven by a moment $M(t)$ acting on bar AB. Bar AB is connected to bar BC by a pivot joint at B, while bar BC is connected to a piston by a pinned joint at C. We analyzed this assembly in Section 4.5.2 as a mechanism, with an input rotation angle without considering the

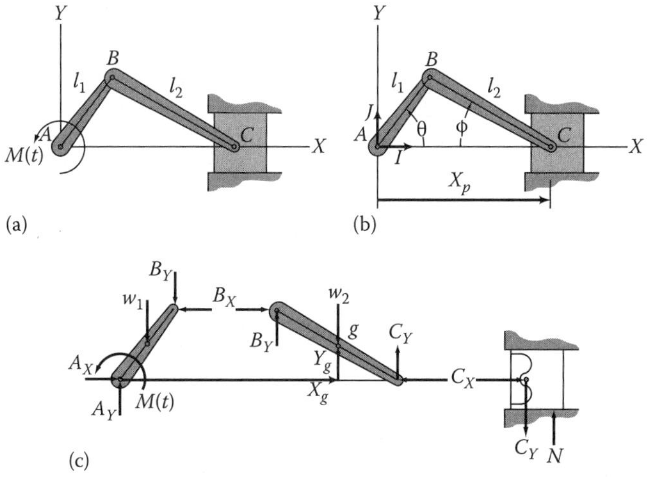

FIGURE 5.59
(a) Slider-crank mechanism driven by moment $M(t)$ acting on bar AB. (b) Coordinate choices. (c) Free-body diagram.

forces or moments that are either required to produce motion or result as reactions due to motion. We now want to develop a simulation model to define the motion and reaction forces resulting from $M(t)$. In developing the equations of motion, we will neglect friction at joints A, B, and C plus friction forces due to the sliding motion of the piston within its bore.

The first step in developing the model is the selection of coordinates. Our choices are shown in Figure 5.59b and coincides with the coordinates used earlier in Figure 4.15. Figure 5.59c provides a free-body diagram for the assembly. We can start the model development by taking moments for bar AB about A to obtain

$$I_A\ddot{\theta} = \sum M_A = M(t) - w_1\frac{l_1}{2}\cos\theta + B_X l_1 \sin\theta - B_Y l_1\cos\theta, \quad I_A = m_1\frac{l_1^2}{3}. \tag{5.245}$$

We could have taken moments about bar AB's mass center, but that approach would have pulled the unknown reaction forces A_X, A_Y into the model. This is the only equation needed for bar AB, and it includes the three unknowns $\ddot{\theta}$, B_X, B_Y.

Bar BC is in general motion, so we will need force equations for its mass center and a moment equation. Applying $\sum f = m\ddot{\boldsymbol{R}}_g$ gives the component equations:

$$\begin{aligned} m_2\ddot{X}_g &= \sum f_X = B_X - C_X \\ m_2\ddot{Y}_g &= \sum f_Y = B_Y + C_Y - w_2. \end{aligned} \tag{5.246}$$

These two equations have the unknowns: $\ddot{X}_g$, $\ddot{Y}_g$, B_X, B_Y, C_X, C_Y. Taking moments about bar BC's mass center gives

$$I_{2g}\ddot{\phi} = \sum M_g = B_X\frac{l_2}{2}\sin\phi + B_Y\frac{l_2}{2}\cos\phi + C_X\frac{l_2}{2}\sin\phi - C_Y\frac{l_2}{2}\cos\phi: \; I_{2g} = m_2\frac{l_2^2}{12}, \tag{5.247}$$

and includes the new unknown $\ddot{\phi}$.

The force equation for the piston is

$$m_p\ddot{X}_p = \sum f_X = C_X, \tag{5.248}$$

and provides the last independent equation that we can obtain using force and moment equations.

We now have five equations for the nine unknowns $\ddot{\theta}$, $\ddot{\phi}$, B_X, B_Y, C_X, C_Y, $\ddot{X}_g$, $\ddot{Y}_g$, $\ddot{X}_p$. Kinematics equations are required to complete the model, and we can start by defining $\ddot{X}_g$ in terms of φ, θ and their derivatives, via

$$X_g = l_1 \cos\theta + \frac{l_2}{2}\cos\phi \Rightarrow \dot{X}_g = -l_1 \sin\theta\dot{\theta} - \frac{l_2}{2}\sin\phi\dot{\phi}$$
$$\ddot{X}_g = -l_1 \sin\theta\ddot{\theta} - l_1\cos\theta\dot{\theta}^2 - \frac{l_2}{2}\sin\phi\ddot{\phi} - \frac{l_2}{2}\cos\phi\dot{\phi}^2. \quad (5.249)$$

Similarly for $\ddot{Y}_g$,

$$Y_g = l_1\sin\theta \Rightarrow \dot{Y}_g = l_1\cos\theta\dot{\theta} \Rightarrow \ddot{Y}_g = l_1\cos\theta\ddot{\theta} - l_1\sin\theta\dot{\theta}^2 \quad (5.250)$$

Substituting for $\ddot{X}_g$ and $\ddot{Y}_g$ into Equations 5.246 gives

$$m_2\left(-l_1\sin\theta\ddot{\theta} - l_1\cos\theta\dot{\theta}^2 - \frac{l_2}{2}\sin\phi\ddot{\phi} - \frac{l_2}{2}\cos\phi\dot{\phi}^2\right) = B_X - C_X$$
$$m_2\left(\frac{l_1}{2}\cos\theta\ddot{\theta} - \frac{l_1}{2}\sin\theta\dot{\theta}^2\right) = B_Y + C_Y - w_2. \quad (5.251)$$

The definition for $\ddot{X}_p$ follows from

$$X_p = l_1\cos\theta + l_2\cos\phi \Rightarrow \dot{X}_p = -l_1\sin\theta\dot{\theta} - l_2\sin\phi\dot{\phi},$$
$$\ddot{X}_p = l_1\sin\theta\ddot{\theta} - l_1\cos\theta\dot{\theta}^2 - l_2\sin\phi\ddot{\phi} - l_2\cos\phi\dot{\phi}^2. \quad (5.252)$$

Substituting $\ddot{X}_p$ into Equation 5.248 gives

$$m_p(-l_1\sin\theta\ddot{\theta} - l_1\cos\theta\dot{\theta}^2 - l_2\sin\phi\ddot{\phi} - l_2\cos\phi\dot{\phi}^2) = C_X \quad (5.253)$$

Having eliminated $\ddot{X}_g$, $\ddot{Y}_g$, $\ddot{X}_p$ Equations 5.245, 5.247, 5.251, and 5.253 provide five equations for the remaining six unknowns $\ddot{\theta}$, $\ddot{\phi}$, B_X, B_Y, C_X, C_Y.

From Figure 5.59b, the kinematic constraint equation relating θ and ϕ is

$$l_1\sin\theta = l_2\sin\phi, \quad \phi = \sin^{-1}\left(\frac{l_1}{l_2}\sin\theta\right) \quad (5.254)$$

This is the starting point for the remaining equation. Differentiating with respect to time gives

$$l_1\cos\theta\dot{\theta} = l_2\cos\phi\dot{\phi} \Rightarrow \dot{\phi} = \left(\frac{l_1\cos\theta}{l_2\cos\phi}\right)\dot{\theta} \quad (5.255)$$

Differentiating again gives

$$l_1\cos\theta\ddot{\theta} - l_1\sin\theta\dot{\theta}^2 = l_2\cos\phi\ddot{\phi} - l_2\sin\phi\dot{\phi}^2. \quad (5.256)$$

This equation completes the model.

Stating Equations 5.245, 5.247, 5.251, 5.253, and 5.256 in matrix format for the unknowns $\ddot{\theta}$, $\ddot{\phi}$, B_X, B_Y, C_X, C_Y gives

$$\begin{bmatrix} I_A & 0 & -l_1\sin\theta & l_1\cos\theta & 0 & 0 \\ 0 & I_{2g} & -\frac{l_2}{2}\sin\phi & -\frac{l_2}{2}\cos\phi & -\frac{l_2}{2}\sin\phi & -\frac{l_2}{2}\cos\phi \\ -m_2 l_1\sin\theta & -m_2\frac{l_2}{2}\sin\phi & -1 & 0 & 1 & 0 \\ -m_2\frac{l_1}{2}\sin\theta & 0 & 0 & -1 & 0 & -1 \\ -m_p l_1\sin\theta & -m_p l_2\sin\phi & 0 & 0 & -1 & 0 \\ l_1\cos\theta & -l_2\cos\phi & 0 & 0 & 0 & 0 \end{bmatrix} \times \begin{Bmatrix} \ddot{\theta} \\ \ddot{\phi} \\ B_X \\ B_Y \\ C_X \\ C_Y \end{Bmatrix} = \begin{Bmatrix} M(t) - w_1\frac{l_1}{2}\cos\theta \\ 0 \\ m_2 l_1\cos\theta\dot{\theta}^2 + m_2\frac{l_2}{2}\cos\phi\dot{\phi}^2 \\ m_1\frac{l_2}{2}\sin\theta\dot{\theta}^2 - w^2 \\ m_p l_1\dot{\theta}^2\cos\theta + m_p l_2\dot{\phi}^2\cos\phi \\ l_1\sin\theta\dot{\theta}^2 - l_2\dot{\phi}^2\sin\phi \end{Bmatrix}. \quad (5.257)$$

The moment $M(t)$ is the input that drives this model with $\theta(t)$, $\phi(t)$ as output coordinates. The piston position $X_p(t) = l_1\cos\theta(t) + l_2\cos\phi(t)$ is also an output coordinate, but is developed in an "open-loop" fashion, outside of Equation 5.257.

Despite having $\theta(t)$, $\phi(t)$ as output coordinates, this model has only one degree of freedom. Choosing θ to be "the" coordinate for the model, Equation 5.257 represents a second-order differential equation in $\theta(t)$. Starting with initial conditions $\theta(0) = \theta(t=0)$, Equation 5.254 is used to evaluate $\phi(0)$. Next, with known values for $\theta(0)$, $\phi(0)$, $\dot{\theta}(0)$ Equation 5.255 is used to solve for $\dot{\phi}(0)$. Plugging these values into Equation 5.257 and solving gives $\ddot{\theta}(0)$, $\ddot{\phi}(0)$, $B_X(0)$, $B_Y(0)$, $C_X(0)$, $C_Y(0)$. This same procedure holds for any value of time and can be used to integrate $\ddot{\theta}(t)$ to obtain $\dot{\theta}(t)$, $\theta(t)$. Think about the steps involved in completing one integration step for a numerical solution of Equation 5.257. The model has only one degree of freedom, and we could choose $x_1 = \theta$, $x_2 = \dot{\theta}$ to be the variables in a state-variable model that looks like

$$\dot{x}_1 = G_1(x_1, x_2, t) = x_2, \quad \dot{x}_2 = G_2(x_1, x_2, t)$$

We cannot reduce Equations 5.257 to a single second-order differential equation in θ. If we could, the corresponding second-order differential equation would be $\ddot{\theta} = G_2(\theta, \dot{\theta}, t)$. We only need θ and $\dot{\theta}$ to evaluate the coefficient matrix and the right-hand side of Equation 5.257 because—given θ—we can use

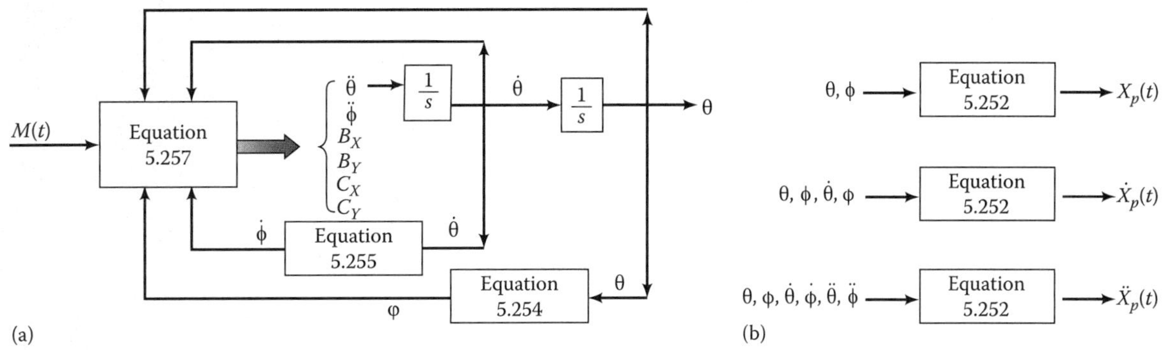

FIGURE 5.60
Block diagram for solving Equation 5.257. (a) Model for Equation 5.257. (b) Solution for $X_p(t)$, $\dot{X}_p(t)$, $\ddot{X}_p(t)$.

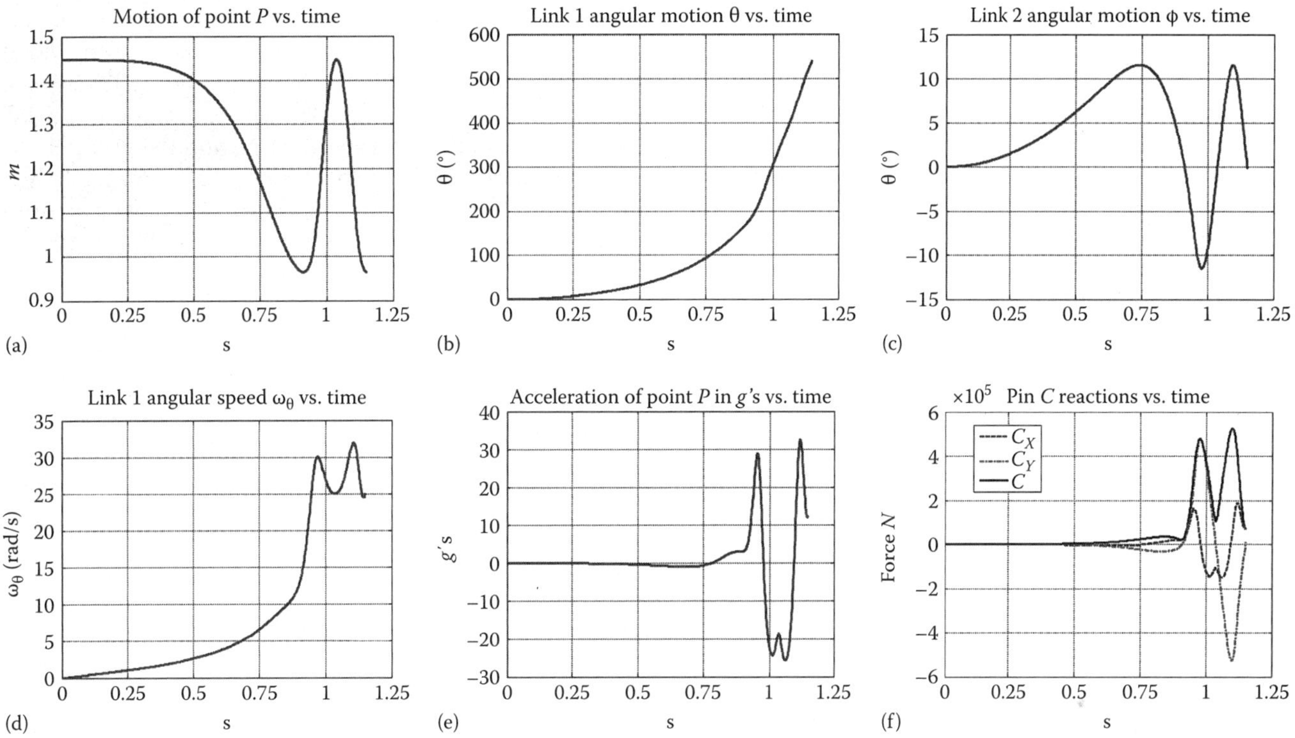

FIGURE 5.61
Solution for Equation 5.257: (a) $X_p(t)$, (b) $\theta(t)$, (c) $\phi(t)$, (d) $\dot{\theta}(t)$, (e) $\ddot{X}(t)$, (f) $C_X(t)$, $C_Y(t) = \sqrt{C_X^2 + C_Y^2}$.

Equations 5.254 to solve for ϕ. Similarly, given θ, ϕ, $\dot{\theta}$, we can use Equations 5.255 to solve for $\dot{\phi}$. Solving Equation 5.257 gives numerical values for the unknowns ($\ddot{\theta}$, $\ddot{\phi}$, B_X, B_y, C_X, C_Y). The numerical evaluation for $\ddot{\theta}$ allows us to calculate $G(x_1, x_2, t)$. A fourth-order Runge–Kutta integration scheme would require four evaluations of this function [and the other derivative $G_1(x_1, x_2, t) = x_2$]. Figure 5.60 illustrates this solution format.

Gardner (2001) has an attractive book that covers solutions of mechanisms using MATLAB® and Simulink®.

Reciprocating Compressor Example* Reciprocating compressors used to compress natural gas weigh up to 668 kN (150,000 lbs), stand around 7.62 m (25 ft) tall, 9.24 m (30 ft) long, and 6.2 m (20 ft) wide. The compressors can have several pistons on the same drive shaft. A work piston is typically oriented in the horizontal plane, can weigh up to 4086 N (900 lbs), and be close to 0.9 m (3 ft) in diameter. Figure 5.61 provides a solution for the data set given below.

* This example problem and solution was developed by Andrew Conkey.

Model Parameters for One Compressor Stage:

- **Link 1: Crank**: $m_1 = 908$ kg ($w_1 = 2000$ lbs), $L1 = 0.241$ m (0.791 ft)
- **Link 2: Connecting Rod**: $m_2 = 245$ kg ($w_2 = 490$ lbs), $L2 = 1.206$ m (3.75 ft)
- **Slider P: Work Piston**: $m_p = 590$ kg ($w_p = 900$ lbs)
- **Steady Operational Speed for** $\dot{\theta}$: 377 rpm (39.5 rad/s)
- **Applied Moment**: **1400** N m

The results shown in Figure 5.61 shows the monotonic increase in the rotation angle $\theta(t)$, an oscillation in $X_p(t)$, and the high acceleration levels in $\ddot{X}_p(t)$ that result from the applied moment. The reaction components and magnitude at C are also illustrated.

Consider the question: What moment $M(\theta)$ is required to drive the mechanism at the constant rate $\dot{\theta} = \omega$? To address this question, we will restate Equation 5.257 to make $\theta = \theta_0 + \omega t \Rightarrow \dot{\theta} = \omega$, $\ddot{\theta} = 0$ the input, with M, $\ddot{\theta}$, $\ddot{\phi}$, B_X, B_Y, C_X, C_Y output functions of θ. For this requirement, Equation 5.257 stops being a differential-equation model and becomes an algebraic model with θ as the input. To answer the question, we bring M to the left-hand side as an unknown and set $\dot{\theta} = \omega$, $\ddot{\theta} = 0$ to produce

$$\begin{bmatrix} 1 & l_1\sin\theta_i & l_1\cos\theta_i & 0 & 0 \\ 0 & -\frac{l_2}{2}\sin\phi_i & -\frac{l_2}{2}\cos\phi_i & -\frac{l_2}{2}\sin\phi_i & \frac{l_2}{2}\cos\phi_i \\ 0 & -1 & 0 & 1 & 0 \\ 0 & 0 & -1 & 0 & -1 \\ 0 & 0 & 0 & -1 & 0 \end{bmatrix} \begin{Bmatrix} M(\theta_i) \\ B_X(\theta_i) \\ B_Y(\theta_i) \\ C_X(\theta_i) \\ C_Y(\theta_i) \end{Bmatrix} = \begin{Bmatrix} w_1\frac{l_1}{2}\cos\theta_i \\ -I_{2g}\ddot{\phi}_i \\ m_2 l_1\omega^2\cos\theta_i + m_2\frac{l_2}{2}\cos\phi_i\dot{\phi}_i^2 + m_2\frac{l_2}{2}\sin\phi_i\ddot{\phi}_i \\ m_1\frac{l_2}{2}\omega^2\sin\theta_i - w_2 \\ m_p l_1\omega^2\cos\theta_i + m_p l_2\dot{\phi}_i^2\cos\phi_i + m_p l_2\sin\phi_i\ddot{\phi}_i \end{Bmatrix} \tag{5.258}$$

This model contains the first five equations from Equation 5.257. We can evaluate Equation 5.258 over the range $\theta \varepsilon\ [0, 2\pi]$. For a specified value of θ_i. we will first use Equation 5.254 to solve for ϕ_i, then use Equation 5.255 with θ_i, ϕ_i, and $\dot{\theta}_i = \omega$ to solve for $\dot{\phi}_i$, and then plug θ_i, ϕ_i, $\dot{\phi}_i$ into Equation 5.256 to solve for $\ddot{\phi}_i$. Equation 5.258 then produces M_i, B_{Xi}, B_{Yi}, C_{Xi}, C_{Yi}. The input power requirement is $\text{Power}_i = M(\theta_i)\omega$. Figure 5.62 illustrates one cycle of rotation from Equation 5.258 for $\omega = 100$

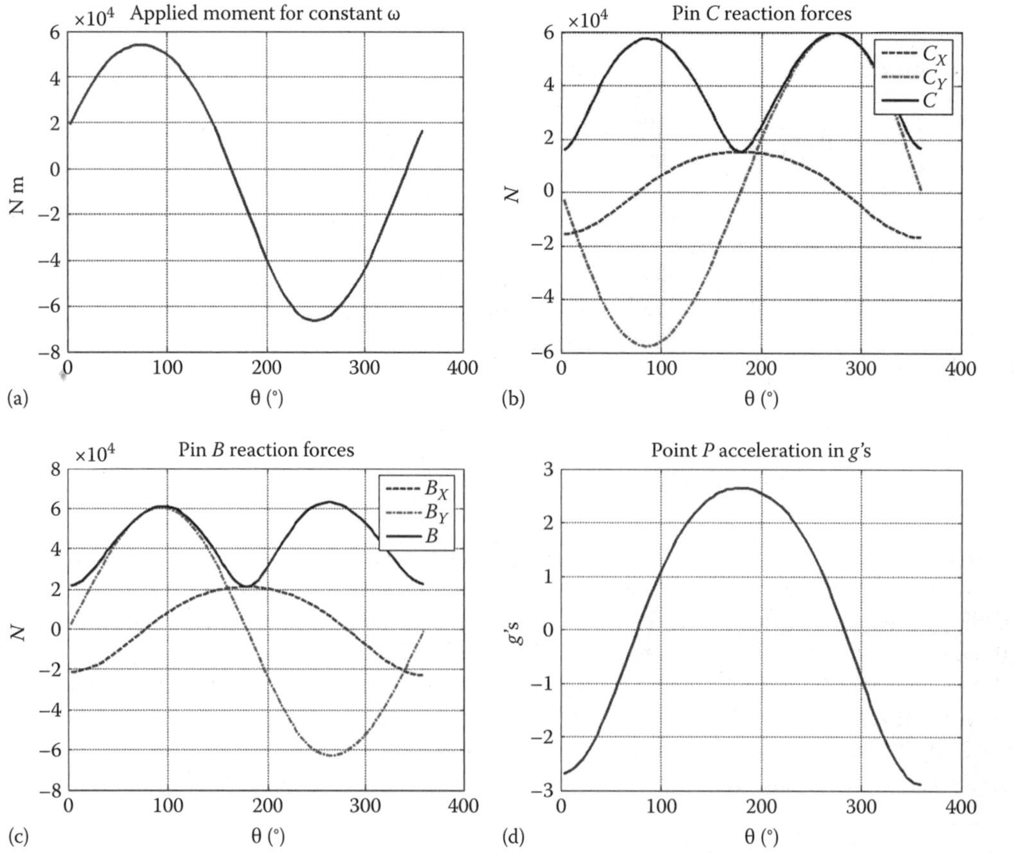

FIGURE 5.62
Solution to Equation 5.258: (a) applied moment $M(\theta)$, reaction force components, (b) $C_X(\theta)$, $C_Y(\theta)$, (c) $B_X(\theta)$, $B_Y(\theta)$, and (d) piston acceleration $\ddot{X}_p(\theta)$.

rpm = 10.5 rad/s. In comparing Figures 5.62 and 5.61, note in Figure 5.61, that $\omega_1 = \dot{\theta} = 10$ rad/s at approximately $t = 0.88$ s. The pin reaction forces and acceleration levels are comparable in both figures at this time.

An alternative statement of this example could have a force $f(t)$ acting on the piston as the input instead of $M(t)$ acting on bar AB. For this option, Equation 5.248 becomes

$$m_p\ddot{X}_p = \sum f_X = f(t) + C_X,$$

We would continue using Equation 5.252 to eliminate $\ddot{X}_p$, obtaining

$$m_p(-l_1 \sin\theta\ddot{\theta} - l_1\cos\theta\dot{\theta}^2 - l_2\sin\phi\ddot{\phi} - l_2\cos\phi\dot{\phi}^2) = f(t) + C_X.$$

The model is changed by dropping $M(t)$ in the first entry and adding $f(t)$ to the fifth entry on the right-hand side vector of Equation 5.257.

5.7.5.2 A Four-Bar-Linkage Example

Figure 5.63a and b illustrate, respectively, a four-bar-linkage assembly in (i) an upright position and (ii) a general position. Linkage dynamics problems are commonly stated in the type of specific configuration provided by Figure 5.63a; however, to begin deriving equations of motion, a general-configuration illustration, as provided in Figure 5.63b, must be developed. Drawing a general-configuration figure is an essential, active part of the problem solution. This figure shows the three angular coordinates α_1, α_2, α_3 that define the orientation of the assembly. Remember, you choose the coordinates as the first step in working the problem, and you could have chosen differently and still had an entirely valid model; e.g., the complements to these angles would work fine. Although this problem has three coordinates, it has only one degree of freedom; specifically, if one of the three angles is specified, the remaining two can be determined.

Our first task is: Assuming that the bar AB is driven by a clockwise moment $M(t)$, find governing differential equations of motion defining $\alpha_1(t)$, $\alpha_2(t)$, $\alpha_3(t)$. The free-body diagram of Figure 5.63c is the starting point. Surveying the three separated bodies, note that equations of motion for bars AB and CD would normally be obtained by stating moment equations about the fixed pivot points at A and D, respectively. Because bar BC is in general motion involving displacement and rotation, you should expect that it will require both a moment equation for some point on the body and an application of a vector force equation $\sum f = m\ddot{\mathbf{R}}_g$ for the mass center. As with the slider-crank mechanism, you should expect to have more unknowns than equations when all the independent force and moment equations have been stated, requiring the development of additional kinematic equations, relating the angles α_1, α_2, α_3.

The appropriate moment equation for bar AB is about point A, avoiding the unwanted reaction components A_X, A_Y, yielding

$$I_{1A}\ddot{\alpha} = \sum M_A = M(t) + w_1\frac{l_1}{2}\sin\alpha_1 + B_X l_1\cos\alpha_1 - B_Y l_1 \sin\alpha_1; \quad I_{1A} = \frac{m_1 l_1^2}{3}. \tag{5.259}$$

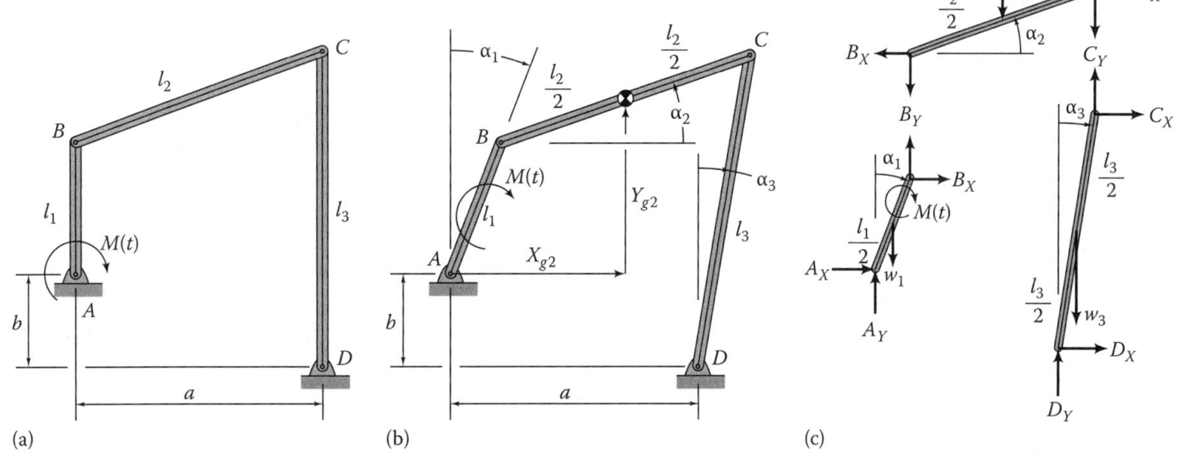

FIGURE 5.63
(a) Four-bar-linkage problem. (b) The four-bar-linkage problem in a general position. (c) Free-body diagram.

The positive sign convention for moments follows from the positive $+\alpha_1$, counterclockwise direction. This equation has the three unknowns $\ddot{\alpha}$, B_X, B_Y, and no additional independent equation of motion can be stated for this body. A moment equation about any other point will bring A_X, A_Y into the governing equations as will a force equation for the bar's mass center.

The moment equation for bar CD about D provides

$$I_{3D}\ddot{\alpha}_3 = \sum M_D = w_3\frac{l_3}{2}\sin\alpha_3 + C_X l_3\cos\alpha_3 - C_Y l_3\sin\alpha_3; \quad I_{3D} = \frac{ml_3^2}{3}. \tag{5.260}$$

The positive signs for moments in this equation are established by the positive clockwise rotation direction for $+\alpha_3$.

For bar BC, we could correctly state moments about any number of points on the body; however, the most direct statement is about the mass center to obtain

$$I_{2g}\ddot{\alpha}_2 = \sum M_g = C_X\frac{l_2}{2}\sin\alpha_2 - C_Y\frac{l_2}{2}\cos\alpha_2 - B_X\frac{l_2}{2}\sin\alpha_2 + B_Y\frac{l_2}{2}\cos\alpha_2; \quad I_{2g} = \frac{m_2 l_2^2}{12} \tag{5.261}$$

The positive signs for moments in this equation are established by the positive counterclockwise rotation direction for $+\alpha_2$.

Writing $\sum \boldsymbol{f} = m\ddot{\boldsymbol{R}}_g$ for the mass center of bar BC gives the two component equations

$$\begin{aligned} m_2\ddot{X}_{2g} &= \sum f_X = -B_X - C_X, \\ m_2\ddot{Y}_{2g} &= \sum f_Y = -w_2 - B_Y - C_Y \end{aligned} \tag{5.262}$$

Equations 5.259 and 5.262 comprise five equations in the nine unknowns $\ddot{\alpha}_1, \ddot{\alpha}_2, \ddot{\alpha}_3, \ddot{X}_{2g}, \ddot{Y}_{2g}, B_X, B_Y, C_X, C_Y$, which should tell you that now is the time to seek kinematic equations.

Returning to Figure 5.63b, we state X_{2g}, Y_{2g} as

$$X_{2g} = l_1\sin\alpha_1 + \frac{l_2}{2}\cos\alpha_2, \quad Y_{2g} = l_1\cos\alpha_1 + \frac{l_2}{2}\sin\alpha_2.$$

Differentiating once with respect to time gives

$$\begin{aligned} \dot{X}_{2g} &= l_1\cos\alpha_1\dot{\alpha}_1 - \frac{l_2}{2}\sin\alpha_2\dot{\alpha}_2, \\ \dot{Y}_{2g} &= -l_1\sin\alpha_1\dot{\alpha}_1 + \frac{l_2}{2}\cos\alpha_2\dot{\alpha}_2. \end{aligned}$$

A second differentiation gives

$$\begin{aligned} \ddot{X}_{2g} &= l_1\cos\alpha_1\ddot{\alpha}_1 - l_1\sin\alpha_1\dot{\alpha}_1^2 - \frac{l_2}{2}\sin\alpha_2\ddot{\alpha}_2 - \frac{l_2}{2}\cos\alpha_2\dot{\alpha}_2^2 \\ \ddot{Y}_{2g} &= -l_1\sin\alpha_1\ddot{\alpha}_1 - l_1\cos\alpha_1\dot{\alpha}_1^2 + \frac{l_2}{2}\cos\alpha_2\ddot{\alpha}_2 - \frac{l_2}{2}\sin\alpha_2\dot{\alpha}_2^2. \end{aligned}$$

Substitution from this equation into Equation 5.262 eliminates $\ddot{X}_{2g}$, $\ddot{Y}_{2g}$ and produces

$$\begin{aligned} & m_2\left(l_1\cos\alpha_1\ddot{\alpha}_1 - l_1\sin\alpha_1\dot{\alpha}_1^2 - \frac{l_2}{2}\sin\alpha_2\ddot{\alpha}_2 - \frac{l_2}{2}\cos\alpha_2\dot{\alpha}_2^2\right) \\ &\quad = -B_X - C_X \\ & m_2\left(-l_1\sin\alpha_1\ddot{\alpha}_1 - l_1\cos\alpha_1\dot{\alpha}_1^2 + \frac{l_2}{2}\cos\alpha_2\ddot{\alpha}_2 - \frac{l_2}{2}\sin\alpha_2\dot{\alpha}_2^2\right) \\ &\quad = -w_2 - B_Y - C_Y. \end{aligned} \tag{5.263}$$

We now have five equations and seven unknowns.

The additionally required kinematic equations are obtained from the following geometric relations from Figure 5.63b:

$$\begin{aligned} &X \text{ component: } a = l_1\sin\alpha_1 + l_2\cos\alpha_2 - l_3\sin\alpha_3 \\ &Y \text{ component: } b = -l_1\cos\alpha_1 - l_2\sin\alpha_2 + l_3\cos\alpha_3. \end{aligned} \tag{5.264}$$

Differentiating with respect to time gives

$$\begin{aligned} l_2\sin\alpha_2\dot{\alpha}_2 + l_3\cos\alpha_3\dot{\alpha}_3 &= l_1\cos\alpha_1\dot{\alpha}_1, \\ l_2\cos\alpha_2\dot{\alpha}_2 + l_3\sin\alpha_3\dot{\alpha}_3 &= l_1\sin\alpha_1\dot{\alpha}_1. \end{aligned} \tag{5.265}$$

An additional differentiation gives

$$\begin{aligned} & l_2\sin\alpha_2\ddot{\alpha}_2 + l_2\cos\alpha_2\dot{\alpha}_2^2 + l_3\cos\alpha_3\ddot{\alpha}_3 - l_3\sin\alpha_3\ddot{\alpha}_3^2 \\ &\quad = l_1\cos\alpha_1\ddot{\alpha}_1 - l_1\sin\alpha_1\dot{\alpha}_1^2 \\ & l_2\cos\alpha_2\ddot{\alpha}_2 - l_2\sin\alpha_2\dot{\alpha}_2^2 + l_3\sin\alpha_3\ddot{\alpha}_3 + l_3\cos\alpha_3\dot{\alpha}_3^2 \\ &\quad = l_1\sin\alpha_1\ddot{\alpha}_1 + l_1\cos\alpha_1\dot{\alpha}_1^2. \end{aligned} \tag{5.266}$$

Equations 5.259, 5.260, 5.261, 5.263, and 5.266 comprise seven equations in the seven remaining unknowns $\ddot{\alpha}_1, \ddot{\alpha}_2, \ddot{\alpha}_3, B_X, B_Y, C_X, C_Y$. Their matrix statement is

$$\begin{bmatrix} I_{1A} & 0 & 0 & -l_1\cos\alpha_1 & +l_1\sin\alpha_1 & 0 & 0 \\ 0 & I_{2g} & 0 & \frac{l_2}{2}\sin\alpha_2 & -\frac{l_2}{2}\cos\alpha_2 & -\frac{l_2}{2}\sin\alpha_2 & \frac{l_2}{2}\cos\alpha_2 \\ 0 & 0 & I_{3D} & 0 & 0 & -l_3\cos\alpha_3 & l_3\sin\alpha_3 \\ -m_2l_1\cos\alpha_1 & \frac{m_2l_2}{2}\sin\alpha_2 & 0 & 0 & 1 & 0 & 1 \\ m_2l_1\cos\alpha_1 & -\frac{m_2l_2}{2}\sin\alpha_2 & 0 & 1 & 0 & 1 & 0 \\ -l_1\cos\alpha_1 & l_2\sin\alpha_2 & l_3\cos\alpha_3 & 0 & 0 & 0 & 0 \\ -l_1\sin\alpha_1 & l_2\cos\alpha_2 & l_3\sin\alpha_3 & 0 & 0 & 0 & 0 \end{bmatrix}$$
$$\times \begin{Bmatrix} \ddot{\alpha}_1 \\ \ddot{\alpha}_2 \\ \ddot{\alpha}_3 \\ B_X \\ B_Y \\ C_X \\ C_Y \end{Bmatrix} = \begin{Bmatrix} M(t) + w_1\frac{l_1}{2}\sin\alpha_1 \\ 0 \\ w_3\frac{l_3}{2}\sin\alpha_3 \\ -w_2 + m_2\left(l_1\cos\alpha_1\dot{\alpha}_1^2 + \frac{l_2}{2}\sin\alpha_2\dot{\alpha}_2^2\right) \\ m_2\left(l_1\sin\alpha_1\dot{\alpha}_1^2 + \frac{l_2}{2}\cos\alpha_2\dot{\alpha}_2^2\right) \\ -l_1\sin\alpha_1\dot{\alpha}_1^2 - l_2\cos\alpha_2\dot{\alpha}_2^2 + l_3\sin\alpha_3\dot{\alpha}_3^2 \\ l_1\cos\alpha_1\dot{\alpha}_1^2 + l_2\sin\alpha_2\dot{\alpha}_2^2 - l_3\cos\alpha_3\dot{\alpha}_3^2 \end{Bmatrix}. \quad (5.267)$$

Equations 5.267 and 5.264 and 5.265 define the model and complete the engineering task. Despite having three coordinates, this model has one degree of freedom. As with the slider-crank mechanism, we could choose "the" coordinate, for example, α_1. With this choice, we can use Equations 5.267, 5.241, and 5.242 to systematically solve for $\ddot{\alpha}_1$. and then proceed from initial conditions to integrate and obtain $\alpha_1(t)$, $\dot{\alpha}_1(t)$.

When we speak of specifying a known values for α_1, and solving for α_2, α_3, recall from Section 4.5.2 that the nonlinear Equations 5.264 can have two solutions, one solution, or no solutions depending on values for l_1, l_2, l_3. Equation 4.27 defines the multiple solution possibilities for the four-bar mechanism of Figure 4.19.

As with the slider-crank mechanism, we can think about driving this mechanism by prescribing the angle $\dot{\alpha}_{1i} = \omega$, $\ddot{\alpha}_i = 0$, $\alpha_{1i} = \omega t_i$. For any α_{1i}, we can (successively) solve (i) Equations 5.264 for the α_{2i}, α_{3i}, (ii) Equations 5.265 for α_{2i}, $\dot{\alpha}_{3i}$, and (iii) Equations 5.266 for $\ddot{\alpha}_{2i}$, $\ddot{\alpha}_{3i}$. We need Equation 5.267 to then solve (algebraically) for B_{Xi}, B_{Yi}, C_{Xi}, C_{Yi} and M_i, the moment required to drive the mechanism at the prescribed motion defined by $\dot{\alpha}_{1i} = \omega$, $\ddot{\alpha}_i = 0$, $\alpha_{1i} = \omega t_i$. The rearranged matrix equation is

$$\begin{bmatrix} 1 & l_1\cos\alpha_1 & -l_1\sin\alpha_1 & 0 & 0 \\ 0 & \frac{l_2}{2}\sin\alpha_2 & -\frac{l_2}{2}\cos\alpha_2 & -\frac{l_2}{2}\sin\alpha_2 & \frac{l_2}{2}\cos\alpha_2 \\ 0 & 0 & 0 & -l_3\cos\alpha_3 & l_3\sin\alpha_3 \\ 0 & 1 & 0 & 1 & 0 \\ 0 & 1 & 0 & 1 & 0 \end{bmatrix} \begin{Bmatrix} M(\alpha_{1i}) \\ B_X(\alpha_{1i}) \\ B_Y(\alpha_{1i}) \\ C_X(\alpha_{1i}) \\ C_Y(\alpha_{1i}) \end{Bmatrix}$$
$$= \begin{Bmatrix} -w_1\frac{l_1}{2}\sin\alpha_{1i} \\ -I_{2g}\ddot{\alpha}_{2i} \\ w_3\frac{l_3}{2}\sin\alpha_{3i} - I_{3D}\ddot{\alpha}_{3i} \\ -w_2 + m_2\left(l_1\cos\alpha_{1i}\omega^2 + \frac{l_2}{2}\sin\alpha_{2i}\dot{\alpha}_{2i}^2\right) - m_2\frac{l_2}{2}\cos\alpha_{2i}\ddot{\alpha}_{2i} \\ m_2\left(l_1\sin\alpha_{1i}\omega^2 + \frac{l_2}{2}\cos\alpha_{2i}\dot{\alpha}_{2i}^2\right) + m_2\frac{l_2}{2}\sin\alpha_{2i}\ddot{\alpha}_{2i} \end{Bmatrix}. \quad (5.268)$$

This model includes the first five equations from Equation 5.267. Again, this is an algebraic model to determine the required driving moment and resulting reaction-force components for prescribed motion $\dot{\alpha}_{1i} = \omega$, $\ddot{\alpha}_i = 0$, $\alpha_{1i} = \omega t_i$. The required power is $\text{Power}_i = M_i \times \omega$.

5.7.5.3 Alternative Slider-Crank Mechanism

Figure 5.64a shows an alternative slider-crank mechanism being driven by a moment $M(t)$ applied to bar AB. Bar AB is connected by a joint at B to bar BD. Bar BD slides in a bushing (having negligible inertia) at C. We will be neglecting friction due to the rotation at joints A, B, and the sliding motion at C. Bars AB and BC have mass m_1 and m_2, respectively. Their lengths are illustrated. Out task is to derive governing equations of motion for this mechanism.

The coordinate choices are shown in Figure 5.64b and repeats the variables of Figure 4.24 considered earlier. Note that S defines the distance from the slider at C to B.

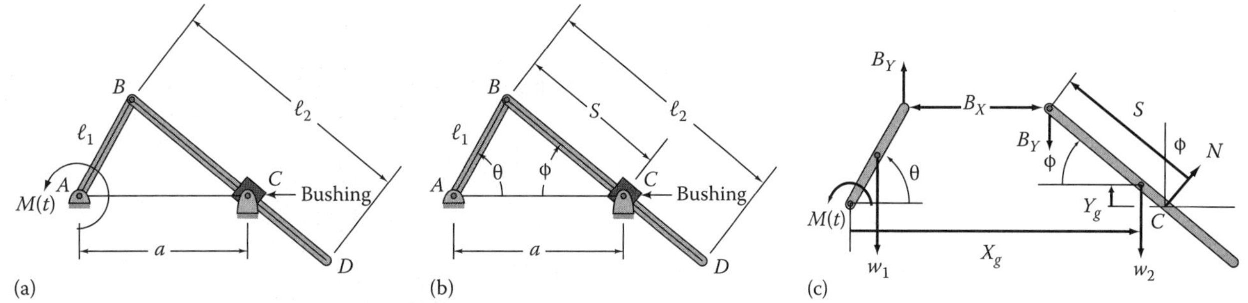

FIGURE 5.64
(a) Alternate slider-crank mechanism driven by a moment $M(t)$ acting on bar AB. (b) Coordinate choices. (c) Free-body diagram.

Since bar AB is rotating about A, a moment equation about A is appropriate and gives

$$I_A\ddot{\theta} = \sum M_A = M(t) - w_1\frac{l_1}{2}\cos\theta + B_X l_1 \sin\theta + B_Y l_1\cos\theta, \quad I_A = m_1\frac{l_1^2}{3}. \tag{5.269}$$

The plus and minus signs for the moment terms follow from the positive counterclockwise direction selected for θ. This equation has the three unknowns $\ddot{\theta}$, B_X, B_Y.

Bar BC is in general motion, and we will need to apply $\sum \boldsymbol{f} = m\ddot{\boldsymbol{R}}_g$ plus a moment equation. The force equation gives the component equations:

$$\begin{aligned} m_2\ddot{X}_g &= \sum f_X = B_X + N\sin\phi \\ m_2\ddot{Y}_g &= \sum f_Y = -B_Y - w_2 + N\cos\phi. \end{aligned} \tag{5.270}$$

These two equations introduce $\ddot{X}_g$, $\ddot{Y}_g$ as new unknowns. Taking moments about bar BC's mass center gives

$$I_{2g}\ddot{\phi} = \sum M_g = B_X\frac{l_2}{2}\sin\phi - B_Y\frac{l_2}{2}\cos\phi - N\left(S - \frac{l}{2}\right), \quad I_{2g} = m_2\frac{l_2^2}{12}. \tag{5.271}$$

and introduces the new unknown $\ddot{\phi}$. The plus and minus signs for the moments follow from the assumed positive clockwise rotation direction for ϕ. We now have four equations for the seven unknowns $\ddot{\theta}$, $\ddot{\phi}$, B_X, B_Y, N, $\ddot{X}_g$, $\ddot{Y}_g$.

The $\ddot{X}_g$ kinematic constraint equation follow from

$$X_g = l_1\cos\theta + \frac{l_2}{2}\cos\phi \Rightarrow \dot{X}_g = -l_1\sin\theta\dot{\theta} - \frac{l_2}{2}\sin\phi\dot{\phi}$$

$$\ddot{X}_g = -l_1\sin\theta\ddot{\theta} - l_1\cos\theta\dot{\theta}^2 - \frac{l_2}{2}\sin\phi\ddot{\phi} - \frac{l_2}{2}\cos\phi\dot{\phi}^2.$$

For $\ddot{Y}_g$,

$$Y_g = l_1\sin\theta - \frac{l_2}{2}\sin\phi \Rightarrow \dot{Y}_g = l_1\cos\theta\dot{\theta} - \frac{l_2}{2}\cos\phi\dot{\phi}$$

$$\ddot{Y}_g = l_1\cos\theta\ddot{\theta} - l_1\sin\theta\dot{\theta}^2 - \frac{l_2}{2}\cos\phi\ddot{\phi} + \frac{l_2}{2}\sin\phi\dot{\phi}^2.$$

Substituting $\ddot{X}_g$ and $\ddot{Y}_g$ into Equations 5.270 gives

$$\begin{aligned} &m_2\left(-l_1\sin\theta\ddot{\theta} - l_1\cos\theta\dot{\theta}^2 - \frac{l_2}{2}\sin\phi\ddot{\phi} - \frac{l_2}{2}\cos\phi\dot{\phi}^2\right) \\ &\quad = B_X + N\sin\phi \\ &m_2\left(l_1\cos\theta\ddot{\theta} - l_1\sin\theta\dot{\theta}^2 - \frac{l_2}{2}\cos\phi\ddot{\phi} + \frac{l_2}{2}\sin\phi\dot{\phi}^2\right) \\ &\quad = -B_Y - w_2 + N\cos\phi. \end{aligned} \tag{5.272}$$

Equations 5.269, 5.271, and 5.272 now comprise four equations in the five unknowns $\ddot{\theta}$, $\ddot{\phi}$, B_X, B_Y, N. The additional required kinematic equations are obtained from the X and Y components for the mechanism. From Figure 5.64b,

$$\begin{aligned} &X \text{ component: } l_1\cos\theta + S\cos\phi = a \\ &Y \text{ component: } l_1\sin\theta = S\sin\phi \end{aligned} \tag{5.273}$$

We are going to choose θ to be "the coordinate." With this decision, we can solve Equation 5.273 for S, ϕ as

$$S = \sqrt{a^2 - 2al_1\cos\theta + l_1^2}, \quad \tan\phi = \frac{\sin\phi}{\cos\phi} = \frac{l_1\sin\theta}{a - l_1\cos\theta} \tag{5.274}$$

Differentiating Equation 5.273 once with respect to time gives

$$\begin{aligned} -l_1\sin\theta\dot{\theta} + \dot{S}\cos\phi - S\dot{\phi}\sin\phi &= 0 \\ l_1\cos\theta\dot{\theta} &= \dot{S}\sin\phi + S\cos\phi\dot{\phi} \end{aligned} \tag{5.275}$$

These equations can be solved for $\dot{S}$, $\dot{\phi}$ as

$$\dot{S} = l_1\dot{\theta}\sin(\theta + \phi), \quad S\dot{\phi} = l_1\dot{\theta}\cos(\theta + \phi). \tag{5.276}$$

Differentiating Equations 5.276 again gives

$$\begin{aligned} \ddot{S} &= l_1\ddot{\theta}\cos(\theta - \phi) + l_1\dot{\theta}(\dot{\theta} + \dot{\phi})\cos(\theta + \phi) \\ S\ddot{\phi} &= -\dot{S}\dot{\phi} + l_1\ddot{\theta}\cos(\theta + \phi) - l_1\dot{\theta}(\dot{\theta} + \dot{\phi})\sin(\theta + \phi). \end{aligned} \tag{5.277}$$

The second of Equation 5.277 defines $\ddot{\phi}$, which completes the model. Putting Equations 5.269, 5.271, and 5.272 and the last of Equation 5.277 into matrix format gives

$$\begin{bmatrix} I_A & 0 & -l_1\sin\theta & -l_1\cos\theta & 0 \\ 0 & I_{g^2} & -\frac{l_2}{2}\sin\phi & \frac{l_2}{2}\cos\phi & \left(S - \frac{l_2}{2}\right) \\ -m_2 l_1\sin\theta & -m_2\frac{l_2}{2}\sin\phi & -1 & 0 & -\sin\phi \\ -m_2 l_1\cos\theta & -m_2\frac{l_2}{2}\cos\phi & 0 & 1 & -\cos\phi \\ l_1\cos(\theta+\phi) & -S & 0 & 0 & 0 \end{bmatrix} \times \begin{Bmatrix} \ddot{\theta} \\ \ddot{\phi} \\ B_X \\ B_Y \\ N \end{Bmatrix} = \begin{Bmatrix} M(t) - w_1\frac{l_1}{2}\cos\theta \\ 0 \\ m_2 l_1\cos\theta\dot{\theta}^2 + m_2\frac{l_2}{2}\cos\phi\dot{\phi}^2 \\ -w_2 + m_2 l_1\sin\theta\dot{\theta}^2 - m_2\frac{l_2}{2}\sin\phi\dot{\phi}^2 \\ \dot{S}\dot{\phi} + l_1\dot{\theta}(\dot{\theta}+\dot{\phi})\sin(\theta+\phi) \end{Bmatrix}. \tag{5.278}$$

The coefficient matrix and the right-hand-side vector are all defined in terms of θ and $\dot{\theta}$. S and ϕ are defined by Equation 5.274. The last of Equation 5.276 defines $\dot{\phi}$ in terms of θ, S, ϕ and $\dot{\theta}$. As with the preceding examples, we could restate the model to solve for the required driving torque $M(\theta)$ to rotate bar AB at a constant rate $\dot{\theta} = \omega$.

5.7.5.4 Closing Comments

Having worked through the material of this section, your "success" would be measured to the extent that you are thinking," Gee, these problems all seem to be worked in about the same way." The examples that have been worked through in detail all have more coordinates than degrees of freedom. Developing the equations of motion employs the following ordered steps:

a. The linkage assembly is drawn in a general configuration, with all coordinates identified.
b. A free-body diagram is drawn for each body in the assembly. In this diagram, the assembly is drawn in an exploded (disjointed) configuration showing the reaction-force components acting at each connection joint of the assembly. The free-body diagram includes all forces acting on a body (or particle) including reaction forces, weight, and external forces and moments.
c. Equations of motion are stated for each body. For a body that is rotating about a pivot point, stating a moment equation about the pivot point is best. A moment equation statement about any other point on the body will pull the reaction-force components at the pivot into the problem as unknowns. In stating the moment equation, remember, positive directions for moments coincide with positive directions for the rotation angle of the body. For a body that is both translating and rotating, the following equations of motion are generally recommended: (i) a moment equation stated for moments taken about the mass center, (ii) the components of $\sum f = m\ddot{R}_g$ for the mass center. Occasionally, taking moments about a point other than the mass center can be very helpful. Stating the moment equation about a pivot point will eliminate the reaction components at the pivot point (but not from the $\sum f = m\ddot{R}_g$ equation). Positive signs for moments in the moment equation are established by the designated positive direction of rotation for the body. Positive signs for forces in the $\sum f = m\ddot{R}_g$ equation follow from the assumed positive directions of mass center coordinates; that is, $+X_g$ and $+Y_g$. Hence, three differential equations of motion are obtained for each body in general motion (two translation and one rotation).
d. At this stage of development, having developed all independent equations from force and moment equations, the unknowns will include angular accelerations of the rigid bodies, mass center acceleration components, and internal reaction-force components. There will be more unknowns than equations, and the remaining equations to be developed are kinematic. Kinematic equations are used to define the mass center acceleration components in terms of the rigid-body angles and their derivatives, eliminating $\ddot{X}_g$, $\ddot{Y}_g$ unknown terms. The remaining kinematic equations follow from geometric constraint equations in the horizontal and vertical directions. Differentiating these equations twice will yield the required equations in the angular-acceleration unknowns. The geometric-differentiation approach was used exclusively in this subsection. The alternative vector-equation approach is presented (at length) in Section 4.5.
e. Having arrived at a sufficient number of equations to define the unknowns, a matrix equation is written for the unknown reaction forces and angular accelerations. This equation can be used as either (i) a differential-equation model with time-varying forces or moments as inputs, and accelerations and reaction forces as unknowns, or (ii) an algebraic model with a constant specified rotation rate as the input and reaction force and required driving moments as the output.

5.8 Moment of Momentum for Planar Motion

5.8.1 Developing Moment-of-Momentum Equations for Planar Motion of Rigid Bodies

The notion of moment of momentum for a particle was previously introduced in Section 3.8. Reviewing the moment-of-momentum concept for a particle may be helpful before taking on the idea for a rigid body. Figure 5.65 illustrates an inertial X, Y, Z coordinate system with origin at O.

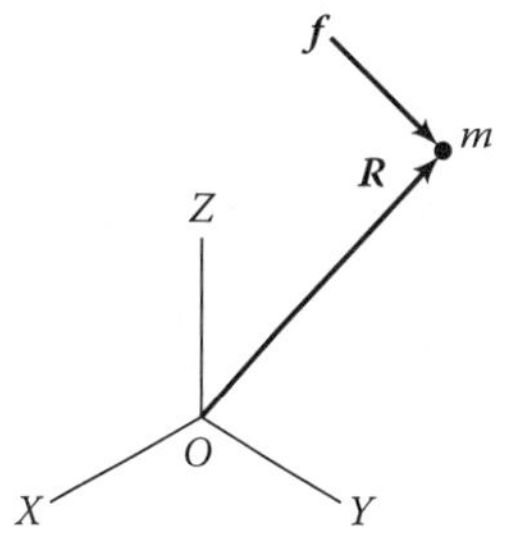

FIGURE 5.65
An inertial X, Y, Z coordinate with the particle m located by the vector $\boldsymbol{R} = \boldsymbol{I}R_X + \boldsymbol{J}R_Y + \boldsymbol{K}R_Z$.

The force f acts on the particle of mass m that is located by the vector $\boldsymbol{R}=\boldsymbol{I}R_X+\boldsymbol{J}R_Y+\boldsymbol{K}R_Z$. The moment of momentum of particle m about O is

$$\boldsymbol{H}_O=\boldsymbol{R}\times m\dot{\boldsymbol{R}}.$$

We demonstrated in Section 3.8 that $\boldsymbol{M}_o=\boldsymbol{R}\times\boldsymbol{f}=\dot{\boldsymbol{H}}_o$, where $\dot{\boldsymbol{H}}_o=d\boldsymbol{H}_o/dt|_{X,Y,Z}$ is the time derivative of $\boldsymbol{H}_O$ with respect to the X, Y, Z system. The proof started with the vector relationship

$$\dot{\boldsymbol{H}}_o=\left.\frac{d}{dt}\right|_{X,Y,Z}(\boldsymbol{R}\times m\dot{\boldsymbol{R}})=m(\dot{\boldsymbol{R}}\times\dot{\boldsymbol{R}}+\boldsymbol{R}\times\ddot{\boldsymbol{R}})=\boldsymbol{R}\times m\ddot{\boldsymbol{R}}.$$

By comparison, taking moments about O for both sides of $\boldsymbol{f}=m\ddot{\boldsymbol{R}}$ gives

$$\boldsymbol{M}_O=\boldsymbol{R}\times\boldsymbol{f}=\boldsymbol{R}\times m\ddot{\boldsymbol{R}}=\dot{\boldsymbol{H}}_O, \tag{5.279}$$

and proves our desired result. Most useful applications for this equation involve recognizing situations for which the moment is (always) zero about some axis. In such a case, moment-of-momentum about the zero-moment axis is constant or "conserved." For example, a constant zero moment about the Z-axis would yield from Equation 5.279 the component equation

$$M_{oz}=0\Rightarrow H_{oz}=\text{constant}=m(R_X\dot{R}_Y-R_Y\dot{R}_X)=mr^2\dot{\theta}.$$

For particles, $H_{OZ}=mr^2\dot{\theta}$ is generally more useful than the Cartesian-coordinate version.

Identifying situations for which moment of momentum is conserved is also the central thrust of moment-of-momentum applications in planar kinetics. First, we need an appropriate definition for a rigid body's moment of momentum. Figure 5.66 illustrates a rigid body with an imbedded, body-fixed x, y, z coordinate system, whose origin o is located in the inertial X, Y coordinate system by $\boldsymbol{R}_o$. Two position vectors locating a particle of mass dm at position P in the rigid body are (i) $\boldsymbol{\rho}=\boldsymbol{i}x+\boldsymbol{j}y+\boldsymbol{k}z$ in the x, y, z system and (ii) $\boldsymbol{r}=\boldsymbol{R}_o+\boldsymbol{\rho}$ in the X, Y system. The body's mass center is located in the x, y, z system by the position vector $\boldsymbol{b}_{og}$ defined earlier in Section 5.2 by

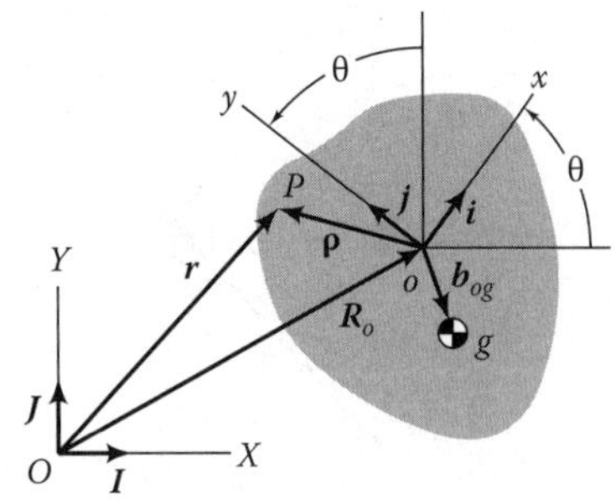

FIGURE 5.66
Body-fixed x, y, z system with origin o located in an inertial X, Y, Z coordinate system by $\boldsymbol{R}_o$. The body's mass center is located in the x, y, z system by $\boldsymbol{b}_{og}$.

$$m\boldsymbol{b}_{og}=m(\boldsymbol{i}b_{ogx}+\boldsymbol{j}b_{ogy}+\boldsymbol{k}b_{ogz})=\int_V\boldsymbol{\rho}\gamma\,dx\,dy\,dz\int_m\boldsymbol{\rho}\,dm, \tag{5.280}$$

where $\gamma(x,y,z)$ is the mass density of the body at point P.

Based on the earlier developments of this chapter leading to the moment equation (Equation 5.24), taking moments of momentum about o, the origin of the x, y, z system, would appear to be a reasonable choice for defining $\boldsymbol{H}_o$ and gives

$$\boldsymbol{H}_o=\int_m\boldsymbol{\rho}\times\dot{\boldsymbol{r}}\,dm=\int_m\boldsymbol{\rho}\times(\dot{\boldsymbol{R}}_o+\dot{\boldsymbol{\rho}})dm$$
$$=\int_m\boldsymbol{\rho}\times(\dot{\boldsymbol{R}}_o+\boldsymbol{\omega}\times\boldsymbol{\rho})dm. \tag{5.281}$$

In this equation, $\dot{\boldsymbol{\rho}}$ is the time derivative of $\boldsymbol{\rho}$ with respect to the X, Y coordinate system and is defined from Equation 2.52 as

$$\dot{\boldsymbol{\rho}}=\left.\frac{d\boldsymbol{\rho}}{dt}\right|_{X,Y}=\hat{\dot{\boldsymbol{\rho}}}+\boldsymbol{\omega}\times\boldsymbol{\rho},$$

where

$$\hat{\dot{\boldsymbol{\rho}}}=\left.\frac{d\boldsymbol{\rho}}{dt}\right|_{x,y,z}=\boldsymbol{i}\dot{x}+\boldsymbol{j}\dot{y}+\boldsymbol{k}\dot{z}=0.$$

The vector $\hat{\dot{\boldsymbol{\rho}}}$ is zero because ρ is fixed in the x, y, z system. Proceeding with Equation 5.281,

$$\boldsymbol{H}_o=m\boldsymbol{b}_{og}\times\dot{\boldsymbol{R}}_o+\int_m\boldsymbol{\rho}\times(\boldsymbol{\omega}\times\boldsymbol{\rho})\,dm.$$

Substituting for ρ and $\boldsymbol{\omega}=\boldsymbol{k}\dot{\boldsymbol{\theta}}$ gives

$$\boldsymbol{H}_o=m\boldsymbol{b}_{og}\times\dot{\boldsymbol{R}}_o+\boldsymbol{k}\dot{\boldsymbol{\theta}}\int_m(x^2+y^2)dm=m\boldsymbol{b}_{og}\times\dot{\boldsymbol{R}}_o+\boldsymbol{k}I_o\dot{\boldsymbol{\theta}}, \tag{5.282}$$

and completes the definition. Differentiating Equation 5.282 gives

$$\dot{\boldsymbol{H}}_o=\left.\frac{d\boldsymbol{H}_o}{dt}\right|_{X,Y}=(m\dot{\boldsymbol{b}}_{og}\times\dot{\boldsymbol{R}})+(m\boldsymbol{b}_{og}\times\ddot{\boldsymbol{R}})+\boldsymbol{k}I_o\ddot{\boldsymbol{\theta}},$$

with the z component,

$$H_{oz} = I_o\ddot{\theta} + (m\dot{\boldsymbol{b}}_{og} \times \dot{\boldsymbol{R}}_o)_z + (m\boldsymbol{b}_{og} \times \ddot{\boldsymbol{R}}_o)_z. \quad (5.283)$$

Comparing this result with the moment equation,

$$\sum M_{oz} = I_o\ddot{\theta} + m(\boldsymbol{b}_{og} \times \ddot{\boldsymbol{R}}_o)_z, \quad (5.24)$$

shows that, generally speaking, $\sum \boldsymbol{M}_o \neq \dot{\boldsymbol{H}}_o$. In words, the net moment about o, the origin of the body-fixed x, y, z system does *not* generally equal $d\boldsymbol{H}_o/dt|_{X,Y}$, the time derivative of $\boldsymbol{H}_o$ with respect to the X, Y system.

The presence of the second term in Equation 5.283 causes $\sum \boldsymbol{M}_o \neq \dot{\boldsymbol{H}}_{oz}$. This term disappears if $\boldsymbol{b}_{og} = 0$ in both equations; that is, taking moments about the mass center yields,

$$\sum M_{gz} = \dot{H}_{gz} = I_g\ddot{\theta}, \quad H_{gz} = I_g\dot{\theta}.$$

Similarly, if point o is fixed in space, such that $\dot{\boldsymbol{R}}_o = \ddot{\boldsymbol{R}}_o = 0$, Equations 5.283 and 5.24 provide

$$\sum M_{oz} = \dot{H}_{oz} = I_o\ddot{\theta}, \quad H_{oz} = I_o\dot{\theta},$$

which corresponds physically to taking moments about a point o, that is *both fixed in the body* and *fixed in space*. These are the obvious (and useful) circumstances for which the objectionable term disappears, and a zero moment implies conservation of momentum. However, these results were previously developed in Section 5.3 and do not provide any new moment-of-momentum information or tools.

Looking for a new and more useful definition, suppose we go back to the particle-dynamics precedent and define the moment-of-momentum vector about an arbitrary point A that is fixed in the inertial X, Y coordinate system. To simplify our algebra, we will use the arrangement of Figure 5.67 with the mass center located at the origin of the x, y, z system, causing

$$\int_m \boldsymbol{\rho}\, dm = m\boldsymbol{b}_{og} = 0.$$

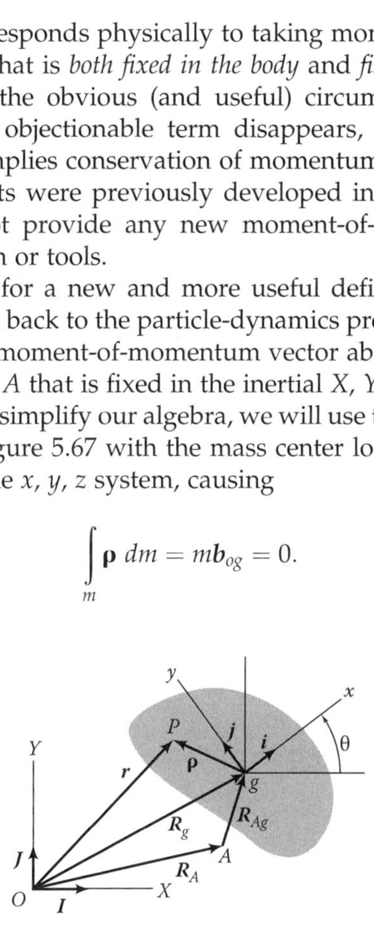

FIGURE 5.67
Body-fixed x, y, z system with origin o. The body's mass center g is located in the inertial X, Y system by $\boldsymbol{R}_g$.

The vector $\boldsymbol{R}_A$ locates an arbitrary fixed point A in the X, Y system. The vector $\boldsymbol{R}_{Ag}$ extends from A to the body's mass center at g, which is located in the X, Y, Z system by $\boldsymbol{R} = \boldsymbol{R}_A + \boldsymbol{R}_{Ag}$; hence, $\dot{\boldsymbol{R}}_g = \dot{\boldsymbol{R}}_{Ag}$ and $\ddot{\boldsymbol{R}}_g = \ddot{\boldsymbol{R}}_{Ag}$ since $\dot{\boldsymbol{R}}_A = \ddot{\boldsymbol{R}}_A = 0$. A point P in the rigid body is located by $\boldsymbol{r}$ in the X, Y system and $\boldsymbol{\rho} = \boldsymbol{i}x + \boldsymbol{j}y + \boldsymbol{k}z$ in the x, y, z system. From $\boldsymbol{r} = \boldsymbol{R}_A + \boldsymbol{R}_{Ag} + \boldsymbol{\rho}$, we obtain $\dot{\boldsymbol{r}} = \dot{\boldsymbol{R}}_g + \dot{\boldsymbol{\rho}}$ and $\ddot{\boldsymbol{r}} = \ddot{\boldsymbol{R}}_g + \ddot{\boldsymbol{\rho}}$. Given that points o and P are fixed in the rigid body, we can use Equations 4.3 to state the velocity and acceleration of point P as

$$\dot{\boldsymbol{r}} = \dot{\boldsymbol{R}}_g + \boldsymbol{\omega} \times \boldsymbol{\rho}, \quad \ddot{\boldsymbol{r}} = \ddot{\boldsymbol{R}}_g + \dot{\boldsymbol{\omega}} \times \boldsymbol{\rho} + \boldsymbol{\omega} \times (\boldsymbol{\omega} \times \boldsymbol{\rho}). \quad (5.284)$$

For moments taken about A, the moment of momentum is defined by

$$\begin{aligned} \boldsymbol{H}_A &= \int_m (\boldsymbol{R}_{Ag} + \boldsymbol{\rho}) \times \dot{\boldsymbol{r}}\, dm \\ &= \int_m (\boldsymbol{R}_{Ag} + \boldsymbol{\rho}) \times (\dot{\boldsymbol{R}}_g + \boldsymbol{\omega} \times \boldsymbol{\rho}) dm, \end{aligned}$$

with $\dot{\boldsymbol{r}} = \dot{\boldsymbol{R}}_g + \boldsymbol{\omega} \times \boldsymbol{\rho}$. Expanding terms gives

$$\begin{aligned} \boldsymbol{H}_A &= m(\boldsymbol{R}_{Ag} \times \dot{\boldsymbol{R}}_g) + \boldsymbol{R}_{Ag} \times \left(\boldsymbol{\omega} \times \int_m \boldsymbol{\rho}\, dm\right) \\ &\quad + \int_m \boldsymbol{\rho}\, dm \times \dot{\boldsymbol{R}}_g + \int_m \boldsymbol{\rho} \times (\boldsymbol{\omega} \times \boldsymbol{\rho})\, dm \\ &= m(\boldsymbol{R}_{Ag} \times \dot{\boldsymbol{R}}_g) + \int_m \boldsymbol{\rho} \times (\boldsymbol{\omega} \times \boldsymbol{\rho})\, dm, \end{aligned}$$

since $\int_m \boldsymbol{\rho}\, dm = m\boldsymbol{b}_{og} = 0$. Substituting $\boldsymbol{\rho} = \boldsymbol{i}x + \boldsymbol{j}y + \boldsymbol{k}z$ and $\boldsymbol{\omega} = \boldsymbol{k}\dot{\boldsymbol{\theta}}$ gives

$$H_{Az} = I_g\dot{\theta} + m(\boldsymbol{R}_{Ag} \times \dot{\boldsymbol{R}}_g)_z. \quad (5.285)$$

Differentiating this equation nets

$$\begin{aligned} \dot{H}_{Az} &= I_g\ddot{\theta} + m(\dot{\boldsymbol{R}}_{Ag} \times \dot{\boldsymbol{R}}_g)_z + m(\boldsymbol{R}_{Ag} \times \ddot{\boldsymbol{R}}_g)_z \\ &= I_g\ddot{\theta} + m(\boldsymbol{R}_{Ag} \times \ddot{\boldsymbol{R}}_g)_z. \end{aligned}$$

The term $\dot{\boldsymbol{R}}_{Ag} \times \dot{\boldsymbol{R}}_g$ is zero because $\dot{\boldsymbol{R}}_g = \dot{\boldsymbol{R}}_{Ag} + \dot{\boldsymbol{R}}_A = \dot{\boldsymbol{R}}_{Ag}$, since $\dot{\boldsymbol{R}}_A = 0$.

Now the question is: Does $\sum M_{Az} = \dot{H}_{Az}$? Taking moments about A of both sides of $\boldsymbol{f} = dm\, \ddot{\boldsymbol{r}}$ for a particle of mass dm gives $(\boldsymbol{R}_{Ag} + \boldsymbol{\rho}) \times \boldsymbol{f} = (\boldsymbol{R}_{Ag} + \boldsymbol{\rho}) \times dm\, \ddot{\boldsymbol{r}}$, and integrating over the mass of the body gives

$$\begin{aligned} \boldsymbol{M}_{Ar} &= \int_m (\boldsymbol{R}_{Ag} + \boldsymbol{\rho}) \times \{\ddot{\boldsymbol{R}}_g + (\dot{\boldsymbol{\omega}} \times \boldsymbol{\rho}) + \boldsymbol{\omega} \times [\boldsymbol{\omega} \times (\boldsymbol{\omega} \times \boldsymbol{\rho})]\} dm \\ &= \boldsymbol{R}_{Ag} \times m\ddot{\boldsymbol{R}}_g + \int_m \boldsymbol{\rho}\, dm \times \ddot{\boldsymbol{R}}_g + \boldsymbol{R}_{Ag} \times \left(\dot{\boldsymbol{\omega}} \times \int_m \boldsymbol{\rho}\, dm\right) \\ &\quad + \boldsymbol{R}_{Ag} \times \left[\boldsymbol{\omega} \times \left(\boldsymbol{\omega} \times \int_m \boldsymbol{\rho}\, dm\right)\right] + \int_m \boldsymbol{\rho} \times (\dot{\boldsymbol{\omega}} \times \boldsymbol{\rho}) dm \\ &\quad + \boldsymbol{\omega} \times \int_m \boldsymbol{\rho} \times (\boldsymbol{\omega} \times \boldsymbol{\rho}) dm \\ &= \boldsymbol{R}_{Ag} \times m\ddot{\boldsymbol{R}}_g + \int_m \boldsymbol{\rho} \times (\dot{\boldsymbol{\omega}} \times \boldsymbol{\rho})\, dm + \boldsymbol{\omega} \times \int_m \boldsymbol{\rho} \times (\boldsymbol{\omega} \times \boldsymbol{\rho}) dm, \end{aligned}$$

where

$\boldsymbol{M}_{Ar}$ is the resultant moment acting on the rigid body about A

$\ddot{r}$ is defined by the second of Equation 5.284

$b_{og} = 0$

Substituting for $\boldsymbol{\omega}$, $\dot{\boldsymbol{\omega}}$ and $\boldsymbol{\rho}$ and integrating yields

$$\begin{aligned} M_{Arz} &= (\boldsymbol{R}_{Ag} \times m\ddot{\boldsymbol{R}}_g)_z + I_g\ddot{\theta} + \boldsymbol{k}\dot{\boldsymbol{\theta}} \times \boldsymbol{k} I_g \dot{\boldsymbol{\theta}} \\ &= (\boldsymbol{R}_{Ag} \times m\ddot{\boldsymbol{R}}_g)_z + I_g\ddot{\theta}. \end{aligned}$$

Comparing this result with Equation 5.285 confirms the result: For moments and moments of momentum about a point A that is fixed in inertial space,

$$\begin{aligned} M_{Azr} &= \dot{H}_{Az} = I_g\ddot{\theta} + (\boldsymbol{R}_{Ag} \times m\ddot{\boldsymbol{R}}_g)_z, \\ \text{where } H_{Az} &= I_g\dot{\theta} + (\boldsymbol{R}_{Ag} \times m\dot{\boldsymbol{R}}_g). \end{aligned} \tag{5.286}$$

This equation completes the required moment-of-momentum equations. In the balance of this section, Equation 5.286 will be applied only to examples where the moment about a z-axis through point A (fixed in space) is zero, implying that the z moment-of-momentum component, $H_{Az} = m(\boldsymbol{R}_{Az} \times \dot{\boldsymbol{R}}_g)_z + I_g\dot{\theta}$, is constant. Most applications of Equation 5.286 are concerned with the collisions of rigid bodies in planar motion. After spending most of your time deriving equations of motion or seeking to integrate equations of motion, you may have trouble "shifting gears" into the proper application of Equation 5.286.

5.8.2 Applying Moment-of-Momentum Equations in Planar Dynamics

We will be applying the following moment-of momentum results in this section:

Moments about the mass center g of a rigid body, implying

$$\sum M_{gz} = \dot{H}_{gz} = I_g\ddot{\theta}; \quad H_{gz} = I_{gz}\dot{\theta}.$$

Moments about a point o, fixed in a rigid body and fixed in inertial space, implying

$$\sum M_{oz} = \dot{H}_{oz} = I_o\ddot{\theta}; \quad H_{oz} = I_o\dot{\theta}.$$

Moments taken about point A, fixed in inertial space, implying

$$\begin{aligned} M_{Azr} &= \dot{H}_{Az} = I_g\ddot{\theta} + (\boldsymbol{R}_{Ag} \times m\ddot{\boldsymbol{R}}_g)_z, \\ H_{Az} &= I_g\dot{\theta} + (\boldsymbol{R}_{Ag} \times m\dot{\boldsymbol{R}}_g). \end{aligned} \tag{5.287}$$

We will start with simple demonstrations of $\sum M_{gz} = I_g\ddot{\theta}$ and $\sum M_{oz} = I_o\ddot{\theta}$, but spend most of our time applying the more general (and less obvious) Equations 5.287.

5.8.2.1 Two Spinning Wheels Connected by an Adjustable-Tension Belt

Figure 5.68 illustrates two pulleys connected by an adjustable tension belt. In Figure 5.68a, the idler pulley at C has been moved away from the belt, relieving it from tension and allowing the belt to slip loosely on

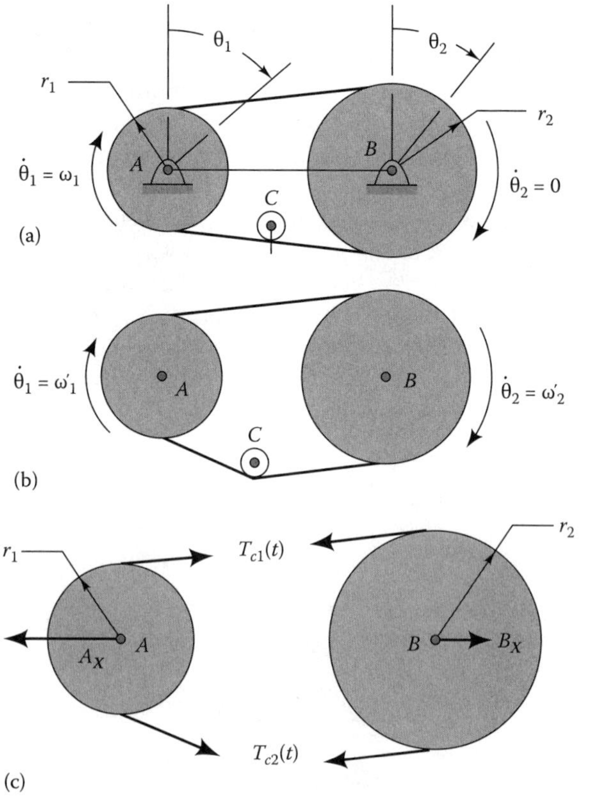

FIGURE 5.68
Two-wheel belt assembly. (a) Idler pulley at C is disengaged with zero belt tension and the belt and driven wheel at rest. (b) Idler wheel engaged with no slipping between the wheels and the belt. (c) Free-body diagram for the wheels during slipping.

both the drive pulley to the left and the driven pulley at the right. In this initial circumstance, the angular velocities of the two disks are $\dot{\theta}_1(t_1) = \omega_1$, $\dot{\theta}_2(t_1) = 0$. In Figure 5.68b, the idler pulley has been moved down, increasing the belt tension and causing the right wheel to accelerate and the left wheel to decelerate before reaching their final nonslipping angular velocities $\dot{\theta}_1(t_1 + \Delta t) = \omega_1'$, $\dot{\theta}_2(t_1 + \Delta t) = \omega_2'$. This type of arrangement might be used in a riding lawn mower where the driven wheel is powered by the torque of the motor plus the effective inertia of the motor's drive assembly. For the present example, no external moments act on the pulleys. The mass, radius, and radius of gyration for the left and right wheels are (m_1, r_1, k_{g1}) and (m_2, r_2, k_{g2}), respectively.

The following tasks apply:

a. Determine the final, nonslipping angular velocities for the two wheels, ω_1', ω_2'.
b. What fraction of kinetic energy is dissipated in transferring from the initial slipping condition to the final nonslipping condition?

First, suppose that we work through Task a without thinking about conservation of momentum. Figure 5.68c provides free-body diagrams that are appropriate for the time interval Δt during which the belt is engaged, and the two wheel's angular velocities are changing. Note that tension $T_{c1}(t)$ applies in the upper portion of the belt, while tension $T_{c2}(t)$ holds for the lower belt segment (neglecting the inertia of the idler pulley). The difference in the two tension values causes wheel 1 to decelerate while wheel 2 accelerates. Writing moment equations for the two pulleys about their mass centers gives

$$\sum M_A = T_{c1}(t)r_1 - T_{c2}(t)r_1 = I_A\ddot{\theta}_1 = m_1k_{g1}^2\ddot{\theta}_1$$
$$\sum M_B = T_{c2}(t)r_2 - T_{c1}(t)r_2 = I_B\ddot{\theta}_2 = m_2k_{g2}^2\ddot{\theta}_2.$$

Moments are positive in a clockwise direction corresponding to $+\theta_1$ direction and $+\theta_2$ direction. Integrating the equations over the Δt time interval provides

$$\int_{t_1}^{t_1+\Delta t} [T_{c1}(t)r_1 - T_{c2}(t)r_2]dt$$
$$= m_1k_{g1}^2\dot{\theta}_1(t_1 + \Delta t) - m_1k_{g1}^2\dot{\theta}_1(t_1) = m_1k_{g1}^2\omega_1' - m_1k_{g1}^2\omega_1$$
$$-\int_{t_1}^{t_1+\Delta t} [T_{c1}(t)r_1 - T_{c2}(t)r_2]dt$$
$$= m_2k_{g2}^2\dot{\theta}_2(t_1 + \Delta t) - m_2k_{g2}^2\dot{\theta}_2(t_1) = m_2k_{g2}^2\omega_2' - 0.$$

As with particle dynamics, the integrals on the left of these equations are called "angular impulse." Each equation states: *Angular Impulse* $= \Delta H$; that is, angular impulse equals the change in moment of momentum. Adding the two equations gives

$$0 = m_1k_{g1}^2\omega_1' - m_1k_{g1}^2\omega_1 + m_2k_{g2}^2\omega_2'$$
$$\Rightarrow m_1k_{g1}^2\omega_1' + m_2k_{g2}^2\omega_2' = m_1k_{g1}^2\omega_1, \tag{5.288a}$$

which provides one equation in the two unknowns ω_1', ω_2'. To complete engineering-analysis Task a, we need the kinematic equation $r_1\omega_1' = r_2\omega_2'$ that applies when the belt has stopped slipping. Substituting this result into Equation 5.288a gives

$$\omega_1' = \frac{m_1k_{g1}^2\omega_1}{m_1k_{g1}^2 + m_2k_{g2}^2(r_2/r_1)}, \quad \omega_2' \frac{m_1k_{g1}^2\omega_1}{m_1k_{g1}^2(r_1/r_2) + m_2k_{g2}^2}, \tag{5.288b}$$

and concludes Task a.

Suppose that we rethink the task in terms of conservation of moment of momentum. No external moment acts on the *system* composed of the two wheels and belt. Hence, assuming counterclockwise momenta to be positive, the z component of the system moment-of-momentum vector is $H_A(t_1) + H_B(t_1) = mk_{g1}^2\omega_1 + 0$ before the belt is engaged, and is $H_A(t_1 + \Delta t) + H_B(t_1 + \Delta t) = mk_{g1}^2\omega_1' + mk_{g2}^2\omega_2'$ after slipping has ended. Equating these results gives

$$m_1k_{g1}^2\omega_1' + m_2k_{g2}^2\omega_2' = m_1k_{g1}^2\omega_1,$$

which repeats Equation 5.287. Recognizing that system moment of momentum is conserved saves a great deal of time and effort, given that the result can be stated by inspection.

There is a direct parallel in the present example to some of the one-dimensional, particle-collision examples in Section 3.7.1. In particular, examples in which two particles "stick together" and have a common velocity following a collision are similar to the present example in which the kinematic condition $r_1\omega_1' = r_2\omega_2'$ applies following the belt engagement.

Moving to Task b, the initial kinetic energy is $T_i = I_1\omega_1^2/2$, and the final kinetic energy is $T_f = I_1\omega_1'^2/2 + I_2\omega_2'^2/2$. Substituting from Equation 5.288 for ω_1', ω_2' (plus algebra) gives

$$\frac{T_f - T_i}{T_i} = \frac{-2r_1r_2I_1I_2}{r_1I_1^2 + 2r_1r_2I_1I_2 + r_2^2I_2^2}.$$

While concluding Task b, this result shows that energy is dissipated, but is otherwise not particularly motivating. To get some idea of the energy dissipated, if the wheels are identical ($I_1 = I_2$; $r_1 = r_2$), 50% of the initial kinetic energy will be dissipated during the transition to the final angular velocities. As with collisions between particles, interactions between bodies such that moment of momentum is conserved are almost always accompanied by a loss in energy.

5.8.2.2 A Particle Impacting a Compound Pendulum

Figure 5.69a illustrates a particle of mass m with a horizontal velocity $\boldsymbol{v} = \boldsymbol{I}v_i$. As shown in Figure 5.69b, it will impact a vertical compound pendulum of length L at a distance b down from the pivot point O. The pendulum is at rest before the collision and has mass M. The bullet lodges in the pendulum following the collision.

The following tasks apply:

a. Find the angular velocity of the pendulum $\dot{\theta}_2$ immediately following the collision.
b. Determine the horizontal linear impulse due to the reaction force at the support point O.

We will follow the same approach with this example as the preceding; namely, first develop a solution without regard for conservation of momentum and then develop the same result (more quickly) via conservation of *system* moment of momentum.

The free-body diagram of Figure 5.69b gives the following equations of motion for the particle and pendulum:

$$\sum f_X = -f(t) = m\dot{v} \Rightarrow -\int_0^{\Delta t} f(t)dt = mv_2 - mv_1$$

$$\sum M_O = f(t)b = I_O\ddot{\theta} \Rightarrow \int_0^{\Delta t} f(t)b\,dt = I_O\dot{\theta}_2 - I_O\dot{\theta}_1 = I_o\dot{\theta}_2. \tag{5.290}$$

The first equation starts with the X component of $\sum \boldsymbol{f} = m\ddot{\boldsymbol{r}}$ for the particle. The second equation starts with the moment equation for the pendulum about the fixed pivot-point O, with positive moments in the counterclockwise $+\theta$ direction. The duration of the collision Δt is very short, so short that the pendulum's position does not change. The velocity changes, but the position does not.

Eliminating the integral in Equations 5.290 gives

$$bmv_1 = I_O\dot{\theta}_2 + bmv_2, \tag{5.291}$$

which is one equation for the two unknowns v_2, $\dot{\theta}_2$. The concluding kinematic equation is $v_2 = b\dot{\theta}_2$, and substitution provides the requested answer

$$\dot{\theta}_2 = \frac{bmv_1}{(I_O + mb^2)}. \tag{5.292}$$

Now, suppose we think about applying conservation of moment of momentum to this example. Clearly, the momenta of the particle and the pendulum change individually. However, the force $\boldsymbol{I}f(t)$ and its reaction $-\boldsymbol{I}f(t)$ are internal to the *system* composed of the particle and the pendulum. The reaction forces O_X, O_Y of Figure 5.69b act through the pivot point O and create no moments about O. Hence, $\sum M_{Oz}$, the *system* moment of momentum

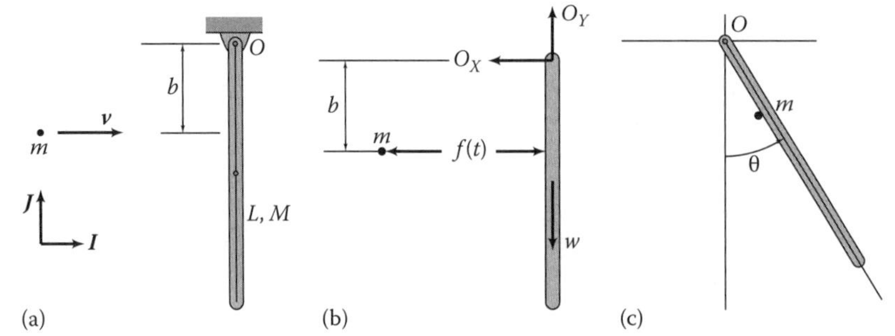

FIGURE 5.69
(a) Particle of mass m impacting an initially motionless, vertical, compound pendulum. (b) Free-body diagram for the particle and pendulum during collision. (c) Post-collision motion of the pendulum with imbedded particle.

about the fixed point O, is conserved; that is, $\sum M_{Oz} = 0 = \sum \dot{H}_{Oz} \Rightarrow \sum H_{Oz} =$ constant before and after the collision.

This is precisely what Equation 5.291 states. Immediately before the collision, the moment-of-momentum elements for the system components are as follows:

a. The pendulum is motionless and has zero moment of momentum.
b. The particle's moment of momentum with respect to O is $(-\boldsymbol{J}b \times \boldsymbol{I}mv_1 = \boldsymbol{K}bmv_1)$.

The left-hand side of Equation 5.291 is the sum of these two terms with counterclockwise moments taken as positive.

Immediately following the collision, the system's moment of momentum with respect to O is the sum of

a. The pendulum's moment of momentum with respect to O, $I_O\dot{\theta}_2$
b. The particle's moment of momentum with respect to O, $(-\boldsymbol{J}b \times \boldsymbol{I}mv_2 = \boldsymbol{K}bmv_2)$.

The right-hand side of Equation 5.291 is the sum of these two terms, and Equation 5.291 can be written by inspection, stating that the system moment of momentum about O is conserved, that is, $\sum H_{O1} = \sum H_{O2}$.

The request in Task b is not particularly motivating, but it leads to an interesting concept. The horizontal reaction force $O_X(t)$ in Figure 5.69b can be obtained from the horizontal component of $\sum \boldsymbol{f} = m\ddot{\boldsymbol{R}}_g$ as

$$f(t) - O_X(t) = m\ddot{X}_g = m\frac{L\ddot{\theta}}{2}$$

$$\Rightarrow \int_0^{\Delta t} f(t)dt - \int_0^{\Delta t} O_X(t)dt = \frac{mL}{2}(\dot{\theta}_2 - \dot{\theta}_1) = \frac{mL}{2}\dot{\theta}_2.$$

Hence, the requested linear impulse is

$$\int_0^{\Delta t} O_X(t)dt = -\int_0^{\Delta t} f(t)dt - \frac{mL}{2}\dot{\theta}_2 = m(v_1 - v_2) - \frac{mL}{2}\dot{\theta}_2.$$

The first of Equations 5.290 was used to eliminate the right-hand integral. Substituting $v_2 = b\dot{\theta}_2$ and then substituting for $\dot{\theta}_2$ from Equation 5.292 gives

$$\int_0^{\Delta t} O_X(t)\, dt = mv_1\left[1 - \frac{mb(b + l/2)}{(I_O + mb^2)}\right], \tag{5.293}$$

which concludes Task b.

Now comes the interesting part, Equation 5.293 suggests that the distance b can be selected, such that there will be no horizontal impulse at O, hence no horizontal reaction force at O either. Solving Equation 5.293 for this b value gives

$$mv_1\left[1 - \frac{m\bar{b}(\bar{b} + l/2)}{(I_O + m\bar{b}^2)}\right] = 0 \Rightarrow \bar{b} = \frac{2l}{3}.$$

The point located by $\bar{b} = 2l/3$ is called the "center of percussion." In referring to baseball bats or tennis racquets, it is called the "sweet spot." Hitting a ball at the center of percussion in a baseball bat will not cause "stinging" or unwanted vibrations in your hands.

5.8.2.3 A Spinning Baton Striking the Ground

The two examples above have been concerned with fixed-axis rotation, and have accordingly not used the general moment-of-momentum definition provided by Equation 5.287. The spinning baton of Figure 5.70a

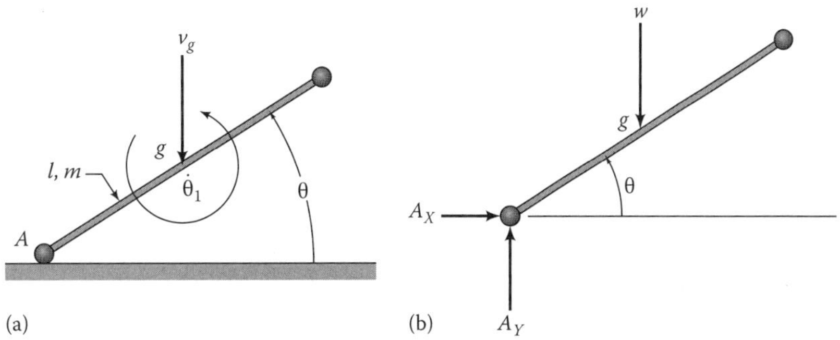

FIGURE 5.70
(a) Falling baton with vertical velocity v_g, angular velocity $\dot{\theta}_1$, length l, and mass m. (b) Free-body diagram for forces during collision.

illustrates the first of several examples that are most easily resolved via the general equation. At impact, the mass center of the baton in Figure 5.70a has the vertical velocity $v_g = -Jv_{gl}$ and the counterclockwise angular velocity $\dot{\theta}_1$. The baton has mass m, length l, and a radius of gyration about the mass center of k_g. The engineering-analysis task is to *determine the baton's angular velocity* $\dot{\theta}_2$ *immediately following impact*.

In this example, we are going to continue our practice of first working the problem the "hard way," followed by a more direct solution from moment-of-momentum conservation. The free-body diagram of Figure 5.70b applies during the impact process and provides, from $\sum f = m\ddot{\boldsymbol{R}}_g$, the two linear impulse-momentum component equations:

$$\sum f_X = A_X(t) = m\ddot{X}_g$$

$$\Rightarrow \int_0^{\Delta t} A_X(t)dt = m\dot{X}_{g2} - m\dot{X}_{g1} = m\dot{X}_{g2} - m(0)$$

$$\sum f_Y = A_Y(t) - w = m\ddot{Y}_g \Rightarrow \int_0^{\Delta t} A_Y(t)\,dt - \int_0^{\Delta t} w\,dt$$

$$= m\dot{Y}_{g2} - m\dot{Y}_{g1} = m\dot{Y}_{g2} - mv_{g1}. \quad (5.294)$$

The moment equation about g provides

$$\sum M_g = A_X\frac{l}{2}\sin\theta - A_Y\frac{l}{2}\cos\theta = I_g\ddot{\theta}$$

$$\Rightarrow \int_0^{\Delta t}\left[A_X(t)\frac{l}{2}\sin\theta(t) - A_Y(t)\frac{l}{2}\cos\theta(t)\right]dt = I_g\dot{\theta}_2 - I_g\dot{\theta}_1. \quad (5.295)$$

We are going to assume that $\theta = \bar{\theta}$ and does not change over the short time interval of the collision. This assumption reduces Equation 5.295 to

$$\frac{l}{2}\sin\bar{\theta}\int_0^{\Delta t} A_X(t)dt - \frac{l}{2}\cos\bar{\theta}\int_0^{\Delta t} A_Y(t)dt = I_g\dot{\theta}_2 - I_g\dot{\theta}_1 \quad (5.296)$$

We need one additional assumption in the second of Equations 5.294; namely, the impulse due to weight, $\int_0^{\Delta t} w(t)dt$, is negligible in comparison to the remaining impulse due to $A_Y(t)$. This assumption is reasonable at face value since w is not the force that causes a change in momenta. In a sense, it is also consistent with the assumption that $\theta = \bar{\theta}$ throughout the collision.

Following the collision, when the baton pivots about A, and g changes position, w will do work and change the kinetic energy of the bar. However, during the collision, with no change in position, w does no work.

These assumptions still leave us with three equations for the five unknowns: $\dot{X}_{g2}$, $\dot{Y}_{g2}$, $\dot{\theta}_2$, $\int_0^{\Delta t} A_X(t)dt$, $\int_0^{\Delta t} A_Y(t)dt$. The required kinematic condition following impact is

$$\dot{\boldsymbol{R}}_{gf} = \boldsymbol{I}\dot{X}_{g2} + \boldsymbol{J}\dot{Y}_{g2} = \dot{\boldsymbol{R}}_A + \boldsymbol{K}\dot{\theta}_2 \times \frac{l}{2}(\boldsymbol{I}\cos\bar{\theta} + \boldsymbol{J}\sin\bar{\theta})$$

$$= 0 + \frac{l\dot{\theta}_2}{2}(\boldsymbol{J}\cos\bar{\theta} - \boldsymbol{I}\,\sin\bar{\theta})$$

$$\therefore \dot{X}_{g2} = -\frac{l\dot{\theta}_2}{2}\sin\bar{\theta}, \quad \dot{Y}_{g2} = -\frac{l\dot{\theta}_2}{2}\cos\bar{\theta}. \quad (5.297)$$

Substituting $\dot{X}_{g2}$, $\dot{Y}_{g2}$ into Equation 5.294 defines the impulse integrals as

$$\int_0^{\Delta t} A_X(t)dt = m\dot{X}_{g2} = -m\frac{l\dot{\theta}_2}{2}\sin\bar{\theta},$$

$$\int_0^{\Delta t} A_Y(t)dt = m(\dot{Y}_{g2} - v_{g1}) = m\left(\frac{l\dot{\theta}_2}{2}\cos\bar{\theta} - v_{gl}\right).$$

Substituting these results into Equation 5.296 gives

$$\frac{l}{2}\sin\bar{\theta}\left(-m\frac{l\dot{\theta}_2}{2}\sin\bar{\theta}\right) - \frac{l}{2}\cos\bar{\theta}\,m\left(\frac{l\dot{\theta}_2}{2}\cos\bar{\theta} + v_{g1}\right)$$

$$= -\frac{ml^2}{4}\dot{\theta}_2 - \frac{mlv_{gl}}{2}\cos\bar{\theta} = I_g\dot{\theta}_2 - I_g\dot{\theta}_1. \quad (5.298)$$

Gathering terms and substituting $I_g = mk_g^2$ nets

$$\left(I_g + \frac{ml^2}{4}\right)\dot{\theta}_2 = I_g\dot{\theta}_1 - mv_{g1}\frac{1}{2}\cos\bar{\theta}$$

$$\Rightarrow \left(mk_g^2 + \frac{ml^2}{4}\right)\dot{\theta}_2 = mk_g^2\dot{\theta}_1 - mv_{gt}\frac{l}{2}\cos\bar{\theta}, \quad (5.299)$$

$$\therefore \dot{\theta}_2 = \frac{4k_g^2\dot{\theta}_1 - 2v_g l\cos\bar{\theta}}{4k_g^2 + l^2},$$

which provides the requested solution for $\dot{\theta}_2$

Now, think about the same task in terms of conservation of moment of momentum. The point A is fixed in space and the reaction force acts at A. Hence, $\sum M_{Az} = 0 \Rightarrow H_{Az} = \text{constant} \Rightarrow H_{Az1} = H_{Az2}$. Note that we have (again) neglected the moment due to weight to reach this result. The initial required moment of momentum is

$$\begin{aligned} H_{Az1} &= I_g\dot{\theta}_1 + (\boldsymbol{R}_{g1} \times m\dot{\boldsymbol{R}}_{g1})_z \\ &= I_g\dot{\theta}_1 + \frac{lm}{2}(\boldsymbol{I}\cos\overline{\theta} + \boldsymbol{J}\sin\overline{\theta}) \times (\boldsymbol{I}\dot{X}_{g1} + \boldsymbol{J}\dot{Y}_{g1}) \\ &= I_g\dot{\theta}_1 + \frac{lm}{2}(\boldsymbol{I}\cos\overline{\theta} + \boldsymbol{J}\sin\overline{\theta}) \times -\boldsymbol{J}v_{g1} \\ &= I_g\dot{\theta}_1 - mv_{g1}\frac{l}{2}\cos\overline{\theta} = mk_g^2\dot{\theta}_1 - mv_{g1}\frac{l}{2}\cos\overline{\theta} \end{aligned}$$

Immediately after the collision, the baton is in pure rotation about A; hence, the moment of momentum is

$$H_{Az2} = I_A\dot{\theta}_2 = \left(m_1k_g^2 + m\frac{l^2}{4}\right)\dot{\theta}_2.$$

Equating these two results gives

$$\begin{aligned} H_{Az1} = H_{Az2} &\Rightarrow mk_g^2\dot{\theta}_1 - mv_{g1}\frac{l\cos\overline{\theta}}{2} = \left(mk_g^2 + m\frac{l^2}{4}\right)\dot{\theta}_2 \\ &\Rightarrow \dot{\theta}_2 = \frac{4kg_g^2\dot{\theta}_1 - 2v_gl\cos\overline{\theta}}{4k_g^2 + l^2}, \end{aligned}$$

which coincides with our earlier results of Equation 5.299.

5.8.2.4 *A Rolling Cylinder That Encounters an Inclined Plane*

Figure 5.71a illustrates a cylinder of mass m and radius r that is rolling without slipping along a horizontal plane. At some point, the cylinder encounters and rolls without slipping up an intersecting plane that is inclined at the angle α to the horizontal. The initial velocity of the cylinder's center is v_1. The following tasks apply:

a. Find the cylinder's angular velocity immediately after it begins moving up the inclined plane.
b. Determine the fraction of energy dissipated during the collision of the cylinder with the inclined plane.

As noted previously, the key to application of conservation of moment of momentum for this type of example concerns finding a point that is fixed in space for which the net moment acting on the rigid body is zero. Figure 5.71b provides a free-body diagram that applies during the collision process. Note that the reaction force components that cause the change in momenta act on the cylinder at A. Hence, neglecting the weight and the reaction-force components at point B, the moment about A is zero, and $\sum M_A = 0 \Rightarrow H_{Az} =$ constant. Taking clockwise moments as positive, the moment of momentum about A before contacting the wall is

$$H_{Az1} = mv_1 \times r\cos\alpha + I_g\dot{\theta}_1 = \frac{mr^2}{2}(1 + 2\cos\alpha)\dot{\theta}_1; \tag{5.300a}$$

where the rolling-without-slipping kinematic condition is $v_1 = r\dot{\theta}_1$, and $I_g = mr^2/2$. The first term in Equation 5.300a is the initial linear momentum of the cylinder mv_1 times the perpendicular distance between its line of action and point A, $r\cos\alpha$. The cylinder's moment of momentum following the collision is

$$H_{Az2} = mv_2 \times r + I_g\dot{\theta}_2 = \left(mr^2 + \frac{mr^2}{2}\right)\dot{\theta}_2 = \frac{3mr^2}{2}\dot{\theta}_2, \tag{5.300b}$$

where $v_2 = r\dot{\theta}_2$, the post-collision, rolling-without-slipping kinematic condition has been used. Equating the results of Equations 5.300 gives

$$\frac{mr^2}{2}(1 + 2\cos\alpha)\dot{\theta}_1 = \frac{3mr^2}{2}\dot{\theta}_2.$$

Hence,

$$\begin{aligned} \dot{\theta}_2 &= \frac{\dot{\theta}_1}{3}(1 + 2\cos\alpha) = \frac{v_1}{3r}(1 + 2\cos\alpha) \\ v_2 &= \frac{v_1}{3}(1 + 2\cos\alpha), \end{aligned} \tag{5.301}$$

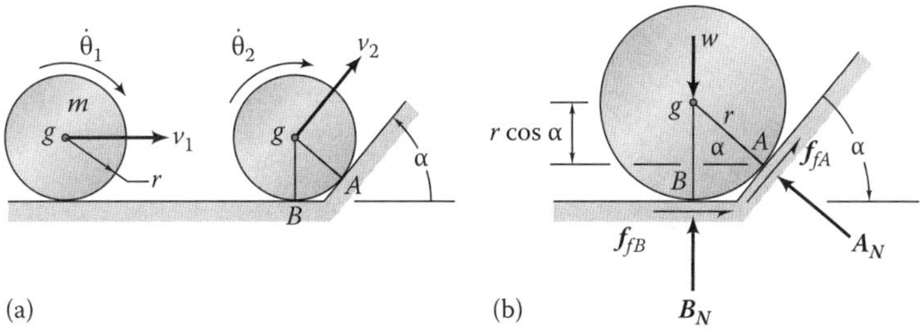

FIGURE 5.71
(a) Cylinder rolling without slipping on a horizontal plane and then shifting to rolling without slipping up an inclined plane. (b) Free-body diagram during the collision process.

which concludes Task a. The results of Equations 5.301 are reasonable since for $\alpha = 0$ (no inclined plane) there is no change in the angular or linear velocity terms.

The kinetic energies before and after colliding with the inclined plane are

$$T_1 = \frac{mv_1^2}{2} + \frac{I_g\dot{\theta}_1^2}{2} = \left(mr^2 + \frac{mr^2}{2}\right)\frac{\dot{\theta}_1^2}{2} = \frac{3mr^2\dot{\theta}_1^2}{4}$$

$$T_2 = \frac{mv_2^2}{2} + \frac{I_g\dot{\theta}_2^2}{2} = \frac{mr^2}{12}(1 + 2\cos\alpha)^2\dot{\theta}_1^2.$$

Hence, the fraction of kinetic energy lost during the initial contact with the inclined plane is

$$\frac{T_2 - T_1}{T_1} = -\frac{4}{9}(1 + \sin^2\alpha - \cos\alpha). \quad (5.302)$$

As would be expected, for $\alpha = 0$ there is no energy loss. Equation 5.302 serves as a continuing reminder that, despite conservation of moment of momentum, kinetic energy is always lost during a real collision.

5.9 Summary and Discussion

We shifted in this chapter from particle kinetics to rigid body kinetics. Initially, the shift required the development of inertia-property definitions including the moment of inertia and the mass center location developments of Section 5.2. With these definitions in hand, in Section 5.3 we proceeded from $\sum f = m\ddot{r}$ for a particle to develop equations of motion for planar motion of a rigid body. The rigid-body force equation was

$$\sum f_i = m\ddot{\boldsymbol{R}}_g. \quad (5.14)$$

This (vital) equation states that the acceleration of a rigid body's mass center $\ddot{\boldsymbol{R}}_g$ can be defined with the same basic equation as a particle.

The general moment equation developed for a rigid body was

$$\sum M_{oz} = I_o\ddot{\theta} + m(\boldsymbol{b}_{og} \times \ddot{\boldsymbol{R}}_o)_z, \quad (5.24)$$

where the moments are taken about the origin o of the body-fixed x, y, z coordinate system, and the x, y plane of the body remains in the inertial X, Y plane. The vector b_{og} locates the mass center g in the x, y, z coordinate system, I_O is the moment of inertia about the z-axis through o, and $\ddot{\boldsymbol{R}}_o$ is the acceleration of point o in the X, Y, Z system. Equation 5.24 is useful in analyzing some systems, for example, the equation of motion for a pendulum with an accelerating pivot point (Section 5.6.4) can be efficiently developed using this general equation. This equation also provides a speedy approach for some rolling-without-slipping applications.

However, reduced forms of Equation 5.24 are used more frequently. Specifically, taking moments about the mass center [$\boldsymbol{b}_{og} = 0$ in Equation 5.24] reduces the moment equation to

$$M_{gz} = I_g\ddot{\theta}. \quad (5.25)$$

The equations of motion for general planar motion (involving translation and rotation) are normally formulated using (i) $\sum f_i = m\ddot{\boldsymbol{R}}_g$ from Equation 5.14 and (ii) $M_{gz} = I_g\ddot{\theta}$ from Equation 5.25. $\sum f_i = m\ddot{\boldsymbol{R}}_g$ is a vector equation with two components. Stating this equation in Cartesian coordinates gives

$$\sum f_{iX} = m\ddot{R}_{gX}, \ \sum f_{iY} = m\ddot{R}_{gY}. \quad (5.15)$$

Hence, there are at most three independent equations of motion for general planar motion of a rigid body. Kinematic constraint equations are required to complete the model if there are more than three unknowns.

For planar rigid-body systems involving rotation about a point o that is fixed in space ($\ddot{\boldsymbol{R}}_o = 0$ in Equation 5.24), the moment equation is reduced to

$$M_{oz} = I_o\ddot{\theta}. \quad (5.26)$$

This equation is very useful in deriving the EOM for rigid bodies that are rotating about an axis that is fixed in inertial space. Multiple fixed-axis-rotation examples were presented in Sections 5.5 and 5.6, including the equation of motion for a compound pendulum. The compound pendulum displayed the same basic nonlinear and linearized differential equation as the simple pendulum of Section 3.4.2. The idea and techniques of linearizing a nonlinear equation for motion about an equilibrium position continued to be a vital tool for engineering analysis. The linearized equation defines the natural frequency for stable systems and can be used to determine whether a system will be statically stable or unstable. The procedure was used in Section 5.6.1 to demonstrate (mathematically) that an inverted pendulum is unstable.

In Chapter 3, we found that many 1DOF problems could be analyzed efficiently using energy concepts, and the same idea applies to planar motion of rigid bodies. The particle work–energy equation, $\text{Work}_{n.c.} = (T_2 + V_2) - (T_1 + V_1)$, also applies for a rigid body, since a body is simply a connection of an infinite number of particles. Hence, the question that needs to be resolved in applying this equation is simply, "How is the kinetic energy of a rigid body to be defined?" Starting from the particle kinetic energy definition $T = mv^2/2$, and integrating over the mass of a body netted

$$T = \frac{m|\dot{\boldsymbol{R}}_g|^2}{2} + \frac{I_g\dot{\theta}^2}{2}. \tag{5.33}$$

This definition applies for general planar motion. For fixed-axis rotation, the following reduced equation applies

$$T = \frac{I_o\dot{\theta}^2}{2}. \tag{5.34}$$

Most of the 1DOF examples in Sections 5.5, 5.6.1 through 5.6.3, 5.7.1, and 5.7.2 are worked in parallel using both Newtonian approaches and the energy equation. The Newtonian approach starts with free-body diagrams and the force and moment equations. Deriving the equation of motion using the energy equation starts by stating the work–energy equation in terms of the 1DOF coordinate, for example θ, and then differentiating with respect to θ using the energy-integral substitution $d(\dot{\theta}/2)/d\theta = \ddot{\theta}$. The examples demonstrate that the equation of motion for many 1DOF, one-coordinate systems can be developed more quickly from the work–energy equation than from a free-body/Newtonian approach. Examples involving Coulomb friction provide a notable exception to this outcome. Having only one degree of freedom is the key to success in developing an EOM by applying the work–energy equation. Systems with one degree of freedom but multiple coordinates that are related through nonlinear constraint equations are generally hard to model using the work–energy equation.

Section 5.7 considered general planar motion involving translation and rotation, starting in Section 5.7.1 with rolling-without-slipping examples. A body that is rolling without slipping has one degree of freedom, since the translation and rotation degrees of freedom are connected by a kinematic constraint equation. The examples of Section 5.7.2 involve one rigid body and one degree of freedom but multiple coordinates. The model for these systems start from either (i) the force and moment equations plus the kinematic constraint equations or (ii) the work–energy equation plus kinematic constraint equations.

Section 5.7.3 examines a limited class of multi-body, 1DOF, one-coordinate examples. Equations of motion for this type of problem can be obtained much more speedily from the work–energy equation than Newtonian approaches. As noted above, this direct application of the work–energy equation to obtain an equation of motion is restricted to 1DOF examples that are readily reduced to one coordinate. Lagrange's equations of motion are the needed "variational" extension of the energy approach for models with more than one degree of freedom, but they are beyond the scope of this book.

Section 5.7.4 concerns MDOF planar motion with 2DOF examples, including torsional vibrations, beams as springs, and double compound pendulums. Models and analysis for these systems generally resemble the earlier particle-dynamics developments of Section 3.5. The pendulum examples were nonlinear, and linearization led to 2DOF linear vibration problems with two natural frequencies, two mode shapes, etc.

Section 5.7.5 considered planar mechanisms that involve multiple bodies and multiple coordinates but having one degree of freedom. Equations of motion were derived for the individual bodies using (i) $\sum f = m\ddot{\boldsymbol{R}}_{\mathbf{g}}$ and $M_{gz} = I_g\ddot{\theta}$ for general motion involving displacements and rotations and (ii) $M_{oz} = I_o\,\ddot{\theta}$ for fixed-axis rotation. With mechanisms, a complete model always requires kinematic constraint equations, as developed in Section 4.5. They cannot generally be reduced to one coordinate—coupled equations for unknown accelerations and reaction forces must be solved simultaneously.

Moment of momentum for planar motion of rigid bodies is the subject of Section 5.8. Moment of momentum and moment of momentum conservation have nothing to do with the development of models or equations of motion. The moment-of-momentum equation is an integral of a moment equation and is primarily of value for circumstances involving planar collisions of rigid bodies or particles. Section 5.8.1 was used to develop the appropriate moment and moment-of-momentum equations:

$$\begin{aligned} M_{Azr} &= \dot{H}_{Az} = I_g\ddot{\theta} + (\boldsymbol{R}_{Ag} \times m\ddot{\boldsymbol{R}}_g)_z, \\ H_{Az} &= I_g\dot{\theta} + (\boldsymbol{R}_{Ag} \times m\dot{\boldsymbol{R}}_g), \end{aligned} \tag{5.286}$$

for moments taken about a point A that is fixed in inertial space. These equations were used in Section 5.8.2 to relate the angular velocities of a rigid body before and after a collision. With the point of contact at A, the moment about A is zero, and H_{Az}, the moment of momentum about A, is conserved (constant) across the collision.

If you have mastered these chapters, you know far more than most practicing engineers. You will understand most of the essential concepts and strategies for developing general kinematic equations and general equations of motion. You should understand that a mathematical model consists of enough (correct) algebraic and differential equations to define the unknown accelerations and reaction forces. You should be comfortable with the idea of numerical solution of nonlinear equations of motion.

You should also be comfortable with the idea of equilibrium and the techniques of linearization for small motion about an equilibrium position. Linearized equations about an equilibrium position can be statically stable or unstable, and we have considered examples with both types of motion. Small stable motion about an equilibrium position normally involves vibration equations, and you should understand and be comfortable with vibration definitions and concepts such as natural frequencies, damping factors, mode shapes, and eigenvalues. Vibrations deal with a particular type of dynamic model, and most practicing engineers working in dynamics need to understand linear vibration concepts and terminology.

Notable topics that are not covered in these chapters include 3D kinematics, 3D kinetics of rigid bodies and systems of rigid bodies, and Lagrange's equations for MDOF systems. Many good texts are available to cover these topics.

Problems

1DOF Torsion Problems

5.1 The system shown in Figure P5.1 consists of a solid circular disk supported at its center by two uniform bars with circular cross sections. The bars' ends are fixed, and they have the following properties: $d_1 = 1$ in., $d_2 = 0.75$ in., $L_1 = 20$ in., $L_2 = 15$ in., $G_1 = 15\text{E}6$ psi, $G_2 = 17\text{E}6$ psi, $D = 20$ in., $h = 1$ in., $\rho = 0.283$ *lb*/in.^{3}.

Tasks:

(a) Draw a free-body diagram for the disk and obtain its equation of motion (EOM) for torsional motion.

(b) Determine the system's natural frequency.

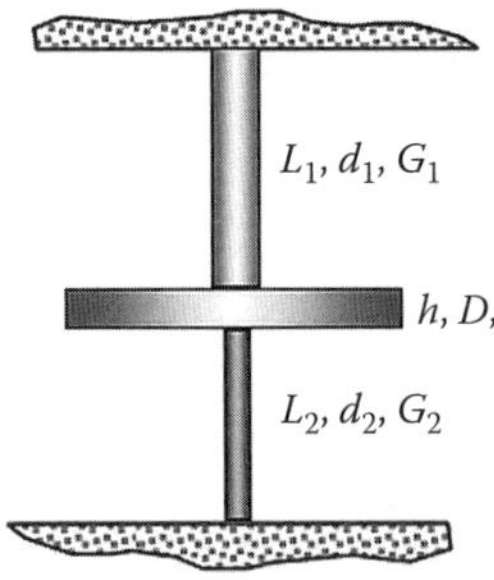

FIGURE P5.1

5.2 Figure P5.2 illustrates a large rectangular block of height h, width w, breadth b, and density ρ, is supported by the hollow torsional shaft of length L, outside diameter d_O, inside diameter d_I, and modulus of rigidity G. The bottom of the block has an oil film between it and ground. The oil film generates a viscous resistance torque on the block with damping coefficient c_θ. Properties: $L = 1750$ mm, $d_O = 150$ mm, $d_I = 145$ mm, $G = 750$ GPa, $h = 500$ mm, $b = 550$ mm, $w = 750$ mm, $\rho = 7.85(10^3)$ kg/m^3, $c_\theta = 2100$ N m s/rad.

Tasks:

(a) Draw a free-body diagram and obtain the EOM for the system.

(b) Determine the system's undamped natural frequency, damping factor, and damped natural frequency.

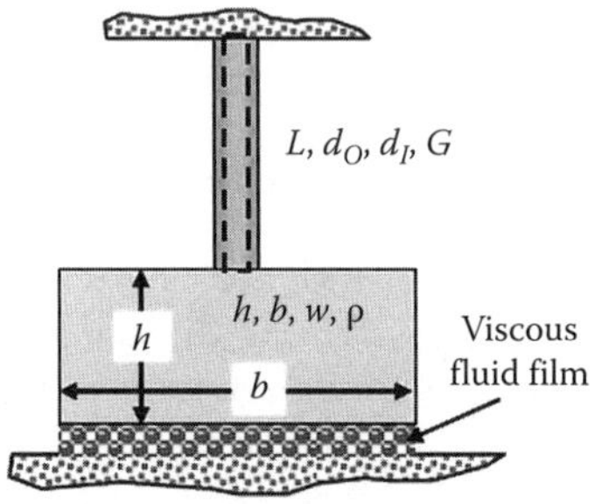

FIGURE P5.2

Motion with Constraints

5.3 As shown in Figure P5.3, gear A has an applied external moment $M(t)$. Mass M_3 is attached to gear B by an inextensible cable. Gears A and B are meshed.

Tasks:

(a) Select coordinates and obtain the kinematic constraint relations.

(b) Draw free-body diagrams and obtain the EOM using Newtonian principles.

(c) Obtain the EOM using energy principles.

(d) Determine the speed and elevation change of mass M_3 at $t = 5$ s.

System properties: $M_1 = 50$ kg, $k_{1g} = 35$ cm, $R_1 = 50$ cm, $M_2 = 30$ kg, $k_{2g} = 25$ cm, $R_2 = 35$ cm, $R_3 = 20$ cm, $M_3 = 40$ kg, $M(t) = 0.5t$ N m; *k_{ig}-radius of gyrations.*

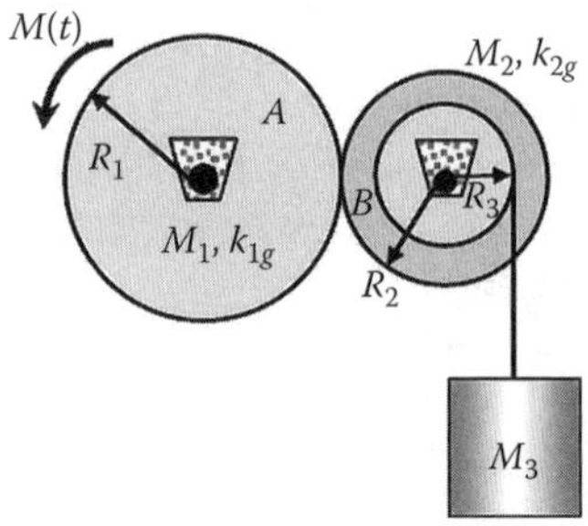

FIGURE P5.3

5.4 The bodies shown in Figure P5.4 are released from rest from the position shown. The cord connecting the bodies is inextensible, and a no-slip condition applies at its contact with the pulleys.

Tasks:

(a) Select coordinates, obtain the kinematic constraint relations, and draw appropriate free-body diagrams.

(b) Obtain the EOM for block A by applying Newtonian principles.

(c) Develop the EOM for block A using energy methods including the disk inertias.

(d) Determine the velocities of bodies A, B, and D after body B has moved 1 foot. Also determine the angular displacement of pulley C. Assume the system is released from rest. Properties: $w_A = 10$ lb, pulleys: $w_C = w_D = 5$ lb, $k_{gC} = k_{gD} =$ 12 in., $r = 5$ in.

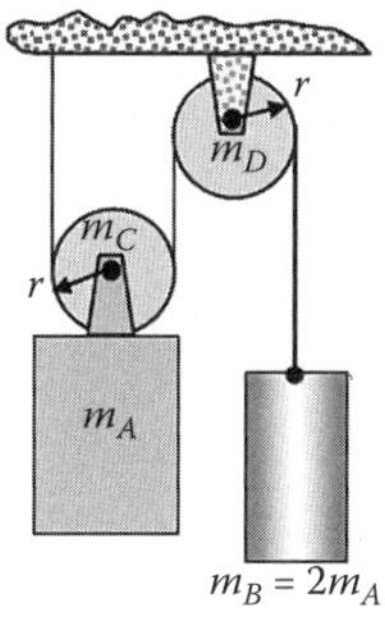

FIGURE P5.4

5.5 Disk 1 in Figure P5.5 has mass M_1 and radius R_1 and has a linear torsional spring with a stiffness k_θ attached at point O_1. Disk 2 has mass M_2 and radius R_2 and is driven by a constant torque $\boldsymbol{M}$ in the counterclockwise direction. The inextensible belt connecting the two pulleys does not slip on either pulley. System properties: $R_1 = 0.375$ ft, $W_1 = 20$ lb, $R_2 = 0.275$ ft, $W_2 = 12$ lb, $k_\theta = 2.5$ ft lb/rad, $M = 5$ ft lb.

Tasks:

(a) Select coordinates and state the kinematic constraint relationships.

(b) Obtain the EOM by applying work–energy relationships. Find the undamped natural frequency.

(c) Assuming the system starts from rest, determine the speed of disk 1 when disk 2 reaches an angle of 75°.

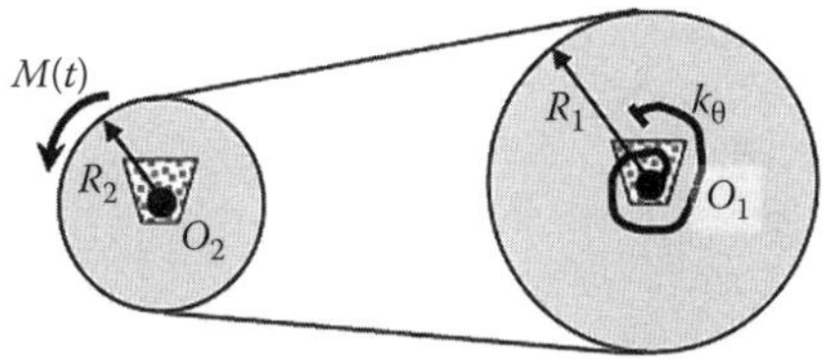

FIGURE P5.5

5.6 The sun gear shown in Figure P5.6 has the constant applied torque, $T = 1.5$ N m. The planet gears' centers are fixed but the gears rotate freely. The ring gear is free to rotate, and a no-slip condition exists at all contact points.

System properties: $m_S = 1.5$ kg, $r_2 = 1.25$ cm, $k_{Sg} = 0.9$ cm, $m_p = 1$ kg, $r_1 = 1$ cm, $k_{pg} = 0.7$ cm, $m_R = 3$ kg, $k_{Rg} = 2.3$ cm.

Note: k_{ig} *is the radius of gyration for corresponding part about the element's center of gravity: S—sun gear, P—planetary gear, R—Ring gear.*

Tasks:

(a) Select coordinates and obtain the required kinematic constraint relationships.

(b) Obtain the EOM for the ring gear using energy principles.

(c) Determine the speed and number of revolutions of the ring gear at $t = 5$ s.

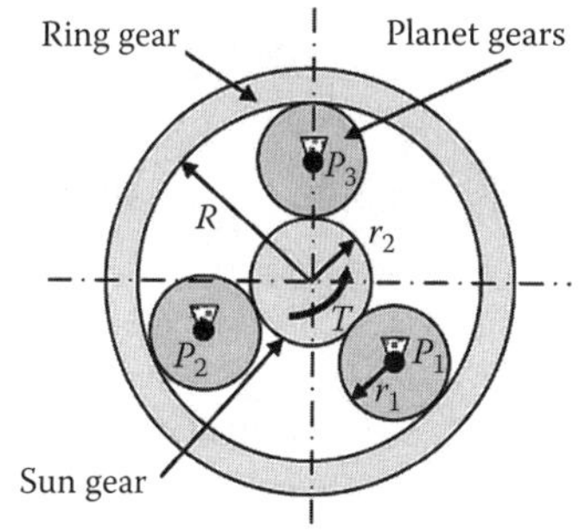

FIGURE P5.6

Pendulum Problems

5.7 The inverted pendulum shown in Figure P5.7 has been released from rest at $\theta = 0°$. Neglect friction.

Tasks:

(a) Draw a free-body diagram and develop a general EOM for θ.

(b) Apply energy principles to obtain the EOM.

(c) Determine the reaction forces of the support pin as a function of θ (only).

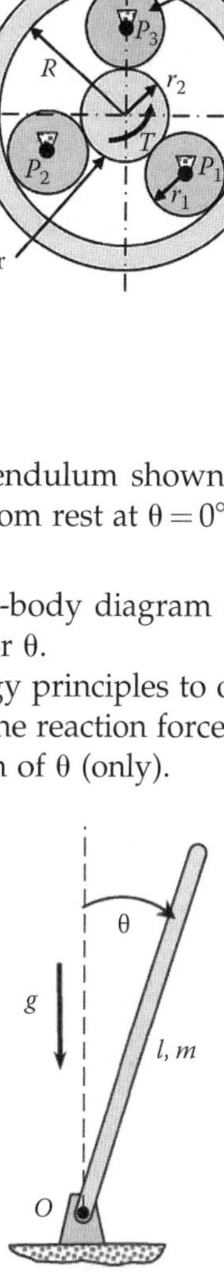

FIGURE P5.7

5.8 The uniform bar shown in Figure P5.8 is supported by a frictionless pin at O and released from rest at $\theta = 0$.

Tasks:

(a) Develop the EOM applying Newtonian principles. (Free-body diagram required.)
(b) Develop the EOM using work–energy principles.
(c) Determine the equilibrium positions, and develop the EOM for small motion about the equilibrium positions.
(d) Define the reaction components at O as a function of θ (only).

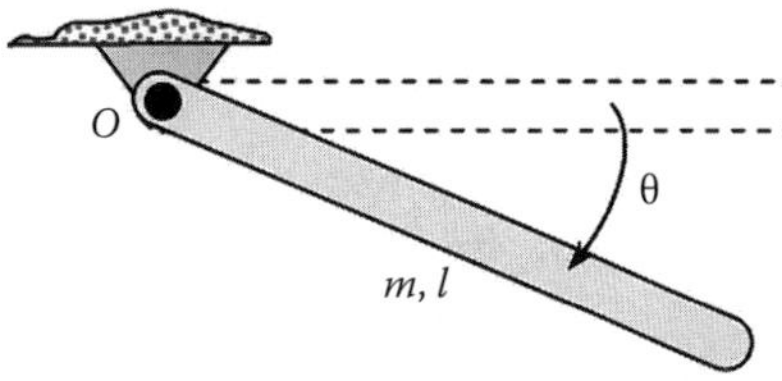

FIGURE P5.8

5.9 The system illustrated in Figure P5.9 is a composite pendulum made from two uniform bars. The top is supported by a frictionless pivot at point O and has mass $2m$ and length $2l$. The cross bar has mass m and length l. Assume that system is released from rest from the inverted, vertical position.

Tasks:

(a) Select a coordinate and develop the EOM using (1) a free-body diagram and (2) energy principles.
(b) Determine the reaction force components at O as a function of the rotation angle only.
(c) Determine the equilibrium positions. Determine from the perturbed EOM whether small-angle motion about the equilibrium positions is stable.

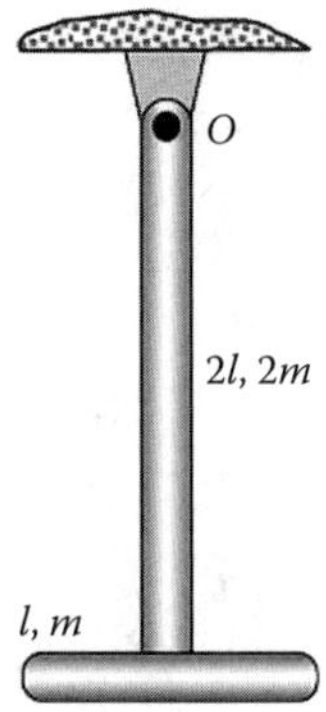

FIGURE P5.9

5.10 The composite pendulum shown in Figure P5.10 is made of a uniform bar (length l and mass m) that is rigidly attached to a disk (mass M and radius r). The pendulum is released from rest in the position shown. Neglect friction.

Tasks:

(a) Select a coordinate, draw a free-body diagram, and develop the EOM.
(b) Develop the EOM using conservation of energy.
(c) Define the pivot reaction-force components as a function of the rotation angle only.
(d) Solve for the stable equilibrium position, and develop the EOM for small motion about it.

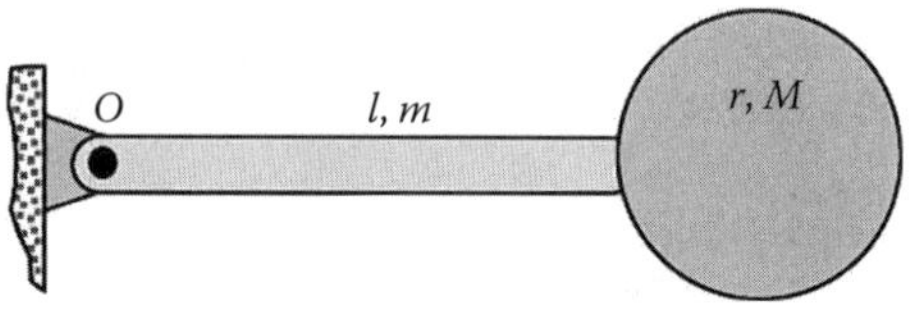

FIGURE P5.10

5.11 The system shown in Figure P5.11 is released from rest at $\theta = 0$. Assume the support pin at O to be frictionless. Mass M is connected to O by two rigid bars with negligible mass.

Tasks:

(a) Draw a free-body diagram and develop the EOM.
(b) Develop the EOM using work–energy principles.
(c) Define the reaction-force components at O as a function of θ only.

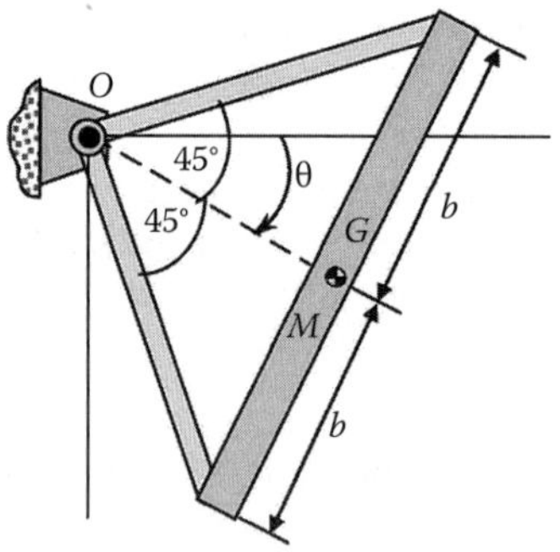

FIGURE P5.11

5.12 The plate shown in Figure P5.12 is released from rest in the position shown. Neglect friction at the support pin at O.

Tasks:

(a) Select a coordinate, draw a free-body diagram, and develop the EOM using Newtonian principles.
(b) Develop the EOM using work–energy principles.
(c) Determine the reaction components at O as a function of the selected coordinate only.

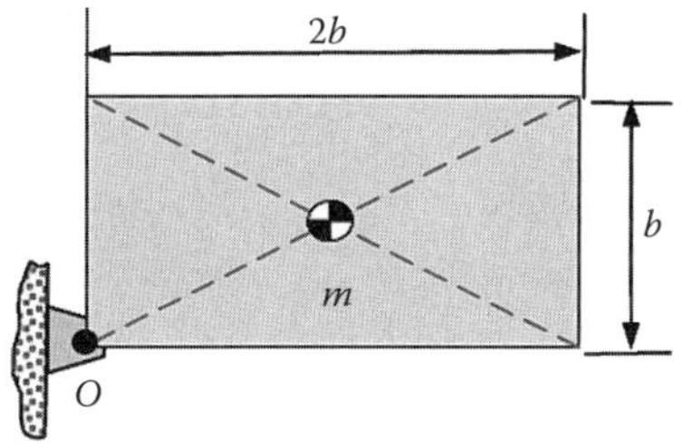

FIGURE P5.12

5.13 The plate shown in Figure P5.13 is released from rest at $\theta = 0$. The support pin at O is frictionless.

Tasks:

(a) Draw a free-body diagram and develop the EOM using Newtonian principles.
(b) Develop the EOM using work–energy principles.
(c) Determine the reaction components at O as a function of θ (only).

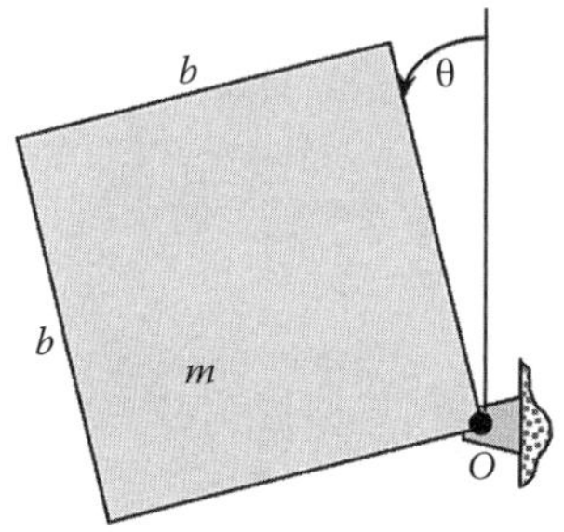

FIGURE P5.13

5.14 The circular plate illustrated in Figure P5.14 is released from rest at $\theta = 0$. The support pin at O is frictionless.

Tasks:

(a) Draw a free-body diagram and develop the EOM using Newtonian principles.
(b) Develop the EOM using work–energy principles.
(c) Determine the reaction components at O as a function of θ (only).

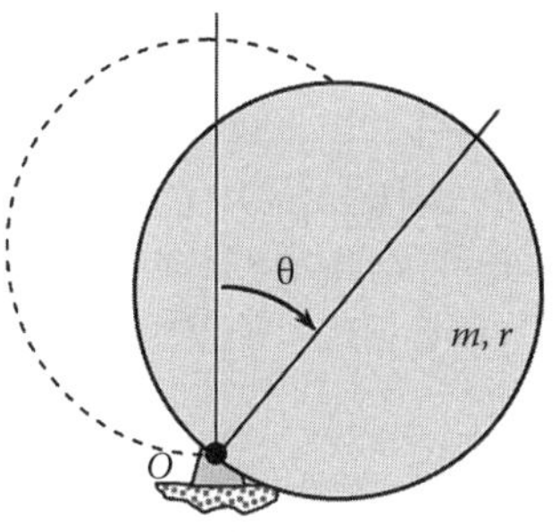

FIGURE P5.14

5.15 The semicircular plate illustrated in Figure P5.15 is released from rest at $\theta = 0$. The support pin at O is frictionless.

Tasks:

(a) Draw a free-body diagram and develop the EOM using Newtonian principles.
(b) Develop the EOM using work–energy principles.
(c) Determine the reaction components at O as a function of θ (only).

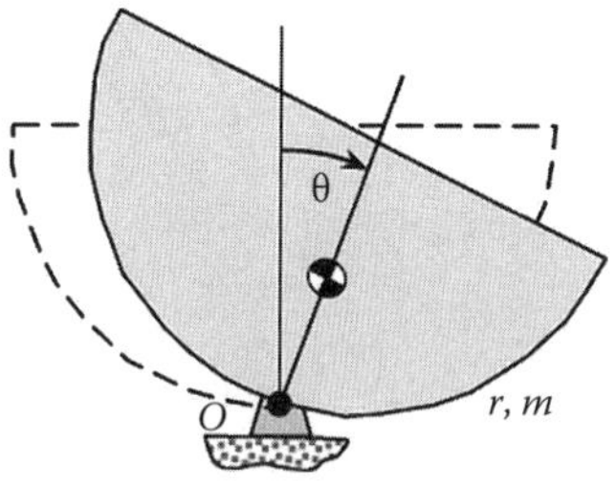

FIGURE P5.15

5.16 The triangular plate shown in Figure P5.16 is pivoted at O. The pin support at O is frictionless.

Tasks:

(a) Draw a free-body diagram and develop the EOM using Newtonian principles.
(b) Develop the EOM using work–energy principles.
(c) Define the body's equilibrium positions. State the EOM for small motion about the stable equilibrium position and determine its natural frequency.
(d) Assuming that the plate is released from rest in the inverted vertical position, state the reaction force components at O as a function of the plate's rotation angle (only).

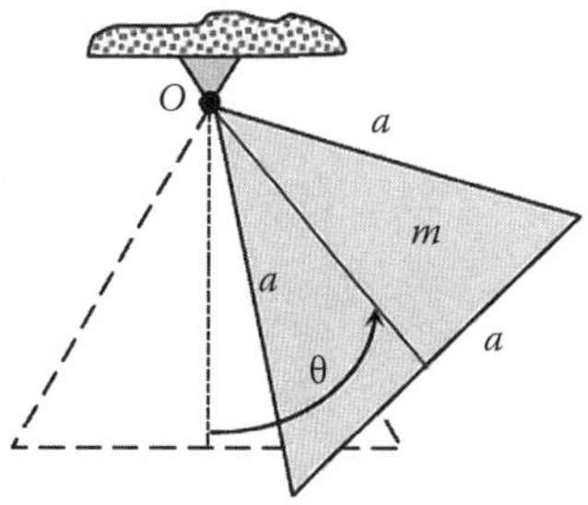

FIGURE P5.16

Bars with Support Springs and Dampers

5.17 The bar illustrated in Figure P5.17 is released from rest in the position shown and rotates to the right. The spring has an undeflected length of $L/3$. The pin support is frictionless.

Tasks:
(a) Draw a free-body diagram and develop the EOM using large angles.
(b) Apply conservation of energy and obtain the EOM.
(c) Determine the bar's equilibrium position and develop an EOM for small motion about it.

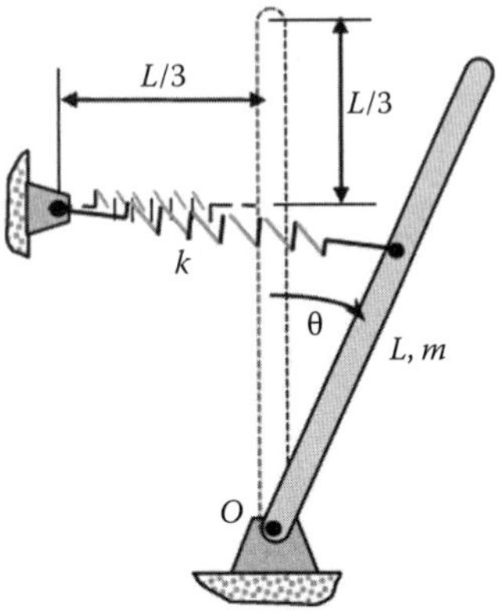

FIGURE P5.17

5.18 The slender bar in Figure P5.18 pivots about the frictionless pivot O.

Tasks:
Assume small θ, and complete the following steps:
(a) Draw a free-body diagram and develop the EOM by applying Newtonian principles.
(b) Develop the EOM by applying conservation of energy.
(c) Determine the natural frequency.
(d) Determine the limiting value of k so that the system remains statically stable.

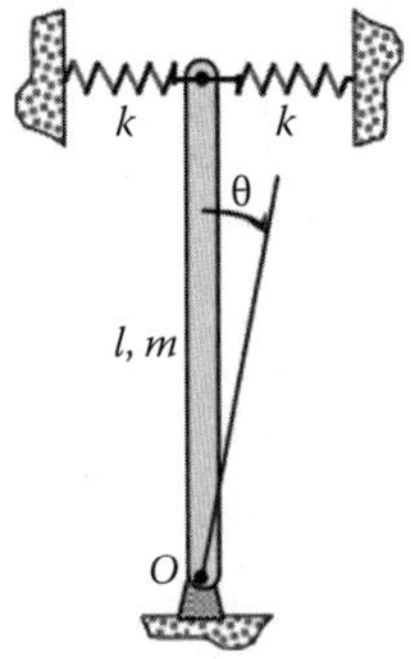

FIGURE P5.18

5.19 The bar shown in Figure P5.19 has mass m and length l and is supported vertically at its top by two springs with spring constants k. Each spring has an initial tension T_0. Two linear dampers are attached at the middle of the bar with damping coefficients c. The pin support at O is frictionless.

Tasks:
Assume small angular motion and complete the following steps:
(a) Draw the system in a general position, select a coordinate, draw a free-body diagram, and develop the EOM.
(b) What is the undamped natural frequency and damping factor?
(c) Drop the damping and use conservation of energy to develop the EOM.
(d) Determine the minimum value of the spring coefficient k at which the bar becomes unstable.

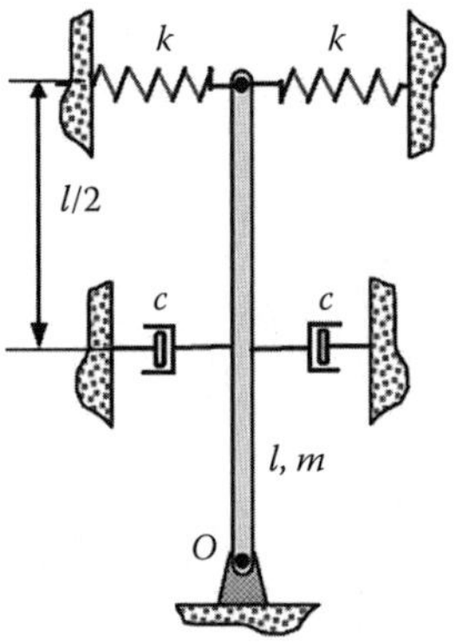

FIGURE P5.19

5.20 The uniform bar illustrated in Figure P5.20 has mass m, length l, and is supported by a frictionless pivot at O. A linear spring with spring coefficient k is connected to the bar at a distance $2l/3$ below the pivot point. A linear damper with damping coefficient c is connected to the bar the distance $l/3$ below the pivot.

Tasks:
For small-rotation angle of the bar
(a) Select a coordinate, and draw the system in a general position.
(b) Draw a free-body diagram, and develop the EOM.
(c) For $k = 350$ N/m, $m = 20$ kg, and $l = 1$ m, determine the undamped natural frequency. Free response test results produced a *log dec* value of 0.637, determine the damping coefficient value c.
(d) Drop the damping and develop the EOM using conservation of energy.

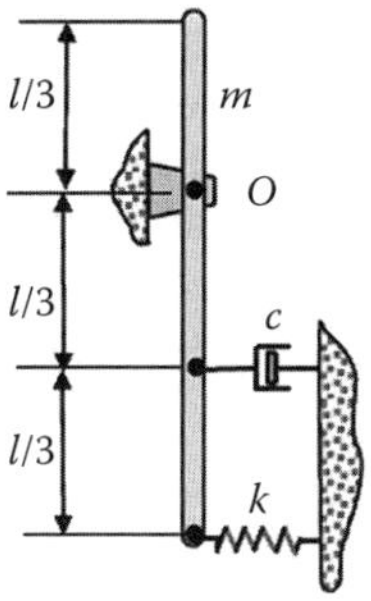

FIGURE P5.20

5.21 The bar illustrated in Figure P5.21 has a mass m, length b, and is supported by a frictionless pivot at O. A linear spring with spring coefficient k is connected to the bar at a distance b below the pivot point. A linear damper with damping coefficient c is connected to the bar a distance a below the pivot.

Tasks:

For a small-rotation angle of the bar

(a) Select a coordinate, and draw the system in a general position.

(b) Draw a free-body diagram, and develop the EOM.

(c) For $k = 225$ N/m, $m = 20$ kg, $a = 0.5$ m, and $b = 1$ m determine the undamped natural frequency. A free response test result is shown. The first labeled peak has values of (0.9, 0.0936), and the second labeled peak has values (2.7, 0.0265) where the values are (time, amplitude). Determine the value for c.

(d) Drop the damping and develop the EOM using conservation of energy.

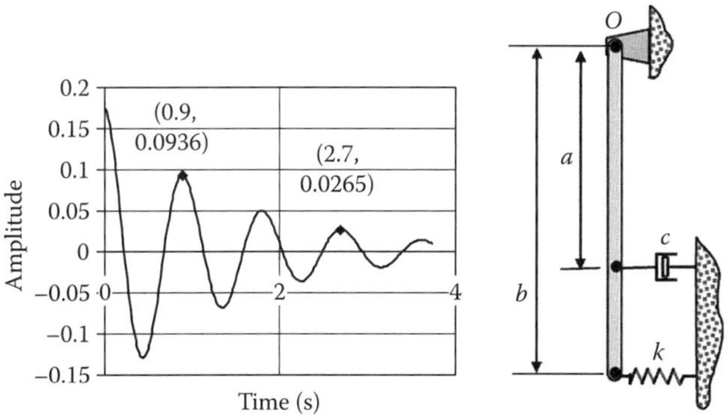

FIGURE P5.21

5.22 The assembly shown in Figure P5.22 is pivoted about the frictionless pivot O, has mass m with mass center at G, and a radius of gyration about O of k_{Og}. The springs (spring constant k) attached to the cross-member ends are in compression at $\theta = 0$, and remain in compression for all θ values. The linear dampers attached to the ends of the cross member have linear damping coefficients c.

Tasks:

(a) For small θ, draw a free-body diagram and develop the EOM. What is the natural frequency? What is the damping factor?

(b) Drop the damping and develop the EOM from conservation of energy.

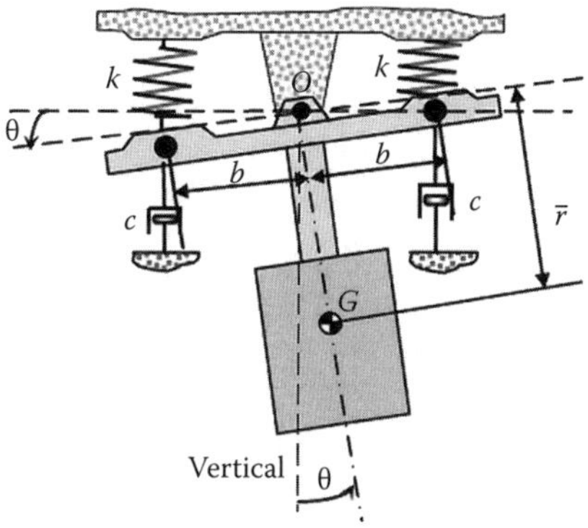

FIGURE P5.22

5.23 The cylinder shown in Figure P5.23a is supported by two inextensible wires and is referred to as a bifilar pendulum. This arrangement is often used to estimate the mass moment of inertia of a body. Assume small motion.

Tasks:

(a) Select a coordinate and draw the system in a general position for rotational motion about the Z-axis.

(b) Draw a free-body diagram of the system and develop the EOM for rotation of the system about the Z-axis.

(c) Determine the undamped natural frequency of the system.

(d) Repeat (b) through (c) for motion swinging motion into the plane of paper.

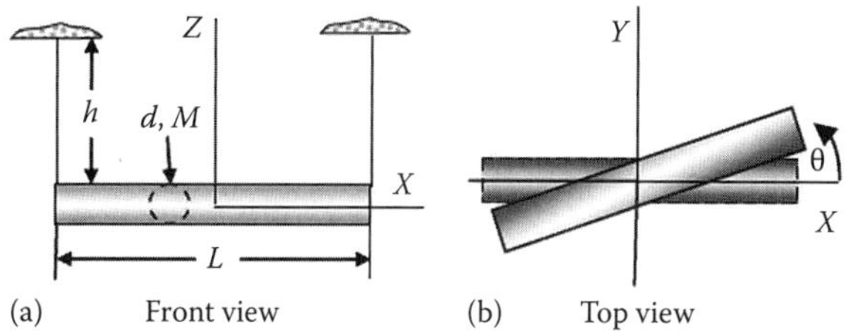

FIGURE P5.23

5.24 As shown in Figure P5.24, the bar (length l and mass m) and square plate (mass M and sides b) assembly is supported by a frictionless pivot at O. A linear spring with coefficient k and a linear damper with coefficient c are connected at the midpoint of the bar. The spring is undeflected in the position shown.

Tasks:

For small angular motion of the assembly

(a) Select a coordinate, and draw the system in a general position.
(b) Draw a free-body diagram, and develop the EOM.
(c) Drop the damping and develop the EOM from conservation of energy.
(d) Determine the limiting value of the spring coefficient k so that the bar becomes statically unstable.

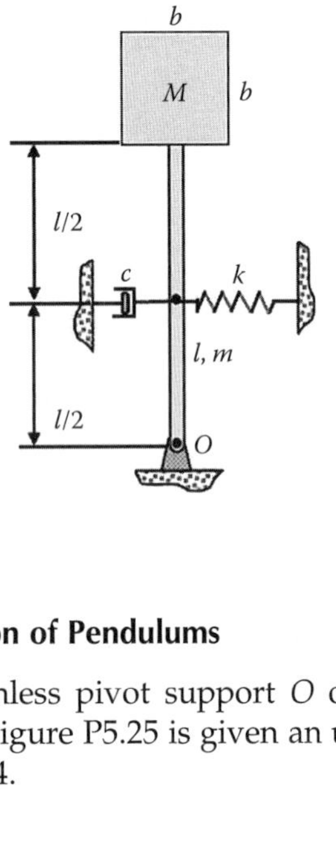

FIGURE P5.24

Base Acceleration of Pendulums

5.25 The frictionless pivot support O of the assembly shown in Figure P5.25 is given an upward acceleration of $g/4$.

Tasks:

(a) Select a coordinate, and draw the system in a general position.
(b) Draw a free-body diagram for the assembly, and develop the EOM for the body.
(c) Determine the support reactions as a function of the selected coordinate only.

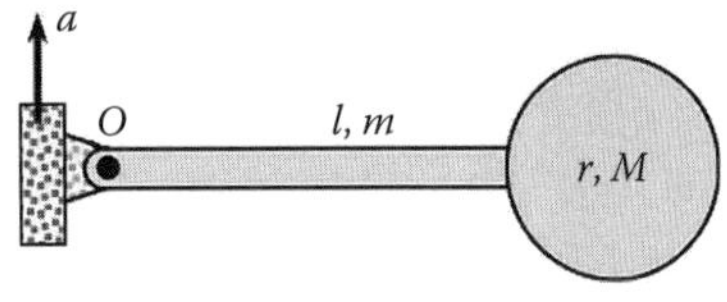

FIGURE P5.25

5.26 The truck shown in Figure P5.26 is accelerating to the right with acceleration a ($a = \ddot{X}$). The plate of length L and mass m is pivoted at O (frictionless pivot) and released from rest at $\theta = 0$.

Tasks:

(a) Draw a free-body diagram and develop the EOM defining $\ddot{\theta}$.
(b) Determine the velocity of the plate at $\theta = 90°$.

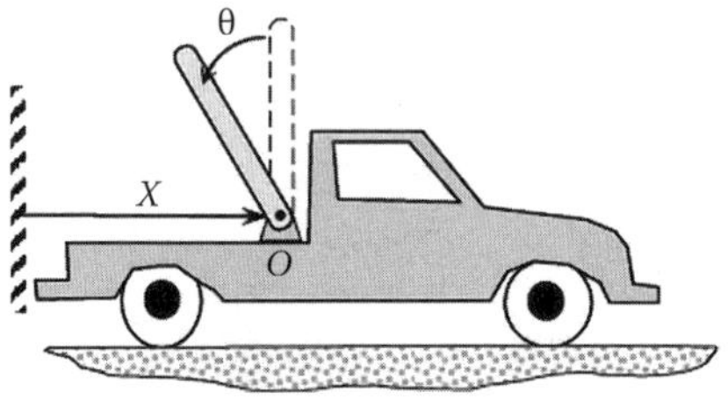

FIGURE P5.26

5.27 The bar illustrated in Figure P5.27 is attached to a vehicle that has a constant horizontal acceleration $a = g/3$. Assume the pivot support of the bar to the plate to be frictionless.

Tasks:

(a) Draw a free-body diagram and develop the EOM for θ.
(b) If the pendulum is released from rest at $\theta = 0$, determine the peak angle for θ, θ_{max}.
(c) Define the radial and circumferential force components at O as a function of θ.
(d) Determine the bar's equilibrium position.

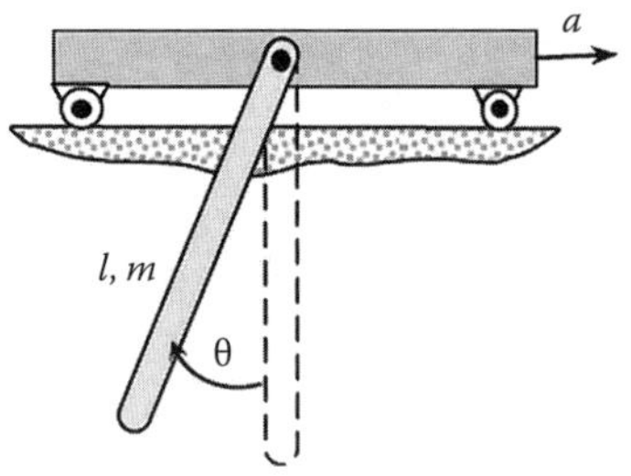

FIGURE P5.27

5.28 The support pin for the bar illustrated in Figure P5.28 is accelerating up the inclined plane at the constant acceleration rate a.

Tasks:

(a) Draw a free-body diagram and develop the governing EOM for θ.
(b) If the bar is released from rest with $\theta = 0$, determine the peak angle for θ, θ_{max}.
(c) Define the radial and circumferential force components at O as a function of θ.
(d) Determine the bar's equilibrium position.

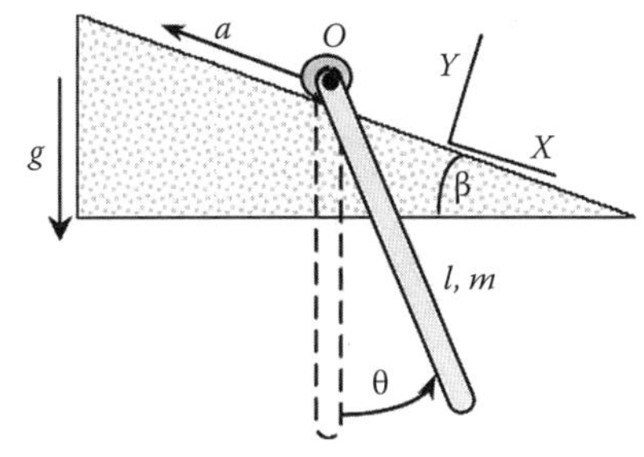

FIGURE P5.28

5.29 The balloon shown in Figure P5.29 is rising with an acceleration of $g/8$. A pendulum (length l, mass m) is attached to the bottom of the balloon by a frictionless pivot pin.

Tasks:

(a) Draw a free-body diagram and develop a general EOM for θ.

(b) Develop general expressions for the components of the reaction force at A.

(c) If the pendulum is released from rest with $\theta = 45°$, find the radial reaction force component (along the axis of the pendulum) as a function of θ only.

(d) For small θ, what is the pendulum's natural frequency?

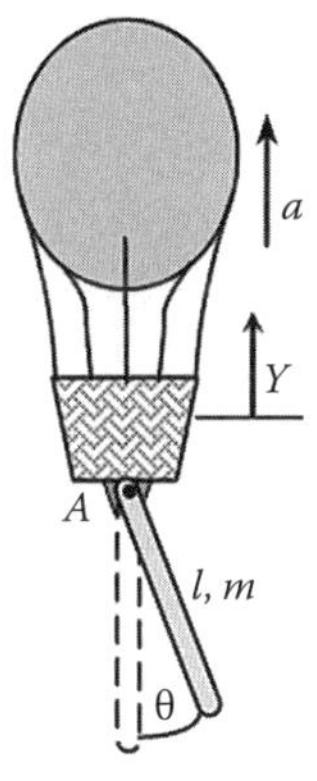

FIGURE P5.29

Rolling-Without-Slipping Examples

5.30 The eccentric disk shown in Figure P5.30 is released from the position shown and rolls without slipping. The mass center is displaced a distance d from the center of the disk. The radius of gyration for the disk about its mass center is k_g.

Tasks:

(a) Select coordinates and draw the cylinder in a general position.

(b) Draw a free-body diagram and develop the equations of motion and the required kinematics equation to reduce the problem to a single EOM.

(c) Obtain the EOM by applying energy principles.

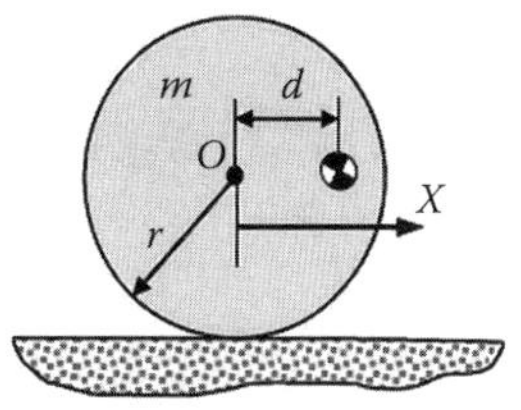

FIGURE P5.30

5.31 The small cylinder of mass m and radius r rolls without slipping on a large cylinder of radius R contained in the vertical plane. Use the coordinates shown on Figure P5.31.

Tasks:

(a) Draw a free-body diagram and develop the EOM for θ.

(b) Use conservation of energy and develop the EOM.

(c) Show that the equation is unstable for small θ.

(d) From the free-body diagram, determine the normal reaction force as a function of θ alone. If $r = R/4$, and the cylinder starts from rest at $\theta = 0$, for what value of θ will the cylinders lose contact.

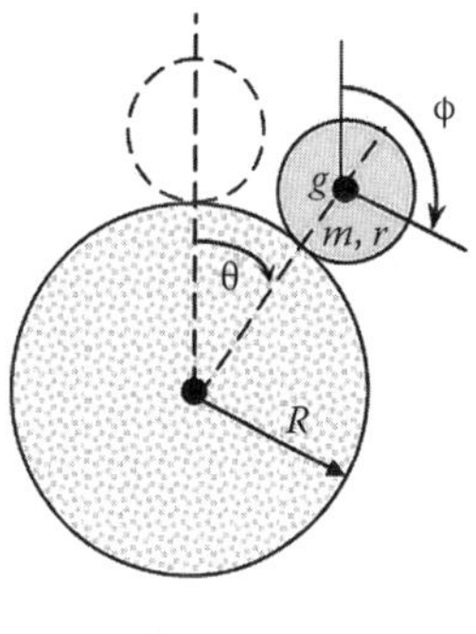

FIGURE P5.31

5.32 As shown in Figure P5.32, the cylinder of mass m is rolling without slipping down the inclined plane. The cylinder's mass center is displaced from the cylinder's geometric center a distance e. The cylinder has a radius of gyration about the geometric center of k_g.

Tasks:

(a) Draw a free-body diagram and develop the EOM for θ and the reaction force components at C.

(b) Develop EOM for θ by applying conservation of energy.

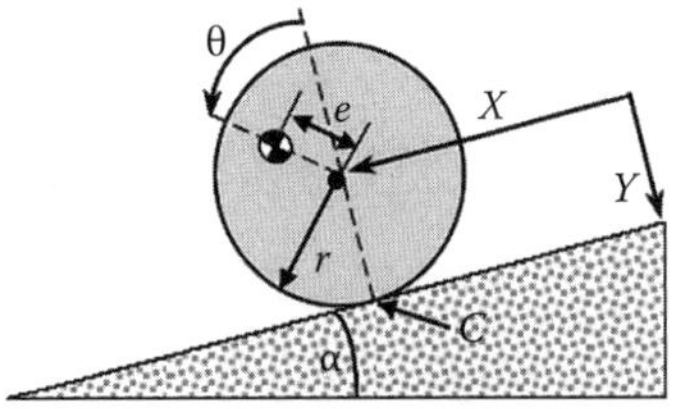

FIGURE P5.32

5.33 In Figure P5.33, the solid cylinder of mass m and radius r is rolling without slipping on a smooth plane that is tilted at the angle α from the horizontal. Two inextensible cords are wrapped around the cylinder and attached to a solid wall through the springs with stiffness coefficient k.

Tasks:

(a) Select coordinates and draw the cylinder in a general position.

(b) Draw a free-body diagram and develop the EOM. What is the natural frequency?

(c) Use conservation of energy to develop the EOM.

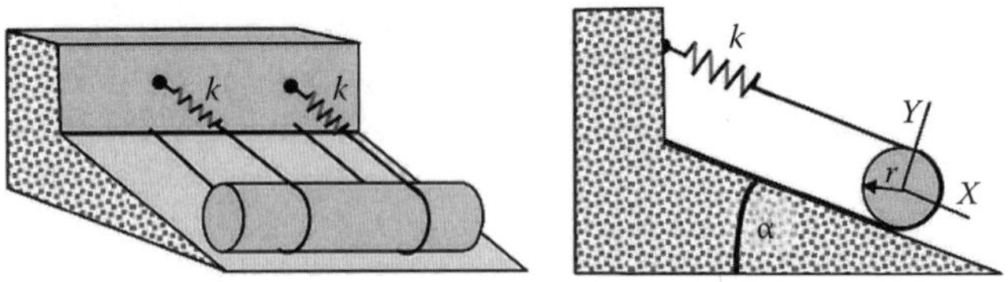

FIGURE P5.33

5.34 The pulley system shown in Figure P5.34 (mass m and radius r) is supported by an inextensible cord. The cord does not slip on the pulley. One end of the cord is attached to a support at A. The other end is connected to a linear spring (spring coefficient k). The other end of the spring is attached to the support at B. A linear damper with damping coefficient c is attached between the support at D and O, the center of the pulley. A frictionless pivot point at O supports a tray of mass M. Treat the pulley as a thin disk.

Tasks:

(a) Select coordinates and draw the system in a general position.

(b) Draw free-body diagrams for the bodies, and develop a single governing EOM.

(c) Drop the damper and obtain the EOM using conservation of energy.

(d) For $k = 175$ N/m, $m = 20$ kg, $M = 40$ kg, and $r = 45$ cm determine the undamped natural frequency. Determine the damping value c such that the damping factor is $\zeta = 0.1$.

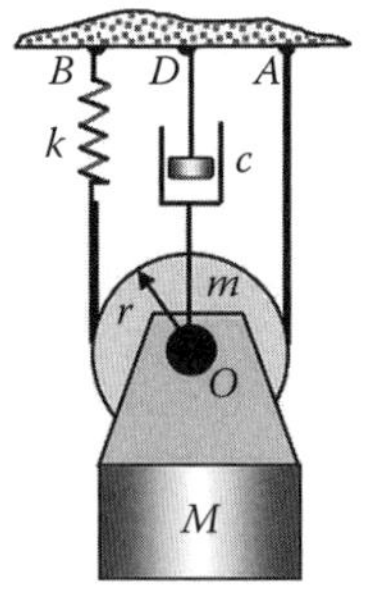

FIGURE P5.34

5.35 The cylinder shown in Figure P5.35 has mass m and rolls without slipping on the inclined surface. It is restrained by a linear spring of stiffness k and a linear damper with damping coefficient c. The cylinder is initially in equilibrium with the spring in tension.

Tasks:

(a) Select coordinates and draw the system in a general position.

(b) Draw a free-body diagram for the cylinder and develop a single EOM.

(c) Drop the damping and develop the EOM using conservation of energy.

(d) For $k = 10$ lb/in., $w = 50$ lb, and $r = 18$ in. Determine the undamped natural frequency. Select the damping value c such that the damping factor is $\zeta = 0.1$.

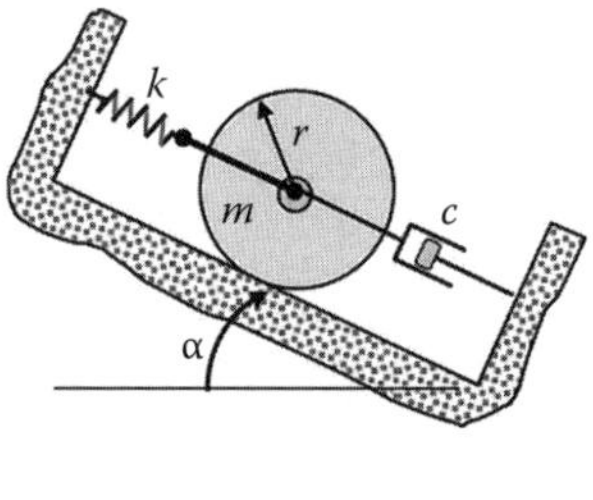

FIGURE P5.35

5.36 The cylinder illustrated in Figure P5.36 (mass m and radius r) is rolling without slipping on the horizontal plane. Its mass center is displaced the distance e from the center O. The disk has a radius of gyration of k_g about O. A linear spring with stiffness coefficient k connects O to ground and it is undeflected in the position shown.

Tasks:

(a) Select coordinates and draw the system in a general position.

(b) Draw a free-body diagram, and develop the EOM.

(c) Use conservation of energy to develop the EOM.

(d) For small motion, determine the natural frequency.

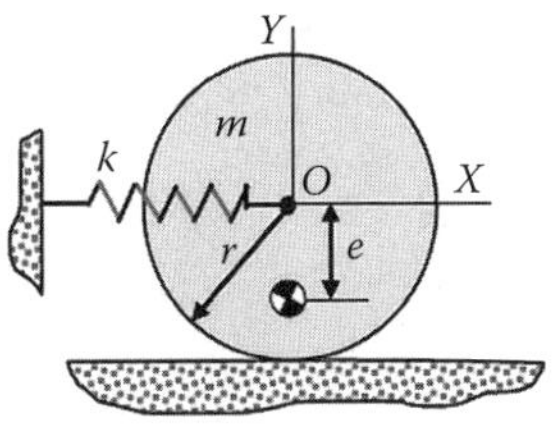

FIGURE P5.36

Planar Motion: One Degree of Freedom

5.37 The bar illustrated in Figure P5.37 is released from rest at $\theta = 90°$. Neglect friction and the mass and inertia of the rollers.

Tasks:

(a) Draw a free-body diagram and develop an EOM for θ.
(b) Use conservation of energy to develop the EOM.
(c) Find the reaction force at the wall as a function of θ.
(d) For what value of θ does the bar lose contact with the vertical wall?

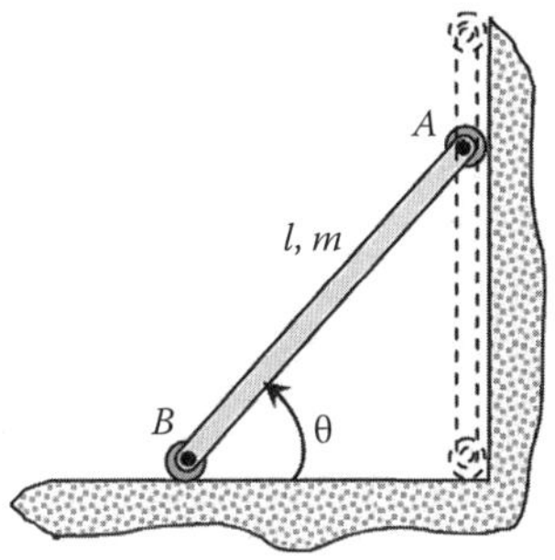

FIGURE P5.37

5.38 The uniform bar shown in Figure P5.38 has length l and mass m. Both masses in the frictionless slots have mass M. The spring constant is k, and the spring is undeflected at $\theta = 0$. The horizontal force $\boldsymbol{F}$ acts on the lower mass A.

Tasks:

(a) Draw free-body diagrams and develop an EOM for θ.
(b) Develop a general work–energy equation and develop the EOM.

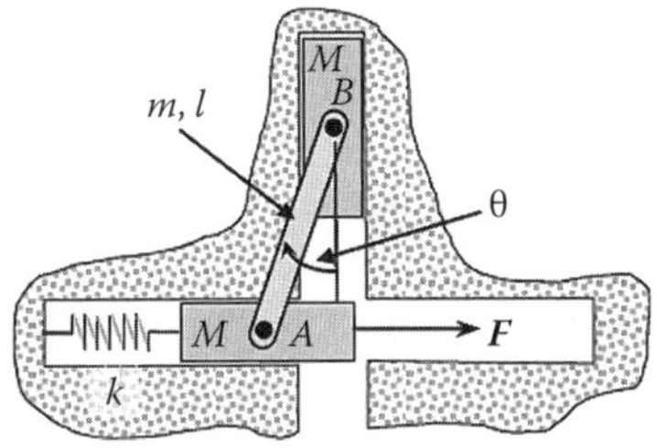

FIGURE P5.38

5.39 For the system illustrated in Figure P5.39, the spring is undeflected for $\theta = 45°$. Neglect the mass and inertia of the rollers.

Tasks:

(a) Draw the free-body diagram for the system and obtain the EOM.
(b) Develop the EOM by applying work–energy principles.

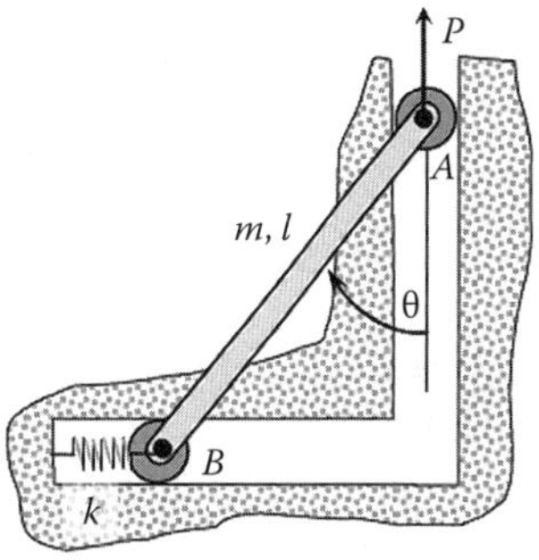

FIGURE P5.39

5.40 In Figure P5.40, the uniform bar of length l and mass m is being elevated by a force P at the bar's mass center that acts perpendicular to the bar. Springs connect points A and B to ground. The springs are undeflected when $\theta = 0$. Neglect friction and the mass and inertia of the rollers.

Tasks:

(a) Draw the free-body diagram and obtain the EOM for θ.
(b) Apply work–energy principles and develop the EOM.

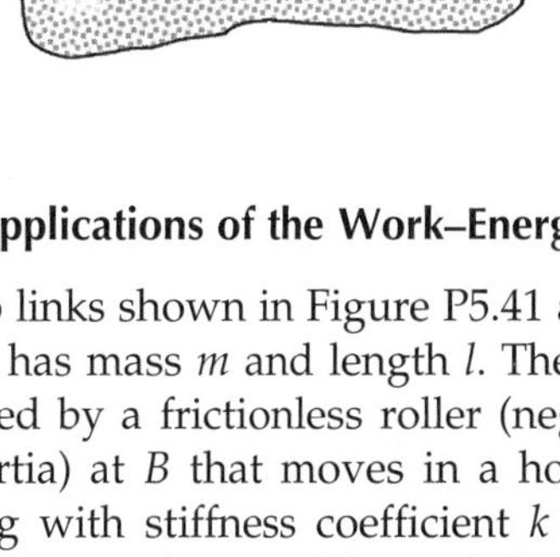

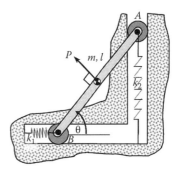

FIGURE P5.40

Multi-Body Applications of the Work–Energy Equation

5.41 The two links shown in Figure P5.41 are pinned at A. Each has mass m and length l. The right link is supported by a frictionless roller (negligible mass and inertia) at B that moves in a horizontal slot. A spring with stiffness coefficient k connects the links at their midpoints. The undeflected length of the spring is $l/2$. A force $\boldsymbol{F}$ acts at the midpoint (and perpendicular to) the right-hand link.

Tasks: State the work–energy equation for the system and develop the EOM.

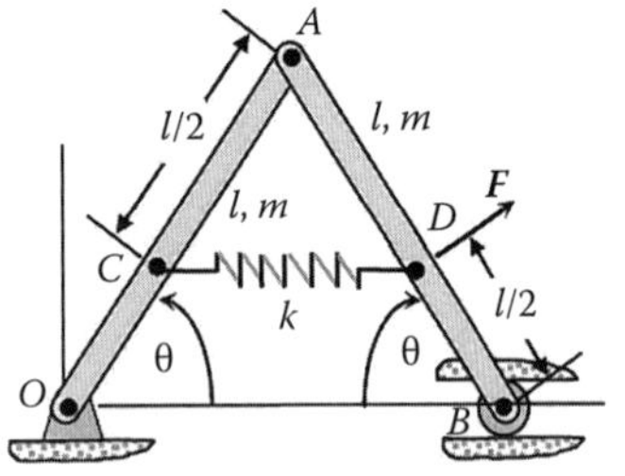

FIGURE P5.41

5.42 The two links shown in Figure P5.42 are pinned at A. Each has mass m and length l. The frictionless roller (negligible mass and inertia) at B moves in a horizontal slot and supports the right link. A spring with stiffness coefficient k connects the links at points O and B. The spring is undeflected when $\theta = 30°$. A horizontal force $\boldsymbol{F}$ acts at A.

Tasks: State the work–energy equation for the system and develop the EOM.

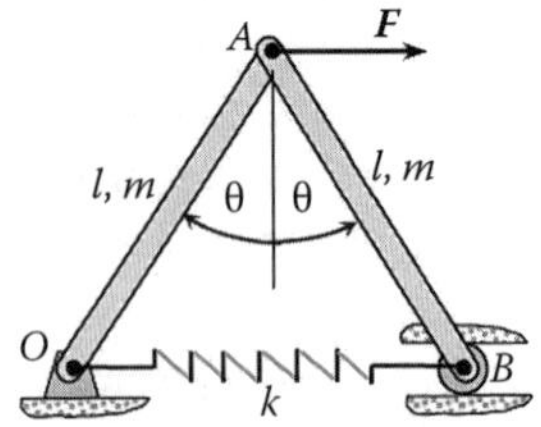

FIGURE P5.42

5.43 The bar on the left of Figure P5.43 is pivoted about its center O and has mass M and length $L = 2l$. The force $\boldsymbol{F}$ acts perpendicular to the lower left end of the bar. The bar on the right has mass m and length l and is supported by a frictionless roller (negligible mass and inertia) at B. A linear spring with stiffness coefficient k connects B to pivot point O. The spring has an undeflected length l_0. The two bars are connected at A by a frictionless pivot joint.

Tasks: State the work–energy equation for the system and develop the EOM.

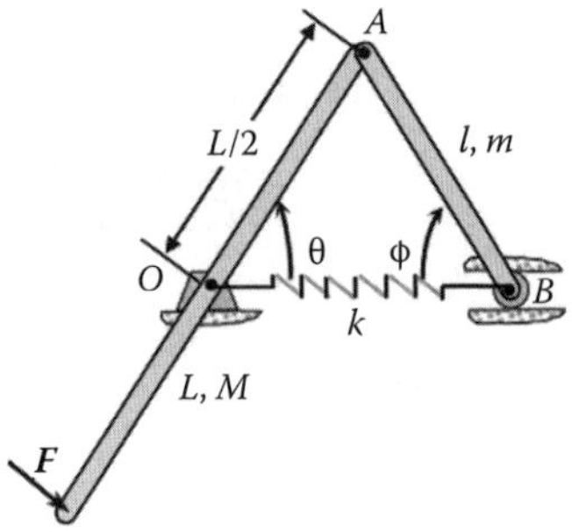

FIGURE P5.43

5.44 As shown in Figure P5.44, four identical bars of length l and mass m are connected together as shown and support mass M. The linear spring shown is undeflected at $\theta = 30°$ and has a stiffness coefficient k. The force $\boldsymbol{F}$ acts on mass M.

Tasks: State the work–energy equation for the system and develop the EOM.

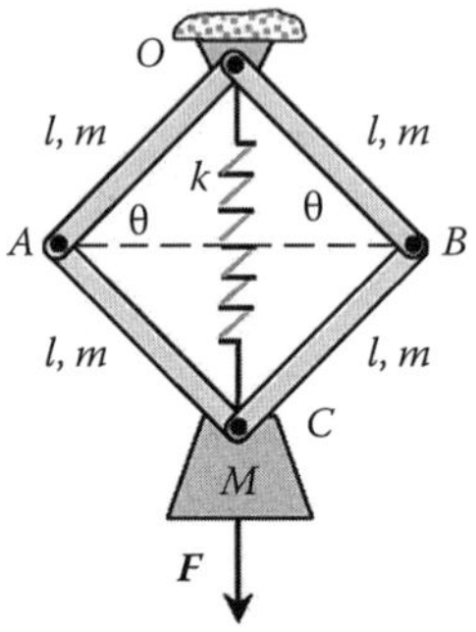

FIGURE P5.44

5.45 The bar assembly shown in Figure P5.45 is being pulled upward by a vertical force F at A. The spring connecting O and A is undeflected when the roller at A is resting on O.

Tasks: State the work–energy equation for the system and develop the EOM.

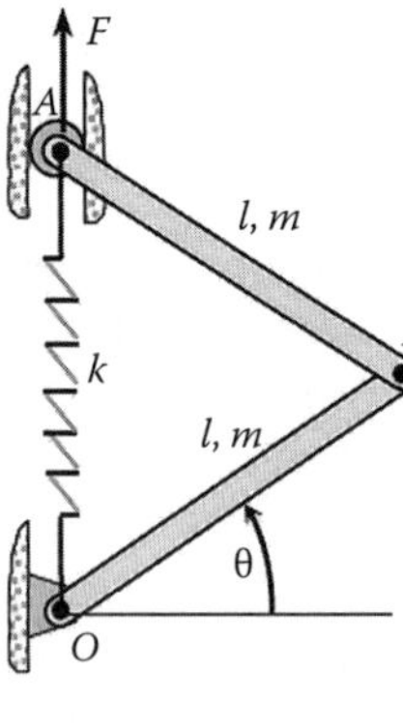

FIGURE P5.45

5.46 The assembly illustrated in Figure P5.46 consists of two bars and a cylinder. The left bar is pivoted at O. The cylinder rolls without slipping on the horizontal plane. A linear spring with stiffness coefficient k connects points O and A. The spring is undeflected at $\theta = \theta_0$. A force with magnitude P acts perpendicular to the right-hand end of the second bar at point D.

Tasks: State the general work–energy equation for this system and develop the EOM.

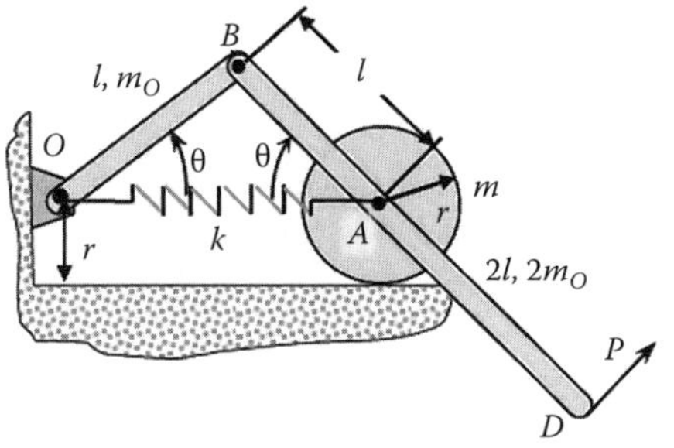

FIGURE P5.46

5.47 Shown in Figure P5.47 is a disk that rolls without slipping on the horizontal plane. The spring has an undeflected length of r. Assume the collar at C has no friction along guide rod. A constant horizontal force $\boldsymbol{F}$ acts at point O.

Tasks: State a general work–energy relationship for the system and develop the EOM.

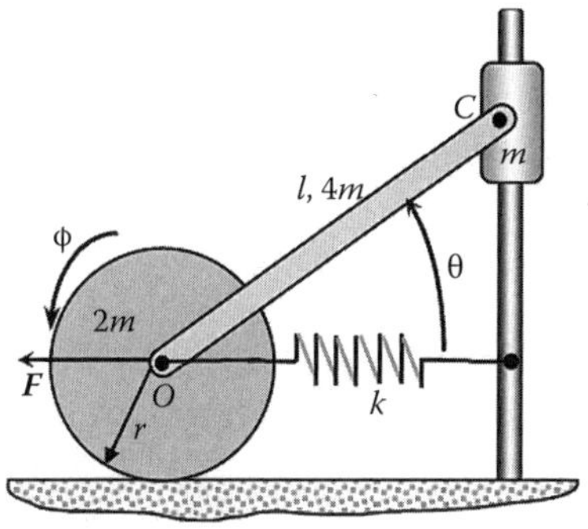

FIGURE P5.47

5.48 As shown in Figure P5.48, the disk of mass M and radius r rolls without slipping on the horizontal plane. Links OA and AB, each of mass m and length l, are released from rest when $\theta = 0$. The spring connecting point B to the wall is undeflected at $\theta = 0$.

Tasks: State the general work–energy expression for the system and develop the EOM.

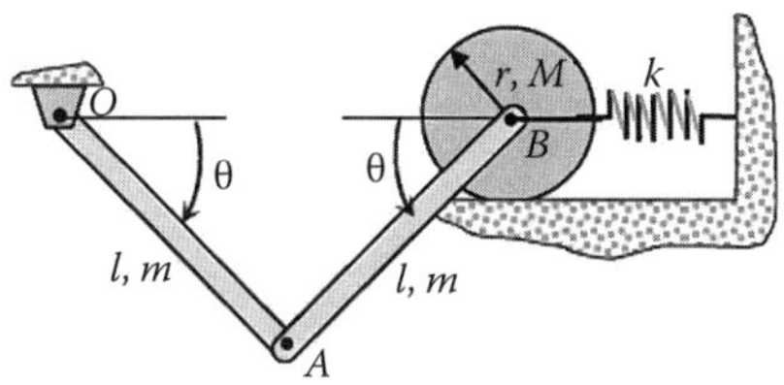

FIGURE P5.48

5.49 For the plate assembly illustrated in Figure P5.49, the plate is connected to ground by a frictionless pivot at O and is connected to the bar by a frictionless pivot at A. The right end of the bar moves in a frictionless horizontal guide at B. The roller at B has negligible mass and inertia. Points O and B are connected together by a linear spring with spring coefficient k and the undeflected length l.

Task: State the general work–energy expression and develop the EOM for θ.

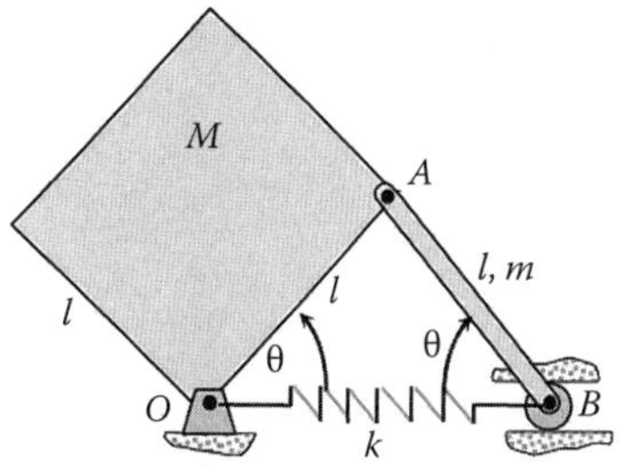

FIGURE P5.49

5.50 The bar-plate assembly illustrated in Figure P5.50 is acted by a horizontal force $\boldsymbol{F}$. The system starts from rest at $\theta = 0$. The spring connecting points O to A is undeflected when $\theta = 0$. Neglect friction and the mass and inertia of the roller at A.

Tasks: State the general work–energy expression and develop the EOM.

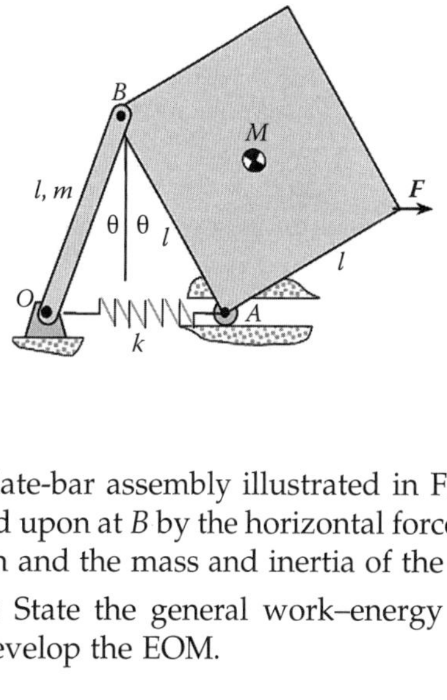

FIGURE P5.50

5.51 The plate-bar assembly illustrated in Figure P5.51 is acted upon at B by the horizontal force $\boldsymbol{F}$. Neglect friction and the mass and inertia of the roller at B.

Tasks: State the general work–energy expression and develop the EOM.

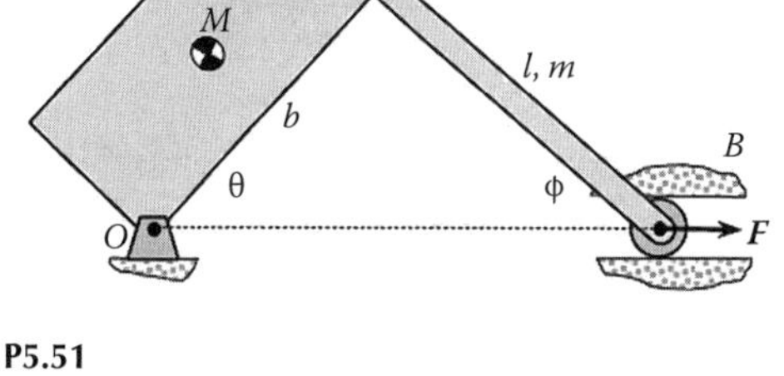

FIGURE P5.51

5.52 The two wheels roll without slipping on the horizontal plane, as shown in Figure P5.52. Bar AB of mass m and length l is connected to each disk by frictionless pivots. The system is released from rest when $\theta = 0°$.

Tasks:

(a) Develop the general work–energy expression for the system in terms of θ and $\dot{\theta}$.

(b) From (a) develop the EOM.

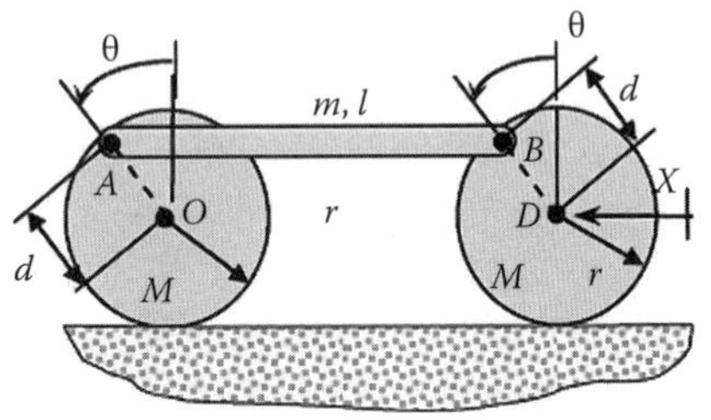

FIGURE P5.52

Multi-Degree-of-Freedom Torsion Examples

5.53 In Figure P5.53, the ends of the elastic circular shafts that support gears 1 and 2 are rigidly attached to ground. The gears are rigidly engaged. The shaft for gear 1 has a length of 20 cm, a diameter of 5 cm, and shear modulus of 80 MPa. Gear 1 has a 15 cm radius and a 1.5 kg-m^2 polar moment of inertia. The shaft for gear 2 has length 10 cm, diameter 7.5 cm, and shear modulus 75 MPa. Gear 2 has a 22.5 cm radius. The pulley attached to gear 2 has a radius 10 cm. The weight attached to the cable from the pulley has 10 kg mass. The combined polar moment of inertia for the pulley and gear 2 is 1.75 kg-m^2.

Tasks:

(a) Select coordinates and draw free-body diagrams for torsional motion of the system.

(b) Obtain an EOM for the system using both Newtonian principles and conservation of energy.

(c) For the properties given, determine the natural frequency.

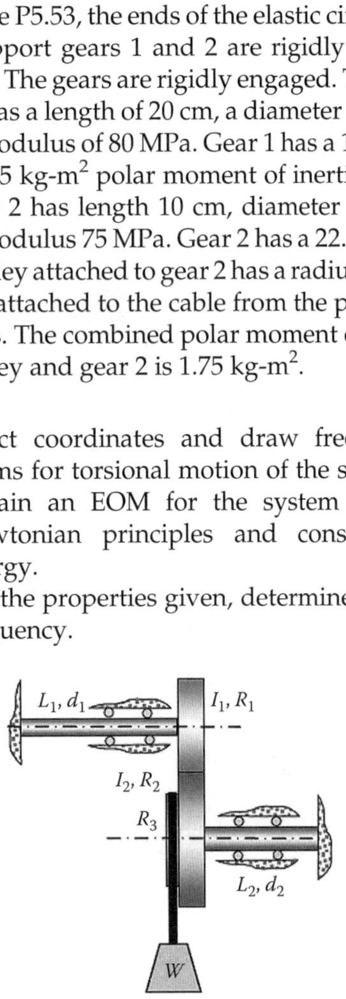

FIGURE P5.53

5.54 In Figure P5.54, for torsional motion, complete the following tasks:

Tasks:

(a) Select coordinates, draw a free-body diagram for each disk, and develop the equations of motion.

(b) State the equations of motion in matrix form.

(c) Determine the eigenvalues, natural frequencies, and eigenvectors given the following data: $M_1 = 200$ kg, $r_1 = 200$ mm, $L = 3000$ mm, $G = 90$ MPa, $d = 75$ mm, $M_2 = 200$ kg, and $r_2 = 200$ mm.

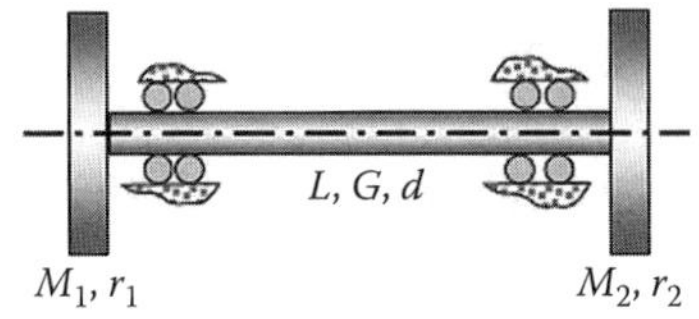

FIGURE P5.54

5.55 Figure P5.55 illustrates two shafts connected at their ends by gears. The gears are rigidly engaged. Neglect torsional drag (friction) at bearings and perform the following tasks:

Tasks:

(a) Select coordinates and draw the free-body diagram for each disk.

(b) Apply the kinematic constraint to eliminate the unneeded coordinate. State the separate independent equations of motion and put them in matrix format.

(c) Determine the eigenvalues and eigenvectors for the following properties: $M_1 = 100$ kg, $r_1 = 200$ mm, $L_1 = 3000$ mm, $d_1 = 50$ mm, $M_2 = 50$ kg, $r_2 = 150$ mm, $M_3 = 25$ kg, $r_3 = 75$ mm, $L_2 = 1500$ mm, $d_2 = 75$ mm, $M_4 = 250$ kg, $r_4 = 200$ mm, $G_1 = G_2 = 80$ GPa.

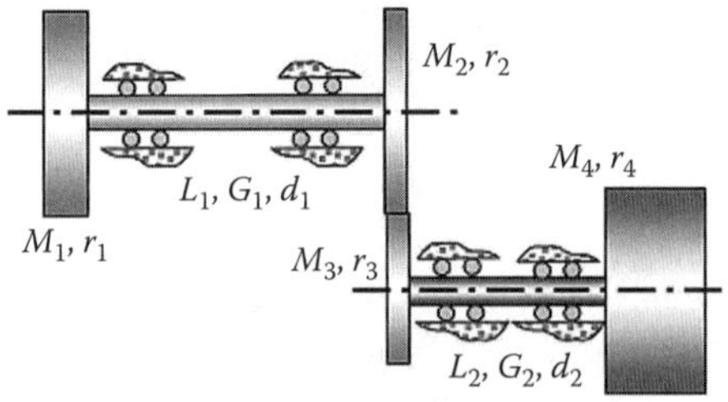

FIGURE P5.55

5.56 In Figure P5.56, the shaft ends that support gears 1 and 2 are rigidly attached to ground. The gears are rigidly engaged. The shaft for gear 1 has a 20 cm length, a 5 cm diameter, and a 95 GPa shear modulus. Gear 1 has a 15 cm radius and a 1.5 kg-m^2 mass moment of inertia. The shaft for gear 2 has a 10 cm length, a 7.5 cm diameter, and a 85 GPa shear modulus. Gear 2 has a 22.5 cm radius and a 1.75 kg-m^2 mass moment of inertia. The pulley attached to gear 2 has a radius of 10 cm. The weight attached to the cable from the pulley has a mass of 10 kg, and the cable has a stiffness of 90 N/mm. The pulley attached to gear 2 has negligible mass. Neglect torsional drag at the bearings.

Tasks:

(a) Select coordinates and draw free-body diagrams for each mass/inertia element.

(b) Apply the kinematic constraint to eliminate an unneeded coordinate. Separately state the independent equations of motion and then put them in matrix format.

(c) For the properties given, determine the natural frequencies.

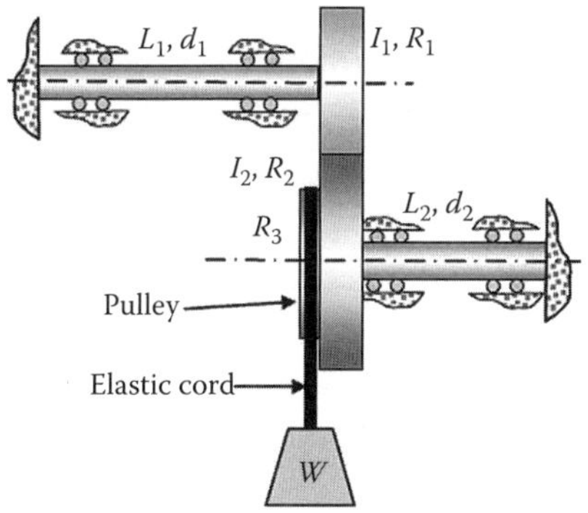

FIGURE P5.56

5.57 The rotor train shown in Figure P5.57 is driven from disk 1, disk 6 has a resisting load, and disks 2, 3, and 4 are a set of gears. The gears are rigidly engaged. Neglect torsional drag of bearings.

Tasks:

(a) Select coordinates and draw free-body diagrams for each mass/inertia element.

(b) Apply the kinematic constraints to eliminate unneeded coordinates. State the separate independent equations of motion and put them in matrix format.

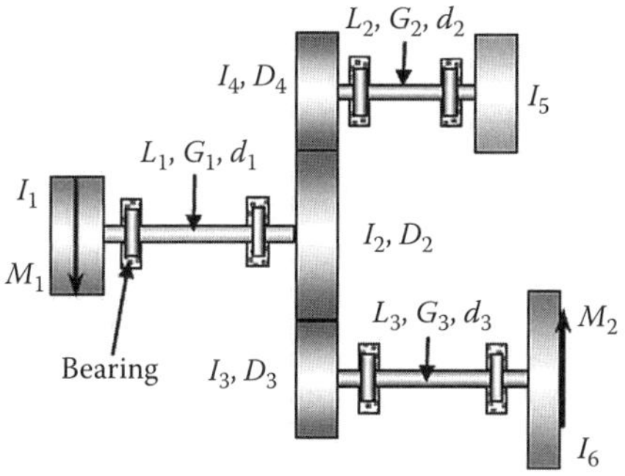

FIGURE P5.57

Beams as Springs Examples

5.58 The cantilever beam shown in Figure P5.58 has length L and a rectangular cross section with width a, height b. The rectangular thin plate mounted on the end of the beam has weight w and dimensions $A = 16$ in. and $B = 18$ in.

Tasks:

(a) For motion within the plane of the paper, draw a free-body diagram for the plate and state the equations of motion. Determine the inertia and stiffness matrices, and state the matrix EOM.

(b) Determine the eigenvalues, natural frequencies, and eigenvectors for the following data: $a = 2$ in., $b = 1$ in., $L = 10$ in., $w = 200$ lb s. The beam is made from steel, $E = 30\text{E}6$ psi.

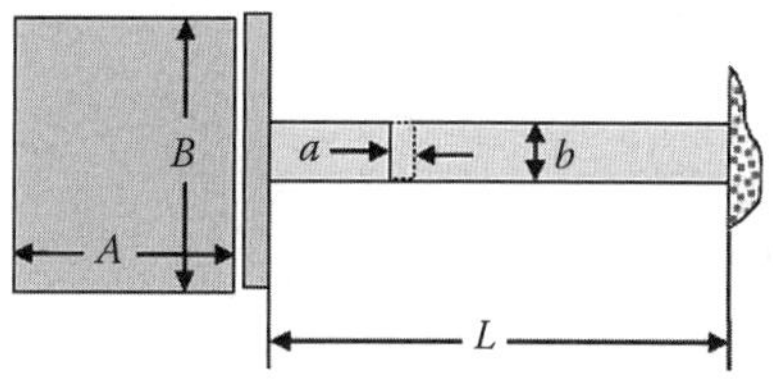

FIGURE P5.58

5.59 Figure P5.59 shows a plate with mass $m = 2000$ kg that is supported by four identical columns that are cantilevered from the ground. Each column has length $l = 2$ m and a square cross section with side dimension $b = 75$ mm. The columns are made from steel with modulus of elasticity of $E = 210$ GPa. The plate is attached to the tops of the columns with pin connections.

Tasks:

(a) Develop the plate's EOM in the horizontal direction, using X to define its position.

(b) Determine the plate's natural frequency.

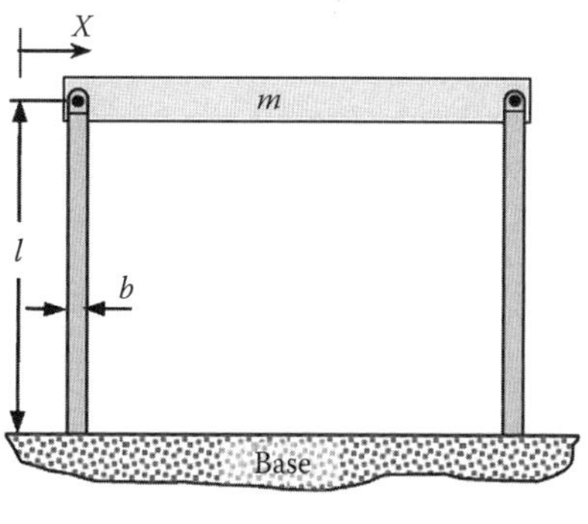

FIGURE P5.59

5.60 Figure P5.60 illustrates a plate with mass $m = 2000$ kg that is supported by four identical columns that are cantilevered from the ground. Each column has length $l = 2$ m and a square cross section with dimension $b = 75$ mm. The columns are made from steel with modulus of elasticity of $E = 210$ GPa. The plate is welded to the tops of the columns.

Tasks:

(a) Develop the plate's EOM in the horizontal direction, using X to define its position.

(b) Determine the plate's natural frequency.

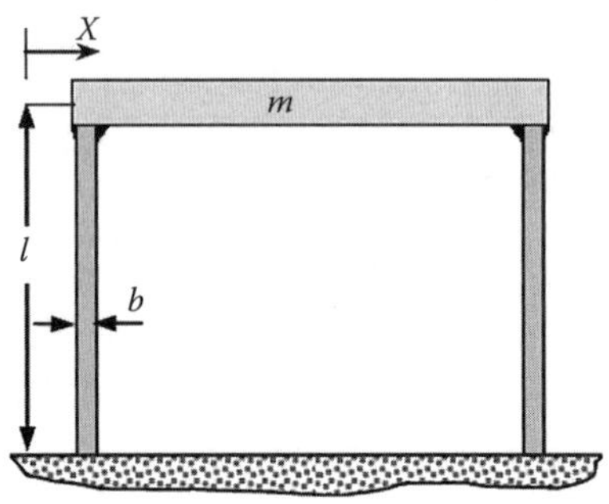

FIGURE P5.60

5.61 Figure P5.61 illustrates a plate with mass $m = 2000$ kg that is supported by four columns that are cantilevered from the ground. The columns have length $l = 2$ m. The two columns on the left have a square cross section with dimension $b = 75$ mm while the two columns on the right have a circular cross section with a diameter of 50 mm. The columns are made from steel with modulus of elasticity of $E = 210$ GPa. The plate is welded to the tops of the left columns and pinned to the columns on the right.

Tasks:

(a) Develop the plate's EOM in the horizontal direction, using X to define its position.

(b) Determine the plate's natural frequency.

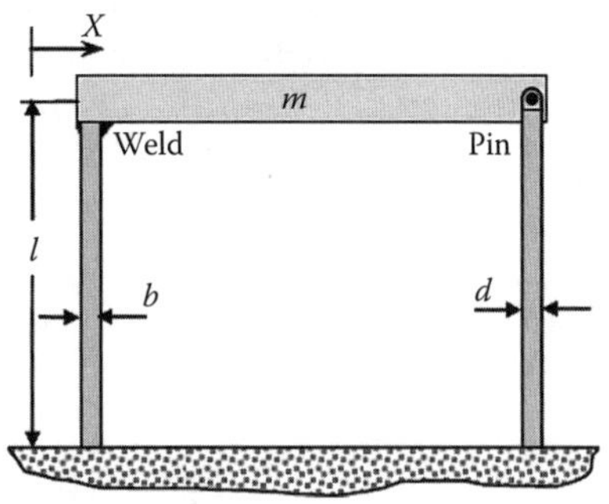

FIGURE P5.61

5.62 Develop the EOM for Figure P5.59 if the base is excited horizontally by a motion of the form $X_B(t) = 5\cos(29t)$ cm. Assume that the support columns produce a damping factor for the slab of 0.075.

Tasks:

(a) Determine the EOM for the system.

(b) Determine the plate's steady-state response.

(c) Determine the peak response of the system.

5.63 The first floor of the two-story structure shown in Figure P5.63 is a 12 ft by 15 ft slab that weighs 75 lb/ft^2 and is supported by four columns sized at HP10×42. The second story is a 12 ft by 15 ft slab that weighs 2/3 of the first floor that is supported from the first floor by four HP10×42 columns. All columns are 12 ft in length.

Tasks:

(a) Develop the equations of motion and present them in matrix format.

(b) For the given properties, determine the eigenvalues, natural frequencies, and eigenvectors. Draw a sketch showing the mode shapes (i.e., eigenvector plots). *Note*: There are two directions of motion. Some additional properties: Columns are made from steel: $E = 30(10^6)$ psi, HP10×42: $Ix = 394$ in.^{4}, $Iy = 127$ in.^{4}.

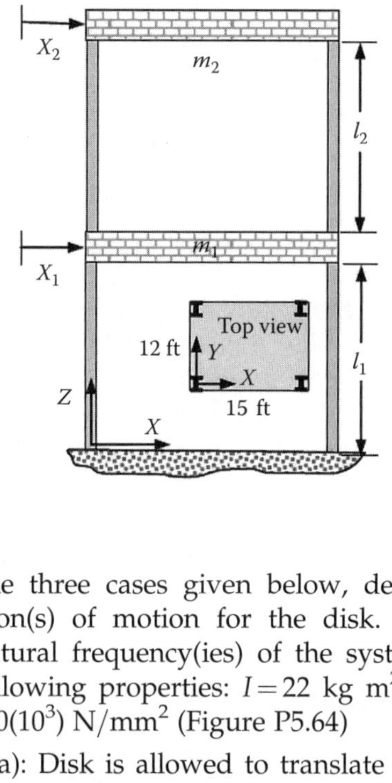

FIGURE P5.63

5.64 For the three cases given below, determine the equation(s) of motion for the disk. Also, solve the natural frequency(ies) of the systems. Apply the following properties: $I = 22$ kg m^2, $M = 2$ kg, $E = 210(10^3)$ N/mm^2 (Figure P5.64)

Case (a): Disk is allowed to translate only. $d = 25$ mm, $L = 2500$ mm.

Case (b): Disk is allowed to pivot only. $d = 25$ mm, $L = 2500$ mm.

Case (c): Disk at end of beam has linear displacement spring and a torsional spring attached to the disk. $K_\theta = 55$ N m/rad, $K = 2000$ N/mm, $L = 2500$ mm.

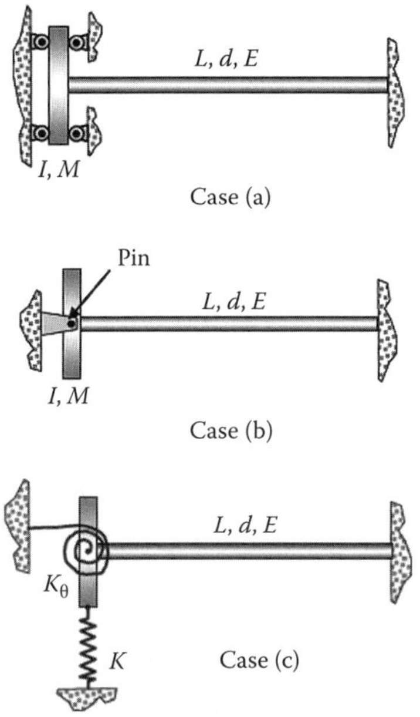

FIGURE P5.64

More Multi-Degree-of-Freedom Examples

5.65 The composite rotor shown in Figure P5.65 is supported on bearings that are modeled as linear displacement springs. Assume small motion (pitch and bounce) about the rotor's equilibrium position. System properties: $d_1 = 6$ in., $D_3 = 2$ in., $L_1 = 20$ in., $D_1 = 7.5$ in., $L_2 = 18$ in., $D_2 = 10.75$ in., $d_2 = 6$ in., $k_A = 750{,}000$ lb/in., $k_B = 900{,}000$ lb/in. The shaft D_3 goes through entire body and is made of steel ($\rho_S = 0.283$ lb/in.^{3}). Disk D_1 is made of steel, and disk D_2 is made from brass ($\rho_B = 0.318$ lb/in.^{3}).

Tasks:

(a) Select coordinates, develop the equations of motion for the system, and present them in matrix format.

(b) Obtain the eigenvalues and eigenvectors for the system.

(c) Draw mode-shape plots for the eigenvectors.

(d) What are the natural frequencies?

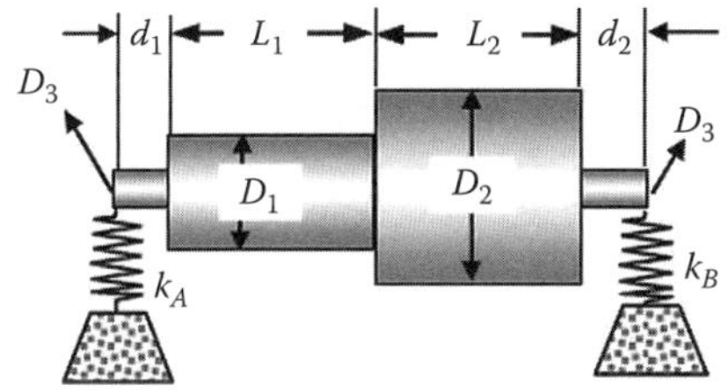

FIGURE P5.65

5.66 As shown in Figure P5.66, disk 1 is driven by an external moment $\boldsymbol{M}(t)$. Disk 2 is connected torsionally to disk 1 by a shaft and rolls without slipping on the cart.

The following system properties apply: $I_1 = 1.5$ (10^5) kg-mm^2, $D_1 = 200$ mm, $L = 1250$ mm, $G = 79$ GPa, $d = 50$ mm, $I_2 = 0.5(10^3)$ kg mm^2, $D_2 = 175$ mm, $M = 75$ kg, $k_A = 2$ kN/mm, $k_B = 1.5$ kN/mm.

Tasks:

(a) Select coordinates and draw free-body diagram for motion of the two disks and the cart.

(b) Obtain the EOM for each body.

(c) Present the coupled equations of motion in matrix format.

(d) Obtain the eigenvalues, natural frequencies, and normalized eigenvectors.

(e) State the uncoupled modal equations of motion and the transformation from modal to physical coordinates.

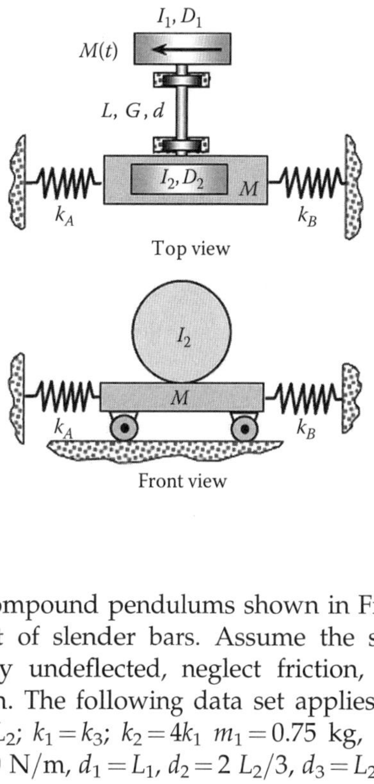

FIGURE P5.66

5.67 The compound pendulums shown in Figure P5.67 consist of slender bars. Assume the springs are initially undeflected, neglect friction, and small motion. The following data set applies: $2m_1 = m_2$; $2L_1 = L_2$; $k_1 = k_3$; $k_2 = 4k_1$ $m_1 = 0.75$ kg, $L_1 = 0.5$ m, $k_1 = 20$ N/m, $d_1 = L_1$, $d_2 = 2\ L_2/3$, $d_3 = L_2$.

Tasks:

(a) Select coordinates and draw the free-body diagrams for the two bodies.

(b) Obtain the equations of motion for the system and present them in matrix format.

(c) Solve for the normalized eigenvalues and eigenvectors.

(d) State the uncoupled modal equations of motion and the transformation from modal to physical coordinates.

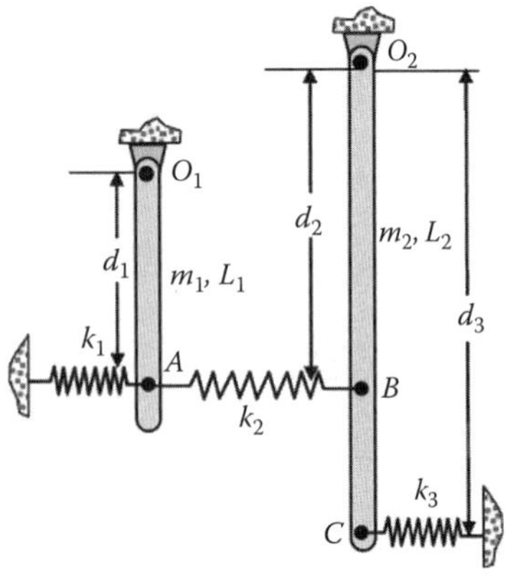

FIGURE P5.67

5.68 In Figure P5.68, two disks of equal mass m and radius R are joined by spring of stiffness k_2 and roll without slipping. Disk A is attached to ground by spring k_1 and disk B is attached to ground by spring k_3. The following data set applies: $k_1 = k_3$; $k_2 = 2k_1$ $m = 5$ kg, $k_1 = 50$ N/m, $R = 0.25$ m.

Tasks:

(a) Select coordinates and draw a free-body diagram for each body.

(b) Obtain the equations of motion for the system and present them in matrix format.

(c) Solve for the eigenvalues and normalized eigenvectors.

(d) State the uncoupled modal equations of motion and the transformation from modal to physical coordinates.

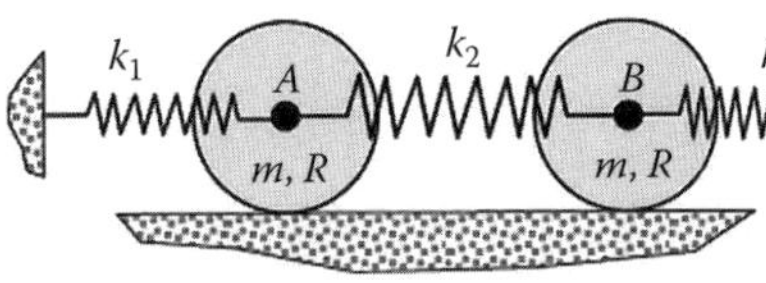

FIGURE P5.68

5.69 The cart shown in Figure P5.69 is attached to the wall by a spring of stiffness k. As shown, the spring is undeflected. The pendulum is attached to the cart by a frictionless pin and a torsional spring of stiffness k_θ. The following data set applies:

$$m_c = 5 \text{ kg}, \quad m_p = 1 \text{ kg}, \quad L = 0.5 \text{ m}, \quad k_\theta = 3.5 \text{ N m/rad}, \quad k = 200 \text{ N/m}$$

Tasks:

(a) Select coordinates and draw the free-body diagram for each body.

(b) Obtain the EOMs for the system using Newtonian principles. Present in matrix format. Apply small angle assumption.

(c) Determine the required conditions for stable equilibrium.

(d) Obtain the eigenvalues and normalized eigenvectors for the system.

(e) State the uncoupled modal equations of motion and the transformation from modal to physical coordinates.

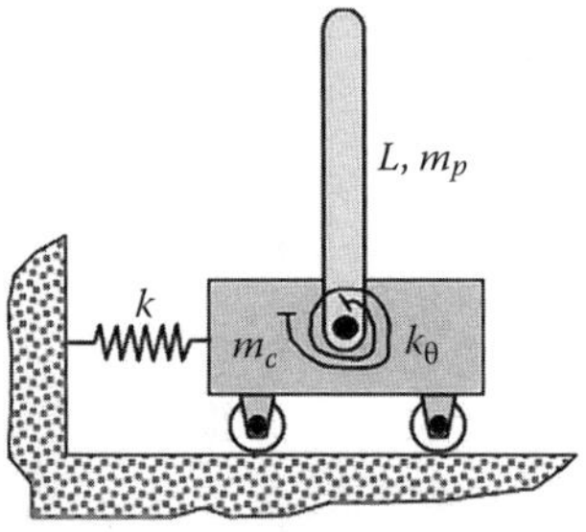

FIGURE P5.69

5.70 As shown in Figure P5.70, the cylinder of mass M and radius R rolls without slipping on the plane and is attached to the cart of mass m by spring k_B. Spring k_A connects the cart to ground and spring k_C connects the cylinder to ground. The following data set applies: $M = 25$ kg, $R = 50$ mm, $m = 10$ kg, $k_A = 1000$ N/m, $k_B = 2500$ N/m, $k_C = 1000$ N/m.

Tasks:

(a) Select coordinates and draw free-body diagrams for each body.

(b) Obtain the EOM for each body. Present the equations in matrix format.

(c) For the given properties, obtain the eigenvalues, normalized eigenvectors, and natural frequencies of the system.

(d) State the uncoupled modal equations of motion and the transformation from modal to physical coordinates.

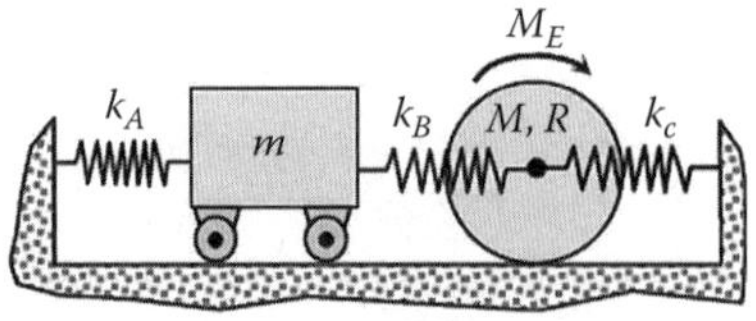

FIGURE P5.70

5.71 As shown in Figure P5.71, the cylinder of mass M and radius R rolls with out slipping on the cart of mass m. The following data set applies: $M = 50$ kg, $R = 150$ mm, $m = 150$ kg, $k_A = 5500$ N/m, $k_B = 2000$ N/m.

Tasks:

(a) Select coordinates and draw a free-body diagram for each body.
(b) Obtain the EOM for each body. Present the equations in matrix format.
(c) For the data provided below, obtain the eigenvalues, normalized eigenvectors, and natural frequencies of the system.
(d) State the uncoupled modal equations of motion and the transformation from modal to physical coordinates.

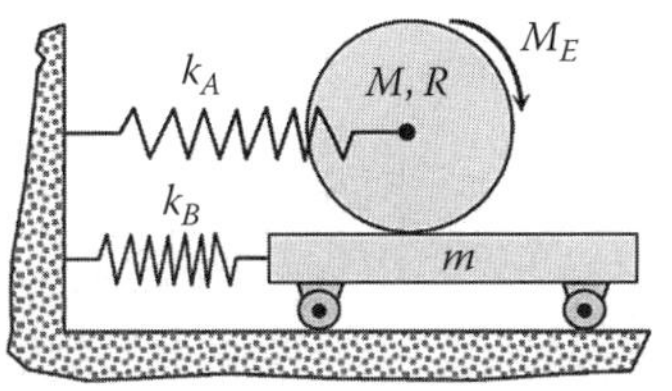

FIGURE P5.71

5.72 As shown in Figure P5.72, the two slender rigid bars are connected to the midpoint of the plate and the disk and rigidly attached to the ends of shaft *AB*. Shaft *AB* does not bend and cannot rotate freely, but can twist. Bars *AC* and *BD* are rigid but have appreciable mass. The system properties are as follows: shaft: $L_s = 16$ in., $d_s = 0.75$ in., $G_s = 15(10^6)$ psi, bar 1: $L_{b1} = 20$ in., $w_{b1} = 2$ lb; bar 2: $L_{b2} = 15$ in., $w_{b2} = 1.5$ lb; plate: $h = 5$ in., $w = 2$ in., $b = 10$ in., $\rho_p = 0.283$ lb/in.^{3}; disk: $R = 3$ in., $t = 2$ in., $\rho_d = 0.318$ lb/in^3.

Tasks:

(a) Select coordinates and draw a free-body diagram for each body.
(b) Obtain the EOM for each body.
(c) Present the equations of motion in matrix format.
(d) Obtain the eigenvalues, normalized eigenvectors, natural frequencies, and uncoupled equations of motion for the system.
(e) State the uncoupled modal equations of motion and the transformation from modal to physical coordinates.

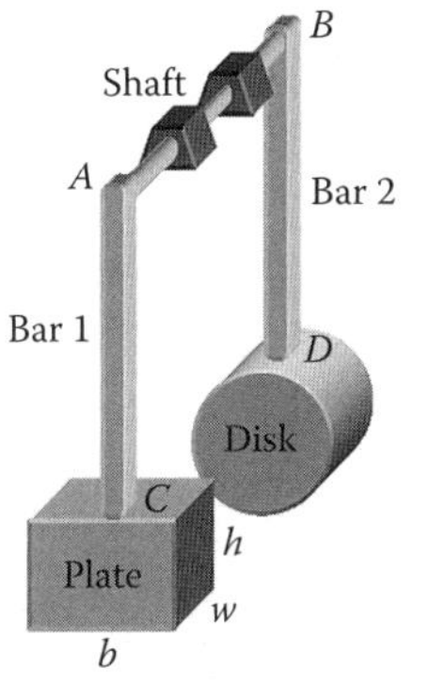

FIGURE P5.72

Mechanism Examples

5.73 As shown in Figure P5.73, the end of link *OB* is acted upon by a constant horizontal force ***F***. Friction is negligible at the pinned connections.

Tasks:

(a) Draw the system in a general orientation, select coordinates, and define the required kinematic constraint relationships.
(b) Draw the required free-body diagrams and develop the equations of motion.
(c) Present the equations for your unknowns in matrix format.

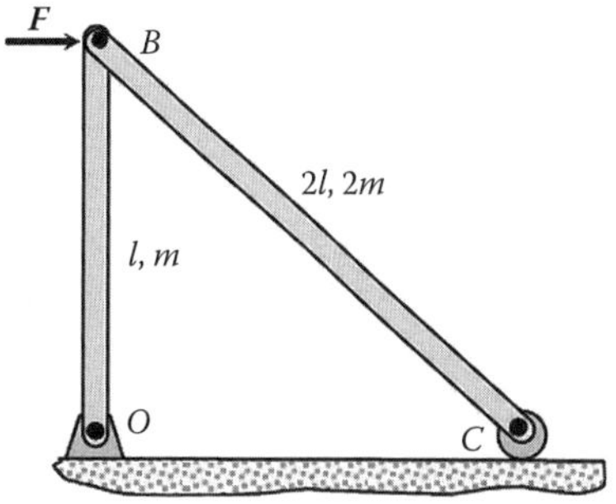

FIGURE P5.73

5.74 As shown in Figure P5.74, a frictionless pin at *A* connects the two bars. Bar *OA* is supported by a frictionless pin at *O*. A small, mass roller at *B* (negligible mass and inertia) supports the right end of bar *AB*. The system starts from rest when *OA* is parallel to the *Y*-axis and is acted on by the horizontal force ***F*** at *B*. Pin friction is negligible.

Tasks:

(a) Select coordinates and define the required kinematic constraint relationships.
(b) Draw the required free-body diagrams and develop the equations of motion.
(c) Put the equations for your unknowns in matrix form.

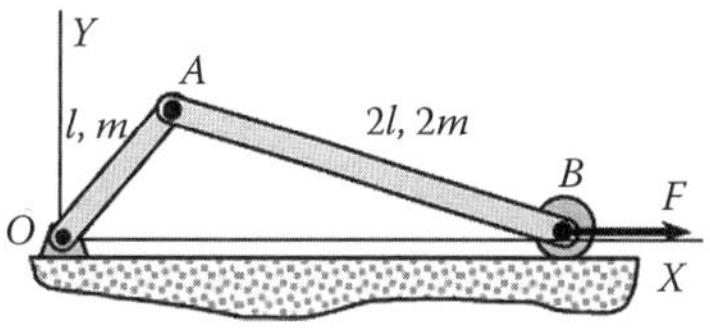

FIGURE P5.74

5.75 The force ***F*** in Figure P5.75 is applied at point *B*, the top of the two-bar assembly. Point *B* is on a guide and can only move vertically. There is no friction, and the mass and inertia of the roller are negligible.

Tasks:
(a) Select coordinates and define the required kinematic constraint relationships.
(b) Draw the necessary free-body diagrams and develop the equations of motion.
(c) State the equations for your unknowns in matrix format.

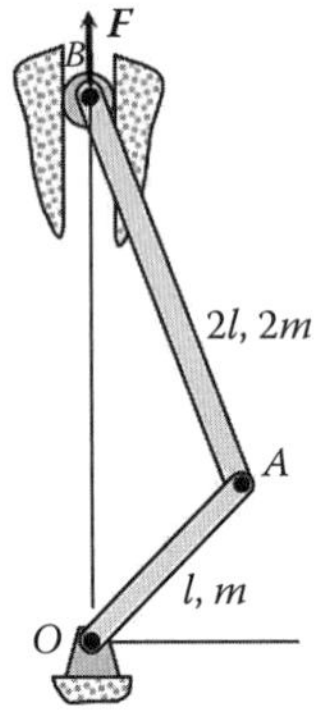

FIGURE P5.75

5.76 A frictionless pin at A connects the two bars shown in Figure P5.76. Bar OA of length l_1 and mass m_1 is supported by a frictionless pin at O. Bar AB has length l_2 and mass m_2, and its right end is supported by a roller at B (negligible mass and inertia). The system starts from rest when OA is vertical and is acted on by a horizontal force $\boldsymbol{F}$ at B. Friction is negligible at the pins, and there is no slipping at the roller B.

Tasks:
(a) Select coordinates and define the required kinematic constraint relationships.
(b) Draw the required free-body diagrams, state the unknown variables, and develop the equations of motion.
(c) State the equations for your unknowns in matrix format.

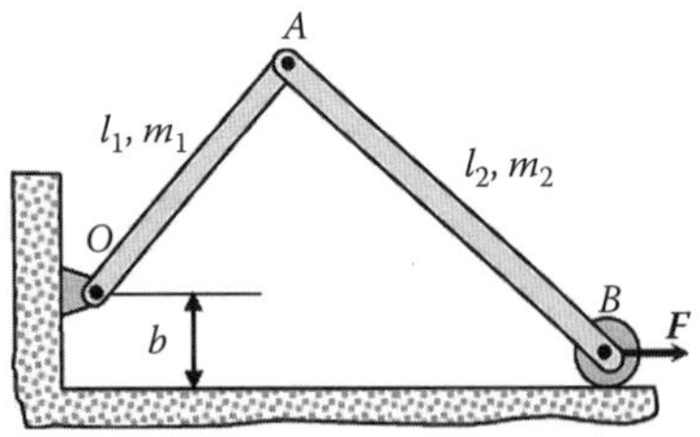

FIGURE P5.76

5.77 The two bars in Figure P5.77 are connected by a frictionless pin at A. Bar OA has length l_1, mass m_1, and is supported by a frictionless pin at O. Bar AB has length l_2 and mass m_2, and its right end is supported by a small frictionless roller at B. The roller has negligible mass and inertia, and friction at the connecting pins is negligible.

Tasks:
(a) Select coordinates and state the required kinematic constraint relationships.
(b) Draw the required free-body diagrams, state the unknown variables, and develop the equations of motion.
(c) State the equations for your unknowns in matrix format.

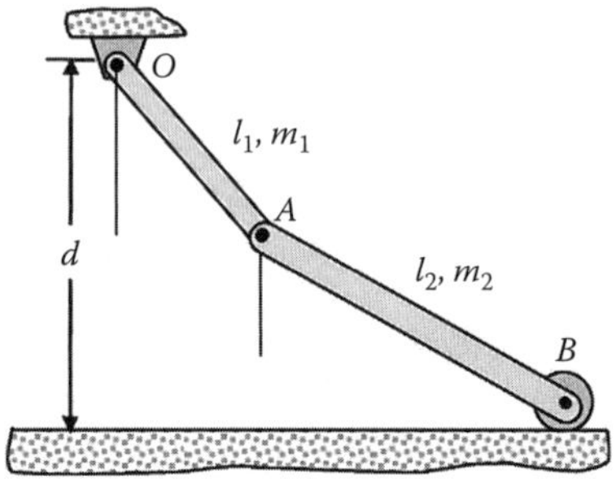

FIGURE P5.77

5.78 In the four-bar linkage illustrated in Figure P5.78, the left bar is driven by a clockwise torque $\boldsymbol{T}$.

Tasks:
(a) Draw the system in a general position, select coordinates, and state the kinematic constraint relationships.
(b) Draw the necessary free-body diagrams and define the unknown variables.
(c) Develop governing equations of motion for the system and state the equations for the unknowns in matrix format (you should have seven unknowns).

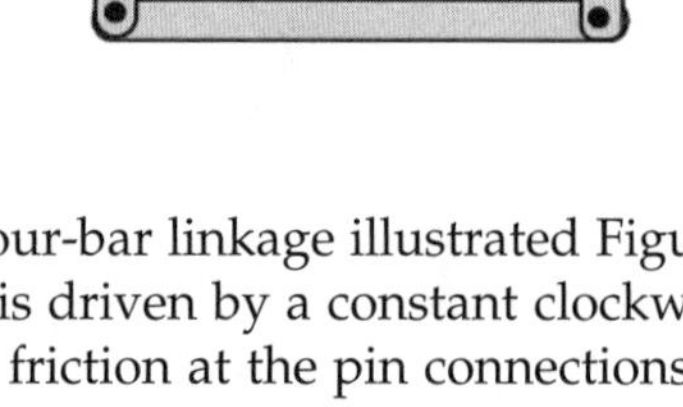

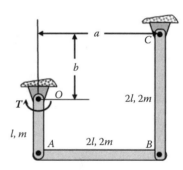

FIGURE P5.78

5.79 In the four-bar linkage illustrated Figure P5.79, the left bar is driven by a constant clockwise torque $\boldsymbol{T}$. Neglect friction at the pin connections.

Tasks:
(a) Draw the system in a general position, select coordinates, and state the required kinematic constraint relationships.
(b) Draw the necessary free-body diagrams and define the unknown variables.

(c) Develop governing equations of motion for the system, and state the equations for the unknowns in matrix format (you should have seven unknowns).

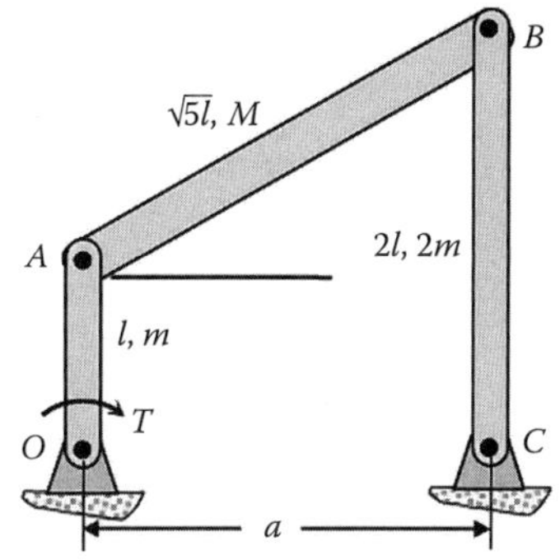

FIGURE P5.79

5.80 In Figure P5.80, the bar of mass m and length $2l$ is supported by two massless cords and is acted on by a known horizontal force $f(t)$. The cords on the left and right have lengths l and $2l$, respectively.

Tasks:

(a) Draw the system in a general position, select coordinates, and state the required kinematic constraint relationships.
(b) Draw a free-body diagram for the bar in a general position and develop governing equations of motion.
(c) State the equations for your unknowns in matrix format.

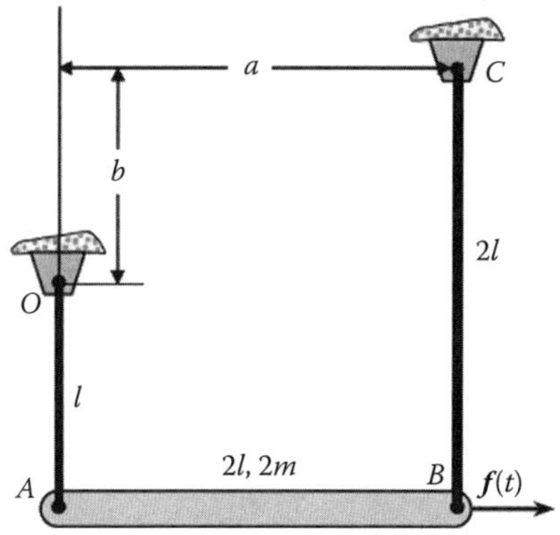

FIGURE P5.80

5.81 The winch shown at C in Figure P5.81 is reeling in cable at a constant rate $\dot{S}$. Bar AB of length L and mass M initially starts from rest in contact with the horizontal surface. The Coulomb dynamic friction coefficient between the bar and ground is μ_d.

Tasks:

(a) Select coordinates and state the required kinematic constraint relationships.
(b) Draw the required free-body diagram and identify the unknowns.
(c) Obtain the equations of motion, identify the unknowns.
(d) State the equations for your unknowns in matrix format.

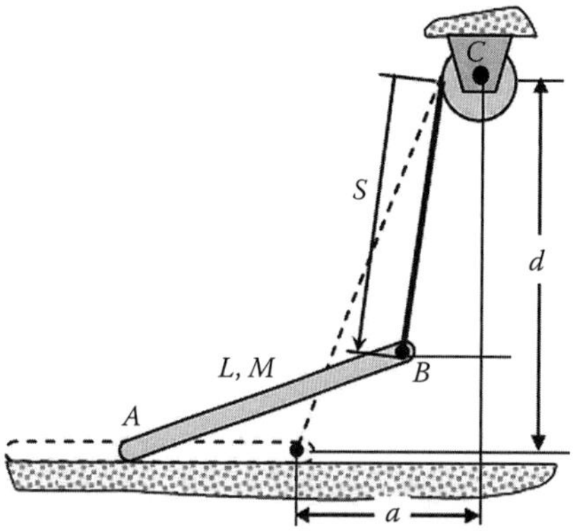

FIGURE P5.81

5.82 The cylinder shown in Figure P5.82 has mass M and radius r and rolls without slipping on the horizontal plane. At O, it is connected via a frictionless joint to bar OA, having mass m and length l. The right end of the bar is connected to point B by an inextensible cord of length l and negligible mass.

Tasks:

(a) Select coordinates and state the required kinematic constraint equations.
(b) Draw the required free-body diagrams for the two bodies and develop the governing equations of motion.
(c) State the governing equations for unknowns in matrix format.

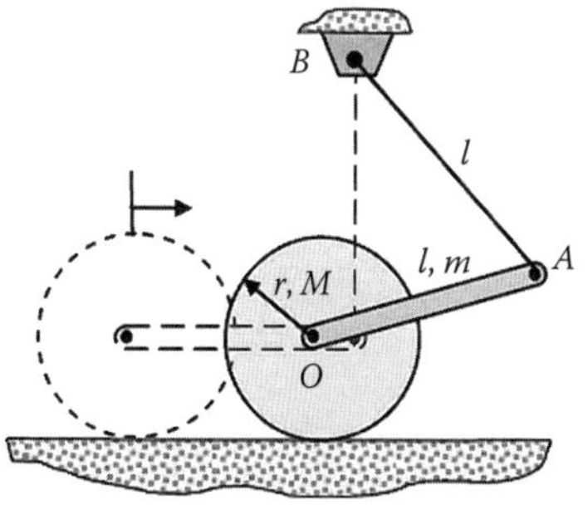

FIGURE P5.82

5.83 The cylinder in Figure P5.83 has mass M and radius r and rolls without slipping on the horizontal plane. It is connected at A via a frictionless joint to a bar OA of mass m and length l. The right end of the bar slides without friction on a horizontal plane.

Tasks:

(a) Select coordinates and state the required kinematic constraint relationships.
(b) Draw the required free-body diagrams for the two bodies and develop the governing equations of motion.
(c) State the governing equations for the unknowns in matrix format.

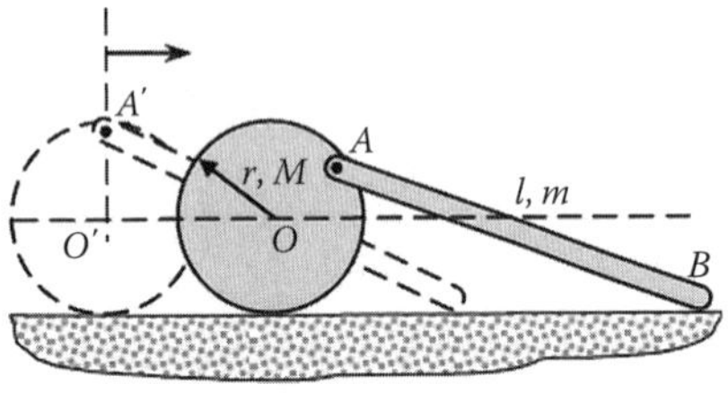

FIGURE P5.83

5.84 In Figure P5.84, the slender bar of length $2l$ and mass m is attached to a circular smooth bar at A by a collar of negligible mass. Point B of bar AB is in contact with the horizontal plane. The coefficient of dynamic friction at the contact point B is μ_d.

Tasks:

(a) Select coordinates and state the kinematic constraint relationships.
(b) Draw the required free-body diagram and state your unknowns.
(c) Obtain the equations of motion for each body.
(d) Develop the system governing equations of motion and place the equations for your unknowns in matrix format.

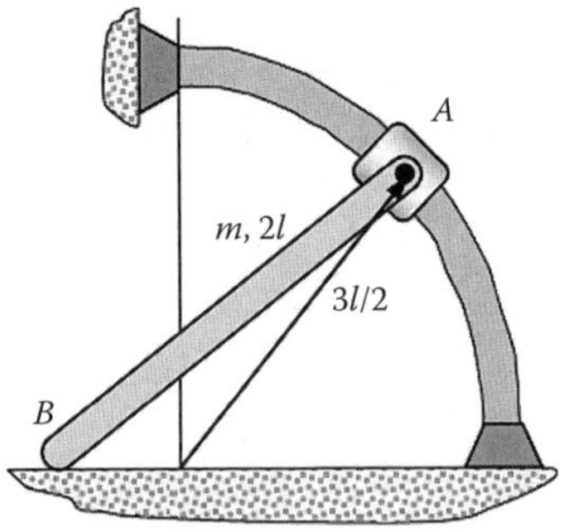

FIGURE P5.84

5.85 Link AC in Figure P5.85 is driven by an external moment $\boldsymbol{M}$. Link OA's mass center is located a distance $\bar{r}$ from O. Bar OB has a radius of gyration k_{Og} and mass of m_2. Link AC is a uniform bar of length r and mass m_1. Friction is negligible.

Tasks:

(a) Select coordinates and state the kinematic constraint relationships.
(b) Draw the required free-body diagrams.
(c) Obtain the equations of motion for each body.
(d) Develop the system governing equations of motion and place the equations for your unknowns in matrix format.

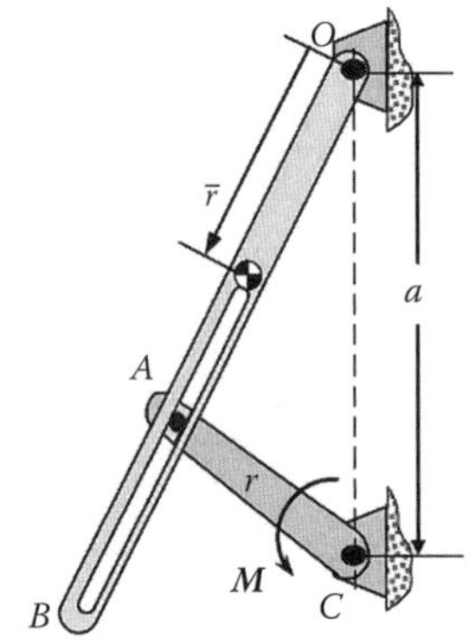

FIGURE P5.85

5.86 The mechanism shown in Figure P5.86 causes the element B to move intermittently as disk A rotates due to an applied torque $\boldsymbol{M}$. The mass center of wheel B is at O_2. Pin C is fixed to disk A and is located a distance r from pivot O_1. Disk A's mass center is located the distance $\bar{r}$ from point O_1. Disk A has mass m_1 and the radius of gyration k_{1g} about point O_2. Friction is negligible.

Tasks:

(a) Draw the system in a general position, select coordinates, and state the kinematic constraint relationships.
(b) Draw the required free-body diagrams.
(c) Obtain the EOM for each body.
(d) Develop the system governing equations of motion and place the equations for your unknowns in matrix format.

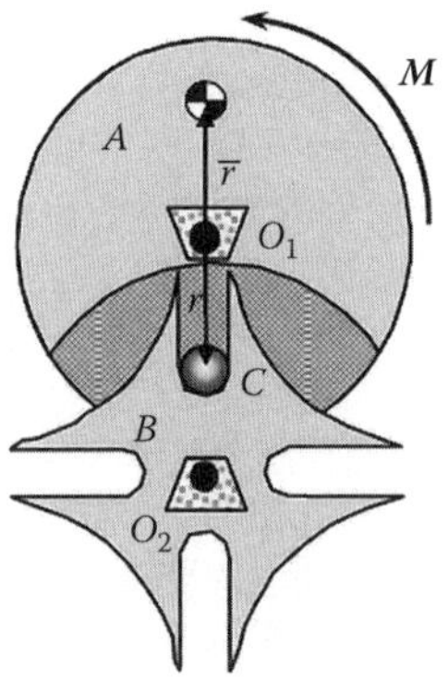

FIGURE P5.86

5.87 In Figure P5.87, link AB is driven by a moment $\boldsymbol{M}$. Cylinder C has mass m_C and is driven by the force $\boldsymbol{F}$. Link AB is a uniform bar of mass m_1 and length l_1; uniform link BC has mass m_2 and length l_2. The mechanism operates in the plane of the paper. Friction is negligible.

Tasks:

(a) Select coordinates and state the kinematic constraint relationships.
(b) Draw the required free-body diagrams.
(c) Obtain the EOM for each body.
(d) Develop the system governing equations of motion and place the equations for your unknowns in matrix format.

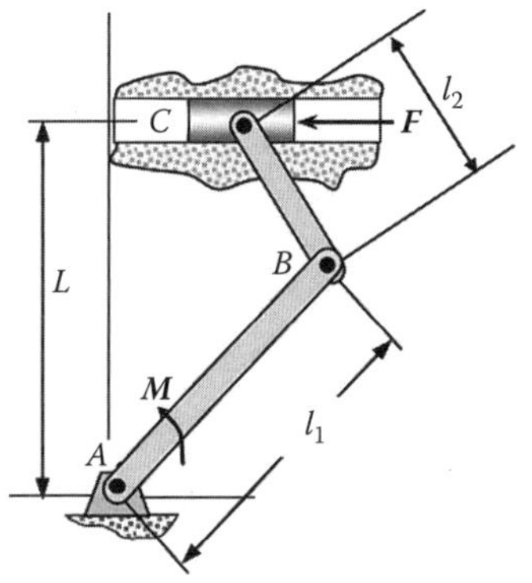

FIGURE P5.87

5.88 In Figure P5.88, link AB a uniform slender link of mass m_1, and length L_1. It is driven by an external moment M. Link CD has mass m_2 and radius of gyration k_g about point C. Its mass center is located at position $\bar{r}$ from C. The slider at point B has a mass of m_3. Friction is negligible.

Tasks:

(a) Select coordinates and state the kinematic constraint relationships.
(b) Draw the required free-body diagrams.
(c) Obtain the equations of motion for each body.
(d) Develop the system governing equations of motion and place the equations for your unknowns in matrix format.

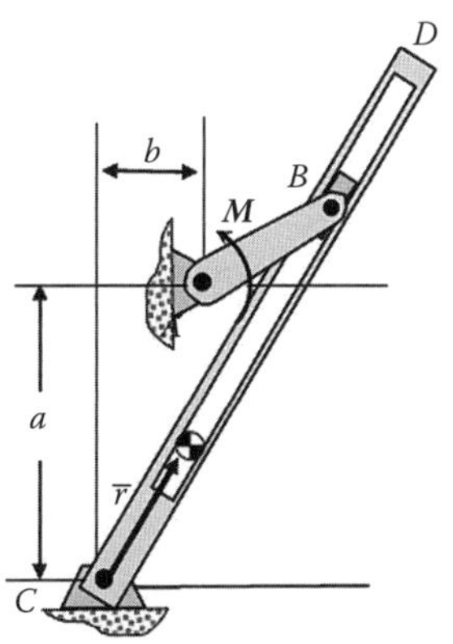

FIGURE P5.88

5.89 The links in the four-bar linkage illustrated in Figure P5.89 are uniform slender bars. For the interval of interest, bar OA has an applied constant input torque M. Friction is negligible.

Tasks:

(a) Draw the mechanism in a general orientation, select coordinates, and state the kinematic constraint relationships.
(b) Draw the required free-body diagrams.
(c) Obtain the equations of motion for each body.
(d) Develop the system governing equations of motion and place the equations for your unknowns in matrix format.

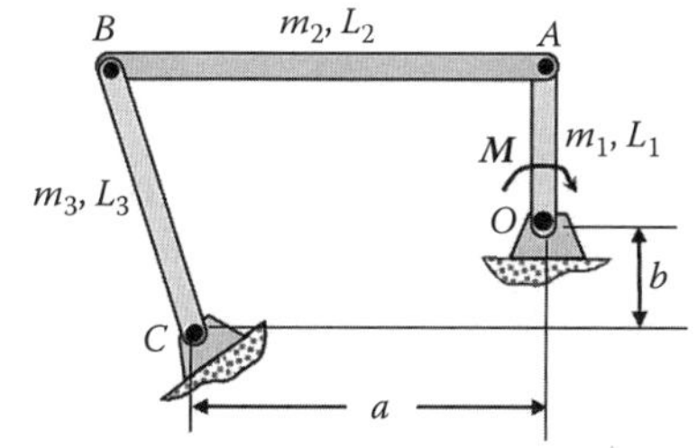

FIGURE P5.89

5.90 The rotation of link OA in Figure P5.90 is controlled by the piston rod of hydraulic cylinder BC. For the time of interest, the hydraulic cylinder exerts a constant force $\boldsymbol{F}$ to point B on the bar OA (parallel to CB). Bar OA is a slender uniform bar of mass m_2 and length L_2. The hydraulic cylinder has a mass of m_1 and a radius of gyration about C of k_g. Assume that the extension of the cylinder does not cause the center of gravity of the hydraulic cylinder to change or appreciably change the cylinder's moment of inertia.

Tasks:

(a) Select coordinates and state the kinematics relationships.
(b) Draw the required free-body diagrams.
(c) Obtain the equations of motion for each body.
(d) Develop the system governing equations of motion and place the equations for your unknowns in matrix format.

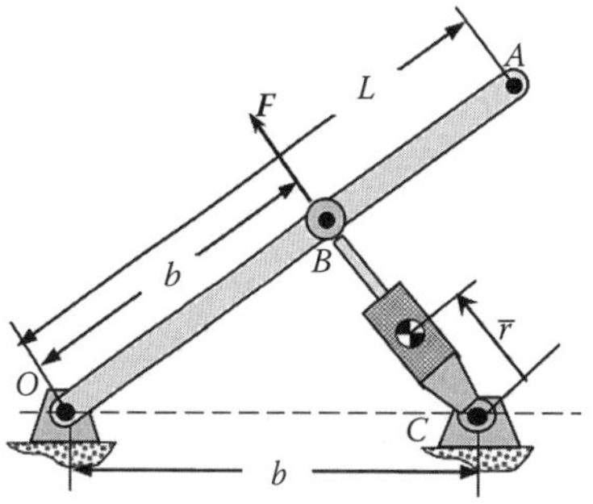

FIGURE P5.90

5.91 In Figure P5.91, link AB is a uniform bar of mass m, and link BC is uniform of mass M. The moment of inertia for the collar at D is negligible. Link AB's rotational speed is constant in the counterclockwise direction. Friction is negligible.

Tasks:

(a) Select coordinates and state the kinematic constraint relationships.
(b) Draw the required free-body diagrams.
(c) Obtain the equations of motion for each body.
(d) Develop the system governing equations of motion and place the equations for your unknowns in matrix format.

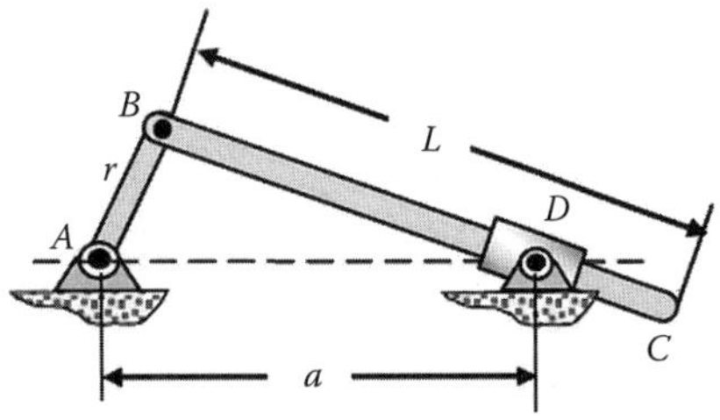

FIGURE P5.91

5.92 The disk at O in Figure P5.92 is rotating at a constant angular speed ω_O in the clockwise direction. Rolling without slipping exists between disks O and A. The pin at B is rigidly attached to the disk at B and slides with negligible friction in the slot on the yoke. Treat the two disks as thin disks and assume that the assembly operates in the plane of the paper. Friction is negligible.

Tasks:

(a) Select coordinates and state the kinematics relationships.
(b) Draw the required free-body diagrams.
(c) Obtain the EOM for each body.
(d) Develop a single EOM for the rotation of the disk supported at A.

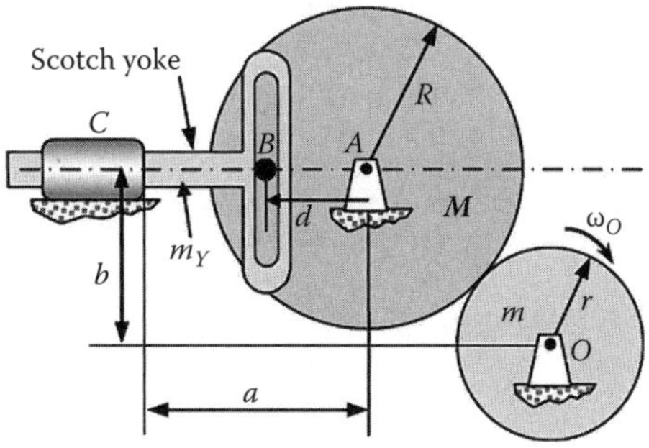

FIGURE P5.92

Appendix A: Essentials of Matrix Algebra

Matrix algebra is an essential part of engineering analysis. The material presented here is intended as a quick review for students who have previously been exposed to the topic.

The simplest matrix example involves two simultaneous equations, for example, $5x+3y=3$, $2x+4y=7$. The matrix statement of these equations is

$$\begin{bmatrix} 5 & 3 \\ 2 & 4 \end{bmatrix} \begin{Bmatrix} x \\ y \end{Bmatrix} = \begin{Bmatrix} 3 \\ 7 \end{Bmatrix}. \tag{A.1}$$

The first matrix in this equation is a "square" matrix with two rows and two columns, and it is referred to as a 2×2 matrix. The second and third matrices are called column matrices. They have two rows and one column; hence, they are referred to as 2×1 matrices. Matrix multiplication is being performed by the first matrix on the second matrix, producing the third matrix.

In general terms, Equation A.1 is stated

$$\begin{bmatrix} a_{11} & a_{12} \\ a_{21} & a_{22} \end{bmatrix} \begin{Bmatrix} x_1 \\ x_2 \end{Bmatrix} = \begin{Bmatrix} b_1 \\ b_2 \end{Bmatrix},$$

where a_{ij} is the entry in the ith row and jth column. In general, a matrix can have m rows and n columns, for example,

$$[A] = \begin{bmatrix} a_{11} & a_{12} & . & a_{1n} \\ a_{21} & a_{22} & & \\ . & & & \\ . & & & \\ a_{m1} & a_{m2} & & a_{mm} \end{bmatrix}$$

The initial example of Equation A.1 provides an intuitive example of matrix multiplication. Consider the following slightly more advanced example involving the multiplication of a 3×3 matrix times a 3×2 matrix producing a 3×2 matrix:

$$\underset{3\times 3}{\begin{bmatrix} a_{11} & a_{12} & a_{13} \\ a_{21} & a_{22} & a_{23} \\ a_{31} & a_{32} & a_{33} \end{bmatrix}} \underset{3\times 2}{\begin{bmatrix} b_{11} & b_{12} \\ b_{21} & b_{22} \\ b_{31} & b_{32} \end{bmatrix}} = [c]$$

$$= \underset{3\times 2}{\begin{bmatrix} a_{11}b_{11} + a_{12}b_{21} + a_{13}b_{31} & a_{11}b_{12} + a_{12}b_{22} + a_{13}b_{32} \\ a_{21}b_{11} + a_{22}b_{21} + a_{23}b_{31} & a_{21}b_{12} + a_{22}b_{22} + a_{23}b_{32} \\ b_{31}b_{11} + a_{32}b_{21} + a_{33}b_{31} & b_{31}b_{12} + a_{32}b_{22} + a_{33}b_{32} \end{bmatrix}}$$

An entry in $[c]$ is defined by $c_{ij} = \sum_{k=1,3} a_{ik}b_{kj}$. Matrix algebra requires that the number of rows in the first matrix equal the number of columns in the second matrix. Note from the two examples considered so far that—unlike customary algebra—$[b]\ [a] \neq [a]\ [b]$. In fact, for the example above, $[b]\ [a]$ does not exist since $[b]$ has two columns and $[a]$ has three rows.

A *diagonal matrix* is a square matrix for which all entries except the diagonal entries are zero. The *identity matrix* $[I]$ is a diagonal matrix with unity diagonal entries ($I_{jj}=1$). For a square matrix $[C]$, the identity matrix has the property $[C]=[C]\ [I]=[I]\ [C]$. The identity matrix is the matrix equivalent to unity in conventional algebra, $A \times 1 = 1 \times A = A$.

Square matrices may have "inverses." The *inverse* of the matrix $[A]$ is denoted as $[A]^{-1}$ and defined by

$$[A]^{-1}\ [A] = [I] = \begin{bmatrix} 1 & 0 & 0 & . & . \\ 0 & 1 & 0 & & \\ 0 & 0 & 1 & & \\ . & & & & \\ . & & & & 1 \end{bmatrix}$$

$[A]^{-1}$ exists if the determinant of $[A]$ is nonzero, that is, $|A| \neq 0$. For some square matrices, the matrix determinant $|A|$ is zero, that is, $|A|=0$, and the inverse is undefined. In terms of a problem with n unknowns defined by n linear simultaneous equations of the form $[A](x)=(b)$, $|A|=0$ implies that the n simultaneous equations are not linearly independent. Multi-degree-of-freedom vibration problems regularly generate "eigenvalue–eigenvector" problems of the form $[A]\ (x)=(0)$ for which $|A|=0$ leads to the only nonzero solution for the unknown vector (x).

Consider the matrix product, $[C]=[A][B]$, where all the matrices are square, $n \times n$, matrices. $[C]$'s inverse, $[C]^{-1}$, is defined by $[C]^{-1}=[B]^{-1}[A]^{-1}$, as can be confirmed from $[C]^{-1}[C]=[B]^{-1}[A]^{-1}[A][B]=[B]^{-1}[I][B]=[B]^{-1}[B]=[I]$. This is the "reverse-order rule" for matrix products.

The matrix $[A]$ has an associated *transpose matrix* $[A]^T$ obtained by interchanging $[A]$'s rows and columns. For a 3×3 matrix,

$$[A] = \begin{bmatrix} a_{11} & a_{12} & a_{13} \\ a_{21} & a_{22} & a_{23} \\ a_{31} & a_{32} & a_{33} \end{bmatrix} \Rightarrow [A]^T = \begin{bmatrix} a_{11} & a_{21} & a_{31} \\ a_{12} & a_{22} & a_{32} \\ a_{13} & a_{23} & a_{33} \end{bmatrix},$$

$$a_{ij} = a_{ji}$$

While this example matrix is square, any matrix can have a transpose. The transpose of an $n \times 1$ column matrix is a $1 \times n$ row matrix, the transpose of a $1 \times n$ row matrix is an $n \times 1$ column matrix, etc. For the matrix, $[C]=[A][B]$, $[C]$'s transpose is $[C]^T=[B]^T[A]^T$, defining the "reverse-order rule" for matrix transpose operations of products.

A matrix whose transpose equals its inverse ($[A]^T=[A]^{-1}$) is said to be orthogonal. Coordinate transformations in kinematics regularly involve orthogonal matrices.

"Cramer's rule" is an efficient approach for solving two simultaneous equations. The solution for the unknowns x_1, x_2 in the equation,

$$\begin{bmatrix} a_{11} & a_{12} \\ a_{21} & a_{22} \end{bmatrix} \begin{Bmatrix} x_1 \\ x_2 \end{Bmatrix} = \begin{Bmatrix} b_1 \\ b_2 \end{Bmatrix},$$

starts with calculation of the determinant D,

$$D = |A| = \begin{vmatrix} a_{11} & a_{12} \\ a_{21} & a_{22} \end{vmatrix} = a_{11}a_{22} - a_{12}a_{21},$$

The solution for the first unknown x_1 starts with developing a new matrix by substituting the right-hand column (b) into the first column of $[A]$, and then performing

$$Dx_1 = \begin{vmatrix} b_1 & a_{12} \\ b_2 & a_{22} \end{vmatrix} = b_1a_{22} - b_2a_{12} \Rightarrow x_1 = \frac{b_1a_{22} - b_2a_{12}}{a_{11}a_{22} - a_{12}a_{21}}.$$

The solution for the second unknown x_2 starts with forming a new matrix by substituting the right-hand column (b) into the second column of $[A]$, and then performing

$$Dx_2 = \begin{vmatrix} a_{11} & b_1 \\ a_{21} & b_2 \end{vmatrix} = b_2a_{11} - b_1a_{21} \Rightarrow x_2 = \frac{b_2a_{11} - b_1a_{21}}{a_{11}a_{22} - a_{12}a_{21}}.$$

For Equation A.1, this procedure nets, $D=|A|=5 \times 4 - 3 \times 2 = 14$;

$$Dx_1 = 14x_1 = \begin{vmatrix} 3 & 3 \\ 7 & 4 \end{vmatrix} = 3 \times 4 - 7 \times 3 = 12 - 21 = -9$$

$$\Rightarrow x_1 = \frac{-9}{14} = -0.643,$$

$$Dx_1 = 14x_2 = \begin{vmatrix} 5 & 3 \\ 2 & 7 \end{vmatrix} = 5 \times 7 - 2 \times 3 = 35 - 6 = 29$$

$$\Rightarrow x_2 = \frac{29}{14} = 2.071.$$

Back substitution confirms (to three significant figures) the correctness of these answers. Cramer's rule is very effective for solving two linear simultaneous equations either numerically or analytically. It has no value as a numerical approach for solving more than two linear simultaneous equations.

Appendix B: Essentials of Differential Equations

B.1 Introduction

Differential equations in dynamics normally arise from Newton's second law of motion, $\Sigma f = m\ddot{r}$. Accordingly, in dynamics, systems of coupled second-order differential equations are the norm. Occasionally, the equation of motion for a particle or rigid body has the form, $\ddot{y} = d^2y/dt^2 = g(y)$, and the energy-integral substitution,

$$\ddot{y} = \frac{d\dot{y}}{dt} = \frac{d\dot{y}}{dy}\frac{dy}{dt} = \dot{y}\frac{d\dot{y}}{dy} = \frac{d}{dy}\left(\frac{\dot{y}^2}{2}\right), \tag{B.1}$$

reduces the second-order equation with time t as the independent variable to the first-order differential equation

$$\frac{d}{dy}\left(\frac{\dot{y}^2}{2}\right) = g(y), \tag{B.2}$$

with displacement y as the independent variable and $\dot{y}^2/2$ as the dependent variable.

B.2 Linear Differential Equations

Most differential equations arising in dynamics and vibrations are nonlinear. The simple pendulum of Section 3.4.2 provides the following common nonlinear-differential-equation example:

$$\ddot{\theta} + \frac{g}{l}\sin\theta = 0. \tag{3.105}$$

The Taylor series for $\sin\theta$,

$$\sin\theta = \theta - \frac{\theta^3}{6} + \frac{\theta^5}{120} + \cdots, \tag{3.106}$$

provides the following "small-angle" approximation for Equation 3.105:

$$\ddot{\theta} + \frac{g}{l}\theta \cong 0 \Rightarrow \ddot{\theta} + \omega_n^2\theta = 0. \tag{3.107}$$

Equation 3.107 is a linear constant-coefficient differential equation. Observe that θ and $\ddot{\theta}$ appear linearly in Equation 3.107. In Equation 3.105, the second derivative $\ddot{\theta}$ appears linearly; however, $\sin\theta$ is a distinctly nonlinear function of θ.

Imagine a child standing in a play yard swing. The child's harmonic up and down motion varies the length of the swing. For small-angle motion of the swing, the equation of motion has the form

$$\ddot{\theta} + \frac{g}{l}(1 + \varepsilon\cos\omega t)\theta = 0.$$

This equation is linear, but the coefficient of θ is now a function of time.

Most differential equations are nonlinear, cannot be solved analytically, and must be solved numerically. Linear differential equations with variable coefficients are also difficult to solve analytically and must generally be solved numerically. The numerical integration of differential equations is introduced briefly in Section B.3. Fortunately, many very important differential equations of vibrations are linear and have very useful analytical solutions. We will briefly review some of these analytical solutions below.

B.2.1 Linear First-Order Equation

A linear first-order equation has the form

$$\frac{dy}{dx} + P(x)y = Q(x). \tag{B.3}$$

The homogeneous equation is obtained by setting the right-hand side to zero and can be solved via

$$\frac{dy_h}{dx} + P(x)y_h = 0 \Rightarrow \frac{dy_h}{y_h} = -P(x)dx.$$

Integrating both sides yields

$$\ln[y_h(x)] = \ln C - \int P(x)dx \Rightarrow y_h(x) = Ce^{-\int P(x)dx}. \tag{B.4}$$

Frequently, the particular solution can be obtained via inspection. The solution to Equation 3.102 is obtained via this approach. The particular solution is formally defined as

$$y_p(x) = e^{-\int P(x)dx}\int Q(x)e^{\int P(x)dx}dx. \tag{B.5a}$$

(Note that the particular solution is linearly proportional to $Q(x)$, the right-hand side of Equation B.3, reflecting the linearity of Equation B.3.) Differentiating Equation B.5a gives

$$\frac{dy_p}{dx} = e^{-\int P(x)dx}P(x)\int Q(x)e^{\int P(x)dx}dx + e^{-\int P(x)dx}Q(x)e^{\int P(x)dx}$$
$$= e^{-\int P(x)dx}P(x)\int Q(x)e^{\int P(x)dx}dx + Q(x). \quad \text{(B.5b)}$$

Substituting these results back into the differential Equation B.3 confirms the correctness of the particular solution. The complete solution is

$$y(x) = y_h(x) + y_p(x)$$
$$= Ce^{-\int P(x)dx} + e^{-\int P(x)dx}\int Q(x)e^{\int P(x)dx}dx. \quad \text{(B.6)}$$

Note that the complete solution must satisfy the single initial condition for $y(x)$.

A falling particle acted on by gravity and aerodynamic damping provides a simple example for this solution formula. The model of Equation 3.57 can be stated

$$\frac{dU}{dY} + bU = g, \quad U = \frac{v^2}{2}. \quad \text{(B.7)}$$

By comparison to Equation B.3, Y is the distance the particle has fallen and replaces x as the independent variable. Hence, $P(Y) = b \Rightarrow \int P(y)dY = bY$, and $U_h = Ce^{-bY}$. By comparison to Equation B.5a, the particular solution is

$$U_P(Y) = e^{-bY}\int ge^{bY}dY = e^{-bY}\frac{g}{b}e^{bY} = \frac{g}{b},$$

and the complete solution is

$$U = U_h + U_p = Ce^{-bY} + \frac{g}{b}.$$

Assuming that the particle starts from rest at $Y = 0$, the constant C is obtained via

$$U(0) = 0 = C + \frac{g}{b} \Rightarrow C = -\frac{g}{b},$$

netting the final solution

$$U(Y) = \frac{g}{b}(1 - e^{-bY}).$$

Assuming that the integrals can be completed, Equation B.6 provides some comfort in assuring that the solution can be developed. However, in many cases, the homogeneous and particular solutions can be obtained more easily by simple inspection.

B.2.2 Constant Coefficient, Linear, Second-Order Differential Equations

B.2.2.1 Undamped Spring-Mass Model

The equation of motion for a spring-mass system can be stated

$$\ddot{Y} + \omega_n^2 Y = g, \quad \text{(B.8)}$$

where
Y is the dependent variable, $\ddot{Y} = d^2Y/dt^2$
t is the independent variable

This equation is in "standard" form with the dependent variable Y and its derivatives on the left-hand side of the equal sign. Functions that are a function only of the independent variable t are on the right-hand side.

The homogeneous differential equation is obtained by setting the right-hand side to zero, netting in this case $\ddot{Y}_h + \omega_n^2 Y_h = 0$. In solving any linear second-order differential equation, we use the following steps: (i) solve the homogeneous equation for $Y_h(t)$, which will include two arbitrary constants A and B; (ii) solve for the particular solution $Y_p(t)$ that satisfies the right-hand side of the equation; (iii) form the complete solution $Y(t) = Y_h(t) + Y_p(t)$; and (iv) use the complete solution to solve for the two unknown constants A and B that satisfy the problem's initial conditions. This sequence works for "initial-value" problems where the initial (at $t = 0$) position $Y(0)$ and velocity $\dot{Y}(0)$ are given. Initial value problems are the norm. Sometimes, as in Example Problem 3.11, "boundary-value" problems arise with coordinates or velocities stated at two times or locations.

The formal solution to the homogeneous differential equation, $\ddot{Y}_h + \omega_n^2 Y_h = 0$, is obtained by assuming a solution of the form $Y_h = Ae^{st} \Rightarrow \ddot{Y}_h = s^2 Ae^{st}$. Substituting into the homogeneous differential equation nets

$$(s^2 + \omega_n^2)Ae^{st} = 0 \Rightarrow s^2 = -\omega_n^2 \Rightarrow s = \pm j\omega_n, \quad j = \sqrt{-1}$$

This result holds, since neither A nor e^{st} are zero. The solution corresponding to $s_1 = j\omega_n$ and $s_2 = -j\omega_n$ is

$$Y_h = \boldsymbol{A}_1 e^{j\omega_n t} + \boldsymbol{A}_2 e^{-j\omega_n t},$$

where $\boldsymbol{A}_1 = A_{1r} + jA_{1i}$, $\boldsymbol{A}_2 = A_{2r} + jA_{2i}$ are complex constants. From the identities,

$$e^{j\omega_n t} = \cos\omega_n t + j\sin\omega_n t, \quad e^{-j\omega_n t} = \cos\omega_n t - j\sin\omega_n t,$$

this solution becomes

$$Y_h = (\boldsymbol{A}_1 + \boldsymbol{A}_2)\cos\omega_n t + j(\boldsymbol{A}_1 - \boldsymbol{A}_2)\sin\omega_n t$$
$$= A\cos\omega_n t + B\sin\omega_n t, \quad \text{(B.9)}$$

where $A = A_{1r} + A_{2r}$; $A_{1r} = A_{2r}$, and $B = -A_{1i} + A_{2i}$; $A_{1i} = -A_{2i}$ are real constants.

The equation $\ddot{Y}_h + \omega_n^2 Y_h = 0$ applies to a spring-mass system. Based on experience, you might decide to simply guess a harmonic solution of the form $Y_h = A\cos\omega t \Rightarrow \ddot{Y}_h = -A\omega^2\cos\omega_n t$. Substituting this result into $\ddot{Y}_h + \omega_n^2 Y_h = 0$ gives

$$(\omega^2 - \omega_n^2)A = 0 \Rightarrow \omega = \pm\omega_n \Rightarrow Y_h = A\cos\omega_n t.$$

The guessed solution $Y_h = B\sin\omega t$ nets a similar result, and the complete homogeneous solution is (again) Equation B.9.

Any constant-coefficient linear equation can be solved by Laplace transforms including the particular solution. However, most particular solutions can be obtained by an inspection of the right-hand side terms and the differential equation itself. For Equation B.8, the right-hand side is constant, and a guessed constant solution of the form $Y_p = c \Rightarrow \dot{Y}_p = \ddot{Y}_p = 0$ yields

$$0 + c\omega_n^2 = g \Rightarrow c = \frac{g}{\omega_n^2} \Rightarrow Y_p = \frac{g}{\omega_n^2}.$$

Note that this particular solution is linearly proportional to g, the excitation on the right-hand side of Equation B.8.

The complete solution to Equation B.8 is

$$Y = Y_h + Y_p = A\cos\omega_n t + B\sin\omega_n t + \frac{g}{\omega_n^2}. \quad \text{(B.10)}$$

Assuming that the initial conditions are $Y(0) = Y_0$, $\dot{Y}(0) = \dot{Y}_0$, we can first solve for the constant A via

$$Y_0 = A + \frac{g}{\omega_n^2} \Rightarrow A = Y_0 - \frac{g}{\omega_n^2}.$$

Similarly, $\dot{Y} = -A\omega_n\sin\omega_n t + B\omega_n\cos\omega_n t$, nets

$$\dot{Y}_0 = B\omega_n \Rightarrow B = \frac{\dot{Y}_0}{\omega_n},$$

and the complete solution (satisfying the initial conditions) is

$$Y = Y_0\cos\omega_n t + \frac{g}{\omega_n^2}(1 - \cos\omega_n t) + \frac{\dot{Y}_0}{\omega_n}\sin\omega_n t. \quad \text{(B.11)}$$

Suppose the spring-mass system is acted on by an external force that increases linearly with time, netting the differential equation of motion

$$\ddot{Y} + \omega_n^2 Y = at, \quad \text{(B.12)}$$

This equation has the same homogeneous differential equation and solution; however, a different particular solution is required. By inspection, a solution of the form $Y_p = At \Rightarrow \dot{Y}_p = A \Rightarrow \ddot{Y}_p = 0$ will work. Substituting this guessed solution into Equation B.12 produces

$$(0 + \omega_n^2 At) = at \Rightarrow A = \frac{a}{\omega_n^2}, \quad Y_p = \frac{at}{\omega_n^2}.$$

Equation B.12's complete solution is now

$$Y = A\cos\omega_n t + B\sin\omega_n t + \frac{at}{\omega_n^2}. \quad \text{(B.13)}$$

Solving this equation in terms of the initial conditions $Y(0) = Y_0$, $\dot{Y}(0) = \dot{Y}_0$, first provides $Y_0 = A$. B is obtained by differentiating Y from Equation B.13, obtaining $\dot{Y} = -\omega_n A\sin\omega_n t + B\omega_n\cos\omega_n t + a/\omega_n^2$. Evaluating $\dot{Y}$ at $t = 0$ gives

$$\dot{Y}_0 = B\omega_n + \frac{a}{\omega_n^2} \Rightarrow B = \frac{\dot{Y}_0}{\omega_n} - \frac{a}{\omega_n^3}.$$

The complete solution for Equation B.12, satisfying the given initial conditions, is

$$Y = Y_0\cos\omega_n t + \left(\frac{\dot{Y}_0}{\omega_n} - \frac{a}{\omega_n^3}\right)\sin\omega_n t + \frac{at}{\omega_n^2}. \quad \text{(B.14)}$$

As a final example, suppose the spring-mass system is acted on by a force that is proportional to t^2, producing an equation of the form

$$\ddot{Y} + \omega_n^3 Y = bt^2, \quad \text{(B.15)}$$

The particular solution for this excitation has the form $Y_p = A + Ct^2 \Rightarrow \ddot{Y}_p = 2C$. Substituting into Equation B.12 gives

$$2C + \omega_n^2(A + Ct^2) = bt^2 \Rightarrow (2C + \omega_n^2 A) + (\omega_n^2 C - b)t^2 = 0. \quad \text{(B.16)}$$

The solution of this equation requires that the constant term and the coefficient of t^2 vanish. Hence,

$$C = \frac{b}{\omega_n^2}, \quad A = -\frac{2C}{\omega_n^2} = -\frac{2b}{\omega_n^4},$$

and the complete solution is

$$Y = A\cos\omega_n t + B\sin\omega_n t - \frac{2b}{\omega_n^4} + \frac{2bt^2}{\omega_n^2}. \quad \text{(B.17)}$$

Table B.1 summarizes the three particular solutions that we have developed.

Note that the following basic steps are taken to produce a complete solution for $\ddot{Y} + \omega_n^2 Y = u(t)$ with initial conditions $Y(0) = Y_0, \dot{Y}(0) = \dot{Y}_0$:

a. The homogeneous solution $Y_h(t) = A \cos \omega_n t + B \sin \omega_n t$ is developed for $\ddot{Y} + \omega_n^2 Y_h = 0$ and involves the two constants A and B.
b. The particular solution $Y_p(t)$ is developed to satisfy the right-hand side of the equation; that is, the particular solution $Y_p(t)$ satisfies $\ddot{Y}_p + \omega_n^2 Y_p = u(t)$.
c. The complete solution, $Y(t) = Y_h(t) + Y_p(t)$, is formed, and the arbitrary constants are determined such that the complete solution satisfies the initial conditions.

Consider the following version of Equation B.8:

$$\ddot{Y} + \omega_n^2 Y = g + at + bt^2.$$

Since this equation is linear, Table B.1 provides the three particular solutions, and the homogeneous solution is defined by Equation B.9. The complete solution is

$$Y = A \cos \omega_n t + B \sin \omega_n t + \frac{g}{\omega_n^2} + \frac{at}{\omega_n^2} + \left(-\frac{2b}{\omega_n^4} + \frac{bt^2}{\omega_n^2} \right).$$

To obtain this result, we have simply added the separate particular solutions corresponding to the separate right-hand sides of the governing equation. This additive nature is a distinct characteristic of linear differential equations.

B.2.2.2 Spring-Mass-Damper Model

The equation of motion for a spring-mass-damper system can be stated

$$\ddot{Y} + 2\zeta\omega_n \dot{Y} + \omega_n^2 Y = g. \tag{B.18}$$

TABLE B.1

Particular Solutions for $\ddot{Y}_p + \omega_n^2 Y_p = u(t)$

Excitation, $u(t)$	$Y_p(t)$
h = constant	$\frac{h}{\omega_n^2}$
at	$\frac{at}{\omega_n^2}$
bt^2	$-\frac{2b}{\omega_n^4} + \frac{bt^2}{\omega_n^2}$

This equation is solved in the same fashion as the undamped spring-mass model of Equation B.8; namely, we first solve the homogeneous equation, $\ddot{Y}_h + 2\zeta\omega_n \dot{Y}_h + \omega_n^2 Y_h = 0$. The solution is obtained (again) by substituting a solution of the form $Y_h = Ae^{st} \Rightarrow \dot{Y}_h = sAe^{st} \Rightarrow \ddot{Y}_h = s^2 Ae^{st}$ into the homogeneous differential equation to obtain

$$(s^2 + 2\zeta\omega_n s + \omega_n^2)Ae^{st} = 0 \Rightarrow (s^2 + 2\zeta\omega_n s + \omega_n^2) = 0. \tag{B.19}$$

The right-hand side result holds, since neither A nor e^{st} is zero.

The following three solution possibilities exist for Equation B.19: (i) $\zeta = 1$, critically damped motion; (ii) $\zeta > 1$, over-damped motion; and (iii) $\zeta < 1$, under-damped motion. Most vibration and dynamics models of interest involve under-damped motion. For completeness, we will first look at homogeneous solutions for the over-damped and critically damped cases. For $\zeta > 1$, the roots to the characteristic Equation B.19 are

$$s = -\zeta\omega_n \pm \omega_n \sqrt{\zeta^2 - 1} \Rightarrow s_1 = \zeta\omega_n + \omega_n \sqrt{\zeta^2 - 1} < 0, \quad s_2 = -\zeta\omega_n - \omega_n \sqrt{\zeta^2 - 1}. \tag{B.20}$$

Note that two real, negative roots are obtained, netting the homogeneous solution

$$Y_h = A_1 e^{-|s_1|t} + A_2 e^{-|s_2|t}, \tag{B.21}$$

where A_1, A_2 are real constants. The over-damped solution is the sum of two exponentially decaying terms.

For $\zeta = 1$ (critical damping), Equation B.20 produces the single root $s = -\omega_n$, and single solution, $Y_h = Ae^{-\omega_n t}$. This homogeneous solution (containing only one constant) is not adequate to satisfy two initial conditions (position and velocity). For less than obvious reasons, the complete homogeneous solution is

$$Y_h = A_1 e^{-\omega_n t} + A_2 t e^{-\omega_n t}, \tag{B.22}$$

where A_1, A_2 are real constants. The second term in this solution also satisfies the differential equation as can be confirmed by substituting for Y_h and

$$\dot{Y}_h = -\omega_n A_1 e^{-\omega_n t} + A_2 e^{-\omega_n t} - \omega_n A_2 t e^{-\omega_n t}$$
$$\ddot{Y}_h = -\omega_n^2 A_1 e^{-\omega_n t} - \omega_n A_2 e^{-\omega_n t} - \omega_n A_2 e^{-\omega_n t} + \omega_n^2 A_2 t e^{-\omega_n t},$$

into $\ddot{Y}_h + 2\zeta\omega_n \dot{Y} + \omega_n^2 Y_h = 0$ to obtain

$$(\omega_n^2 - 2\omega_n^2 + \omega_n^2)A_1 e^{-\omega_n t} + [(-2\omega_n + 2\omega_n) + t(\omega_n^2 - 2\omega_n^2 + \omega_n^2)]A_2 e^{-\omega_n t} = 0.$$

In looking at Equation B.22, you might wonder whether the term $A_2te^{-\omega_n t}$ would decay with increasing time. However, using L'Hospital's rule,

$$te^{-\omega_n t} \Rightarrow \underset{t\to\infty}{\text{limit}}\left|\left(\frac{t}{e^{\omega_n t}}\right) = \frac{\infty}{\infty} \Rightarrow \underset{t\to\infty}{\text{limit}}\right|\left(\frac{1}{\omega_n e^{\omega_n t}}\right) = 0.$$

Hence, the solution of Equation B.22 clearly vanishes with increasing time. The critically damped solution is interesting from a mathematical viewpoint, as a limiting condition, but has minimal direct engineering value.

By contrast, under-damped motion, characterized by $\zeta < 1$, is of very considerable interest. In this case, Equation B.21 defines the complex conjugate roots:

$$\begin{aligned} s &= -\zeta\omega_n \pm j\omega_n\sqrt{1-\zeta^2} \\ &= -\zeta\omega_n + j\omega_d, \quad -\zeta\omega_n - j\omega_d, \end{aligned} \tag{B.23}$$

where $\omega_d = \omega_n\sqrt{1-\zeta^2}$. These two roots correspond to the $\pm$ signs in the square root of Equation B.23; hence, the homogeneous solution looks like

$$Y_h = \boldsymbol{A}_1 e^{(-\zeta\omega_n + j\omega_d)t} + \boldsymbol{A}_2 e^{(-\zeta\omega_n - j\omega_d)t}, \tag{B.24}$$

where $\boldsymbol{A}_1$, $\boldsymbol{A}_2$ are *complex* coefficients. Substituting the identities,

$$e^{j\omega_d t} = \cos\omega_d t + j\sin\omega_d t, \quad e^{-j\omega_d t} = \cos\omega_d t - j\sin\omega_d t,$$

into Equation B.24 yields a final homogeneous solution of the form

$$Y_h = e^{-\zeta\omega_n t}(A\cos\omega_d t + B\sin\omega_d t). \tag{B.25}$$

where A, B are *real* constants.

The particular solution for $\ddot{Y} + 2\zeta\omega_n\dot{Y} + \omega_n^2 Y = g$ is $Y_p = g/\omega_n^2$, and the complete solution is

$$Y = e^{-\zeta\omega_n t}(A\cos\omega_d t + B\sin\omega_d t) + \frac{g}{\omega_n^2}. \tag{B.26}$$

For the initial condition, $Y(0) = Y_0$, Equation B.26 yields

$$Y(0) = Y_0 = A + \frac{g}{\omega_n^2} \Rightarrow A = Y_0 - \frac{g}{\omega_n^2}. \tag{B.27}$$

Differentiating Equation B.26 gives

$$\begin{aligned} \dot{Y} = &- \zeta\omega_n e^{-\zeta\omega_n t}(A\cos\omega_d t + B\sin\omega_d t) \\ &+ e^{-\zeta\omega_n t}(-\omega_d A\sin\omega_d t + \omega_d B\cos\omega_d t). \end{aligned}$$

Hence, the initial condition, $\dot{Y}(0) = \dot{Y}_0$, defines B via,

$$\begin{aligned} \dot{Y}_0 &= -\zeta\omega_n A + \omega_d B \Rightarrow B = \frac{\dot{Y}_0}{\omega_d} + \frac{\zeta\omega_n A}{\omega_d} \\ &= \frac{\dot{Y}_0}{\omega_d} + \frac{\zeta}{\sqrt{1-\omega_n^2}}\left(Y_0 - \frac{g}{\omega_n^2}\right), \end{aligned} \tag{B.26$'$}$$

and the complete solution is

$$\begin{aligned} Y = e^{-\zeta\omega_n t}\Bigg\{&\left(Y_0 - \frac{g}{\omega_n^2}\right)\cos\omega_d t \\ &+ \left[\frac{\dot{Y}_0}{\omega_d} + \frac{\zeta}{\sqrt{1-\omega_n^2}}\left(Y_0 - \frac{g}{\omega_n^2}\right)\right]\sin\omega_d t\Bigg\} + \frac{g}{\omega_n^2}. \end{aligned}$$

If the spring-mass-damper system is excited by a force that is proportional to time, the differential equation can be stated

$$\ddot{Y} + 2\zeta\omega_n\dot{Y} + \omega_n^2 Y = at. \tag{B.27$'$}$$

Substituting the assumed particular solution $Y_p = Bt + A \Rightarrow \dot{Y}_p = B, \ddot{Y}_p = 0$ into Equation B.27 gives

$$\begin{aligned} &2\zeta\omega_n B + \omega_n^2(Bt + A) = at \\ &\Rightarrow \omega_n(2\zeta B + \omega_n A) + (\omega_n^2 B - a)t = 0. \end{aligned}$$

Both the constant coefficient (coefficient of $t^0 = 1$) and the coefficient of t must be zero to satisfy this equation, yielding

$$B = \frac{a}{\omega_n^2} \Rightarrow A = -\frac{2\zeta B}{\omega_n} = -\frac{2\zeta a}{\omega_n^3}.$$

Hence, the particular solution for Equation B.27 is

$$Y_p = \frac{a}{\omega_n^2}\left(t - \frac{2\zeta}{\omega_n}\right).$$

Combining the homogeneous solution from Equation B.25, the complete solution is

$$\begin{aligned} Y &= Y_h + Y_p \\ &= e^{-\zeta\omega_n t}(A\cos\omega_d t + B\sin\omega_d t) + \frac{a}{\omega_n^2}\left(t - \frac{2\zeta}{\omega_n}\right). \end{aligned} \tag{B.28}$$

The solution would be completed by solving for A and B to satisfy the initial conditions.

Continuing, suppose the force acting on the spring-mass-damper system is proportional to t^2, producing the model

$$\ddot{Y} + 2\zeta\omega_n\dot{Y} + \omega_n^2 Y = bt^2. \quad \text{(B.29)}$$

Substituting the assumed particular solution $Y_p = Ct^2 + Bt + A \Rightarrow \dot{Y}_p = 2Ct + B, \ddot{Y}_p = 2C$ gives

$$2C + 2\zeta\omega_n(2Ct + B) + \omega_n^2(Ct^2 + Bt + A) = bt^2 \quad \text{or}$$
$$(2C + 2\zeta\omega_n B + \omega_n^2 A) + \omega_n(4\zeta C + \omega_n B)t + (\omega_n^2 C - b)t^2.$$

This equation is satisfied if the coefficients of t^0, t, t^2 are all zero, netting

$$C = \frac{b}{\omega_n^2} \Rightarrow B = -\frac{4\zeta C}{\omega_n} = -\frac{4\zeta b}{\omega_n^3} \Rightarrow A = -\frac{2C}{\omega_n^2} - \frac{2\zeta B}{\omega_n}$$
$$= -\frac{2b}{\omega_n^4}(1 - 4\zeta^2),$$

and the particular solution is

$$Y_p = \frac{b}{\omega_n^2}\left[t^2 - \frac{4\zeta t}{\omega_n} - \frac{2}{\omega_n^2}(1 - 4\zeta^2)\right].$$

Table B.2 summarizes the three particular solutions that we have developed.

As with the undamped equation, the following basic steps are taken to produce a complete solution for $\ddot{Y} + 2\zeta\omega_n\dot{Y} + \omega_n^2 Y = u(t)$:

a. The homogeneous solution $Y_h(t) = e^{-\zeta\omega_n t}(A\cos\omega_d t + B\sin\omega_d t)$ is developed for $\ddot{Y} + 2\zeta\omega_n\dot{Y} + \omega_n^2 Y = 0$ and involves the two constants A and B.
b. The particular solution $Y_p(t)$ is developed to satisfy the right-hand side of the equation $\ddot{Y} + 2\zeta\omega_n\dot{Y} + \omega_n^2 Y = u(t)$.
c. The complete solution $Y(t) = Y_h(t) + Y_p(t)$ is formed, and the arbitrary constants are determined such that the complete solution satisfies the specified initial conditions.

TABLE B.2

Particular Solutions for $\ddot{Y}_p + 2\zeta\omega_n\dot{Y}_p + \omega_n^2 Y_p = u(t)$

Excitation, $u(t)$	$Y_p(t)$
h = constant	$\frac{h}{\omega_n^2}$
at	$\frac{a}{\omega_n^2}\left(t - \frac{2\zeta}{\omega_n}\right)$
bt^2	$\frac{b}{\omega_n^2}\left[t^2 - \frac{4\zeta t}{\omega_n} - \frac{2}{\omega_n^2}(1 - 4\zeta^2)\right]$

B.3 An Introduction to Numerical Integration of Differential Equations

Today, numerical solutions can be readily developed for systems of differential equations via "canned" numerical integration packages. The implicit message from these developments is that you don't have to know anything about numerical methods to develop a solution to a differential equation, and the message is generally true. However, knowing something about numerical integration procedures is important in setting up differential equation models for dynamics.

Euler's method is the simplest procedure for integrating differential equations. It is not used in commercial codes but is a good starting point for understanding how an integrator might work. Consider the simple first-order differential equation

$$\frac{dy}{dt} = y, \quad y(t = 0) = y_0.$$

This homogeneous differential equation has the solution, $y(t) = y_0 e^t$. Think about starting with the initial condition y_0 and estimating the solution $y_1 = y(\delta t)$ at a small time increment later, $t = \delta t$. From the approximation,

$$\frac{dy}{dt}(t = 0) \cong \frac{y(\delta t) - y_0}{\delta t}$$

we can solve for

$$y(\delta t) = y_1 \cong \frac{dy}{dt}(t = 0) + y_0\delta t = y_0 + y_0\delta t.$$

Having arrived at an estimate for the solution at $t = \delta t$, we can proceed to the next time increment $t = 2\delta t$ and calculate $y_2 = y(2\delta t)$, etc. Thinking about the differential equation stated generally as $dy/dt = f(y, t)$, *note that Euler's method requires one evaluation of* $f(y, t)$ *for each time increment*. Accordingly, the Euler method is called a first-order integrator.

Starting with the initial condition, $y(t = 0) = 1$, Figure B.1 illustrates the numerical and exact solutions over the time interval $t \in [0, 1]$ for $\delta t = 1.0, 0.5, 0.1$. For $\delta t = 1.0$, the solution starts at the initial position, and the initial slope is used to extrapolate forward linearly to $t = 1$. The numerical solution would only consist of the (t_i, y_i) points, (0, 1), (1, 2), and the error at $t = 1$ is about 26%. For $\delta t = 0.5$, the initial slope is used to predict $y(t = 0.5)$. At $t = 0.5$, the slope is recalculated, and the solution is projected forward to $t = 1$. The solution consists of the points (0, 1), (0.5, 1.5), (1., 2.25), and the error at $t = 1$ is about −17%. Shifting to $t = 0.1$, the solution has 11 points, and the error at $t = 1$ is about −4.6%.

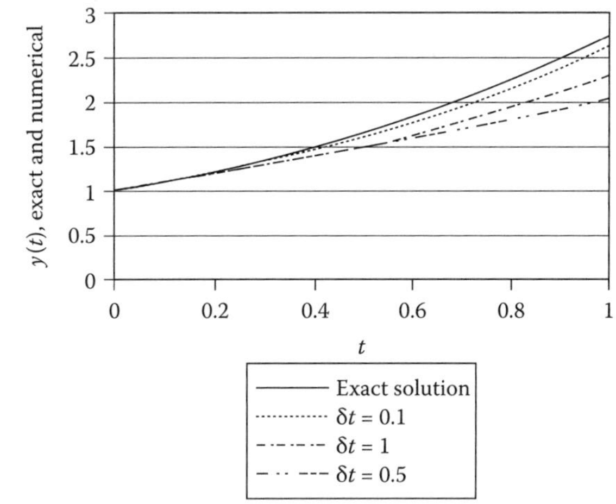

FIGURE B.1
Exact and approximate solutions for $dy/dt = y$ using Euler's method.

We could reduce the error further by reducing the integration step size δt; however, eventually, round-off error from the finite length of the numbers used in the calculations would limit this strategy.

In seeking numerical solutions, systems of second-order differential equations are regularly restated as equivalent systems of first-order equations—the "state-variable" model form. For example, a simple pendulum of length l is modeled by the second-order differential equation, $\ddot{\theta} + (g/l)\sin\theta = 0$, where g is the acceleration of gravity. The substitution $\theta = x_1$, $\dot{\theta} = x_2$, converts this equation into the following state-variable format:

$$\dot{x}_1 = x_2, \quad \dot{x}_2 = -\left(\frac{g}{l}\right)\sin x_1 \Rightarrow \begin{Bmatrix} \dot{x}_1 \\ \dot{x}_2 \end{Bmatrix} = \begin{Bmatrix} x_2 \\ -(g/l)\sin x_1 \end{Bmatrix} = \begin{Bmatrix} f_1 \\ f_2 \end{Bmatrix}. \tag{B.30}$$

There is no unique procedure for converting a model into state-variable form. We could have used, $\theta = x_1, \dot{\theta}\sqrt{g/l} = x_2$ to obtain

$$\dot{\omega} = -\sqrt{\frac{g}{l}}\sin\theta, \quad \dot{\theta} = \sqrt{\frac{g}{l}}\omega \Rightarrow \begin{Bmatrix} \dot{\omega} \\ \dot{\theta} \end{Bmatrix} = \sqrt{\frac{g}{l}}\begin{Bmatrix} -\sin\theta \\ \omega \end{Bmatrix}.$$

Either model is a valid state-variable model for the initial differential equation.

The point is that you will have to convert your original model (consisting—at a minimum of complexity—of second-order differential equations) to the state-variable format to use most numerical integrator packages. The integrator code will expect your model to be of the form

$$\dot{x}_i = f_i(x_i, t); \quad i = 1, \ldots, n.$$

The integrator works by repeatedly evaluating the right-hand function in this model. (We will talk about specific numerical integrators later.) At the initial time, the functions can be evaluated in terms of initial conditions. For the pendulum example, any solution—numerical or analytical—would start with the initial conditions, $\theta(t = 0) = \theta_0$, $\dot{\theta}(0) = \dot{\theta}_0$.

Many models that arise in dynamics resist conversion to the state-variable form. For example, consider the model for a two-degree-of-freedom (2DOF) mass-spring-damper system

$$\begin{bmatrix} m_1 & 0 \\ 0 & m_2 \end{bmatrix}\begin{Bmatrix} \ddot{x}_1 \\ \ddot{x}_2 \end{Bmatrix} + \begin{bmatrix} (c_1 + c_2) & -c_2 \\ -c_2 & (c_2 + c_3) \end{bmatrix}\begin{Bmatrix} \dot{x}_1 \\ \dot{x}_2 \end{Bmatrix} + \begin{bmatrix} (k_1 + k_2) & -k_2 \\ -k_2 & (k_2 + k_3) \end{bmatrix}\begin{Bmatrix} x_1 \\ x_2 \end{Bmatrix} = \begin{Bmatrix} F_1(t) \\ F_2(t) \end{Bmatrix}.$$

We can use the substitution, $\dot{x}_1 = v_1$, $\dot{x}_2 = v_2$ to obtain

$$\begin{bmatrix} m_1 & 0 \\ 0 & m_2 \end{bmatrix}\begin{Bmatrix} \dot{v}_1 \\ \dot{v}_2 \end{Bmatrix} = \begin{Bmatrix} F_1(t) \\ F_2(t) \end{Bmatrix} - \begin{bmatrix} (c_1 + c_2) & -c_2 \\ -c_2 & (c_2 + c_3) \end{bmatrix}\begin{Bmatrix} v_1 \\ v_2 \end{Bmatrix} - \begin{bmatrix} (k_1 + k_2) & -k_2 \\ -k_2 & (k_2 + k_3) \end{bmatrix}\begin{Bmatrix} x_1 \\ x_2 \end{Bmatrix}\begin{Bmatrix} \dot{x}_1 \\ \dot{x}_2 \end{Bmatrix} = \begin{Bmatrix} v_1 \\ v_2 \end{Bmatrix}.$$

Because of the mass-matrix coupling in the first two component equations, this is not quite the state-variable format that the numerical-integrator code is expecting. However, the inverse of the mass matrix $[M]^{-1}$ can be easily determined and used to create the model

$$\begin{Bmatrix} \dot{x}_1 \\ \dot{x}_2 \end{Bmatrix} = \begin{Bmatrix} x_3 \\ x_4 \end{Bmatrix}$$

$$\begin{Bmatrix} \dot{x}_3 \\ \dot{x}_4 \end{Bmatrix} = \begin{bmatrix} m_1 & 0 \\ 0 & m_2 \end{bmatrix}^{-1}\left\{\begin{Bmatrix} F_1(t) \\ F_2(t) \end{Bmatrix} - \begin{bmatrix} (c_1 + c_2) & -c_2 \\ -c_2 & (c_2 + c_3) \end{bmatrix}\begin{Bmatrix} x_3 \\ x_4 \end{Bmatrix} - \begin{bmatrix} (k_1 + k_2) & -k_2 \\ -k_2 & (k_2 + k_3) \end{bmatrix}\begin{Bmatrix} x_1 \\ x_2 \end{Bmatrix}\right\},$$

where $\dot{x}_1 = v_1 = x_3$, $\dot{x}_2 = v_2 = x_4$.

Consider the following model arising from a double pendulum:

$$\begin{bmatrix} l_1(m_1 + m_2) & m_2 l_2 \cos(\theta_2 - \theta_1) \\ m_2 l_1 \cos(\theta_2 - \theta_1) & m_2 l_2 \end{bmatrix}\begin{Bmatrix} \ddot{\theta}_1 \\ \ddot{\theta}_2 \end{Bmatrix} = \begin{Bmatrix} -(w_1 + w_2)\sin\theta_1 + m_2 l_2 \dot{\theta}_2^2 \sin(\theta_2 - \theta_1) \\ -w_2 \sin\theta_2 - m_1 l_1 \dot{\theta}_1^2 \sin(\theta_2 - \theta_1) \end{Bmatrix}$$

Unlike the previous example, the inertia matrix for this model is a function of the coordinates, not constant. For this small 2DOF example, we could use Cramer's rule and solve for $\ddot{\theta}_1$, $\ddot{\theta}_2$ analytically. Larger-dimension models arise where this option does not exist, and the coordinate-dependent coefficient matrix must be dealt with numerically. For the present example, suppose we choose the coordinates, $x_1 = \theta_1$, $x_2 = \theta_2$, $x_3 = \dot{\theta}_1, x_4 = \dot{\theta}_2$ as state-variable coordinates. With this choice, the state-variable model becomes

$$\begin{Bmatrix} \dot{x}_1 \\ \dot{x}_2 \end{Bmatrix} = \begin{Bmatrix} x_3 \\ x_4 \end{Bmatrix} = \begin{Bmatrix} f_1 \\ f_2 \end{Bmatrix}$$

$$\begin{bmatrix} l_1(m_1 + m_2) & m_2 l_2 \cos(x_2 - x_1) \\ m_2 l_1 \cos(x_2 - x_1) & m_2 l_2 \end{bmatrix} \begin{Bmatrix} \dot{x}_3 \\ \dot{x}_4 \end{Bmatrix}$$

$$= \begin{Bmatrix} -(w_1 + w_2)\sin x_1 + m_2 l_2 x_4^2 \sin(x_2 - x_1) \\ -w_2 \sin x_2 - m_1 l_1 x_3^2 \sin(x_2 - x_1) \end{Bmatrix} = \begin{Bmatrix} g_3 \\ g_4 \end{Bmatrix}.$$

The last two equations provide two linear equations in the two unknowns $\dot{x}_3 = f_3, \dot{x}_4 = f_4$, and they will need to be solved *numerically* each time the numerical integrator code needs to evaluate the right-hand side of the state-variable model. Remember, the numerical integrator expects to find numerical evaluations for the right-hand side of the model

$$\begin{Bmatrix} \dot{x}_1 \\ \dot{x}_2 \\ \dot{x}_3 \\ \dot{x}_4 \end{Bmatrix} = \begin{Bmatrix} f_1 \\ f_2 \\ f_3 \\ f_4 \end{Bmatrix}$$

based on numerical values for the present state-variable vector. Stated differently, starting from known initial conditions for the state vector, trying to ''intrude'' into a ''canned'' numerical integrator code to solve simultaneous linear equations can be difficult.

A ''fourth-order Runge–Kutta (RK)'' integrator is commonly used in numerical integrators. This integrator requires four evaluations of the right-hand side functions. Some RK integrators use variable step sizes to limit the errors produced by integration.

Appendix C: Mass Properties of Common Solid Bodies

Mass Moments of Inertia

Geometry	Inertia Properties	Diagram of Geometry
Slender bar	$I_{XX} = I_{YY} = \frac{1}{12}ml^2$ $I_{X'X'} = I_{Y'Y'} = \frac{1}{3}ml^2$	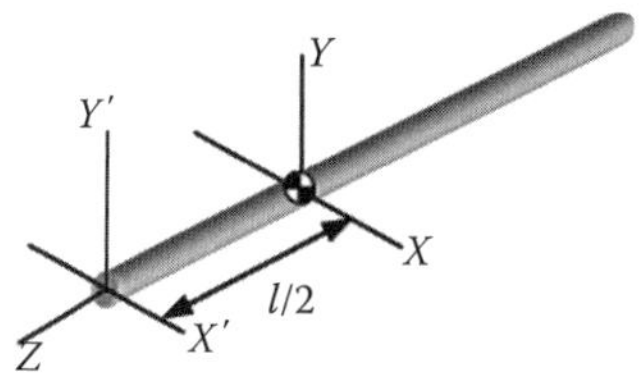
Cylinder (to apply for thin disk, let L go to zero)	$I_{XX} = I_{YY} = \frac{1}{12}m(3R^2 + L^2)$ $I_{ZZ} = \frac{1}{2}mR^2$	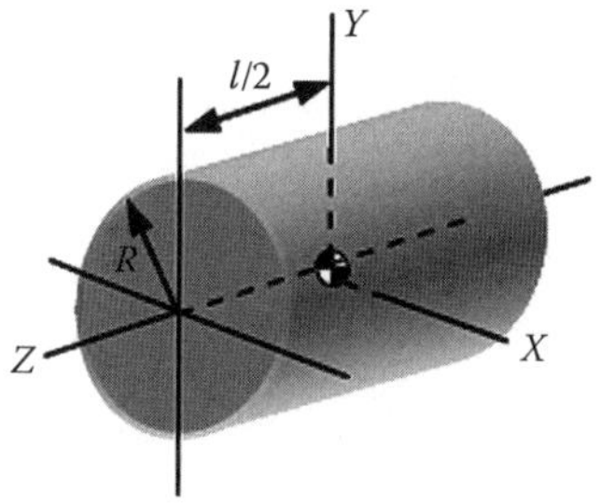
Rectangular prism (to apply for thin plate, let L go to zero)	$I_{XX} = \frac{1}{12}m(L^2 + a^2)$ $I_{YY} = \frac{1}{12}m(L^2 + b^2)$ $I_{ZZ} = \frac{1}{12}m(a^2 + b^2)$	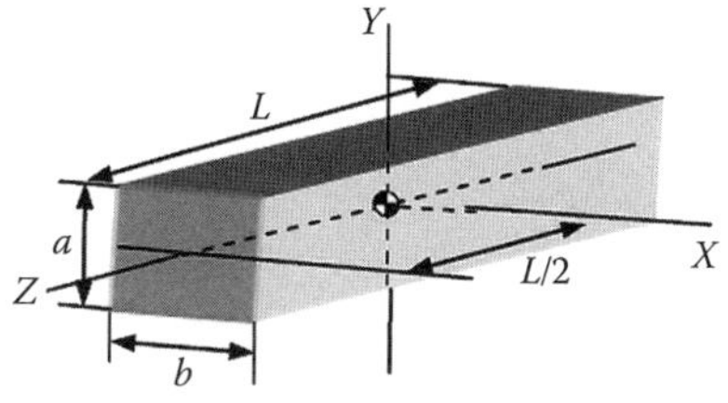
Semi-cylinder	$I_{XX} = \frac{9\pi^2 - 64}{36\pi^2}mR^2 + \frac{1}{12}mL^2$ $I_{YY} = \frac{1}{12}m(3R^2 + L^2)$ $I_{ZZ} = \frac{9\pi^2 - 32}{18\pi^2}mR^2; \quad I_{Z'Z'} = \frac{mR^2}{2}; \quad \bar{r} = \frac{4R}{3\pi}$	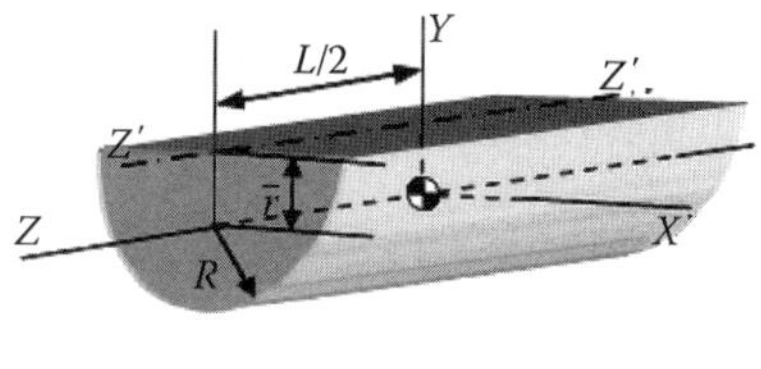
Thin half cylindrical shell	$I_{XX} = I_{YY} = \frac{1}{2}mR^2 + \frac{1}{12}mL^2$ $I_{ZZ} = \left(1 - \frac{4}{\pi^2}\right)mR^2$ $\bar{r} = \frac{2R}{\pi}$	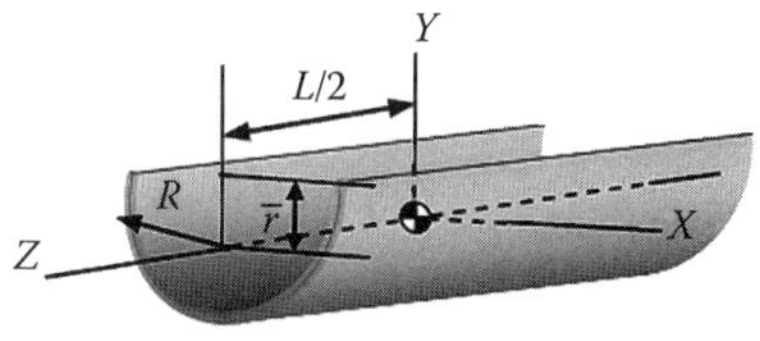
Thin circular cylindrical shell	$I_{XX} = I_{YY} = \frac{1}{12}m(6R^2 + L^2)$ $I_{ZZ} = mR^2$	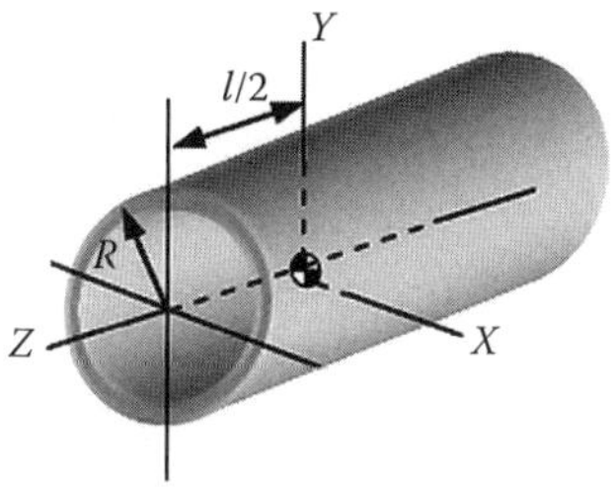

(continued)

(continued)

Mass Moments of Inertia

Geometry	Inertia Properties	Diagram of Geometry
Isosceles triangle of unit thickness	$I_G = \frac{1}{18} ma^2(3 \sin^2 \theta + \cos^2 \theta)$ $\bar{r} = \frac{2}{3} a \cos(\theta)$	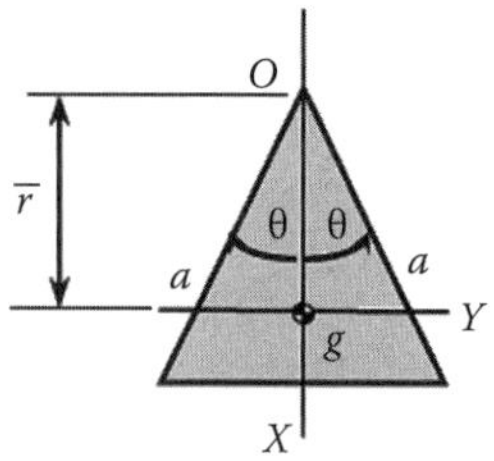
Thick cylinder	$I_{ZZ} = \frac{1}{2} m(R^2 + r^2)$ About mid-length axis perpendicular to ZZ $I_{BB} = \frac{1}{12} m(3R^2 + 3r^2 + h^2)$	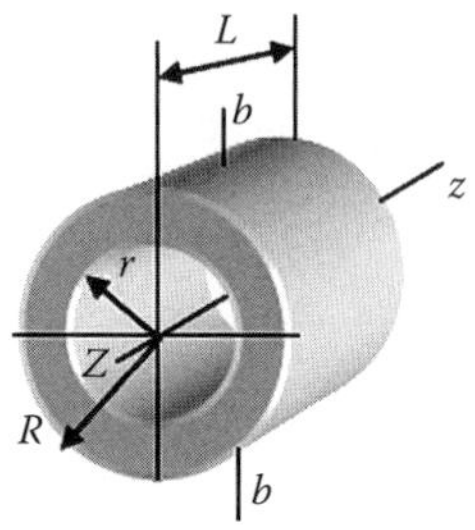

General expression for mass moment of inertia: $I_O = \int r_O^2\, dm$, where r_O is the distance from a z-axis through O to the differential mass element dm.

Parallel axis theorem: $I_A = I_g + mb_{gA}^2$, where I_g is the mass moment of inertia about a z-axis through the center of mass, and b_{gA} is the distance from a z-axis through the mass center g to a z-axis through point A.

Radius of gyration: $I_O = mk_{Og}^2$, where k_{Og} is the radius of gyration about point O and I_O is the mass moment of inertia about point O.

: Indicates mass center of body.

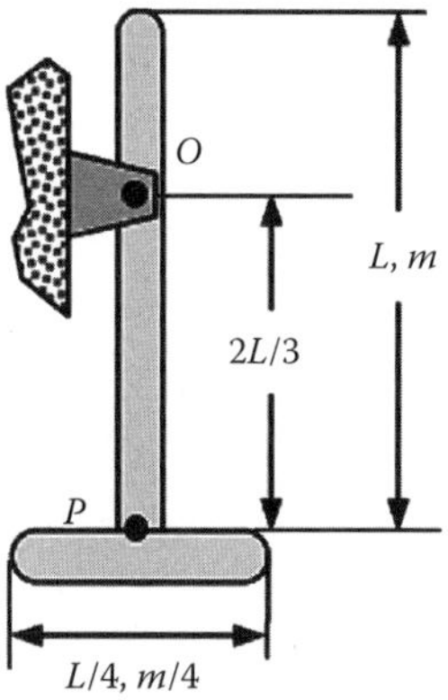

FIGURE PC.1
Thin bars.

Problems

Center of Mass and Mass Moment of Inertia

C.1–C.4 Perform the following tasks for the assemblies shown in Figures PC.1 through PC.4.

Tasks: (Reference frames are required)

(a) Determine the mass center g for the system shown relative to point O.
(b) Determine the mass moment of inertia about point O.
(c) Determine the mass moment of inertia about point P.
(d) Determine the radius of gyrations about point O and point P.

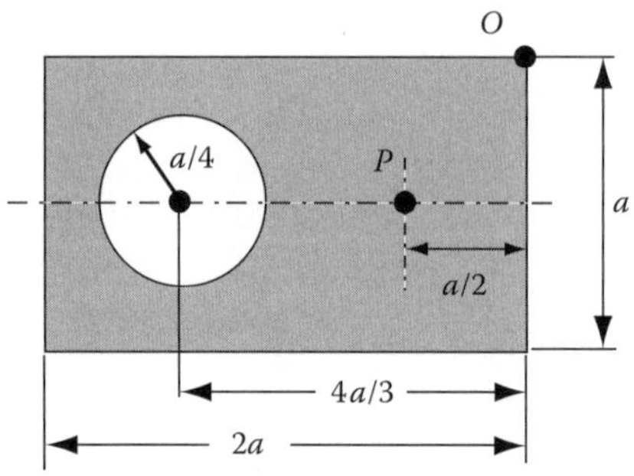

FIGURE PC.2
Rectangular plate of mass M with hole of removed mass $m_c = 0.2M$.

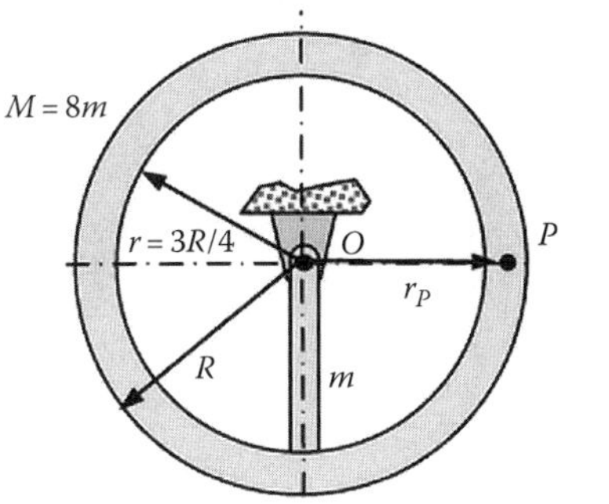

FIGURE PC.3
Thick ring of mass M with pivot bar of mass m r_P midway between r and R.

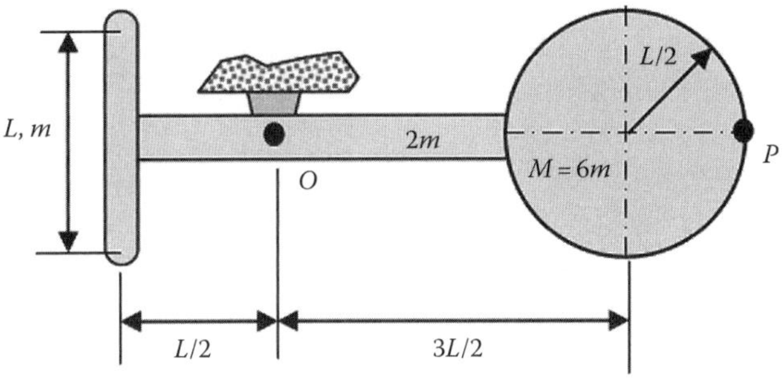

FIGURE PC.4
Composite slender bars and disk.

C.5 The body shown in Figure PC.5 has a mass of 10 kg and a radius of gyration about P of 1.25 m. The mass center of the assembly is located at $g\{0.5, 0.25\}$ m. Determine the mass moment of inertia about point O $\{1.25, 0.75\}$ m.

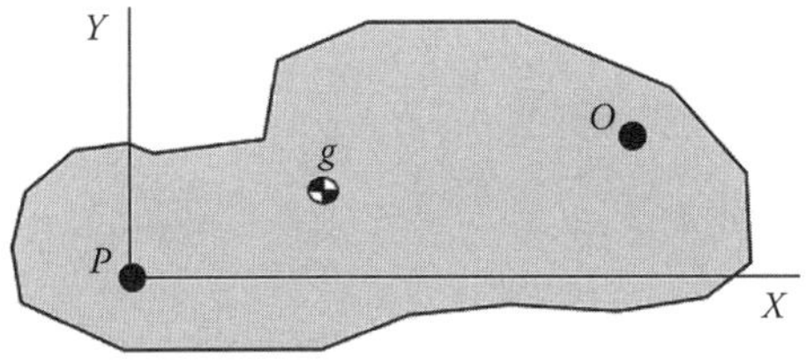

FIGURE PC.5
Rigid body

C.6 The two steel rods are welded together to form a T assembly, as shown in Figure PC.6. Locate the mass center g of the assembly and determine the mass moment of inertia about the mass center.

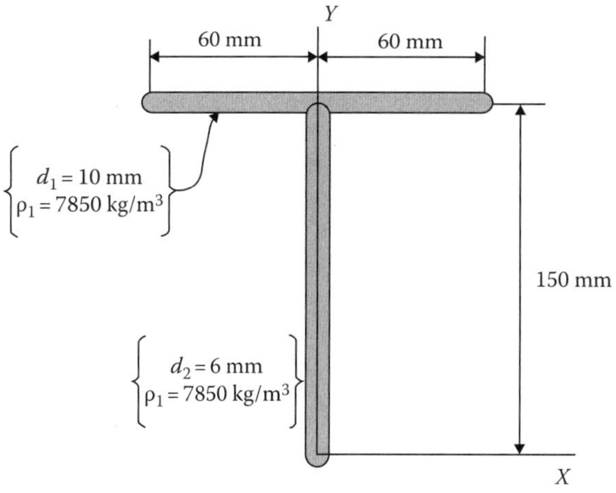

FIGURE PC.6
Two-bar assembly

C.7 The machine part shown in Figure PC.7 is made from steel and has the specific weight $\gamma_w = 0.283$ lb/in.^{3}.

Tasks:

(a) Determine the location of the mass center g.
(b) Determine the mass moment of inertia about the mass center Z-axis.
(c) Determine the mass moment of inertia about point P about the Z-axis.
(d) Determine the radius of gyration about the Z-axis through point P.

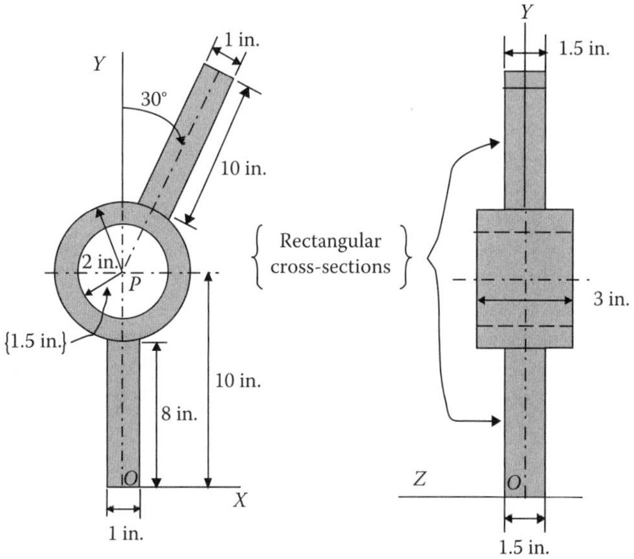

FIGURE PC.7
Machine part

Answers

PC.1 $\bar{r}_{Og} = \frac{4L}{15}$; $I_P = 0.3346mL^2$; $I_O = 0.2235mL^2$; $k_{Og} = 0.4228L$; $k_{Pg} = 0.51747L$

PC.2 $\bar{\mathbf{r}}_{Og} = -0.917a\mathbf{I} - 0.5a\mathbf{J} \rightarrow \bar{r}_{Og} = 1.044a$; $I_P = 0.739Ma^2$; $I_O = 1.292Ma^2$; $k_{Og} = 1.271a$; $k_P = 0.961a$

PC.3 $\bar{r}_{Og} = \frac{3R}{72}$; $I_P = 13.72mR^2$; $I_O = 6.46mR^2$; $k_{Og} = 0.847R$; $k_{Pg} = 1.238R$

PC.4 $\bar{r}_{Og} = L$; $I_P = 15.08mL^2$; $I_O = 15.08mL^2$; $k_{Og} = 1.294L$; $k_{Pg} = 1.294L$

PC.5 $I_O = 20.63$ kg-m^2

PC.6 $\bar{r}_Y = 0.1267$ m; $I_g = 28.03(10^{-5})$ kg m^2

PC.7 $\bar{\mathbf{r}}_{Og} = (1.207\mathbf{I} + 10.53\mathbf{J} + 0\mathbf{K})$ in.; $I_g = 0.8741$ lb in. s^2; $I_P = 0.9295$ lb in. s^2; $k_{Pg} = 5.4$ in.

Appendix D: Answers to Selected Problems

Chapter 2

2.1

$$\dot{\mathbf{r}} = \dot{r}\boldsymbol{\varepsilon}_r + r\dot{\theta}\boldsymbol{\varepsilon}_\theta \quad \ddot{\mathbf{r}} = (\ddot{r} - r\dot{\theta}^2)\boldsymbol{\varepsilon}_r + (r\ddot{\theta} + 2\dot{r}\dot{\theta})\boldsymbol{\varepsilon}_\theta$$

2.2

$$\mathbf{a} = \dot{v}\boldsymbol{\varepsilon}_t + \frac{v^2}{\rho}\boldsymbol{\varepsilon}_n = \ddot{s}\boldsymbol{\varepsilon}_t + \frac{v^2}{\rho}\boldsymbol{\varepsilon}_n$$

2.3

$$\mathbf{a} = \ddot{s}\boldsymbol{\varepsilon}_t + \frac{v^2}{\rho}\boldsymbol{\varepsilon}_n$$

2.4

r–θ and n–t are not mutually perpendicular to each other

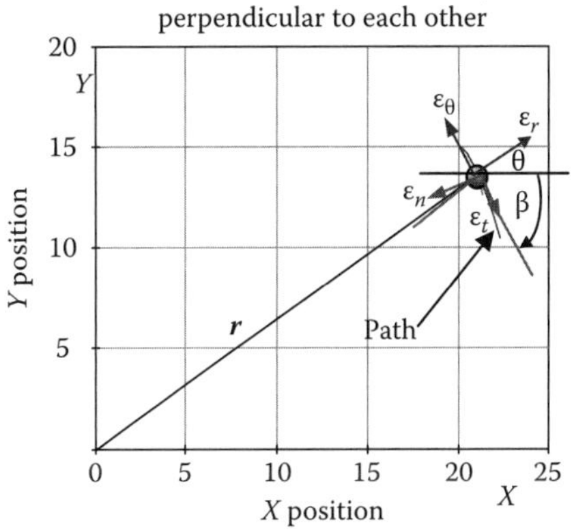

Path and coordinates diagram

$$\begin{Bmatrix} v_X \\ v_Y \end{Bmatrix} = \begin{Bmatrix} 1 \\ -2 \end{Bmatrix} \text{mm/s};$$

$$\begin{Bmatrix} a_X \\ a_Y \end{Bmatrix} = \begin{Bmatrix} 0.5 \\ -2 \end{Bmatrix} \text{mm/s}^2;$$

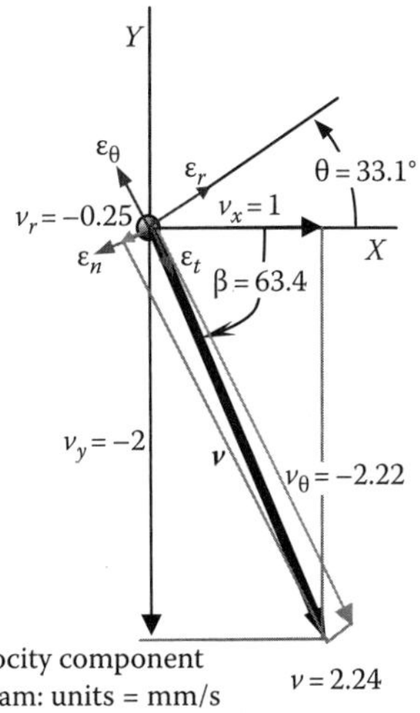

Velocity component diagram: units = mm/s

$$\begin{Bmatrix} v_r \\ v_\theta \end{Bmatrix} = \begin{Bmatrix} -0.255 \\ -2.22 \end{Bmatrix} \text{mm/s};$$

$$\begin{Bmatrix} a_r \\ a_\theta \end{Bmatrix} = \begin{Bmatrix} -0.673 \\ -1.948 \end{Bmatrix} \text{mm/s}^2;$$

Acceleration component diagram: units = mm/s^2

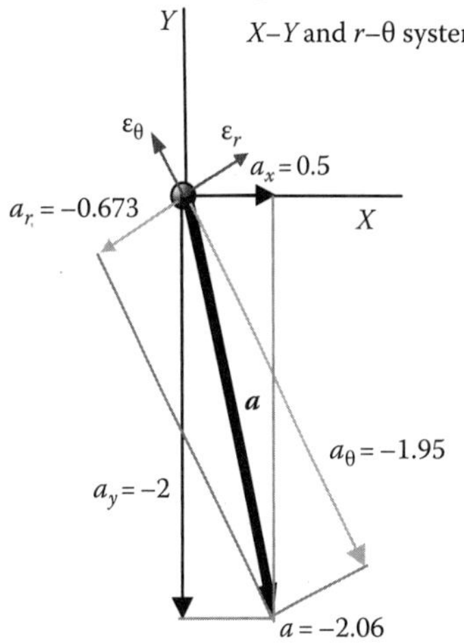

$$\begin{Bmatrix} v_t \\ v_n \end{Bmatrix} = \begin{Bmatrix} 2.24 \\ 0 \end{Bmatrix} \text{mm/s};$$

$$\begin{Bmatrix} a_t \\ a_n \end{Bmatrix} = \begin{Bmatrix} 2.01 \\ 0.447 \end{Bmatrix} \text{mm/s}^2$$

Acceleration component diagram: units = mm/s^2

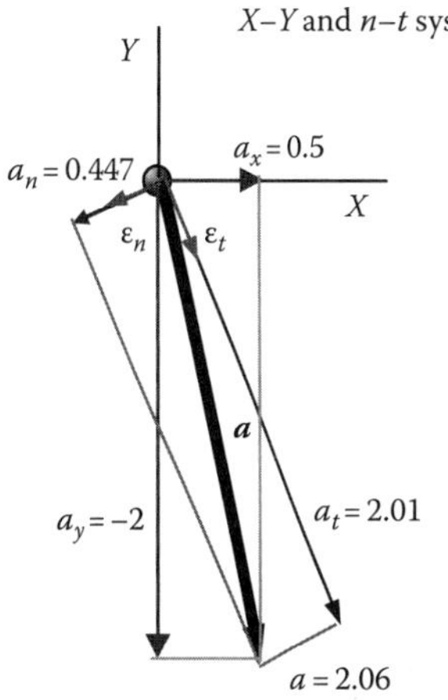

2.5

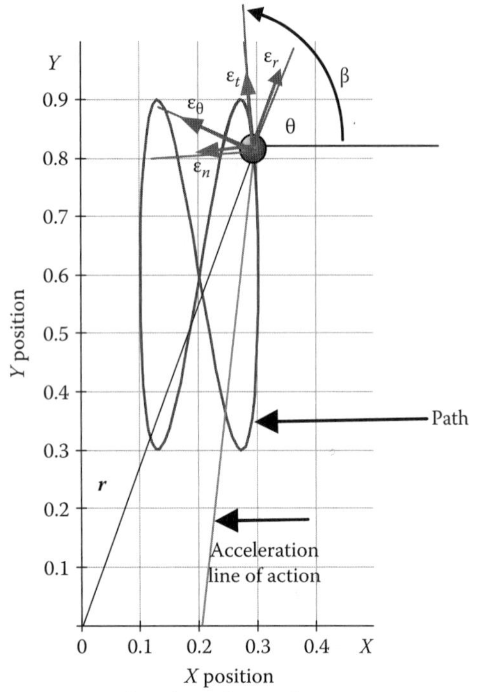

Path and coordinates diagram

$$\begin{Bmatrix} v_X \\ v_Y \end{Bmatrix} = \begin{Bmatrix} -0.383 \\ 4.24 \end{Bmatrix} \text{ m/s;}$$

$$\begin{Bmatrix} a_X \\ a_Y \end{Bmatrix} = \begin{Bmatrix} -9.24 \\ -84.9 \end{Bmatrix} \text{ m/s}^2$$

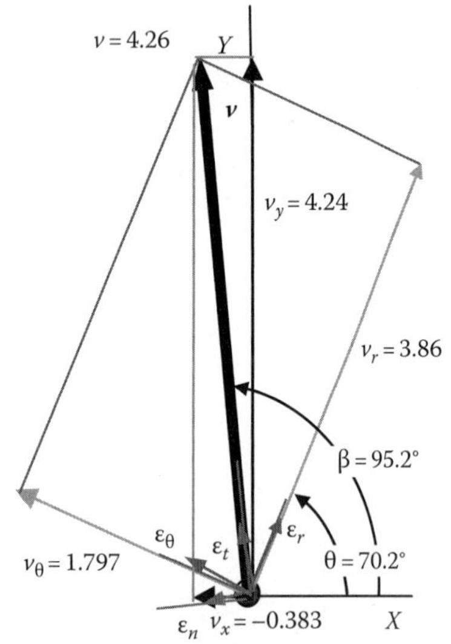

Velocity component diagram: units = in./s

$$\begin{Bmatrix} v_r \\ v_\theta \end{Bmatrix} = \begin{Bmatrix} 3.86 \\ 1.797 \end{Bmatrix} \text{ m/s;}$$

$$\begin{Bmatrix} a_r \\ a_\theta \end{Bmatrix} = \begin{Bmatrix} -83.0 \\ -20.1 \end{Bmatrix} \text{ m/s}^2$$

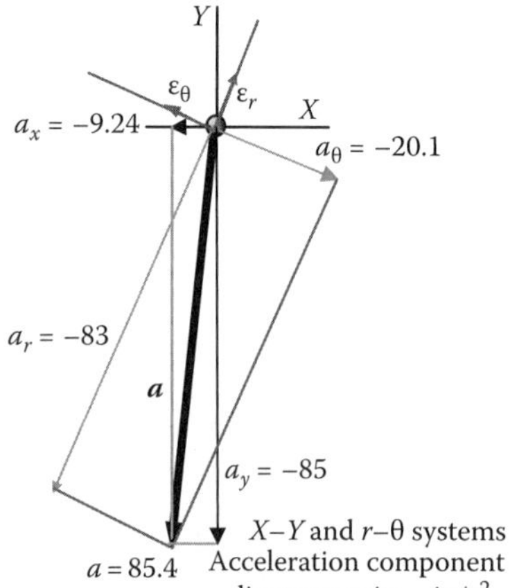

X–Y and r–θ systems
Acceleration component diagram: units = in/s²

$$\begin{Bmatrix} v_t \\ v_n \end{Bmatrix} = \begin{Bmatrix} 4.26 \\ 0 \end{Bmatrix} \text{ m/s;}$$

$$\begin{Bmatrix} a_t \\ a_n \end{Bmatrix} = \begin{Bmatrix} -83.7 \\ 16.84 \end{Bmatrix} \text{ m/s}^2$$

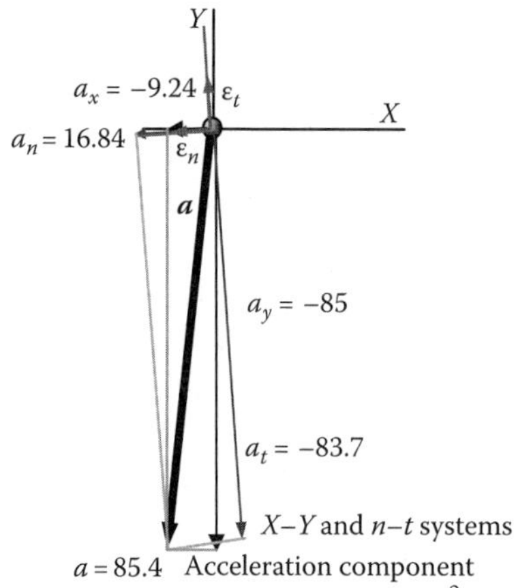

X–Y and n–t systems
Acceleration component diagram: units = in/s²

2.6

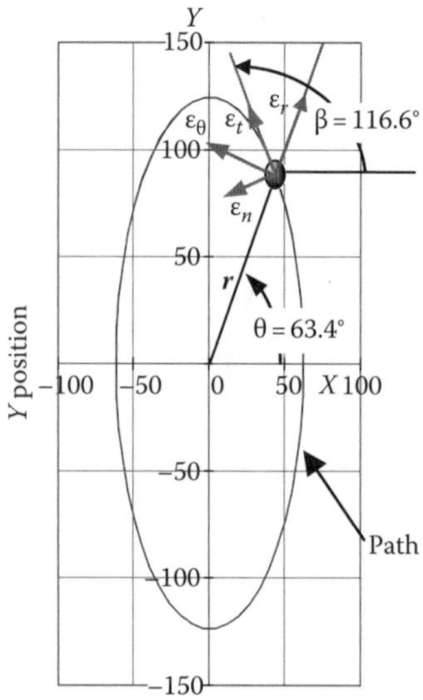

X position
Path and coordinates diagram

2.9

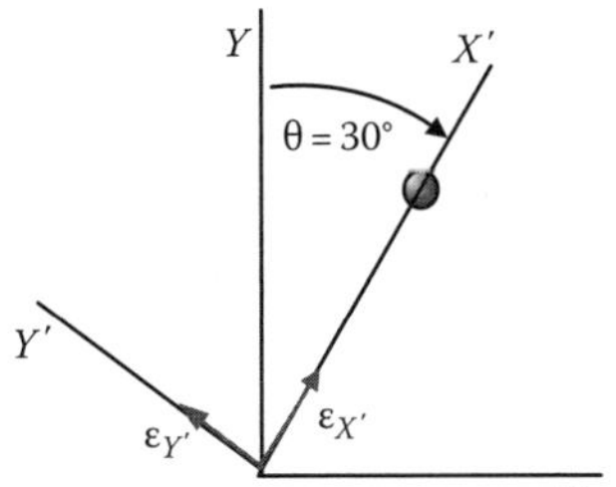

Path and coordinates diagram

$$\text{(a)}\begin{Bmatrix} v_X \\ v_Y \end{Bmatrix} = \begin{bmatrix} 180.2 \\ -118.3 \end{bmatrix} \text{ ft/s;}$$

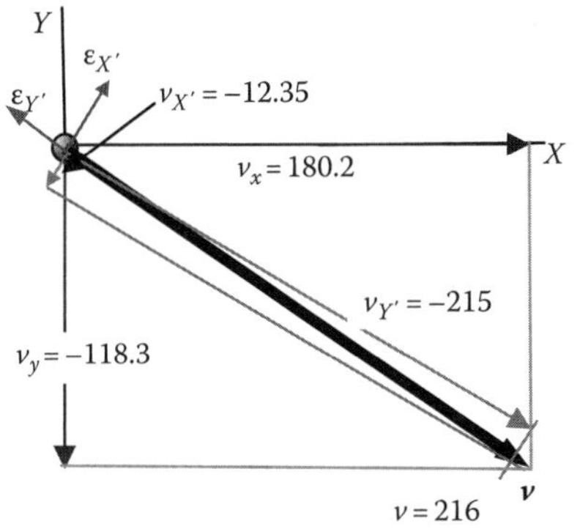

Velocity component diagram: units = ft/s

$$\text{(b)}\begin{Bmatrix} v_{X'} \\ v_{Y'} \end{Bmatrix} = \begin{bmatrix} -12.35 \\ -215.22 \end{bmatrix} \text{ ft/s}$$

2.10

$$\begin{Bmatrix} v_r \\ v_\theta \end{Bmatrix} = \begin{Bmatrix} -r_0\omega/2 \\ 0 \end{Bmatrix} \quad \begin{Bmatrix} a_r \\ a_\theta \end{Bmatrix} = \begin{Bmatrix} 0 \\ -\pi r_0\omega^2/6 \end{Bmatrix};$$

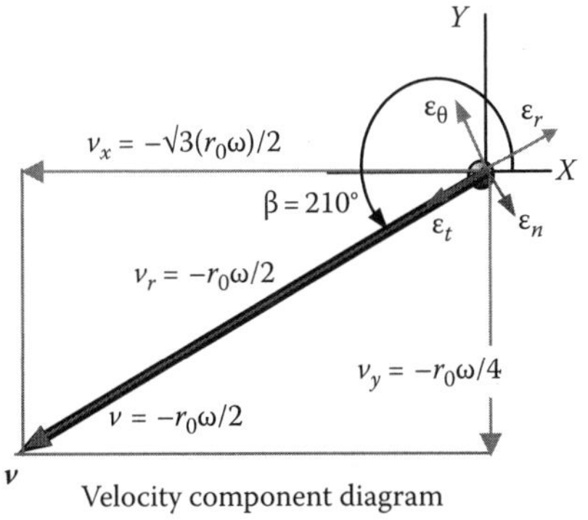

Velocity component diagram

$$\begin{Bmatrix} v_X \\ v_Y \end{Bmatrix} = \begin{Bmatrix} -\sqrt{3}r_0\omega/4 \\ -r_0\omega/4 \end{Bmatrix}$$

$$\begin{Bmatrix} a_X \\ a_Y \end{Bmatrix} = \begin{Bmatrix} r_0\pi\omega^2/12 \\ -\sqrt{3}r_0\pi\omega^2/12 \end{Bmatrix};$$

X–Y and *r–θ* systems

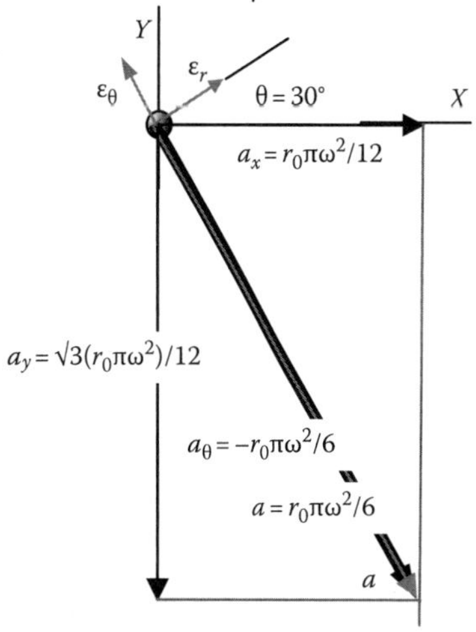

Acceleration component diagram: units = NA

$$\begin{Bmatrix} v_t \\ v_n \end{Bmatrix} = \begin{Bmatrix} r_0\omega/2 \\ 0 \end{Bmatrix} \quad \begin{Bmatrix} a_t \\ a_n \end{Bmatrix} = \begin{Bmatrix} 0 \\ r_0\pi\omega^2/6 \end{Bmatrix}$$

X–Y and *n–t* systems

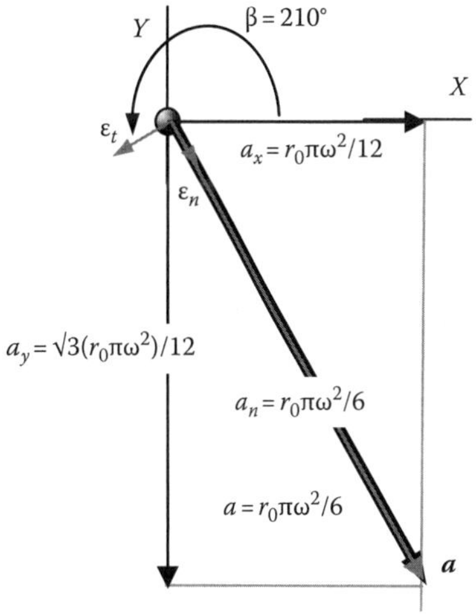

Acceleration component diagram: units = NA

2.11

$$\begin{Bmatrix} v_r \\ v_\theta \end{Bmatrix} = \begin{Bmatrix} -300 \\ 1600 \end{Bmatrix} \text{ mm/s;}$$

$$\begin{Bmatrix} a_r \\ a_\theta \end{Bmatrix} = \begin{Bmatrix} -12800 \\ -8800 \end{Bmatrix} \text{ mm/s}^2$$

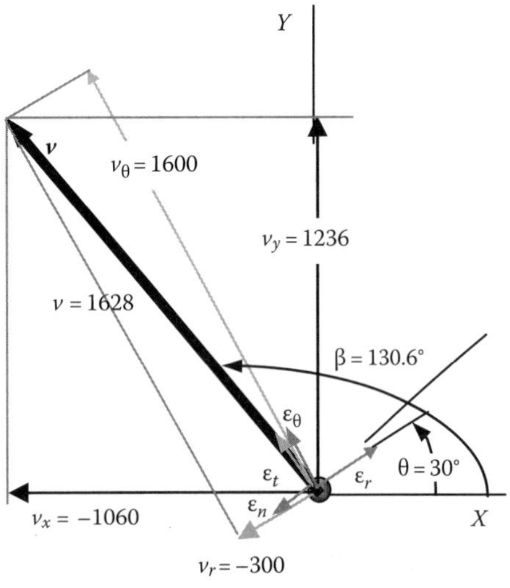

Velocity component diagram: units = mm/s

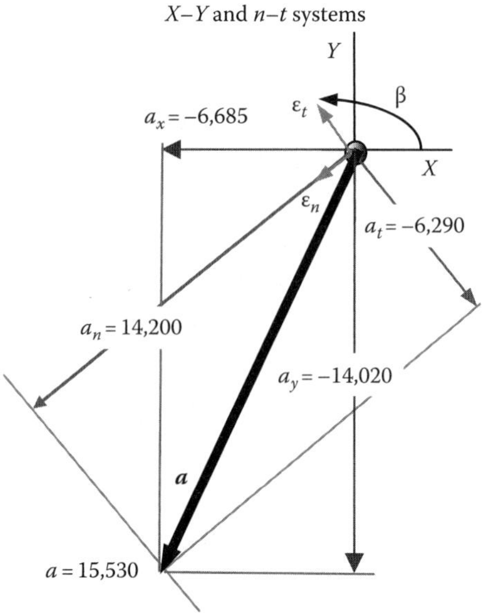

Acceleration component diagram: units = mm/s²

$$\begin{Bmatrix} v_X \\ v_Y \end{Bmatrix} = \begin{Bmatrix} -1060 \\ 1236 \end{Bmatrix} \text{mm/s};$$

$$\begin{Bmatrix} a_X \\ a_Y \end{Bmatrix} = \begin{Bmatrix} -6685 \\ -14021 \end{Bmatrix} \text{mm/s}^2$$

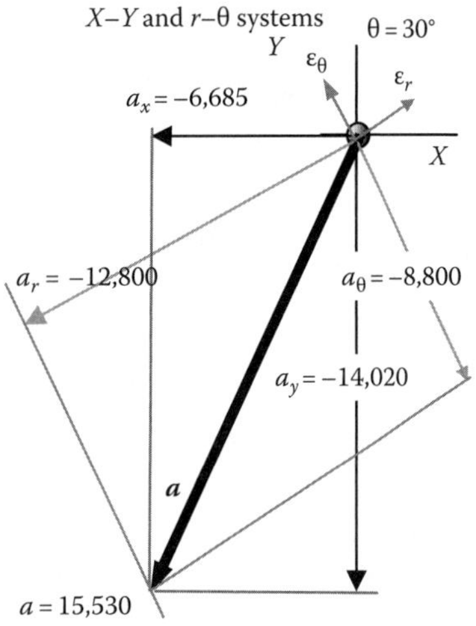

Acceleration component diagram: units = mm/s²

$$\begin{Bmatrix} v_t \\ v_n \end{Bmatrix} = \begin{Bmatrix} 1628 \\ 0 \end{Bmatrix} \text{mm/s};$$

$$\begin{Bmatrix} a_t \\ a_n \end{Bmatrix} = \begin{Bmatrix} -6295 \\ 14200 \end{Bmatrix} \text{mm/s}^2$$

2.12

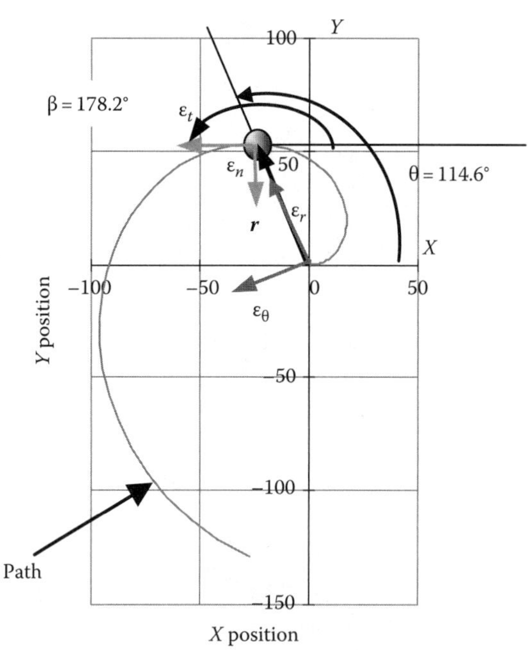

Path and coordinates diagram

$$\begin{Bmatrix} v_r \\ v_\theta \end{Bmatrix} = \begin{Bmatrix} 117.3 \\ 236.4 \end{Bmatrix} \text{m/s};$$

$$\begin{Bmatrix} a_r \\ a_\theta \end{Bmatrix} = \begin{Bmatrix} -831 \\ 1175 \end{Bmatrix} \text{m/s}^2$$

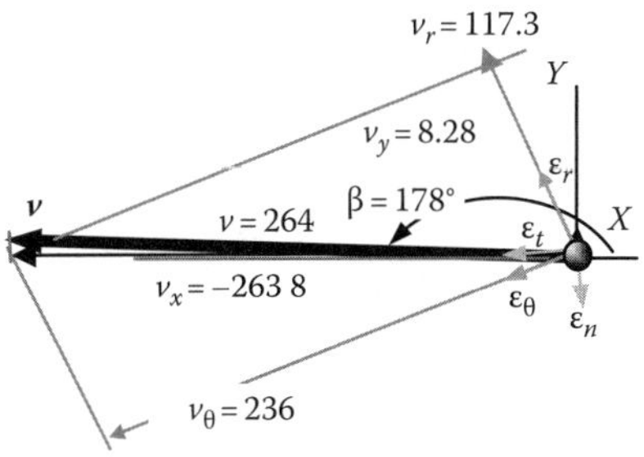

Velocity component diagram: units = m/s

$$\begin{Bmatrix} v_X \\ v_Y \end{Bmatrix} = \begin{Bmatrix} -264 \\ 8.28 \end{Bmatrix} \text{ m/s};$$
$$\begin{Bmatrix} a_X \\ a_Y \end{Bmatrix} = \begin{Bmatrix} -722. \\ -1245 \end{Bmatrix} \text{ m/s}^2$$

X–Y and r–θ systems

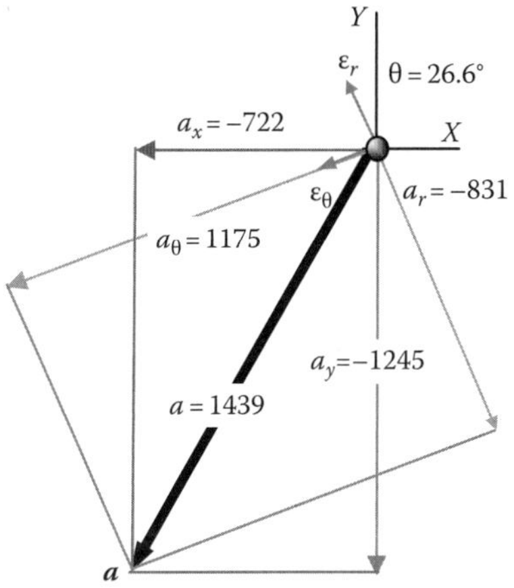

Acceleration component diagram: units = m/s^2

$$\begin{Bmatrix} v_t \\ v_n \end{Bmatrix} = \begin{Bmatrix} 264 \\ 0 \end{Bmatrix} \text{ m/s};$$
$$\begin{Bmatrix} a_t \\ a_n \end{Bmatrix} = \begin{Bmatrix} 683 \\ 1267 \end{Bmatrix} \text{ mm/s}^2$$

X–Y and n–t systems

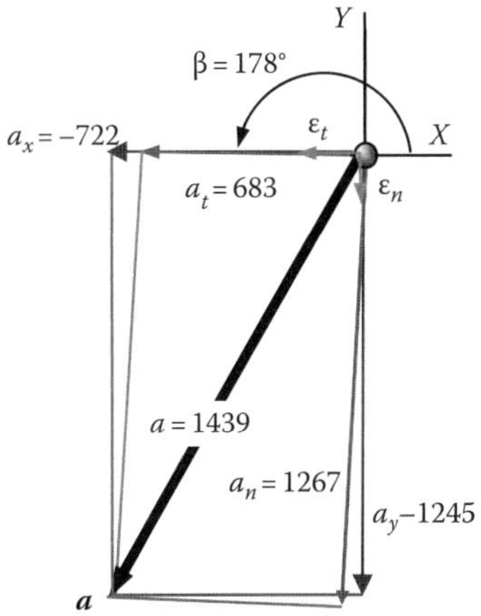

Acceleration component diagram: units = m/s^2

2.13

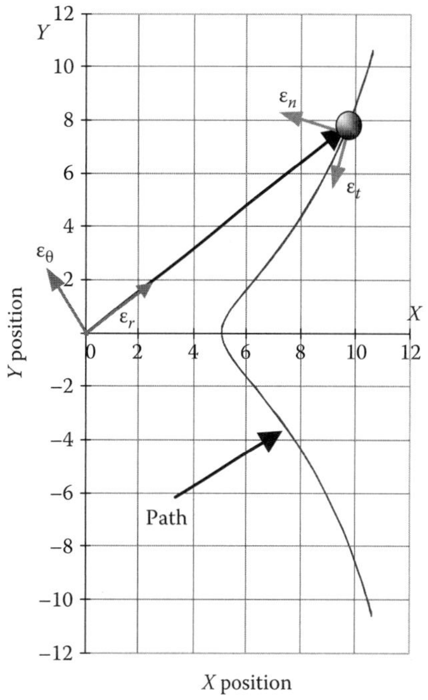

Path and coordinates diagram

$$\begin{Bmatrix} v_r \\ v_\theta \end{Bmatrix} = \begin{Bmatrix} -3265 \\ -1850 \end{Bmatrix} \text{ mm/s};$$
$$\begin{Bmatrix} a_r \\ a_\theta \end{Bmatrix} = \begin{Bmatrix} -169.5 \\ -242 \end{Bmatrix} \times 10^3 \text{ mm/s}^2$$

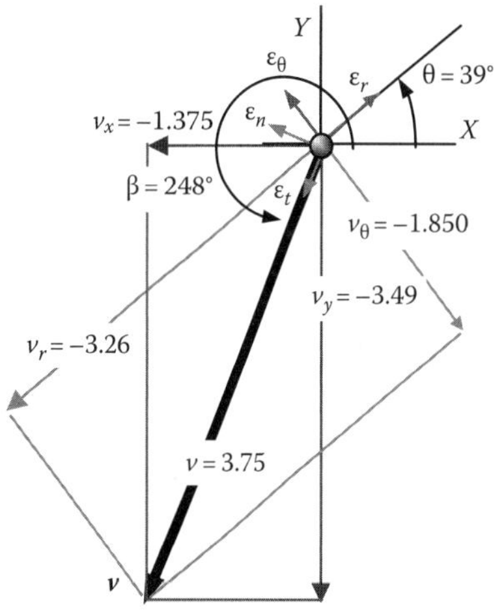

Velocity component diagram: units = 10^3 mm/s

$$\begin{Bmatrix} v_X \\ v_Y \end{Bmatrix} = \begin{Bmatrix} -1375 \\ -3492 \end{Bmatrix} \text{ mm/s};$$
$$\begin{Bmatrix} a_X \\ a_Y \end{Bmatrix} = \begin{Bmatrix} -1.166 \\ -1.254 \end{Bmatrix} \times 10^6 \text{ mm/s}^2$$

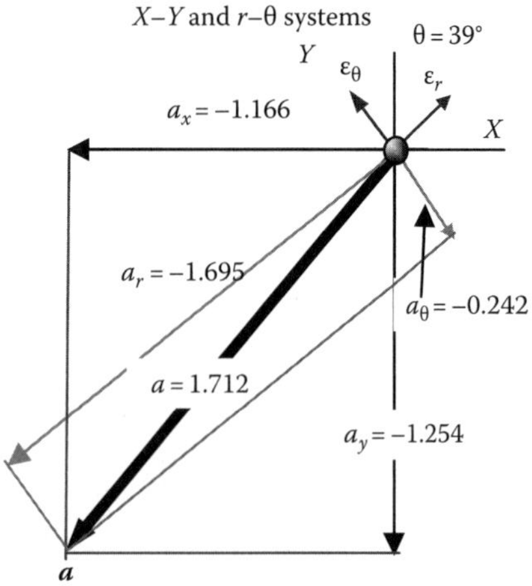

Acceleration component diagram: units = 10^6 mm/s^2

$$\begin{Bmatrix} v_t \\ v_n \end{Bmatrix} = \begin{Bmatrix} 3753 \\ 0 \end{Bmatrix} \text{ mm/s;}$$

$$\begin{Bmatrix} a_t \\ a_n \end{Bmatrix} = \begin{Bmatrix} 1.594 \\ 0.625 \end{Bmatrix} \times 10^6 \text{ mm/s}^2$$

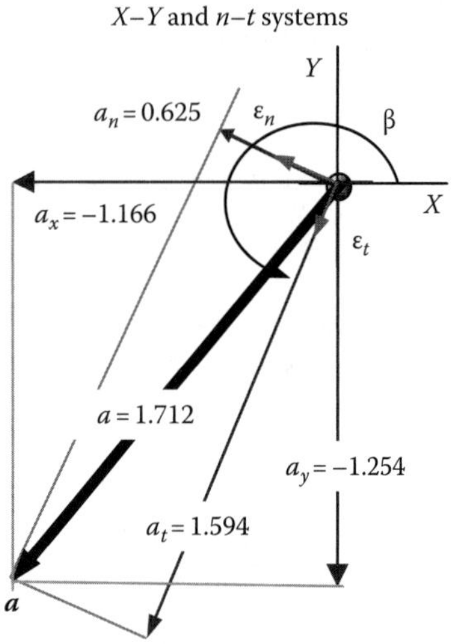

Acceleration component diagram: units = 10^6 mm/s^2

2.14

$$\begin{Bmatrix} v_r \\ v_\theta \end{Bmatrix} = \begin{Bmatrix} 500 \\ 220 \end{Bmatrix} \text{ m/s;}$$

$$\begin{Bmatrix} a_r \\ a_\theta \end{Bmatrix} = \begin{Bmatrix} -17 \\ 78 \end{Bmatrix} \text{ m/s}^2$$

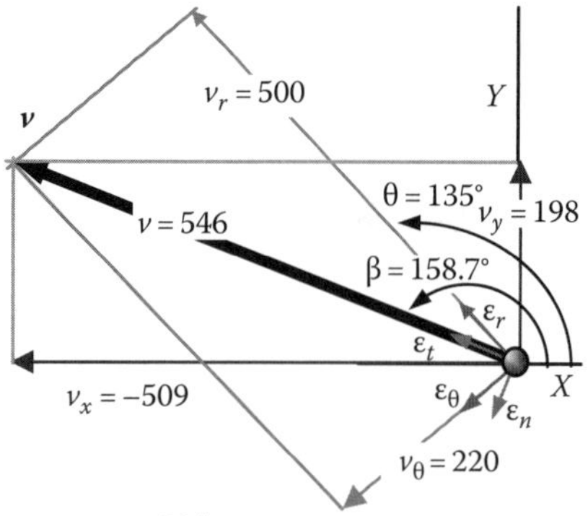

Velocity component diagram: units = m/s

$$\begin{Bmatrix} v_X \\ v_Y \end{Bmatrix} = \begin{Bmatrix} -509 \\ 198 \end{Bmatrix} \text{ m/s;}$$

$$\begin{Bmatrix} a_X \\ a_Y \end{Bmatrix} = \begin{Bmatrix} -43.1 \\ -67.2 \end{Bmatrix} \text{ m/s}^2$$

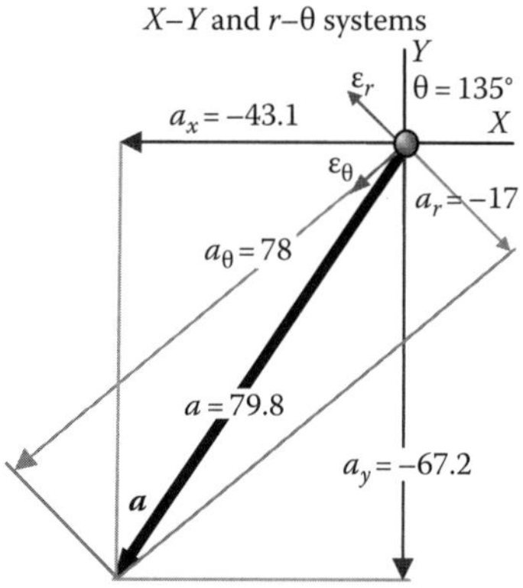

Acceleration component diagram: units = m/s^2

$$\begin{Bmatrix} v_t \\ v_n \end{Bmatrix} = \begin{Bmatrix} 546 \\ 0 \end{Bmatrix} \text{ m/s;}$$

$$\begin{Bmatrix} a_t \\ a_n \end{Bmatrix} = \begin{Bmatrix} 15.78 \\ 78.3 \end{Bmatrix} \text{ m/s}^2$$

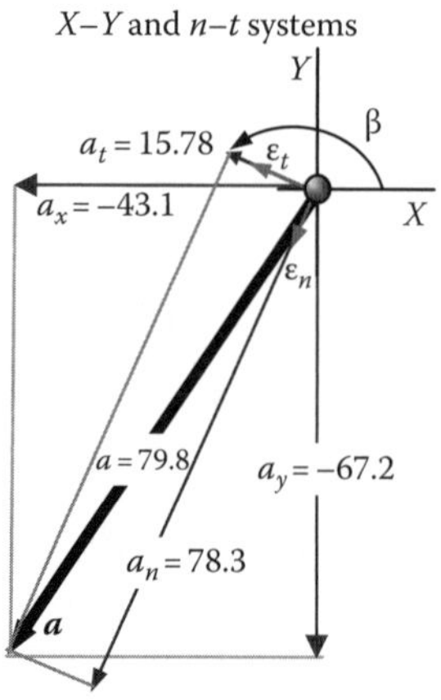

Acceleration component diagram: units = m/s^2

2.15

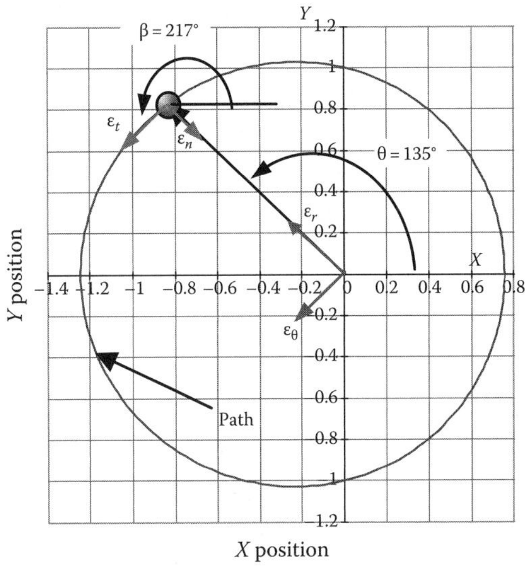

Path and coordinates diagram

$$\begin{Bmatrix} v_r \\ v_\theta \end{Bmatrix} = \begin{Bmatrix} 1.102 \\ 7.38 \end{Bmatrix} \text{ m/s;}$$

$$\begin{Bmatrix} a_r \\ a_\theta \end{Bmatrix} = \begin{Bmatrix} -53.3 \\ 13.85 \end{Bmatrix} \text{ m/s}^2$$

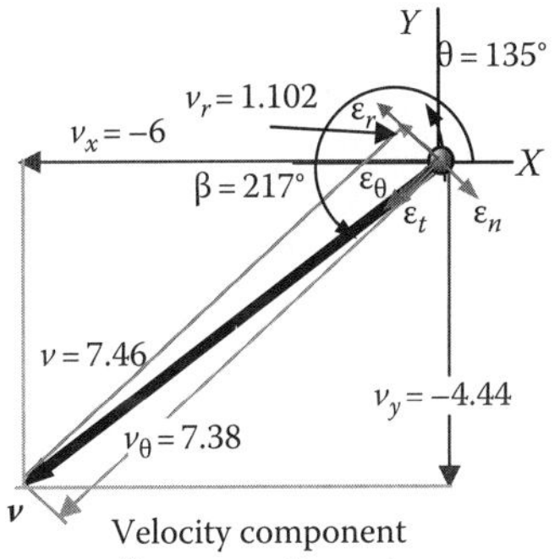

Velocity component diagram: units = m/s

$$\begin{Bmatrix} v_X \\ v_Y \end{Bmatrix} = \begin{Bmatrix} -6.00 \\ -4.44 \end{Bmatrix} \text{ m/s;}$$

$$\begin{Bmatrix} a_X \\ a_Y \end{Bmatrix} = \begin{Bmatrix} 27.9 \\ -47.5 \end{Bmatrix} \text{ m/s}^2$$

X–*Y* and *r*–θ systems

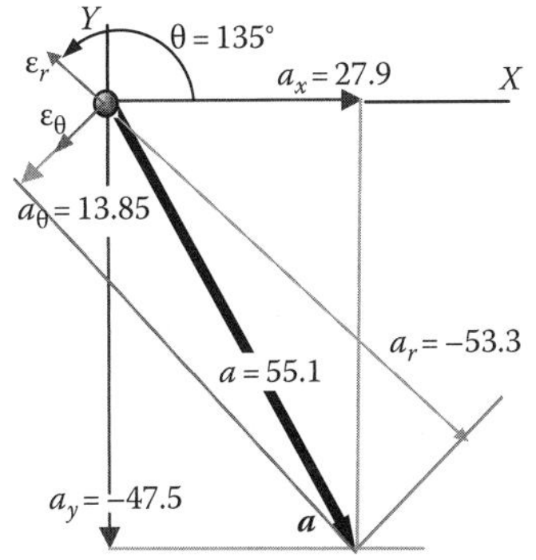

Acceleration component diagram: units = m/s^2

$$\begin{Bmatrix} v_t \\ v_n \end{Bmatrix} = \begin{Bmatrix} 7.46 \\ 0 \end{Bmatrix} \text{ m/s;}$$

$$\begin{Bmatrix} a_t \\ a_n \end{Bmatrix} = \begin{Bmatrix} 5.88 \\ 54.9 \end{Bmatrix} \text{ m/s}^2$$

X–*Y* and *n*–*t* systems

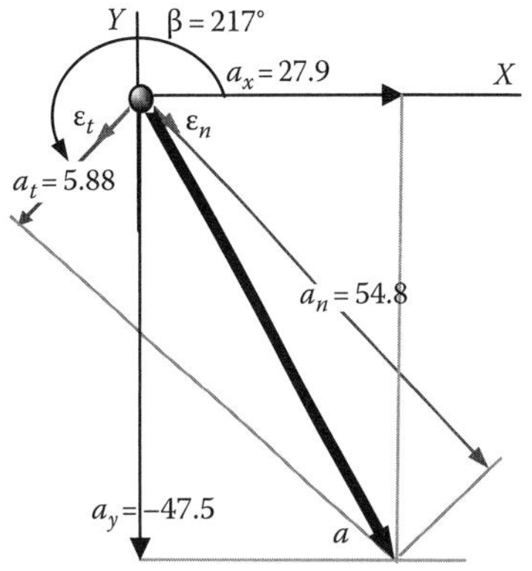

Acceleration component diagram: units = m/s^2

2.16

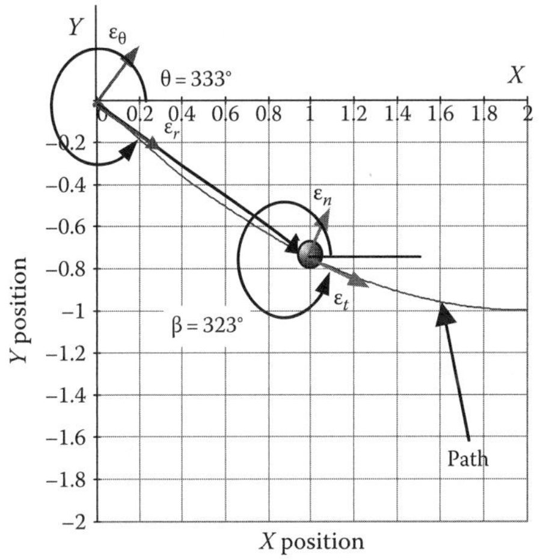

Path and coordinates diagram

$$\begin{Bmatrix} v_t \\ v_n \end{Bmatrix} = \begin{Bmatrix} 6.9498 \\ 0 \end{Bmatrix} \text{ ft/s}$$

$$\begin{Bmatrix} a_t \\ a_n \end{Bmatrix} = \begin{Bmatrix} 14.398 \\ 17.28 \end{Bmatrix} \text{ ft/s}^2\text{;}$$

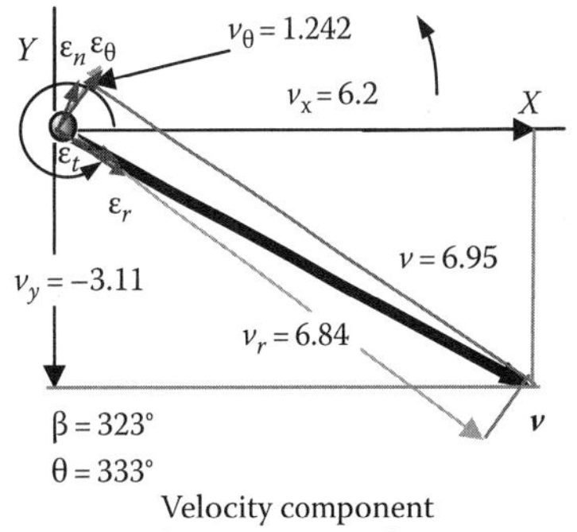

Velocity component diagram: units = ft/s

$$\begin{Bmatrix} v_X \\ v_Y \end{Bmatrix} = \begin{Bmatrix} 6.214 \\ -3.112 \end{Bmatrix} \text{ ft/s}$$

$$\begin{Bmatrix} a_X \\ a_Y \end{Bmatrix} = \begin{Bmatrix} 20.61 \\ 9.00 \end{Bmatrix} \text{ ft/s}^2;$$

X–Y and *r–θ* systems

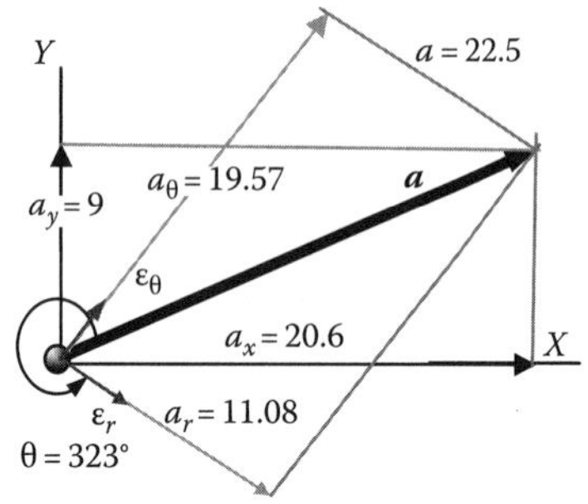

Acceleration component diagram: units = ft/s²

$$\begin{Bmatrix} v_r \\ v_\theta \end{Bmatrix} = \begin{Bmatrix} 6.838 \\ 1.242 \end{Bmatrix} \text{ ft/s}$$

$$\begin{Bmatrix} a_r \\ a_\theta \end{Bmatrix} = \begin{Bmatrix} 11.08 \\ 19.57 \end{Bmatrix} \text{ ft/s}^2$$

X–Y and *n–t* systems

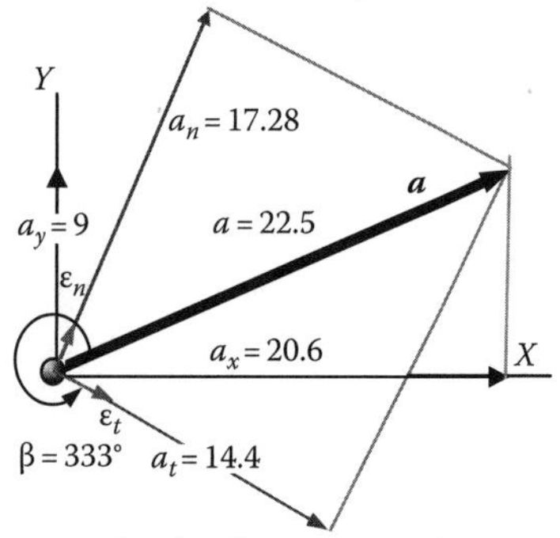

Acceleration component diagram: units = ft/s²

$$\begin{Bmatrix} v_t \\ v_n \end{Bmatrix} = \begin{Bmatrix} 27.78 \\ 0 \end{Bmatrix} \text{ m/s};$$

$$\begin{Bmatrix} a_t \\ a_n \end{Bmatrix} = \begin{Bmatrix} 2 \\ 0.7617 \end{Bmatrix} \text{ m/s}^2;$$

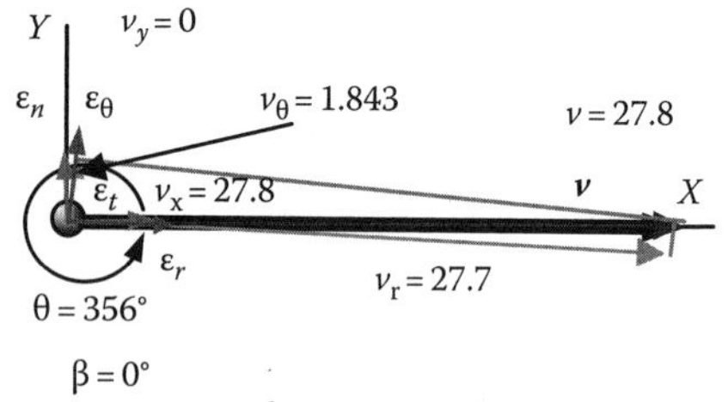

Velocity component diagram: units = m/s

$$\begin{Bmatrix} v_X \\ v_Y \end{Bmatrix} = \begin{Bmatrix} 27.78 \\ 0 \end{Bmatrix} \text{ m/s};$$

$$\begin{Bmatrix} a_X \\ a_Y \end{Bmatrix} = \begin{Bmatrix} 2 \\ 0.7617 \end{Bmatrix} \text{ m/s}^2;$$

X–Y and *r–θ* systems

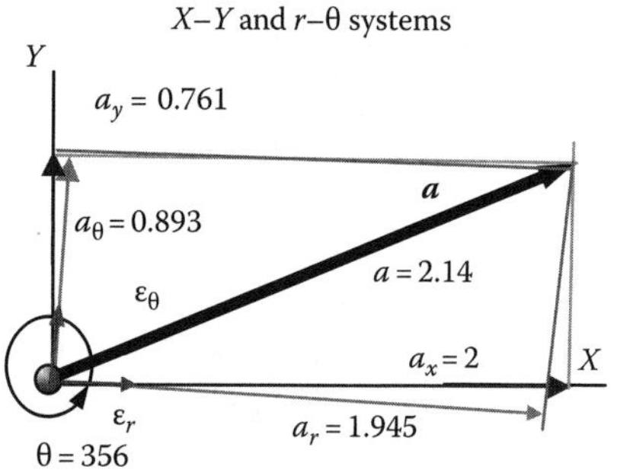

Acceleration component diagram: units = ms²

$$\begin{Bmatrix} v_r \\ v_\theta \end{Bmatrix} = \begin{Bmatrix} 27.72 \\ 1.846 \end{Bmatrix} \text{ m/s};$$

$$\begin{Bmatrix} a_r \\ a_\theta \end{Bmatrix} = \begin{Bmatrix} 1.945 \\ 0.893 \end{Bmatrix} \text{ m/s}^2$$

2.17 Let $\theta = 3.81°$ in the CCW direction

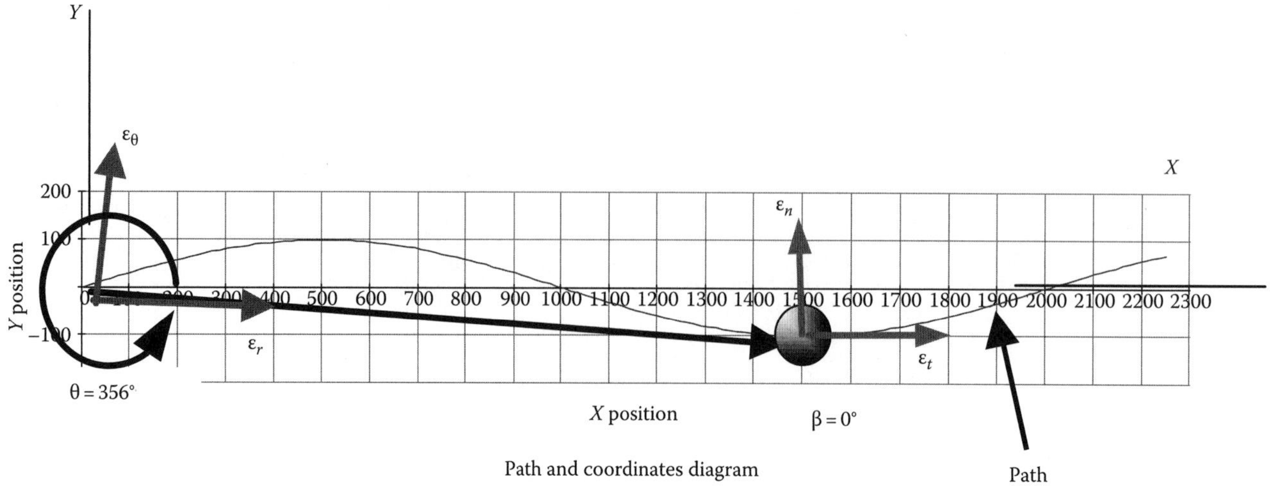

Path and coordinates diagram

X–Y and n–t systems

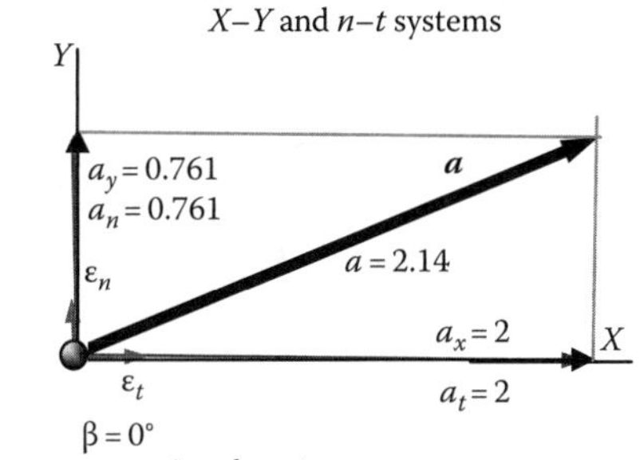

Acceleration component diagram: units = m/s²

2.18

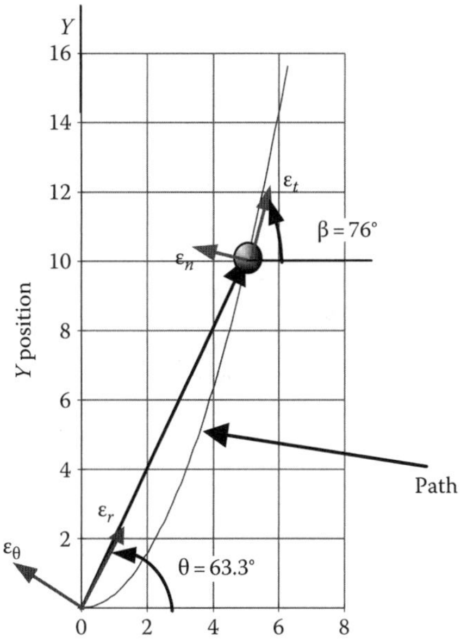

Path and coordinates diagram

$$\begin{Bmatrix} v_t \\ v_n \end{Bmatrix} = \begin{Bmatrix} 200 \\ 0 \end{Bmatrix} \text{ m/s}$$

$$\begin{Bmatrix} a_t \\ a_n \end{Bmatrix} = \begin{Bmatrix} -1 \\ 0.4565 \end{Bmatrix} \text{ m/s}^2;$$

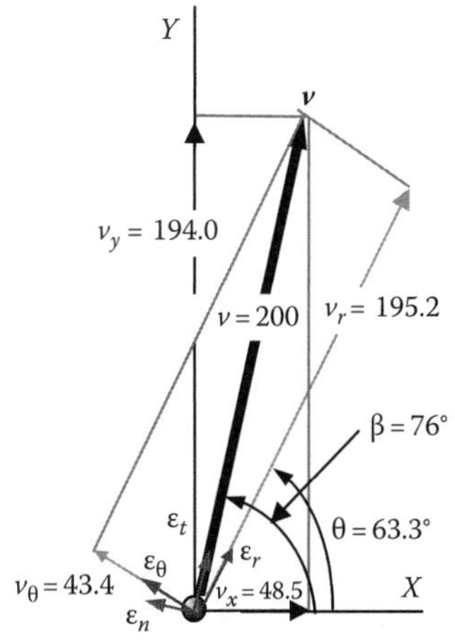

Velocity component diagram: units = m/s

$$\begin{Bmatrix} v_X \\ v_Y \end{Bmatrix} = \begin{Bmatrix} 48.51 \\ 194.03 \end{Bmatrix} \text{ m/s}$$

$$\begin{Bmatrix} a_X \\ a_Y \end{Bmatrix} = \begin{Bmatrix} -0.685 \\ -0.859 \end{Bmatrix} \text{ m/s}^2;$$

X–Y and r–θ systems

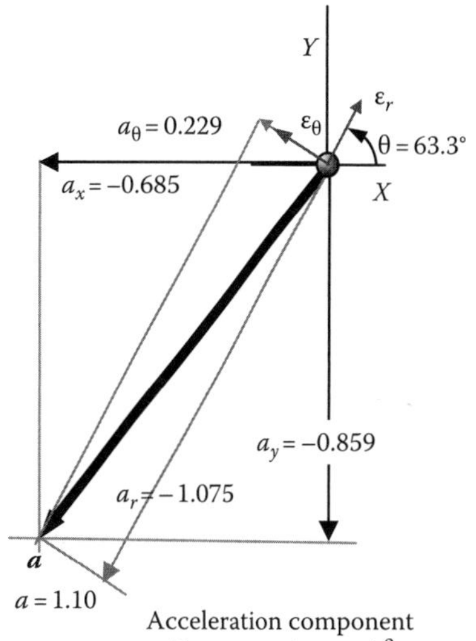

Acceleration component diagram: units = m/s²

$$\begin{Bmatrix} v_r \\ v_\theta \end{Bmatrix} = \begin{Bmatrix} 195.24 \\ 43.39 \end{Bmatrix} \text{ m/s}$$

$$\begin{Bmatrix} a_r \\ a_\theta \end{Bmatrix} = \begin{Bmatrix} -1.075 \\ 0.229 \end{Bmatrix} \text{ m/s}^2$$

X–Y and n–t systems

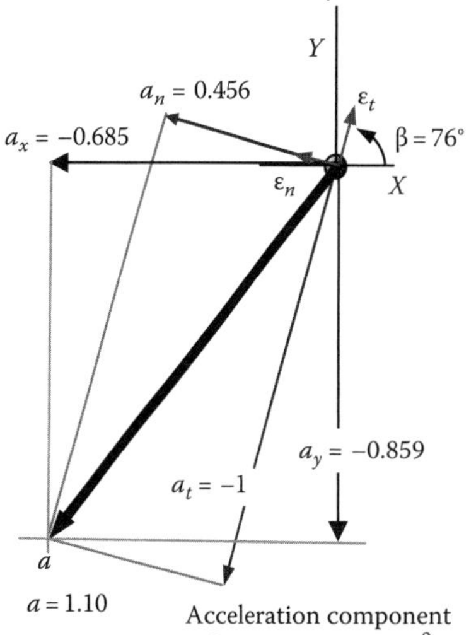

Acceleration component diagram: units = m/s²

2.19

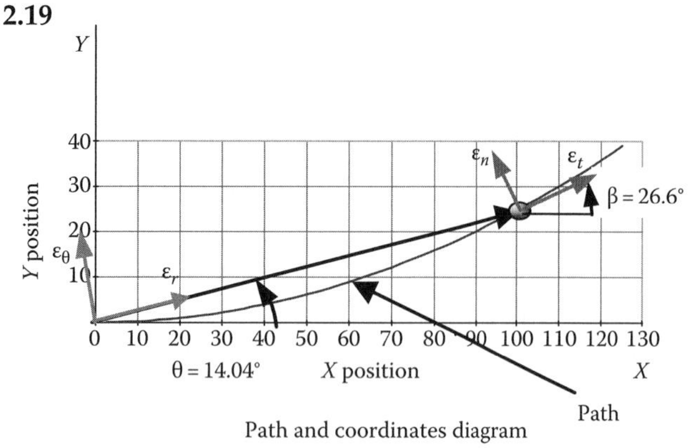

Path and coordinates diagram

$$\begin{Bmatrix} v_t \\ v_n \end{Bmatrix} = \begin{Bmatrix} 20 \\ 0 \end{Bmatrix} \text{ m/s;}$$

$$\begin{Bmatrix} a_t \\ a_n \end{Bmatrix} = \begin{Bmatrix} 2 \\ 1.432 \end{Bmatrix} \text{ m/s}^2$$

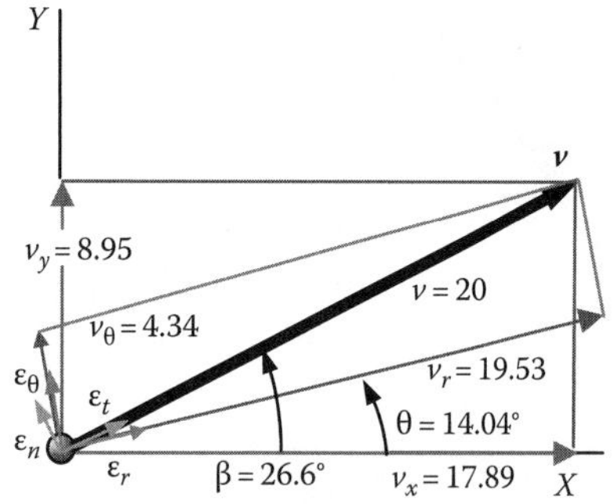

Velocity component diagram: units = m/s

$$\begin{Bmatrix} v_X \\ v_Y \end{Bmatrix} = \begin{Bmatrix} 17.89 \\ 8.95 \end{Bmatrix} \text{ m/s;}$$

$$\begin{Bmatrix} a_X \\ a_Y \end{Bmatrix} = \begin{Bmatrix} 1.148 \\ 2.175 \end{Bmatrix} \text{ m/s}^2$$

X–Y and r–θ systems

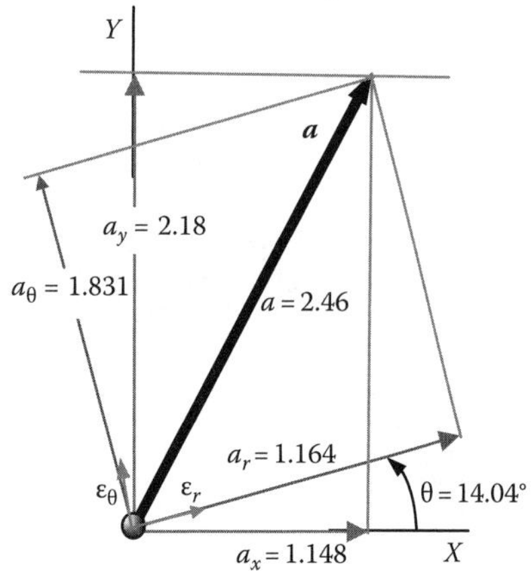

Acceleration component diagram: units = m/s²

$$\begin{Bmatrix} v_r \\ v_\theta \end{Bmatrix} = \begin{Bmatrix} 19.53 \\ 4.34 \end{Bmatrix} \text{ m/s;}$$

$$\begin{Bmatrix} a_r \\ a_\theta \end{Bmatrix} = \begin{Bmatrix} 1.164 \\ 1.831 \end{Bmatrix} \text{ m/s}^2$$

X–Y and n–t systems

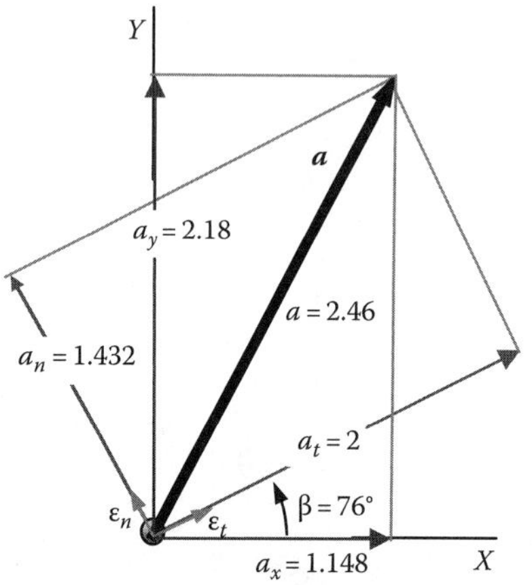

Acceleration component diagram: units = m/s²

2.20

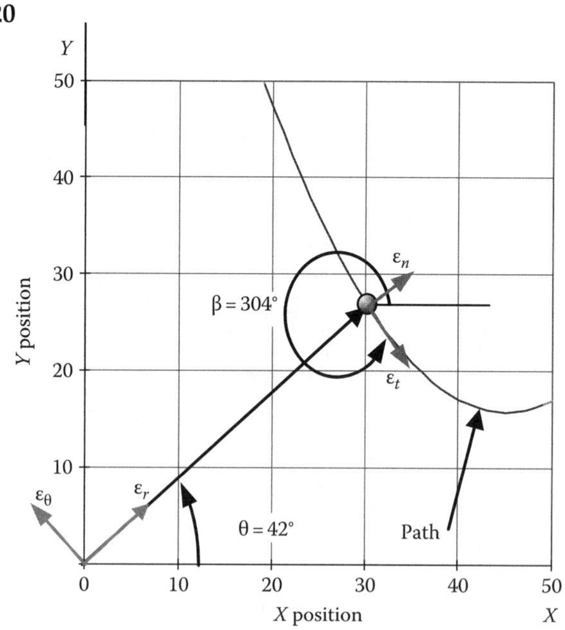

Path and coordinates diagram

$$\begin{Bmatrix} v_t \\ v_n \end{Bmatrix} = \begin{Bmatrix} 40 \\ 0 \end{Bmatrix} \text{ ft/s;}$$

$$\begin{Bmatrix} a_t \\ a_n \end{Bmatrix} = \begin{Bmatrix} 0.32 \\ 27.3 \end{Bmatrix} \text{ ft/s}^2$$

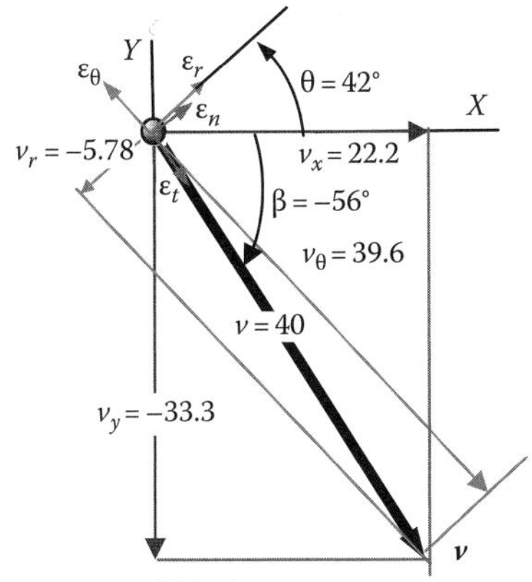

Velocity component diagram: units = ft/s

$$\begin{Bmatrix} v_X \\ v_Y \end{Bmatrix} = \begin{Bmatrix} 22.2 \\ -33.3 \end{Bmatrix} \text{ft/s};$$

$$\begin{Bmatrix} a_X \\ a_Y \end{Bmatrix} = \begin{Bmatrix} 22.9 \\ 14.89 \end{Bmatrix} \text{ft/s}^2$$

X–Y and r–θ systems

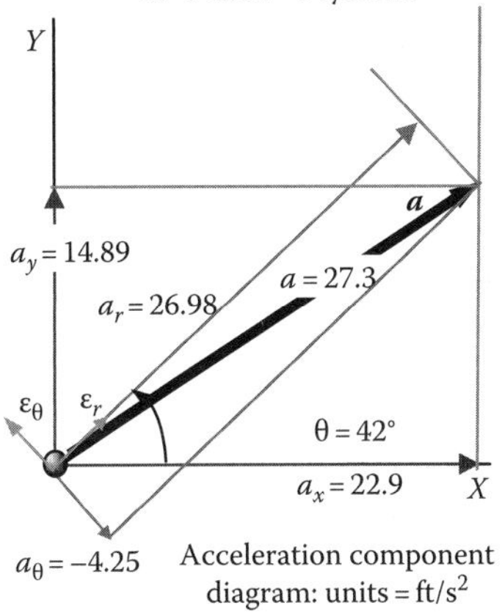

Acceleration component diagram: units = ft/s²

$$\begin{Bmatrix} v_r \\ v_\theta \end{Bmatrix} = \begin{Bmatrix} -5.78 \\ 39.6 \end{Bmatrix} \text{ft/s};$$

$$\begin{Bmatrix} a_r \\ a_\theta \end{Bmatrix} = \begin{Bmatrix} 26.98 \\ 4.25 \end{Bmatrix} \text{ft/s}^2$$

X–Y and n–t systems

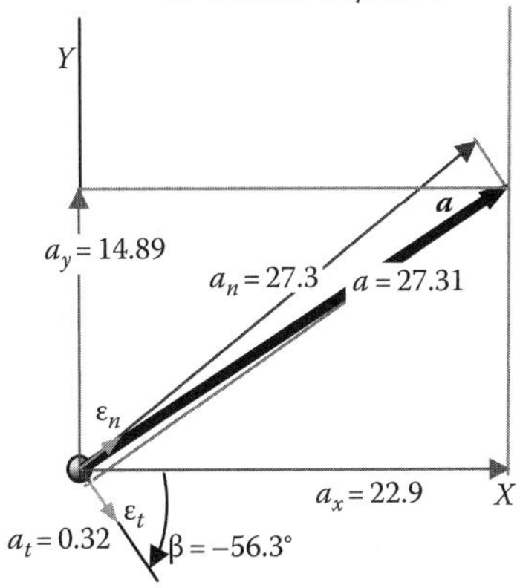

Acceleration component diagram: units = ft/s²

2.21

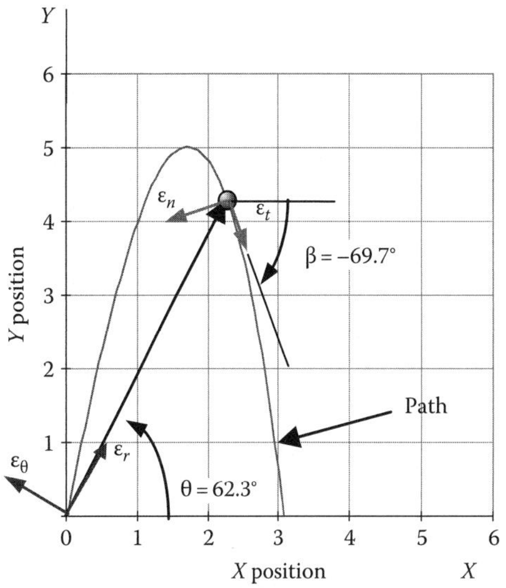

Path and coordinates diagram

$$\begin{Bmatrix} v_t \\ v_n \end{Bmatrix} = \begin{Bmatrix} 5 \\ 0 \end{Bmatrix} \text{m/s};$$

$$\begin{Bmatrix} a_t \\ a_n \end{Bmatrix} = \begin{Bmatrix} -1.5 \\ 5.67 \end{Bmatrix} \text{m/s}^2$$

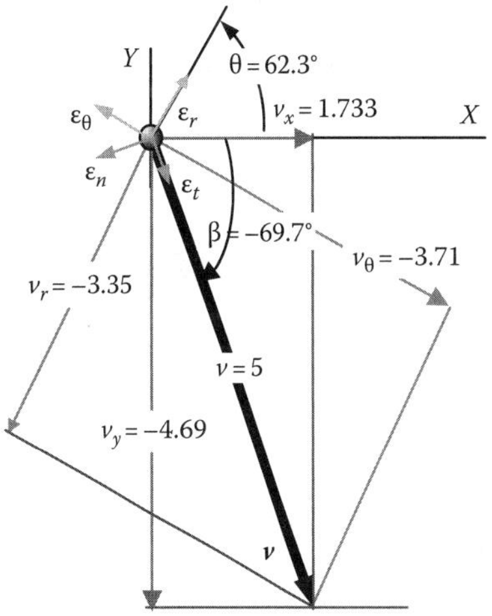

Velocity component diagram: units = m/s

$$\begin{Bmatrix} v_X \\ v_Y \end{Bmatrix} = \begin{Bmatrix} 1.733 \\ -4.69 \end{Bmatrix} \text{m/s};$$

$$\begin{Bmatrix} a_X \\ a_Y \end{Bmatrix} = \begin{Bmatrix} -5.842 \\ -0.559 \end{Bmatrix} \text{m/s}^2$$

X–Y and *r–θ* systems

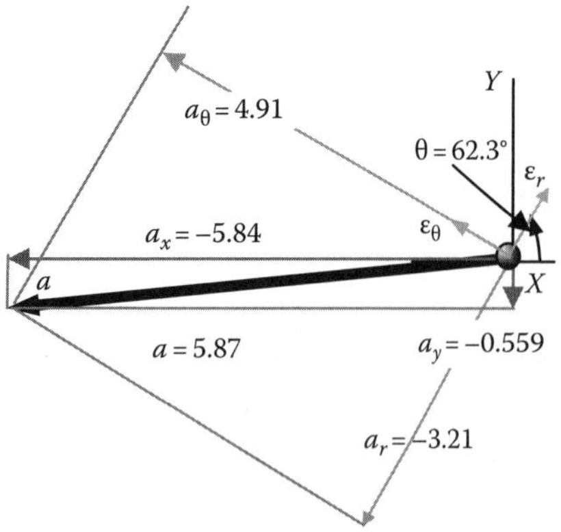

Acceleration component diagram: units = m/s²

$$\begin{Bmatrix} v_r \\ v_\theta \end{Bmatrix} = \begin{Bmatrix} -3.35 \\ -3.71 \end{Bmatrix} \frac{\text{m}}{\text{s}};$$

$$\begin{Bmatrix} a_r \\ a_\theta \end{Bmatrix} = \begin{Bmatrix} -3.21 \\ 4.91 \end{Bmatrix} \frac{\text{m}}{\text{s}^2}$$

X–Y and *n–t* systems

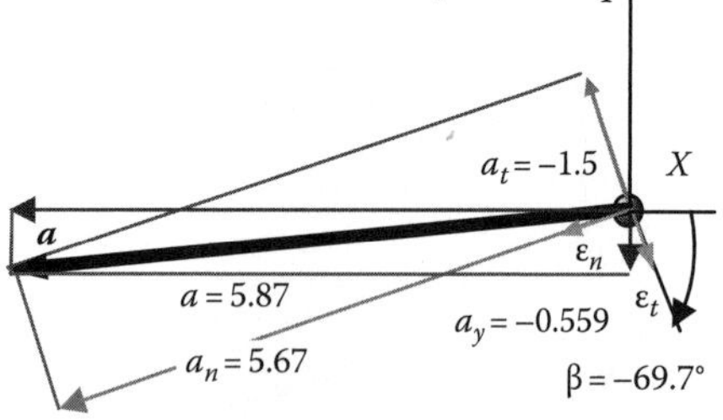

Acceleration component diagram: units = ft/s²

Chapter 3

3.1

$$m\ddot{Y} + \gamma_f\left(\frac{\pi d^2}{4}\right)Y = w; \quad \overline{Y} = \frac{4w}{\pi d^2 \gamma_f}$$

3.2

$$m\ddot{Y} + \frac{3EI}{L^3}Y = -w; \quad \omega_n = \left(\frac{3EIg}{L^3 w}\right)^{1/2}$$

3.3

$$m\ddot{Y} + \frac{AE}{L}Y = w; \quad \omega_n = \left(\frac{AEg}{Lw}\right)^{1/2}$$

3.4

(a) $m\ddot{X} + \left(\dfrac{k_3 k_2}{k_2 + k_3} + k_1\right)X = 0;$

(b) $m\ddot{X} + (k_1 + k_2 + k_3)X = 0;$

(c) $m\ddot{X} + \left(\dfrac{(k_3 + k_2)k_1}{k_3 + k_2 + k_1}\right)X = 0;$

(a) $f_n = 1.950$ Hz, (b) $f_n = 2.76$ Hz, (c) $f_n = 1.3$ Hz

3.5

$$m\ddot{\delta} + (k_1 + k_2)\delta = 0; \quad f_n = 3.56 \text{ Hz}$$

3.6

$$m\ddot{\delta} + (k_1 + k_2)\delta = 0; \quad f_n = 8.28 \text{ Hz}$$

3.7

$$m\ddot{Y} + 2c\dot{Y} + 3kY = -w; \quad f_n = 7.2943 \text{ Hz};$$
$$\zeta = 0.175; \quad f_d = 7.184 \text{ Hz}$$

3.8

$$m\ddot{X} + 2c\dot{X} + kX = -w \sin\alpha;$$
$$f_n = 1.9784 \text{ Hz}; \quad \zeta = 0.1937; \quad f_d = 1.9401 \text{ Hz};$$
$$\overline{X} = \frac{-w \sin\alpha}{k}; \quad m\ddot{\delta} + 2c\dot{\delta} + k\delta = 0$$

3.9 *Note*: The time that the load is applied be $t = 0.15\tau_n$

(a) $\ddot{X} + 500X = 2.5$ {transient EOM}

$$\text{(b)}\ X(t) = \begin{cases} (-0.005\cos(22.36t) + 0.005)\text{m} & 0 \le t \le 0.0422 \text{ s} \\ (0.00206\cos(22.36(t - 0.0422)) + 0.00404\sin(22.36(t - 0.0422)))\text{m} & 0.0422 \le t \end{cases};$$

(c) $t_L = 0.0492$ s.

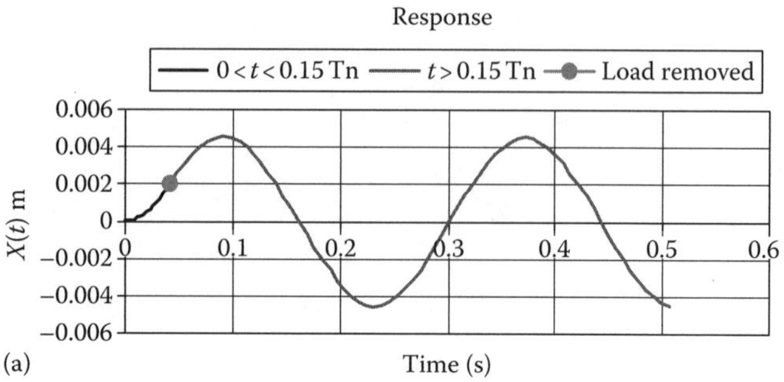

(a)

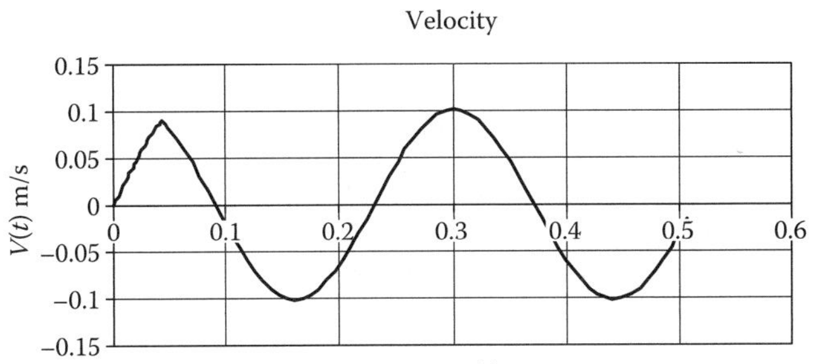

(b)

3.10

(a) $\ddot{X} + 3864X = 48.3t$;

(b)

$$X(t)=\begin{cases}0.000202\sin(62.16t)+0.0125t & 0\le t\le 0.5\,\text{s}\\ 0.00618\cos(62.16(t-0.5))+0.0001907\sin(62.16(t-0.5)) & 0.5\le t\end{cases}$$

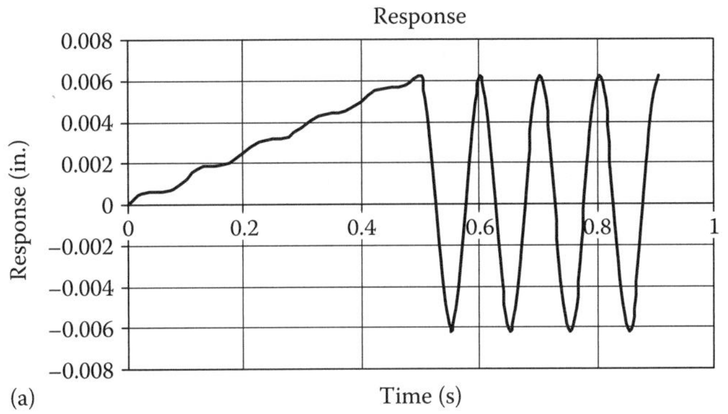

(a)

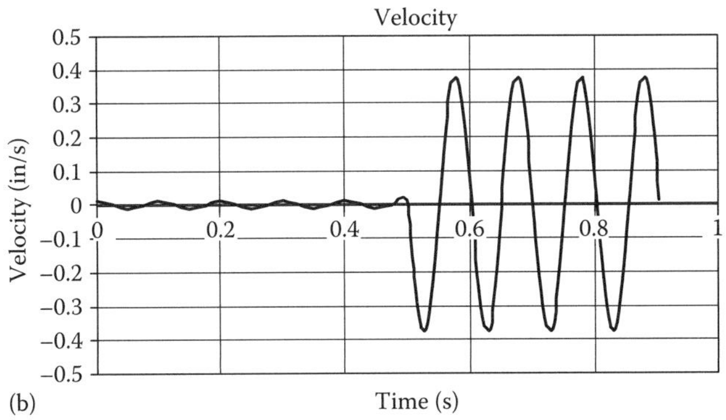

(b)

3.11

(a) $\ddot{X} + 952.2X = 7.5t - 3.05t^2$ transient EOM;

(b)

$$X(t)=\begin{cases}-0.2553\sin(30.86t)+0.00635t^2-7.875t\ \text{mm for } 0\le t\le 1\ \text{s}\\ -1.605\cos(30.86(t-1))+0.373\sin(30.86(t-1))\ \text{mm for } 1\le t\ \text{s}\end{cases}$$

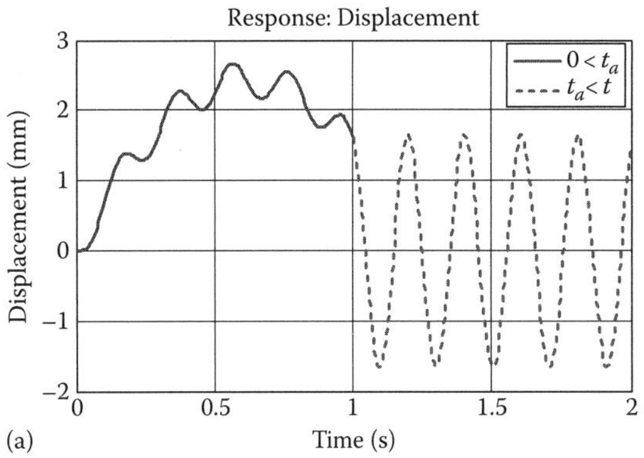

(a)

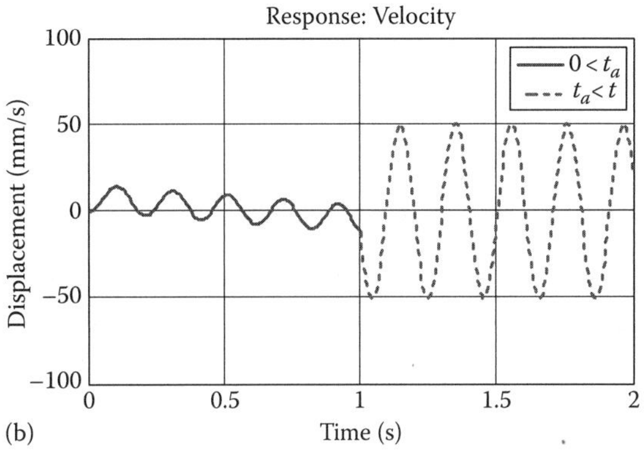

(b)

3.12 *Note*: Time for applied force, $t_a = 2.5\tau_d$, where τ_d is the damped period.

(a) $\ddot{X} + 5.196\dot{X} + 300X = 7$ {transient EOM}

(b) $$X(t)=\begin{cases}0.0233+e^{-2.598t}(-0.0233\cos(17.12t)-0.00354\sin(17.12t))\ \text{m}\ 0\le t\le 0.917\ \text{s}\\ e^{-2.598(t-0.917)}(0.025\cos(17.12(t-0.917))+0.00379\sin(17.12(t-0.917)))\ \text{m}\ 0.917\le t\ \text{s}\end{cases};$$

(c) $t_{lag} = 0.1837$ s

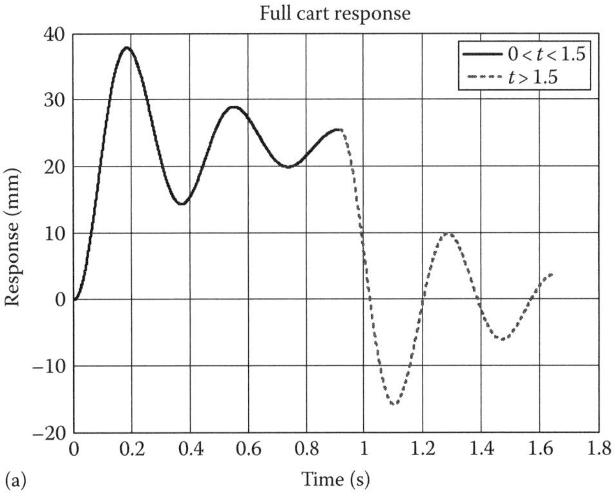

(a)

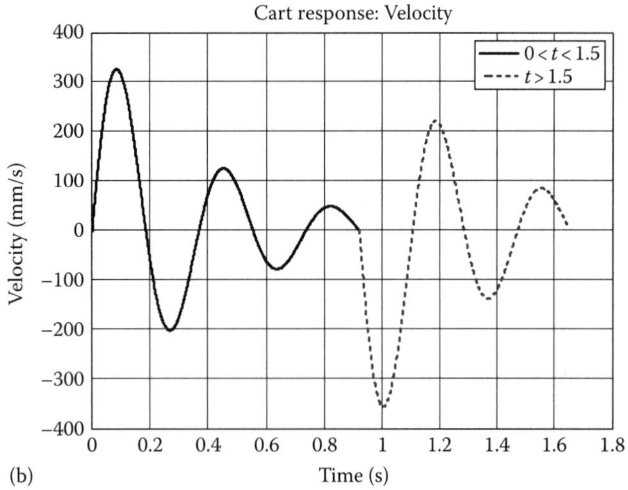

(b)

(c) $t_P = 1.54$ s

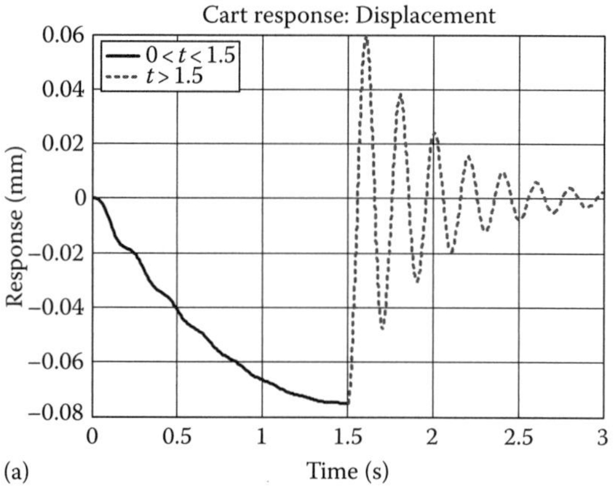

(a)

3.13

(a) $\ddot{X} + 15.46\dot{X} + 5152X = 5.15t$

(b) $X(t) = \begin{cases} e^{-7.68t}[-1.401(10^{-5})\sin(71.36t)] + 0.001t \text{ in. } 0 \le t \le 0.25 \text{ s} \\ e^{-7.68(t-0.25)}[2.489(10^{-4})\cos(71.36(t-0.25)) + 4.005(10^{-5})\sin(71.36(t-0.25))] \text{ in. } 0.25 \le t \end{cases}$;

(c) $t_P = 0.295$ s.

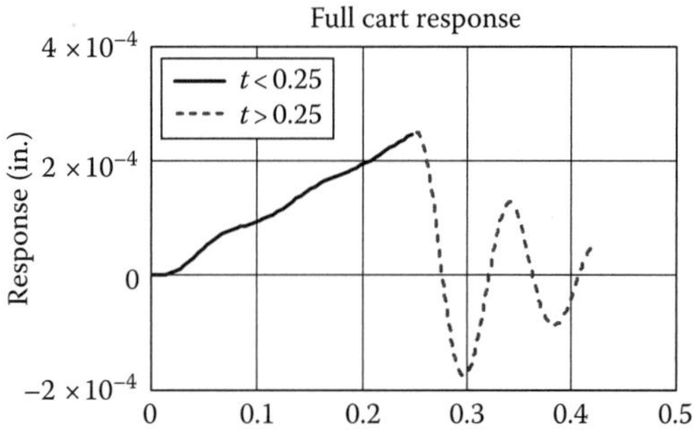

(a)

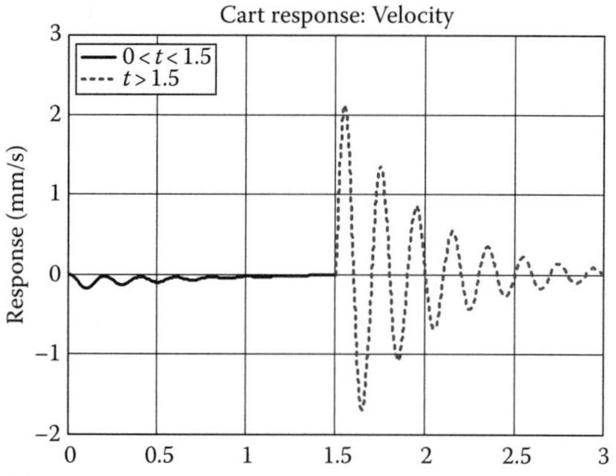

(b)

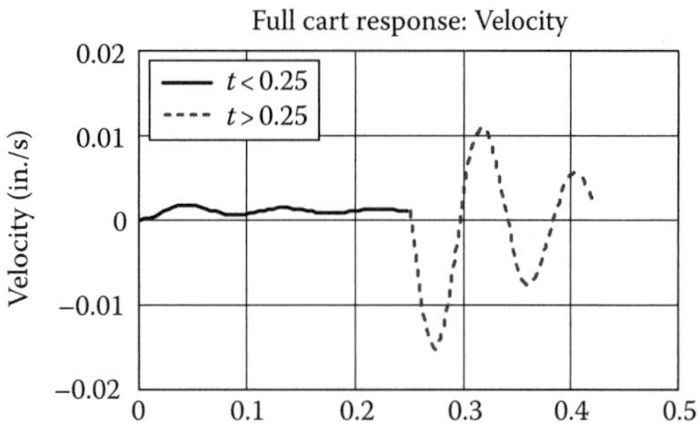

(b)

3.15

$$\delta(t) = e^{-\zeta\omega_n t}\left[\frac{D}{\omega_n^2}\cos(\omega_d t) + \frac{\zeta D}{\omega_n \omega_d}\sin(\omega_d t)\right] - \frac{D}{\omega_n^2}$$

3.14

(a) $\ddot{X} + 5.028\dot{X} + 31.62X = 0.0333t^2 - 0.1t$

(b)

$$X(t) = \begin{cases} e^{-2.265t}[3.88(10^{-7})\cos(31.52t) + 3.152(10^{-6})\sin(31.52t)] + 3.33(10^{-5})t^2 - 10.03(10^{-5})t + 0.0388(10^{-5}) \text{ m } 0 \le t \le 1.5 \text{ s} \\ e^{-2.265(t-1.5)}[-7.51(10^{-5})\cos(31.54(t-1.5)) - 5.056(10^{-6})\sin(31.54(t-1.5))] \text{ m } 1.5 \le t \end{cases}$$

3.16

(a) $\ddot{X} + 18.85\dot{X} + 35520X = 18540 + 30850t$
{transient EOM}

(b) $X(t) = \begin{cases} e^{-9.424t}(-0.521\cos(188.24t) - 0.0307\sin(188.24t)) + 0.521 + 0.868t \text{ in. } 0 \le t \le 0.25 \\ e^{-3.864(t-0.25)}(0.788\cos(152.21(t-0.25)) + 0.0301\sin(152.21(t-0.25))) \text{ in. } t \ge 0.25 \end{cases};$

(c) $t_P = 0.01$ s.

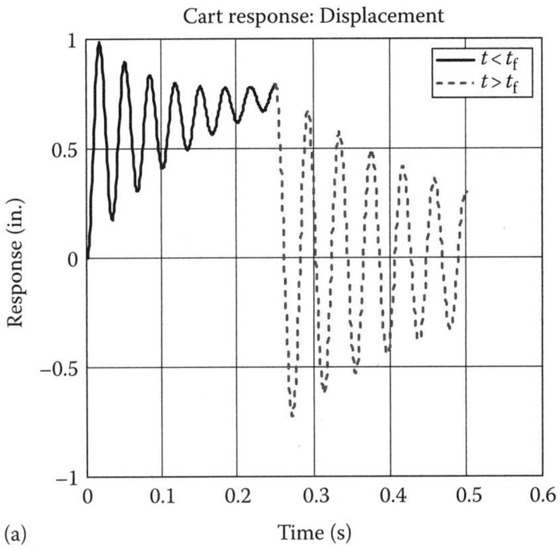

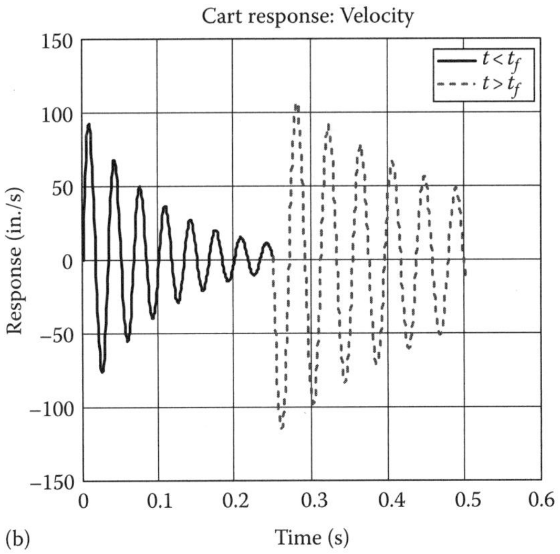

(c) Peak response occurs at $t = 3.245$ s.

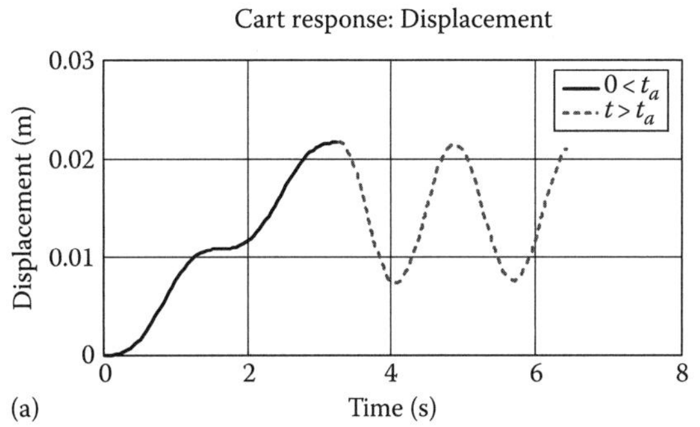

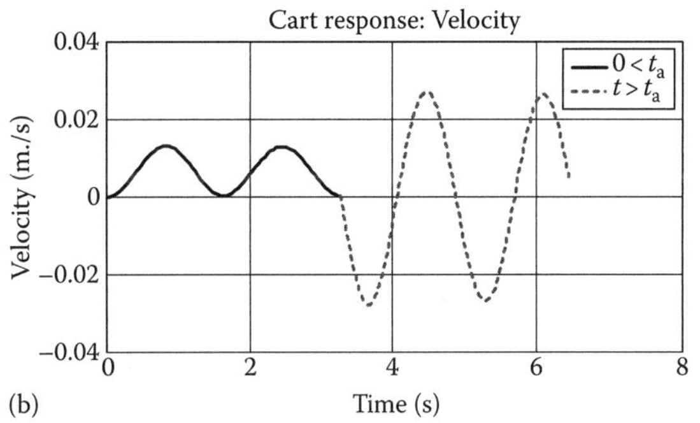

3.17

(a) For $t < t_a$ $\ddot{X}_1 + 0.0375\dot{X}_1 + 15X_1 = 0.1t + 0.00025$ and for $t > t_a$ $m\ddot{X}_2 + (c_A + c_B)\dot{X}_2 + (k_A + k_B)X_2 = k_B(x(t_a) - X_1(t_a)) = f_0$

(b) $X(t) = \begin{cases} e^{-0.01875t}[1.721(10^{-3})\sin(3.872t)] + 6.667(10^{-3})t \text{ m } 0 \le 3.245 \text{ s} \\ e^{-0.01875(t-3.245)}[0.00723\cos(3.872(t-3.245)) + 1.366(10^{-4})\sin(3.872(t-3.245)) + 0.0144] \text{ m } 3.245 \le t \end{cases}$

3.18

(a) $\ddot{X} + 50.23\dot{X} + 1700X = 463.7t^2 + 69.55t$

(b) $X(t) = \{e^{-25.1t}(1.0504(10^{-3})\cos(32.70t) + 4.895(10^{-5})\sin(32.70t)) + 0.2728t^2 + 0.02478t - 0.001054\}$ in.

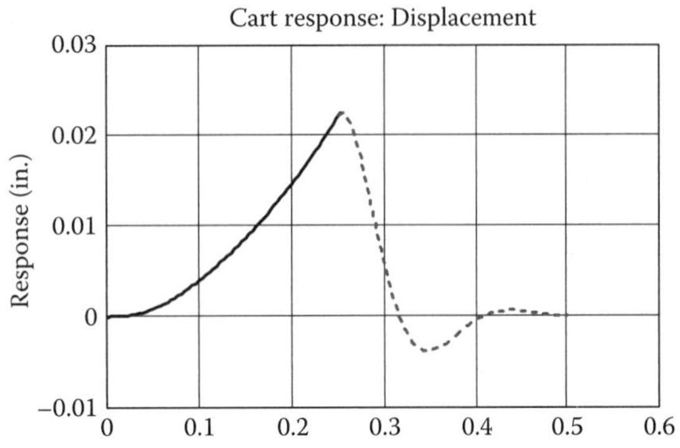

(a)

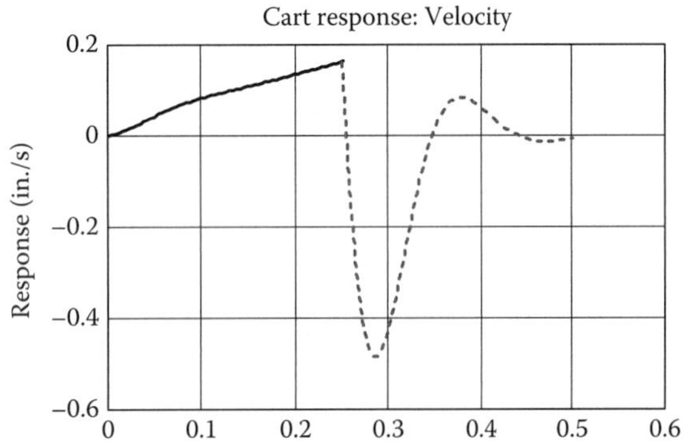

(b)

3.19

For case (a) $\zeta_1 = 0.0833$ $Y_{OPa} = 26.9$ mm
for case (b) $\zeta_2 = 0.0$ $Y_{OPb} = 51.3$ mm;
(c) $\omega < \omega_1 = 13.40$ Hz; $\omega > \omega_2 = 17.25$ Hz

3.20

(a) $Y_{OP} = 0.0691$ in.; (b) $\omega_1 = 27.62$ Hz; $\omega_2 = 34.46$ Hz; (c) $c = 8.49$ lbs/in.

3.21

(a) $c = 0.018$ lbs/in.; (b) $Y = 1.593$ in.

3.22

$c = 22.11$ N s/m

3.23

(a) $v_0 = 10.79$ km/h; (b) $c = 6084$ N s/m

3.24

(a) $v_0 = 10.23$ mi/h; (b) $c = 2{,}279$ lbs/ft

3.25

(a) $m\ddot{X} + c\dot{X} + (k_1 + k_2)X = k_2 A \cos \omega t$;
(b) $Y_{OP} = 35.6$ mm; (c) $\omega_1 = 15.33$ Hz; $\omega_2 = 20.46$ Hz

3.26

(a) $Yopss = 4.89$ mils, (b) $Yopmax = 11.2$ mils,
(c) $a = 11.11$ mils

3.27

(a) $a = 1.835$ mm, (b) $a = 0.916$ mm,
(c) $527 < \omega < 554$ rpm

3.28

(a) $a = 8.33$ mils, (b) $Y_{OP} = 11.27$ mils,
(c) $a_{new} = 6.26$ mils, 40% change in stiffness.

3.29

$$T = \frac{1}{2}m\left(g + \frac{v_0^2}{2h\sqrt{2}}\right)$$

3.30 (X (↓+) corresponds to block B, Y (↓+) corresponds to block C)

(a) $\ddot{X} = \frac{-5}{19}g$; $\ddot{Y} = \frac{15}{19}g$; (c) $\dot{X} = \frac{-5}{19}g$; $\dot{Y} = \frac{15}{19}g$;

(d) $\dot{X} = \sqrt{\frac{10g}{57}}$ and $\dot{Y} = \sqrt{\frac{30g}{19}}$

3.31

(b) $\ddot{X} = g\left[\frac{(3m_B/2) - m_A(\sin\alpha + \mu\cos\alpha)}{(m_A + (9/4)m_B)}\right]$;

(c) $\dot{X}_A = 2.68\,\frac{\text{m}}{\text{s}}$ moving up incline;

(d) $\dot{X}_A = 2.313\,\frac{\text{m}}{\text{s}}$ moving up incline

3.32

(a) $\left(m_A + \frac{m_B}{4}\right)\ddot{X}_A = \left(W_A + \frac{\mu_d W_B}{2}\right)\sin\theta - \left(\mu_d W_A + \frac{W_B}{2}\right)\cos\theta$

$\ddot{X}_A = 9.233\,\frac{\text{ft}}{\text{s}^2}$ (up incline);

$\ddot{X}_B = 4.61\,\frac{\text{ft}}{\text{s}^2}$ (down incline);

(b) $\dot{X}_A = 12.27\,\frac{\text{ft}}{\text{s}}$ (up incline),

$\dot{X}_B = 6.14\,\frac{\text{ft}}{\text{s}}$ (down incline);

(c) $\dot{X}_A = 4.95\,\frac{\text{ft}}{\text{s}}$ (up incline);

$\dot{X}_B = 2.47\,\frac{\text{ft}}{\text{s}}$ (down incline)

3.33

(a) $(m_A + m_B + m_C)\ddot{X}_C = W_A \sin\theta - \mu_d W_A \cos\theta - \mu_d W_B + W_C$;

(b) $\dot{X}_A = \dot{X}_B = \dot{Y}_C = 8.425$ in./s;

(c) $\dot{X}_A = \dot{X}_B = \dot{Y}_C = 26.62$ in./s

3.34

(a) $(m_B + 4m_A)\ddot{Y} = W_B - 2\mu_d W_A$;

(b) $\dot{X}_A = 1.058\ \frac{\text{m}}{\text{s}}\quad \dot{Y}_B = 0.529$ m/s;

(c) $\dot{X}_A = 0.280\ \frac{\text{m}}{\text{s}}\quad \dot{Y}_B = 0.1401$ m/s

3.35

$$(m_B + 9m_C)\ddot{X} + c_B\dot{X} + (k_B + 9k_C)X = w_B - w_C;\quad \omega_n = \sqrt{\frac{k_B + 9k_C}{m_B + 9m_C}};$$

$$\zeta = \frac{c_B}{\sqrt{(k_B + 9k_C)(m_B + 9m_C)}}$$

3.36 Block A taken to be moving to the right.

$$\ddot{X}_A = \frac{-2kX_A}{3m_A} + \frac{2g}{3}\quad \text{and}\quad \ddot{X}_B = \frac{1}{3}\left(\frac{-kX_A}{m_A} + g\right)$$

$$\dot{X}_A = \left(\frac{-2k}{3m_A}X_A^2 + \frac{4gX_A}{3}\right)^{1/2}\quad \text{and}$$

$$\dot{X}_B = 2\left(\frac{-2k}{3m_A}X_A^2 + \frac{4gX_A}{3}\right)^{1/2}$$

3.37

$$m\ddot{X} + 4kX = 0;\quad \omega_n = 2\sqrt{\frac{k}{m}}$$

3.38

$$(m_1 + 4m_2)\ddot{X} + c\dot{X} + 16kX = 0;\quad \omega_n = 4\sqrt{\frac{k}{m_1 + 4m_2}}$$

3.39

$$(m_1 + 0.25m_2)\ddot{X} + 0.25c\dot{X} + kX = m_1 g \sin\alpha - 0.5m_2 g;$$

$$\omega_n = \sqrt{\frac{k}{m_1 + 0.25m_2}}$$

3.40

$$(m_1 + 2.25m_2)\ddot{X} + 2.25c\dot{X} + kX = 0;$$

$$\omega_n = \sqrt{\frac{k}{m_1 + 2.25m_2}}$$

3.41

$$(m_1 + m_2)\ddot{X}_1 + c\dot{X}_1 + 4kX_1 = 0;$$

$$\omega_n = 2\sqrt{\frac{k}{m_1 + m_2}};\quad \zeta = \frac{c}{4\sqrt{(m_1 + m_2)k}}$$

3.42

$$\begin{bmatrix} m_1 & 0 & 0 \\ 0 & m_2 & 0 \\ 0 & 0 & m_3 \end{bmatrix}\begin{Bmatrix} \ddot{X}_1 \\ \ddot{X}_2 \\ \ddot{X}_3 \end{Bmatrix} + \begin{bmatrix} k_1 & -k_1 & 0 \\ -k_1 & k_1 + k_2 & -k_2 \\ 0 & -k_2 & k_2 \end{bmatrix}\begin{Bmatrix} X_1 \\ X_2 \\ X_3 \end{Bmatrix} = \begin{Bmatrix} 0 \\ 0 \\ 0 \end{Bmatrix}$$

3.43

(a) $v1' = -5$ m/s, $v2' = 20$ m/s, $v3' = 0$;

(b) $X_1(t) = -5t$ m,
$X_2(t) = 1.333t + 0.385 \sin(17.32t)$ m;
$X_3(t) = 1.333t - 0.769 \sin(17.32t)$ m

(b) $v1' = 0.25$ m/s, $v2' = 16.5$ m/s, $v3' = 0$

(b) $X_1(t) = 0.25t$ m,
$X_2(t) = 10.997t + 0.317 \sin(17.32t)$ m;
$X_3(t) = 10.997t - 0.635 \sin(17.32t)$ m

(c) $v1' = 10$ m/s, $v2' = 10$ m/s, $v3' = 0$

(b) $X_1(t) = X_2(t) = 7.69t + 0.1431 \sin(16.13t)$ m;
$X_3(t) = -7.69t + 0.476 \sin(16.13t)$ m

3.44

$$\begin{bmatrix} m_1 & 0 & 0 & 0 \\ 0 & m_2 & 0 & 0 \\ 0 & 0 & m_3 & 0 \\ 0 & 0 & 0 & m_4 \end{bmatrix}\begin{Bmatrix} \ddot{X}_1 \\ \ddot{X}_2 \\ \ddot{X}_3 \\ \ddot{X}_4 \end{Bmatrix} + \begin{bmatrix} k_1 & -k_1 & 0 & 0 \\ -k_1 & k_1 + k_2 + k_3 & -k_2 & -k_3 \\ 0 & -k_2 & k_2 + k_4 & 0 \\ 0 & -k_3 & 0 & k_3 + k_5 \end{bmatrix}\begin{Bmatrix} X_1 \\ X_2 \\ X_3 \\ X_4 \end{Bmatrix} = \begin{Bmatrix} 0 \\ f_0 \sin 2\omega t \\ 0 \\ k_5 A \cos\omega t \end{Bmatrix}$$

3.45

$$\begin{bmatrix} m_1 & 0 & 0 \\ 0 & m_2 & 0 \\ 0 & 0 & m_3 \end{bmatrix}\begin{Bmatrix} \ddot{X}_1 \\ \ddot{X}_2 \\ \ddot{X}_3 \end{Bmatrix} + \begin{bmatrix} k_1+k_2+k_5 & -k_2 & -k_5 \\ -k_2 & k_2+k_3 & -k_3 \\ -k_5 & -k_3 & k_3+k_4+k_5 \end{bmatrix}\begin{Bmatrix} X_1 \\ X_2 \\ X_3 \end{Bmatrix} = \begin{Bmatrix} k_1 A \cos\omega t + f_1(t) \\ f_2(t) \\ f_3(t) \end{Bmatrix}$$

3.46

$$\begin{bmatrix} m_1 & 0 & 0 \\ 0 & m_2 & 0 \\ 0 & 0 & m_3 \end{bmatrix}\begin{Bmatrix} \ddot{X}_1 \\ \ddot{X}_2 \\ \ddot{X}_3 \end{Bmatrix} + \begin{bmatrix} c_1 & -c_1 & 0 \\ -c_1 & c_1+c_2 & -c_2 \\ 0 & -c_2 & c_2 \end{bmatrix}\begin{Bmatrix} \dot{X}_1 \\ \dot{X}_2 \\ \dot{X}_3 \end{Bmatrix} + \begin{bmatrix} k_1 & -k_1 & 0 \\ -k_1 & k_1+k_2 & -k_2 \\ 0 & -k_2 & k_2 \end{bmatrix}\begin{Bmatrix} X_1 \\ X_2 \\ X_3 \end{Bmatrix} = \begin{Bmatrix} 0 \\ 0 \\ 0 \end{Bmatrix}$$

3.47

$$\begin{bmatrix} m_1 & 0 & 0 \\ 0 & m_2 & 0 \\ 0 & 0 & m_3 \end{bmatrix}\begin{Bmatrix} \ddot{X}_1 \\ \ddot{X}_2 \\ \ddot{X}_3 \end{Bmatrix} + \begin{bmatrix} c_1+c_2+c_5 & -c_2 & -c_5 \\ -c_2 & c_2 & 0 \\ -c_5 & 0 & c_4+c_5 \end{bmatrix}\begin{Bmatrix} \dot{X}_1 \\ \dot{X}_2 \\ \dot{X}_3 \end{Bmatrix} + \begin{bmatrix} k_1+k_2+k_5 & -k_2 & -k_5 \\ -k_2 & k_2+k_3 & -k_3 \\ -k_5 & -k_3 & k_3+k_4+k_5 \end{bmatrix}\begin{Bmatrix} X_1 \\ X_2 \\ X_3 \end{Bmatrix} = \begin{Bmatrix} k_1 A \cos\omega t - c_1 A\omega \sin\omega t + f_1(t) \\ f_2(t) \\ f_3(t) \end{Bmatrix}$$

3.48

$$\begin{bmatrix} m_1 & 0 & 0 & 0 \\ 0 & m_2 & 0 & 0 \\ 0 & 0 & m_3 & 0 \\ 0 & 0 & 0 & m_4 \end{bmatrix}\begin{Bmatrix} \ddot{X}_1 \\ \ddot{X}_2 \\ \ddot{X}_3 \\ \ddot{X}_4 \end{Bmatrix} + \begin{bmatrix} c_1 & -c_1 & 0 & 0 \\ -c_1 & c_1+c_2+c_3 & -c_2 & -c_3 \\ 0 & -c_2 & c_2+c_4 & 0 \\ 0 & -c_3 & 0 & c_3+c_5 \end{bmatrix}\begin{Bmatrix} \dot{X}_1 \\ \dot{X}_2 \\ \dot{X}_3 \\ \dot{X}_4 \end{Bmatrix} + \begin{bmatrix} k_1 & -k_1 & 0 & 0 \\ -k_1 & k_1+k_2+k_3 & -k_2 & -k_3 \\ 0 & -k_2 & k_2+k_4 & 0 \\ 0 & -k_3 & 0 & k_3+k_5 \end{bmatrix}\begin{Bmatrix} X_1 \\ X_2 \\ X_3 \\ X_4 \end{Bmatrix} = \begin{Bmatrix} 0 \\ -f_0 \sin 2\omega t \\ 0 \\ k_5 D\cos\omega t - c_5 D\omega\sin\omega t \end{Bmatrix}$$

3.49

(a) $\lambda = \{2597 \quad 15403\}^T (\text{rad/s})^2$;

$f_n = \{8.11 \quad 19.75\}^T$ Hz;

(b) $A = \begin{bmatrix} 1 & 1 \\ 1.175 & -0.4254 \end{bmatrix}$;

(c) $A^* = \begin{bmatrix} 0.4611 & 0.7664 \\ 0.5419 & -0.3260 \end{bmatrix}$

(d) $\ddot{q}_1 + 2597 q_1 = 0$; $\ddot{q}_2 + 15403 q_2 = 0$;

(e) $\{q(t)\} = \begin{Bmatrix} 0.1304 \cos(50.96t) \\ -0.0132 \cos(124.1t) \end{Bmatrix}$

(f) $\{x(t)\} = \begin{Bmatrix} 0.06012 \cos(50.96t) - 0.01012 \cos(124.1t) \\ 0.07067 \cos(50.96t) + 0.004304 \cos(124.1t) \end{Bmatrix}$ m

3.50

(a) $\lambda = \begin{bmatrix} 7.984 \\ 36.16 \end{bmatrix}$ rad^2/s^2 $f_n = \begin{bmatrix} 0.45 \\ 0.957 \end{bmatrix}$ Hz;

(b) $A = \begin{bmatrix} 1 & 1 \\ 1.372 & -4.372 \end{bmatrix}$;

(c) $A^* = \begin{bmatrix} 0.9087 & 1.0836 \\ 1.2468 & -4.7376 \end{bmatrix}$

(d) $\ddot{q}_1 + 7.984 q_1 = 0$ and $\ddot{q}_2 + 36.161 q_2 = 0$

(e) $q_1(t) = 0.01672\cos(2.83t)$

and $q_2(t) = -0.01402\cos(6.013t)$

(f)
$$\theta_1(t) = 0.01519\cos(2.83t) - 0.01519\cos(6.013t)\ \text{rad}$$
$$\theta_2(t) = 0.02085\cos(2.83t) + 0.06642\cos(6.013t)\ \text{rad}$$

3.51

(a) $\lambda = \begin{bmatrix} 1222 \\ 2469 \end{bmatrix} \text{rad}^2/\text{s}^2 \quad f_n = \begin{bmatrix} 5.65 \\ 7.912 \end{bmatrix}$ Hz;

(b) $A = \begin{bmatrix} 1 & 1 \\ -0.1715 & 0.1924 \end{bmatrix}$;

(c) $A^* = \begin{bmatrix} 2.632 & 1.4885 \\ -0.4514 & 0.2863 \end{bmatrix}$

3.52

(a) $\lambda = \begin{bmatrix} 0 \\ 58.33 \\ 150 \end{bmatrix} \text{rad}^2/\text{s}^2; \quad f_n = \begin{bmatrix} 0 \\ 1.216 \\ 1.949 \end{bmatrix}$ Hz,

$$A = \begin{bmatrix} 1 & 1 & 1 \\ 1 & -0.1665 & -2 \\ 1 & -0.75 & 2 \end{bmatrix};$$

(b) $q_1(t) = 0.12284 + 3.1525t - 1.3207\sin(2.307t)$

$$q_2(t) = 0.3266\sin(2.307t) - 0.03489\cos(7.637t) - 0.1201\sin(7.637t)$$
$$q_3(t) = 1.4967\sin(2.307t) - 0.07462\cos(12.247t) - 0.2918\sin(12.247t)$$

(c)

$$\begin{Bmatrix} X_1(t) \\ X_2(t) \\ X_3(t) \end{Bmatrix} = \begin{Bmatrix} 0.04643 + 1.192t - 0.01948\cos(7.638t) - 0.057\sin(7.638t) - 0.01591\cos(12.25t) - 0.0622\sin(12.25t) + 0.00226\sin(2.39t) \\ 0.04643 + 1.192t + 0.00325\cos(7.638t) - 0.0095\sin(7.638t) + 0.0318\cos(12.25t) + 0.1244\sin(12.25t) - 1.168\sin(2.39t) \\ 0.04643 + 1.192t + 0.01461\cos(7.638t) + 0.0428\sin(7.638t) - 0.0318\cos(12.25t) - 0.1244\sin(12.25t) + 0.00226\sin(2.39t) \end{Bmatrix} \text{m}$$

3.53

$$\ddot{q}_1 + 1.711\dot{q}_1 + 292.9q_1 = 1000t^2$$
$$\ddot{q}_2 + 4.132\dot{q}_2 + 1701q_2 = 1000t^2$$

$$q_1(t) = e^{-0.856t}[0.0231\cos(17.09t) + 0.00349\sin(17.09t)] + 3.414t^2 - 0.0399t - 0.0231$$

$$q_2(t) = e^{-2.07t}[6.749\cos(41.27t) + 1.027\sin(41.27t)]10^{-4} + 0.586t^2 - 2.836(10^{-3})t - 6.79(10^{-4})$$

$$x_1(t) = 0.325[e^{-1.259t}(3.202\cos(25.15t) + 0.4841\sin(25.15t))10^{-3}] + 0.3667t^2 - 1.678t - 0.001043\ \text{m}$$

$$x_2(t) = 0.444[e^{-1.259t}(3.202\cos(25.15t) + 0.4841\sin(25.15t))10^{-3}] + 0.443t^2 - 2.292t - 0.001421\ \text{m}$$

3.54

$$\begin{Bmatrix} x_1(t) \\ x_2(t) \end{Bmatrix} = \begin{Bmatrix} 0.485\cos(0.662t) - 0.486\cos(2.14t) \\ 8.649\cos(0.662t) + 1.364\cos(2.14t) \end{Bmatrix} \text{mm}$$

3.55

$$\begin{Bmatrix} x_1(t) \\ x_2(t) \end{Bmatrix} = \begin{Bmatrix} 6.67t - 3.85\sin(1.732t) \\ 6.67t + 1.925\sin(1.732t) \end{Bmatrix} \text{mm}$$

3.56

$$\begin{Bmatrix} x_1(t) \\ x_2(t) \end{Bmatrix} = \begin{Bmatrix} -0.6213\cos(0.662t) - 0.3788\cos(2.14t) + 1 \\ -1.105\cos(0.662t) + 0.1062\cos(2.14t) + 1 \end{Bmatrix} \text{mm}$$

3.57

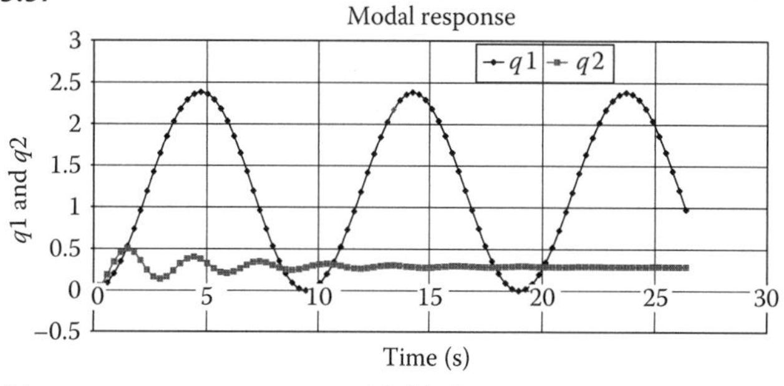

(a) Modal plots

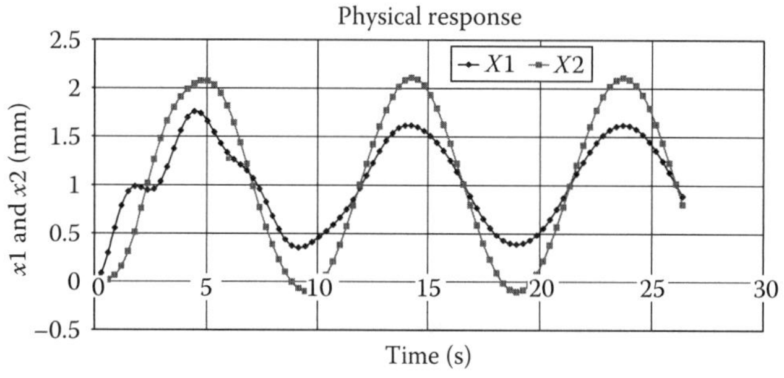

(b) Physical plots

$q_1(t) = -1.1905\cos(0.662t) + 1.1905$ and

$$q_2(t) = e^{-0.214t}[-0.2882\cos(2.13t) - 0.02896\sin(2.13t)] + 0.2882$$

$$\begin{Bmatrix} x_1(t) \\ x_2(t) \end{Bmatrix} = \begin{Bmatrix} -0.6213\cos(0.662t) - e^{-0.214t}[0.3788\cos(2.13t) + 0.03806\sin(2.13t)] + 1 \\ -1.105\cos(0.662t) + e^{-0.214t}[0.1062\cos(2.13t) + 0.01113\sin(2.13t)] + 1 \end{Bmatrix} \text{mm}$$

3.58

$$q_1(t) = e^{-0.0993t}[-1.1905\cos(0.655t) - 0.1805\sin(0.655t)] + 1.1905$$

$$q_2(t) = e^{-0.321t}[-0.2882\cos(2.12t) - 0.04364\sin(2.12t)] + 0.2882$$

$$\begin{Bmatrix} x_1(t) \\ x_2(t) \end{Bmatrix} = \begin{Bmatrix} e^{-0.0993t}[-0.6213\cos(0.655t) - 0.0942\sin(0.655t)] - e^{-0.321t}[0.3788\cos(2.12t) + 0.05735\sin(2.12t)] + 1 \\ e^{-0.0993t}[-1.105\cos(0.655t) - 0.1679\sin(0.655t)] + e^{-0.321t}[0.1062\cos(2.12t) + 0.01609\sin(2.12t)] + 1 \end{Bmatrix} \text{mm}$$

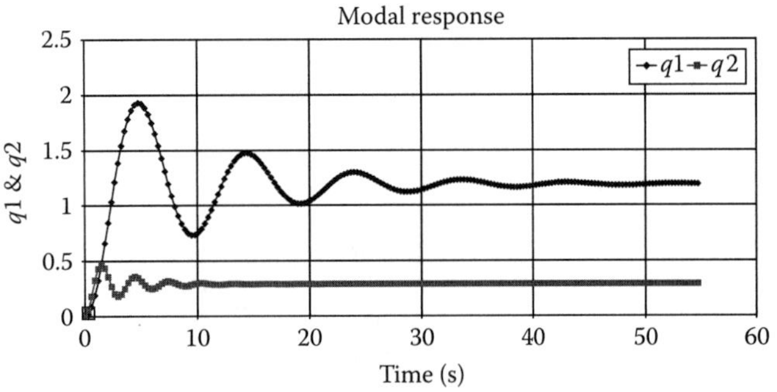

(a) Modal plots

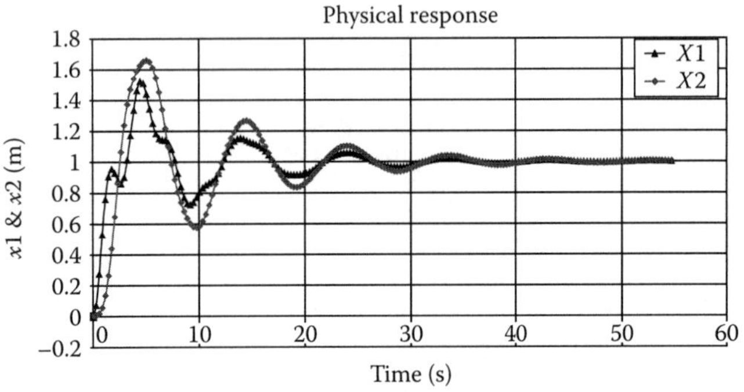

(b) Physical plots

3.59

(a) $\lambda = \begin{bmatrix} 0.382 \\ 2.618 \end{bmatrix}$ rad²/s² $f_n = \begin{bmatrix} 0.0984 \\ 0.258 \end{bmatrix}$ Hz,

(b) $A = \begin{bmatrix} 1 & 1 \\ 1.618 & -0.618 \end{bmatrix}$,

(c) $A^* = \begin{bmatrix} 0.3718 & 0.6015 \\ 0.6015 & -0.3717 \end{bmatrix}$

(d) $\ddot{q}_1 + 0.382q_1 = 0.3718f_1(t) + 0.6015f_2(t)$
$\ddot{q}_2 + 2.618q_2 = 0.6015f_1(t) - 0.3718f_2(t)$,

(e) $q_1(t) = 1.267\cos(0.618t) - 1.140\cos(2t)$
$q_2(t) = -0.03504\cos(1.618t) + 0.0394\cos(2t)$

(f) $\begin{Bmatrix} x_1(t) \\ x_2(t) \end{Bmatrix}$

$$= \begin{Bmatrix} 0.4711\cos(0.618t) - 0.0211\cos(1.618t) - 0.4\cos(2t) \\ 0.7621\cos(0.618t) + 0.01303\cos(1.618t) - 0.7\cos(2t) \end{Bmatrix} \text{m}$$

3.60 (*Note*: Both masses assumed to move to the right)

(a)

$$\begin{bmatrix} m_1 & 0 \\ 0 & m_2 \end{bmatrix}\begin{Bmatrix} \ddot{X}_1 \\ \ddot{X}_2 \end{Bmatrix} + \begin{bmatrix} k_1 + k_2 + k_4 & -k_2 \\ -k_2 & k_2 + k_3 \end{bmatrix}\begin{Bmatrix} X_1 \\ X_2 \end{Bmatrix} = \begin{Bmatrix} 0 \\ -f(t) \end{Bmatrix};$$

(b) $\lambda = \begin{bmatrix} 2454 \\ 4791 \end{bmatrix}$ rad²/s² $f_n = \begin{bmatrix} 7.922 \\ 11.03 \end{bmatrix}$ Hz,

(c) $A = \begin{bmatrix} 1 & 1 \\ 0.7656 & -3.265 \end{bmatrix}$,

(e) $q_1(t) = 0.1657\cos(49.54t) + 0.001270(188.5t)$
$q_2(t) = 0.0272\cos(69.21t) - 0.002877(188.5t)$;

(f)

$$\begin{Bmatrix} X_1(t) \\ X_2(t) \end{Bmatrix} = \begin{Bmatrix} 0.927\cos(49.54t) + 0.07369\cos(69.21t) - 0.000689\cos(188.5t) \\ 0.7097\cos(49.54t) - 0.2406\cos(69.21t) + 0.03089\cos(188.5t) \end{Bmatrix} \text{in.}$$

3.61 (*Note*: Both masses assumed to move to the left)

(a)

$$\begin{bmatrix} m_1 & 0 \\ 0 & m_2 \end{bmatrix}\begin{Bmatrix} \ddot{X}_1 \\ \ddot{X}_2 \end{Bmatrix} + \begin{bmatrix} k_1 + k_2 + k_3 + k_4 + k_5 & -(k_2 + k_3) \\ -(k_2 + k_3) & k_2 + k_3 \end{bmatrix}\begin{Bmatrix} X_1 \\ X_2 \end{Bmatrix} = \begin{Bmatrix} k_1X_0(t \\ 0 \end{Bmatrix};$$

(b) $\lambda = \begin{bmatrix} 7.5 \\ 20 \end{bmatrix}$ rad²/s²; $f_n = \begin{bmatrix} 0.4359 \\ 0.7118 \end{bmatrix}$ Hz,

(c) $A = \begin{bmatrix} 1 & 1 \\ 2.667 & -1.5 \end{bmatrix}$;

(d) $\ddot{q}_1 + 7.5q_1 = 4.5\cos(62.9t)$
$\ddot{q}_2 + 20q_2 = 6\cos(62.9t)$

(e) $\begin{Bmatrix} x_1(t) \\ x_2(t) \end{Bmatrix}$

$$= \begin{Bmatrix} 0.0849 \cos(2.74t) + 0.0119 \cos(4.47t) - 9.296(10^{-2}) \cos(62.9t) \\ 0.1133 \cos(2.74t) - 0.0229 \cos(4.47t) - 9.12(10^{-2}) \cos(62.9t) \end{Bmatrix} \text{mm}$$

3.62

$$\dot{X}_{11} = \sqrt{2gX_{10} \sin\alpha}; \quad \dot{X}_{22} = \dot{X}_{11}\frac{(1+e)m_1}{m_1+m_2};$$

$$X_2(t) = e^{-\zeta\omega_n t}\left(\frac{-w_2 \sin\alpha}{k} \cos\omega_d t + \frac{\dot{X}_{22}}{\omega_d} \sin\omega_d t\right) + \frac{w_2 \sin\alpha}{k}$$

$f_d = 2.24\,\text{Hz} \quad \zeta = 0.106;$

$\dot{X}_{21} = -0.315\ \text{m/s} \quad \dot{X}_{22} = 2.55\ \frac{\text{m}}{\text{s}}; \quad \Delta_{max} = 0.185\ \text{m}$

3.63

$$X(t) = e^{-\zeta\omega_n t}\left(\frac{v_0}{\omega_n} \sin\omega_n t\right); \; X(t) = 3.411e^{-3.868ct} \sin(87.91t)$$

3.64

$$\ddot{\theta} - \sin^4\theta_0 \dot{\phi}_0^2 \frac{\cos\theta}{\sin\theta} = -\frac{g \sin\theta}{R}$$

3.65

$$\ddot{S} - \sin^2\alpha\left(\frac{S_0^4}{S^3}\right)\dot{\theta}_0^2 = -g \cos\alpha$$

3.66

$$(m+M)\ddot{S} - m\sin^2\alpha\left(\frac{S_0^4}{S^3}\right) = mg\cos\alpha - Mg$$

Chapter 4

Some useful trigonometry identities:

$$\sin(A \pm B) = \sin A \cos B \pm \cos A \sin B;$$
$$\cos(A \pm B) = \cos A \cos B \mp \sin A \sin B;$$
$$\sin 2A = 2 \sin A \cos A;$$
$$\cos 2A = \cos^2 A - \sin^2 A$$

4.1

(a) $\boldsymbol{\omega} = 247.98\ \mathbf{K}$ rad/s;

(b) $v_P = \mathbf{I}(-15) - \mathbf{J}(6.26)$ m/s;

(c) $\boldsymbol{\alpha} = -395\ \mathbf{K}$ rad/s^2;

(d) $\mathbf{a}_P = (8167)\ \mathbf{I} - (48.65)\mathbf{J}$ m/s^2

4.2

(a) $v_P = (-78.07)$ I – (104.7) J ft/s;

(b) $a_P = (6585)$ I – (4281.1) J ft/s^2

4.3

$\mathbf{v}_A = (5.12)\mathbf{I} + (2.12)\mathbf{J}\ \frac{\text{m}}{\text{s}}; \quad \mathbf{v}_B = (3)\mathbf{I} - (1.5)\mathbf{J}$ m/s;

$\mathbf{a}_A = (150.6)\mathbf{I} + (-162.1)\mathbf{J}$ m/s^2;

$\mathbf{a}_B = [(-117.5)\mathbf{I} + (2.5)\mathbf{J}]$ m/s^2

4.4

$$\mathbf{v}_E = (v_O + \omega R \sin\phi)\mathbf{I} + (\omega R \cos\phi)\mathbf{J};$$
$$\mathbf{a}_E = (a_O + R\alpha \sin\phi + R\omega^2 \cos\phi)\mathbf{I} + (R\alpha \cos\phi - R\omega^2 \sin\phi)\mathbf{J}$$

4.5

$\boldsymbol{\omega} = -1\ \mathbf{K}$ rad/s; $\mathbf{v}_D = 1.5\,\mathbf{I}$ m/s; $\mathbf{v}_B = (0.5\,\mathbf{I} - 1\,\mathbf{J})$ m/s;

$\mathbf{v}_A = -0.5\mathbf{I}$ m/s; $\mathbf{a}_D = (7.35\,\mathbf{I} - 1\,\mathbf{J})$ m/s^2;

$\mathbf{a}_B = (1.45\,\mathbf{I} - 4.9\,\mathbf{J})$ m/s^2; $\mathbf{a}_A = -2.45\,\mathbf{I} + 1\,\mathbf{J}$ m/s^2

4.6

$\mathbf{v}_O = 1.565\ \mathbf{I}$ m/s; $\mathbf{v}_A = -1.565\ \mathbf{I}$ m/s;

$\mathbf{a}_O = -4.905\ \mathbf{I}$ m/s^2; $\mathbf{a}_A = (4.905\ \mathbf{I} + 9.797\ \mathbf{J})$ m/s^2

4.7

$\mathbf{v}_G = -2.11\ \mathbf{J}$ ft/s; $\mathbf{v}_B = (2.18\ \mathbf{I} - 4.296\ \mathbf{J})$ ft/s;

$\mathbf{a}_G = -2.76\ \mathbf{J}$ ft/s^2; $\mathbf{a}_B = (-4.24\ \mathbf{I} - 12.71\ \mathbf{J})$ ft/s^2

4.8

(a) $\boldsymbol{\omega} = -3.333\,\mathbf{K}$ rad/s; (b) $\boldsymbol{\alpha} = -0.111\,\mathbf{K}$ rad/s^2;

(c) $v_O = 10$ ft/s;

(d) $\mathbf{v}_B = (15.07\ \mathbf{I} - 7.075\ \mathbf{J})$ ft/s,

$\mathbf{a}_B = (-23.47\ \mathbf{I} - 23.36\ \mathbf{J})$ ft/s^2

4.9

(a) $\omega_O = \frac{v_G}{2r}, \quad \omega_D = \frac{v_G}{2R},$ (b) $\mathbf{v_E} = -1\mathbf{I} + 1\mathbf{J}$ ft/s,

(c) $\boldsymbol{\alpha}_\mathbf{D} = -0.24\ \mathbf{K}$ rad/s^2

4.10

(a) $\omega = \frac{v_a - v}{r+R};$

(b) $\mathbf{v}_O = \mathbf{I}\left(v_A - (v_A - v)\frac{R}{r+R}\right)$ (taken wrt pt C) or,

$\mathbf{v}_O = \mathbf{I}\left(v + \frac{r}{R+r}(v_A - v)\right)$ (taken wrt pt E);

(c) $\mathbf{v}_B = \left[v_A + (v_A - v)\frac{R}{r+R}(\sin\theta - 1)\right]\mathbf{I} + \left[(v_A - v)\frac{R}{r+R}\cos\phi\right]\mathbf{J}$ taken wrt pt C.

(e) $\alpha = \frac{-a_A}{R+r}$

4.11 $\mathbf{v}_P = \dfrac{(v_B + v_A)R}{R+r}\mathbf{I} + \left(v_A + \dfrac{-(v_B+v_A)R}{R+r}\right)\mathbf{J};$

$$\mathbf{a}_P = \left(\frac{(a_B + a_A)R}{R+r}\right)\mathbf{I} + \left(a_A + \frac{-(a_B + a_A)R}{R+r} - \left(\frac{(v_B+v_A)}{R+r}\right)^2 R\right)\mathbf{J}$$

4.12

$\boldsymbol{\omega} = -4.94\,\mathbf{K}$ rad/s; $\boldsymbol{\alpha} = -23.53\,\mathbf{K}$ rad/s^2;

$\mathbf{a}_B = (-5\mathbf{I} + 10.19\mathbf{J})$ m/s^2

4.13

$\boldsymbol{\omega} = -2\ \mathbf{K}$ rad/s; $\boldsymbol{\alpha} = 8\,\mathbf{K}$ rad/s^2;

$v_{\mathrm{A}} = 0$; $a_{\mathrm{Ax}} = 0$; $\mathbf{a}_D = (2\mathbf{I} - 4\,\mathbf{J})$ m/s^2

4.14

$\mathbf{a}_C = (-1.577\,\mathbf{I} + 30.67\,\mathbf{J})$ in./s^2;

$\mathbf{a}_L = -2.664\,\mathbf{J}$ in./s^2; $\mathbf{a}_D = (-33.34\,\mathbf{I} - 4.241\,\mathbf{J})$ in./s^2

4.15

(a) $\boldsymbol{\omega} = -\left(\dfrac{v}{R+r}\right)\mathbf{K}$,

(b) $\mathbf{v}_B = [v - \omega(r + R\sin\phi)\,\mathbf{I} + (-\omega R\cos\phi)\,\mathbf{J}]$,

(d) $\dot{\boldsymbol{\omega}} = \left(\dfrac{a_O}{R}\right)\mathbf{K}$, $\mathbf{a}_C = \left[a_O\left(1 - \frac{r}{R}\right)\right]\mathbf{I} - r\omega^2\,\mathbf{J}$

4.16

(a) $\theta = \dfrac{X}{r}, \dot{\theta} = \dfrac{v_f}{r}\sqrt{\dfrac{X}{X_f}}$ (a_T: acceleration of truck, X_f: final position of truck, v_f: final velocity of truck), (b) $\theta = 57.3$ revolutions, $\omega = 1471$ rpm.

4.17

(a) $\omega_O = \dfrac{b\omega_1}{R}$, (b) $v_O = 0$, (c) $\mathbf{v_C} = -b\omega_1\mathbf{J}$

4.18

(a) $\omega_S = \dfrac{d}{D}\omega_m$; (b) $D = 29.57$ in.;

(c) $\mathbf{a}_A = (-27{,}740\mathbf{e}_r + 81.6\mathbf{e}_\theta)$ in./s^2

4.19

(a) $\boldsymbol{\omega} = 6000\,\mathbf{K}$ rpm; (b) $v_C = 1727$ in./s;

(c) $\mathbf{a}_C = (275t^4\,\mathbf{I} + 55t\,\mathbf{J})$ in./s^2 and $\mathbf{a}_C(0.75) = 87\,\mathbf{I} - 41.25\,\mathbf{J}$ in./s^2

4.20

(a) $\boldsymbol{\omega}_A = \frac{R\omega_O}{2r}\,\mathbf{K}$; (b) $\boldsymbol{\omega}_{OA} = \frac{R\omega_O}{2(R-r)}\,\mathbf{K}$;

(c) $v_A = \frac{R\omega_O}{2} = (R - r)\omega_{OA}$

For kinematic diagrams used for mechanisms, see respective figures.

4.21

(a) $v_B = 344.4$ in./s; (b) $a_B = 212087$ in./s^2;

(c) $\mathbf{v}_G = (-313.6\,\mathbf{I} + 244.9\,\mathbf{J})$ in./s;

$\mathbf{a}_G = (-198{,}372\,\mathbf{I} - 53{,}296\,\mathbf{J})$ in./s^2

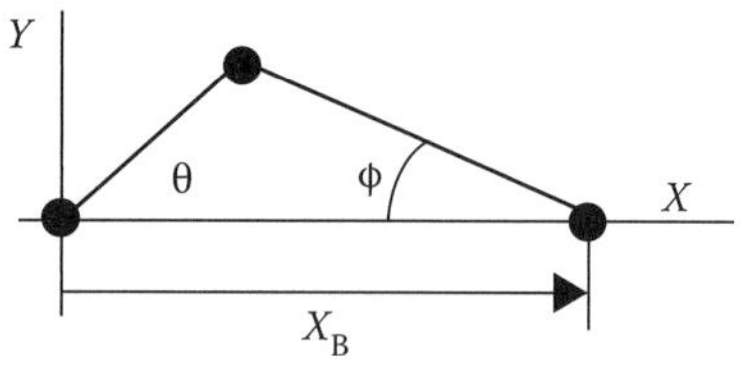

4.22

$$\dot{\theta} = \frac{\dot{X}\cos\beta}{L_1\sin(\theta+\beta)} \quad \dot{\beta} = \frac{\dot{X}\cos\theta}{L_3\sin(\theta+\beta)};$$

$$\ddot{\beta} = \frac{-L_1\dot{\theta}^2}{L_3\sin(\theta+\beta)} - \frac{\dot{\beta}^2\cos(\beta+\theta)}{\sin(\theta+\beta)} + \frac{\ddot{X}\cos\theta}{L_3\sin(\theta+\beta)}$$

$$\ddot{\theta} = \frac{\ddot{X}\cos\beta}{L_1\sin(\theta+\beta)} - \frac{\dot{\theta}^2}{\tan(\theta+\beta)} - \frac{L_3\dot{\beta}^2}{L_1\sin(\theta+\beta)};$$

$$\mathbf{v}_C = [\dot{X} - (L_1 + L_2)\dot{\theta}\sin\theta]\,\mathbf{I} + [(L_1 + L_2)\dot{\theta}\cos\theta]\,\mathbf{J}$$

$$\mathbf{a}_C = [\ddot{X} - (L_1 + L_2)\ddot{\theta}\sin\theta - \dot{\theta}^2(L_1 + L_2)\cos\theta]\,\mathbf{I} + [(L_1 + L_2)\ddot{\theta}\cos\theta - \dot{\theta}^2(L_1 + L_2)\sin\theta]\,\mathbf{J}$$

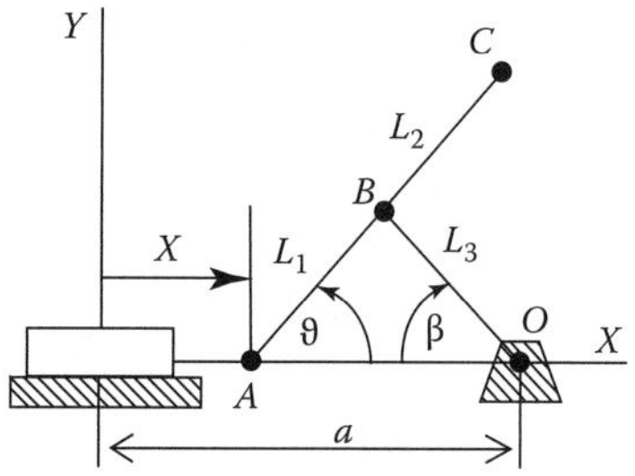

4.23

$$\dot{S} = r\dot{\theta}\sin(\theta+\phi)\ \dot{\phi} = \frac{r\dot{\theta}}{S}\cos(\theta+\phi);$$

$$\ddot{S} = r\ddot{\theta}\sin(\theta+\phi) + r\dot{\theta}^2\cos(\theta+\phi) + S\dot{\phi}^2;$$

$$\ddot{\phi} = \frac{r\ddot{\theta}\cos(\theta+\phi)}{S} - \frac{r\dot{\theta}^2\sin(\theta+\phi)}{S} - \frac{2\dot{S}\dot{\phi}}{S}$$

$\mathbf{v}_C = (-r\dot{\theta}\sin\theta - L\dot{\phi}\sin\phi)\,\mathbf{I} + (r\dot{\theta}\cos\theta - L\dot{\phi}\cos\phi)\,\mathbf{J}$;

$\mathbf{v}_C = (-2.54\ \mathbf{I} + 0.627\ \mathbf{J})$ m/s, $\mathbf{a}_C = (44.3\ \mathbf{I} - 11.09\ \mathbf{J})$ m/s^2

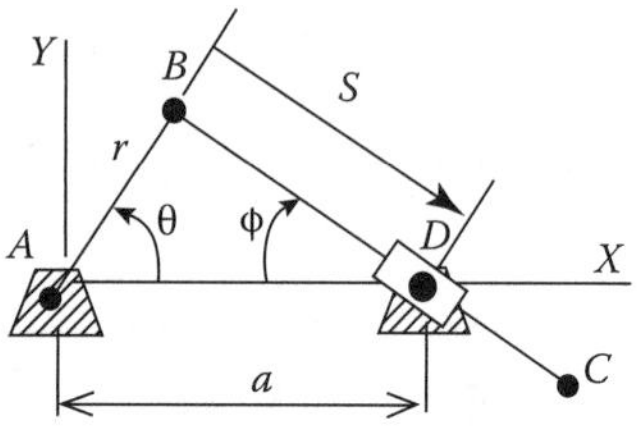

4.24

$$\dot{S} = -L_1\dot{\theta}\cos(\theta-\phi); \dot{\phi} = \frac{-L_1\dot{\theta}}{S}\sin(\theta-\phi);$$

$$\ddot{S} = -L_1\ddot{\theta}\cos(\theta-\phi) + L_1\dot{\theta}^2\sin(\theta-\phi) - \dot{S}\dot{\phi}\sin 2\phi + S\dot{\phi}^2$$

$$\ddot{\phi} = \frac{-L_1\ddot{\theta}\sin(\theta-\phi)}{S} - \frac{L_1\dot{\theta}^2\cos(\theta-\phi)}{S} - \frac{2\dot{S}\dot{\phi}\cos 2\phi}{S};$$

$$\mathbf{v}_B = (L_2\dot{\phi}\sin\phi)\,\mathbf{I} + (L_2\dot{\phi}\cos\phi)\,\mathbf{J}$$

$$\mathbf{a}_B = (L_2\ddot{\phi}\sin\phi + \dot{\phi}^2 L_2\cos\phi)\,\mathbf{I} + (L_2\ddot{\phi}\cos\phi - \dot{\phi}^2 L_2\sin\phi)\,\mathbf{J};$$

$$\mathbf{v}_B = (4.21\ \mathbf{I} + 4.03\ \mathbf{J})\ \text{ft/s},$$

$$\mathbf{a}_B = (-142.8\ \mathbf{I} - 164.8\ \mathbf{J})\ \text{ft/s}^2$$

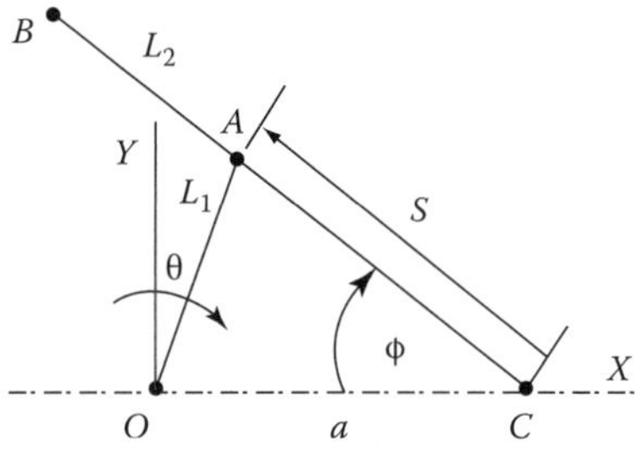

4.25

$$\mathbf{v}_A = \frac{\dot{S}}{b\sin(\beta+\theta)}(-\sin\theta\ \mathbf{I} + \cos\theta\ \mathbf{J})$$

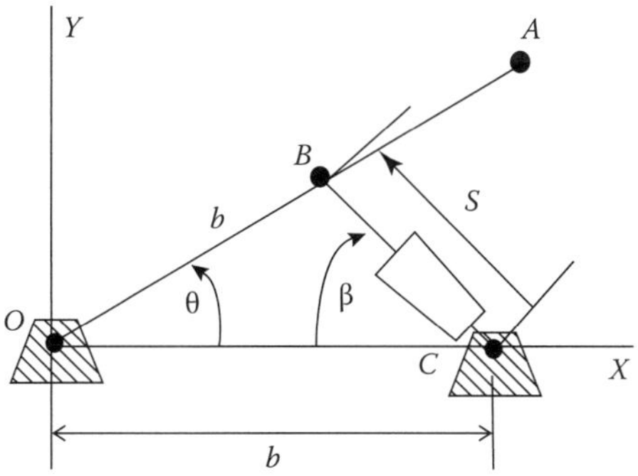

4.26

$$\dot{S} = L_1\dot{\theta}\cos(\phi+\theta); \quad \dot{\phi} = \frac{L_1\dot{\theta}}{S}\sin(\phi+\theta);$$

$$\ddot{S} = L_1\ddot{\theta}\cos(\phi+\theta) - L_1\dot{\theta}^2\sin(\phi+\theta) - L_1\dot{\theta}\dot{\phi}\sin(\phi+\theta)$$

$$\ddot{\phi} = \frac{2\dot{S}\dot{\theta}}{S} + \frac{L_1\dot{\theta}^2}{S}\sin(\phi-\theta);$$

$$\mathbf{v}_B = (L_2\dot{\phi}\cos\phi)\,\mathbf{I} + (L_2\dot{\phi}\sin\phi)\,\mathbf{J};$$

$$\mathbf{a}_B = (L_2\ddot{\phi}\cos\phi - \dot{\phi}^2 L_2\sin\phi)\,\mathbf{I} + (L_2\ddot{\phi}\sin\phi + \dot{\phi}^2 L_2\cos\phi)\,\mathbf{J}$$

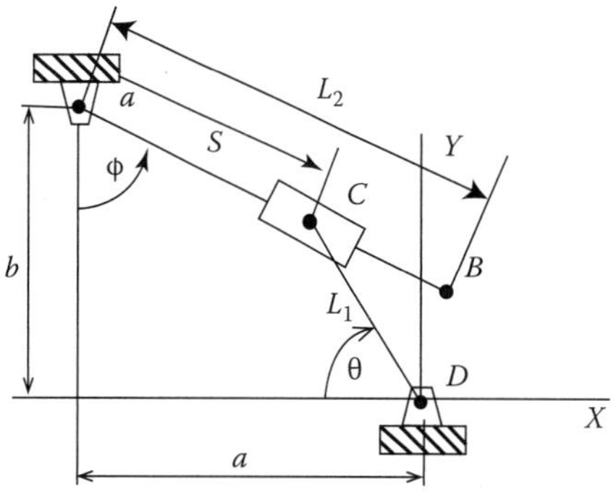

4.27

$$\dot{S} = r\dot{\theta}\sin(\gamma-\theta);$$

$$\dot{\gamma} = \frac{r\dot{\theta}\cos(\theta-\gamma)}{S};$$

$$\ddot{S} = r\ddot{\theta}\sin(\theta-\gamma) - r\dot{\theta}^2\cos(\gamma-\theta) + S\dot{\gamma}^2;$$

$$\ddot{\gamma} = \frac{r\ddot{\theta}\cos(\theta-\gamma) - r\dot{\theta}^2\sin(\theta-\gamma) - 2\dot{S}\dot{\gamma}}{S}$$

$$\mathbf{v}_B = (r\dot{\theta}\sin\theta - L\dot{\gamma}\sin\gamma)\,\mathbf{I} + (r\dot{\theta}\cos\theta - L\dot{\gamma}\cos\gamma)\,\mathbf{J}$$

$$\mathbf{a}_B = (r\ddot{\theta}\sin\theta + r\dot{\theta}^2\cos\theta - L\ddot{\gamma}\sin\gamma - L\dot{\gamma}^2\cos\gamma)\,\mathbf{I} + (r\ddot{\theta}\cos\theta - r\dot{\theta}^2\cos\theta - L\ddot{\gamma}\cos\gamma + L\dot{\gamma}^2\sin\gamma)\,\mathbf{J}$$

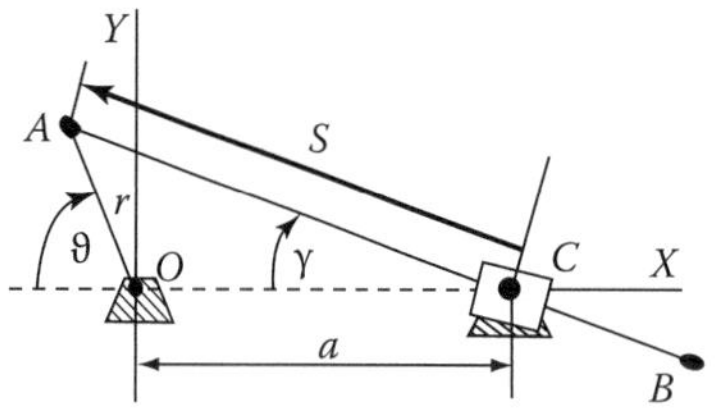

4.28

(a) $\boldsymbol{R} = \boldsymbol{OA}$, (b) $\dot{\theta} = \dfrac{X_0\omega\cos\omega t}{\sqrt{R^2 - X_0^2\sin^2\omega t}}$,

$$\ddot{\theta} = \frac{X_0\omega^2\sin(\omega t)}{\sqrt{R^2 - X_0^2\sin^2(\omega t)}}\left[\frac{X_0^2\cos(\omega t)}{R^2 - X_0^2\sin^2(\omega t)} - 1\right],$$

and

(d) $\mathbf{v}_B = -137.3\mathbf{I} + 96.85\ \mathbf{J}$ in./s,
$\mathbf{a}_B = -4.117\ \mathbf{I} + 114.4\ \mathbf{J}$ in./s^2

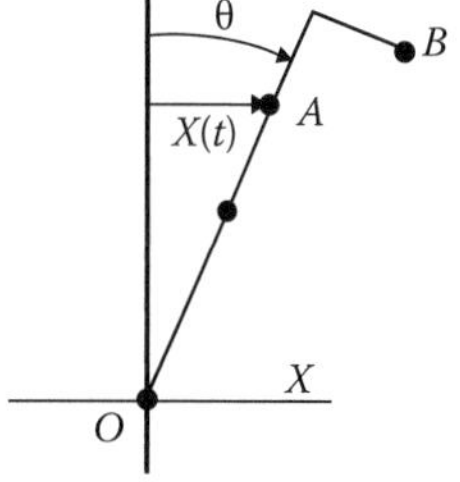

4.29

$$\dot{\beta} = \frac{-L_1\dot{\theta}\,\sin(\theta+\phi)}{L_2\sin(\beta+\phi)};\quad \dot{\phi} = \frac{L_1\dot{\theta}\,\sin(\beta-\theta)}{L_3\sin(\beta+\phi)};$$

$$\ddot{\beta} = \frac{-L_1\dot{\theta}^2\cos(\theta+\phi)}{L_2\sin(\beta+\phi)} - \frac{\dot{\beta}^2\cos(\beta+\phi)}{\sin(\beta+\phi)} - \frac{-L_3\dot{\phi}^2}{L_2\sin(\beta+\phi)}$$

$$\ddot{\phi} = \frac{-L_1\dot{\theta}^2\cos(\theta-\beta)}{L_3\sin(\beta+\phi)} + \frac{-L_2\dot{\beta}^2}{L_3\sin(\beta+\phi)} + \frac{-\dot{\phi}^2\cos(\phi+\beta)}{\sin(\beta+\phi)}$$

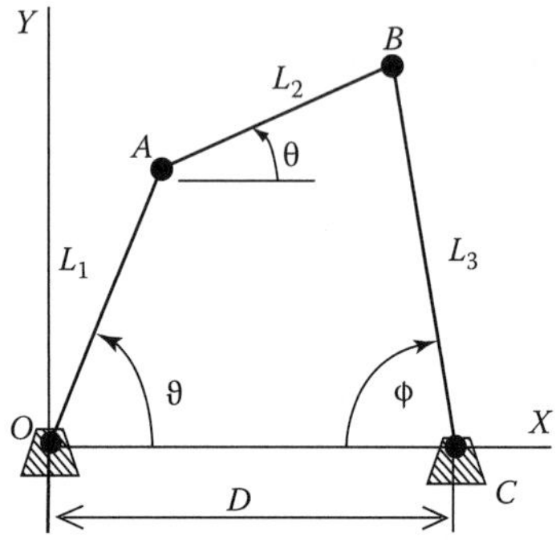

4.30

$$\dot{\beta} = \frac{L_1\cos(\theta-\phi)}{L_2\sin(\beta-\phi)};\quad \dot{\phi} = \frac{-L_1\sin(\beta+\theta)}{L_3\sin(\beta-\phi)};$$

$$\ddot{\beta} = \frac{L_1\dot{\theta}^2\cos(\theta+\phi)}{L_2\sin(\beta-\phi)} - \frac{\dot{\beta}^2\cos(\beta-\phi)}{\sin(\beta-\phi)} + \frac{L_3\dot{\phi}^2(2\sin 2\phi)}{L_2\sin(\beta-\phi)}$$

$$\ddot{\phi} = \frac{-L_1\dot{\theta}^2\sin(\beta+\theta)}{L_3\sin(\beta-\phi)} + \frac{L_2\dot{\beta}^2}{L_3\sin(\beta-\phi)} - \frac{\dot{\phi}^2\sin(\beta+\phi)}{\sin(\beta-\phi)}$$

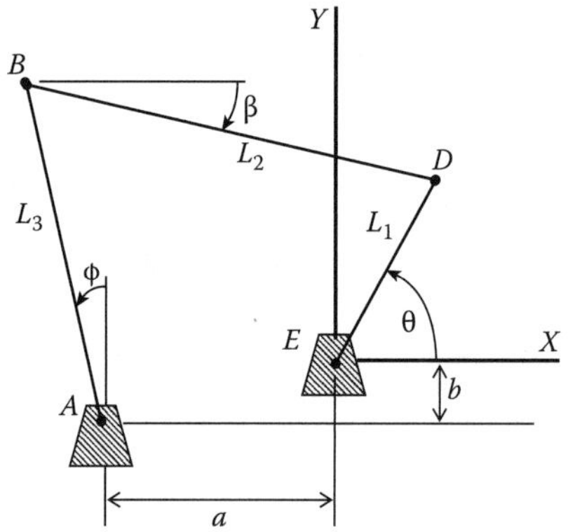

4.31

$$\dot{\beta} = \frac{r\,\sin(\theta+\phi)\omega_0}{L_1\cos(\beta-\phi)};$$

$$\dot{\phi} = \frac{r\,\cos(\theta-\beta)\omega_0}{L_2\cos(\beta-\phi)};$$

$$\ddot{\beta} = \frac{r\omega_0^2\cos(\theta-\phi)}{L_1\cos(\beta-\phi)} + \frac{\dot{\beta}^2\sin(\beta-\phi)}{\cos(\beta-\phi)} - \frac{L_2\dot{\phi}^2}{L_1\cos(\beta-\phi)}$$

$$\ddot{\phi} = \frac{-r\omega_0^2\sin(\beta-\theta)}{L_2\cos(\beta-\phi)} + \frac{L_1\dot{\beta}^2}{L_2\cos(\beta-\phi)} - \frac{L_2\dot{\phi}^2\sin(\beta-\phi)}{L_1\cos(\beta-\phi)}$$

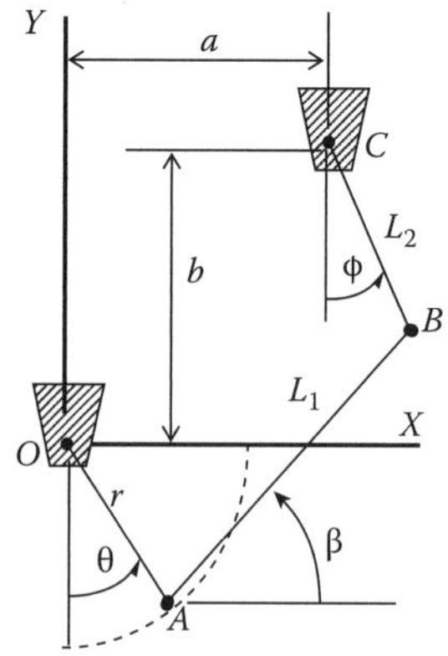

4.32

$$\dot{\beta} = \frac{-l_1\dot{\theta}\cos(\theta+\phi)}{l_2\cos(\beta-\phi)};\quad \dot{\phi} = \frac{l_1\dot{\theta}\sin(\theta+\beta)}{l_3\cos(\beta-\phi)};$$

$$\ddot{\beta} = \frac{l_1\ddot{\theta}\cos(\phi-\theta) + l_1\dot{\theta}^2\sin(\phi+\theta) + l_2\dot{\beta}^2\sin(\phi+\beta) + l_3\dot{\phi}^2\cos 2\phi}{l_2\cos(\beta-\phi)}$$

$$\ddot{\phi} = \frac{l_1\ddot{\theta}\sin(\beta+\theta) + l_1\dot{\theta}^2\cos(\beta+\theta) + l_2\dot{\beta}^2\cos(2\beta) + l_3\dot{\phi}^2\sin(\phi+\beta)}{l_3\cos(\beta-\phi)}$$

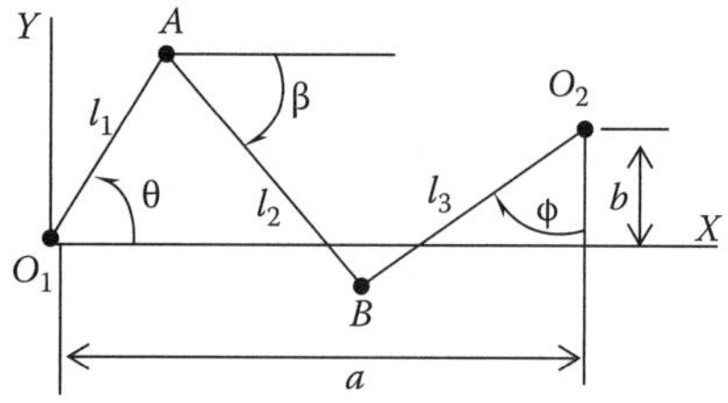

4.33

(b) $\begin{bmatrix} L_2\sin\beta & L_1\sin\theta \\ L_2\cos\beta & L_1\cos\theta \end{bmatrix}\begin{Bmatrix}\dot{\beta} \\ \dot{\theta}\end{Bmatrix} = \begin{bmatrix} 0 \\ \dot{S}\end{bmatrix};$

(c) $\begin{bmatrix} L_2\sin\beta & L_1\sin\theta \\ L_2\cos\beta & L_1\cos\theta \end{bmatrix}\begin{Bmatrix}\ddot{\beta} \\ \ddot{\theta}\end{Bmatrix}$

$$= \begin{bmatrix} -L_2\dot{\beta}^2\cos\beta - L_1\dot{\theta}^2\cos\theta \\ L_2\dot{\beta}^2\sin\beta + L_1\dot{\theta}^2\sin\theta \end{bmatrix};$$

(d) $\mathbf{v}_D = (-L_3\dot{\beta}\,\sin\beta)\,\mathbf{I} + (\dot{S} + L_3\dot{\beta}\,\cos\beta)\,\mathbf{J};$

$\mathbf{a}_D = (-L_3\ddot{\beta}\,\sin\beta - L_3\dot{\beta}^2\cos\beta)\,\mathbf{I} + (L_3\ddot{\beta}\,\cos\beta - L_3\dot{\beta}^2\sin\beta)\,\mathbf{J}$

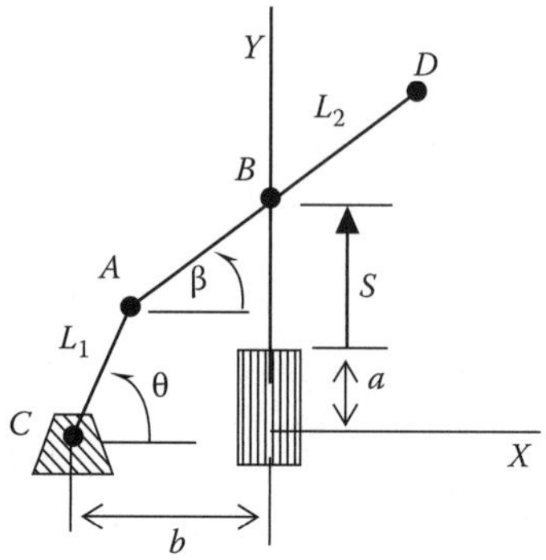

4.34

$$\dot{S} = b\dot{\theta}_1 \sin(\theta_1 - \phi) + a\dot{\theta}_2 \sin(\theta_2 - \phi); \quad \dot{\phi} = \frac{-b\dot{\theta}_1}{S}\cos(\theta_1 - \phi) - \frac{a\dot{\theta}_2}{S}\cos(\theta_2 - \phi);$$

$$\ddot{S} = b\ddot{\theta}_1 \sin(\theta_1 - \phi) + b\dot{\theta}_1^2 \cos(\theta_1 - \phi) + S\dot{\phi}^2$$

$$\ddot{\phi} = \frac{-b\ddot{\theta}_1 \cos(\phi - \theta_1) + b\dot{\theta}_1^2 \sin(\theta_1 - \phi) - 2\dot{S}\dot{\phi} - a\ddot{\theta}_2 \cos(\phi - \theta_2) + a\dot{\theta}_2^2 \sin(\theta_2 - \phi)}{S}$$

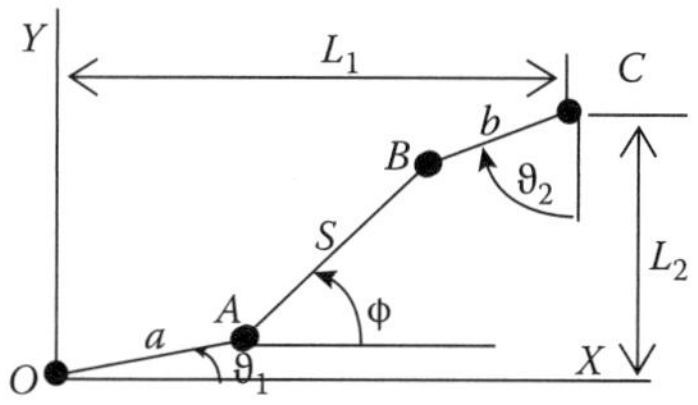

4.35

$$\dot{S} = \frac{-l_1\dot{\theta}\cos(\theta + \phi)}{\cos\phi}; \quad \dot{\phi} = \frac{-l_1\dot{\theta}\sin\theta}{l_2\cos\phi};$$

$$\ddot{S} = \frac{l_1\ddot{\theta}\sin(\phi - \theta) - l_1\dot{\theta}^2\cos(\theta + \phi) - l_2\dot{\phi}^2}{\cos\varphi};$$

$$\ddot{\phi} = \frac{-l_1\ddot{\theta}\cos\theta + l_1\dot{\theta}^2\sin\theta + l_2\dot{\phi}^2\sin\phi}{l_2\cos\phi}$$

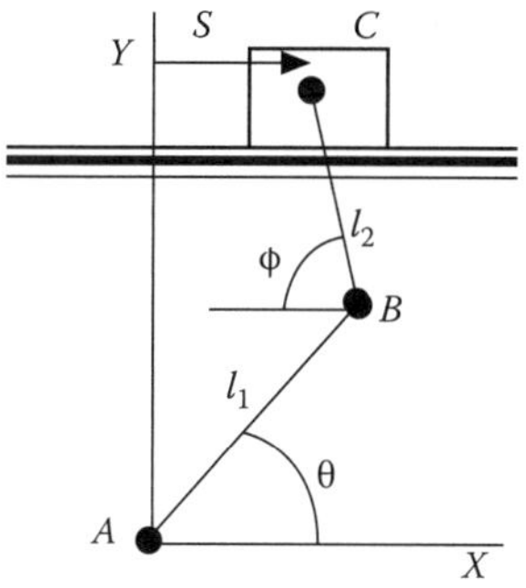

4.36

$$\dot{S} = -R\dot{\theta}\sin(\theta + \phi); \quad \dot{\phi} = \frac{R\dot{\theta}}{S}\cos(\theta - \phi);$$

$$\ddot{S} = -R\dot{\theta}^2\cos(\theta - \phi) + S\dot{\phi}^2;$$

$$\ddot{\phi} = \frac{R}{S}\dot{\theta}^2\sin(\theta + \phi) - \frac{2\dot{S}\dot{\phi}}{S}$$

$$\mathbf{v}_D = (-L\dot{\phi}\sin\phi)\,\mathbf{I} + (L\dot{\phi}\cos\phi)\,\mathbf{J};$$

$$\mathbf{a}_D = \left(-L\ddot{\phi}\sin\phi - \omega_{CD}^2 L\sin\phi\right)\mathbf{I} + \left(L\ddot{\phi}\cos\phi - \omega_{CD}^2 L\cos\phi\right)\mathbf{J}$$

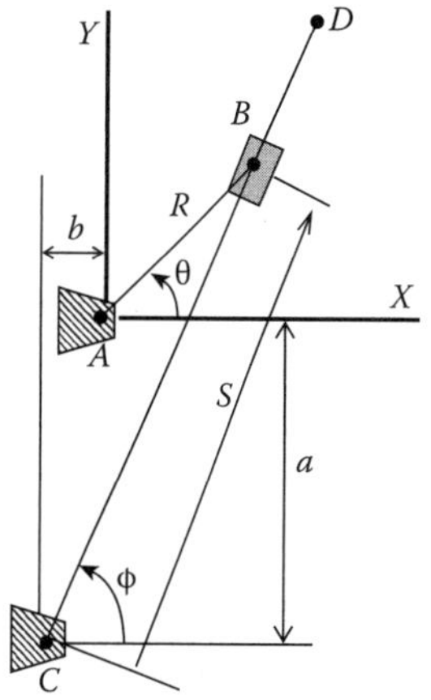

4.37

$$\dot{S} = r\dot{\theta}\sin(\theta - \phi); \quad \dot{\phi} = \frac{-r\dot{\theta}}{S}\cos(\theta - \phi);$$

$$\ddot{S} = r\ddot{\theta}\sin(\theta + \phi) + r\dot{\theta}^2\cos(\theta - \phi) + \dot{S}\dot{\phi}\sin 2\phi + S\dot{\phi}^2$$

$$\ddot{\phi} = \frac{-r\ddot{\theta}}{S}\cos(\theta - \phi) + \frac{r\dot{\theta}^2}{S}\sin(\theta - \phi) - \frac{2\dot{S}\dot{\phi}}{S};$$

$$\mathbf{v}_C = (-r\dot{\theta}\cos\theta - L\dot{\phi}\sin\phi)\,\mathbf{I} + (-r\dot{\theta}\sin\theta - L\dot{\phi}\cos\phi)\,\mathbf{J}$$

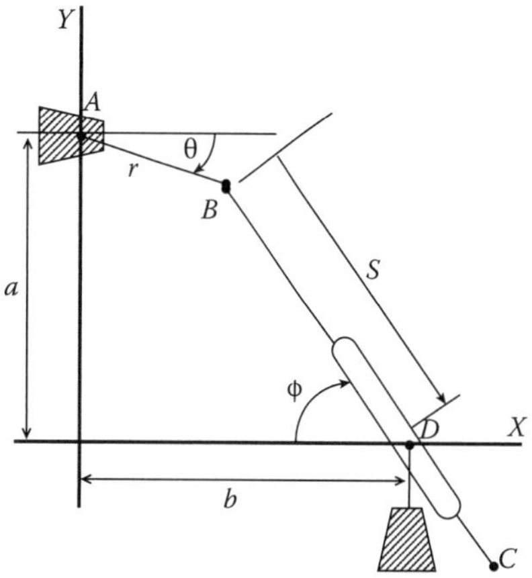

4.38

(a) $a = X_C + d\cos\phi$; (b) $\dot{X}_C = \frac{dr}{R}\omega_O \sin\left(\frac{r}{R}\omega_O t\right)$,

$\ddot{X}_C = \frac{-dr^2}{R^2}\omega_O^2 \cos\left(\frac{r}{R}\omega_O t\right)$;

(c) $\dot{X}_C = 574\frac{\text{in.}}{\text{s}}$, $\ddot{X}_C = 14054\,\text{in.}/\text{s}^2$

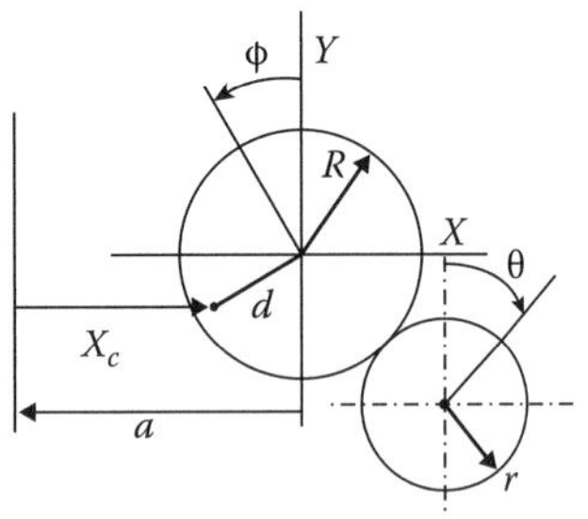

4.39

(a) $$\begin{bmatrix} \sin\phi & S\cos\phi \\ \cos\phi & S\sin\phi \end{bmatrix} \begin{Bmatrix} \dot{S} \\ \dot{\phi} \end{Bmatrix} = \begin{bmatrix} -r\dot{\theta}\cos\theta \\ -r\dot{\theta}\sin\theta \end{bmatrix};$$

(b) $$\begin{bmatrix} \sin\phi & S\cos\phi \\ \cos\phi & S\sin\phi \end{bmatrix} \begin{Bmatrix} \ddot{S} \\ \ddot{\phi} \end{Bmatrix} = \begin{bmatrix} -r\dot{\theta}^2\sin\theta - 2\dot{S}\dot{\phi}\cos\phi + S\dot{\phi}^2\sin\phi \\ -r\dot{\theta}^2\cos\theta + 2\dot{S}\dot{\phi}\sin\phi + S\dot{\phi}^2\cos\phi \end{bmatrix};$$

(c) $$\mathbf{v}_B = (-L\dot{\phi}\sin\phi)\,\mathbf{I} + (L\dot{\phi}\cos\phi)\,\mathbf{J}$$

$$\mathbf{a}_D = (-L\ddot{\phi}\sin\phi + L\dot{\phi}^2\cos\phi)\,\mathbf{I} + (L\ddot{\phi}\cos\phi + L\dot{\phi}^2\sin\phi)\,\mathbf{J}$$

$$\begin{bmatrix} \sin\phi & S\cos\phi \\ \cos\phi & -S\sin\phi \end{bmatrix} \begin{Bmatrix} \dot{S} \\ \dot{\phi} \end{Bmatrix} = \begin{bmatrix} r\dot{\theta}\cos\theta \\ r\dot{\theta}\cos\theta \end{bmatrix};$$

$$\begin{bmatrix} \sin\phi & S\cos\phi \\ \cos\phi & -S\sin\phi \end{bmatrix} \begin{Bmatrix} \ddot{S} \\ \ddot{\phi} \end{Bmatrix} = \begin{bmatrix} -r\dot{\theta}^2\sin\theta - 2\dot{S}\dot{\phi}\cos\phi + S\dot{\phi}^2\sin\phi \\ r\dot{\theta}^2\sin\theta + 2\dot{S}\dot{\phi}\sin\phi + S\dot{\phi}^2\cos\phi \end{bmatrix}$$

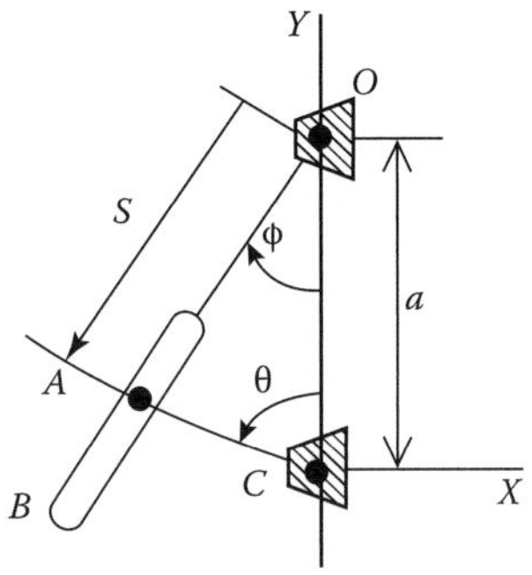

4.40

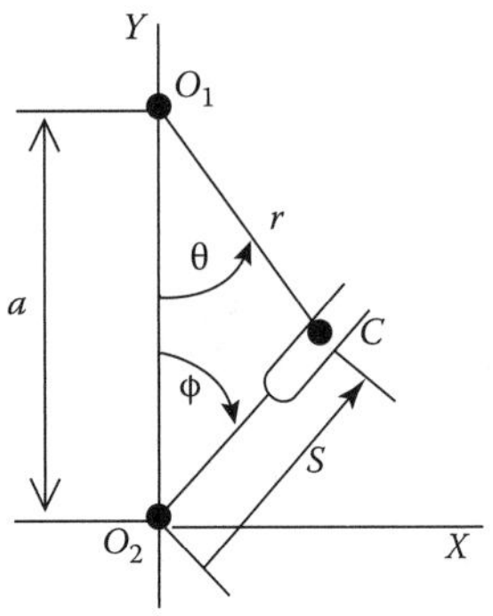

4.41

$$\dot{\phi} = \frac{e\dot{\theta}\sin(\theta-\phi)}{S};$$

$$\dot{S} = \frac{e\dot{\theta}(S\cos(\theta-\phi) - R\sin(\phi-\theta))}{S};$$

$$\ddot{\phi} = \frac{e\ddot{\theta}\sin(\theta-\phi) + e\dot{\theta}(\dot{\theta}-\dot{\phi})\cos\theta - \dot{\phi}\dot{S}}{S}$$

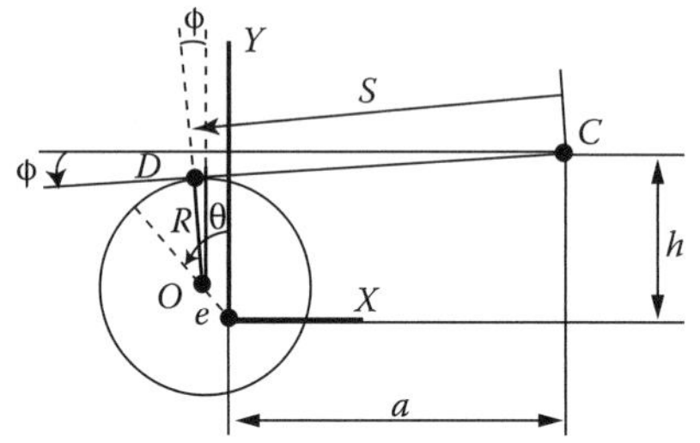

4.42

$$\begin{bmatrix} \sin\beta & S_1\cos\beta \\ \cos\beta & -S_1\sin\beta \end{bmatrix} \begin{Bmatrix} \dot{S}_1 \\ \dot{\beta} \end{Bmatrix} = \begin{bmatrix} R\dot{\theta}\cos\theta \\ R\dot{\theta}\sin\theta \end{bmatrix};$$

$$\begin{bmatrix} 1 & L_2\cos\phi \\ 0 & L_2\sin\phi \end{bmatrix} \begin{Bmatrix} \dot{S}_2 \\ \dot{\phi} \end{Bmatrix} = \begin{bmatrix} L_1\dot{\beta}\cos\beta \\ L_1\dot{\beta}\sin\beta \end{bmatrix};$$

$$\begin{bmatrix} \sin\beta & S_1\cos\beta \\ \cos\beta & -S_1\sin\beta \end{bmatrix} \begin{Bmatrix} \ddot{S}_1 \\ \ddot{\beta} \end{Bmatrix} = \begin{bmatrix} -R\dot{\theta}^2\sin\theta - 2\dot{S}_1\dot{\beta}\cos\beta + S_1\dot{\beta}^2\sin\beta \\ R\dot{\theta}^2\cos\theta + 2\dot{S}_1\dot{\beta}\sin\beta + S_1\dot{\beta}^2\cos\beta \end{bmatrix};$$

$$\begin{bmatrix} 1 & L_2\cos\phi \\ 0 & L_2\sin\phi \end{bmatrix} \begin{Bmatrix} \ddot{S}_2 \\ \ddot{\phi} \end{Bmatrix} = \begin{bmatrix} L_1\ddot{\beta}\cos\beta - L_1\dot{\beta}^2\sin\beta + L_2\dot{\phi}^2\sin\phi \\ L_1\ddot{\beta}\sin\beta + L_1\dot{\beta}^2\cos\beta + L_2\dot{\phi}^2\cos\phi \end{bmatrix}$$

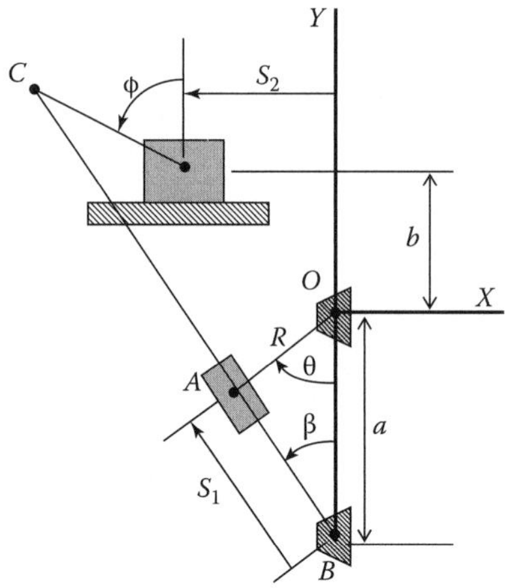

Chapter 5

5.1

(a) $I_O\ddot{\theta} + (k_{\theta 1} + k_{\theta 2})\theta = 0$, (b) $f_n = 15.48$ Hz

5.2

(a) $\ddot{\theta} + 18.00\,\dot{\theta} + 23144\,\theta = 0$,

(b) $f_n = 24.2$ Hz, $f_d = 12.16$ Hz

5.3

(b) $$\left(M_1 k_1^2 + M_2\left(\frac{k_2 R_1}{R_2}\right)^2 + M_3\left(\frac{R_1 R_3}{R_2}\right)^2\right)\ddot{\theta} = M + \frac{R_1 R_3}{R_2} w_3;$$

(d) $\theta(5) = 131.09$ rad, $Y(5) = 37.45$ m, $\dot{Y}(5) = 15.04$ m/s

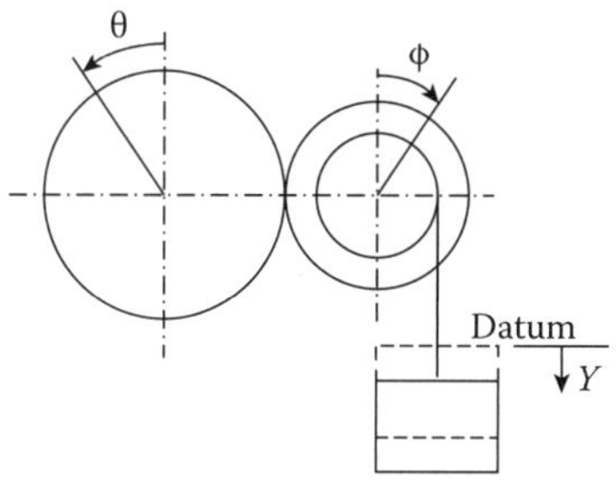

Assumed kinematics

5.4

(b) $$\left(m_A + \frac{I_E}{r^2} + \frac{4I_D}{r^2} + 4m_B\right)\ddot{X} = w_A + w_C - 2w_B;$$

(d) $$\dot{X} = \sqrt{\frac{2(-7w_A + w_C)gX}{9w_A + 5w_C(k_g^2/r^2)}};$$

$\dot{X} = 2.99$ ft/s; $\dot{Y} = 5.98$ ft/s; $\dot{\theta}_1 = 7.18$ rad/s; $\dot{\theta}_2 = 14.35$ rad/s $\theta_C = 1.2$ rad s

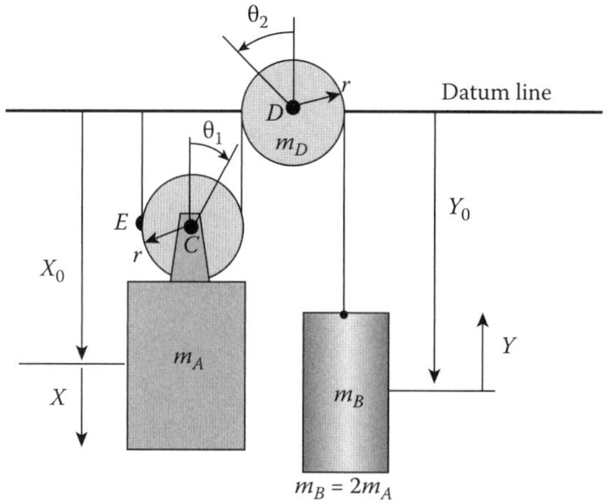

5.5

(b) $$\left(I_1 + \left(\frac{R_1}{R_2}\right)^2 I_2\right)(\ddot{\theta}_1) + k_\theta \theta_1 = M\left(\frac{R_1}{R_2}\right),$$

$f_n = 0.952$ Hz;

(b) $\dot{\theta}_1 = 10.21$ rad/s

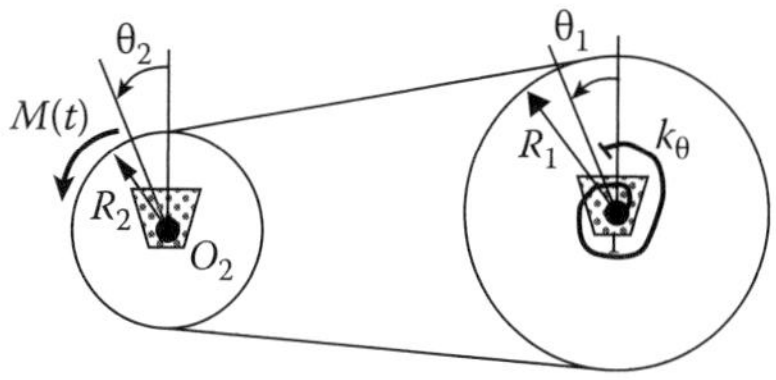

5.6

(b) $$\left(I_R + 3I_p\left(\frac{R}{r_p}\right)^2 + I_S\left(\frac{R}{r_2}\right)^2\right)\ddot{\theta}_R = \frac{R}{r_2}T;$$

(c) $\theta_R(5) = 24,614.8$ rad s or $3,917.6$ revs

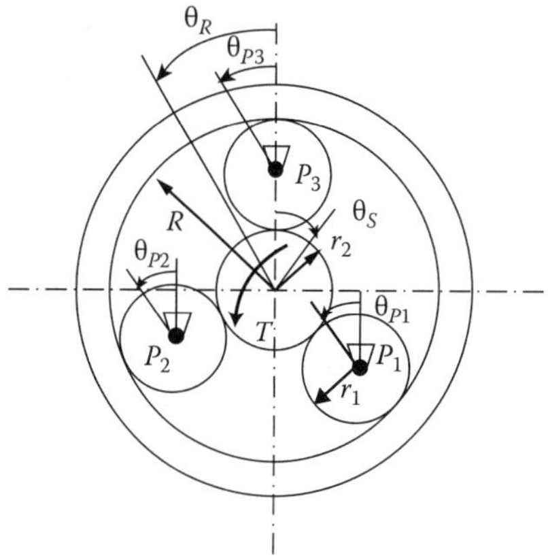

5.7

(a) $\ddot{\theta} = \frac{3}{2}\frac{g\sin\theta}{L}$;

(c) $O_r = \frac{W}{2}(5W\cos\theta - 3)$, $O_\theta = \frac{1}{4}W\sin\theta$

5.8

(a) $\ddot{\theta} = \dfrac{3g}{2L}\cos(\theta)$; (c) $\bar{\theta} = \dfrac{\pi}{2}, \dfrac{3\pi}{2}$ $\delta\ddot{\theta} + \dfrac{3g}{2l}\delta\theta = 0$;

(d)

$$O_r = \frac{5}{2}W\sin(\theta) - \frac{3}{2}W\sin\theta_0, \quad O_\theta = \frac{W\cos(\theta)}{4}$$

5.9

(a) $\ddot{\theta} = \dfrac{16g}{81L}\sin(\theta)$;

(b)

$$O_r = \frac{115W}{243}\cos(\theta) + \frac{128W}{243}, \quad O_\theta = \frac{179W\sin(\theta)}{243},$$

(c) $\bar{\theta} = 0, \pi$, and (d) $\omega_n = \dfrac{4}{9}\sqrt{\dfrac{g}{L}}$

5.10

(a)
$$\left(\left(\frac{1}{3}m+M\right)l^2+\frac{3}{2}Mr^2+2Mrl\right)\ddot{\theta}=\frac{(m+2M)l+2MR}{2(m+M)}W\cos\theta;$$

(c) $O_r = W\sin\theta+(m+M)\bar{r}\left(\frac{2\bar{r}W\sin\theta}{I_O}\right)$,

$O_\theta = W\cos\theta-(m+M)\bar{r}^2\frac{W\cos\theta}{I_O}$;

where $W=(m+M)g$;

(d) $\bar{\theta}=\frac{\pi}{2},\frac{3\pi}{2}$; (d) $\delta\ddot{\theta}+\frac{\bar{r}W}{I_O}\delta\theta=0$

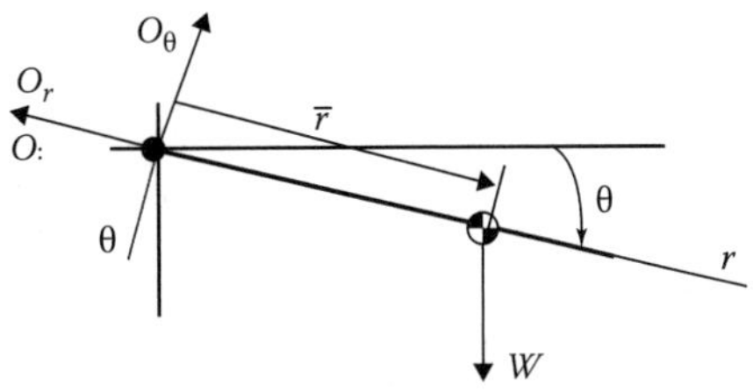

Free-body diagram

5.11

(a) $\ddot{\theta}=\frac{3}{4}\frac{g}{b}\cos\theta$;

(c) $O_r=\frac{5}{2}W\sin\theta,\quad O_\theta=\frac{W\cos\theta}{4}$

5.12

(a) $\ddot{\theta}=\frac{3g}{2\sqrt{5}b}\cos(\alpha-\theta)$;

(c) $O_r=\frac{3}{2}W\sin(\alpha)-\frac{1}{2}W\sin(\alpha-\theta)$,

$O_\theta=\frac{W\cos(\alpha-\theta)}{4}$

5.13

(a) $\ddot{\theta}=\frac{3g}{b2\sqrt{2}}\sin(\theta+\alpha)$,

(c) $O_r=2mg\cos(\theta+\alpha)-3mg\cos(\alpha)$,

$O_\theta=\frac{mg}{4}\sin(\theta+\alpha)$

5.14

(a) $\ddot{\theta}=\frac{2g}{3r}\sin(\theta)$,

(c) $O_r=\frac{5mg}{3}\cos\theta-\frac{2mg}{3}\cos\theta_0,\quad O_\theta=\frac{mg}{3}\sin\theta$

5.15

(a) $\ddot{\theta}=\frac{2(3\pi-4)g}{(9\pi-16)r}\sin(\theta)$,

(c) $O_r=\mathrm{mg}\left(1+\frac{(3\pi-4)^2}{9\pi-16}(\cos\theta-\cos\theta_0)\right)$,

$O_\theta=mg\left(1-\frac{2(3\pi-4)^2}{3\pi(9\pi-16)}\right)\sin\theta$

5.16

(a) $\ddot{\theta}+\frac{4\sqrt{3}}{5}\frac{g}{a}\sin\theta=0$,

(c) $\delta\ddot{\theta}+\frac{4\sqrt{3}}{5}\frac{g}{a}\sin\delta\theta=0$,

(d) $O_r=\frac{mg}{5}[13\cos\theta-8\cos\theta_0],\quad O_\theta=\frac{mg}{5}\sin\theta$

5.17

(a)
$$\ddot{\theta}+\frac{2k}{m}\left(\left[1+\frac{4}{9}(\sin\theta-2\cos\theta)\right]^{1/2}-\frac{1}{3}\right)\times\cos(\theta-\phi)-\frac{3g}{2L}\sin\theta=0$$

5.18

(a) $\ddot{\theta}+\left(\frac{6k}{m}-\frac{3g}{2l}\right)\theta=0$, (c) $\omega_n=\sqrt{\frac{6k}{m}-\frac{3g}{2l}}$,

(d) $k\geq\frac{W}{4l}$

5.19

(a)
$$\ddot{\theta}+\frac{3c}{2m}\dot{\theta}+\left(\frac{6k}{m}-\frac{3g}{2l}\right)\theta=0,$$

$\omega_n=\sqrt{\frac{6k}{m}-\frac{3g}{2l}}$,

$\zeta=\frac{3c}{4m\sqrt{(6k/m)-(3g/2l)}}$, (c) $k<\frac{W}{4l}$

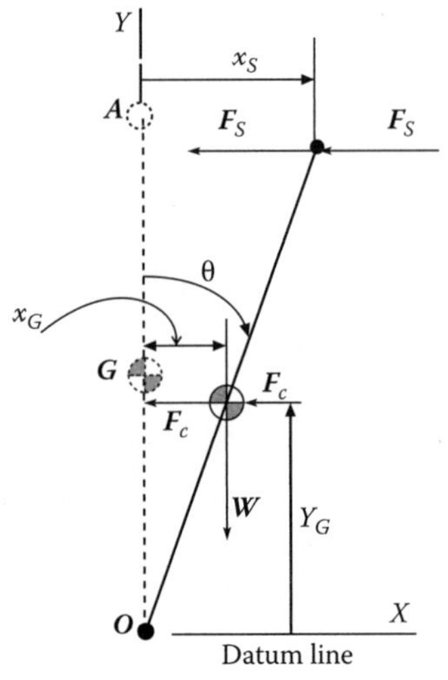

5.20

(a) $\ddot{\theta} + \dfrac{c}{m}\dot{\theta} + \left(\dfrac{4k}{m} + \dfrac{3g}{2l}\right)\theta = 0;$

(c) $\omega_n = 3.89$ rad/s, $c = 0.925$ N s/m

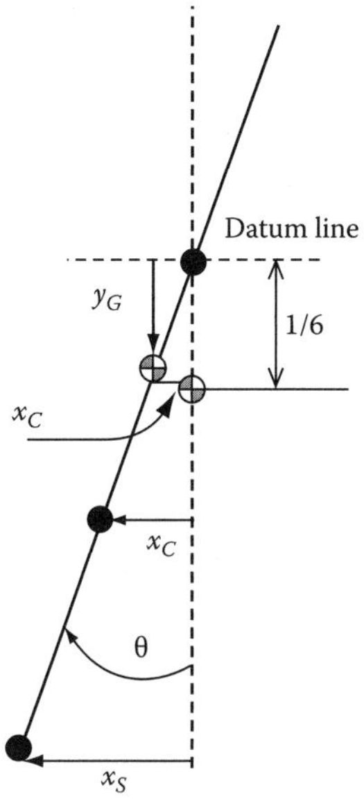

5.21

(a) $\ddot{\theta} + \dfrac{3c}{4m}\dot{\theta} + \left(\dfrac{3k}{m} + \dfrac{3g}{2l}\right)\theta = 0;$

(b) $\omega_n = 6.96$ rad/s, $c = 37.12$ N s/m

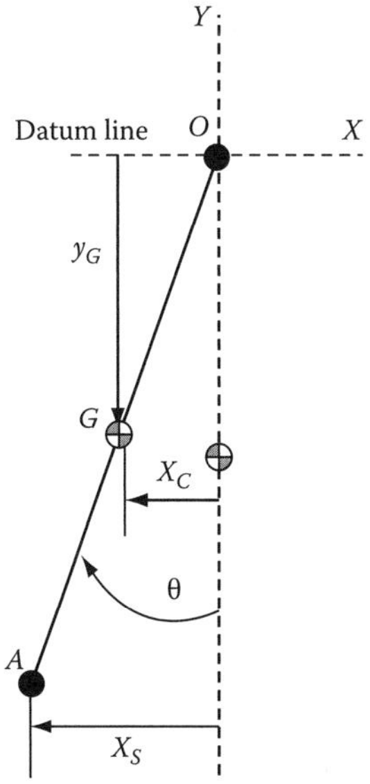

5.22

(a) $I_O\ddot{\theta} + 2b^2c\dot{\theta} + (2kb^2 + \bar{r}W)\theta = 0,$

$$\omega_n = \sqrt{\frac{2kb^2 + \bar{r}W}{mk_O^2}}, \quad \zeta = \frac{cb^2}{\sqrt{(mk_O^2)(2kb^2 + \bar{r}W)}}$$

5.23

(a) $\ddot{\theta} + \dfrac{g}{3h}\theta = 0$ (*note*: θ angle about *Z*-axis);

(d) $\ddot{\psi} + \dfrac{g}{h + (d/2)}\sin\psi = 0$ (*Note*: ψ angle of swing in *Y*–*Z* plane)

5.24

(b) $I_O\ddot{\theta} + \dfrac{cl^2}{4}\dot{\theta} + \left(\dfrac{kl^2}{4} - w_M\left(l + \dfrac{b}{2}\right) - w_m\dfrac{l}{2}\right)\theta = 0,$

(d) $\bar{k} = \dfrac{4(w_M(l + (b/2)) + w_m(l/2))}{l^2}$

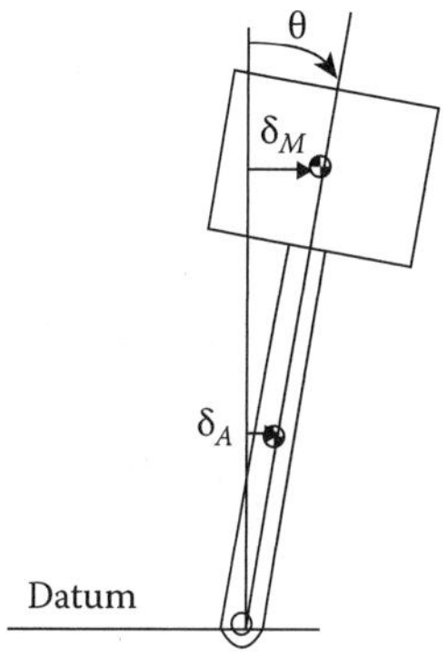

5.25

(b) $$\left(\frac{1}{3}ml^2 + \frac{1}{2}Mr^2 + M(l+r)^2\right)\ddot{\theta} = \left(\frac{ml}{2} + M(l+r)\right)\frac{3g}{4}\cos\theta,$$

(c)

$$O_r = \frac{3}{4}g\left[(m+M)\sin\theta - \frac{2((ml/2)+M(l+r))^2}{((1/3)ml^2+(1/2)Mr^2+M(l+r)^2)}(\sin\theta - \sin\theta_0)\right],$$

$$O_\theta = \frac{g}{4}\left[5(m+M) + 3\left(\frac{((ml/2)+M(l+r))^2}{(1/3)ml^2+(1/2)Mr^2+M(l+r)^2}\right)\right]\cos\theta$$

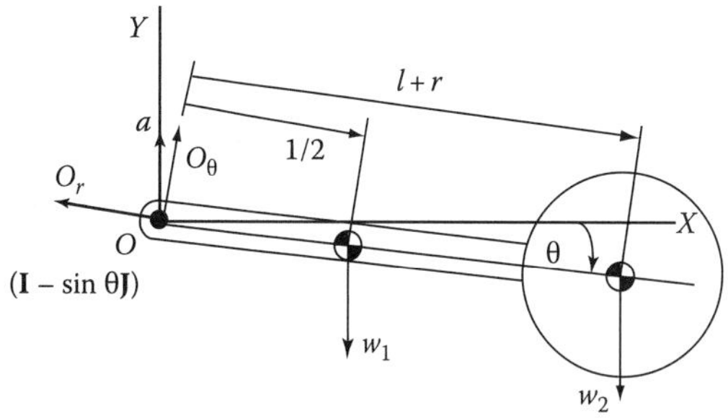

5.26

(a) $\ddot{\theta} = \dfrac{3}{2L}(a_O + g\sin\theta),$

(b) $\dot{\theta} = \left[\dfrac{3g}{L}\left(\dfrac{a_0}{g} + 1 - \cos\theta\right)\right]^{1/2}$

5.27

(a) $\ddot{\theta} = \frac{g}{2l}(\cos\theta - 3\sin\theta)$,

(b) $\theta = 36.9^\circ$, (c) $O_r = \frac{mg}{2}\left(\frac{17}{3}\cos\theta + \sin\theta - 3\right)$,
$O_\theta = \frac{mg}{4}\left(\sin\theta + \frac{7}{3}\cos\theta\right)$

5.28

(a) $\ddot{\theta} = \dfrac{3}{2}\left(\dfrac{a}{l}\cos(\theta+\beta) - \dfrac{g}{l}\sin(\theta)\right)$,

(b) $\tan\left(\dfrac{\theta}{2}\right) = \dfrac{a\cos\beta}{a\sin\beta + g}$,

(c) $O_r = mg\left(\dfrac{5}{2}\cos\theta - \dfrac{3}{2}\right) + \dfrac{3ma}{2}(\sin(\theta+\beta) - \sin\beta)$,

$\theta_\theta = \dfrac{mg}{4}\sin\theta + \dfrac{3ma}{4}\cos(\theta+\beta)$

5.29

(a) $\ddot{\theta} = \dfrac{-27g}{16l}\sin\theta$,

(b) $A_r = \dfrac{mg}{16}(45\cos\theta - 27\cos\theta_0)$, $A_\theta = \dfrac{9}{32}mg\sin\theta$

5.30

(b) $\ddot{\theta} = \dfrac{gd\cos\theta + rd\dot{\theta}^2\cos\theta}{(k^2 + r^2 + d^2 - 2rd\sin\theta)}$

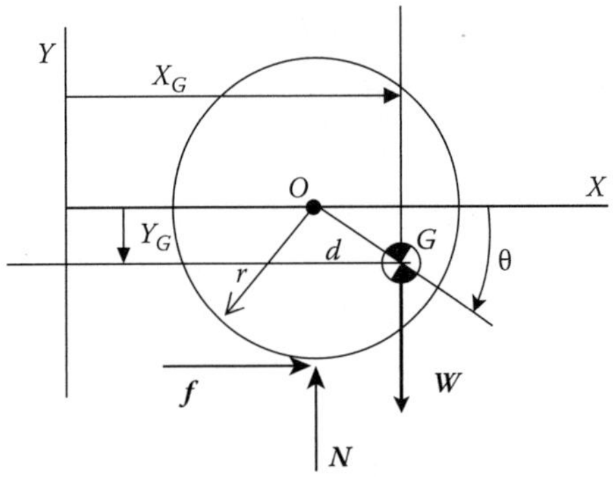

5.31

(a) $\ddot{\theta} = \frac{2g\sin\theta}{3(R+r)}$, (d) $\theta_L = 55.15^\circ$

5.32

(a) $(I_G + 2erm\cos\theta + (r^2 + e^2)m)\ddot{\theta}$
$= w(e\sin(\theta+\alpha) + r\cos\alpha) + erm\dot{\theta}^2\sin\theta$

5.33

(b) $\dfrac{3mr}{4}\ddot{\theta} + 4rk\theta = W\cos\alpha$, $\omega_n = \sqrt{\dfrac{16k}{3m}}$

5.34

(b) $\left(M + \dfrac{3m}{2}\right)r\ddot{\theta} + cr\dot{\theta} + 4kr\theta = W + w$,

(d) $\omega_n = 3.16$ rad/s, $c = 44.3$ N s/m

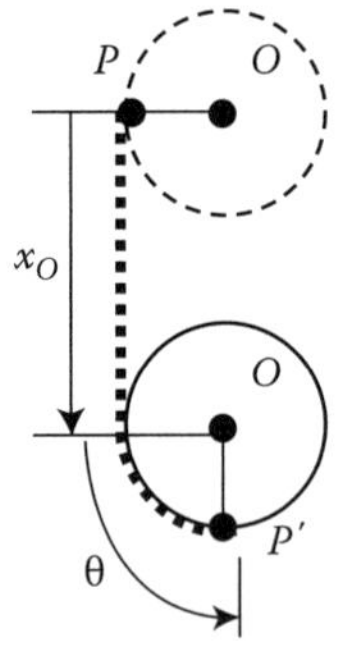

5.35

(b) $\dfrac{3m}{2}r\ddot{\theta} + cr\dot{\theta} + kr\theta = -w\sin\alpha$,

(d) $f_n = 1.143$ Hz, $c = 0.279$ lbs/in.

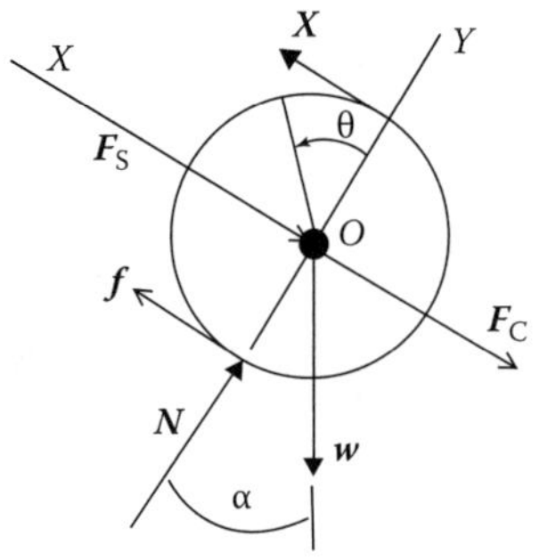

5.36

(b) $\left[k_g^2 + r^2 - 2er\cos\theta\right]\ddot{\theta} + er\dot{\theta}^2\sin\theta$
$+ eg\sin\theta + \dfrac{k}{m}r^2\theta = 0$;

(d) $\omega_n = \sqrt{\dfrac{(k/m)r^2 + ge}{r^2 - 2re + k_g^2}}$

5.37

(a) $\ddot{\theta} = \dfrac{-3}{2}\dfrac{g}{l}\cos\theta$,

(c) $N_1 = \dfrac{3mg}{2}\left(\dfrac{3}{2}\sin\theta\cos\theta - \cos\theta\right)$,

$N_2 = m\left(1 + \dfrac{3g}{l}\left(\dfrac{-\cos^2\theta}{2} + \sin^2\theta\right) - \dfrac{3g}{l}\sin\theta\right)$,

(d) $\theta_{LC} = 41.8^\circ$

5.38

(a)

$$\left(\frac{1}{3}m_{AB} + [m_B\sin^2\theta + m_A\cos^2\theta]\right)l^2\ddot{\theta}$$
$$= l((w_B + w_{AB})\sin\theta - (F + kl\sin\theta)\cos\theta)$$
$$+ [-m_B + m_A]l^2\dot{\theta}^2\cos\theta\sin\theta$$

5.39

$$\text{(a)}\ \frac{1}{3}ml^2\ddot{\theta} = (w - 2P)\frac{l}{2}\sin\theta - kl^2(\sin\theta - \sin\theta_0)(\cos\theta)$$

5.40

$$\text{(a)}\ \frac{1}{3}ml^2\ddot{\theta} = -w\frac{l}{2}\cos\theta + P\frac{l}{2}\cos 2\theta - k_2l^2\cos\theta\sin\theta - k_1l^2\cos\theta\sin\theta$$

5.41

$$\left(\frac{2}{3} - 2\sin^2\theta\right)ml\ddot{\theta} - ml\dot{\theta}^2\sin 2\theta + mg\cos\theta + kl(\sin\theta - \sin 2\theta) = F(-3\sin^2\theta + \cos^2\theta)$$

5.42

$$2ml\left(\frac{1}{3} + \cos^2\theta\right)\ddot{\theta} + ml\sin 2\theta(\dot{\theta}^2) - mg\sin\theta + kl(\sin 2\theta - \cos\theta) = F\cos\theta$$

5.43

$$2ml\left(\frac{1}{3} + \sin^2\theta\right)\ddot{\theta} + \frac{ml\dot{\theta}^2}{2}\sin 2\theta + \frac{mg}{2}\cos\theta + k(l_0\sin\theta - \sin 2\theta) = F$$

5.44

$$\left[\left(\frac{1}{3} + \frac{7}{2}\cos^2\theta\right)m + 4M\cos^2\theta\right]l\ddot{\theta} + [2M - 1m]l\dot{\theta}^2 + 2k(l\sin 2\theta - l_0\cos\theta) = 2F\cos\theta$$

5.45

$$\left(\frac{2}{3} + 2\cos^2\theta\right)ml\ddot{\theta} - ml\dot{\theta}^2\sin 2\theta + 2mg\cos\theta + 2kl\sin 2\theta = 0$$

5.46

$$(m_ol^2 + 4(2m_o + m)l^2\sin^2\theta + 4ml^2\sin^2\theta)\ddot{\theta} + ((4(2m_o + m))l^2\sin\theta\cos\theta + 2ml^2\sin\theta\cos\theta)\dot{\theta}^2 + k4l^2(\cos\theta - \cos\theta_0)(-\sin\theta) + \frac{m_ogl}{2}\cos\theta = Pl[1 + 2\sin^2\theta]$$

5.47

$$ml^2\left(\frac{7}{3} + 2\sin^2\theta\right)\ddot{\theta} + 2ml^2\cos\theta\sin\theta(\dot{\theta}^2) + 3mgl\cos\theta - kl^2(\cos\theta - \cos\theta_0)\sin\theta = -Fl\sin\theta$$

5.48

$$\left(ml^2\left(\frac{2}{3} + 2\sin^2\theta\right) + 6Ml^2\sin^2\theta\right)\ddot{\theta} + (2ml^2 + 6Ml^2)\sin\theta\cos\theta\dot{\theta}^2 - lmg\cos\theta + 4kl^2(\cos\theta - \cos\theta_0)(-\sin\theta) = 0$$

5.49

$$ml\left[\frac{1}{3} + 2\sin^2\theta\right]\ddot{\theta} + ml\dot{\theta}^2\sin 2\theta + \frac{Mg}{\sqrt{2}}\cos(\theta + 45) + mg\cos\theta + kl(\sin\theta(1 - \cos\theta)) = 0$$

5.50

$$Ml^2\left[\cos^2\theta + \frac{2}{\sqrt{2}}\cos\theta\cos(\theta+\alpha) + \cos^2(\theta+\alpha) + \frac{1}{3}\right]\ddot{\theta} - \frac{mgl}{2}\sin\theta + \frac{Mgl}{\sqrt{2}}\cos(\theta+\alpha) + 4kl^2(\sin\theta - \sin\theta_0)\cos\theta$$
$$- Ml^2\left[\cos\theta\sin\theta + \frac{1}{\sqrt{2}}(\sin\theta\cos(\theta+\alpha) + \cos\theta\sin(\theta+\alpha)) + \cos(\theta+\alpha)\sin(\theta+\alpha)\right]\dot{\theta}^2 = Fl(2\cos\theta - \sin\theta)$$

5.52

$$\text{(b)}\ \left[Mr^2\left(\frac{3}{2} + 1 - 2\cos\theta + \cos^2\theta\right) + 2mr^2(1 + \cos\theta)\right]\ddot{\theta} + (-mr^2\sin\theta + Mr^2(\sin\theta - \sin\theta\cos\theta))\dot{\theta}^2 - mgr\sin\theta = 0$$

5.53

$$\text{(b)}\ \left(I_{O1} + \left(\frac{R_1}{R_2}\right)^2 I_{O2} + \left(\frac{R_3R_1}{R_2}\right)^2 m\right)\ddot{\theta}_1 + \left(K_1 + \left(\frac{R_1}{R_2}\right)^2 K_2\right)\theta_1 = \frac{-R_1R_3}{R_2}W,$$

(c) $f_n = 4.64$ Hz

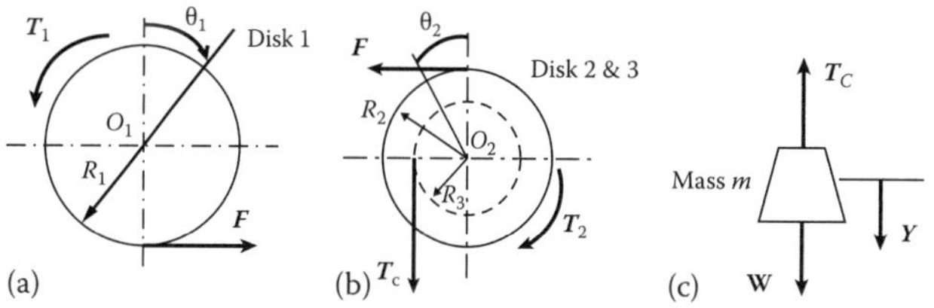

5.54

$$\text{(b)}\ \begin{bmatrix} I_{O1} & 0 \\ 0 & I_{O2} \end{bmatrix}\begin{Bmatrix} \ddot{\theta}_1 \\ \ddot{\theta}_2 \end{Bmatrix} + \begin{bmatrix} k_\theta & -k_\theta \\ -k_\theta & k_\theta \end{bmatrix}\begin{Bmatrix} \theta_1 \\ \theta_2 \end{Bmatrix} = \begin{Bmatrix} 0 \\ 0 \end{Bmatrix},$$

(c) $f_{n1} = 0$ (rigid body mode);

$$f_{n2} = 1.086\ \text{Hz},\quad A = \begin{bmatrix} 1 & 1 \\ 1 & -1 \end{bmatrix}$$

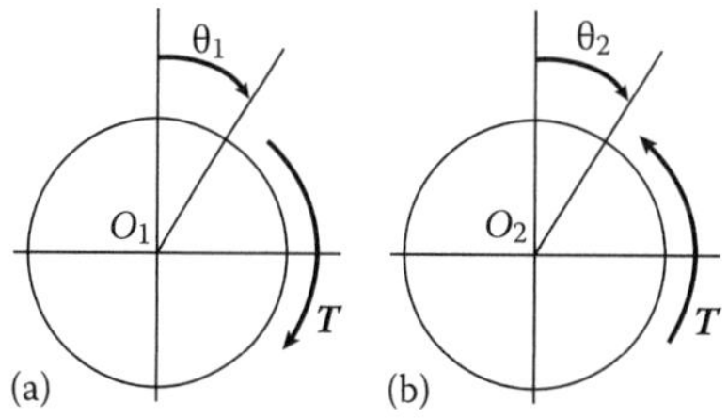

5.55

(b)
$$\begin{bmatrix} I_{O1} & 0 & 0 \\ 0 & I_{O2}+\left(\frac{r_2}{r_3}\right)^2 I_{O3} & 0 \\ 0 & 0 & I_{O4} \end{bmatrix}\begin{Bmatrix} \ddot{\theta}_1 \\ \ddot{\theta}_2 \\ \ddot{\theta}_4 \end{Bmatrix} + \begin{bmatrix} k_{\theta 1} & -k_{\theta 1} & 0 \\ -k_{\theta 1} & k_{\theta 1}+\left(\frac{r_2}{r_3}\right)^2 k_{\theta 2} & -\frac{r_2}{r_3}k_{\theta 2} \\ 0 & -\frac{r_2}{r_3}k_{\theta 2} & k_{\theta 2} \end{bmatrix}\begin{Bmatrix} \theta_1 \\ \theta_2 \\ \theta_4 \end{Bmatrix} = \begin{Bmatrix} 0 \\ 0 \\ 0 \end{Bmatrix},$$

(c)
$\lambda_1 = 0,\quad \lambda_2 = 8765\ \text{rad/s}^2;\quad \lambda_3 = 837228\ \text{rad/s}^2,$

$$A = \begin{bmatrix} 1 & 1 & 1 \\ 1 & -9.741 & -1025 \\ 2 & -26.49 & 84.47 \end{bmatrix}$$

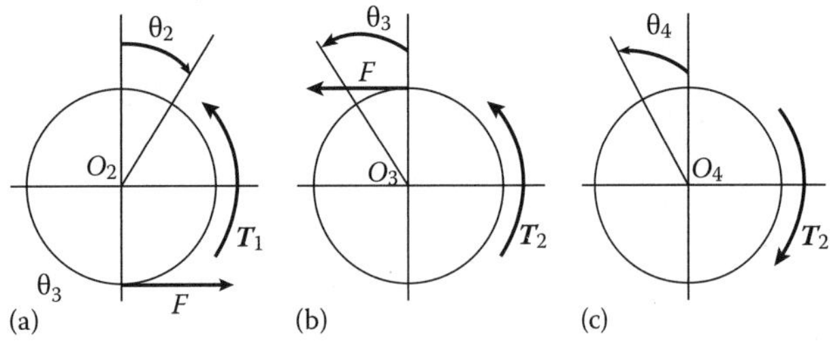

5.56

(b)
$$\left[\begin{array}{c:c} I_{O1}+\left(\frac{R_1}{R_2}\right)^2 I_{O2} & 0 \\ \hdashline 0 & m \end{array}\right]\begin{Bmatrix} \ddot{\theta}_1 \\ \ddot{Y} \end{Bmatrix} + \left[\begin{array}{c:c} K_1+\left(\frac{R_1}{R_2}\right)^2 K_2+\left(\frac{R_1R_3}{R_2}\right)^2 k_c & -\frac{R_1R_3}{R_2}k_c \\ \hdashline -\frac{R_1R_3}{R_2}k_c & k_c \end{array}\right]\begin{Bmatrix} \theta_1 \\ Y \end{Bmatrix} = \begin{Bmatrix} 0 \\ W \end{Bmatrix}$$

(c) $\lambda_1 = 8998(\text{rad/s})^2 \rightarrow f_{n1} = 15.10$ Hz and

$\lambda_2 = 643328(\text{rad/s})^2 \rightarrow f_{n2} = 127.7$ Hz,

$$A = \begin{bmatrix} 1 & 1 \\ 240.8 & -0.02486 \end{bmatrix}$$

5.57

(b)
$$\left[\begin{array}{c:c:c:c} I_{O1} & 0 & 0 & 0 \\ \hdashline 0 & I_{O2}+\left(\frac{r_2}{r_3}\right)^2 I_{O3}+\left(\frac{r_2}{r_4}\right) I_{O4} & 0 & 0 \\ \hdashline 0 & 0 & I_{O5} & 0 \\ \hdashline 0 & 0 & 0 & I_{O6} \end{array}\right]\begin{Bmatrix} \ddot{\theta}_1 \\ \ddot{\theta}_2 \\ \ddot{\theta}_5 \\ \ddot{\theta}_6 \end{Bmatrix}$$
$$+ \left[\begin{array}{c:c:c:c} K_1 & -K_1 & 0 & 0 \\ \hdashline -K_1 & K_1+\left(\frac{r_2}{r_3}\right)^2 K_3+\left(\frac{r_2}{r_4}\right)^2 K_2 & -\frac{r_2}{r_4}K_2 & -\frac{r_2}{r_3}K_3 \\ \hdashline 0 & -\frac{r_2}{r_4}K_2 & K_2 & 0 \\ \hdashline 0 & \frac{-r_2}{r_3}K_3 & 0 & K_3 \end{array}\right]\begin{Bmatrix} \theta_1 \\ \theta_2 \\ \theta_5 \\ \theta_6 \end{Bmatrix}$$
$$= \begin{bmatrix} -M_1 \\ 0 \\ 0 \\ -M_2 \end{bmatrix}$$

5.58

$$\begin{bmatrix} m & 0 \\ 0 & I_Z \end{bmatrix}\begin{Bmatrix} \ddot{Y} \\ \ddot{\beta} \end{Bmatrix} + \begin{bmatrix} \frac{12EI_b}{L^3} & \frac{-6EI_b}{L^2} \\ \frac{-6EI_b}{L^2} & \frac{4EI_b}{L} \end{bmatrix}\begin{Bmatrix} Y \\ \beta \end{Bmatrix}$$
$$= \begin{Bmatrix} 0 \\ 0 \end{Bmatrix} \rightarrow \begin{bmatrix} 0.518 & 0 \\ 0 & 13.98 \end{bmatrix}\begin{Bmatrix} \ddot{Y} \\ \ddot{\beta} \end{Bmatrix}$$
$$+ \begin{bmatrix} 60 & -300 \\ -300 & 2000 \end{bmatrix}(10^3)\begin{Bmatrix} Y \\ \beta \end{Bmatrix} = \begin{Bmatrix} 0 \\ 0 \end{Bmatrix}$$
$$\lambda = \begin{Bmatrix} 17.15 \\ 241.9 \end{Bmatrix} 10^3\ \frac{\text{rad}^2}{\text{s}^2};$$
$$A = \begin{bmatrix} 1 & 1 \\ -0.1704 & -0.217 \end{bmatrix};\quad f_n = \begin{Bmatrix} 20.8 \\ 78.3 \end{Bmatrix}\text{Hz}$$

5.59

(a) $m\ddot{X} + 4\left(\frac{3EI}{l^3}\right)X = 0$, (b) $f_n = 3.245$ Hz

5.60

(a) $m\ddot{X} + 4\left(\frac{12EI}{l^3}\right)X = 0$, (b) $f_n = 6.49$ Hz

5.61

(a) $m\ddot{X} + \left(\frac{E}{l^3}\left(2b^4 + \frac{3\pi d^4}{32}\right)\right)X = 0$, (b) $f_n = 4.656$ Hz

5.62

(a) $m\ddot{X} + 4c\dot{X} + 4\left(\frac{3EI}{l^3}\right)X = 4\left(\frac{3EI}{l^3}\right)X_B + 4c\dot{X}_B$,

(b) $zX_{SS} = 4.883$ cm, $X_{PK} = 33.71$ cm

5.63

x-direction: $$\begin{bmatrix} m_1 & 0 \\ 0 & m_2 \end{bmatrix}\begin{Bmatrix} \ddot{x}_1 \\ \ddot{x}_2 \end{Bmatrix} + \begin{bmatrix} 8k_x & -4k_x \\ -4k_x & 4k_x \end{bmatrix}\begin{Bmatrix} x_1 \\ x_2 \end{Bmatrix} = \begin{Bmatrix} 0 \\ 0 \end{Bmatrix};$$

$$k_x = \frac{12EI_x}{L^3}; \quad A_x = \begin{bmatrix} 1 & 1 \\ -1 & 0.667 \end{bmatrix} \quad \lambda_x = \begin{Bmatrix} 226 \\ 1358 \end{Bmatrix} \text{rad}^2/\text{s}^2$$

$$f_{nx} = \begin{Bmatrix} 2.39 \\ 5.87 \end{Bmatrix} \text{Hz}$$

y-direction: $$\begin{bmatrix} m_1 & 0 \\ 0 & m_2 \end{bmatrix}\begin{Bmatrix} \ddot{y}_1 \\ \ddot{y}_2 \end{Bmatrix} + \begin{bmatrix} 8k_y & -4k_y \\ -4k_y & 4k_y \end{bmatrix}\begin{Bmatrix} y_1 \\ y_2 \end{Bmatrix} = \begin{Bmatrix} 0 \\ 0 \end{Bmatrix};$$

$$k_y = \frac{12EI_y}{L^3}; \quad A_y = \begin{bmatrix} 1 & 1 \\ -1 & 0.667 \end{bmatrix}$$

$$\lambda_y = \begin{Bmatrix} 73 \\ 438 \end{Bmatrix} \frac{\text{rad}^2}{\text{s}^2} \quad f_{ny} = \begin{Bmatrix} 1.36 \\ 3.33 \end{Bmatrix} \text{Hz}$$

5.64

(a) $m\ddot{Y} + \dfrac{12EI_b}{l^3} Y = 0, \quad f_n = 6.26\,Hz$

(b) $I_G\ddot{\beta} + \dfrac{4EI_b}{l}\beta = 0, \quad f_n = 2.725\,Hz$

(c)
$$\begin{bmatrix} m & 0 \\ 0 & I \end{bmatrix}\begin{Bmatrix} \ddot{Y} \\ \ddot{\beta} \end{Bmatrix} + \begin{bmatrix} \dfrac{12EI_b}{l^3} + K & -\dfrac{6EI_b}{l^2} \\ -\dfrac{6EI_b}{l^2} & \dfrac{4EI}{l} + K_\theta \end{bmatrix}\begin{Bmatrix} Y \\ \beta \end{Bmatrix} = 0,$$

$$\lambda_1 = 295.01\left(\frac{\text{rad}}{\text{s}}\right)^2 \to \underline{\omega_{1n} = 17.17\frac{\text{rad}}{\text{s}}\{2.735\,\text{Hz}\}}$$

$$\lambda_2 = 1000898\left(\frac{\text{rad}}{\text{s}}\right)^2 \to \underline{\omega_{2n} = 1000\frac{\text{rad}}{\text{s}}\{159.3\,\text{Hz}\}}$$

5.65

$$\begin{bmatrix} m & 0 \\ 0 & I_G \end{bmatrix}\begin{Bmatrix} \ddot{Y}_G \\ \ddot{\theta} \end{Bmatrix} + \begin{bmatrix} k_A + k_B & -k_A\overline{X} + k_B l_B \\ -k_A\overline{X} + k_B l_B & k_A\overline{X}^2 + k_B l_B^2 \end{bmatrix}\begin{Bmatrix} Y_G \\ \theta \end{Bmatrix} = 0;$$

$\overline{X}$—mass center relative to left end.

$$\lambda_1 = 387184\left(\frac{\text{rad}}{\text{s}}\right)^2 \to f_{n1} = 99.1\text{ Hz}$$

$$\lambda_2 = 816349\left(\frac{\text{rad}}{\text{s}}\right)^2 \to f_{n2} = 143.9\text{ Hz}$$

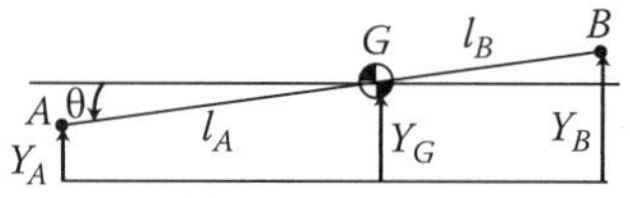

Kinematic diagram

5.66

(c)
$$\left[\begin{array}{c:c} I_1 & 0 \\ \hdashline 0 & I_2 + Mr_2^2 \end{array}\right]\begin{Bmatrix} \ddot{\theta}_1 \\ \ddot{\theta}_2 \end{Bmatrix} + \left[\begin{array}{c:c} k_\theta & -k_\theta \\ \hdashline -k_\theta & k_\theta + r_2^2(k_A + k_B) \end{array}\right]\begin{Bmatrix} \theta_1 \\ \theta_2 \end{Bmatrix} = \begin{bmatrix} M(t) \\ 0 \end{bmatrix},$$

(d) $\lambda_1 = 35785\left(\dfrac{\text{rad}}{\text{s}}\right)^2 \to f_{n1} = 30.12\,\text{Hz}$,

$\lambda_2 = 336822\left(\dfrac{\text{rad}}{\text{s}}\right)^2 \to f_{n2} = 92.41\,\text{Hz}$

$$A = \begin{bmatrix} 1 & 1 \\ 0.8616 & -0.3029 \end{bmatrix};$$

(e) $\ddot{q}_1 + 35779q_1 = 1.734M(t)$
$\ddot{q}_2 + 336887q_2 = 4.933M(t)$

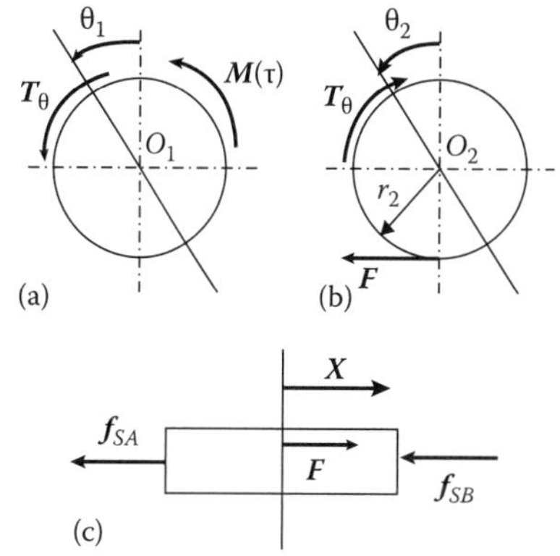

5.67

(b)
$$\begin{bmatrix} I_{O1} & 0 \\ 0 & I_{O2} \end{bmatrix}\begin{Bmatrix} \ddot{\theta}_1 \\ \ddot{\theta}_2 \end{Bmatrix} + \left[\begin{array}{c:c} (k_1 + k_2)d_1^2 + \frac{L_1}{2}w_1 & -k_2 d_1 d_2 \\ \hdashline -k_2 d_1 d_2 & k_2 d_2^2 + k_3 d_3^2 + \dfrac{L_2}{2}w_2 \end{array}\right]\begin{Bmatrix} \theta_1 \\ \theta_2 \end{Bmatrix} = 0,$$

(c) $\lambda_1 = 63.62\left(\dfrac{\text{rad}}{\text{s}}\right)^2 \to f_{n1} = 1.27\text{ Hz}$,

$\lambda_2 = 491.6\left(\dfrac{\text{rad}}{\text{s}}\right)^2 \to f_{n2} = 3.53\text{ Hz}$

$$A = \begin{bmatrix} 1 & 1 \\ 0.8573 & -0.1457 \end{bmatrix};$$

(d) $\ddot{q}_1 + 63.6q_1 = 0 \quad \ddot{q}_2 + 491.79\,q_2 = 0$

5.68

(b)

$$\begin{bmatrix} I_A + mR^2 & 0 \\ 0 & I_B + mR^2 \end{bmatrix} \begin{Bmatrix} \ddot{\theta}_1 \\ \ddot{\theta}_2 \end{Bmatrix} + R^2 \begin{bmatrix} k_1 + k_2 & -k_2 \\ -k_2 & k_2 + k_3 \end{bmatrix} \begin{Bmatrix} \theta_1 \\ \theta_2 \end{Bmatrix} = 0,$$

(c)

$$\lambda_1 = 6.667(\text{rad/s})^2 \rightarrow f_{n1} = 0.4112 \text{ Hz}$$
$$\lambda_2 = 33.33(\text{rad/s})^2 \rightarrow f_{n2} = 0.919 \text{ Hz},$$
$$A = \begin{bmatrix} 1 & 1 \\ 1 & -1 \end{bmatrix};$$

(d) $\ddot{q}_1 + 6.67q_1 = 0 \quad \ddot{q}_2 + 33.3q_2 = 0$

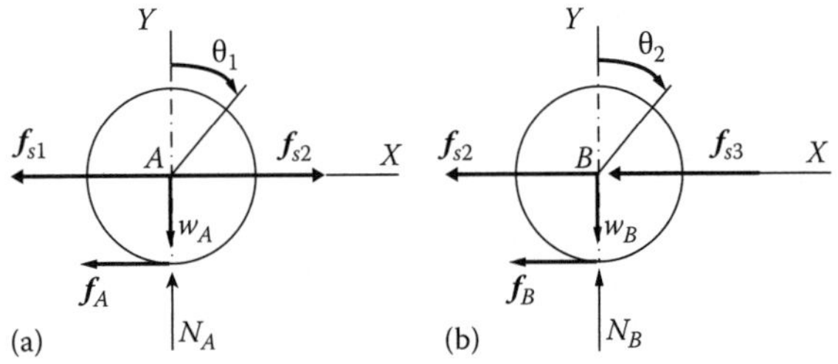

5.69

(b)

$$\begin{bmatrix} m_C + m_P & \dfrac{m_P L}{2} \\ \dfrac{m_P L}{2} & I_G + m_P \dfrac{L^2}{4} \end{bmatrix} \begin{Bmatrix} \ddot{X}_C \\ \ddot{\theta} \end{Bmatrix} + \begin{bmatrix} k & 0 \\ 0 & k_\theta \dfrac{-w_P L}{2} \end{bmatrix} \begin{Bmatrix} X_C \\ \theta \end{Bmatrix} = 0,$$

(c) $k_\theta \geq \dfrac{w_P L}{2}$,

(d)

$$\lambda_1 = 11.77(\text{rad/s})^2 \rightarrow f_{n1} = 0.546 \text{ Hz}$$
$$\lambda_2 = 40.7(\text{rad/s})^2 \rightarrow f_{n2} = 1.016 \text{ Hz},$$
$$A = \begin{bmatrix} 1 & 1 \\ 43.96 & -4.035 \end{bmatrix};$$

(e) $\ddot{q}_1 + 11.76q_1 = 0 \quad \ddot{q}_2 + 40.69q_2 = 0$

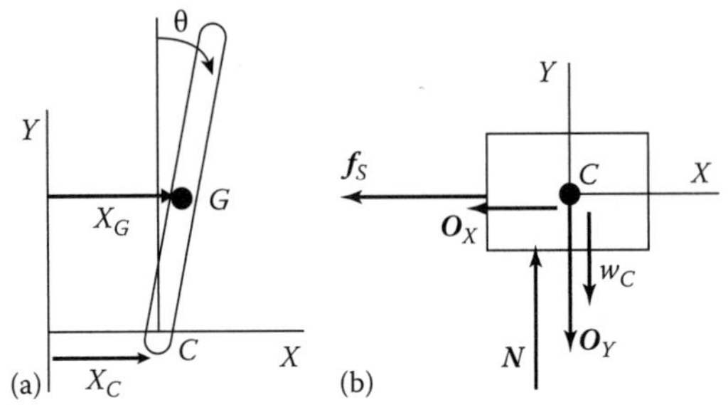

5.70

(b)

$$\begin{bmatrix} m & 0 \\ 0 & \frac{3}{2}MR^2 \end{bmatrix} \begin{Bmatrix} \ddot{X}_1 \\ \ddot{\theta} \end{Bmatrix} + \begin{bmatrix} k_A + k_B & -k_B R \\ -k_B R & (k_B + k_C)R^2 \end{bmatrix} \begin{Bmatrix} X_1 \\ \theta \end{Bmatrix} = \begin{Bmatrix} 0 \\ M(t) \end{Bmatrix},$$

(c)

$$\lambda_1 = 39.63(\text{rad/s})^2 \rightarrow f_{n1} = 1.002 \text{ Hz}$$
$$\lambda_2 = 403.7(\text{rad/s})^2 \rightarrow f_{n2} = 3.2 \text{ Hz},$$
$$A = \begin{bmatrix} 1 & 1 \\ 24.83 & -4.296 \end{bmatrix};$$

(d) $\ddot{q}_1 + 39.63q_1 = 0 \quad \ddot{q}_2 + 403.7q_2 = 0$

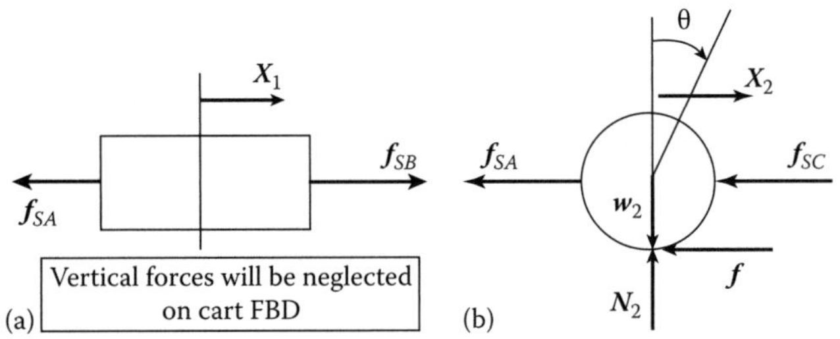

5.71

(b)

$$\begin{bmatrix} \dfrac{m + M}{R} & M \\ M & \dfrac{3MR}{2} \end{bmatrix} \begin{Bmatrix} \ddot{X}_2 \\ \ddot{\theta} \end{Bmatrix} + \begin{bmatrix} \dfrac{k_A + k_B}{R} & k_A \\ k_A & k_A R \end{bmatrix} \begin{Bmatrix} X_2 \\ \theta \end{Bmatrix} = \begin{Bmatrix} 0 \\ \dfrac{M(t)}{R} \end{Bmatrix},$$

(c)

$$\lambda_1 = 11.33(\text{rad/s})^2 \rightarrow f_{n1} = 0.536 \text{ Hz}$$
$$\lambda_2 = 77.67(\text{rad/s})^2 \rightarrow f_{n2} = 1.403 \text{ Hz},$$
$$A = \begin{bmatrix} 1 & 1 \\ -7.073 & 33.13 \end{bmatrix}$$

(d) $\ddot{q}_1 + 11.33q_1 = 0 \quad \ddot{q}_2 + 77.67q_2 = 0$

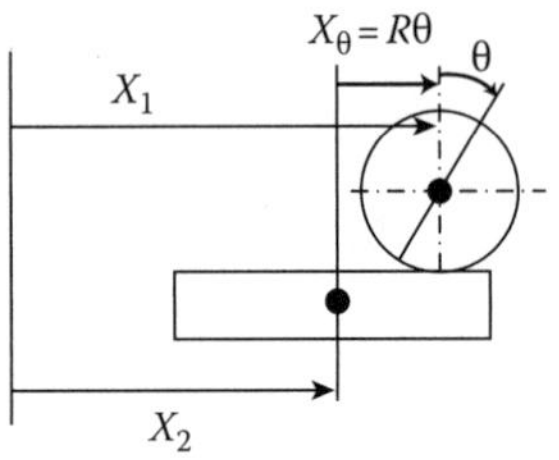

5.72

(c)

$$\begin{bmatrix} I_A & 0 \\ 0 & I_B \end{bmatrix} \begin{Bmatrix} \ddot{\theta}_A \\ \ddot{\theta}_B \end{Bmatrix} + \left[\begin{array}{c:c} k_\theta + w_{1A}\overline{X}_{1A} + w_{2A}\overline{X}_{2A} & -k_\theta \\ \hdashline -k_\theta & k_\theta + w_{1B}\overline{X}_{1B} + w_{2B}\overline{X}_{2B} \end{array}\right] \times \begin{Bmatrix} \theta_A \\ \theta_B \end{Bmatrix} = 0,$$

(d)

$$\lambda_1 = 17.043(\text{rad/s})^2 \rightarrow f_{n1} = 0.657 \text{ Hz}$$
$$\lambda_2 = 2254(\text{rad/s})^2 \rightarrow f_{n2} = 7.56 \text{ Hz},$$
$$A = \begin{bmatrix} 1 & 1 \\ 1 & -1.9604 \end{bmatrix};$$

(e) $\ddot{q}_1 + 17.04q_1 = 0 \quad \ddot{q}_2 + 2254q_2 = 0$

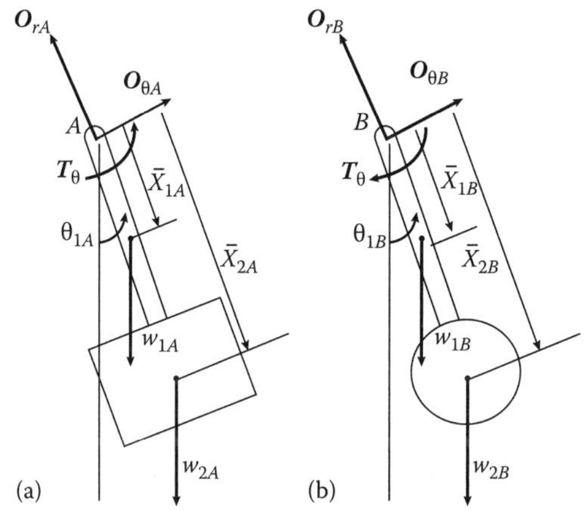

5.73

$$\left[\begin{array}{c:c:c:c:c:c} l\cos\theta & -2l\sin\phi & 0 & 0 & 0 & -1 \\ \hdashline -l\sin\theta & -2l\cos\phi & 0 & 0 & 0 & 0 \\ \hdashline I_o & 0 & -l\cos\theta & l\sin\theta & 0 & 0 \\ \hdashline 0 & I_{g2} & l\sin\phi & l\cos\phi & l\cos\phi & 0 \\ \hdashline 2ml\cos\theta & -2ml\sin\phi & 1 & 0 & 0 & 0 \\ \hdashline -2ml\sin\theta & -2ml\cos\phi & 0 & 1 & -1 & 0 \end{array}\right] \times \begin{bmatrix} \ddot{\theta} \\ \ddot{\phi} \\ B_x \\ B_y \\ R_c \\ \ddot{X}_c \end{bmatrix} = \left[\begin{array}{c} l\sin\theta\dot{\theta}^2 + 2l\cos\phi\dot{\phi}^2 \\ \hdashline l\cos\theta\dot{\theta}^2 - 2l\sin\phi\dot{\phi}^2 \\ \hdashline mg\frac{l}{2}\sin\theta + Fl\cos\theta \\ \hdashline 0 \\ \hdashline 2ml\sin\theta\dot{\theta}^2 + 2ml\cos\phi\dot{\phi}^2 \\ \hdashline 2ml\cos\theta\dot{\theta}^2 - 2ml\sin\phi\dot{\phi}^2 - 2mg \end{array}\right]$$

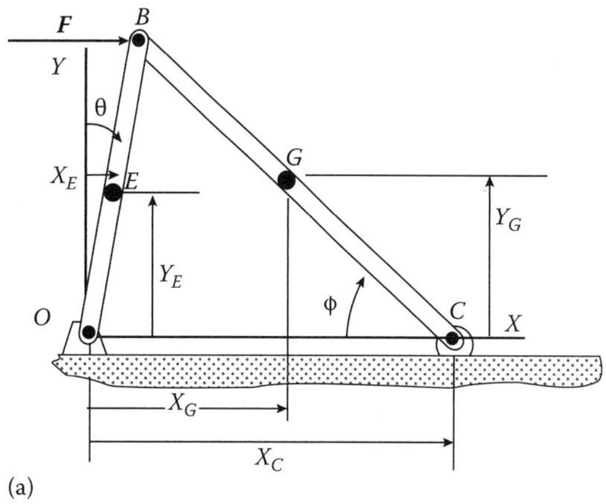

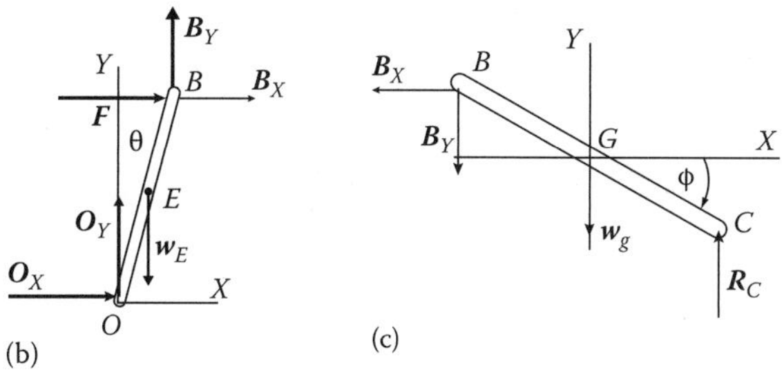

5.74

$$\left[\begin{array}{c:c:c:c:c:c} -l\sin\theta & -2l\sin\phi & 0 & 0 & 0 & -1 \\ \hdashline l\cos\theta & -2l\cos\phi & 0 & 0 & 0 & 0 \\ \hdashline I_o & 0 & l\sin\theta & -l\cos\theta & 0 & 0 \\ \hdashline 0 & I_g & l\sin\phi & l\cos\phi & l\cos\phi & 0 \\ \hdashline -2ml\sin\theta & -2ml\sin\phi & 1 & 0 & 0 & 0 \\ \hdashline 2ml\cos\theta & -2ml\cos\phi & 0 & 1 & -1 & 0 \end{array}\right] \times \begin{bmatrix} \ddot{\theta} \\ \ddot{\phi} \\ A_x \\ A_y \\ R_B \\ \ddot{X}_B \end{bmatrix} = \left[\begin{array}{c} l\cos\theta\dot{\theta}^2 + 2l\cos\phi\dot{\phi}^2 \\ \hdashline l\sin\theta\dot{\theta}^2 - 2l\sin\phi\dot{\phi}^2 \\ \hdashline -mg\frac{l}{2}\cos\theta \\ \hdashline -Fl\sin\phi \\ \hdashline 2ml\cos\theta\dot{\theta}^2 + 2ml\cos\phi\dot{\phi}^2 + F \\ \hdashline 2ml\sin\theta\dot{\theta}^2 - 2ml\sin\phi\dot{\phi}^2 - 2mg \end{array}\right]$$

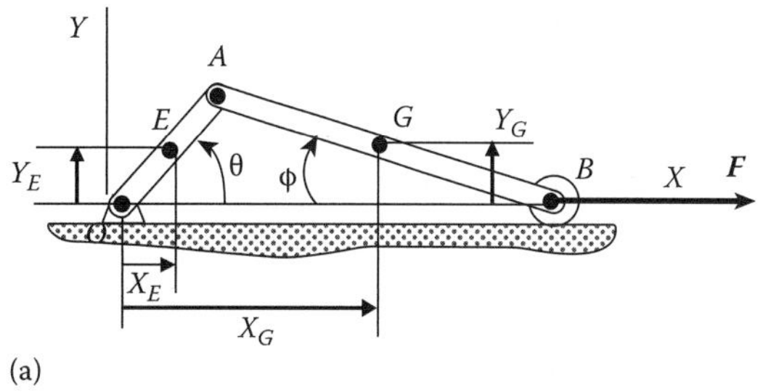

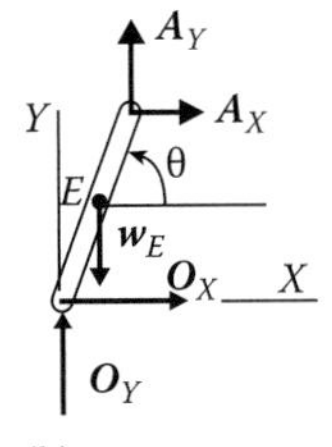

(b)

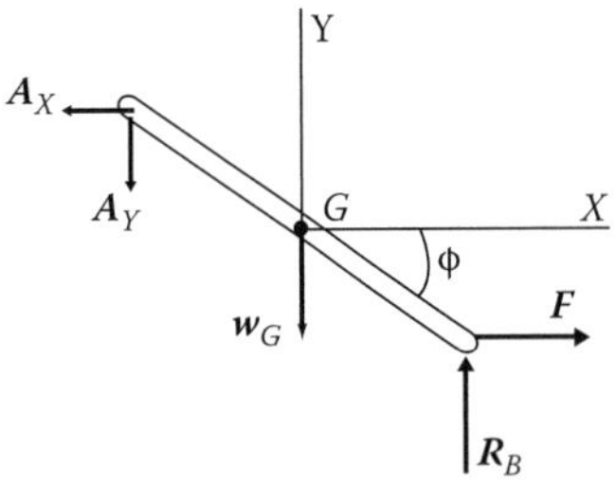

(c)

5.75

$$\begin{bmatrix} l\cos\theta & -2l\sin\phi & 0 & 0 & 0 & -1 \\ -l\sin\theta & -2l\cos\phi & 0 & 0 & 0 & 0 \\ I_o & 0 & -l\sin\theta & l\cos\theta & 0 & 0 \\ 0 & I_g & -l\cos\phi & l\sin\phi & -l\cos\phi & 0 \\ -2ml\sin\theta & -2ml\cos\phi & -1 & 0 & 1 & 0 \\ 2ml\cos\theta & -2ml\sin\phi & 0 & 1 & 0 & 0 \end{bmatrix} \times \begin{bmatrix} \ddot{\theta} \\ \ddot{\phi} \\ A_x \\ A_y \\ R_B \\ \ddot{X}_B \end{bmatrix} = \begin{bmatrix} l\sin\theta\dot{\theta}^2 + 2l\cos\phi\dot{\phi}^2 \\ l\cos\theta\dot{\theta}^2 - 2l\sin\phi\dot{\phi}^2 \\ -mg\frac{l}{2}\sin\theta \\ -Fl\sin\phi \\ 2ml\cos\theta\dot{\theta}^2 - 2ml\sin\varphi\dot{\varphi}^2 \\ 2ml\sin\theta\dot{\theta}^2 + 2ml\cos\phi\dot{\phi}^2 - 2mg + F \end{bmatrix}$$

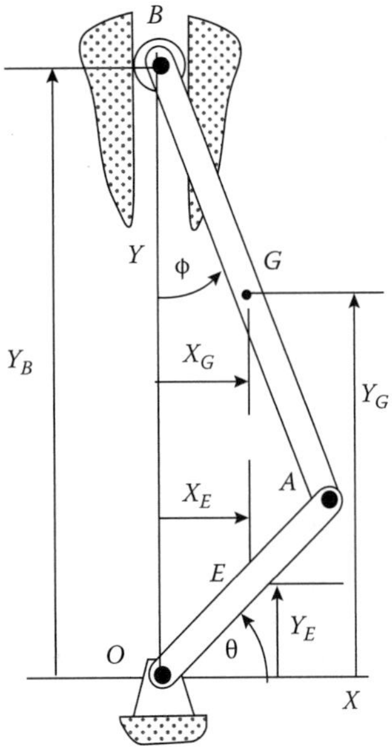

(a)

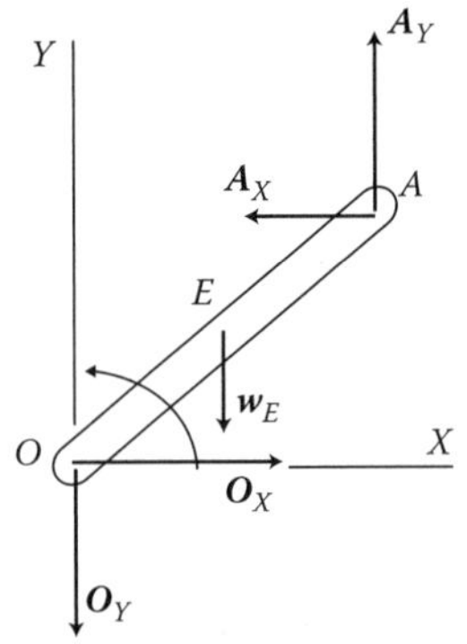

(b)

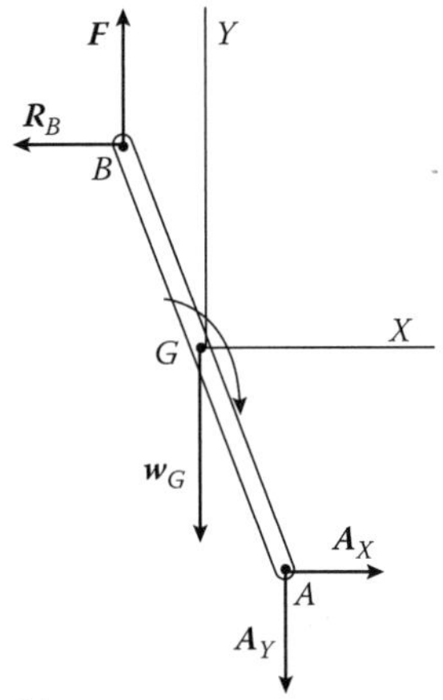

(c)

5.76

$$\begin{bmatrix} -l_1\sin\theta & -l_2\sin\beta & 0 & 0 & 0 & -1 \\ -l_1\cos\theta & -l_2\cos\beta & 0 & 0 & 0 & 0 \\ I_o & 0 & l_1\sin\theta & l_1\cos\theta & 0 & 0 \\ 0 & I_g & -\frac{l_2}{2}\sin\beta & -\frac{l_2}{2}\cos\beta & \frac{l_2}{2}\cos\beta & 0 \\ -m_2l_1\sin\theta & -m_2l_2\sin\beta & 1 & 0 & 0 & 0 \\ m_2l_1\cos\theta & -m_2l_2\cos\beta & 1 & -1 & -1 & 0 \end{bmatrix} \times \begin{bmatrix} \ddot{\theta} \\ \ddot{\beta} \\ A_x \\ A_y \\ R_B \\ \ddot{X}_B \end{bmatrix} = \begin{bmatrix} l_1\cos\theta\dot{\theta}^2 + l_2\cos\beta\dot{\beta}^2 \\ -l_1\sin\theta\dot{\theta}^2 + l_2\sin\beta\dot{\beta}^2 \\ -m_1g\frac{l_1}{2}\sin\theta \\ -F\frac{l_2}{2}\sin\beta \\ F + m_2l\cos\theta\dot{\theta}^2 + m_2\frac{l_2}{2}\cos\beta\dot{\beta}^2 \\ -m_2g + m_2l_1\sin\theta\dot{\theta}^2 - m_2\frac{l_2}{2}\sin\beta\dot{\beta}^2 \end{bmatrix}$$

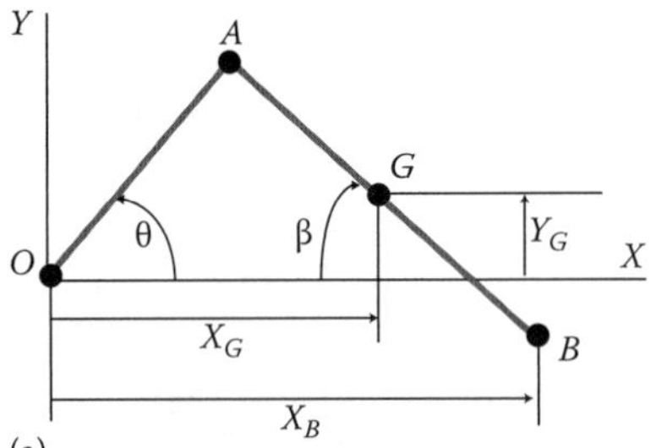

(a)

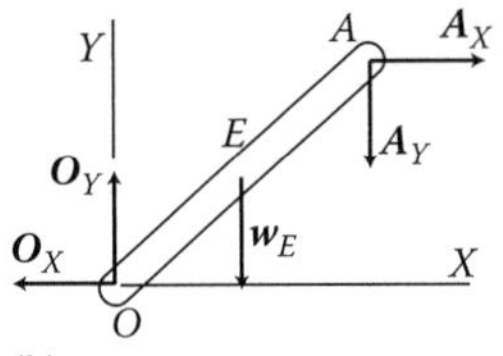

(b)

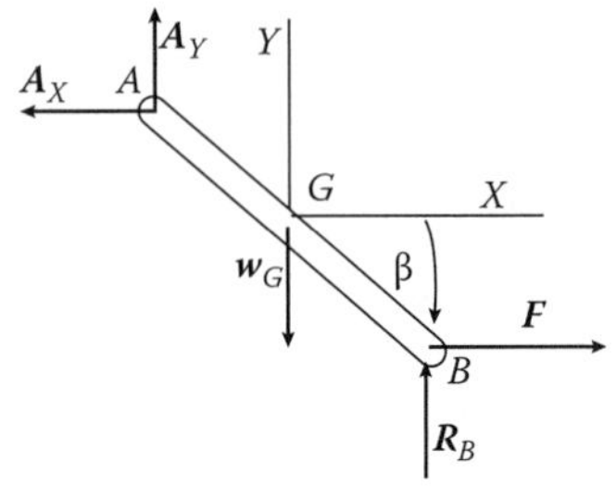

(c)

5.77

$$
\begin{bmatrix}
l_1\cos\theta_1 & l_2\cos\theta_2 & 0 & 0 & 0 & 1\\
-l_1\sin\theta_1 & -l_2\sin\theta_2 & 0 & 0 & 0 & 0\\
I_o & 0 & l_1\cos\theta_1 & l_2\sin\theta_1 & 0 & 0\\
0 & I_g & \frac{l_2}{2}\cos\theta_2 & \frac{l_2}{2}\sin\theta_2 & -\frac{l_2}{2}\sin\theta_2 & 0\\
m_2 l_1\cos\theta_1 & m_2\frac{l_2}{2}\cos\theta_2 & -1 & 0 & 0 & 0\\
0 & -m_2\frac{l_2}{2}\sin\theta_2 & 0 & -1 & -1 & 0
\end{bmatrix}
\times
\begin{bmatrix}
\ddot{\theta}_1\\ \ddot{\theta}_2\\ A_x\\ A_y\\ R_B\\ \ddot{X}_B
\end{bmatrix}
=
\begin{bmatrix}
l_1\sin\theta_1\dot{\theta}_1^2+l_2\sin\theta_2\dot{\theta}_2^2\\
l_1\cos\theta_1\dot{\theta}_1^2+l_2\cos\theta_2\dot{\theta}_2^2\\
-m_1 g\frac{l_1}{2}\sin\theta_1\\
0\\
m_2 l_1\sin\theta_1\dot{\theta}_1^2+m_2\frac{l_2}{2}\sin\theta_2\dot{\theta}_2^2\\
-m_2 g+m_2\frac{l_2}{2}\cos\theta_2\dot{\theta}_2^2
\end{bmatrix}
$$

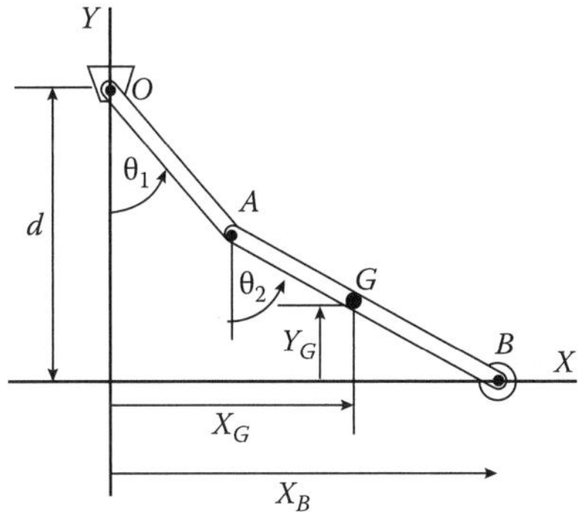

(a)

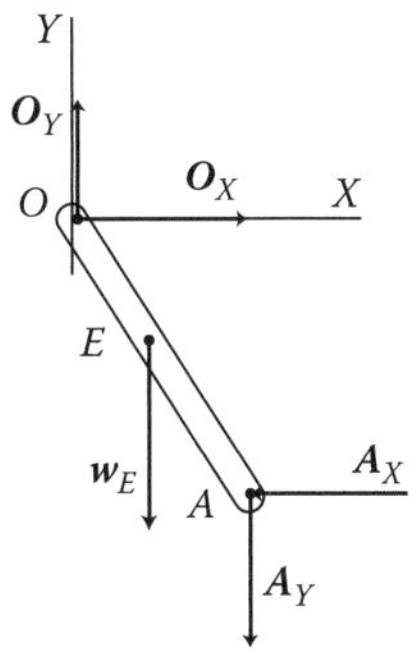

(b)

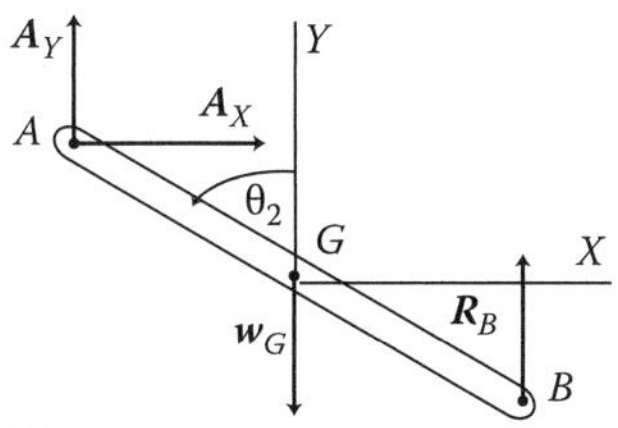

(c)

5.78

$$
\begin{bmatrix}
-l\cos\theta & 2l\sin\phi & 2l\cos\beta & 0 & 0 & 0 & 0\\
l\sin\theta & 2l\cos\phi & -2l\sin\beta & 0 & 0 & 0 & 0\\
I_o & 0 & 0 & l\cos\theta & l\sin\theta & 0 & 0\\
0 & I_g & 0 & l\sin\phi & l\cos\phi & l\sin\phi & -l\cos\phi\\
0 & -2ml\sin\phi & 4ml\cos\beta & -1 & 0 & 1 & 0\\
0 & 2ml\cos\phi & -4ml\sin\beta & 0 & 1 & 0 & 1\\
0 & 0 & I_c & 0 & 0 & -2l\cos\beta & 2l\sin\beta
\end{bmatrix}
\times
\begin{bmatrix}
\ddot{\theta}\\ \ddot{\phi}\\ \ddot{\beta}\\ A_x\\ A_y\\ B_x\\ B_y
\end{bmatrix}
=
\begin{bmatrix}
2l\sin\beta\dot{\beta}^2+2l\cos\phi\dot{\phi}^2-l\sin\theta\dot{\theta}^2\\
2l\cos\beta\dot{\beta}^2+2l\sin\phi\dot{\phi}^2-l\cos\theta\dot{\theta}^2\\
T-mg\frac{l}{2}\sin\theta\\
0\\
4ml\sin\beta\dot{\beta}^2+ml\cos\phi\dot{\phi}^2\\
4ml\cos\beta\dot{\beta}^2+2mg\\
-2mgl\sin\beta
\end{bmatrix}
$$

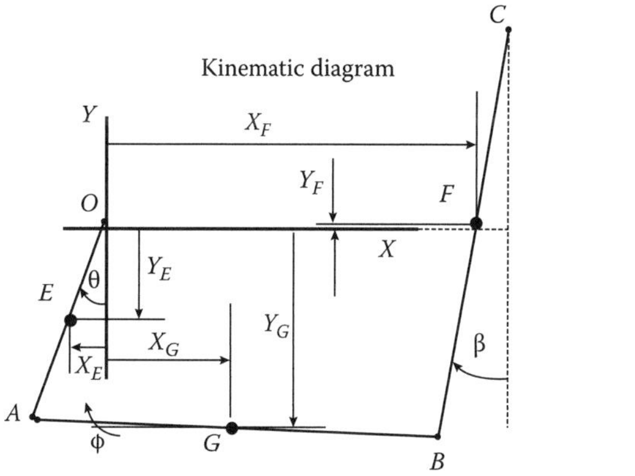

(a)

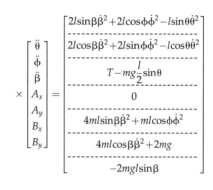

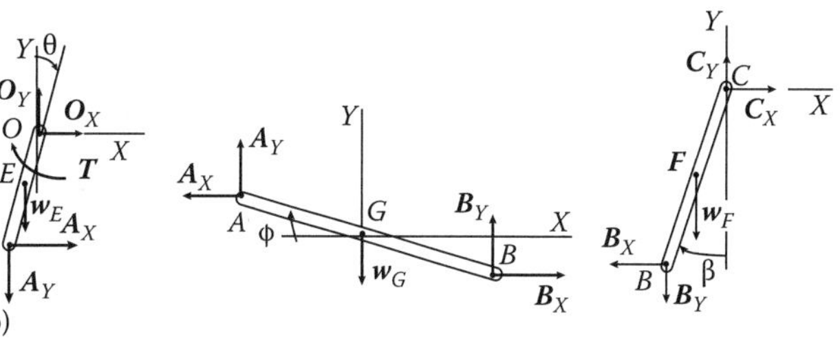

(b)

5.79

$$\begin{bmatrix} l\cos\theta & -\sqrt{5}l\sin\phi & -2l\cos\beta & 0 & 0 & 0 & 0 \\ -l\sin\theta & \sqrt{5}l\cos\phi & 2l\sin\beta & 0 & 0 & 0 & 0 \\ I_o & 0 & 0 & l\cos\theta & -l\sin\theta & 0 & 0 \\ 0 & I_g & 0 & -\frac{\sqrt{5}}{2}l\sin\phi & \frac{\sqrt{5}}{2}l\cos\phi & -\frac{\sqrt{5}}{2}l\sin\phi & -\frac{\sqrt{5}}{2}l\cos\phi \\ Ml\cos\theta & -M\frac{\sqrt{5}}{2}l\sin\phi & 0 & -1 & 0 & 1 & 0 \\ -Ml\sin\theta & M\frac{\sqrt{5}}{2}l\cos\phi & 0 & 0 & -1 & 0 & -1 \\ 0 & 0 & I_c & 0 & 0 & -2l\cos\beta & -2l\sin\beta \end{bmatrix}$$

$$\times \begin{bmatrix} \ddot{\theta} \\ \ddot{\phi} \\ \ddot{\beta} \\ A_x \\ A_y \\ B_x \\ B_y \end{bmatrix} = \begin{bmatrix} -2l\sin\beta\dot{\beta}^2+\sqrt{5}l\cos\phi\dot{\phi}^2+l\sin\theta\dot{\theta}^2 \\ -2l\cos\beta\dot{\beta}^2+\sqrt{5}l\sin\phi\dot{\phi}^2+l\cos\theta\dot{\theta}^2 \\ T+mg\frac{l}{2}\sin\theta \\ 0 \\ Ml\sin\theta\dot{\theta}^2+M\frac{\sqrt{5}}{2}l\cos\phi\dot{\phi}^2 \\ Ml\cos\theta\dot{\theta}^2+M\frac{\sqrt{5}}{2}l\sin\phi\dot{\phi}^2-Mg \\ 2mgl\sin\beta \end{bmatrix}$$

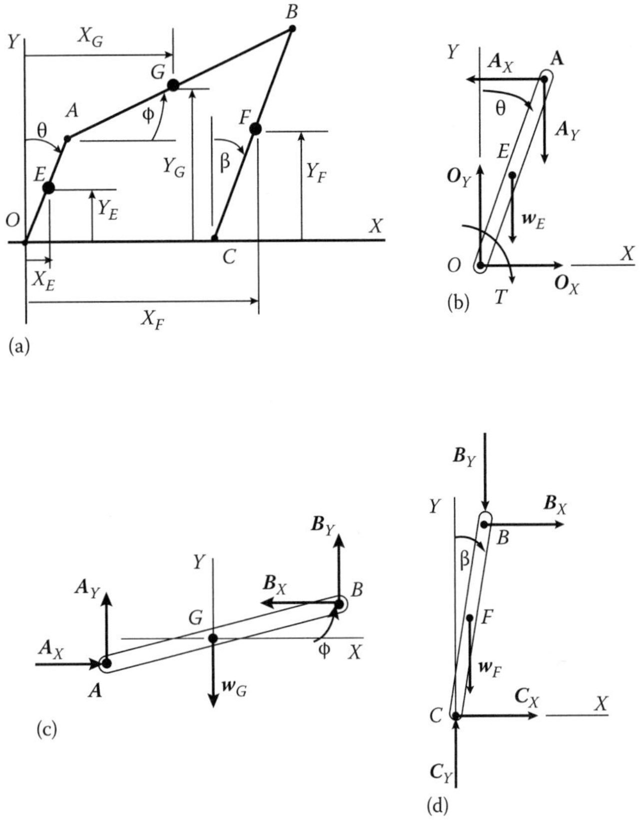

5.80

$$\begin{bmatrix} \cos\theta & -2\sin\phi & -2\cos\beta & 0 & 0 \\ \sin\theta & -2\cos\phi & -2\sin\beta & 0 & 0 \\ 2ml\cos\theta & -2ml\sin\phi & 0 & \sin\theta & \sin\beta \\ -2ml\sin\theta & 2ml\cos\phi & 0 & -\cos\theta & -\cos\beta \\ 0 & \frac{2ml}{3} & 0 & -\cos(\theta-\phi) & \cos(\beta+\phi) \end{bmatrix}$$

$$\times \begin{Bmatrix} \ddot{\theta} \\ \ddot{\phi} \\ \ddot{\beta} \\ T_1 \\ T_2 \end{Bmatrix} = \begin{Bmatrix} \dot{\theta}^2\sin\theta+2\dot{\phi}^2\cos\phi-2\dot{\beta}^2\sin\beta \\ -\dot{\theta}^2\cos\theta-2\dot{\phi}^2\sin\phi+2\dot{\beta}^2\cos\beta \\ f(t)+2ml\dot{\theta}^2\sin\theta+2ml\dot{\phi}^2\cos\phi \\ -w_G+2ml\dot{\theta}^2\cos\theta+2ml\dot{\phi}^2\sin\phi \\ f(t)\sin\phi \end{Bmatrix}$$

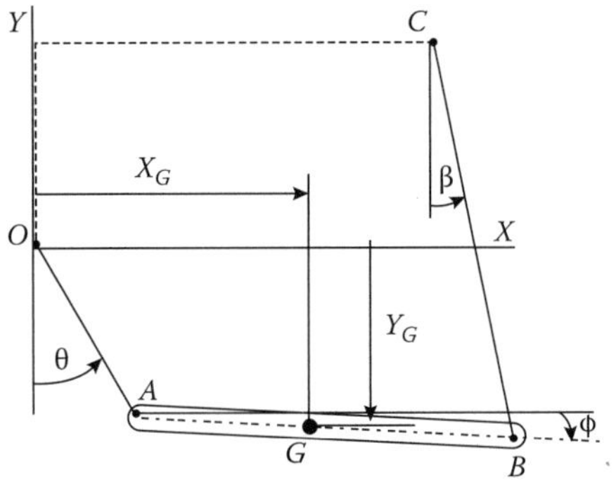

5.81

$$\begin{bmatrix} L\cos\theta & S\cos\phi & 0 & 0 \\ -\frac{ML}{2}\sin\theta & -MS\sin\phi & \cos\phi & -\mu_d \\ \frac{ML}{2}\cos\theta & -MS\cos\phi & \sin\phi & 1 \\ \frac{ML}{6} & 0 & \sin(\theta-\phi) & \mu_d\sin\theta+\cos\theta \end{bmatrix}$$

$$\times \begin{Bmatrix} \ddot{\theta} \\ \ddot{\phi} \\ T \\ N \end{Bmatrix} = \begin{Bmatrix} L\dot{\theta}^2\sin\theta+S\dot{\phi}^2\sin\phi-2\dot{S}\dot{\phi}\cos\phi \\ 2M\dot{S}\dot{\phi}\sin\phi+MS\dot{\phi}^2\cos\phi+\frac{ML}{2}\dot{\theta}^2\cos\theta \\ w_G+\frac{ML}{2}\dot{\theta}^2\sin\theta+MS\dot{\phi}^2\sin\phi-2M\dot{S}\dot{\phi}\cos\phi \\ 0 \end{Bmatrix}$$

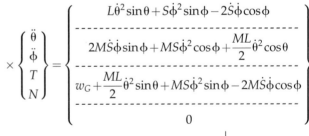

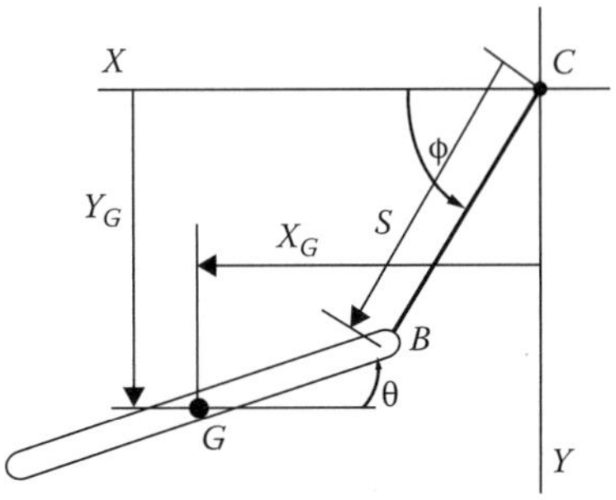

5.82

$$
\begin{bmatrix}
r & -\sin\phi & -\cos\beta & 0 & 0 & 0 \\
0 & \cos\phi & -\sin\beta & 0 & 0 & 0 \\
\frac{3Mr}{2}+mr & \frac{-ml}{2}\sin\phi & 0 & \sin\beta & 0 & 0 \\
-(M+m) & \frac{ml}{2}\sin\phi & 0 & \frac{-\sin\beta}{r} & 1 & 0 \\
0 & \frac{ml}{2}\cos\phi & 0 & -\cos\beta & 0 & -1 \\
\frac{-mr}{2}\sin\phi & \frac{ml}{3} & 0 & -\cos\beta\cos\phi & 0 & 0
\end{bmatrix}
\times
\begin{Bmatrix}
\ddot{\theta} \\ \ddot{\phi} \\ \ddot{\beta} \\ T \\ f \\ N
\end{Bmatrix}
=
\begin{Bmatrix}
\dot{\phi}^2\cos\phi - \dot{\beta}^2\sin\beta \\
\dot{\phi}^2\sin\phi + \dot{\beta}^2\cos\beta \\
\frac{l}{2}\dot{\phi}^2\cos\phi \\
\frac{-ml}{2}\dot{\phi}^2\cos\phi \\
-w_O - w_G + \frac{ml}{2}\dot{\phi}^2\sin\phi \\
\frac{w_G}{2}\cos\phi
\end{Bmatrix}
$$

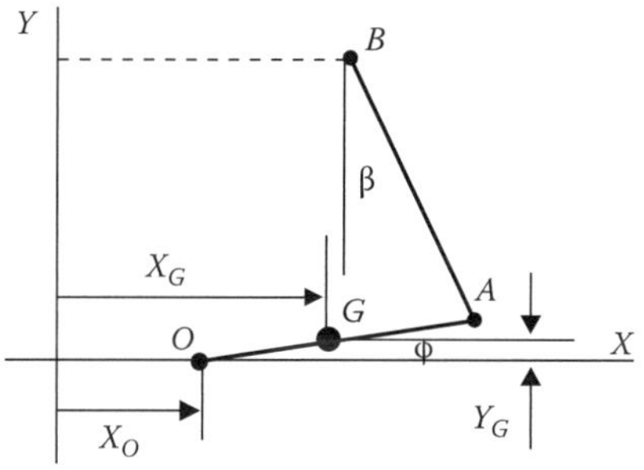

5.83

$$
\begin{bmatrix}
r\sin\theta & l\cos\phi & 0 & 0 & 0 \\
r(m(1+\cos\theta)+M) & \frac{-ml\sin\phi}{2} & 0 & -1 & 0 \\
\left(\frac{Mr}{2} - Mr\cos\theta\right) & 0 & -\sin\theta & 1+\cos\theta & 0 \\
mr\sin\theta & \frac{ml\cos\phi}{2} & 1 & 1 & 0 \\
\frac{-mr}{2}(1+\cos\theta) & \frac{ml}{12}(1+3\sin^2\phi) & \frac{-\cos\phi}{2} & \frac{\cos\phi}{2} & 0
\end{bmatrix}
\times
\begin{Bmatrix}
\ddot{\theta} \\ \ddot{\phi} \\ N \\ f \\ N_B
\end{Bmatrix}
=
\begin{Bmatrix}
r\dot{\theta}^2\cos\theta + l\dot{\phi}^2\sin\phi \\
rm\dot{\theta}^2\sin\theta + \frac{l}{2}\dot{\phi}^2\cos\phi \\
-w_O\sin\theta \\
w_O + w_G - mr\dot{\theta}^2\cos\theta - \frac{ml}{2}\dot{\phi}^2\sin\phi \\
\frac{-\cos\phi}{2}w_O - \frac{mr}{2}\dot{\theta}^2\sin\theta\sin\phi - \frac{ml}{2}\dot{\phi}^2\cos\phi\sin\phi
\end{Bmatrix}
$$

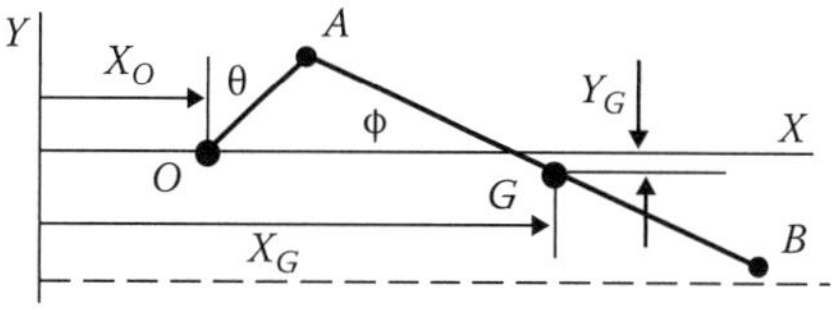

5.84

$$
\begin{bmatrix}
2l\cos\theta & \frac{-3l}{2}\cos\phi & 0 \\
\frac{2ml^2}{3} & \frac{-3ml^2}{2}\cos(\theta+\phi) & 2l\cos\theta \\
(-\sin\theta\tan\phi + \cos\theta) & \frac{3}{2}(\sin\phi\tan\phi - \cos\phi) & \frac{(\mu_d+1)\tan\phi}{ml}
\end{bmatrix}
\times
\begin{Bmatrix}
\ddot{\theta} \\ \ddot{\phi} \\ N
\end{Bmatrix}
=
\begin{Bmatrix}
2l\dot{\theta}^2\sin\theta - \frac{3l}{2}\dot{\phi}^2\sin\phi \\
w_G l\cos\theta + ml^2\sin 2\theta - \frac{3ml^2}{2}\sin(\phi+\theta) \\
\frac{m}{l} + (\cos\theta\tan\phi + \sin\theta)\dot{\theta}^2 + \frac{3}{2}(\tan^2\phi)\dot{\phi}^2
\end{Bmatrix}
$$

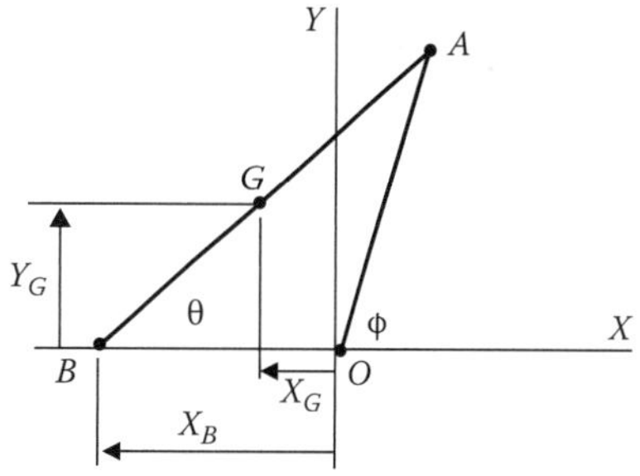

5.85

$$
\begin{bmatrix}
r\cos\theta & -s\cos\phi & -\sin\phi & 0 \\
r\sin\theta & s\sin\phi & -\cos\phi & 0 \\
I_c & 0 & 0 & r\cos(\theta+\phi) \\
0 & I_o & 0 & -S
\end{bmatrix}
\begin{bmatrix}
\ddot{\theta} \\ \ddot{\phi} \\ \ddot{S} \\ N
\end{bmatrix}
=
\begin{bmatrix}
r\dot{\theta}^2\sin\theta - s\dot{\phi}^2\sin\phi + 2\dot{S}\cos\phi\dot{\phi} \\
-r\dot{\theta}^2\cos\theta - s\dot{\phi}^2\cos\phi - 2\dot{S}\sin\phi\dot{\phi} \\
M + \frac{r}{2}W_E\sin\theta \\
-\bar{r}W_g\sin\phi
\end{bmatrix}
$$

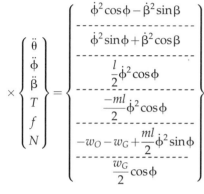

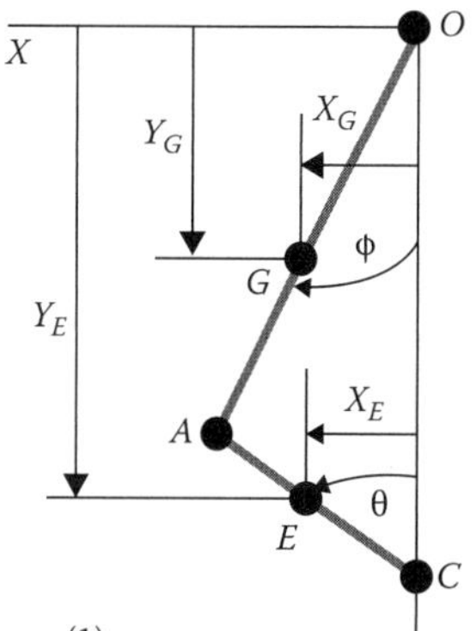

(1)

5.86

$$\begin{bmatrix} r\sin\theta & s\sin\phi & -\cos\phi & 0 \\ r\sin\theta & s\sin\phi & -\cos\phi & 0 \\ I_{o1} & 0 & 0 & r\cos(\theta+\phi) \\ 0 & I_{o2} & 0 & -S \end{bmatrix} \begin{bmatrix} \ddot\theta \\ \ddot\phi \\ \ddot S \\ N \end{bmatrix} = \begin{bmatrix} -r\dot\theta^2\cos\theta - s\dot\phi^2\cos\phi - 2\dot S\sin\phi\dot\phi \\ r\dot\theta^2\sin\theta - s\dot\phi^2\sin\phi + 2\dot S\cos\phi\dot\phi \\ M + \bar r W_A\sin\theta \\ 0 \end{bmatrix}$$

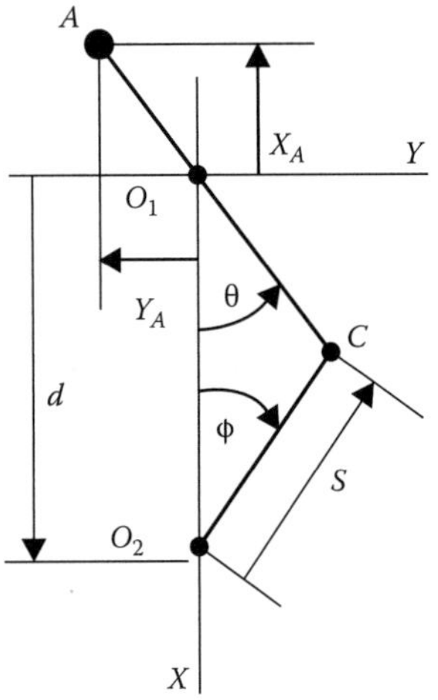

5.87

$$\begin{bmatrix} l_1\sin\theta & -l_2\sin\phi & 1 & 0 & 0 & 0 & 0 \\ l_1\cos\theta & l_2\cos\phi & 0 & 0 & 0 & 0 & 0 \\ I_A & 0 & 0 & -l_1\sin\theta & l_1\cos\theta & 0 & 0 \\ 0 & I_G & 0 & \frac{l_2}{2}\sin\phi & -\frac{l_2}{2}\cos\phi & \frac{l_2}{2}\sin\phi & -\frac{l_2}{2}\cos\phi \\ -m_2l_1\sin\theta & m_2\frac{l_2}{2}\sin\phi & 0 & -1 & 0 & 1 & 0 \\ m_2l_1\cos\theta & m_2\frac{l_2}{2}\cos\phi & 0 & 0 & 1 & 0 & -1 \\ 0 & 0 & m_3 & 0 & 0 & -1 & 0 \end{bmatrix}$$

$$\times \begin{bmatrix} \ddot\theta \\ \ddot\phi \\ \ddot S \\ B_x \\ B_y \\ C_x \\ C_y \end{bmatrix} = \begin{bmatrix} -l_1\dot\theta^2\cos\theta + l_2\dot\phi^2\cos\phi \\ l_1\dot\theta^2\sin\theta + l_2\dot\phi^2\sin\phi \\ 0 \\ 0 \\ m_2l_1\dot\theta^2\cos\theta - m_2\frac{l_2}{2}\dot\phi^2\cos\phi \\ m_2l_1\dot\theta^2\sin\theta + m_2\frac{l_2}{2}\dot\phi^2\sin\phi \\ -F \end{bmatrix}$$

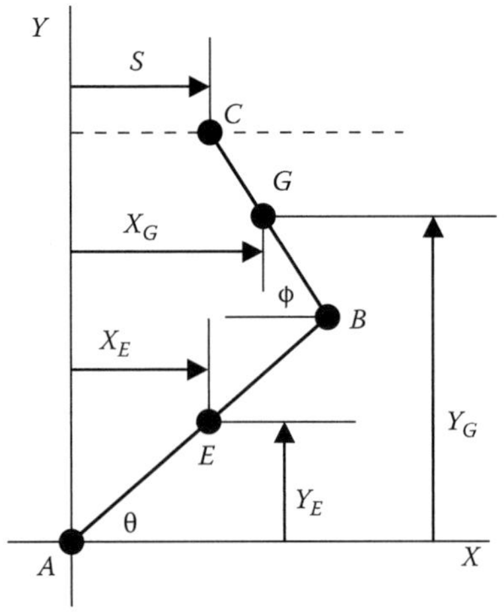

5.88

$$\begin{bmatrix} L_1\sin\theta & -S\sin\phi & \cos\phi & 0 & 0 & 0 \\ L_1\cos\theta & -S\cos\phi & -\sin\phi & 0 & 0 & 0 \\ I_A & 0 & 0 & -L_1\cos\theta & -L_1\sin\theta & 0 \\ 0 & I_C & 0 & 0 & 0 & S \\ -m_3L_1\sin\theta & 0 & 0 & -1 & 0 & \sin\phi \\ m_3L_1\cos\theta & 0 & 0 & 0 & 1 & -\cos\phi \end{bmatrix}$$

$$\times \begin{bmatrix} \ddot\theta \\ \ddot\phi \\ \ddot S \\ B_X \\ B_Y \\ N \end{bmatrix} = \begin{bmatrix} 2\dot S\dot\phi\sin\theta - L_1\dot\theta^2\cos\theta + S\dot\phi^2\cos\phi \\ 2\dot S\dot\phi\cos\phi + L_1\dot\theta^2\sin\theta - S\dot\phi^2\sin\phi \\ -\frac{L_1}{2}W_E\sin\theta \\ -\bar r W_G\cos\phi \\ m_3L_1\dot\theta^2\cos\theta \\ -W_B + m_3L_1\dot\theta^2\cos\theta \end{bmatrix}$$

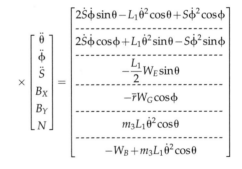

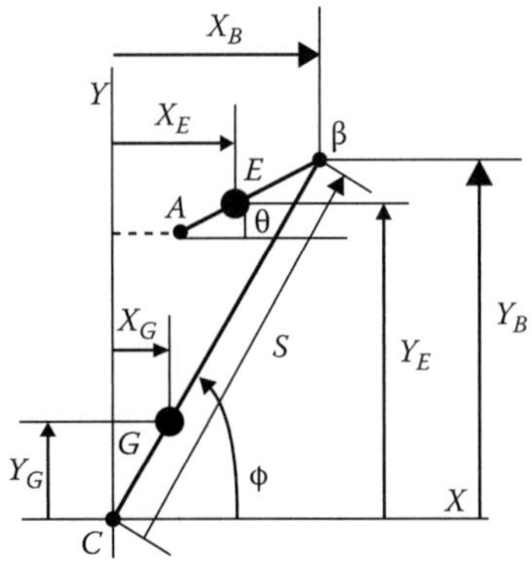

5.89

$$\begin{bmatrix} L_1\cos\theta & -L_2\cos\phi & -L_3\sin\beta & 0 & 0 & 0 & 0 \\ L_1\sin\theta & -L_2\cos\phi & L_3\cos\beta & 0 & 0 & 0 & 0 \\ I_O & 0 & 0 & L_1\cos\theta & L_1\sin\theta & 0 & 0 \\ 0 & I_F & 0 & \frac{L_2}{2}\sin\phi & -\frac{L_2}{2}\cos\phi & \frac{L_2}{2}\sin\phi & -\frac{L_2}{2}\sin\phi \\ 0 & -m_2\frac{L_2}{2}\sin\phi & m_2L_3\sin\beta & -1 & 0 & 1 & 0 \\ 0 & -m_2\frac{L_2}{2}\cos\phi & m_2L_3\cos\beta & 0 & 1 & 0 & -1 \\ 0 & 0 & I_C & 0 & 0 & -L_3\sin\beta & L_3\cos\beta \end{bmatrix}$$

$$\times \begin{bmatrix} \ddot{\theta} \\ \ddot{\phi} \\ \ddot{\beta} \\ A_X \\ A_Y \\ B_X \\ B_Y \end{bmatrix} = \begin{bmatrix} L_1\dot{\theta}^2\cos\theta + L_2\dot{\phi}^2\cos\phi - L_3\dot{\beta}^2\cos\beta \\ -L_1\dot{\theta}^2\cos\theta - L_2\dot{\phi}^2\sin\phi + L_3\dot{\beta}^2\sin\beta \\ M \\ 0 \\ -m_2L_3\dot{\beta}^2\cos\beta + m_2\frac{L_2}{2}\dot{\phi}^2\cos\phi \\ -W_F + m_2L_3\dot{\beta}^2\sin\beta - m_2\frac{L_2}{2}\dot{\phi}^2\sin\phi \\ -W_E\frac{L_3}{2}\cos\beta \end{bmatrix}$$

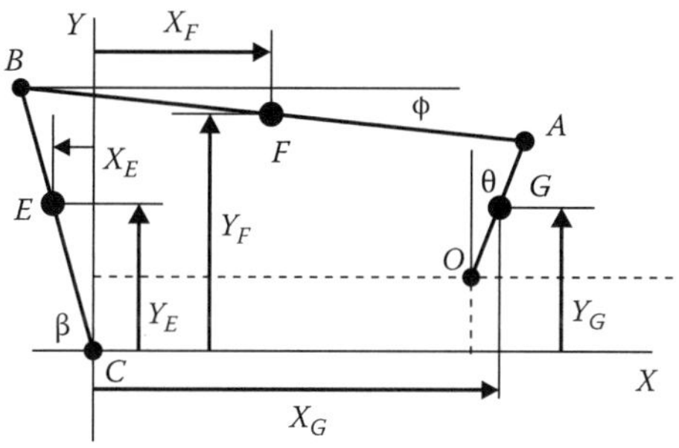

5.90

$$\begin{bmatrix} b\sin\theta & S\sin\phi & -\cos\phi & 0 & 0 \\ -b\cos\theta & S\cos\phi & \sin\phi & 0 & 0 \\ 0 & 0 & S & 0 & 0 \\ I_O & 0 & 0 & b\sin\theta & b\cos\theta \\ 0 & I_C & 0 & S\sin\phi & -S\cos\phi \end{bmatrix}$$

$$\times \begin{bmatrix} \ddot{\theta} \\ \ddot{\phi} \\ \ddot{S} \\ B_X \\ B_y \end{bmatrix} = \begin{bmatrix} -b\dot{\theta}^2\cos\theta - 2\dot{S}\dot{\phi}\sin\phi - S\dot{\phi}^2\cos\phi \\ -b\dot{\theta}^2\sin\theta - 2\dot{S}\dot{\phi}\sin\phi + S\dot{\phi}^2\sin\phi \\ 2b^2(\dot{\theta}^2\cos\theta) - \dot{S}^2 \\ Fb\sin(\theta+\phi) - \frac{L_2}{2}w_G\cos\theta \\ -\frac{w_G}{2}L\cos\phi \end{bmatrix}$$

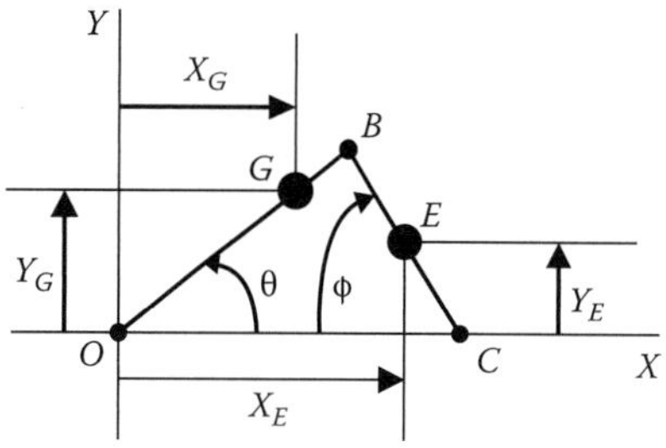

5.91

$$\begin{bmatrix} S\sin\phi & -\cos\phi & 0 & 0 & 0 \\ S\cos\phi & \sin\phi & 0 & 0 & 0 \\ 0 & S & 0 & 0 & 0 \\ 0 & 0 & r\sin\theta & r\cos\theta & 0 \\ I_E & 0 & \frac{L}{2}\sin\phi & -\frac{L}{2}\cos\phi & -\left(S-\frac{1}{2}\right) \end{bmatrix}$$

$$\times \begin{Bmatrix} \ddot{\phi} \\ \ddot{S} \\ B_X \\ B_Y \\ N \end{Bmatrix} = \begin{Bmatrix} -r\dot{\theta}^2\cos\theta - 2\dot{S}\dot{\phi}\sin\phi - S\dot{\phi}^2\cos\phi \\ -r\dot{\theta}^2\sin\theta - 2\dot{S}\dot{\phi}\cos\phi + S\dot{\phi}^2\sin\phi \\ 2ar(\dot{\theta}^2\cos\theta) - \dot{S}^2 \\ -W_G\frac{L}{2}\cos\theta \\ 0 \end{Bmatrix}$$

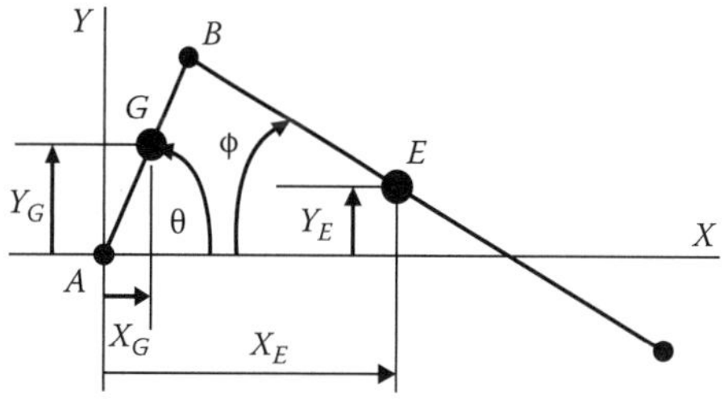

5.92 *Note*: Distance from A to B is d

$$\left(I_A + \frac{R^2}{r^2}I_0 + m_y d^2\sin^2\phi\right)(\ddot{\phi})$$

$$= \frac{R}{r}T - m_y\dot{\phi}^2 d\sin\phi\cos\phi$$

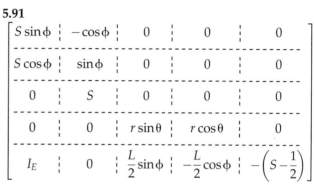

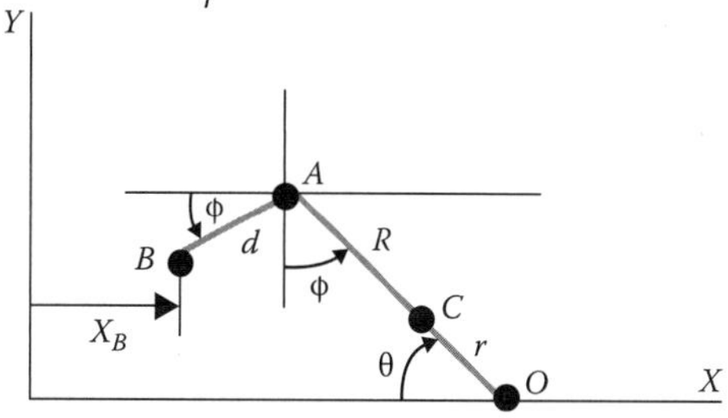

References

Childs, D.W. (1993), *Turbomachinery Rotordynamics*, Wiley, New York.

Dimarogonas, A. (1995), *Vibration for Engineers*, 2nd edn., Prentice Hall, Englewood Cliffs, NJ.

Föppl, A. (1895), Das Problem der Laval'schen Turbinewelle, *Civilengenieur*, **41**, 332–342.

Gardner, J. (2001), *Simulations of Machines Using MATLAB and SIMULINK*, Brooks/Cole, Pacific Grove, CA.

Jeffcott, H.H. (1919), Lateral vibration of loaded shafts in the neighbourhood of a whirling speed—The effect of want of balance, *Philosophical Magazine*, **37**, 304–314.

Mechtly, E.A. (1969), The international system of units, NASA SP-7012, Washington, DC.

Meirovitch, L. (1997), *Fundamentals of Vibrations*, McGraw Hill, New York.

Meriam, J. and Kraige, L. (1992), *Engineering Mechanics Dynamics*, 3rd edn., Vol. 2, Wiley, New York.

Index